Drafting Fundamentals for the Entertainment Classroom

Drafting Fundamentals for the Entertainment Classroom: A Process-Based Introduction to Hand Drafting, Vectorworks, and SketchUp guides students through a syllabus-formatted semester of integrated drafting concepts and skills.

This book links beginner visualization practices with fundamental software knowledge through step-by-step exercises and examples. By presenting hand drafting and Vectorworks through incremental exercises, students not only gain an understanding of the tools used in drafting but also learn why the tools, practices, and standards exist in the first place. SketchUp, a user-friendly 3D modeling program, is integrated into the various exercises to help readers visualize concepts and begin modeling their own ideas. By the end of the book, students will understand drawing construction techniques, United States Institute for Theatre Technology (USITT)-recommended graphic standards, and the typical drawings created for entertainment design, preparing them to dive more deeply into the further complexities and opportunities of Vectorworks and SketchUp.

Drafting Fundamentals for the Entertainment Classroom is written to complement a 14- or 15-week semester of an Entertainment Drafting course. The book's format also provides structure for independent and self-directed study.

Eric Appleton is an associate professor of scenic and lighting design at the University of Wisconsin-Whitewater. He has been a freelance lighting and scenic designer since 1994.

Drafting Fundamentals for the Entertainment Classroom

A Process-Based Introduction Integrating Hand Drafting, Vectorworks, and SketchUp

Eric Appleton

Routledge
Taylor & Francis Group

NEW YORK AND LONDON

First published 2022
by Routledge
605 Third Avenue, New York, NY 10158

and by Routledge
2 Park Square, Milton Park, Abingdon, Oxon, OX14 4RN

Routledge is an imprint of the Taylor & Francis Group, an informa business

Library of Congress Cataloging-in-Publication Data
Names: Appleton, Eric, author.
Title: Drafting fundamentals for the entertainment classroom : a process-based introduction integrating hand drafting, Vectorworks, and SketchUp / Eric Appleton.
Description: New York, NY : Routledge, 2022. | Includes bibliographical references and index.
Identifiers: LCCN 2021018373 (print) | LCCN 2021018374 (ebook) | ISBN 9780367724719 (hbk) | ISBN 9780367724702 (pbk) | ISBN 9781003154921 (ebk)
Subjects: LCSH: Mechanical drawing–Textbooks. | Theaters–Stage-setting and scenery–Drawings–Textbooks. | Interior architecture–Computer-aided design–Textbooks. | SketchUp–Textbooks. | VectorWorks–Textbooks. | Computer-aided design–Textbooks.
Classification: LCC T353 .A639 2022 (print) | LCC T353 (ebook) | DDC 604.2–dc23
LC record available at https://lccn.loc.gov/2021018373
LC ebook record available at https://lccn.loc.gov/2021018374

ISBN: 978-0-367-72471-9 (hbk)
ISBN: 978-0-367-72470-2 (pbk)
ISBN: 978-1-003-15492-1 (ebk)

DOI: 10.4324/9781003154921

Vectorworks images used by permission of Vectorworks. SketchUp images used by permission of SketchUp/Trimble

Typeset in GillSansStd-Light
by KnowledgeWorks Global Ltd.

CONTENTS

Preface   *xiii*
Acknowledgments   *xv*

INTRODUCTION   GETTING READY   1

FOR THE STUDENTS (BUT INSTRUCTORS ARE WELCOME)   1
    What is Drafting?   1
    Who is Drafting For?   5
    Equipment and Materials   5

FOR THE INSTRUCTOR (BUT STUDENTS ARE WELCOME)   11
    In the Classroom   11
    Evaluating Drafting Projects   12

CHAPTER 1   WHERE TO BEGIN   15

MODULE 1: LETTERING   15
    Exercise: Lettering   17
    Homework: Lettering   18

MODULE 2: SCALE AND MEASUREMENTS   18
    Measurement Notation   18
    Exercise: Reading a Ruler and Writing Measurements   19
    Introduction to Commonly Used Scales   19
    Reading the Scale Ruler   20
    Exercise: Scale Measurements   21

MODULE 3: THE DRAWING SPACE   21
    Boundaries and Borders   21
    Setting Up the Drawing Space for Hand Drafting   22
    Exercise: Setting Up the Drawing Space for Hand Drafting   23
    Construction Lines/Guidelines   24
    Exercise: Straightedge Practice #1   26
    Visible Lines   26
    Exercise: Straightedge Practice #2   27
    Exercise: Defining the Drawing Space by Laying Out Borders   27
    Transporting a Drawing   29

Tutorial: Setting up a Base File in Vectorworks   **29**
Page Navigation   **30**
2D or 3D   **31**
Lines: Weights, Types, and Styles   **31**
Exercise: Drawing Lines in Vectorworks   **35**
Geometric Terms and Construction Refresher   **36**

CHAPTER 2   PROJECTION SYSTEMS AND LINE STYLE CONTEXT   39

MODULE 1: LINE STYLE FAMILIARIZATION   39
Tutorial: Vectorworks Line Style Familiarization   **40**
To Print the Drawing   **44**
To Create a PDF of the Drawing   **45**
Homework: Hand Drafting the Line Styles   **45**

MODULE 2: PROJECTION SYSTEMS   46
Exercise: Orthographic Visualization   **51**
Homework: More Orthographic Visualization   **53**

MODULE 3: DERIVING THE MEANING OF LINE WEIGHTS AND TYPES   54
Exercise: Building Four Cubes in SketchUp   **54**
Exercise: Choosing Appropriate Line Styles   **59**
Rules for Hidden Lines   **60**
Homework: Using the Rectangle Tool   **66**
Homework: Geometric Construction in Vectorworks   **66**

MODULE 4: CIRCLES AND ARCS   67
Exercise: Drawing Circles with a Template   **67**
Exercise: Drawing Circles in Vectorworks   **68**
Exercise: Drawing Circles in SketchUp   **69**
When Arcs Meet Lines   **70**
Exercise: Drawing Arcs Meeting Lines   **70**

CHAPTER 3   ORTHOGRAPHIC VISUALIZATION AND AUXILIARY VIEWS   73

MODULE 1: VISUALIZING ORTHOGRAPHIC PROJECTION   73
Exercise: Using SketchUp to Visualize Orthographic Views   **73**
Exercise: Additional SketchUp Visualization   **76**
Homework: Drafting the SketchUp Object   **76**
Homework: Adjacent Views with Missing Information   **77**

MODULE 2: MITER LINES   79
Homework: Diagonal Views with Missing Information   **81**

MODULE 3: AUXILIARY VIEWS   81
The Two-Triangle Technique   **83**
Exercise: Using the Two-Triangle Technique to Draw Angled Shapes   **83**
Exercise: Modeling an Incline in SketchUp   **84**
Drawing Auxiliary Views   **85**

Exercise: Hand Drafting a Simple Auxiliary View   **85**
Homework: Drafting a Simple Incline in Vectorworks   **86**
Irregular Inclines and Section Views   **87**
Exercise: Modeling a Tent-Shaped Object in SketchUp   **88**
Cutting Plane Lines   **90**
Crosshatching the Section View   **93**
Exercise: Adding Crosshatching and Fills to Shapes   **93**
Exercise: Drawing Auxiliary Views   **95**
Homework: Constructing More Auxiliary Views   **96**

CHAPTERS 1–3 DRAFTING REVIEW CHECKLIST   97

CHAPTER 4   DIMENSIONING AND NOTATION   99

MODULE 1: DIMENSIONING   99
Dimensioning Format Systems   **99**
Extension Lines   **100**
Center Lines   **101**
Dimension Lines and Terminators   **102**
Measurement Notation Placement   **102**
Applying Dimensions in Vectorworks   **104**
Leader Lines   **105**
Dimensioning Angles   **106**
Exercise: Hand Drafting Dimensioning Practice   **106**
Determining the Choice of Dimensions   **108**
Discussion and Exercise: Determining Placement of Dimensions   **108**
Exercise: Dimensioning in SketchUp   **111**
Discussion and Exercise: Dimensioning with Auxiliary Views and Hidden Lines   **112**

MODULE 2: DRAWING FULLY DIMENSIONED ORTHOGRAPHIC VIEWS OF FOUR
CUBES IN VECTORWORKS   114
Titles, Labels, and Notes   **125**
Viewports and Sheet Layers   **127**
Title Blocks   **128**

CHAPTER 5   THE SYSTEM OF WORKING   135

INDIVIDUAL DRAWINGS   135
The Drawing Package   **136**
Discussion: The Drawing Package   **138**

CHAPTER 6   DRAFTING FLATS AND PLATFORMS   139

MODULE 1: FLATS   139
What is a Flat?   **139**
Parts of a Flat   **140**
Two Common Styles of Flat Construction   **140**
Lumber Size: Nominal and Actual   **141**

Exercise: Drafting a Standard Style Rectangular Flat   **142**
Adding a Human Figure for Proportion   **144**
Exercise: Modeling a Flat in SketchUp   **146**
Homework: Modeling a Hollywood Style Flat   **151**
Homework: Drafting a Series of Flats   **151**

MODULE 2: REVEALS, DOORS, AND MOLDING   152
Simple Openings with Depth   **152**
Exercise: Draft a Flat with Opening and Reveal   **153**
Homework: Modeling a Flat with Opening and Reveal   **155**
Doors and Molding   **156**
Molding Nomenclature   **157**
Exercise: Drawing a Door Flat with Details   **157**

MODULE 3: PLATFORMS   162
What is a Platform?   **162**
Parts of a Platform   **162**
Exercise: Drawing Platforms (Design)   **164**
Homework: Modeling a Platformed Area   **167**
Exercise: Drawing a Stepped Platform (Design)   **167**
Exercise: Construction Drawing of a Platformed Area   **169**

FLATS AND PLATFORMS DRAFTING CHECKLIST   171

CHAPTER 7   DRAWING AND MODELING THE PLAN OF A PROSCENIUM STYLE VENUE   173

MODULE 1: DRAFTING THE PLAN OF A VENUE   173
Base Drawings   **173**
Vocabulary Terms   **174**
Exercise: Drafting a Venue Plan in Vectorworks   **174**
Scale Bars   **180**
The Line Schedule   **181**
Base Drawing Groundplan Lay Out Checklist   **185**
Homework: Copy an Existing Plan   **185**

MODULE 2: MODELING THE VENUE   186
Exercise: Modeling the Venue in SketchUp   **186**

CHAPTER 8   SCENIC GROUNDPLANS   193

MODULE 1: WHAT IS A GROUNDPLAN?   193
How Are Groundplans Used?   **193**
What's on a Groundplan?   **194**
Discussion: Reading a Groundplan   **198**
Exercise: Drawing Standard Groundplan Graphics   **202**

MODULE 2: CONSTRUCTING THE MODEL OF A BOX SET   209
Exercise: Creating a Rough Groundplan Sketch   **209**
Exercise: Modeling the Box Set in SketchUp   **210**

MODULE 3: LAYING OUT A GROUNDPLAN   218

    Students   **219**

    Homework: Extrapolating a Groundplan from an Image   **225**

REVIEW OF GROUNDPLAN SYSTEM OF WORKING   226

CHAPTER 9   THE CENTER LINE SECTION   227

MODULE 1: WHAT IS A CENTER LINE SECTION?   227

    Important Elements of a Center Line Section   **228**

    Discussion: Looking at Center Line Sections   **229**

MODULE 2: ADDING HEIGHT TO THE SKETCHUP VENUE MODEL   231

    Project: Modeling Battens   **235**

MODULE 3: DRAFTING THE CENTER LINE SECTION   235

    Students   **236**

    Homework: Copying an Existing Section   **247**

REVIEW OF CENTER LINE SECTION SYSTEM OF WORKING   248

CHAPTER 10   THE SCENIC FRONT ELEVATION   249

MODULE 1: HOW IS A FRONT ELEVATION USED?   249

    What is a Scenic Front Elevation?   **249**

    Parts of a Scenic Elevation   **250**

MODULE 2: RETURN TO THE BOX SET MODEL   251

    Exercise: Dressing the Set Model   **251**

MODULE 3: DRAFTING THE ELEVATION   251

    Composite Elevations   **254**

    Exercise: Sketching Individual Flat Elevations   **255**

    Exercise: Composite Elevations   **256**

    Exercise: SketchUp Paint Elevations   **256**

    Homework: A Different Set in the Venue   **257**

REVIEW OF SCENIC FRONT ELEVATION SYSTEM OF WORKING   259

CHAPTER 11   DETAIL PLATES   261

MODULE 1: WHAT ARE DETAIL PLATES?   261

    Discussion: Breaking Apart the Box Set   **265**

    As-Built Drawings   **267**

    Exercise: Layout and Drafting of Detail Plates   **267**

    Homework: Detail Plates for an Existing Scenic Structure   **270**

MODULE 2: MORE ABOUT WINDOWS   271

    Window Vocabulary   **271**

    Exercise: Drafting a Double-Hung Window   **273**

    Exercise: SketchUp Model of the Double-Hung Window   **275**

Exercise: Drafting a Free-Standing Palladian Window with Window Seat   **275**
Exercise: Model the Palladian Window in SketchUp   **278**

MODULE 3: STAIRS, BANISTERS, AND BALUSTRADES   279
Staircase Vocabulary   **279**
Discussion and Exercise: Staircase Math   **280**
Discussion and Exercise: Staircase Layout   **281**
Discussion and Exercise: Stairs for a Unit Set   **281**
Exercise: Stairs for a Unit Set with Curved Platforms   **284**
Exercise: Modeling Platformed Areas and Stairs in SketchUp   **284**
Exercise: Drafting Step Units   **284**
Homework: Modeling Two Step Units in SketchUp   **287**
Exercise: Drafting a Stair Unit with Banister and Landing   **287**
Discussion and Exercise: Laying out the Staircase Drawing   **288**

REVIEW OF DETAIL PLATE SYSTEM OF WORKING   292

CHAPTER 12   DRAWINGS FOR LIGHTING DESIGN   293

MODULE 1: LIGHTING VOCABULARY AND STANDARD GRAPHICS   293
Lighting Vocabulary   **296**
Basic Unit Categories and Their Symbols   **297**
The Legend   **298**
The Symbol as Drawn on the Position   **300**
Essential Elements of a Light Plot   **301**
Discussion and Exercise: Reading a Light Plot   **303**

MODULE 2: DRAFTING THE LIGHT PLOT   303
Exercise: Laying out a Position Using a Rough Plot   **308**
Exercise: Drafting from the Rough Plot in Vectorworks   **310**
Label Legends for Right and Left Facing Units   **317**
Drawing Units Placed below the Grid or Rig   **317**
Exercise: Drafting Vertical Positions   **320**
Exercise: Drafting Booms and Ladders   **322**
Other FOH Positions   **324**
Exercise: Adding Lighting Positions to the Venue Base Drawings   **324**

MODULE 3: LIGHTING SECTIONS   326
Adding Box Boom Positions to the Base Venue Section   **328**
Homework: Drafting a Lighting Section   **330**

REVIEW OF LIGHTING DRAWINGS SYSTEM OF WORKING   332

CHAPTER 13   DRAFTING FOR SOUND DESIGN   335

MODULE 1: TYPES OF DRAWINGS AND SOUND VOCABULARY   335
Sound Vocabulary   **337**
Constructing Symbols with Records in Vectorworks   **338**
Exercise: Constructing a Symbol in Vectorworks   **338**

MODULE 2: SOUND PLOTS AND SECTIONS   343
    Speaker Plots   **343**
    Sections   **349**
    Project: Drafting Speaker Plots and Sections   **349**

MODULE 3: SIGNAL FLOW DIAGRAMS AND RACK DRAWINGS   353
    Signal Flow Diagrams   **353**
    Exercise: Constructing Stock Symbols   **354**
    Homework: Drafting a Signal Flow Diagram   **358**
    Rack Drawings   **359**
    Exercise: Drafting a Rack Drawing   **361**

Bibliography   *363*
Index   *365*

PREFACE

I was introduced to mechanical drawing in the first semester of high school. Wanting to be an engineer, I took drafting every following semester. Embarking on a chemistry major in college, the next drafting course I took was as a lighting design MFA candidate. For me, it was a refresher that re-awakened my enjoyment of drafting. For my classmates, it was a noticeably rockier road. My first take-away from that experience was that one semester of drafting is seldom enough to do much more than instill a dislike of drafting. The second was that drafting is best learned when decoupled from design or technical learning; it's difficult to draft something you don't yet understand.

From my experience in working with students as they learn the fundamentals of drafting, I believe that drafting basics should be taught prior to incorporation into the design process so that students are better able to communicate their design ideas. Adding computer programs (Computer-Aided Drafting/Design, or CAD) to the mix can muddle things further, as learning software adds yet another complex layer of instruction. For a novice, the convenience of a mouse-click may supersede awareness of just what that mouse-click adds to the page.

Unfortunately, small- to mid-sized college theater programs, as well as high school programs, seldom have the curricular space to offer a stand-alone full semester drafting course, much less multi-semester study and practice. Drafting may be introduced as a unit within a scenic design or stagecraft course or touched upon as required by various assignments and projects. Even if a student becomes proficient in CAD and can turn out an intricate 3D model of a lighting rig, they may not understand how to convert that dazzling wireframe structure into working drawings useful to the electricians. Taking a drafting class elsewhere on a campus is often not an option unless the institution also has an engineering program. Students at technical and community colleges may have access to drafting courses, but no entertainment design courses that connect to that content.

In my years of teaching design and drafting, I have found the available textbooks to be excellent providers of information *about* drafting but less satisfactory in addressing *how* to draft. For example, the idea of drawing a line before you can measure out a line is not particularly out intuitive and often a conceptual hurdle for students. Granted, these texts expect the instructor to provide the linking process-information, but for someone who is new to drafting, or new to teaching drafting, it's easy to miss things a textbook assumes the instructor already knows.

Similarly, many texts make a leap between introducing the tools of drafting and a completed drawing. Providing the definition of a scenic groundplan and several examples of Broadway quality drawings is of limited help to the beginner laying out a groundplan for the first time. It's a bit like being trained on tools in a woodshop and then sent off to build a sofa without having been provided the necessary sofa-knowledge. The question remains: How do I use the given tools to create a drawing like *that*?

Because it's my feeling that a draftsperson should understand the objects they draft, there is a fair amount of stagecraft and vocabulary integrated throughout the text. The exercises and explanations of this book are not limited to scenic design students, but also meant to engage technicians, lighting designers, and sound designers.

Drafting is visual description and graphic problem-solving. The draftsperson, whether designer or technician, drafts to understand, describe, and begin the concrete realization of ideas that have, up until then, been mostly in the imagination. Drafting *should* be fiddly. It *should* take time. Paying attention to detail and striving for clarity means that the draftsperson cares about the end product: *I give you this drawing so that, together, we can make this dream a reality.*

Hand drafting is indeed falling by the wayside. Certainly, there are maddeningly repetitive aspects of drafting that computers have made much easier. However, the transition from hand drafting to CAD doesn't mean that drafting itself has become automatically easier to learn or to teach. While CAD knowledge is now an essential component of a career in entertainment technology and design, hand drafting remains an invaluable step in learning what goes on the page and why it goes on the page: after all, the draftsperson is responsible for every single line on a drawing. Besides, computers crash; licenses expire; companies are bought out; Internet connections go down; batteries die.

This book can be used as a classroom textbook, preparation tool for the instructor, or a guide for independent study. Hand drafting provides foundational, conceptual exercises. Vectorworks projects move drafting into the digital realm: same item, different way to draw it. SketchUp is used to facilitate the mental conversion of a 3D object into a 2D description of that object, and vice versa. The emphasis is on process: *how* to build a drawing.

Vectorworks and SketchUp are powerful software programs that are used throughout the entertainment industry. They are likely to remain so for quite some time. Vectorworks features a workspace called Spotlight, which has become a major CAD resource for lighting design. Through Spotlight Vectorworks links with Lightwright, a widely used lighting design database program. Information can be transferred between the graphics of Vectorworks and the spreadsheets of Lightwright, making documentation maintenance far easier than it was in hand drafting days. SketchUp is a modeling program that quickly gets the novice up and running, and aids building the visualization skills that go hand-in-hand with drafting.

This book is not meant to replace the online tutorials and webinars offered by Vectorworks and SketchUp. Indeed, those tutorials should be used to supplement the classroom experience. By the end of this course of study, the budding draftsperson should find those tutorials even *more* useful since fundamentals have been mastered and terminology has become familiar. This book is less interested in exploiting the full potential of these programs than in giving the student a place to begin.

I enjoy drafting, whether by hand or by computer. Just as in playing a musical instrument or painting a picture, there are ah-ha moments, as well as Zen-like passages of time when you look up to discover hours have flown by because the work is so engrossing. To paraphrase Martha Graham, discipline first means freedom later.

Let's get drafting.

ACKNOWLEDGMENTS

I would like to extend my appreciation and thanks to the University of Wisconsin-Whitewater (UW-W) Department of Theatre/Dance, the College of Arts and Communication, and university administration, who approved the sabbatical proposal that provided the time to write this book. A special thanks to the UW-W students who sparked this project, particularly Samuel J. Hess and Nicolas Sole, who also contributed some of their work. Gratitude also goes out to Nancy Mae, who read the first draft and provided copy-editing notes as well as feedback on the book's overall comprehensibility to someone outside the entertainment industry.

I'd also like to thank Jeremy Powell at Vectorworks and Jennifer Pratt, Michelle White, and the SketchUp team at Trimble for facilitating the permission process at their respective companies. At USITT, David Grindle and Richard Dionne made drafts of revised recommendations available and granted permission to reprint selected graphics. Thank you to Noele Stollmack, Thomas C. Umfrid, and Steve Barnes for contributing samples of their lighting and scenic work.

Ben Truppin-Brown, Matthew Tibbs, Josh Schmidt, and Beth Lake generously shared their time and expertise regarding sound design and drafting for sound design. I greatly appreciate the extra effort Ben and Matt took to help me ensure that the content of the sound drafting chapter was reflective of the current state of the field. Thanks also to Matt Hubbs and Mikhail Fiksel for permitting the use of drawings of their sound designs.

Thank you to Lucia Accorsi, who shepherded this book through the publication process at Routledge, and to Thomas C. Umfrid for his insightful, thoughtful, and corrective comments as technical reviewer. Finally, thanks to my husband Rick for his patience and support, because writing a book is like having a houseguest who just doesn't know when to leave.

INTRODUCTION

GETTING READY

TOPICS AND GOALS

For the students (but instructors are welcome):

- What is drafting?
  - Standardized graphic language
  - Architectural and mechanical drawing
- Who is drafting for?
  - Reading drawings
  - Reproduction for the end-user
- Programs, equipment, and materials

For the instructor (but students are welcome):

- In the classroom
- Evaluating drafting projects

FOR THE STUDENTS (BUT INSTRUCTORS ARE WELCOME)

What is Drafting?

Drafting is communication.

Drafting is the practice of constructing drawings that describe items to be fabricated and that guide their fabrication.

Scenic designers draft scenic designs. Lighting designers draft light plots. Technical directors draft construction drawings. Sound designers create signal flow diagrams and speaker plots. Carpenters read drawings. Electricians read light plots. Scenic artists work from paint elevations. Stage managers read groundplans to set up a rehearsal space.

Directors read groundplans to understand the space they're working with the scenic designer to develop. Producers look at drawings to see how their money is being spent. Accurate material and labor budgeting requires detailed drawings. Devices such as computer numerical control (CNC) routers and 3D printers require digital drawings in order to cut and model desired products.

The craft fields and trade unions of entertainment are often classed amongst the Arts. They are, more correctly, an industry that produces art. If a theater practitioner expects to collaborate with craftspeople and artisans in order to put a product in front of an audience, acceptance of the industrial side of things is not only preferable, but essential. Most technicians can offer anecdotes regarding drawings and diagrams delivered by artists that resulted in substantial miscommunication because the artists were speaking their own idiosyncratic art dialect while the technicians were speaking standardized technical theatre

Technical drawing, or drafting, employs a **graphic language** of standardized symbols, page formats, types of lines, and particular views of an object to convey specific information about that object. The rules, symbols, and conventions of this graphic language means that any trained reader will understand what has been drawn, or at least know what questions to ask. This language can be mastered through study, discipline, and practice. If you can draw a straight line along the edge of a ruler, or select an icon and click a mouse, competency in technical drawing is within your reach.

This graphic language emerged as Western architecture and engineering moved through the Industrial Age of the nineteenth century. There was an increasing need to produce copies of clear drawings specific to unique projects. Pictorial forms of reproduction like engraving and lithography

DOI: 10.4324/9781003154921-1

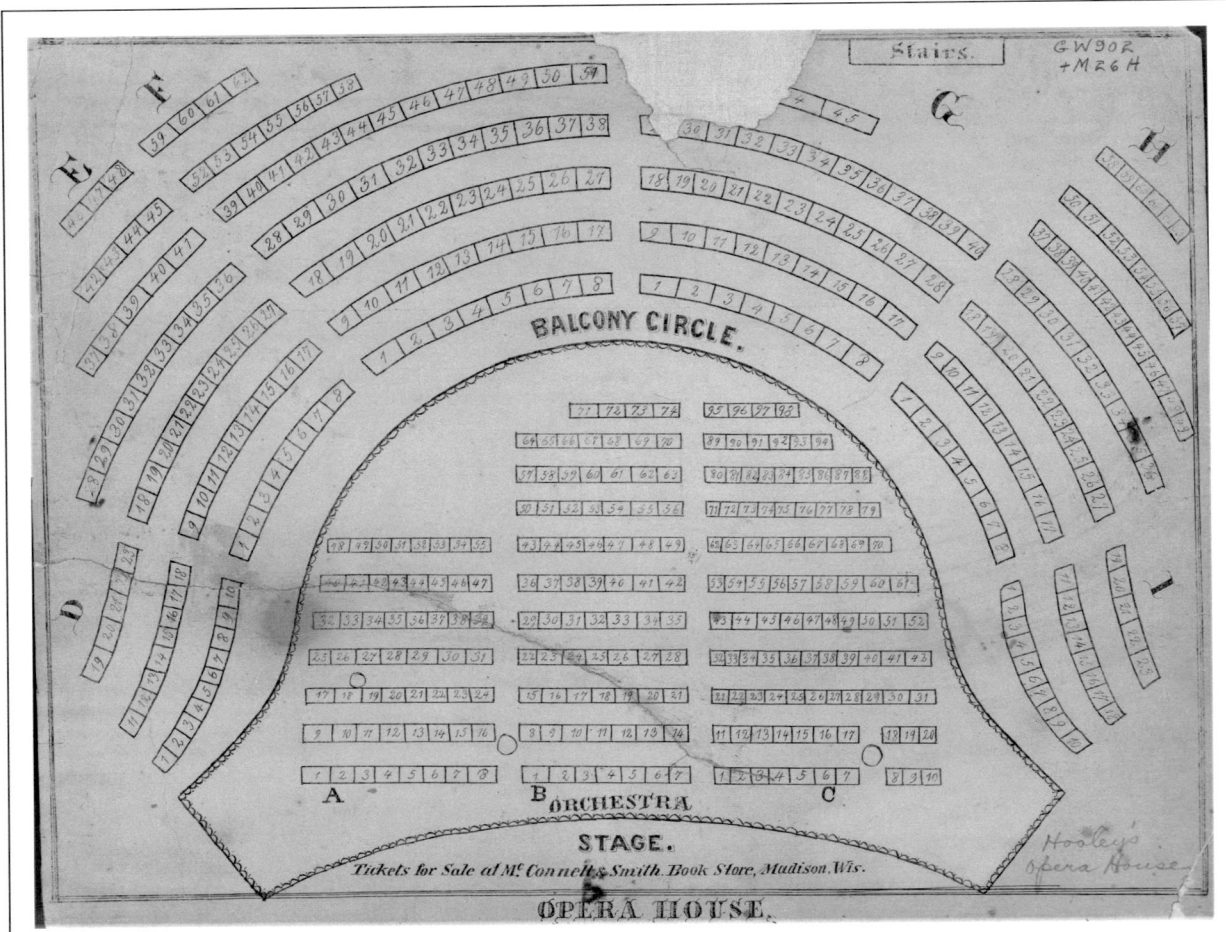

FIGURE I-I   *Lithograph of Hooley's Opera House, 1872. Used by permission: Wisconsin Historical Society, WHS-37794; color image reproduced in grayscale*

were not sufficiently exact, nor adequately time-sensitive (Figure I-1). With the introduction of the blueprinting process in the mid-1800s, artistically detailed drawings gave way to plainer schematic drawings that gave better results when reproduced.[1]

To create a **blueprint**, a drawing made on translucent paper (vellum) was placed atop a sheet of paper covered with a light-sensitive emulsion. Light shined through the vellum exposed the paper beneath. Because the drawing's black graphite lines were opaque, emulsion directly beneath them remained unexposed. When run through a chemical bath, the photosensitive paper's exposed areas turned blue or black while the unexposed lines remained white. Multiple, accurate copies could be easily made and distributed (Figure I-2).

In the twentieth century, blueprinting was superseded by the diazzo-moist and diazzo-dry processes, the dry version becoming more prevalent. These processes produced whiteprints (as opposed to blueprints), or **bluelines**.

The drawing's lines could be black, blue, or red depending on what paper was used (Figure I-3); blue was the most common. Drawings were both exposed and developed in a machine called Ozalid Streamliner, with ammonia vapors as the developing agent. Older drafting studios may still have an old blueline machine tucked away in a corner and a stack of photo-sensitive paper moldering away in a flat file drawer. By the end of the twentieth century, both blueprint and whiteprint reproduction methods were made obsolete by photographic and digital processes.[2]

Because of the need for readily available interchangeable parts, industrial processes themselves required more and more standardization. North American groups like the American Society of Mechanical Engineers (ASME) began developing and disseminating industrial standards, as well as the graphic standards to describe those industrial products.[3] ASME and the American National Standards Institute (ANSI) continue to establish and maintain standards

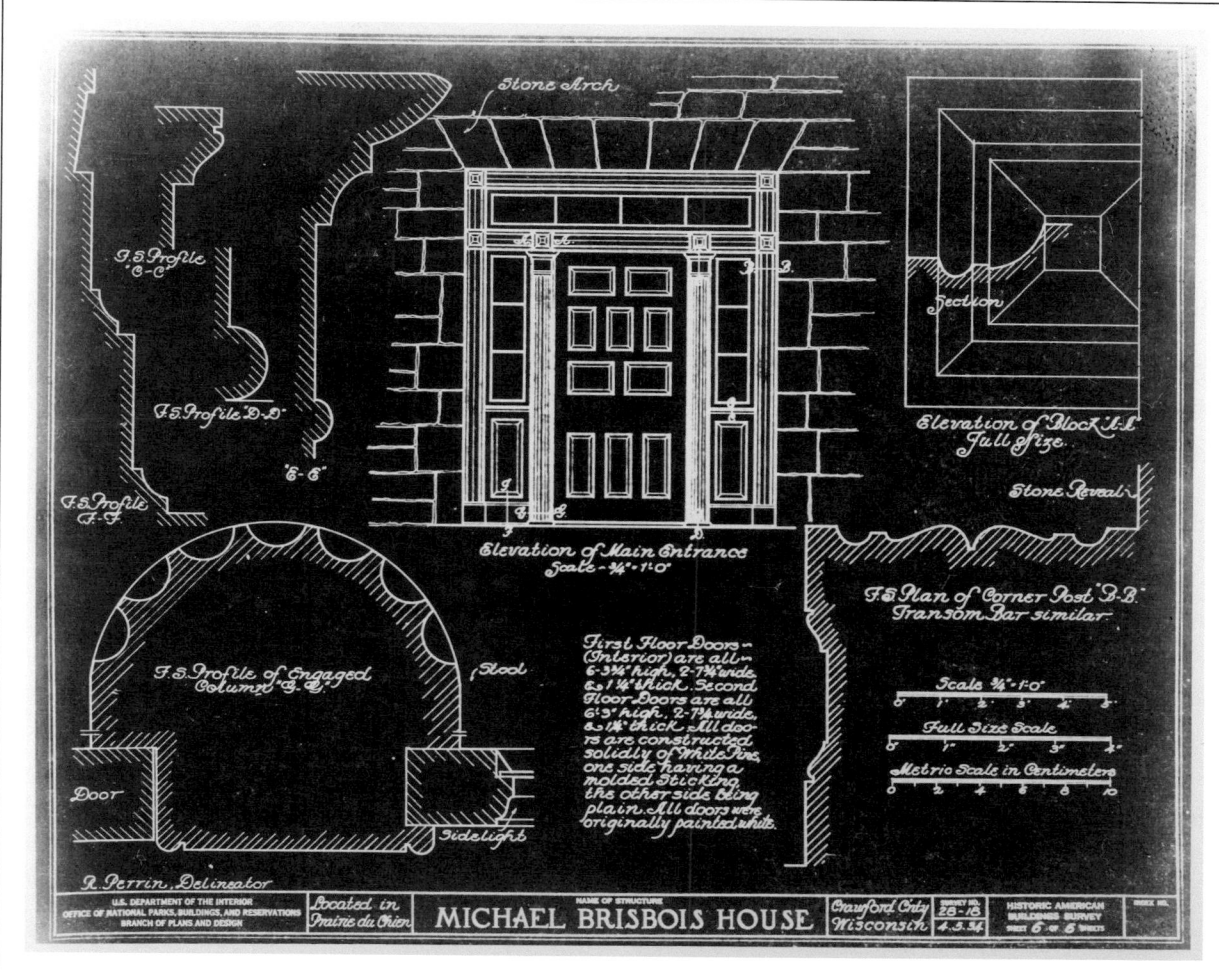

FIGURE I-2   *Blueprint of Brisbois House, 1934. Used by permission: Wisconsin Historical Society, WHS-42449*

throughout business and industry. The International Organization for Standardization (ISO) sets worldwide standards. When making selections in CAD programs, the draftsperson will often run across those organizations' acronyms, among others. Just as theatrical rigging is a descendant of nautical practices, entertainment drafting has evolved and borrowed from mainstream industrial and architectural practices.

These organizations can be found via their websites:

American Society of Mechanical Engineers: www.asme.org

American National Standards Institute: www.ansi.org

United States Institute for Theatre Technology: www.usitt.org

International Organization for Standardization: www.iso.org

The United States Institute for Theatre Technology (USITT) is the organization that first developed standards for use in the North American entertainment industry, often drawing upon practices already established by bodies like ASME and ANSI. For their 2020 updates, USITT moved away from referring to "standards," preferring "recommended best practices" to better embrace the variety of fields within the entertainment industry and be CAD platform-neutral. The current recommendations use as their baseline the *USITT 1992 Graphic Standard* and the *United States National CAD Standard (USNCS_Version4.)*[4] In Vectorworks and other CAD programs, menu selections offer a variety of industrial standards. Choose the ones that most closely align with USITT recommendations.

There are two main, yet interrelated, branches of technical drawing: **mechanical drawing** and **architectural drawing**. In entertainment practice this roughly divides between lighting, sound, video, and technical direction (mechanical), and scenic design and props

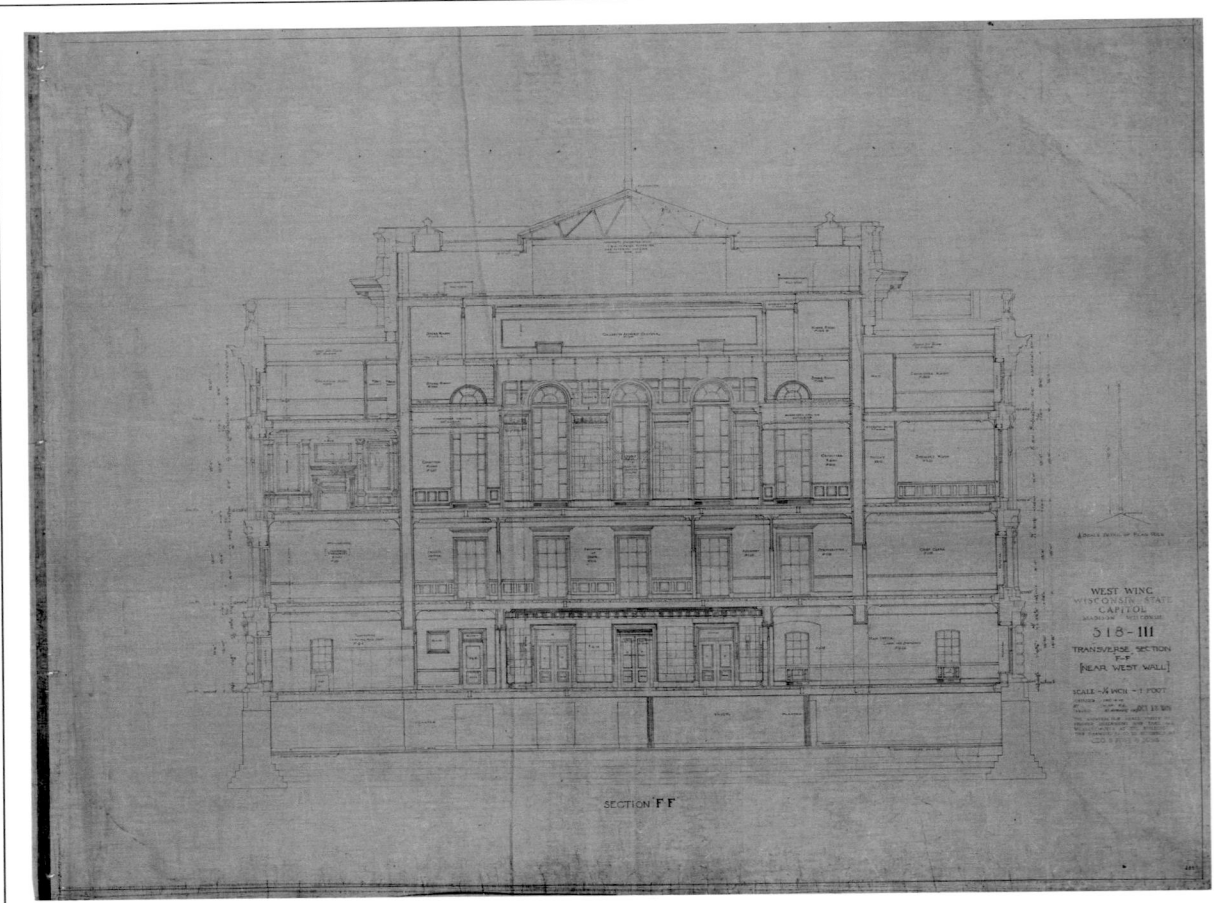

FIGURE I-3   *Blueline of west wing transverse section of Wisconsin State Capitol, 1907. Used by Permission: Wisconsin Historical Society, WHS-1306094; color image reproduced in grayscale*

(architectural). Mechanical drawing is diagrammatic, offering true measurements in scale along with methods of assembly. Architectural drawing leans toward the pictorial, offering images of how things are seen by the eye or creating images with atmosphere and mood. Costume design can also be considered this way: costume renderings are architectural, patterning is mechanical.

While architectural drawings can be attractive and essential in explaining a project, accuracy is often not their strong point. Historically, it has been not uncommon for a scene shop to receive gorgeous plates of pictorial hand drafting from a scenic designer from which it is virtually impossible to lift an accurate measurement with a scale ruler. It works the other way, too—the technical director's aesthetic imagination must enable their construction of schematic drawings that maintain the spirit of the designer's conception.

This book focuses on mechanical drawing, as the skills acquired can readily be applied to future pictorial work. Both SketchUp and Vectorworks allow work to be viewed

and printed in a variety of projection systems, including perspective (see Chapter 2). In SketchUp's default mode, perspective images of the model are automatically generated; the modeler does not need to know how to draw in perspective. This book will leave perspective drawing training to other courses and texts.

Finally, the exercises and discussions in this book break the construction of a mechanical drawing into four broad phases:

- Understanding the object to be drawn, including the object's parts and the names of its parts

- Visualizing the object as both a 3D object described via the system of 2D multi-view orthographic projection (again, more about that in Chapter 2)

- Understanding and applying the graphic standards and recommendations typical to 2D description

- Employing an organized, systematic approach to lay out the drawing

Who is Drafting For?

All drafting is done for the reader.

If the person to whom you hand the drawing cannot decipher the information you have placed there, the drawing has failed. **Clarity** is the force that drives the placement of notes and dimensions, which line-type to use, what views of an object to present, working within the borders on the page, selecting which details to present in their own break-out drawings, even the layout and content of title information. There are always questions to be asked and answered, and in entertainment, there are many unique and uniquely complex items to be designed and built. A maximum effort at clear communication needs to happen on the page *first* so that appropriate follow-up questions can be formulated.

The reader is also expected to have an understanding of drafting's graphic language and be able to mentally convert 2D images into a visualized 3D object. Just as creating a drawing is a learned skill, reading a drawing is a learned skill. When a director responds to a drawing with "I don't understand what's going on here," it may be less concern about the actual design and more about not having the visual training to interpret the drawing. Similarly, a novice carpenter needs to learn to read a construction drawing. It's not automatic.

Readability is not just about what is drawn, but also how a drawing is reproduced. In the days of blueprinting and bluelining, the reproduction of a drawing was automatically the same size as the original. Regardless of whether a photocopier or scanner is set for 100% sized reproduction, it's still taking a photograph that slightly changes the size of the reproduction. How much do you trust the calibration of your home printer?

Printing drawings directly from a digital file may seem like accuracy is assured, but the draftsperson must check settings both in the program and at the printer. "Shrink to Fit" is a dangerous printing option. Because the paper size has changed, the indicated scale of the drawing is no longer accurate. A reader who does not realize there is a slight shift in scale may use their scale ruler to lift inaccurate measurements from the drawing, to the detriment of the project as a whole.

Light plots in particular have suffered at the hands of digital printing. Traditionally, light plots are drafted in ½″ = 1′–0″ scale (more on scale in Chapter 1) and broken into multiple sheets of paper, or plates. Because digital programs make it easier to lay out the plan of an entire venue in the virtual drawing space, there is the temptation to print the whole thing in a size that fits onto a single sheet of paper. By shrinking the drawing, numbers become smaller, symbols become smaller, and there's less space to add notes. The light plot is a working drawing consulted by a group of people in a semi-lit venue and will be marked up with circuiting notes, draped over the top rung of a ladder, or carried into even darker catwalks. Convenience for the draftsperson can be a limiting factor for the technician.

Good drafting is part of being a responsible designer. Don't give technicians impossible tasks because you haven't paid attention to the details. It not only slows down the production and raises costs but also can put technicians and performers in physical danger. Know the venue's ceiling height. Think about escape stairs. Don't place lighting units where there is no viable position or viable way to create a lighting position. Drafting means engaging with the physical aspects of the design and taking the first step of turning theoretical into practical. If it can't be built within the time, budget, and labor allocations of the production, the envelope-pushing dream is of little use to anyone but the designer.

Be sensitive to the needs and limitations of those who will be working from your drawings. It is possible to be both brilliant *and* a joy to work with.

Equipment and Materials

The software discussed and demonstrated in this book are Vectorworks versions from 2019 to 2021 and SketchUp's 2021 version for the PC via Microsoft Windows. While the appearance of the screens and placement of toolbars will differ somewhat from PC to Mac, the overall functioning of the programs remains the same. If Mac is the prevalent classroom technology, the instructor should prepare handouts formatted for platform differences.

Vectorworks is a vector-based drafting program. This means that rather than thinking pixel-by-pixel as a raster-based program would (such as Adobe Photoshop), Vectorworks focuses on the beginnings, ends, and intersections of lines rather than all of the points in between. To over-simplify the matter, as you zoom in and out, the computer's processor has less processing to do since it's keeping track of a smaller number of points.[5] However, this also means that an action such as erasing is not as straightforward as it would be in a Paint-style program. Rather than erasing part of a line with an erase tool, a line segment with terminations (whether endpoint or intersection) needs to be selected and then "trimmed" or "cut." When looking at tool icons in Vectorworks, you won't find one featuring the picture of a small eraser, but rather, scissors. Icon nomenclature reflects how a program approaches the manipulation of data.

Vectorworks also includes powerful 3D modeling and rendering capabilities. In advanced usage, the draftsperson can model an object and then extract the required drawings by importing views onto designated printable sheets. This book, though, focuses on using Vectorworks for 2D drafting. Building the model of a room usually requires generating a floorplan first, so we'll walk before we run.

SketchUp is a modeling program that is closer to drawing than drafting. As a drawing program, it does not readily incorporate the standard graphic terminology of technical drawing. However, its relative ease of use means that designers can quickly construct and manipulate virtual models. This functionality is very handy in production meetings, as a designer can swiftly alter a model as the discussion happens and present alternatives within minutes. SketchUp features an adjacent program called LayOut, which can be used to create 2D drawings. At the time of writing, most large scene shops prefer CAD drawings. LayOut and SketchUp are currently less expensive programs than Vectorworks, and smaller theater companies may be perfectly happy to accept LayOut drawings.

For all models, the level of detail displayed is the level of detail the user puts in. If a doorknob is to look like a doorknob, the user must build or import the desired virtual doorknob. SketchUp offers an online warehouse of objects users around the world have built and uploaded. While handy, it must be remembered that you're more likely to find an object *close* to what you're looking for rather than *exactly* what you're looking for. When sharing a model with the production team, you run the risk of sowing false impressions with placeholder items. Be sure to factor in enough time to model the details. While virtual models have become more predominant and are certainly useful in the design and production process, whenever possible a physical model should be constructed. Interaction with an actual 3D object provides spatial relationship information that a 3D model displayed on a 2D screen still cannot. Use the physical model to inform the virtual model, and vice versa.

For CAD (and for any digital work), saving files frequently and keeping backup copies is imperative, as it may be necessary to revert to a point before an unrectifiable mistake occurred. Have a way to move files from workstation to workstation whether via cloud storage, USB drives, or sending yourself emails with attachments. Digital files may need to be physically transported to a commercial print shop if no large format plotter is readily available.

For both Vectorworks and SketchUp, large monitor screens make it easier to work. A bonus of drafting on paper is that the entirety of a large object can be seen in consistent scale across the page. On a monitor, when you zoom out to see the whole of a large object, detail recedes and is lost. When you zoom in to see detail, a sense of the whole is lost. A large monitor allows the draftsperson to see more of the drawing in a more readable size more of the time. If you are using this text as an e-book, a second screen is recommended—one to follow the text, and one to do the exercise.

Since this book approaches hand drafting more as a way to understand layout and concepts rather than as an end

in itself, there will be less emphasis on finesse when hand drafting. You'll still need to procure items such as triangles and a scale ruler, but using a 2H drawing pencil rather than a lead-holder style mechanical pencil is acceptable. Drafting vellum is expensive; smooth white bond paper is sufficient for most of this book's hand drafting projects. There are times that vellum's transparency is useful; those instances will be pointed out and the instructor can decide whether purchasing vellum is worth the expense. Use graph paper for sketching and layout planning. Lighting students should purchase a lighting symbol drafting template even if they intend to never hand draft a light plot again. Plastic lighting templates remain useful for making on-site revisions to printed paperwork. The time may be near when a plot can be distributed to technicians via tablets and phones—however, paper doesn't shatter when dropped from the top of a ladder, nor is it so easy to lose a paper plot amid a table of equipment.

Art supply companies carry some drafting items; office supply vendors carry others. Specialized items such as lighting templates may require a little more searching; lighting equipment supply companies can usually either order them or direct you toward a source. Allow ample time between ordering an item online and expecting to use it in class.

The following is a list of the basic equipment and materials a student will need to engage with this book's various hand drafting projects.

Architectural Scale Ruler

There are multiple types of scale ruler. An **architectural** scale ruler is required, even though the overall style of drafting in this book leans toward the mechanical. These scale rulers feature the common range of American Standard scales (Figure I-4). There is a *big* difference between architectural scale rulers and other available types.

Mechanical Engineering and Civil Engineering scale rulers divide inches into decimal divisions, such as 10, 20, 30, or more parts per inch, rather than using the standard parts of the inch to represent scale sizes (¼" on the paper representing 1'–0" in reality). Mechanical engineers need a high degree of accuracy when drafting things like machine components. Civil engineers lay out very large objects such as road projects and need rulers that can convert hundreds of feet into the short distances on a page.

The United States uses the Imperial ("Imperial" being a throwback term to the British Empire) system of measurement, where 12 inches equal 1 foot, 3 feet equal 1 yard, 5,280 feet equal 1 mile. Imperial scale rulers note scale as a formula: ½" = 1'–0": one-half inch equals one foot zero inches. The rest of the world uses the decimal-based metric system, where 10 millimeters (mm) equals 1 centimeter (cm), 100 cm equals 1 meter (m), 1000 m equals 1 kilometer (km).

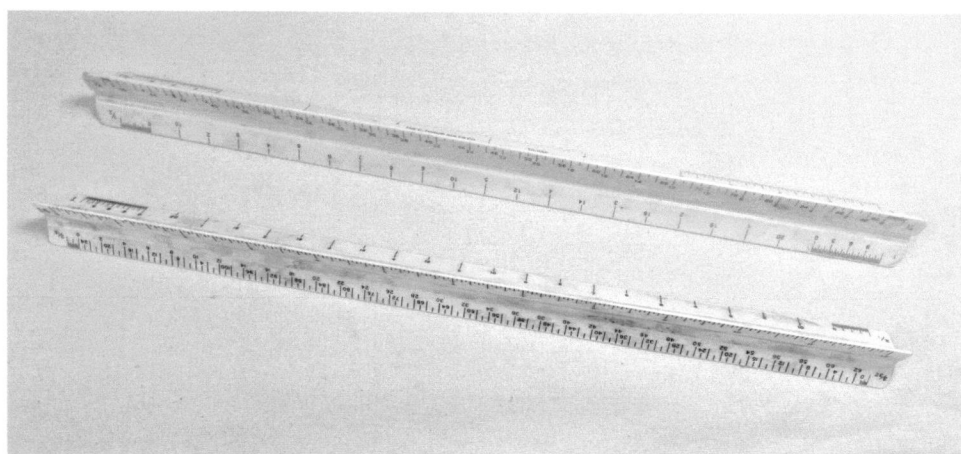

FIGURE I-4   *Two architectural scale rulers, turned to show various available scales*

Metric practices are written as ratios. 1:1 means full size; 1:2 means half size. The metric scale of 1:20 is the most common metric scale used for scenic drawings, and is a bit larger than the Imperial Standard of ½″ = 1′–0″[6] (which appears as 1:24 when converted to a ratio). Vectorworks and SketchUp can convert measurements between Imperial and metric, which comes in handy when shipping drawings beyond US borders (see Figure I-5).

The scale ruler is not a drawing tool. It is a *measuring* tool. To maintain its accuracy over time, *do not draw lines with it!* Wear and tear, not to mention graphite build up, degrades the markings.

Drafting Board or Table

If all work will be completed in a classroom equipped with drawing/drafting tables, a portable drawing board is not necessary. Portable drawing surfaces with attached parallel rules are available, which replace the need for a T-square (see Figure I-6). Affixing a drawing mat or heavyweight piece of paper to the hard surface of the drawing table provides a bit of give, allowing for more pencil control.

T-Square

The T-square (Figure I-7) must be long enough to reach across the whole width of larger-sized sheets of paper when affixed to a drawing surface. A 24″ T-square is a solid choice. Longer T-squares tend to be harder to control, while shorter ones don't allow access to the whole width of the paper. Better T-squares feature strips of transparent plastic along the drawing edges (the blade) so the draftsperson can better see what they're doing.

Triangles

Two triangles are required (Figure I-7): a 45° triangle and a 30°/60° triangle. Both feature a 90° angle in one corner. Triangles come in a range of sizes: too small and they're rather useless, too large and they're difficult to manipulate. Triangles with at least one 8″ side are a good bet. Clear, uncolored plastic is preferable so you can easily see lines beneath.

Circle Template(s)

Several are usually necessary, as the larger the circle, the fewer that fit onto a single template. A compass may still be required for arcs and circles with substantial diameters (Figure I-8).

Bow Compass (Optional)

Bow compasses open and close by rotating a threaded rod that joins its two arms. This means that once set to a distance, the arms will not slide or shift (Figure I-8).

French Curve (Optional)

This tool aids in drawing irregular curves. A curve is lightly drawn freehand, and then segments of the line are matched against parts of the French curve when the line is to be darkened (Figure I-8).

Protractor

While triangles allow the draftsperson to easily draw lines at angles of 90°, 75°, 60°, 45°, 30°, and 15°, for any other angle, a protractor will be needed (Figure I-8).

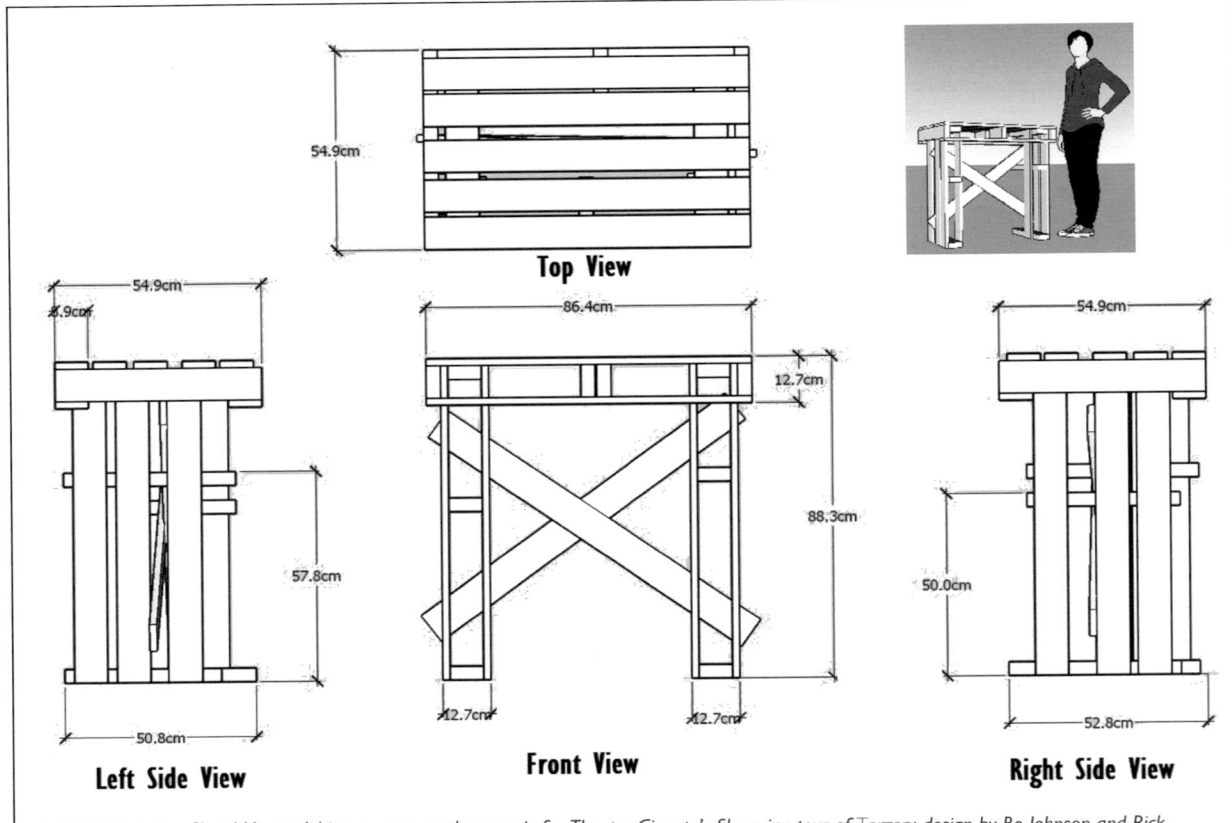

Top View

54.9cm

54.9cm

6.9cm

Left Side View

57.8cm

50.8cm

86.4cm

12.7cm

88.3cm

12.7cm   12.7cm

Front View

54.9cm

50.0cm

52.8cm

Right Side View

FIGURE I-5   *SketchUp model image converted to metric for Theater Gigante's Slovenian tour of* Tarzan*; design by Bo Johnson and Rick Graham, drawing by author*

Lighting Symbol Drafting Template

This is essential for hand drafting a light plot, as well as for making quick revisions on printed plots (Figure I-9). Be sure that the template is in ½″ = 1′–0″. ¼″ = 1′–0″ templates are available, but are less technician-friendly. Hand-me-down templates can save money, but lighting technology changes so quickly that older templates may not include symbols that represent contemporary lighting fixtures.

Pencils

A variety of graphite (or, lead) hardnesses are desirable. 4H, 2H, and H are recommended (see Figure I-10). Most art supply stores carry pencils individually and as sets. Soft leads (H) create thick black lines easily, but also smudge easily. Hard leads (4H) create easily erasable guidelines and thin, tightly controlled black lines. Practice will determine which hardness to choose and how much pressure to apply in order to draw a desired line style. Avoid click-style plastic mechanical pencils,

FIGURE I-6   *Portable drafting board with attached parallel rule*

FIGURE I-7   *24″ T-square with 45° and 30°/60° triangles*

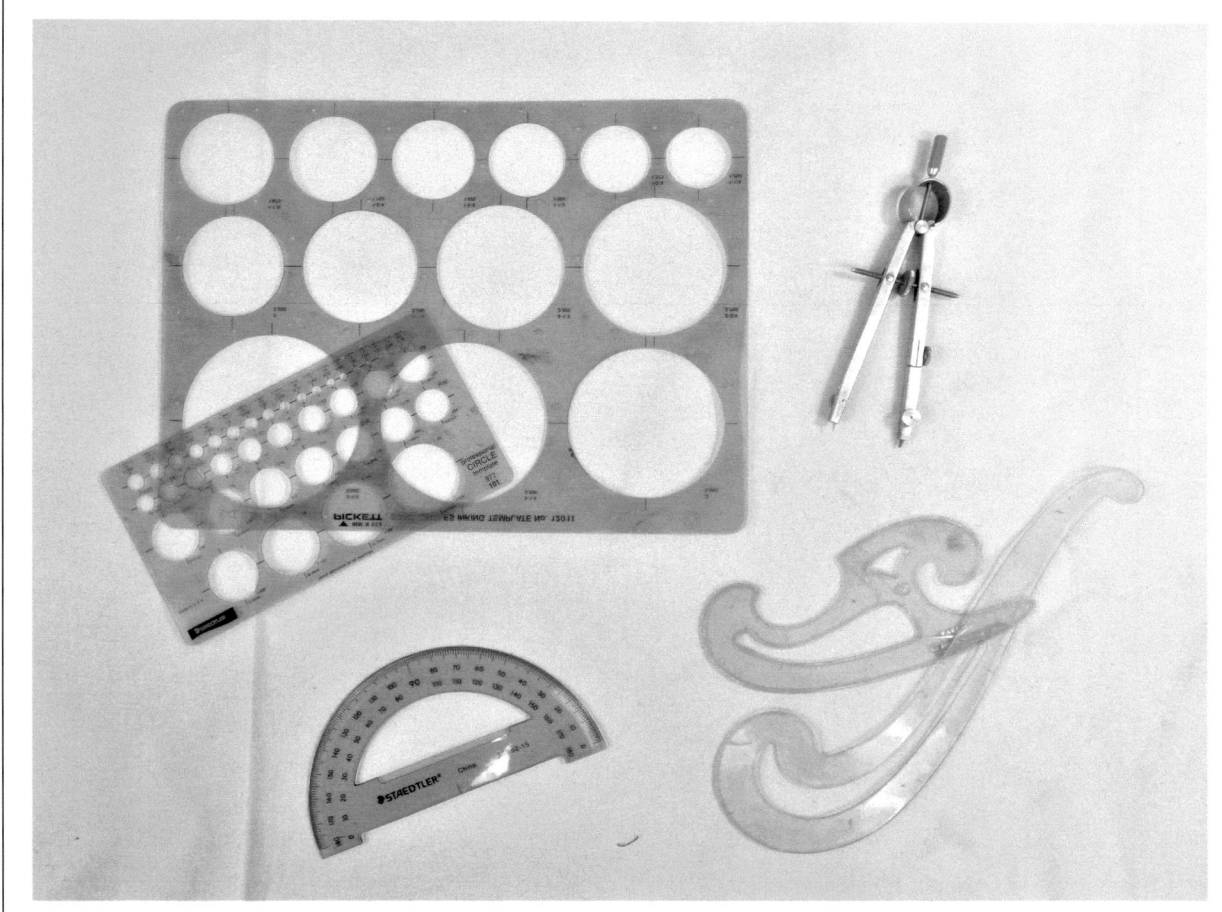

FIGURE I-8 *Circle templates, bow compass, French curves, and protractor*

as the graphite is too slender to survive much pressure. If a pencil can't be pressed hard enough to produce a firm black line, it is of limited use. Hand drafting does not generally use leads softer than H because they smudge too easily.

A Good Pencil Sharpener

It's difficult to draw a clean, precise line without a sharp point. Have a sharpener at your drawing table, since sharpening should be a frequent activity. Electric sharpeners tend to produce finer, more uniform points than small plastic hand-held pointer types. The softer the graphite, the more quickly the point will dull. If you're using a fancy lead-holder style pencil, a special lead-pointer is required (Figure I-10).

Eraser

White vinyl pencil erasers (Figure I-10) are made for drafting, and erase lines without leaving colored gummy residue.

Eraser Shield (Optional)

This s a small metal template with openings of various shapes and sizes that allows highly selective erasing (Figure I-10).

Dry Erase Pad/Canister of Eraser Crumbs (Optional)

The dry erase pad is a loosely woven bag filled with eraser crumbs (Figure I-10). They are sprinkled over the paper before drawing begins and replenished throughout. The granules allow the T-square and triangles to glide over the surface of the paper. This prevents smudging and helps keep the drawing clean. Deliberately covering the paper with eraser crumbs often runs counter to ingrained habits; resist the urge to sweep the paper clean after every erasure. You can also purchase eraser crumbs in canisters.

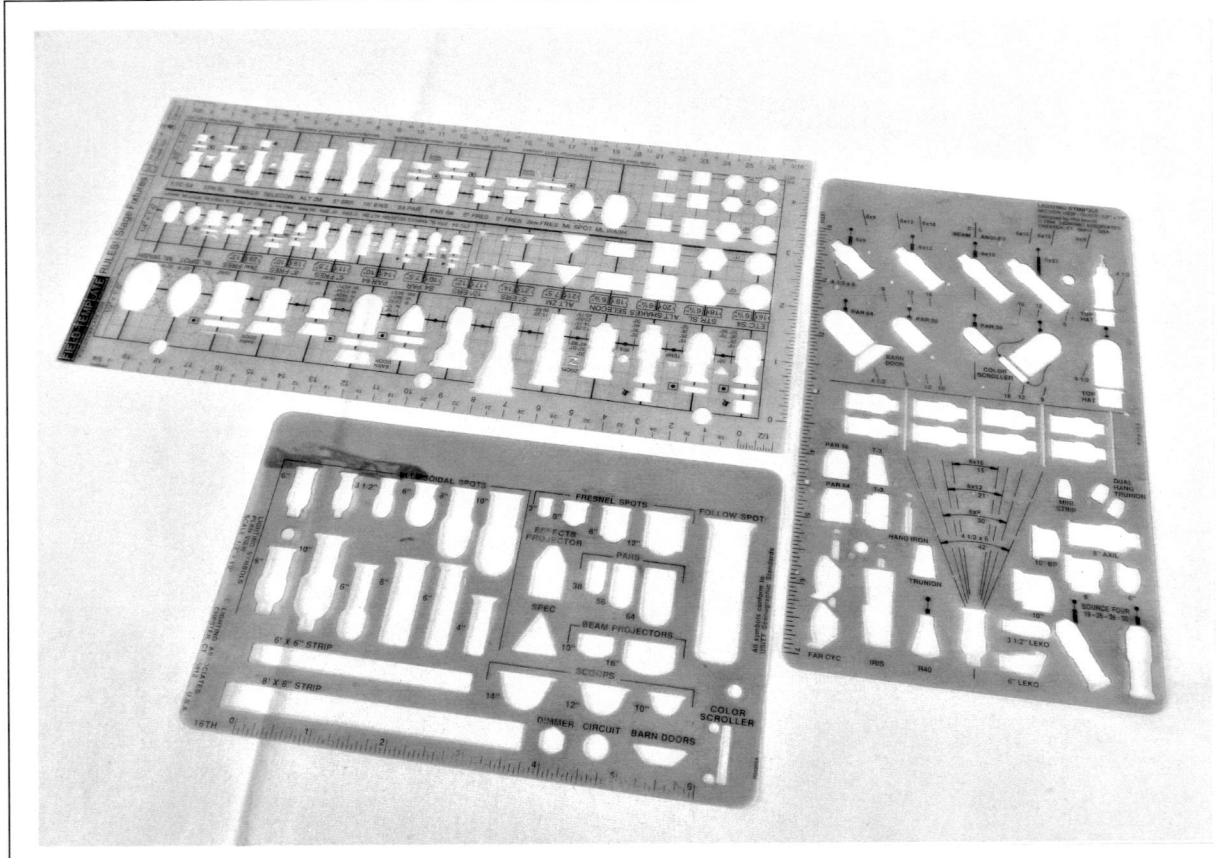

FIGURE I-9 *Lighting symbol drafting templates*

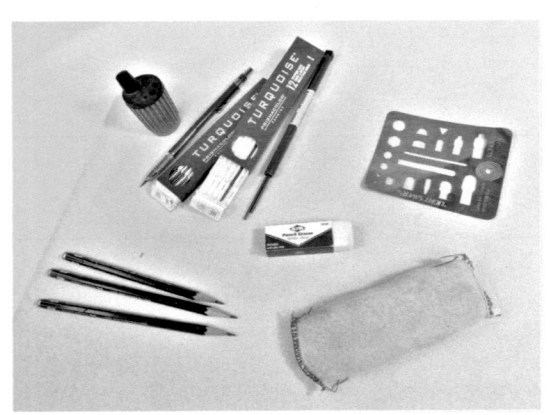

FIGURE I-10 *Lead pointer, lead holder, 2 mm drawing leads, white vinyl eraser, lighting design erasing shield, drawing pencils of various hardnesses, and a dry erase pad*

Brush

Rather than using your hand to sweep detritus from the paper, use a brush to prevent smudging. A cheap, brand-new paint brush is perfectly adequate, though drafting brushes are available (Figure I-11).

Tape

A tape is needed that will pull up easily and not rip corners off of drawings. Drafting tapes (and drafting dots) are available, but tend to be expensive. Blue painter's tape is easier to find and works quite well. If you use ordinary masking tape, stick down and pull up the strip of tape on a not-terribly-clean surface several times before using it to affix the drawing. This dirties the adhesive so that it's less likely to rip the paper when removed (Figure I-11).

FIGURE I-11 *New paint brush, drafting brush, and blue painter's tape*

Paper

Since the transparency of vellum is not a requirement, a smooth bond printer-style paper will work well enough for most projects in this book. The first projects start small and can be accomplished on 8½″ × 11″ copy paper. Later projects will require 11″ × 17″ and 24″ × 36″ sheets. Paper (and vellum), unless it comes on a roll, should be stored flat until used. 18″ × 24″ and 24″ × 30″ bond paper can be found in tablets at most art supply stores.

Vellum (Optional)

Vellum is the optimal drafting surface and allows for finer pencil work. If there is interest in exploring heightened degrees of line control, an in-class comparison between copy paper and vellum can be illuminating. The transparency of vellum is particularly useful when transferring information between drawings. Vellum in sheets or rolls can usually be ordered from art supply companies (Clearprint is a major brand). Sheets of vellum often include preprinted borders and title blocks.

Finally, *write in this book*. Make notes to remember shortcuts, tools, and modes that either make a process easier or go beyond what is covered in this book. Clarify things for yourself based on class discussion. Keep track of discoveries to turn this book into a growing, personalized reference tool as you continue to develop your drafting skills into the future.

FOR THE INSTRUCTOR (BUT STUDENTS ARE WELCOME)

In the Classroom

This book assumes that besides having the requisite number of drafting/drawing surfaces, the instructor also has access to a classroom computer terminal and projection screen also.

The room should have enough terminals for the number of students in the class (unless they're working with their own laptops) as well as space to lay out drawings and other materials next to their terminals. The instructor should be able to circulate amongst the students as they work in class. Bookstands are useful to hold up project materials beside monitors. Depending on a student's motor-control abilities, a digitizing pad and stylus may be preferred to a mouse. Most basic commands in both SketchUp and Vectorworks have keystroke equivalents.

There are other drafting programs—AutoCAD is the earlier major drafting program still favored by many technical directors and manufacturers. User-friendly modeling programs proliferate, especially for use in interior design and home remodeling. If processing capability is an issue, less demanding modeling programs can still be effective visualization tools. Projects and exercises can be translated into different CAD programs, but require the instructor to assemble their own handouts and examples. Regardless of what software is used, keep in mind that it will inevitably be updated and may require a period of refamiliarization before the user gets back up to speed. There is a change in some of the Vectorworks graphics partway through this book because updates occurred during the book's preparation. Both newer and older graphics have been retained to aid the reader in interpreting future icon and screen layout changes.

Should there be limited institutional support for technology, educational versions may be available for student use on personal devices. See Vectorworks's and SketchUp's websites for details. Personal licenses can be a major hit to the wallet; if students intend to use these programs in a freelance career, be sure to discuss the costs of maintaining a license on a home computer, as well as the logistics of printing drawings larger than standard 8½″ × 11″ copy paper.

> Further information on these programs can be found on these websites:
>
> Vectorworks: www.vectorworks.net
> SketchUp: www.sketchup.com
> AutoCAD: www.autodesk.com

The modules of this book will generally engage one topic at a time. Exercises and tutorials immediately put those topics into practice, with explanations integrated throughout. It is up to the instructor whether to review embedded content as part of the day's introductory discussion, or as pauses during the execution of in-class work. Each module is organized for classroom use rather than for classroom reference, and they are designed to be encountered in order. Extra projects are offered as possible homework and can serve as templates for the instructor to create assignments more specific to the needs of that particular group of students.

The first chapters feature exercises and tutorials with explicit step-by-step instructions. As students' familiarity with concepts and tool usage grows, exercises more often state the end result, expecting the draftsperson to figure out how to get there. The "harder" way to do something will be explored first so that students understand why a shortcut is indeed a shortcut.

The steps outlined for Vectorworks exercises purposefully reflect and adapt a traditional hand drafting approach to layout and tool usage. As with most software programs, there are multiple effective ways to achieve a certain result. The deeper a draftsperson explores Vectorworks (or any other CAD program) many highly efficient ways to construct a drawing will be discovered. However, these advanced tools are often not particularly user-friendly or comprehensible unless the draftsperson has a solid idea of what drawing construction goal these tools might accomplish.

Having step-by-step instructions does not mean that demonstrations are any less important. Check in with individual students periodically throughout the class session. There will be need for individual intervention as well as need to pause the whole class to offer reminders or corrections. Instructors should not forget to tell students when something is right on track and looking good.

While engaged with Vectorworks and SketchUp, students *will* find new and novel ways to wander into unfamiliar territory. Sometimes new and interesting tools will be discovered; at other times a drawing may drift into intractable confusion. Troubleshooting during class can be a useful side-topic and provide the encouraging nudge sometimes required to get students to open and navigate help menus.

Students progress at different paces, especially on in-class work. The instructor will discover early which students are slow but solid and which ones jump ahead at jack-rabbit speed. While swiftness is a desirable trait, speed in execution must remain linked to accuracy and neatness. If an exercise is completed quickly but sloppily, the student should be told to redo it from the start, to slow down, to pay closer attention to detail. An additional selection of smaller projects on the day's topic can provide further practice for the speedy student without causing the other students to miss out on new information. Work not completed by the end of a session should be due when the class next meets, so everyone can move forward together.

It is useful to separate in-class projects from homework. The work done in the classroom with the instructor present should serve as the launch pad for students' independent endeavors. Graded assignments should be reflective of the student's level of mastery without immediate supervision. There are occasions where in-class homework workdays are important, especially for large projects, but these should be scheduled judiciously. Discuss obstacles that resist solo problem solving in the next class session. More than one student may have run into it and all might benefit from a group effort to solve the issue.

As the semester progresses and projects become larger and more complex, be sure to check in with the students to make sure that projects can be completed within the expected time frame. Not only will the pace of the individual student come into play, but other course and production work can be a major time-crunch factor. Deadlines are important, especially in entertainment. Opening night remains opening night, and if the light plot is two weeks late, the whole production is in serious trouble.

On the other hand, in an educational environment, being able to complete a paper project and turn it in a little bit late is sometimes better than turning in panicky, unfinished work on time. Speed will come with proficiency; proficiency is fostered through patience. Certainly, it does not help the instructor to have a variety of projects coming in at random times, but canvassing the class to establish regular, reasonable, and reasonably firm due dates (and then applying judicious late penalties) can preserve discipline while preventing the course from turning into a grueling death-march.

Evaluating Drafting Projects

Feedback is vitally important, and the format of the project will channel the form of the feedback. Rubrics are useful, but better still to mark up the drawing, point out errors, make suggestions, and indicate where the student did well. When students first attempt to control their line weights in hand drafting, there's usually a short segment somewhere on the drawing where a consistently thick, consistently black line was produced; highlighting this indicates that this goal is achievable and gives them a visual toward which to aim.

Be sure that grading addresses the primary goal of that project. If the goal is to practice accurate measurement, that is more important than evaluating line weight on that assignment. Comment on everything that needs to be commented upon, but grade the project's primary topic.

Things to expect and look for on any drawing, regardless of method of production, include:

- Evidence of proper tool use, whether by hand or by computer

- Use of guidelines for layout

- Consistent and appropriate use of line styles

- Accuracy of measurement

- Appropriate use and placement of dimension notation

- Appropriate use and placement of labels and other notation

- Good overall layout, with effective use of white space

When setting up the course, decide how digital projects are to be submitted for evaluation.

Trying to evaluate the full Vectorworks file of a project presents the instructor with a certain amount of detective work. Is it the *process* of creating the drawing that is being evaluated, or is it the *drawing*? How does the instructor excavate the multitude of steps it took to create the drawing? The printed product will usually present plenty of clues as to whether confusion occurred. Process is usually better subjected to live observation and intervention in the classroom. Asking the student to demonstrate how they built that curious part of the drawing will then reveal which tools and steps need to be clarified.

For SketchUp, seeing the virtual model in its virtual environment allows the evaluator to move about the space and see the nuts and bolts of the work from various directions and distances. The viewer can zoom in to examine details that would not be evident unless multiple views of the model are printed out in an exceptionally large format.

A nice feature of digital work is that marking up a drawing does not damage the original. Depending on the PDF reader on the instructor's computer, comments may be written directly on an image. Converting a PDF into a JPEG file through a paint or illustrator program allows the instructor to place color-coded notes anywhere on the drawing. The student can then make revisions based on the instructor's comments should the drawing be intended as a portfolio piece, or if the instructor expects it to be redone. For hand drafting, using sticky notes for comments is one way to minimize impact on the original.

NOTES

1. Frederick E. Giesecke, Alva Mitchell, and Henry Cecil Spender, *Technical Drawing*, 4th ed. (New York, NY: The MacMillan Company, 1958), p. 763.
2. Giesecke, Mitchell, and Spencer, p. 766–777.
3. Giesecke, Mitchell, and Spencer, p. 6.
4. USITT Graphic Recommended Best Practices, 2021.
5. John Holloway, *Illustrated Theatre Production Guide*, 2nd ed. (New York, NY and London: Focal Press, 2010), p. 175.
6. Patricia Woodbridge and Hal Tine, *Designer Drafting and Visualizing for the Entertainment World*, 2nd ed. (New York, NY and London: Focal Press, 2013), p. 44.

CHAPTER 1

WHERE TO BEGIN

On the first day of class, after the syllabus and expectations have been reviewed, examples of materials and tools shown, and the initial concepts of drafting as presented in the Introduction have been discussed, there are multiple reasons lettering is a good place to begin the study of drafting.

First, it's unlikely that everyone will have already procured T-squares, triangles, and other required materials. An architectural scale ruler is required for the second module of this chapter. T-squares and triangles are required for the third. Lettering exercises can be done with little more than No. 2 pencils and handouts.

Second, lettering introduces concepts that reverberate throughout the rest of the course: clarity, intention, neatness, standardization, the use of guidelines, and layout.

Finally, while many students will be frustrated in their first attempts at lettering, there is usually enough improvement by the end of the session to demonstrate the principle that information applied to practice bears fruit.

The second module of this chapter addresses scale, the use of the scale ruler, and how to notate measurement. The third module goes through the setup of the drawing space, both for hand drafting and for Vectorworks.

TOPICS AND GOALS

- Introduction to standardized practices
- Lettering
- Using the scale ruler
- Introduction to commonly used scales
- Measurement notation
- Setting up the drawing space

MODULE 1: LETTERING

Lettering is *not* the same as handwriting or penmanship. It is useful to think of lettering as drawings of letters. As letters are symbols with meaning placed on the page, and drafting aims toward universal readability, there is a standard method to draw letters. Lettering has two primary goals:[1]

1. Legibility regardless of drawing size

2. Ease and rapidity of execution

Legibility is accomplished not just through the shapes of the letters but the space between them, the space between words, and the space between lines of letters. The choice of font is not an expression of the draftsperson's personality or the aesthetic qualities of the project being drafted. In computer-aided design (CAD), a well-rounded, sans-serif font such as Arial, Calibri, or Helvetica is a solid choice. CAD means that you do not need to become a superb letterer; still, increased legibility of the handwritten word will serve anyone well.

Architectural drawing has a tradition of stylized lettering that has been adopted by many scenic designers who still hand draft. While attractive, legibility is somewhat reduced. Mechanical drawing lettering uses a style derived from **single-stroke commercial gothic**; the height of letters for most purposes is 1/8″. Adopted as a standard in 1935 by the American Standards Association (ASA), this has been the basis of mechanical drawing lettering ever since.[2] The Vertical American Standard Alphabet (Figure 1-1) is a descendant of this style.[3]

For those with limited manual dexterity, lettering templates of various sizes are available. For the able-handed,

DOI: 10.4324/9781003154921-2

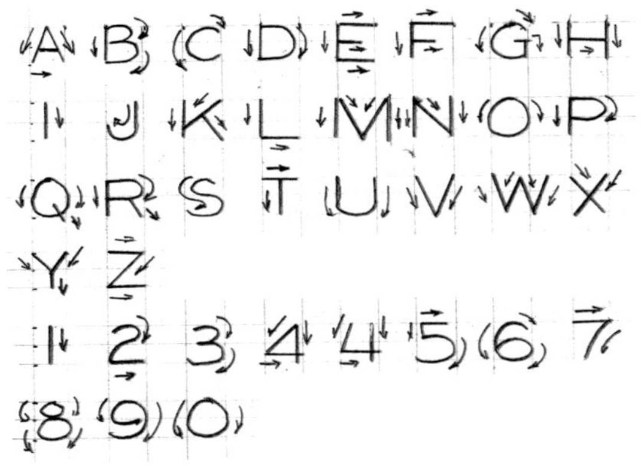

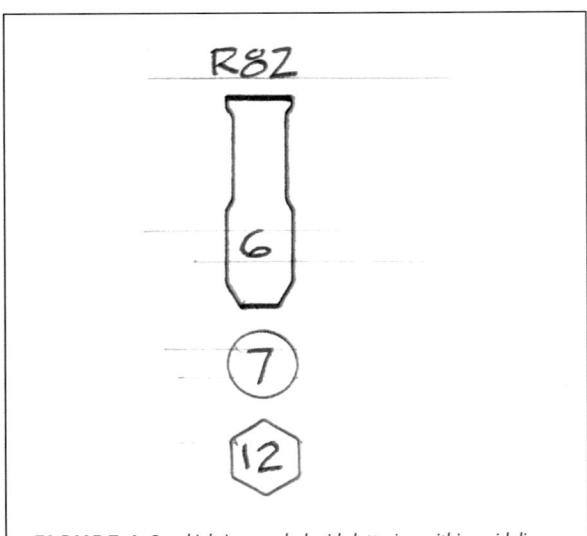

FIGURE 1-1   *Alphabet and numerals hand lettered in Vertical American Standard Alphabet, with direction of pencil strokes; 4 is drawn in both open and closed versions—the closed 4 is standard, the open 4 is often preferred*

though, templates usually slow down lettering and result in figures with a stencil-like appearance.

The goal is not to achieve perfection in lettering, but to develop the ability to draw legible text swiftly and easily. As you develop lettering skill, strokes are consolidated, curves become smoother, and a bit of personal style will creep back in.

There are several things to note in the lettering example in Figure 1-1:

1. Strokes shown are for right-handed draftspersons; left-handers should adapt the strokes as required.

2. All letters are represented by their upper-case forms. To reduce variation that diminishes legibility, lower-case letters are seldom used.

3. All letters and numerals are drawn within upper and lower guidelines to establish uniformity of size. Letters should not float between the guidelines but be anchored by them.

4. Use a guideline to help keep the center bars of A's, B's, E's, F's, etc. level and centered. Vertical guidelines can help keep lettering from slanting to one side or the other.

Guidelines keep lines of words leveled and orderly (Figure 1-2). This is essential for the readability of blocks of text. Notation that slants and drifts across the page diminishes legibility. Sloppy lettering can also diminish the credibility of a drawing by giving the impression that the draftsperson has not paid sufficient attention to detail.

Guidelines become optional in cases where notation consists of one or two letters or numerals; placing the gel color name in front of a lighting unit symbol on a light plot is such a case (Figure 1-3). Still, even accomplished

THE QUICK BROWN
FOX JUMPED OVER
THE LAZY DOGS

FIGURE 1-2   *Lettering with guidelines; letters should touch top and bottom guidelines*

draftspersons use guidelines for their lettering; it's one reason why their drawings look so good.

Letters and numerals are built by individual pencil strokes with defined beginnings and ends (Figure 1-1). An A requires three separate strokes: the two angled sides and the cross-bar. Keep the pencil on the paper until the stroke is complete. Firm, even pressure from beginning to end produces the best result. Lift the pencil from the paper at the end of one line and move it to begin the next line of the letter. For an A, both angled strokes begin at the top of the letter. Starting at the bottom and drawing both angled lines with a single stroke may result in a rounded top, not a point. As, Rs, and 4s can easily look similar when hastily drawn. For letters with round portions, aim for curves that are circular rather than oval.

Most letters and numbers have uniform width. Consider the letter as fitting within and filling out a square. Exceptions are M and W, which are slightly wider, and I and 1, which are both a single vertical stroke. Closed 4s should have sharp points; some designers prefer open 4s to prevent confusion with 9s.

R82

6

7

12

FIGURE 1-3   *Lighting symbol with lettering within guidelines*

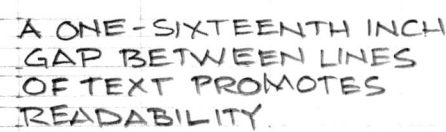

FIGURE 1-4 *Block of lettering with 1/16″ spacing between lines of text*

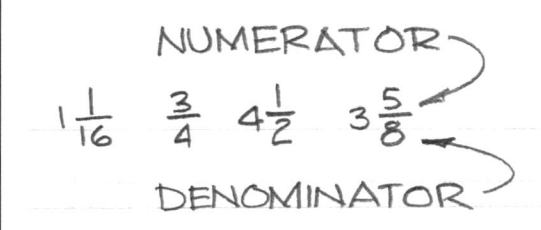

FIGURE 1-6 *Standard hand drafting fraction format*

The general rules for spacing are (Figure 1-4):

- Letters and numerals do not touch each other, nor are they so far apart that a single word breaks apart into sections.

- Leave one letter's width of space between words.

- Leave the width of three letters between the ends of one sentence and the beginning of the next.

- Blocks of text should include space above and below each line of at least 1/16″.

Most lettering is 1/8″ high (12pt for CAD), regardless of the scale of the drawing. One-eighth of an inch is used for most notation, including measurements. One-quarter-inch-high lettering is generally reserved for titles and labels, and important-do-not-miss-this headers. It is more difficult to draw consistent ¼″ high letters and to make letters larger than ¼″ look good, as the width of the pencil stroke does not match the proportion of the letter itself. Tools such as the Ames Lettering Guide can aid the draftsperson in laying out evenly spaced lettering guidelines (Figure 1-5). Affixing a short piece of drafting tape marked with increments to one side of a triangle also works.

Fractions are also drawn with 1/8″ high numerals. The top numeral (numerator) extends above the upper

guideline, and the bottom numeral (denominator) extends below the lower guideline (Figure 1-6). The line between the numerator and denominator (the fraction bar) is horizontal. Numbers should not touch the fraction bar, which is slightly wider than the widest part of the numerals. Center the numerator and denominator along the same vertical axis (Figure 1-6). CAD fonts typically swap the horizontal bar for a diagonal slash.

Typically, a softer lead (H) is used for lettering, allowing the draftsperson to draw smoother, blacker lines more swiftly. 2H also works well, but more pressure must be applied. Since soft leads smudge easily, lettering is among the last elements to be added to a drawing. Repeatedly running a T-square or triangle over a block of lettering reduces the crispness of the letters and leaves graphite tracks elsewhere on the page.

If you are unsure which font and format to select when setting up a CAD drawing, review these lettering principles and choose the options that best serve legibility.

Exercise: Lettering

Instructor

Provide a handout that is guidelined and ready for use. See Figure 1-7 for an example. Guidelines should be 1/8″ apart with 1/16″ separating each tier. They should run across the page, margin to margin. Margins are 1″ from the edge of the paper. Include vertical guidelines from top to bottom, spaced about 1″ apart. Have faster students repeat Steps 4 and 5 until the entire class has completed the full set of letters and numerals.

Review and discuss completed lettering efforts as a class. What was the most difficult aspect of lettering for the students? What was easier? Were there common problems or successes? Do any of the students have tips or discoveries to share with their classmates? What technical suggestions can be made to help each student further improve?

Remember that public examination and comparison of work should not humiliate the student or point out personal inadequacies: the focus should be on practical ways to improve the work. Often, it takes an outside observer

FIGURE 1-5 *Ames lettering guide*

FIGURE 1-7 *Sample lettering worksheet with guidelines*

to identify issues the student cannot see because they are too close to the work. Looking at a project upside down or in a mirror can bring to light issues not normally observed because the viewer can no longer bring prior expectations to the observation.

Students

1. Choose a pencil hardness. Using your normal handwriting, write the alphabet in upper-case letters within the first tier of guidelines on the handout distributed by your instructor.

2. In your normal handwriting, write all ten numerals within the guidelines on the next tier.

3. Consider the graphite hardness and whether you had adequate control. Change hardness to see if something else works better. Write the alphabet and numerals again.

4. Look at the example of gothic standard letters in Figure 1-1. Follow the example and letter the alphabet again. Go slowly. Keep verticals vertical and horizontals horizontal. Don't erase unsatisfactory letters—identify the issue and draw the letter again.

5. Do the same with all ten numerals.

6. Compare and discuss the differences observed.

   – Does your handwriting normally slant one way or another?
   – Do you have trouble with horizontals or verticals?
   – Do you have trouble with curves?
   – When and where do you normally lift the pencil when making a letter?
   – What are the differences between your first and last lines of text?

7. Write THE QUICK BROWN FOX JUMPED OVER THE LAZY DOG using standard lettering. Focus on improving those areas in which you had difficulty. Write the sentence several times. Write numerals from 0 to 20, several times. Practice develops muscle memory.

Homework: Lettering

Instructor

Assign a paragraph of text to letter, both in 1/8″ high letters and in ¼″ high letters. Provide a guidelined worksheet for both. At the next class session, review efforts as a class. Identify areas of improvement and areas still troublesome. If time allows, start the next several class periods with a brief lettering exercise before introducing the day's topic.

Students

Letter the assigned block of text in 1/8″ high letters and again in ¼″ high letters. Pay attention to word and sentence spacing. Which size lettering is easier to control and provides more a more consistent appearance?

When lettering larger blocks of text, you may find that your lettering deteriorates as your hand becomes tired. If that happens, stop, take a break, and then return. This way, constructing the good lettering reinforces muscle memory.

Letter as much as possible in other classes and activities. It will be slower than your normal handwriting, but speed will come with practice. You may even find aspects of lettering creeping into and improving the legibility of your normal handwriting.

MODULE 2: SCALE AND MEASUREMENTS

Measurement Notation

There is a standard way to write measurements. In the United States, measurement is typically expressed in total *feet* and *inches*. The number representing the total number of complete feet is written first. The number of complete inches is written after a dash or hyphen. Remaining fractions of inches are placed to the right of that number. "Two feet plus three and one quarter inches" is written as 2′–3¼″.

The foot measurement is denoted by placing a single mark (3′) after the numeral. Properly, this is not an apostrophe but a straight mark called a **prime.** Inches are denoted with a **double prime**. Prime marks are small; it's easy to mistake 3′0″ for 30″ or 30′, so the dash ensures that the two numbers are read as feet *and* inches (3′–0″).

When a measurement ends with a complete foot, the lack of extra inches is still noted: 3′–0″. This tells the reader that they are not missing any information. It is not typical to write a zero for the foot measurement if the measurement is less than a foot. For four inches, 4″ is preferred to 0′–4″. A lone numeral is typically interpreted to be inches.

Measurements are never expressed in fractions of a foot. "A foot and a half" is never written 1½′ because this looks like it could be either 1′–½″ or 1½″. A foot and a half is written 1′–6″ and *sometimes* as 18″.

Expressing measurements solely in inches for lengths longer than one foot usually requires the reader getting out a calculator to break it down into feet and inches. Every time a reader has to do extra math, there is more opportunity to accumulate error. Still, there are occasions when a measurement is written as a total number of inches rather than a combination of feet and inches. This is most common when noting the height of raised flooring, such as platforming and the treads of staircases. It is typical to write the surface height of a platform as + X″ above the stage floor: e.g., + 48″ rather than + 4′–0″.

Both Vectorworks and SketchUp allow the user to enter fractions of an inch in decimal format: ¾″ as 0.75″. For increments smaller than quarter inches, such as eighths and sixteenths, it's usually easier to enter the fraction rather than the decimal equivalent (5/8″ = 0.625″). For most purposes, fractions are the preferred drawing notation format. If a drawing is generated for use in 3D printing or computer controlled cutting and molding devices, the accuracy of decimal measurements carried out to several places may be required.

In metric, decimals are the default system of notation. The measurement 1.25 meters (m) equals 1 m plus 25 centimeters (cm), as there are 100 cm in a meter. In CAD, line weights are measured in millimeters (mm).

To convert a fraction to decimal format, divide the numerator by the denominator. For example, to convert 5/8 to decimal format, 5 divided by 8 equals 0.625.

Exercise: Reading a Ruler and Writing Measurements

Instructor

For many, a college or high school scene shop is the first time they use tape measures or once-common hand and power tools. A student already knowing how to read a ruler or tape measure cannot be taken for granted. Canvass the class to see if a fractions and ruler-reading refresher is necessary. If so, distribute a handout with lettering guidelines and a list of classroom items to be measured. After completion, compare findings and discuss differences.

Students

Use a tape measure and/or ruler to measure the listed items in the room. Letter the item's name and its measurements on the handout. How many measurements, such as height, depth, width, circumference, diameter, etc., are required to fully describe each object? When you compare findings with your classmates, are there any differences? If there are differences, can you identify their causes?

Introduction to Commonly Used Scales

Scale refers to the ratio or formula used to proportionally reduce or enlarge the measurements of a full-sized object in order to represent it on a page. A drawing that represents an object full-size is in a scale of 1′–0″ = 1′–0″, or, a ratio of 1:1. In ½″ = 1′–0″, a line that is ½″ long on the page represents a line that is 1′–0″ long in reality. The smaller the scale, the larger the object that can fit on the page. When speaking about scales, it's common to refer to the scale by its fraction. To say something is in "half-inch" or "half-inch scale" means it is in ½″ = 1′–0″.

The most common scales for printed drawings are ¼″ = 1′–0″ and ½″ = 1′–0″. Scenery typically uses ¼″ = 1′–0″ for groundplans and large scenic elements. Items requiring more detail are better described in ½″ = 1′–0″ or even 1″ = 1′–0″. Small, highly detailed elements may be drawn full-size. Upon receiving drawings from the scenic designer the technical director interprets, engineers, and produces a second wave of drawings, converting the scenic designer's more architectural images into working drawings, choosing a scale that is most convenient for the carpenters and technicians. Light plots should be printed in ½″ = 1′–0″ and often feature a note indicating that the drawing is in that scale when printed on a certain sized sheet of paper: e.g., "When printed on Arch D sized paper, this drawing is in ½″ = 1′–0″."

Arch D paper is a standard architectural drafting sheet of paper that is 24″ × 36″. Other sizes include Arch A (9″ × 12″), Arch B (12″ × 18″), Arch C 18″ × 24″), etc.

When hand drafting, scale is one of the first decisions that a draftsperson must make. In CAD, the scale in which the drawing is to be printed *can* be one of the last things to be determined, as it is possible to draft in full-size within the virtual drawing environment. For the exercises that follow and to reinforce scale awareness, a scale will be selected during the initial setup process for each drawing, whether hand drafting or using CAD.

Reading the Scale Ruler

The triangular format of the architectural scale ruler allows the draftsperson to select from eleven different scales, from full-size to 3/32″ = 1′–0″. To squeeze eleven different scales onto it, the ruler can be read from either end. Along any edge, the markings for the scale read from the left overlap with those for the scale read from the right. The only place where this is not the case is the edge featuring the full-size (1:1) scale, which looks like a plain old ruler and is sometimes labeled "16" (Figure 1-8).

The name of each scale is located at the end of the ruler from which it is to be read. For the full-size edge, the 16 notation means that each division in the full-size scale is 1/16″ (one inch is broken up into 16 sixteenths of an inch). There is no number at the other end of this edge because overlapping another scale on top of it results in illegibility. The full-size scale markings begin with a zero at the start of the inch increments. Numerals appear at the marks for each full inch increment. Halves, quarters, eighths, and sixteenths of an inch are all marked within each one inch increment.

The 1″ = 1′–0″ edge is denoted by a 1 at one end of the ruler (Figure 1-9). This scale is read *from* the 1. The first section is broken up with very small marks that decrease in value, from 9 to 6 to 3 to 0. This section represents inches. Marks beyond the 0 represent feet.

Beyond 0, the first mark is 20, not 1. The 20 mark is only one-half inch away from the 0 and therefore can't represent a full foot in this scale, much less 20 feet. In 1″ = 1′–0″, a real half-inch equals six inches. The 20 belongs to the scale read from the *other* end of the ruler: ½″ = 1′–0″. The 20 accompanies a mark *shorter* than the next one, which is denoted as 1. This 1 marks the end of the first full foot in 1″ = 1′–0″. Along each edge, the numbers for each scale *increase in value* from the end at which the scale begins.

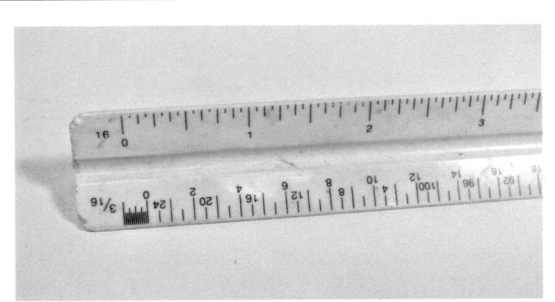

FIGURE 1-8 *Scale ruler turned to show the side with full size scale (16)*

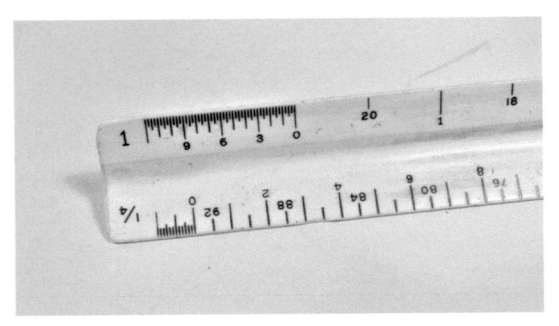

FIGURE 1-9 *Close-up of the 1″ = 1′–0″ end of the scale ruler*

Along the 1″ = 1′–0″ edge, the markings read 0, 20, 1, 18, 2, 16, 3, etc. 0, 1, 2, 3 belong to 1″ = 1′–0″. 20, 18, 16 belong to ½″ = 1′–0″.

Measuring a line is not as simple as setting the scale ruler down with the 0 at one end of the line and reading the mark that the termination of the line aligns with. Reading the scale ruler can be thought of in two steps:

1. Determining how many *full feet* there are

2. Determining how many *full inches* there are

Only in the larger scales, such as 1½″ = 1′–0′ or 3″ = 1′–0″, are fractional inches easily read.

To measure a line (Figure 1-10):

- Draw a line (with a straightedge, not the scale ruler)

- Select the desired scale

- Decide which termination of the line is the beginning, and which is the end. This is arbitrary and only to help with the following steps

- Place the 0 of that scale at the beginning of the line

- See what *last* whole foot mark the line passes (e.g., the end falls between the marks for 7 and 8 feet; therefore, the line is 7 feet plus an unknown number of inches long)

- Slide the ruler so that the last whole foot mark aligns with the end of the line (in this case, the 7). The *beginning* of the line will now fall somewhere among the inch marks

- Look at where the beginning of the line falls within the inch marks. Count *from* the zero to determine how many inches

Reading the scale ruler takes practice. Every draftsperson spends several moments rotating the tool to find the desired scale, flipping it end over end to determine from which direction to read it. If a measurement seems odd and not in keeping with other measurements, double-check it. As you

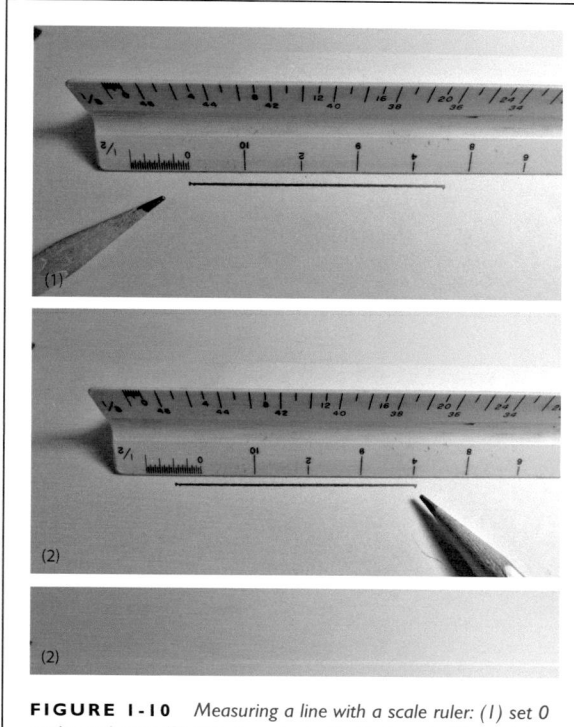

FIGURE 1-10 *Measuring a line with a scale ruler: (1) set 0 at the endpoint; (2) slide the ruler to the foot mark; (3) read the number of full feet and inches*

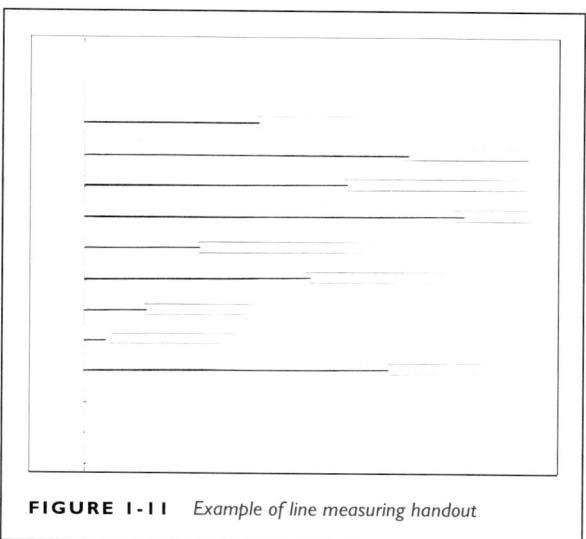

FIGURE 1-11 *Example of line measuring handout*

develop a sense of what things are *supposed* to look like when drawn in a particular scale, you will catch errors more easily.

Exercise: Scale Measurements

Instructor

Provide three copies of the same handout, featuring a number of evenly spaced horizontal lines of various lengths. Include 1/8″ lettering guidelines beside each line. See Figure 1-11 for an example. Have the students measure the sets of lines in a variety of scales. When complete, compare class findings. Were measurements consistent throughout the class? If someone had a different measurement, why? Did they use the wrong scale? Did they forget to slide the ruler back? Which scales seemed to give the most accurate/consistent measurements?

Students

1. On the first handout provided by your instructor, measure each line in full-size (1″ = 1″, or, 1:1). Letter the measurement within the guidelines provided.

2. On the second handout, measure each line in ½″ = 1′–0″. Letter the measurement within the guidelines provided.

3. On the third handout, measure each line in ¼″ = 1′–0″. Letter the measurement within the guidelines provided.

4. When the class discusses the results, identify any problem areas you had and try to identify solutions. Were you in the wrong scale? Did you forget to slide the ruler back to zero? Were the markings too small to read accurately?

MODULE 3: THE DRAWING SPACE

Without a drawing space, there is no drawing.

The drawing space can be a piece of paper or it can be an infinite virtual universe displayed on a computer monitor. Regardless of medium, the drawing space is the first element that shapes what ends up on the page.

Boundaries and Borders

When hand drafting, the edge of the paper is a natural boundary of the drawing. Boundaries in CAD are a little trickier. Programs like Vectorworks and SketchUp offer an infinite space within which to draw, so it's not always possible to know at the start where the edges of the printed drawing will end up. In Vectorworks, there is a blue reference grid that appears when the program is first opened. The appearance of the grid changes depending on the scale and paper size selected (Figure 1-12). For scales such as ½″ = 1′–0″ and ¼″ = 1′–0″, the reference grid can represent the selected paper size and be used to aid page layout. When drafting large objects in 1:1, the reference grid gets smaller and smaller, eventually disappearing into the background since it is also being shown in its real-world size. Depending on software version and whether your computer is PC

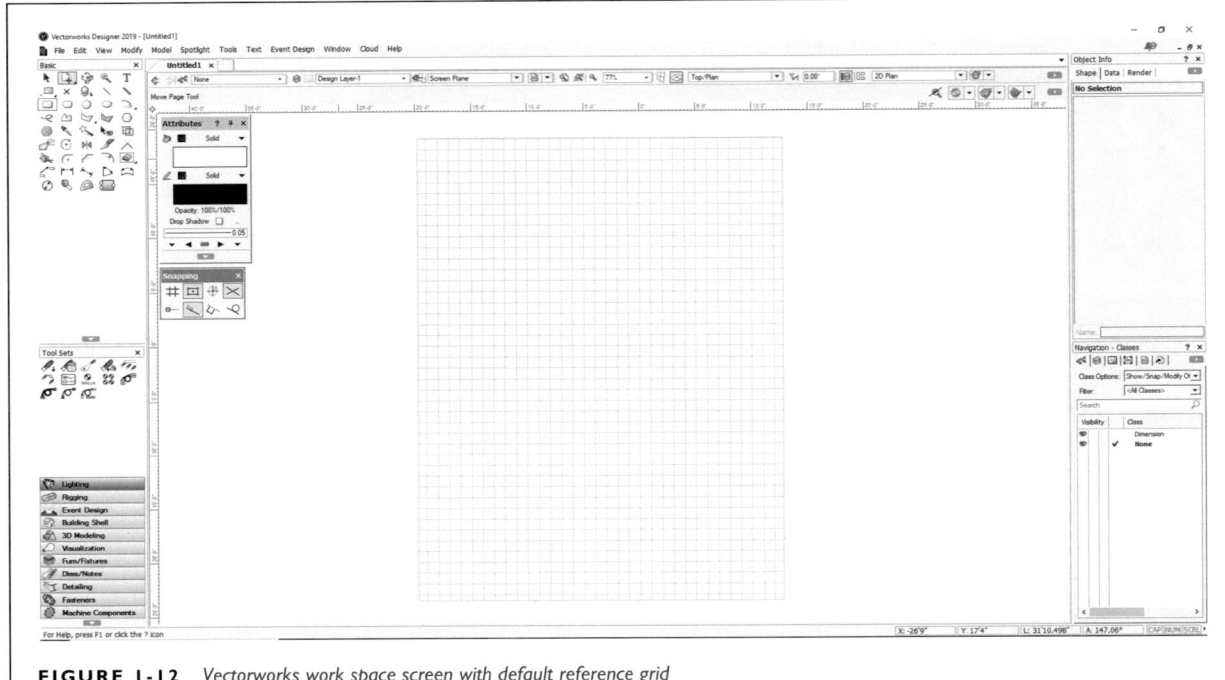

FIGURE 1-12   *Vectorworks work space screen with default reference grid*

or Mac, the appearance of your grid may differ from the examples shown.

In most CAD programs, after the virtual drawing or model is finished, the draftsperson can select portions of the drawing that then are transferred to a separate page space. In Vectorworks, the drawing space is called a **design layer**. The page space to be printed is called a **sheet layer**. It's a bit like cutting an image out of a sketchbook and pasting it onto the center of a fresh, blank sheet of paper. The early Vectorworks exercises in this book use the reference grid to establish the page boundaries and borders at the beginning of the drawing process. This helps develop good layout practices and promotes the mental process of *thinking in scale*. Having a sense of what standard scales look like on the page means that preparatory sketches and doodles can better approximate actual proportions and provide collaborators with a more truthful expectation of the eventual product. A reader also has a better chance of noticing when a reproduced drawing has been printed in a size different from what the scale notation indicates.

It is easy to unintentionally reproduce or display only a portion of a drawing. **Borders** are drawn around the edge of the drawing space to indicate the full extent of the drawing. Seeing uninterrupted borders around the drawing's perimeter tells the reader that they are receiving the whole of the drawing, not just a part of it.

Borders are placed ½″ within and parallel to the top, right, and bottom edges of the paper. The left border is placed 1″ from paper's edge. When multiple sheets

are bound together, the wider left margin ensures that information is not lost in the binding. In CAD, file name information (folder file string) is typically placed along and outside the left border. If multiple CAD drawings are bound together, the left border is increased to 1½″ from the paper's edge to ensure that the file string information is not lost in the crease. When the border is drawn with a single line, it is typically the thickest line on the drawing. It is not uncommon to see a double line used for borders.

> When purchasing sheets of drafting vellum with preprinted borders and title blocks, the border may be ½″ on all sides of the sheet. Leave extra white space on the left side of the drawing so that binding does not obscure information.

Setting Up the Drawing Space for Hand Drafting

In hand drafting, the first step is to affix the paper to the drawing table surface. This is not as straightforward as it might seem.

First, the draftsperson must determine which side of the drawing table will be **the working edge.** For right-handed draftspersons this is the left edge of the table; for left-handers it's the right edge. Those who are ambidextrous get to choose which side will work best for them. The

FIGURE 1-13   *A T-square firmly placed against the table's working edge*

non-dominant hand is used to hold the T-square steady while the dominant hand wields the pencil. Be sure there is clearance to allow free movement of the T-square along the table edge (Figure 1-13).

The working edge of the drawing table is the base reference for every line drawn on the page. Vertical lines are parallel to the working edge. Horizontal lines are perpendicular to the working edge. Angled lines are drawn in relationship to the vertical axis represented by the working edge. Even freehand shapes are drawn within this reference system. When a drawing is removed from the table and re-taped elsewhere, it must be aligned with that table's working edge to ensure that newly drawn lines align with already drawn lines.

Once a drawing is begun, the working edge must remain constant throughout. While consumers can have reasonable expectations that a drawing table has straight edges, 90° corners, and that opposite sides are parallel to each other, this is not a certainty. The T-square should always be used along the established working edge of the table and not moved to the top or bottom edge to draw vertical lines, or to the opposite side to get to the far side of the paper.

> There are always exceptions. A 24″ T-square may not be sufficient to span to the full width of the paper. If you flip the T-square to the non-working edge, be aware that this is a case of breaking the rules, even if it is a necessary breaking of the rules.

Aiming for consistency also guides T-square usage. The draftsperson assumes that the top and bottom edges of the T-square blade are parallel, but that, too, is not a certainty. The top edge of the blade is the only edge that

is drawn along, and against which triangles and other tools are placed.

Exercise: Setting Up the Drawing Space for Hand Drafting

Instructor

This exercise outlines the steps in affixing a piece of paper to the drawing table. Supply 8½″ × 11″ paper as needed. The students should have their own tools, but some may need to be provided. Work through the steps as a class, pausing as needed to ensure everyone keeps up. If time allows, students should leave this paper affixed to the drawing table to use for the Straightedge Practice #1 Exercise (below).

Students

Use a sheet of 8½″ × 11″ copy paper, oriented in landscape position. As well as your drawing table/surface, you will need a T-square, pencil, and drafting tape (Figure 1-14).

1. Determine which side of the drawing table will be the working edge.

2. Determine how far the T-square reaches across the table top. Affix a piece of drafting tape on the table to mark the farthest reach of the T-square.

3. Place the paper on the drawing surface, centered within the T-square's span. If the paper is too close to the working edge, it will be difficult to position and draw along triangles; too close to the far end of the T-square, and it will be difficult to firmly hold the T-square's head against the working edge while drawing.

4. Be sure that the paper is high enough on the drawing table to allow easy access to the bottom of the page. There should be a distance at least equal to the full width of the T-square blade between the bottom of the paper and the table edge; too close to the edge and the top of the T-square blade will not access the bottom of the sheet.

5. Place the T-square head firmly against the table's working edge. While holding the T-square in place, align the bottom edge of the paper against the top edge of the T-square blade.

6. Use tape to affix the two upper corners of the paper.

7. Slide the T-square up onto the paper to permit access to the bottom two corners. This helps keep the paper in place as you tape down the bottom two corners.

Even though the bottom edge of the paper was aligned with the top of the T-square, this does not ensure that

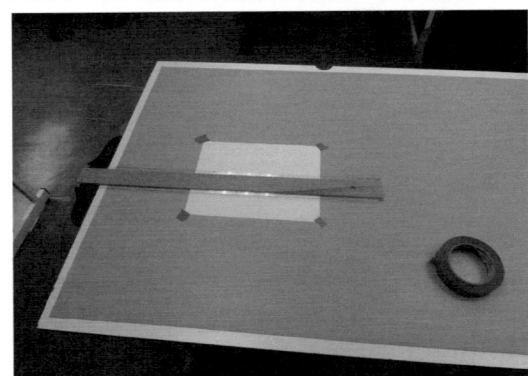

FIGURE 1-14 *Steps to affix paper to the drawing table surface: place paper along T-square; tape upper corners; tape lower corners*

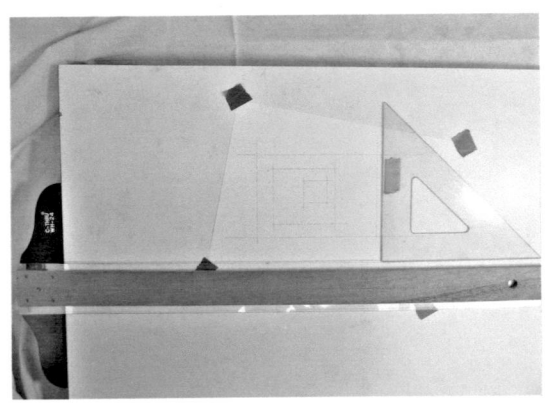

FIGURE 1-15 *Even if the paper is crooked, everything on the drawing still aligns with the working edge*

When re-taping a drawing to a table after transport, before taping it down:

1. Select an already drawn line, preferably horizontal, as a reference.

2. Hold the T-square head against the new working edge and orient the paper so that the reference line aligns with the top edge of the T-square.

3. Tape down the paper.

4. After taping, double-check alignment of the reference line against the T-square.

5. Redo, if necessary.

Construction Lines/Guidelines

Guidelines are the armature of a drawing. No final line of the drawing should be drawn without first drawing a guideline to determine its placement. This is true whether hand drafting or working in CAD.

Guidelines serve two major purposes:

1. Placement of information

2. Transfer of information

The scale ruler is a measuring tool, not a drawing tool. Lines are drawn along drawing tools. T-squares and triangles have edges that are straight, so they are categorized as **straightedges.** Since lines are never drawn along a ruler's edge, how does a measured line get drawn?

Construction lines are drawn first. As the term implies, construction lines help the draftsperson construct the drawing. They also move information around the page. A measurement can be made in one location and transferred elsewhere through

horizontal lines drawn on the paper will be parallel to either the top or bottom edges of the paper. What this does ensure is that all horizontal lines drawn along the top edge of the T-square will be parallel to each other. As long as the paper is firmly affixed to the drawing table and rules of consistency have been maintained in drawing the borders and everything within the borders, the drawing should be fine. This is another function of the border—the paper's edges are not a reference; the borders *are* (Figure 1-15).

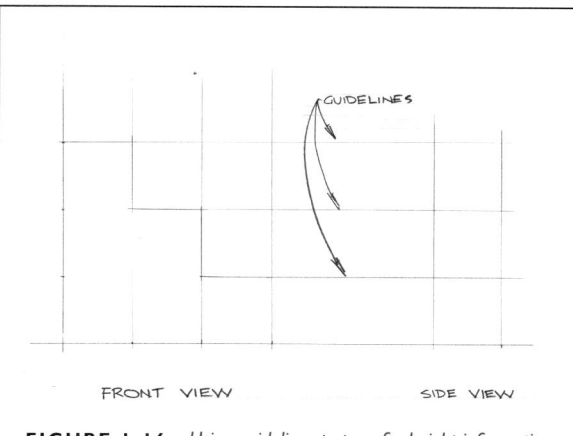

FIGURE 1-16   *Using guidelines to transfer height information from a front view to a side view without re-measuring*

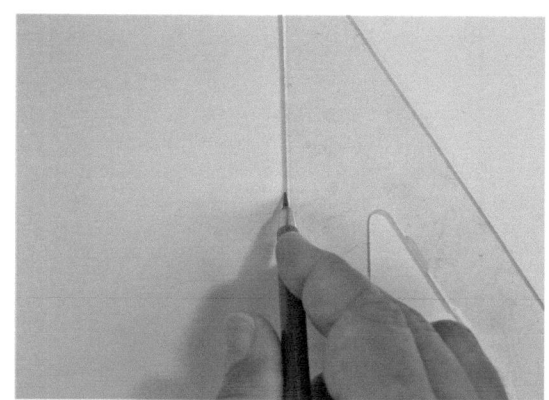

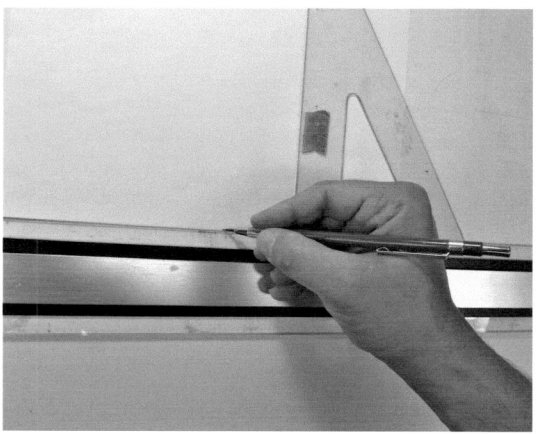

FIGURE 1-18   *Pencil orientation against a straightedge; keep the pencil parallel to the straight-edge*

the use of a construction line (Figure 1-16). The shorter term "guidelines" will be used from here-on throughout this book (SketchUp uses "guidelines," so it is better to settle on the more universal term).

Guidelines are drawn with a thin (narrow), gray, almost translucent line. A hard pencil, such as 4H, is best. They are drawn lightly and can be easily erased, though in most reproduction processes lightly drawn guidelines vanish into the white space making erasure unnecessary. There may be some ghostly residue, but this can be useful as a visual connection between parts of the drawing. The residue should not, however, visually challenge the more important lines of the drawing. The thick (wide) black lines that are the important presentational lines of a drawing are called **visible lines** (Figure 1-17).

When the guideline line style is set up in Vectorworks (see Chapter 2), make them as thin as possible with a color easily differentiated from other lines in the drawing space. In Vectorworks the visibility of guidelines can be turned on and off. Guidelines in SketchUp always appear as dashed lines, and there is a menu choice to delete guidelines within the model.

Be generous with the length of guidelines. They should begin *before* the point the visible line is expected to begin and extend *past* the point where the visible line is expected to terminate. In SketchUp, guidelines can extend into infinity. In Vectorworks, just as in hand drafting, the draftsperson needs to be strategic regarding overall guideline length. As the drawing progresses, extra length aids laying out other portions and views of the object. When placing dimensions (measurements), these extra lengths can help align notation. Avoid frequent erasing, as a guideline that at first might seem to be clutter may be useful later on.

There is a technique to drawing even, consistent lines along a straightedge. This usually requires breaking previously established habits of ruler use (Figure 1-18).

1. Hold the pencil with its length parallel to the straightedge so that the point does not dig beneath or splay away from the edge.

2. Allow the shoulder of the pencil point to ride against the straightedge. The point will be a slight distance away

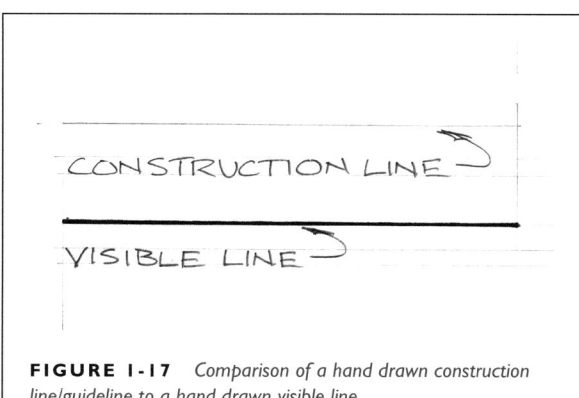

FIGURE 1-17   *Comparison of a hand drawn construction line/guideline to a hand drawn visible line*

from the edge. This allows the draftsperson to better see the line being drawn. It takes practice to maintain consistency.

3. A line should be drawn with a single, well-paced stroke from start to end. This is facilitated by pulling your arm across the length of the line rather than pivoting from the wrist. Short strokes using the wrist as a pivot result in choppy, segmented line work. Gently twirling the pencil as you draw a line maintains the point for a longer time; the hexagonal shape of standard pencils facilitates this motion.

Exercise: Straightedge Practice #1

Students

Spend some time drawing lines with a straightedge. If the paper taped down in the previous exercise is still on the table, use that. Otherwise, tape down a fresh sheet following the steps already described. When drawing, firmly butt the T-square against the table's working edge.

A hard pencil lead, such as 4H, is recommended. Draw vertical, horizontal, and angled guidelines of various lengths and orientations until comfortable.

When using triangles, it takes practice to hold all the tools firmly in place while drawing. For vertical lines, it is best to begin at the top and draw the pencil toward you rather than pushing the pencil away from you.

Visible Lines

Drafting's graphic language employs a range of line weights and line types. Line weight and type together determine a line's **style** (Figure 1-19).

Line **weight** refers to the thickness (width) and opacity of a line. Line **type** refers to the configuration of the line: solid, a series of dashes, undulating, etc. Each weight and type has a particular meaning so that a reader knows whether a line refers to the outline of an object or indicates a surface or edge hidden behind something else.

The term **visible line** refers to the style of thick, solid, unbroken lines used to outline objects and features. The two most used weights are Thick and Medium. USITT recommends 0.50 mm width for thick visible lines, and 0.35 mm for medium visible lines whether by hand or CAD.

Which graphite hardness to use and how much pressure to apply are determined through experimentation and practice. Softer leads, such as H, are generally preferred for visible lines, though 2H also works well. No one actually measures the thickness of the line as they are drawing it; conventional rulers don't have small enough increments to do so. CAD programs allow you to set weights prior to drawing. For hand drafting, it's a case of being visually familiar with the standard weights.

When first learning to control line weight, keep a list of printed examples on your drawing table and compare results frequently. In Vectorworks, setting up line styles is one of the first setup activities. The tutorial in Chapter 2, Module 1 generates a sample sheet of line weights and types.

When hand drawing visible lines:

1. A line should be the same width and opacity from start to finish. Apply even pressure as the line is drawn. If a line must be gone over a second time, the second run should also go from start to finish.

2. If the pencil must be lifted midway through a stroke, ease the pencil off the surface so that the termination fades, rather than stops abruptly. When returning to the line, ease the pencil down at the start of the fade and increase pressure as you continue the line to create a smooth join.

3. The starts and ends of lines must be explicit. Press down a little harder at the ends of lines and bring the pencil to a complete stop before lifting it from the page. Some draftspersons add a minute tick mark to the ends of visible lines to crisp up terminations.

4. The blunter the pencil point, the wider the line. Soft leads like H blunt quickly, and the end of the line may end up wider than the beginning of the line. Slowly twirl the pencil as you draw to help maintain the ideal point over distance. Soft leads with needle-sharp points are also more likely to shatter as you apply pressure. Blunting a hard lead after sharpening can minimize point breakage.

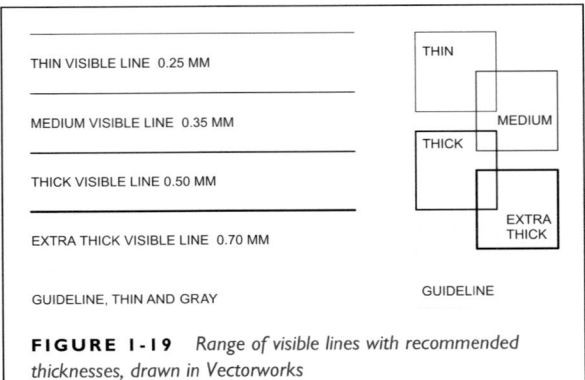

FIGURE 1-19 *Range of visible lines with recommended thicknesses, drawn in Vectorworks*

If using a lead-holder style of drafting pencil, the sharpener, or "lead-pointer," usually has two depth holes in its cover: one for fine points and one for blunt points. Drop the end of the lead into the hole, then bring the holder down to the lid's surface. The lead will extend from the tip the correct amount for sharpening to that type of point.

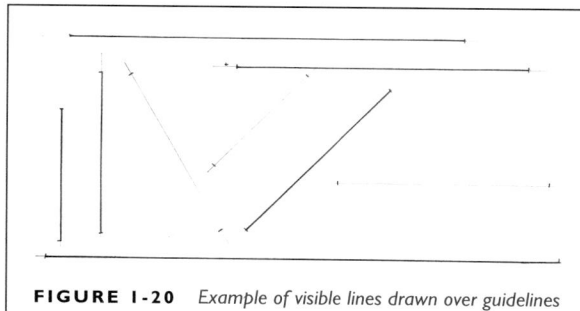

FIGURE 1-20 *Example of visible lines drawn over guidelines*

5. To draw an exceptionally thick line (e.g., for borders), rather than further blunting the point, draw two thick lines side by side that just overlap (no white space between them). Draw the first line, and then shift the straightedge ever so slightly for the second. If you have good pencil control, this can also be done by slightly shifting the angle of the pencil rather than moving the straightedge.

Exercise: Straightedge Practice #2

Students

Use the sheet of paper from Straight-edge Practice #1. Practice drawing thick visible lines *over* the guidelines that were already drawn (Figure 1-20).

1. Make two tick marks along each guideline: one for the start point, the other for the endpoint. Vary the distance between tics for each guideline. Leave excess guideline beyond each tick mark. The visible line will be shorter than the guideline it's drawn over.

2. Practice drawing horizontal, vertical, and angled lines of consistent weight from start tick mark to end tick mark. Aim for crisp terminations.

Exercise: Defining the Drawing Space by Laying Out Borders

Instructor

This exercise combines laying out guidelines, measuring with the scale ruler, and drawing visible lines to lay out borders on an 8½″ × 11″ sheet of paper. Since this is a small sheet of paper that will not be bound with other sheets, all borders will be ½″ from the paper edge.

Students

The following steps outline laying out borders on an 8½″ × 11″ sheet of paper. You will need paper, pencil, tape, scale ruler, T-square, and triangles (Figure 1-21).

1. Tape the paper to the drawing table in portrait orientation, using the steps previously outlined.

2. Using the T-square and a hard pencil (4H), draw a horizontal guideline across the full width of the paper, from side to side. This is an arbitrary guideline used for the placement of other lines. It does not matter how high on the page it is drawn.

3. On the right hand side of the paper, measure ½″ along the guideline from the edge of the paper. Make a tick mark.

4. Since the paper is 8½″ wide in this orientation, ½″ borders will create a drawing space that is 7½″ wide. From the tick mark on the horizontal guideline, measure 7½″ and make another tick mark. This should end up about ½″ from the left side of the paper. You can also measure ½″ in from the left edge of the paper; it's less math and usually sufficient—but let's do it the hard way first.

5. Align a side (not the hypotenuse) of your tallest triangle against the top edge of the T-square blade. Slide it until its vertical edge lines up with the left-hand tick mark. Be sure you can see the guideline and the tick mark. Placing the triangle so that its corner meets the T-square right at the tick mark means you can't extend the vertical line below the tick mark.

6. Draw a vertical guideline the full height of the paper, from edge to edge. If your triangle is not long enough, draw half of the line, move and realign your tools, then complete the line.

7. Line up the triangle with the right hand tick mark. Draw a vertical guideline from the top edge to bottom edge. These two vertical lines are the guidelines for the borders on the right and left sides of the paper.

8. At the top of right-hand vertical guideline, measure ½″ down from the paper edge and make a tick mark. Since the paper is 11″ tall in this orientation, borders placed ½″ from top and bottom will create a 10″ high drawing space. Measure 10″ down from the tick to place another mark. These two marks place the guidelines for top and bottom borders. Since you've measured their placement along one side of the page, there's no need to makes measurements along the other vertical guideline—the T-square will transfer that information across the page.

9. Align the T-square with the bottom tick mark on the right-hand vertical guideline. Draw a guideline from the right-hand vertical line to the left-hand vertical line. Be sure to go a little past it so there are clearly defined intersections.

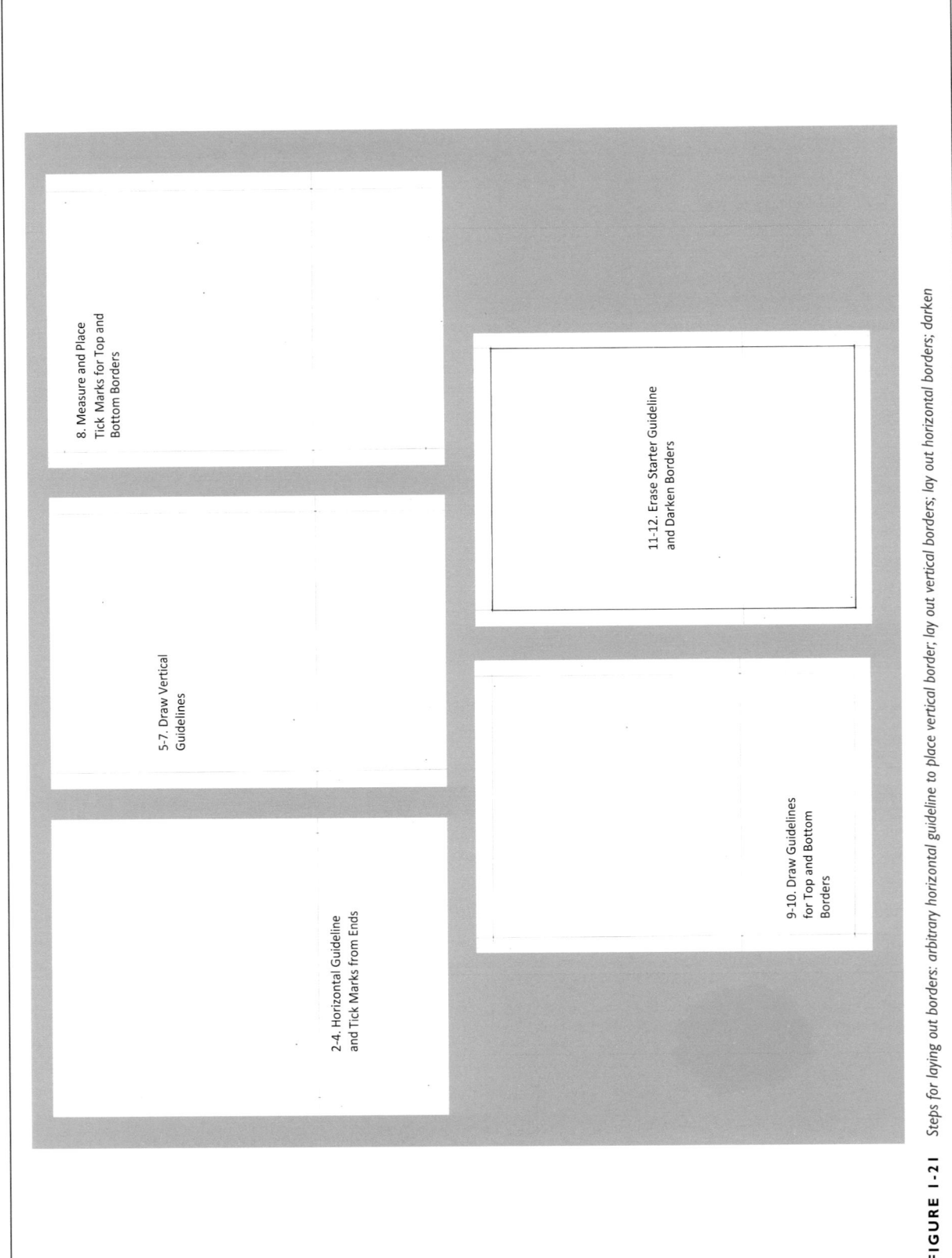

8. Measure and Place
Tick Marks for Top and
Bottom Borders

5-7. Draw Vertical
Guidelines

2-4. Horizontal Guideline
and Tick Marks from Ends

11-12. Erase Starter Guideline
and Darken Borders

9-10. Draw Guidelines
for Top and Bottom
Borders

FIGURE 1-21 *Steps for laying out borders: arbitrary horizontal guideline to place vertical border; lay out vertical borders; lay out horizontal borders; darken*

10. Align the T-square with the top mark on the right-hand vertical line. Draw a horizontal guideline across to the left vertical line. Guidelines for borders on all four sides of the paper are now in place.

11. The original, arbitrarily placed horizontal guideline is no longer required. This may be erased.

12. With a softer pencil (H), use straightedges to draw the border as visible lines. The lines should terminate at each intersection of guidelines, making crisp, clean corners. Don't run visible lines out to the edges of the paper.

You've just laid out and drawn a large rectangle. This is the same basic process used to lay out most objects: begin with a guesstimate to place an initial reference guideline, then build out from there.

If you're not using paper with preprinted borders, darkening the borders is among the very last tasks done on a drawing. As the aim of this exercise was to draw borders, once you have darkened the borders this exercise is complete.

Transporting a Drawing

When hand drafting, you may need to clear off your classroom drawing table and transport your drawing to another location for completion.

Never fold an in-progress drawing, as the resultant creases cause new lines that cross them to have bumps and gaps. Upward fold peaks capture graphite on the underside of tools, leaving undesirable streaks across the drawing.

Small drawings, such as 8½″ × 11″, can be slipped flat into a folder or binder. Be sure the folder's covers are larger than the paper so that the edges of the drawing aren't damaged by other items sharing the pack or bag.

For larger drawings, procure a tube of 3″–4″ in diameter. If the drawing is rolled too tightly, the resultant curling will be troublesome. Plastic tubes can be purchased through most art supply stores. Empty plotter paper tubes also work well; tape paper over one end so the drawing cannot slide out the bottom.

Before sliding the drawing into a tube, roll it after turning it *face down* on the table. When unrolled, gravity and the paper's own weight will work against the curl, making it easier to tape it down to the drawing table.

Tutorial: Setting up a Base File in Vectorworks

The final activity of this chapter is creating a base drawing file in Vectorworks. Beginning a drawing in Vectorworks requires more setup than simply opening a new file and immediately using a line tool to draw a line. The base file created in this tutorial is used as the starting point of later Vectorworks projects.

Instructor

Depending on time available, the following tutorial may be begun during class time, or assigned as homework. If assigned as homework, be sure to discuss discoveries and difficulties during the next class session. This exercise introduces the concepts of layers and classes, and sets up classes for frequently used line styles.

Students

1. Open Vectorworks.

2. Along the Upper Menu, select Tools. At the bottom of the drop-down menu, select Workspaces. From the Workspace menu, select Fundamentals (Figure 1-22).

    When you select Fundamentals, the screen will undergo some changes. Be patient and wait for it to refresh. Vectorworks features several Workspaces, each with features specific to the industry most likely to use that workspace. Spotlight is the workspace that contains many entertainment industry tools. Since these first tutorials focus on learning the basics, Fundamentals is the workspace best suited to beginning efforts.

3. The reference grid in portrait orientation should appear in the center of the page. The default reference grid represents a page that is 8.14″ wide by 10.65″ high. This will fit comfortably within an 8½″ × 11″ sheet, so leave this for now. The wheel on your mouse allows you to zoom in and out. Zoom out until the whole grid fits within the screen space.

> If your computer is connected to a plotter, there may be a default larger paper size already selected. Go into File > Page Setup and change paper sizes (as you would do in Word). Select the various paper sizes offered and see what they do to the reference grid. If your printer does not accommodate larger sheets, the gray lines that appear and break up the reference grid indicate tiles by which the drawing will be divided for printing.

Vectorworks uses a concept called **layers**, which can be thought of as layers of tracing paper. Each layer contains part of the drawing, and when you stack and then look through all of the layers, you see the whole drawing. The program opens into a default layer titled Design Layer-1. This exercise will be drawn in 1:1 (full-size) scale. We'll need to be sure this layer knows we'll be drawing in this scale.

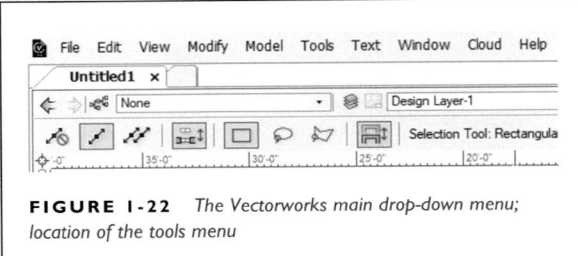

FIGURE 1-22 *The Vectorworks main drop-down menu; location of the tools menu*

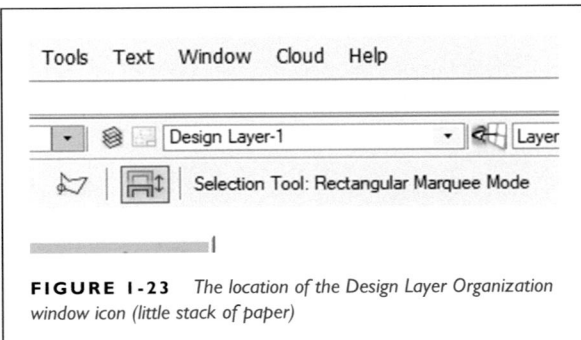

FIGURE 1-23 *The location of the Design Layer Organization window icon (little stack of paper)*

There are several settings that allow layers to interact with each other; when you draw a line in the active layer, for instance, that line can snap to or otherwise have an active intersection with a line in a different layer. Open the View drop-down menu on the main upper toolbar. Choose Layer Options. Checkmark Gray/Snap Others. This setting will mute lines in inactive layers to gray and preserve the ability to interact with them from the active layer.

4. Click on the icon that looks like a little stack of papers (like a stack of tracing paper) to the left of the window that says Design Layer-1. The Organization window will appear (Figures 1-23 and 1-24).

In the Scale column, 1:1 should be noted. If it does, good. If it doesn't, click on the Edit button. The Edit Design Layers window will appear. Even if the scale is noted as 1:1, open the Edit Design Layers window as there are two important selections to consider (Figure 1-25).

Click the button that reads Scale… The Layer Scale dialog window will appear. Select 1:1 on the Metric/Engineering side, as there is no equivalent choice for full-size on the Imperial side.

Checkmark the box that says All Layers so that whenever you add a layer, it automatically uses the scale previously selected. Having this box checked in your base file now will prevent odd things from happening later on. Click OK to back out of the windows and return to the main drawing space. The blue reference grid may have changed its appearance as it now reflects the new chosen scale (Figure 1-26).

The larger the paper size, the more grid lines there will be. There is a snap-to-grid feature that snaps endpoints of lines to the reference grid. If you know the full-size distance between reference grid lines, you can lay out some items with less measuring. Snapping features will be discussed later.

Page Navigation

It's necessary not only to move the cursor around the screen, but to zoom in, zoom out, and shift the drawing area. The wheel on your mouse allows you to zoom in and out. Zooming action is centered on the cursor's location at the moment of zoom. There is also a Zoom selection accessed by the View menu on the main upper toolbar. This selection offers only three zoom sizes, though they are of use when navigating particularly large drawings.

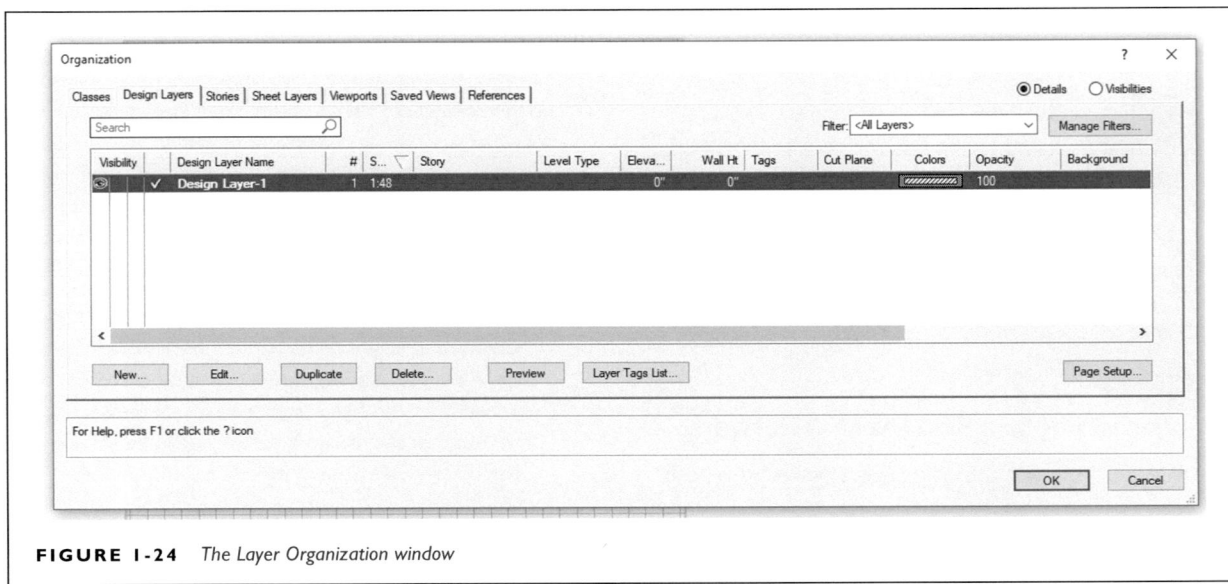

FIGURE 1-24 *The Layer Organization window*

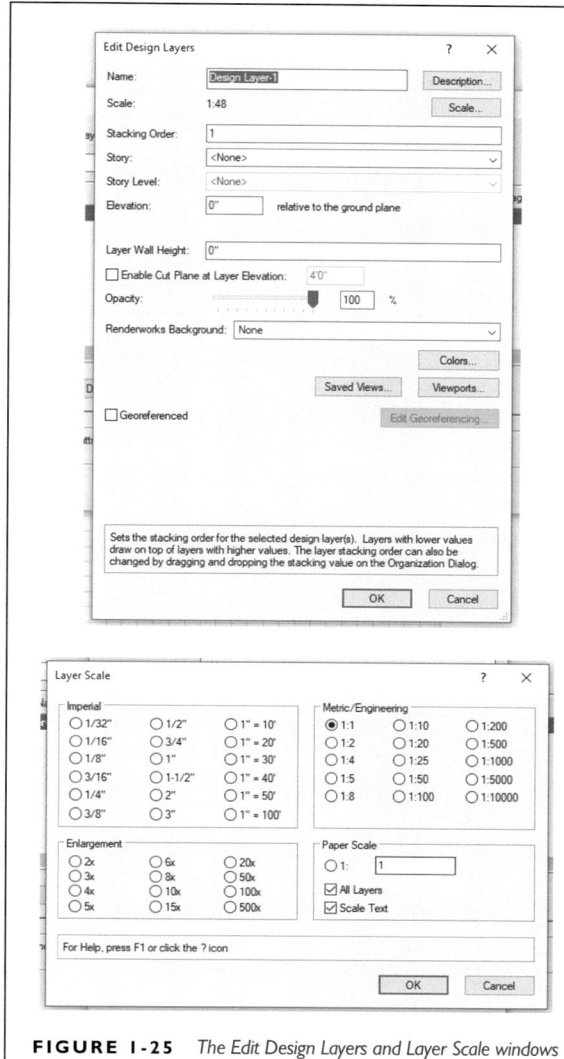

FIGURE 1-25   *The Edit Design Layers and Layer Scale windows*

To move the drawing area itself, find the **Pan** tool (Figure 1-27) in the Basic Toolset. It looks like a small hand. Selecting this tool and holding down the left mouse button allows you to drag the overall view around the screen. Release the mouse to release the cursor. Try zooming and panning a few times to get the hang of it.

Note that the Pan tool icon has a small arrow at the bottom corner. If you select this tool and hold down the left mouse button, an option appears allowing you to change the tool to Move Page. This moves the reference grid in relationship to drawn lines. If your drawing is off-center or runs off the side of the reference grid, you can use this tool to re-center the drawing in the grid.

Finally, the arrow icon is the **Selection** tool (Figure 1-27). This allows you to highlight and activate various lines and objects. Since nothing is drawn yet, there's nothing except the reference grid to select in the drawing space. It's a good habit to select the Select tool after other actions to prevent accidents.

2D or 3D

Since you'll be drafting 2D views of objects in all of this book's Vectorworks exercises, be sure that the program is working in 2D and not 3D.

5. Along the upper main toolbar, it should say **2D Plan** in the box pictured in Figure 1-28. When the mouse hovers over this choice a yellow box appears reading Current Projection (more about projection systems in Chapter 2).

    A nearby text box reads Top/Plan. This indicates that you are looking *down* at the page or object in the viewing space. By changing this view, you could look at the 2D drawing space edge-on (front) and therefore see only the edge of the reference grid, not its surface. If the drawing suddenly shifts into something incomprehensible, check these settings.

Lines: Weights, Types, and Styles

Before drawing a line in Vectorworks you need to tell it what type of line you wish to draw. This means setting up a library of line styles: a pencil box. For now, **thick visible lines**, **medium visible lines**, **dimension lines**, **hidden lines**, and **guidelines** will suffice. More line styles will be added as drawings require them. Thick and medium visible lines are used for object outlines. Dimension lines are used for placing measurements and notation. Hidden lines represent edges behind surfaces and other objects.

Setting Up Classes (The Pencil Box)

Along with layers, the other major organizational tool of Vectorworks is **classes**. Line styles, and therefore objects drawn with a particular line style, can be assigned to classes. When you turn off the visibility of a class, all of the lines and objects drawn with that line style will disappear from view regardless of which layer they are in.

Once a class is created, it can be selected from the class navigation window (Figure 1-29), whose default placement is the lower right corner of the screen. The line style Attributes box (also Figure 1-29) shows you the qualities current for the activated class. You can also use the Attributes box to change the qualities of a selected line.

6. Along the upper tool bar, click on the stack-of-paper icon to open the Organization window. In the upper corner of the Organization window click on the Classes tab (Figure 1-30). The Class Organization window will open. A couple of default classes will be listed (Figure 1-31). In the Mac version, Classes is located on the left side/top of the Organization window.

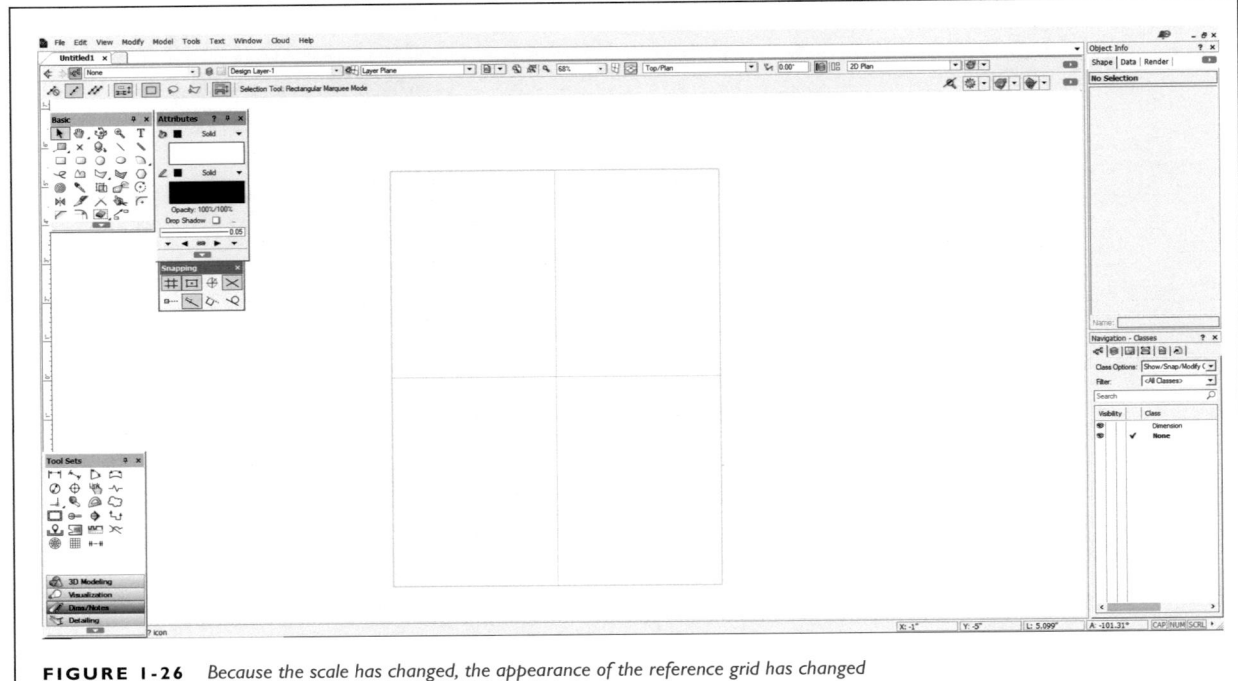

FIGURE 1-26 *Because the scale has changed, the appearance of the reference grid has changed*

Along the column header, **Visibility** refers to whether or not this class of objects is turned on in the drawing. A small eye icon in the far left column indicates that it is. **Class Name** is what you name this class of objects. Each class will have the attributes of **Fill**, **Pen**, **Line**, and **Thickness** assigned to it. Those are the columns of concern at this point.

If you have a default Dimension class and it is highlighted in blue, click on the Edit button. If you do

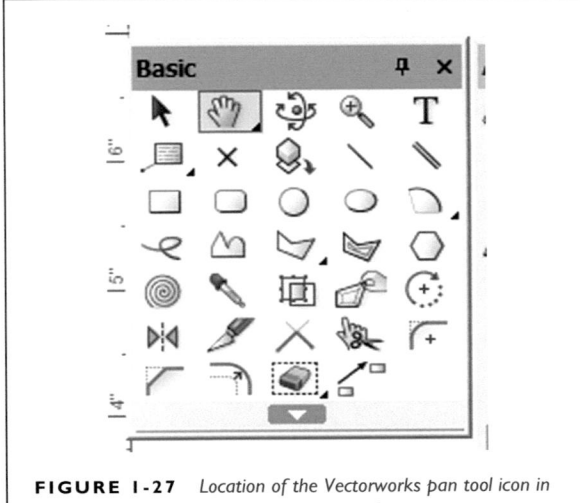

FIGURE 1-27 *Location of the Vectorworks pan tool icon in the basic tool set (it looks like a small hand)*

not, highlight the line that says None, instead. The Edit Classes window will appear. You can also select New to start from scratch (Figure 1-32).

7. Start by assigning desired attributes to the default Dimension line class. If the Name window does not already read Dimension, replace the default name with Dimension.

Check the box at the top of the window that says Use at Creation. This allows you to use that class as soon as you return to the drawing. In general, whenever you create a line class, check this box.

Pen assigns a weight and type (thickness, solid, dashed, etc.) to the class of lines. Select solid for Style. Choose black for Color.

Dimension lines are thinner than visible lines so that they are not mistaken as part of the object itself. USITT recommends a line weight of 0.25 mm for dimension lines in CAD programs. Select a thickness of 0.25 mm. Be sure the Opacity is at 100% (more about dimensioning in Chapter 4).

Fill assigns a pattern or shading to the interior of shapes drawn with this line class. As soon as the shape is completed it is filled with the selected pattern. At this point, let's avoid this happening automatically. Select None for the style of Fill.

Click OK to return to the Class Organization Window. Your changes should be reflected on the line titled Dimension.

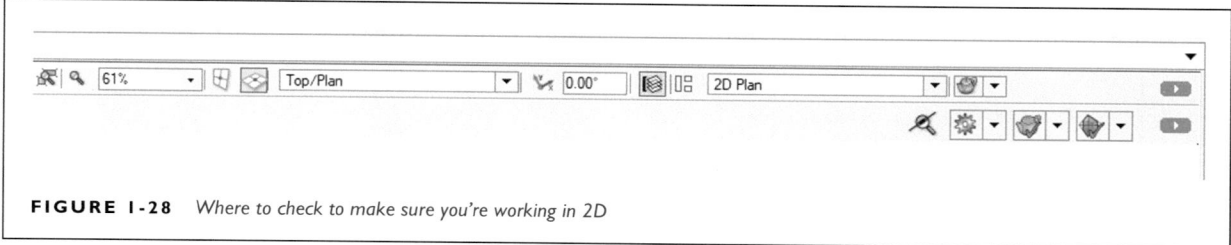

FIGURE I-28 *Where to check to make sure you're working in 2D*

8. Create new classes for the four other standard line styles. Open the Organization dialog box. Click the New button to open the New Class window. Type in the desired class name from the list below. When you click OK, you'll be taken back to the Organization window. Go into the Edit window to enter the line style's attributes. Don't worry about any other settings or options right now. The line weights listed below are the 2021 USITT recommendations for CAD.

For the thick visible line:

Name:	Thick Visible Line
Style:	Solid
Color:	Black
Thickness:	0.50 mm
Fill:	None

For the medium visible line:

Name:	Medium Visible Line
Style:	Solid
Color:	Black
Thickness:	0.35 mm
Fill:	None

For hidden lines:

Name:	Hidden Line
Style:	Line Type (this will open a Line Type selection window)
Line Type:	ISO-02 Dashed
Color:	Black
Thickness:	0.35 mm
Fill:	None

Hidden lines are dashed lines with 1/8″ long dashes and gaps of 1/32″. When the Line Type selection window opens, scroll through the library's selections until you find a line type labeled Hidden Line ISO-02 Dashed. Select it and finish entering the class's information.

Rather than the translucent gray line used for guidelines in hand drafting, use a thin, light-colored line that will show up well on the monitor. Depending on the complexity of the drawing, multiple colors of guideline classes can be created. If your guideline color choice overwhelms the visible lines drawn over them, return to this edit menu and select a more muted color. If you change the color of the guideline class, all of the previously drawn guidelines will reflect this change. There is no recommended weight for guidelines: the thinner, the better.

For guidelines:

Name:	Guidelines
Style:	Solid
Color:	pick a muted red from the selection grid
Thickness:	0.05 mm
Fill:	None

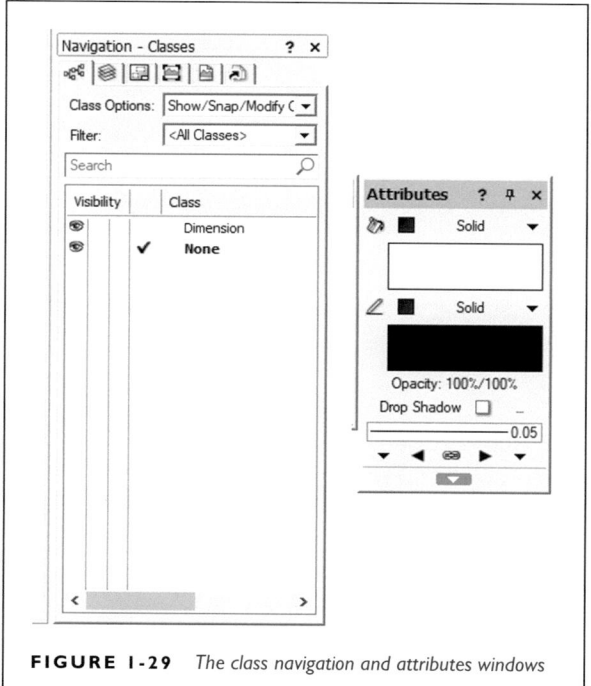

FIGURE I-29 *The class navigation and attributes windows*

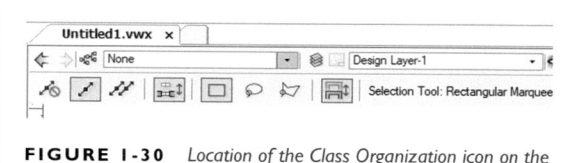

FIGURE I-30 *Location of the Class Organization icon on the main tool bar (looks like a flow chart)*

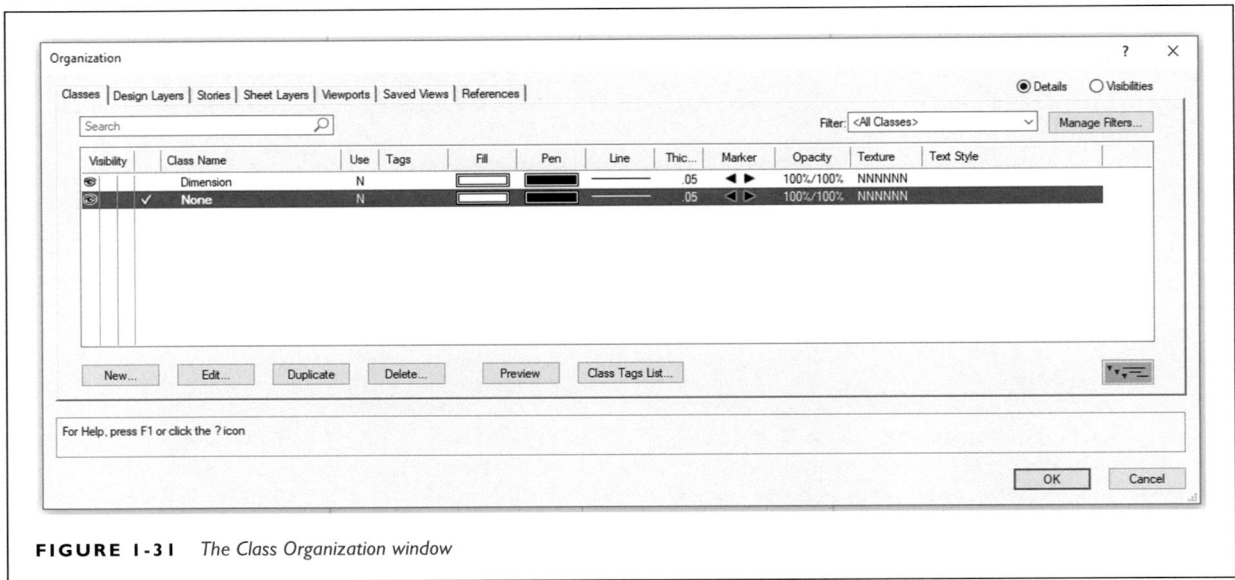

FIGURE 1-31 *The Class Organization window*

When you return to the drawing, if a line weight or type does not appear to be correct when drawn, it may be because Use at Creation was not checked.

After all five classes of line styles are created, the Organization window will look like Figure 1-33. Click OK to return to the drawing. In the lower right hand corner of the screen, the Navigation-Classes window should look like Figure 1-34.

To use a line style, click on its name. A checkmark will appear to indicate that it is active. While that class is active all lines drawn will have those attributes.

9. **Save As**. Save this file and title it [Your Name]'s Vectorworks Base Drawing. This will save the drawing as a .VWX file, which can be placed in any desired folder but only be opened by Vectorworks. As drawings become more complex, the base file will be altered to

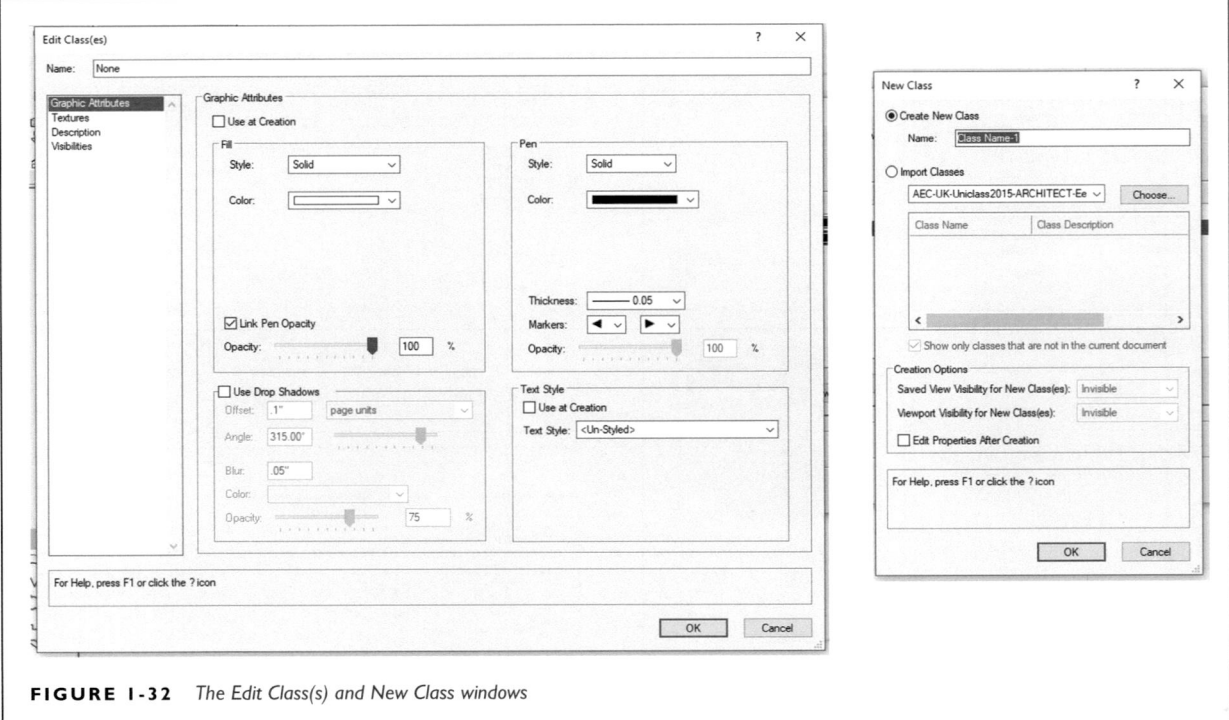

FIGURE 1-32 *The Edit Class(s) and New Class windows*

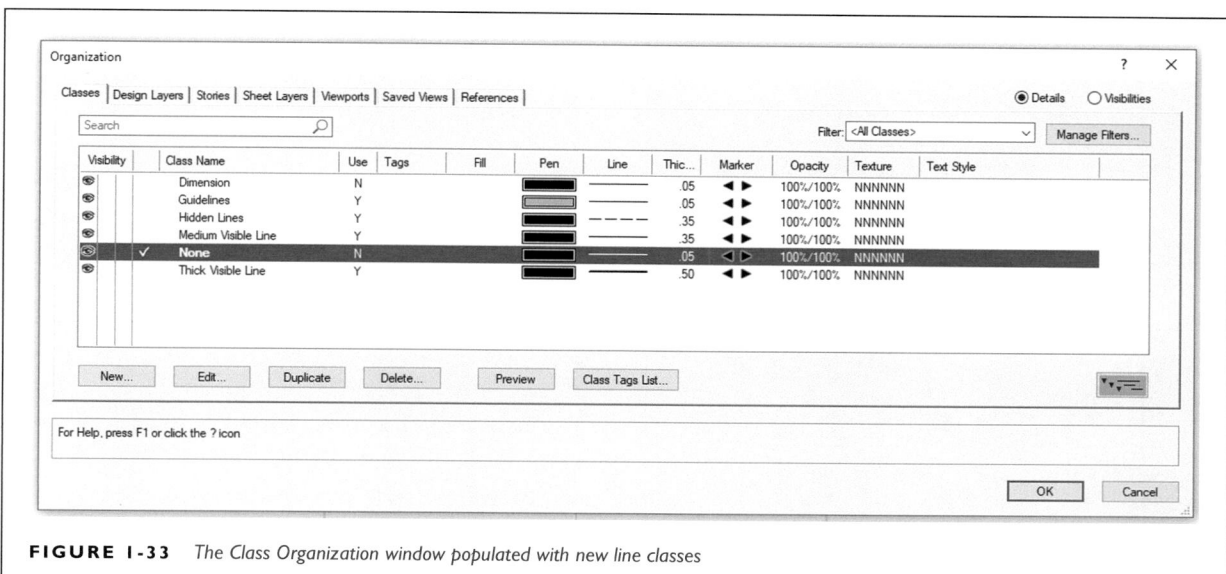

FIGURE 1-33   *The Class Organization window populated with new line classes*

reflect advanced setup needs. Multiple base files can serve as templates for various types of projects. Be sure to name them clearly and intentionally. There is also a Save as Template option in the File drop-down menu, which saves the work as an STA file, which can then be stored in a designated folder in the Vectorworks library.

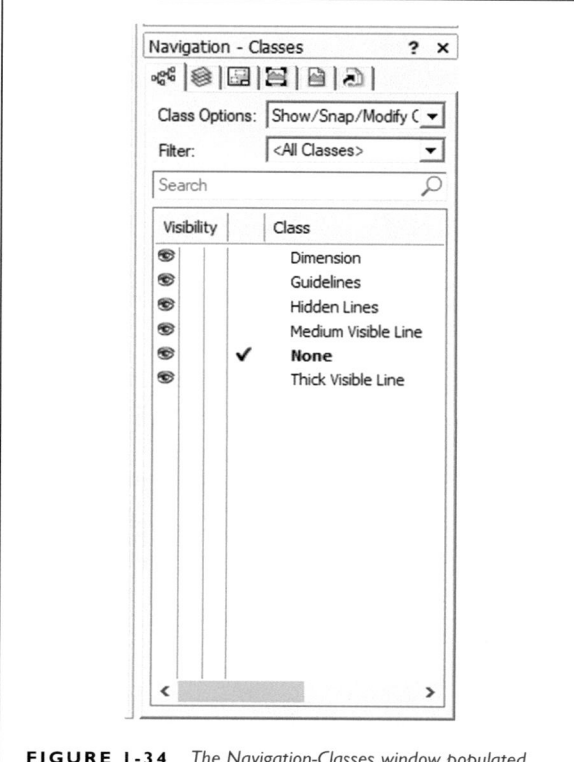

FIGURE 1-34   *The Navigation-Classes window populated with new classes*

Other Line Attributes: Snapping and Constraining

A line's endpoint can be set to "snap" to a particular type of point—the center of a line, a termination, the reference grid, etc. You can also **constrain** the line. Constraining means that the line defaults to angles of 30°, 45°, 90°, or their complements. This prevents accidentally angled lines. You can switch to **unconstrained** when needed.

When the Line tool (the single diagonal line icon in the Basic Tool Set) is selected and active, a number of windows and buttons appear in the upper left corner of the screen that offer various options (Figure 1-35).

Along the options toolbar, hover the cursor over each symbol to see what it does. Click on the Constrained Mode symbol when you find it (it looks like a plus sign). The icon will highlight blue and the choice be noted in text to the right of the buttons.

When the Line tool is active the Snapping window appears below the Attributes box. If it does not appear, enable it by using the **Window** drop down tab in the main toolbar. Select **Palettes** and then select **Snapping**.

Hover the cursor over each Snapping icon to see what it does. Turn on Snap to Object, Snap to Angle, and Snap to Intersection. These three are the most helpful for the coming exercises. Turn the remaining five off. If you find the line you're trying to draw not behaving, chances are a snapping tool is trying to help you. Turn the tools off if they get in the way. You can always turn them back on again.

Save this file again.

Exercise: Drawing Lines in Vectorworks

To draw lines in Vectorworks, tell it what kind of line you wish to draw by selecting the desired class. If the instructor and

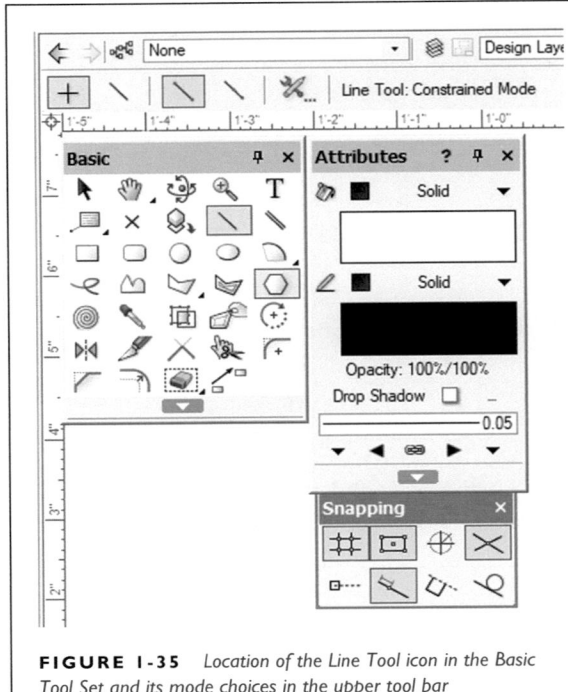

FIGURE 1-35 *Location of the Line Tool icon in the Basic Tool Set and its mode choices in the upper tool bar*

class wish to repeat the straightedge exercises in Vectorworks, it can be done now that classes have been setup.

Students

1. Open your base file. The blue reference grid will appear.

2. Save As "Line Exercise" so that you don't alter the base file.

3. In the Class Navigation box in the lower right corner, click on the Guideline class. A checkmark appears next to the name when it is active.

4. The Line tool is found in the upper left hand Basic Tool Set. Select it (Figure 1-35).

5. Move the cursor to any point within the reference grid. Left-click the mouse to affix one endpoint of the line to the drawing space. Drag the mouse a short distance away and left click again to affix the other end of the line. Lines can be drawn beyond the boundaries of the reference grid.

6. Draw lines. Try snapping. Connect lines to other lines. Practice until you have the hang of it.

7. After enough guidelines have been drawn, select the Thick Visible Line class and trace over them. If the color of the guideline prevents you from seeing the visible line, return to the Class Organization window and edit the

guideline class's color. When enough visible lines have been drawn, turn off the visibility of the Guideline class to see what happens.

Geometric Terms and Construction Refresher

It is useful to review the geometry terms listed below as they will pop up throughout Vectorworks and SketchUp.

*   **Point**: a location in space or on the drawing which has no height, width, or depth. A point can be represented by the intersection of two lines.

*   **Line**: the path between two points. A straight line is the *shortest* distance between two points.

*   **Angle**: formed by two intersecting lines. There are 360 degrees in a circle; angles are described by the number of degrees between the two lines.

*   **Right Angle**: name for any 90° angle.

*   **Perpendicular**: two lines that are at 90° to one another.

*   **Parallel**: lines running alongside each other that maintain a constant distance over their full lengths.

*   **Acute Angle**: any angle less than 90°.

*   **Obtuse Angle**: any angle greater than 90°.

*   **Triangle**: a plane (2D) figure bounded by three straight sides. The sum of the interior angles always equal 180°.

*   **Equilateral Triangle**: a triangle whose three sides are all the same length.

*   **Right Triangle**: any triangle with one corner of 90°. The sum of the other two angles will equal 90°.

*   **Quadrilateral**: any plane (2D) figure bounded by four straight sides.

*   **Parallelogram**: any quadrilateral whose opposite sides are parallel.

*   **Polygon**: any plane (2D) figure bounded by straight sides, such as pentagons, hexagons, and octagons. If the polygon has unequal sides, it is an **irregular polygon**.

*   **Circle**: a plane (2D) figure which is a closed curve, where all points on the curve are an equal distance from the center.

*   **Arc**: a portion of a circle, oval, or ellipse.

*   **Circumference**: the distance around a circle.

*   **Diameter**: the straight line distance across a circle, passing through the center (twice the radius).

*   **Radius**: the straight line distance from the circle's center to the circumference (half the diameter).

Geometric construction is the practice of using geometry to construct shapes on a 2D surface. CAD programs provide shortcuts for placing commonly used shapes such as circles and squares on the drawing. Remember, though, the draftsperson is responsible for every line on the page; don't use a shortcut until you understand *why* it is useful to use a shortcut. Knowing how to lay out shapes by hand fosters a better understanding of how the CAD tools can be used, as well as develops better rough sketching abilities. In the shop, CAD will not help you lay out a hexagon on a sheet of plywood.

There many geometric construction diagrams and guides available in other drafting texts, as well as from online sources. Construction problems such as bisecting an angle, dividing a line into equal segments, and constructing regular polygons are good class-starting exercises that develop measurement and layout skills.

NOTES

1. Frederick E. Giesecke, Alva Mitchell, and Henry Cecil Spencer, *Technical Drawing*, 4th ed. (New York, NY: The MacMillan Company, 1958), p. 66.
2. Giesecke, Mitchell, and Spencer, p. 66.
3. Dennis Dorn and Mark Shanda, *Drafting for the Theatre* (Carbondale, IL: Southern Illinois University Press, 1992), p. 25.

CHAPTER 2

PROJECTION SYSTEMS AND LINE STYLE CONTEXT

The first modules of this chapter help in practicing drawing line styles both by hand and with Vectorworks to gain familiarity with their appearance. Later modules discuss how these line styles derive further meaning through the context in which they are encountered. The various projection systems employed by mechanical drawing are introduced, and SketchUp visualization exercises explore orthographic projection's standard views of an object. The final module addresses the layout of circles and arcs.

TOPICS AND GOALS

*   Systematic drawing construction
*   Basic Vectorworks icons and tools
*   Visual familiarity with standard line styles
*   Using guidelines to place visible lines
*   Commonly used projection systems
*   Multi-view orthographic projection
*   The glass box concept
*   Alignment of views
*   Introduction to SketchUp
*   Selection and use of appropriate line types
*   Adding borders in Vectorworks
*   Drawing circles and arcs

MODULE 1: LINE STYLE FAMILIARIZATION

The five line styles introduced so far are:

*   **Guidelines**: layout and construction of the drawing
*   **Thick Visible Lines**: cutting plane lines, title blocks (more about those in Chapter 4)
*   **Medium Visible Lines**: object lines
*   **Hidden Lines**: edges that are hidden behind surfaces/masses
*   **Dimension Lines**: measurements around a drawn object

There is a certain amount of discretion in choosing between medium and thick visible lines. At 0.35 mm, medium visible lines are close to the 0.25 mm of dimension lines. Also, many draftspersons, especially designers, find it useful to use a thicker visible line to outline major features of an object and medium and thin visible lines for interior details. Using the dimension line weight for object lines is not recommended, as this can cause visual confusion between an object and a measurement. For complex objects, it might be preferable to use thick visible lines (0.50 mm) for major outlines, medium visible lines (0.35 mm) for interior detail, and then create an even thicker line class for features such as the title block and borders that require them. If the line weights are not clearly different when the drawing is printed, adjust them.

A line's full meaning is revealed when style is joined with context. An extra-thick visible (0.70 mm) line ½″ from the edge of the paper is read as a border. This visual information is linked with the knowledge that the border surrounds and

DOI: 10.4324/9781003154921-3

contains the page's information, and a complete and closed border means that the sheet contains all of the information the draftsperson intended to place on that sheet. A dashed line in the same location does not mean the same thing. Since it is not standard to use a dashed line as a border, the reader will be in doubt as to what this dashed line is meant to represent.

Tutorial: Vectorworks Line Style Familiarization

Instructor

This tutorial uses the line style classes created in Chapter 1 to help students get acquainted with what they look like on the page, as well as to practice drawing lines in Vectorworks. A methodical system of drawing layout is also introduced.

Students

When constructing any drawing, there is an overall **system of working** (see Chapter 5). Broadly, guidelines for borders are laid out first, *then* guidelines for all of the objects to be drawn are placed, and *then* lines are darkened. As will be seen in forthcoming exercises, a global approach to layout is one of the drafting's efficiencies: approach the drawing as a whole, rather than attacking and completing a drawing detail by detail.

1. Open your Vectorworks Base Drawing file.

2. Save As: [Your Name] Line Weight Exercise. Close the base drawing file and work only in the new file. **Get into the habit of not only saving frequently, but saving to multiple storage locations and/ or devices. Vectorworks reminds you to save a backup file of your drawing every fifteen minutes. Do it. Crashes and file corruption give little warning. Mistakes and mysterious drawing errors may require reverting to an earlier version of the drawing.**

3. The blue reference grid should be centered on the screen. The line style classes should be visible in the Navigation-Classes window at the lower right. The top toolbar should note that Design Layer-1 is active.

4. The next steps depend on your intentions for the drawing, so planning is required. Whenever you draft something, you should already be reasonably familiar with the object you intend to describe. This line style exercise comprises a column of lines equally spaced within margins. Figure 2-1 shows the hand drafted final product.

5. Borders will not be drawn for this exercise; instead, the margins will serve as terminations of the line segments to be drawn. The margins are 1″ within the perimeter

FIGURE 2-1   *Sketch of hand drafted finished line weight exercise*

of the blue reference grid. Before laying out guidelines for the margins, follow the steps below to create a design layer in which to put them. Having a design layer dedicated solely to guidelines allows their visibility to be turned on and off and permits deletion without fear of accidentally losing a visible line.

Design Layer-1 is the default active design layer. To create a second design layer, open the Design Layer organization window (the stack of paper icon seen in Figure 1-23). After the Organization window opens, click the New button. The New Design Layer window will appear (Figure 2-2). Name this layer Guidelines. Click OK. The new layer will appear in the Organization window.

Since the All Layers box was checked when setting up the base drawing, the scale for this new layer should read as 1:1 (full-size). If it does not, open the scale of this layer and make the change. Be sure that All Layers is indeed checkmarked. Click OK to return to the drawing. The active layer is now the new Guidelines layer (Figure 2-3).

You can move between layers by clicking on the downward pointing arrow at the end of the active layer text box. This opens a drop-down menu listing all of the layers created for the drawing. Click on the name of the

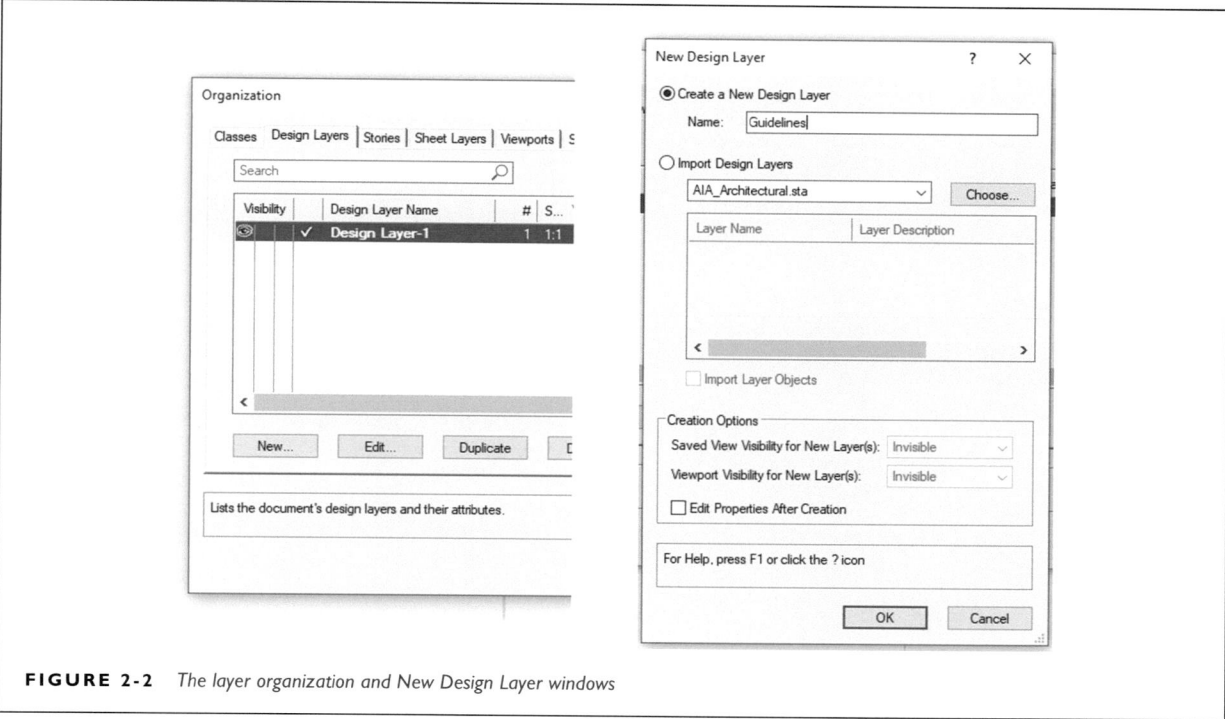

FIGURE 2-2 *The layer organization and New Design Layer windows*

desired layer to make it active. For this exercise, only two layers will be required: Guidelines and Design Layer-1.

6. In the Navigation-Classes window (lower right side of the screen), select the Guidelines class. Work in the Guideline layer and draw with the Guidelines class. To lay out the margins, select the Line tool (Figure 2-4). Choose the constrained mode so that the line doesn't accidently angle off in unexpected directions.

   In hand drafting, draw a guideline first, then measure and mark 1″ along the guideline from the edge of the paper. This approach holds for Vectorworks. Anywhere along the left-hand edge of the blue grid, click the cursor and drag a line horizontally toward the right. Before you click again, type 1″ on the keyboard. A text box will appear and note that L = 1″. When you hit Enter, a dashed circle appears with the dragged line as the radius. Left click the mouse and the circle disappears, leaving a 1″ horizontal line segment (Figure 2-5).

Click on the right endpoint of the line segment and drag a vertical line to the top of the grid. Drag a second vertical line from the same endpoint to the bottom of the grid. Repeat on the right side of the grid to lay out the right-hand margin.

If something goes wrong, you can **Undo** (uppermost toolbar, under the Edit drop-down menu), or use the **Trim Tool** (icon of hand with scissors) to delete the line. You can also Select the line (arrow icon) and then right click the mouse to Cut the line. The line will

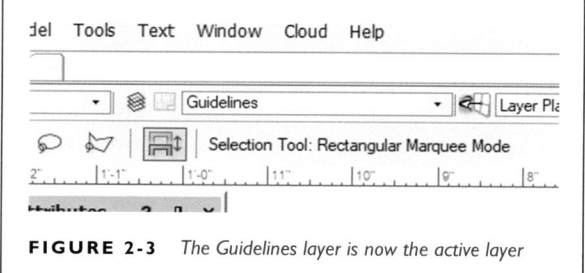

FIGURE 2-3 *The Guidelines layer is now the active layer*

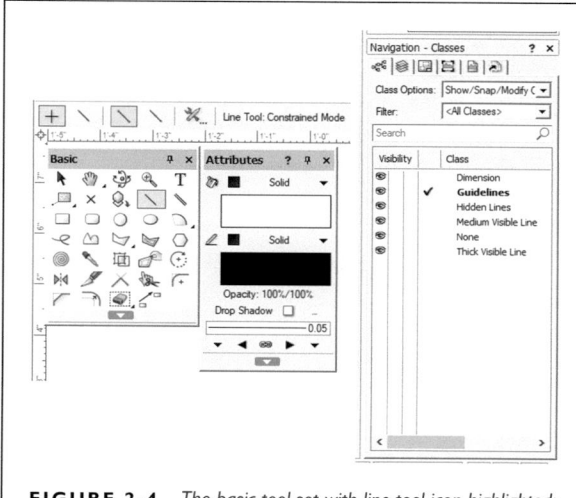

FIGURE 2-4 *The basic tool set with line tool icon highlighted; Navigation-Classes window with Guidelines class checkmarked*

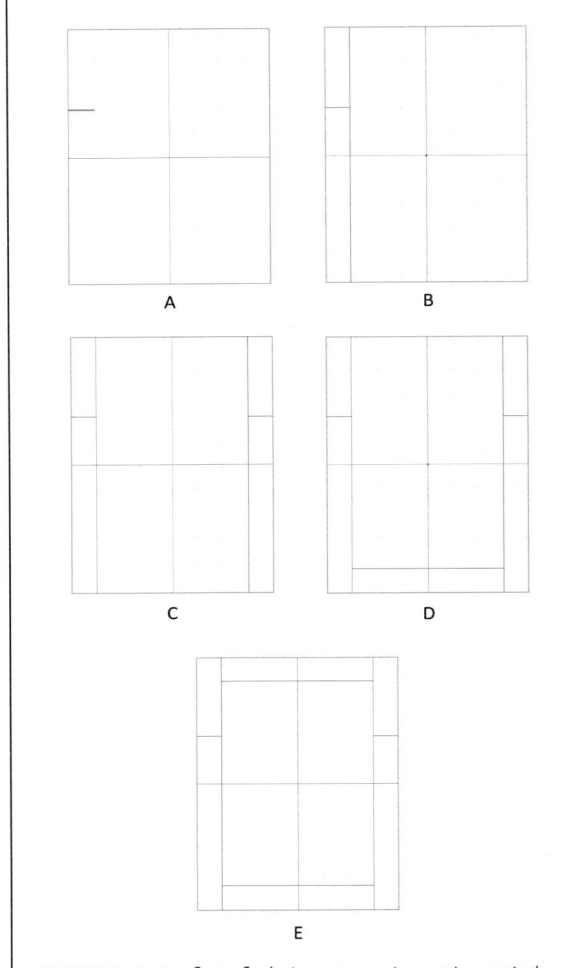

FIGURE 2-5 *Steps for laying out margins: setting vertical border distance from edge of grid; measuring horizontal border placement; laying out horizontal borders*

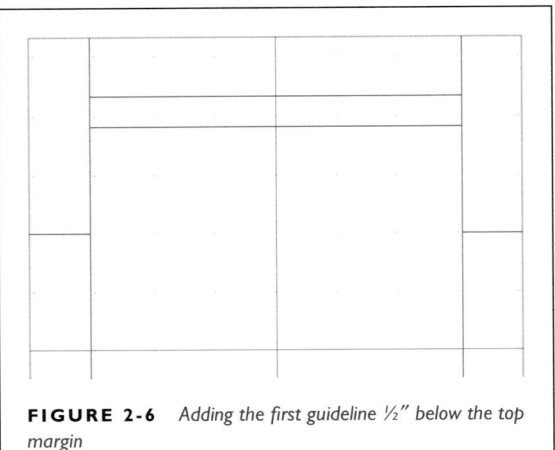

FIGURE 2-6 *Adding the first guideline ½″ below the top margin*

highlight in orange while active (selected). A selected item remains active until another action is chosen. Hitting Escape on the keyboard will deactivate an object.

In Vectorworks, lines can be drawn on top of other lines. The appearance of the drawing may not change, but a stack of lines is created. If lines are stacked, cutting and trimming eliminates only the line on the top of the stack. Depending on how many lines are stacked, it may take several trim/cut actions to get to the bottom and clear the space.

7. To lay out horizontal guidelines, click on the intersection of the bottom of the left vertical line and the reference grid. Drag a 1″ line up vertically over the left line. A new endpoint will be indicated at the end of this short segment. Use that endpoint to drag a horizontal line over to the right-hand vertical line. With the snapping features that have been turned on in the previous

tutorial, the horizontal line should snap to the right-hand vertical line once the cursor gets close to it. Repeat for the top margin.

If you feel the need to delete the extra bits of guideline used to lay out the central rectangle, use the Trim tool.

8. Because a guideline should always be used to place the eventual location of a visible line, the next step is to lay out horizontal guidelines that will be used to place each line style sample. Space the guidelines ½″ apart. Start at the top left corner intersection of the border guidelines. Drag down a vertical line ½″ on top of the already existing vertical line. Drag a horizontal line to the right-hand margin from the new endpoint. This places a horizontal guideline ½″ below and parallel to the top margin guideline (Figure 2-6).

Repeat all the way down the page until you get to the bottom margin (Figure 2-7). Resist the urge to look for short cuts like copying and pasting (which *are* options). Drawing all of these guidelines one-by-one is tedious, but it's also practice that will develop eye/hand/ mouse/screen coordination and muscle memory.

9. The page is now ready for sample line styles to be drawn over the guidelines. Switch the active layer to Design Layer-1. Note that the guidelines become muted because a different layer is active. It is still possible to snap to their intersections. In the Navigation-Classes window, select Thick Visible Line. Using the Line tool, click and drag a thick visible line over the first four horizontal guidelines, endpoint to endpoint (Figure 2-8).

10. Selecting each of the five line classes in turn, draw four lines of each style over the remaining guidelines. If the guideline overwhelms the line class being used, turn off guideline visibility (click on the center column in the Navigation-Classes window), or further adjust the color of the guidelines class. When all of the lines are drawn,

FIGURE 2-7   *Page with all of the guidelines laid out*

turn off guideline visibility. The final product should look like Figure 2-9.

11. Use the Text tool to add your name and course information to the page. The Text Tool icon is the T found in the Basic Tools Set (Figure 2-10).

Select the icon. Move the cursor to the bottom of the page somewhere between your bottom hidden line and the bottom of the page. When you left-click the mouse, an Object Info dialog box appears on the right-hand side of the screen above the Navigation-Classes window (Figure 2-11).

Text Style is about halfway down the dialog box. The default is **Un-Styled**, which means a default text style will appear until a style is selected. In future exercises, a text style will be selected when classes are created so that whenever you draw with that class, text appears automatically in the chosen font and size.

Change the size to 20pt. Find Horiz. Align and use its arrow to open its menu. Select Left so the beginning of each line of text lines up vertically (left justification). A blue box will appear where you click the cursor. As you type, letters appear and expand the box. Type your name, course, assignment title, and date, hitting Enter after each line to create four lines of text (Figure 2-12).

Once entering information is complete, you may wish to move the text box to a better location. Move the cursor away from the text and left click to affix the text box to the drawing. Use Select to select and highlight the text box; this allows you to drag it to a new location. If the box persists in snapping to a place near but not quite exactly where you want it, turn off the snapping tools until you can set the text box in the desired location. If you wish to edit the text, choose the Text tool and click on the block of text to be edited.

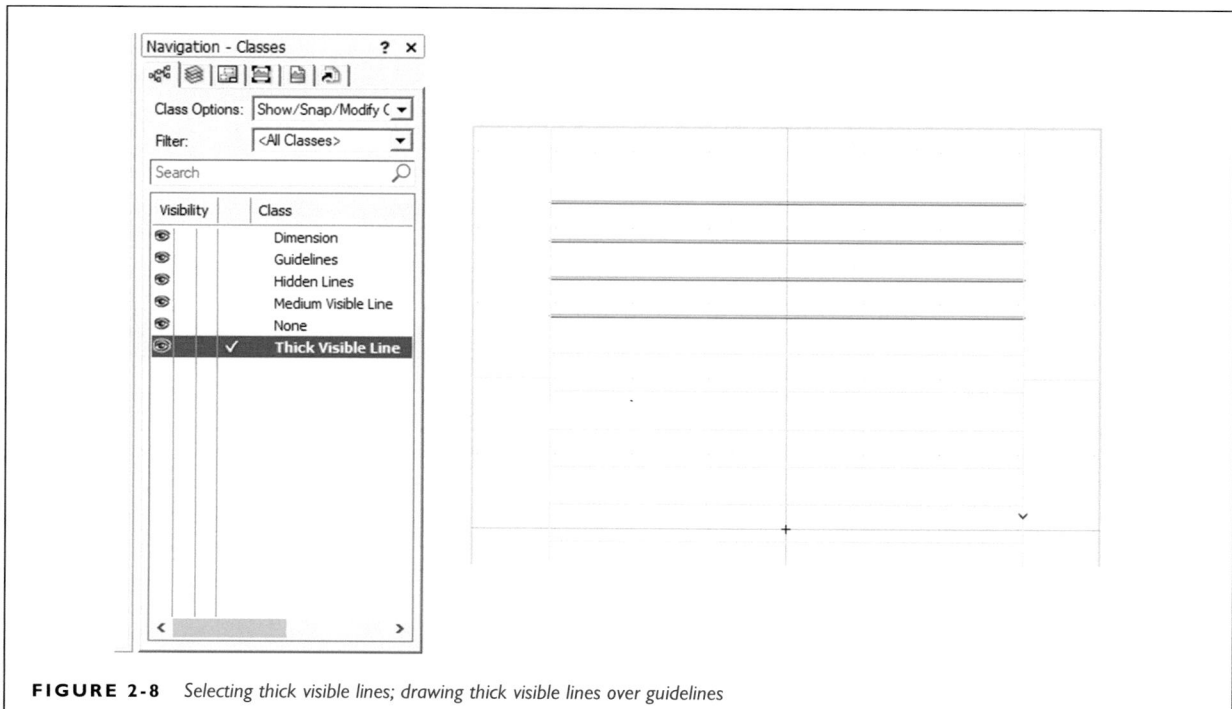

FIGURE 2-8   *Selecting thick visible lines; drawing thick visible lines over guidelines*

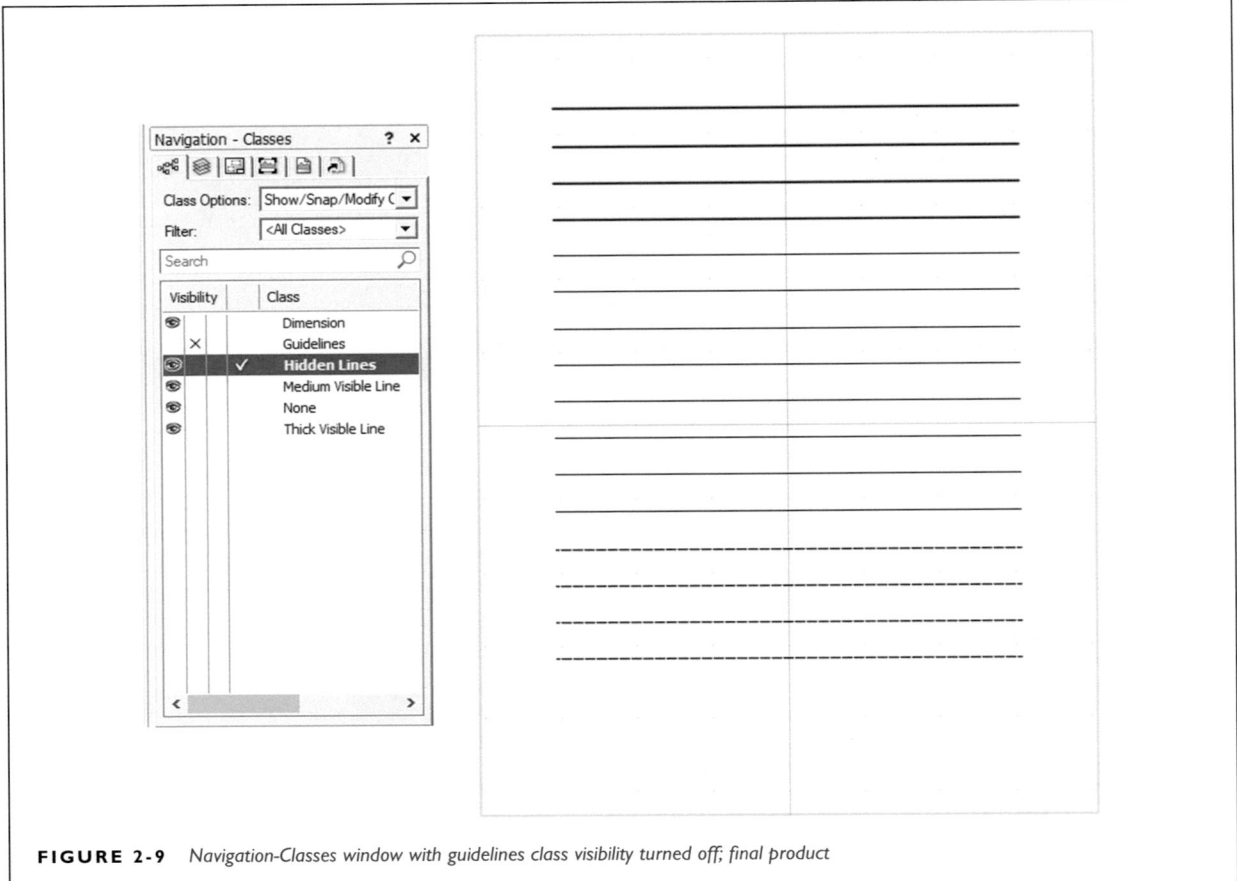

FIGURE 2-9 *Navigation-Classes window with guidelines class visibility turned off; final product*

The text box may have a white background that obscures any lines that it's placed over. This white background is a default Fill set for text. This can be useful later on, but for now let's go with a transparent background. If you have a Fill, Select the text box. In the Attributes dialog box, there is a small paint can icon next to a text box. Use the downward-pointing arrow to open the drop-down menu. Select None to turn off the Fill (Figure 2-13).

12. **Remember to save your work.**

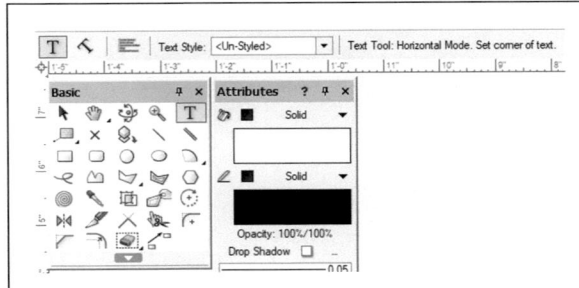

FIGURE 2-10 *Locating the text tool icon in the basic tool set and its modes in the upper tool bar*

To Print the Drawing

On the main toolbar at the top of the screen, select File. Locate Print on the drop-down menu. Be sure that the drawing is sent to the correct printer or plotter. Click Print.

These steps will print whatever is contained within the blue reference grid. If the computer you are using is not connected to a printer or plotter, it will be necessary to transport the file to a different terminal. Converting the drawing to PDF file can expedite this. A PDF is a much smaller size than the Vectorworks file and can be opened on a computer that does not have Vectorworks installed. The drawing can also be saved as a JPEG or PNG file via the Export choice; choose Export Image File and save the file into the desired folder.

The reference grid is not quite as big as an 8½″ × 11″ sheet of standard copy paper and you may get two borders on the hardcopy: the actual border and lines denoting the edge of the reference grid. If you go to Page Setup and change the paper size, the reference grid will shift. Gray tile lines will appear that denote the edges of the new paper size, and when printed, multiple sheets of paper will be used. For this project, using the default page size is fine.

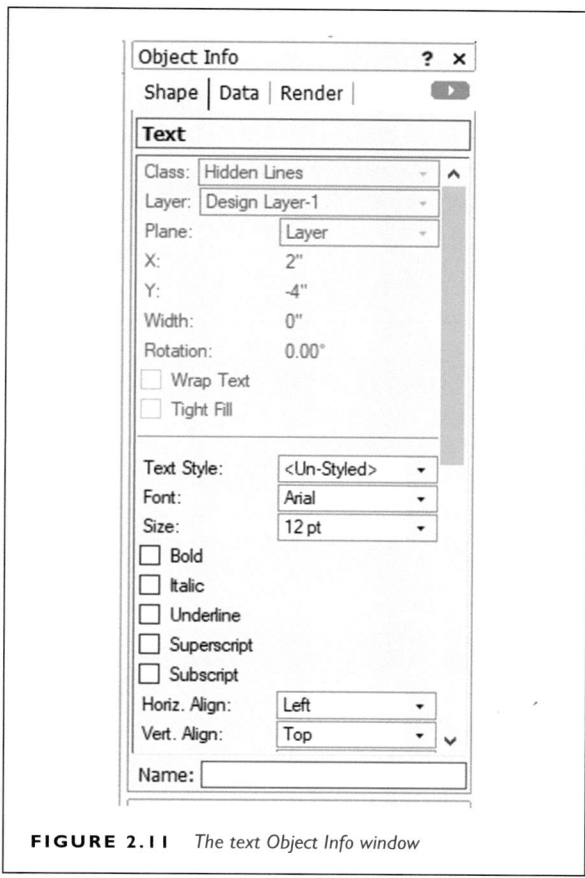

FIGURE 2.11   *The text Object Info window*

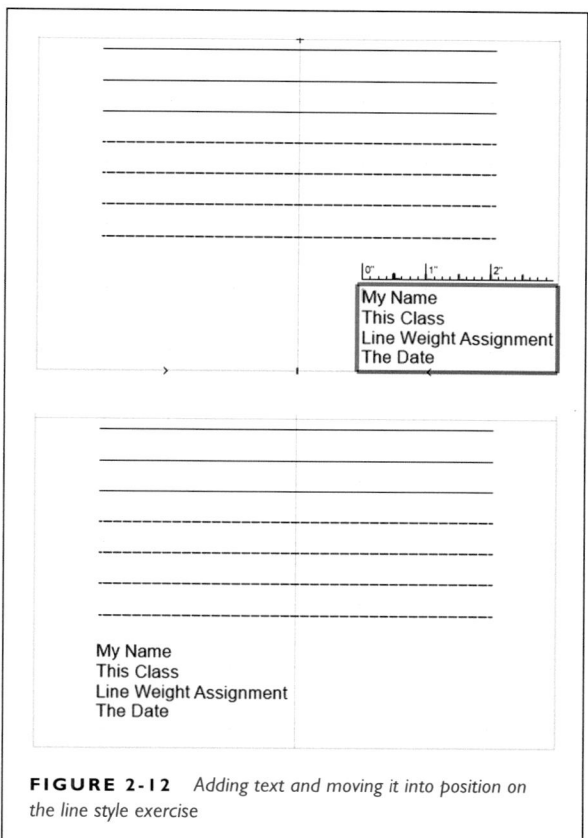

FIGURE 2-12   *Adding text and moving it into position on the line style exercise*

To Create a PDF of the Drawing

1. Under File on the main toolbar, choose the Export option.

2. Choose Export PDF… Don't worry about all the settings in the navigation window that pops up. Default settings are fine for this exercise. Click Export. Whatever PDF creator you have on your computer should appear and allow you to save to the file to the desired hard drive, USB drive, or cloud service.

As with any software program, PDF makers (such as Adobe Acrobat) are regularly updated and reconfigured. If the draftsperson creates the drawing at an institution or company, the PDF maker may require login credentials via the institution's license to create or open a file. Always check with a drawing's end-user to learn what file formats they are able to open and read.

Homework: Hand Drafting the Line Styles

Instructor

In this exercise, students will draw the first five line styles by hand. They should have the hard copy of the previous Vectorworks exercise nearby to refer to. Lines should have consistent opacity and width for their entire length. Hidden lines should have uniform dashes and gaps. When reviewing the exercise, note those places where the line style came closest to the ideal. Ask the student what they did differently for that segment to figure out how to repeat that success.

Students

The print copy of the previous Vectorworks tutorial creates a sample page of ideal line weights for each line style. This assignment asks you to practice pencil control while hand-drawing these line styles.

1. Use 8½″ × 11″ paper, portrait orientation.

2. Use guidelines to lay out margins 1″ from the page edges. Follow the steps shown in Figure 2-5; this time, rather than drawing two horizontal guidelines from left and right, draw one that spans the whole page.

3. Use one of the vertical border guidelines to measure ½″ vertical increments to place horizontal guidelines. Lay out all of the tick marks before drawing any lines.

4. Draw a column of evenly spaced horizontal guidelines.

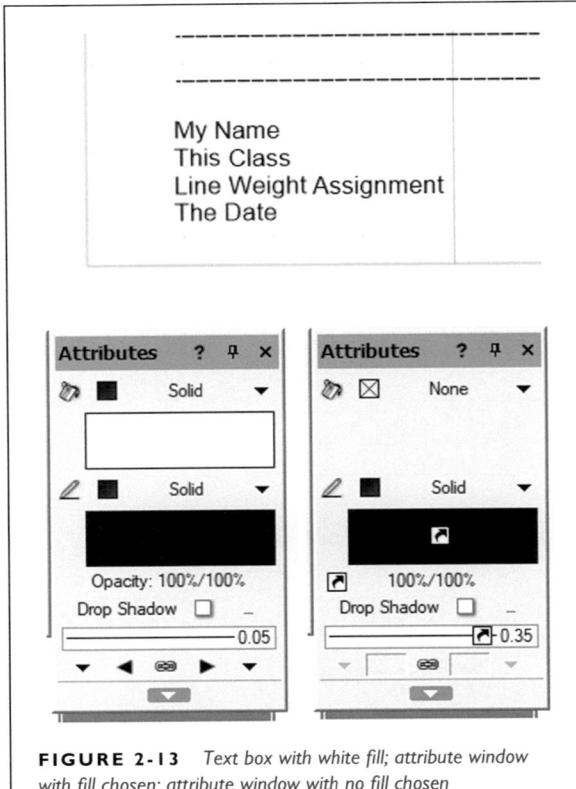

FIGURE 2-13   *Text box with white fill; attribute window with fill chosen; attribute window with no fill chosen*

5. Using the printed Vectorworks tutorial for reference, draw four lines in each line style. Do your best to emulate the thickness of each line of the printout, as well as the length and gaps of dashes. Use pencils of different hardnesses to find the ones that provide the best result for each style.

6. Using guidelines with 1/8″ spacing, letter your name, class, date, and assignment title in the lower left corner of the page. Use a vertical guideline to align the first letter of all lines of text (justify left).

When complete, compare with the Vectorworks sample. Look for consistency in opacity, width, and spacing. Neatness counts. Which portions of the hand-drawn lines were most successful? Can you recall what you did that resulted in a successful line or line portion? What pencil hardness worked best for which line types? Why? If you don't feel the results were all that successful, analyze the problem and draw the line again.

MODULE 2: PROJECTION SYSTEMS

There is minimal interpretive complication when drawing a 2D object on a 2D surface. For a 2D object, information regarding **width** and **depth** is all that is required. Once

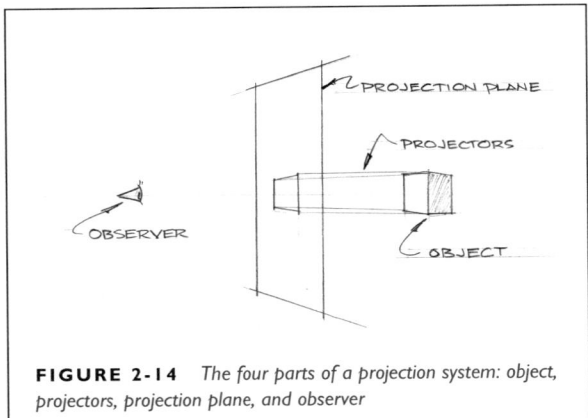

FIGURE 2-14   *The four parts of a projection system: object, projectors, projection plane, and observer*

the third axis, **height,** is added, projection systems are required to fully describe an object. A **projection system** can be thought of as surrounding an object with an array of transparent screens and then looking at each side of the object through its aligned screen.

Each system looks at an object in a particular manner and has its own set of rules. If the viewer knows that a drawing uses Orthographic Projection, one set of rules applies; if the drawing uses Isometric Projection, another set of rules applies. Knowing how you wish to describe an object will determine what system is chosen.

Each projection system involves four elements (Figure 2-14):[1]

1. Observer's eye (station point)

2. Object

3. Plane(s) of projection (picture plane)

4. Projectors (lines of sight, or visual rays)

The observer's eye and the object are the most obvious of these four elements: there is an object, and it is seen by someone.

A **plane of projection** (or picture plane) is an imaginary transparent plane placed between the viewer and the object, much like a pane of glass. The object is viewed *through* the plane of projection. Just as the outline of an object can be traced on the window glass it is viewed through, the image of an object is "traced" onto the plane of projection.

The **projectors**, or visual rays, are imaginary lines that connect the corners (endpoints) of the object to the viewer's eye (station point). This is where projection systems differ. In perspective drawing, projectors emerging from the corners of the object converge. During the Italian Renaissance, the artist and architect Filippo Brunelleschi worked out the graphic rules of **mechanical perspective**, producing drawings that could express architectural geometry in a compositionally cohesive manner (Figure 2-15).

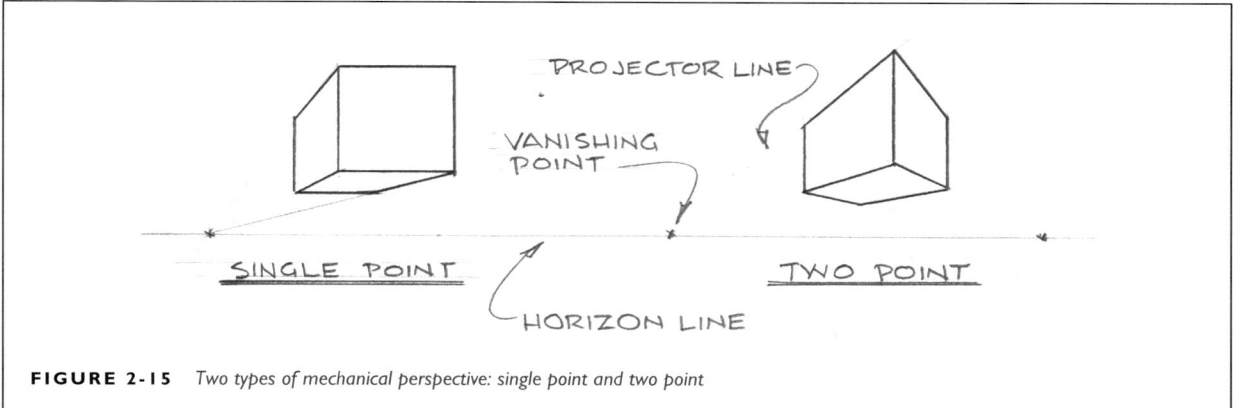

FIGURE 2-15 *Two types of mechanical perspective: single point and two point*

In entertainment drafting, perspective systems are most often used to depict spatial qualities in a pictorial manner, whether a set, venue interior, or theme park building. While perspective drawings offer the overall appearance of the final product, they are not intended to be construction drawings (or, working drawings).

The projection system most used in mechanical drawing is **parallel projection.** In this case, the visual rays connected to the corners of the object remain parallel; they do not converge. This flattens the view of the object, allowing the use of scale since there is no distortion of the object's image.

There are two system branches within parallel projection: **oblique projection** and **orthographic projection**. Oblique means "not parallel or perpendicular."

Within oblique projection, there are two further commonly used branches: **cabinet** and **cavalier**. Both systems describe an object with its front surface parallel to a plane of projection, but with the side and the top views angling off obliquely to that plane (Figure 2-16). Lines that are vertical on the object itself remain vertical on the drawing. Lines representing depth are drawn in a consistent chosen angle, typically 45°, 30°, or 60° depending on the needs of the draftsperson. Cavalier presents the side and top in the same scale as the front

surface; cabinet reduces the top and side to one-half the scale of the front surface.

While cavalier views are easier to measure because of the consistency of scale, cabinet views look more natural. Both systems of projection are useful when a 3D representation is desired but mechanical perspective has yet to be mastered. Be sure that the drawing is labeled so that the reader knows which projection system is used.

There are two systems within the orthographic system heading: **axonomic projection** and **multi-view projection**. Multi-view is the system implied when the term "orthographic" is used on its own. Axonomic has within its family three further systems. The one most commonly used is **isometric projection**. In architectural practice, the axonomic family of projections are also called **paraline** drawings in that sets of lines are infinitely parallel to each other.[2]

Isometric means "equal measure." It is another system, like cabinet and cavalier, that represents all three axes of an object in a single scaled view. Isometric, too, can provide a reasonable 3D representation of an object if mechanical perspective skills are not yet up to the task (Figure 2-17). Many instruction manuals employ exploded isometric drawings to show how individual parts fit together.

In isometric, the object is rotated in relation to the front projection plane so that its principal edges are at equal

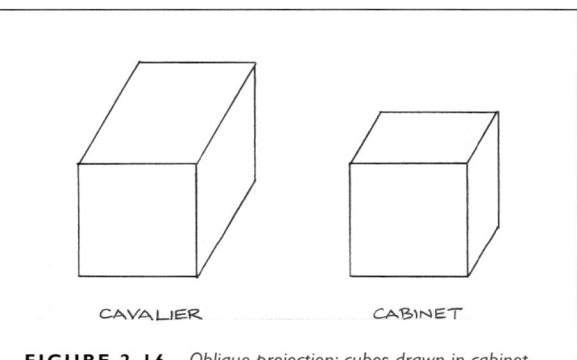

FIGURE 2-16 *Oblique projection: cubes drawn in cabinet and cavalier projection systems*

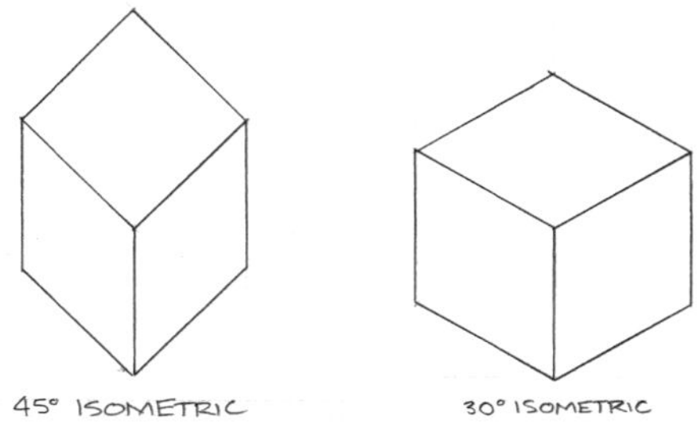

FIGURE 2-17 *Cubes drawn in isometric projection*

angles to that plane, usually 30° or 45°.[3] The object's vertical lines remain vertical; parallel lines remain parallel. Lines representing depth and width are angled and extend away from the plane of projection. All lines, regardless whether height, width, or depth, can be drawn in the same scale. While the view may appear distorted, measurements remain accurate and can be lifted with a typical scale ruler. Circles in an isometric view are seen as ellipses. While there are geometric construction methods to draw ellipse, it's more expedient to purchase ellipse drawing templates. You can buy special isometric rulers that make the object appear more natural by gently distorting the scale as measurements recede from the front, but that's usually overkill.

Multi-view orthographic projection is the prevalent system used in mechanical drawing.

The **glass box concept** is a good way to visualize how multi-view orthographic projection works (Figure 2-18).

Imagine a simple rectilinear object surrounded by a glass cube: a cube of projection planes. There are six sides of the glass cube: front, right side, rear, left side, top, and bottom. Each face of the object within the box is parallel projected onto its respective picture plane. "Orthographic projection" literally means "thrown forward, drawn at right angles."[4] Another way of phrasing this is "straight line projection," as the projectors travel in straight, parallel lines from the object to the projection plane. Mentally unfold the box and flatten its six surfaces into a single continuous plane. This flattened box represents the complete orthographic multi-view projection of the object. Figure 2-19 presents the American National Standard arrangement of views.[5]

Note that the flattened cube presents each view of the object in an aligned visual relationship. The top and bottom views are always aligned above and below the front view. The left and right views are always aligned to either side of the front view. If a rear view is shown, it is placed laterally in line with the front view.

Adjacent views are reciprocal. This means that adjacent views share a common "hinge" of the glass cube. For example, in the top view, the edge that represents the front face of the object (seen from above) is drawn at the bottom of the view, where it is closest to the front view (Figure 2-20). Through visual alignment, each view references information in other views.

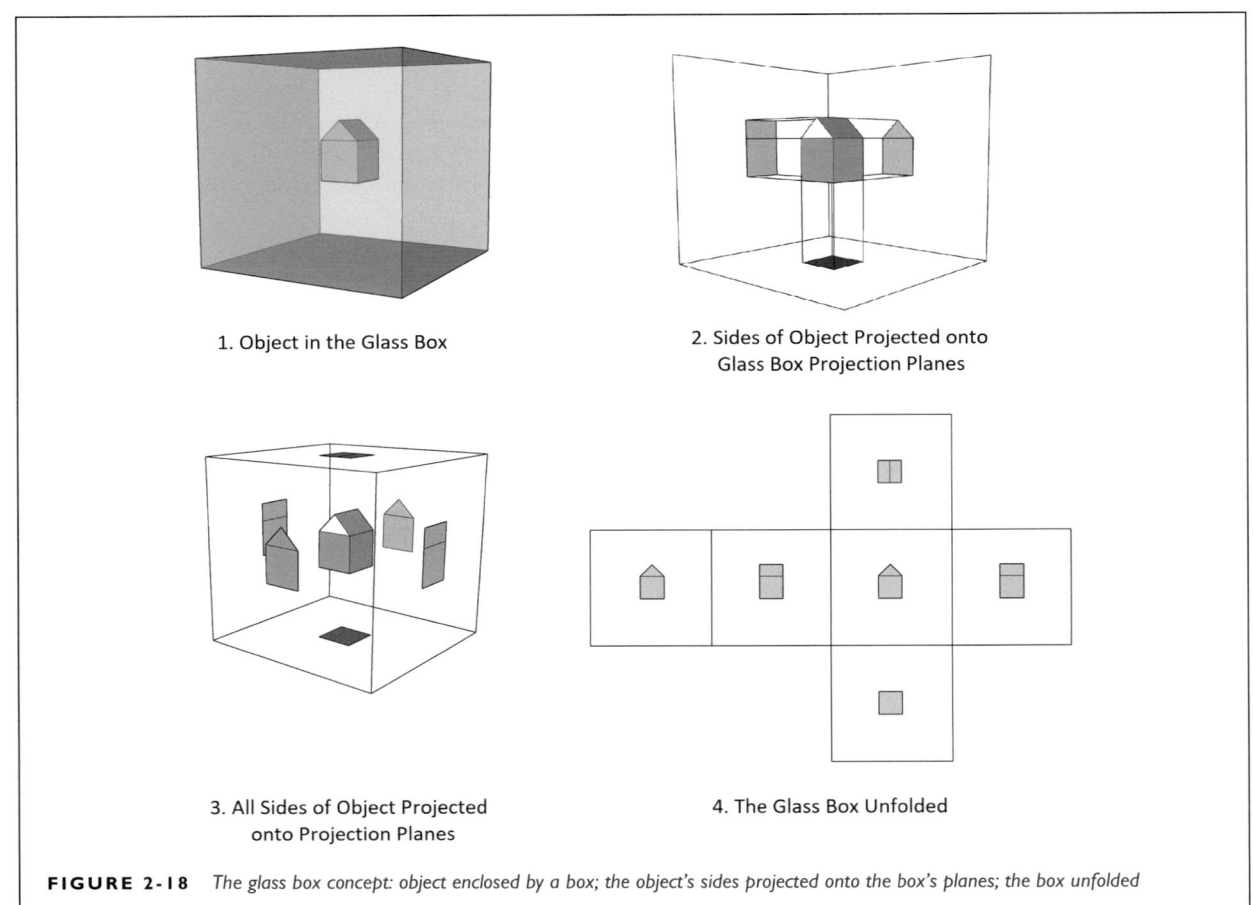

1. Object in the Glass Box

2. Sides of Object Projected onto Glass Box Projection Planes

3. All Sides of Object Projected onto Projection Planes

4. The Glass Box Unfolded

FIGURE 2-18 *The glass box concept: object enclosed by a box; the object's sides projected onto the box's planes; the box unfolded*

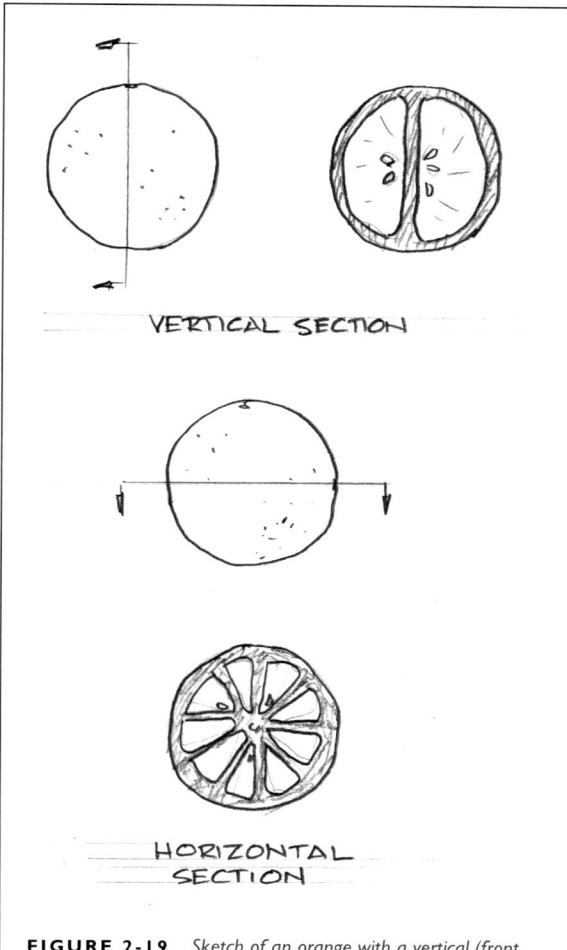

VERTICAL SECTION

HORIZONTAL SECTION

FIGURE 2-19 *Sketch of an orange with a vertical (front view) section and a horizontal (top view) section*

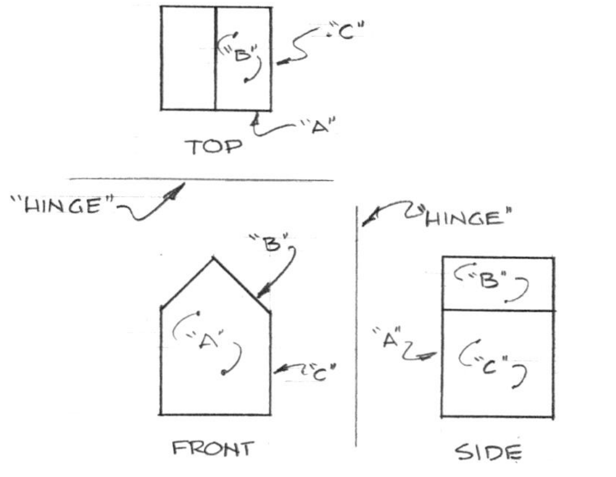

TOP

"HINGE"

"HINGE"

FRONT

SIDE

FIGURE 2-20 *Adjacent view reciprocity: front view plane A is seen as a line in top and side views; plane B in top view is seen as a line in front and side views; plane C is seen as a line in front and top views*

Each view presents information in only two of the three possible dimensions.

Front and Rear:	width and height
Top and Bottom:	width and depth
Left and Right:	depth and height

Not all six views are required for every object. Usually, the front, top, and one of the side views are sufficient.

When a line or surface is parallel to a picture plane, it can be drawn in its true and measurable size in that picture plane. If an object includes an angled surface that is *not* parallel to any of the six standard picture planes, an **auxiliary** view may be required (see Chapter 3). When planning the drawing, choose the views required to fully describe the object most clearly and with the least redundancy.

For large structures such as rooms, buildings, and stages, the term "**plan**" can replace the term "top," as in "plan view" instead of "top view": the groundplan is a plan of the ground. Front, left, and right views are also called **elevations** because they show the height of the object—its elevation. Instead of labeling a view "Front View," it may be labelled "Front Elevation." Elevations that include color and are intended to guide scenic artists in their work are called **paint elevations.**

A **section** view offers the interior view of an object that has been sliced parallel to one picture plane. For example, the standard front, top, and side views of an orange will not reveal the interior structure of the fruit. To see this interior structure, it's necessary to slice the orange; the axis chosen will reveal different information about the interior structure (Figure 2-19).

In addition to the basic meaning of a line's style, in any single orthographic view, a line can be interpreted in one of the three ways (Figure 2-21):

1. The intersection of two surfaces

2. The edge or side view of a surface

3. The contour of a curved surface[6]

In orthographic projection, at least three views are typically required to correctly interpret a line.

Because each line has three possible interpretations, it cannot be over-stressed that orthographic views of an object must be arranged as per the standard glass box concept. The reader gains instant recognition of which views are shown simply from a view's placement on the page. Even if only the front and top views are required to fully describe an object, the top view is always placed directly above the front view. Paper size and paper orientation are to be chosen after determining how many and which views will be used. Placing the top view on a separate sheet because it doesn't fit on a page with the front and side

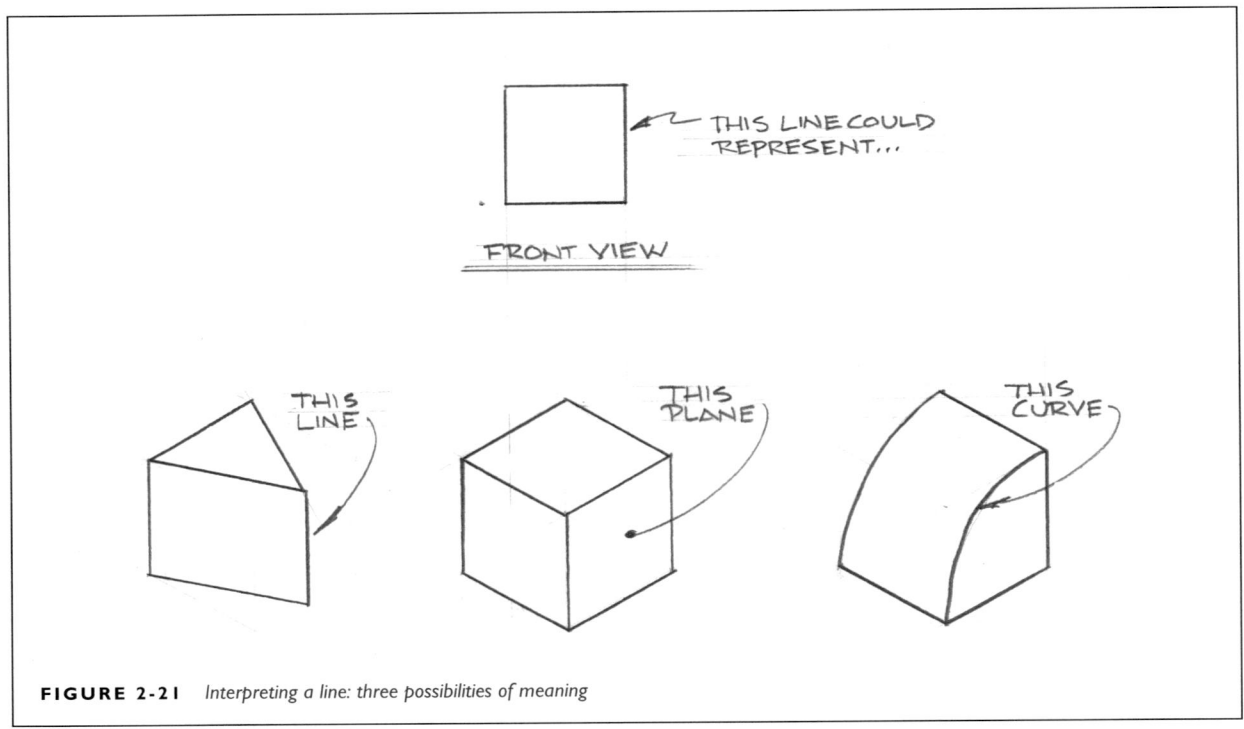

FIGURE 2-21   *Interpreting a line: three possibilities of meaning*

views defeats the integration of information provided by orthographic projection (Figure 2-22).

Keeping in mind the three interpretations of any given line in an orthographic view, knowing which view is being read helps the reader to correctly infer the following information (Figure 2-23):[7]

- A straight line parallel to the picture plane shows its true length.

- A straight line perpendicular to the picture plane shows a single point.

- A straight line inclined to two picture planes (e.g., front/rear and top/bottom) and parallel to a third (right/left sides) only shows its true length on the picture plane to which it is parallel.

- A straight line inclined to all three projection planes (top/bottom, right/left, front/rear) will not show its true length in any of the picture planes. An auxiliary view is required.

- A curved line in a plane parallel to a picture plane shows its true shape on that plane.

- A curved line in a plane that is inclined to two planes of projection (e.g., front and top) and perpendicular to the third (right or left) will not show its true shape on any of the planes; an auxiliary view is required.

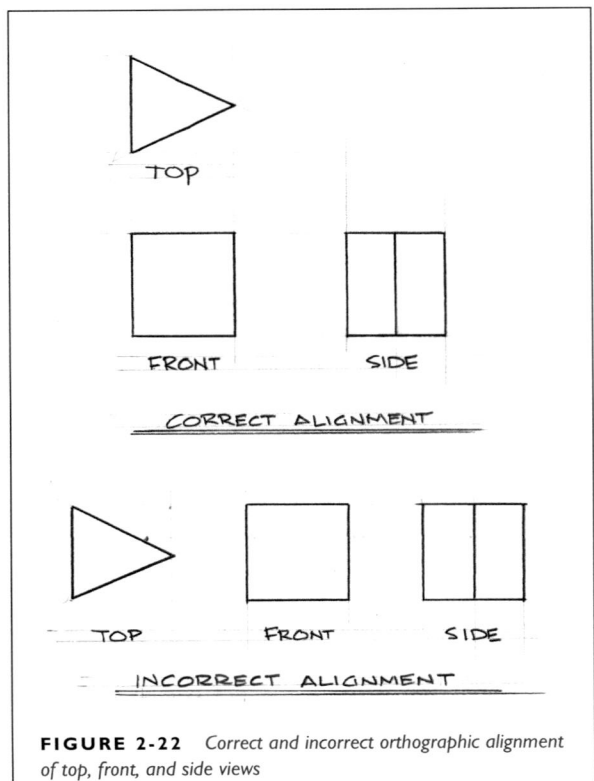

FIGURE 2-22   *Correct and incorrect orthographic alignment of top, front, and side views*

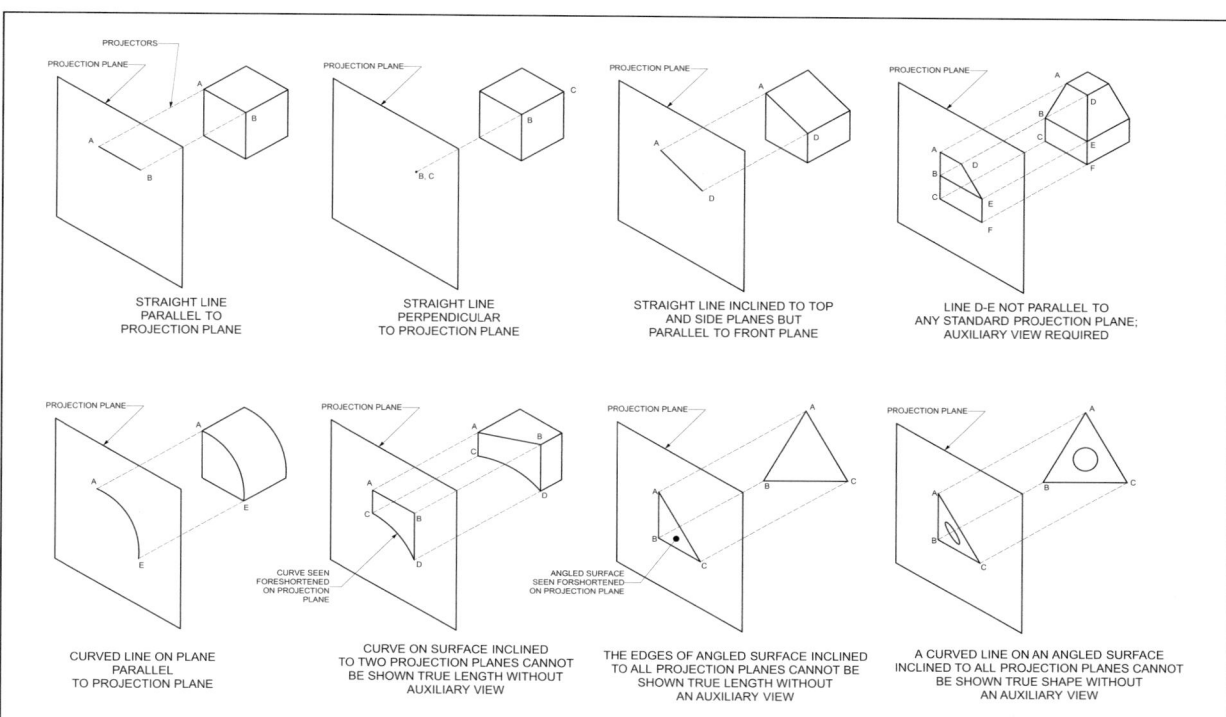

FIGURE 2-23 *Inferring information about a line: top row: line parallel to plane, line perpendicular to plane, edge of incline parallel to plane, edge of incline not parallel to any plane; bottom row: curved line parallel to plane, curved line on incline not parallel to two planes, incline not parallel to any plane, curved line on incline not parallel to any plane*

- A curved line in a plane that is inclined to all three planes (top/bottom, right/left, front/rear) cannot be shown in its true shape in any of them; an auxiliary view is required.

Exercise: Orthographic Visualization

Instructor

Prepare a number of small regular rectilinear 3D shapes (see Figure 2-24 for examples). Make all surfaces whole number increments of measurement. Objects can be fabricated using

FIGURE 2-24 *Sample orthographic visualization practice blocks*

scrap lumber, left over insulation foam, or hot-glued children's blocks. Supply enough so that every student can draw two or three different objects. This exercise is best done on graph paper to provide instant guidelines and regular units of distance.

If time allows, repeat this exercise with different or slightly more complex objects. Sketching objects can also be assigned as homework, and is a good class warm-up. Students should hang onto the sketches, as they will be used in a future exercise.

Students

1. Look at the object(s) provided by the instructor. Determine which side is to be the front. This will be the side that presents the most obvious information, with the fewest features hidden behind other features.

2. Imagine a plane of projection (picture plane) between you and the object. Once the front of the object is determined, use a pencil or small pieces of tape to label all of the surfaces of the object that are parallel to the front picture plane. Label them F for front view.

3. Rotate the object so that the top of the object faces you. Label all of the surfaces parallel to the picture plane with T.

FIGURE 2.25   *Exercise object with surfaces labeled front, top, and side*

4.   Rotate the object so the front faces you. Rotate the object to the right; consider what can be seen from that view. Return to the front and rotate to view the left side of the object. Decide which side view—right or left—gives the clearest information with the fewest features hidden behind other features. Once you have made your choice, label the surfaces on the desired side S (or R, or L). (Figure 2-25)

5.   Sketch the three views (F, T, and S) orthographically. Turn the object so that the front faces you. You don't need to use a straightedge to sketch these objects if graph paper is used. Lay out all three views with guidelines before darkening outlines.

     When laying out the front view, extend vertical guidelines to help place elements of the top view directly above. Do the same with horizontal guidelines to place the side view (whether right or left). In each view, only the surfaces parallel to that view's picture plane will be drawn. Surfaces perpendicular to that picture plane appear as lines (edges).

6.   Rotate the object to the top view. Use the guidelines extended upward from the front view to sketch the top view directly above the front view. Leave adequate white space between the top and front views.

7.   Rotate the object to the preferred side view. Use the guidelines extended from the front view to sketch the side view directly to the side of the front view. Leave an inch or two of white space between the front and side views (Figure 2-26).

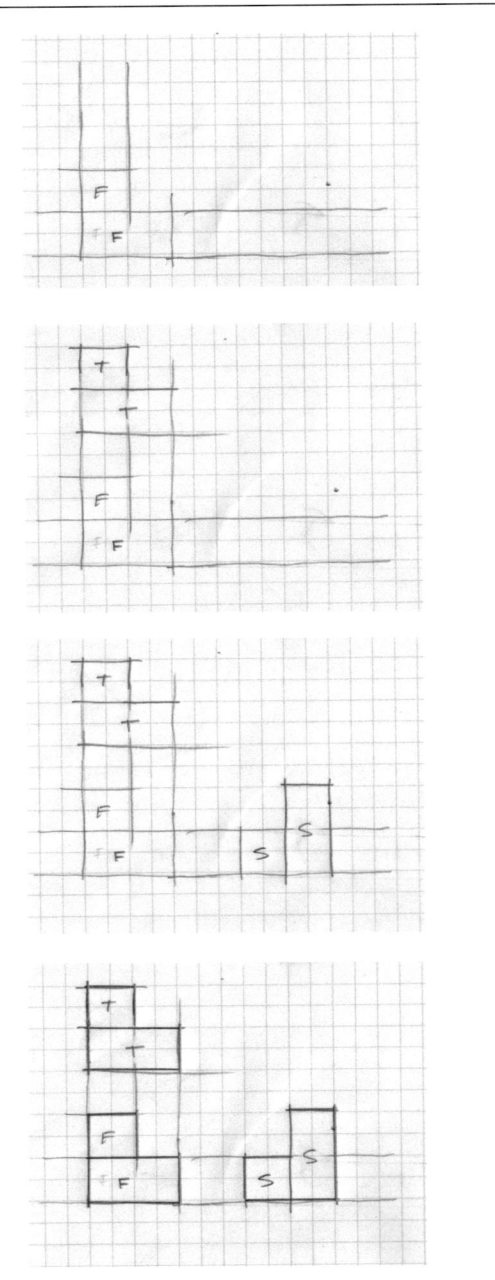

FIGURE 2-26   *Steps of object orthographic layout: lay out front view; extend guidelines for other views; lay out top and side views; darken*

8.   If there are edges hidden behind surfaces in any view, these edges are represented by hidden lines *only in the view in which they are hidden.* An edge visible in the side view may be hidden in the top view. In this case, the line is represented by hidden lines only in the top view. If the hidden edge lies directly behind the edge that obscures it, it is not added to that view (Figure 2-27).

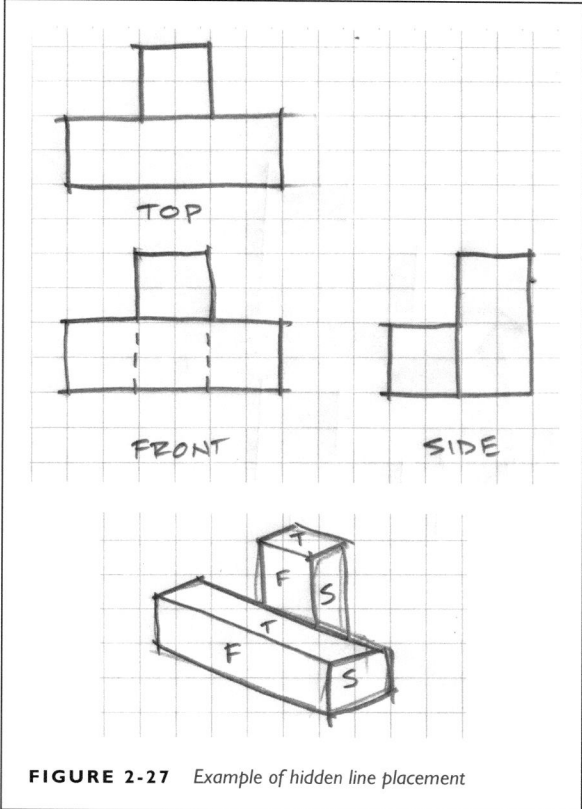

FIGURE 2-27 *Example of hidden line placement*

FIGURE 2.28 *Top, front, and right side photos of small chest of drawers, arranged in standard orthographic relationship*

There are do's and don'ts for hidden lines; these are discussed in Module 3 of this Chapter. For now, a dashed line will suffice.

9. Using guidelines for your lettering, label each view (if the graph paper has increments of 1/8″, use the lines of the paper as guides). Center the label beneath the view, about ½″ to ¾″ below it.

In class, review and discuss the sketches. Why was the right or left view chosen? What feature is clearer in which view? Did you have trouble "flattening" the object out (imagining the object via parallel projection)? Are the views clearly and correctly aligned? Were all of the hidden features indicated with hidden lines? Were there features indicated with hidden lines that were *not* hidden in that view?

Homework: More Orthographic Visualization

Instructor

The students will draw a scale rough sketch of an object in three standard orthographic views: front, side, and top. They should select a generally rectilinear object in their residence, such as the small chest of drawers pictured in Figure 2-28.

Students

Select a small, rectilinear piece of furniture, such as the chest of drawers in Figure 2-28. Sketch this object orthographically in scale. Choose a scale that allows the object to fit on the paper, yet still allows an adequate level of detail to be seen. Use graph paper for the sketch. To help you understand the views, take photos of the object from the front, top, and chosen side.

1. Determine which side of the object provides most of the information about the object's features; use that side view for the sketch.

2. Measure the object. Draw a rough sketch of the object to record the measurements.

3. On graph paper, lay out the perimeter of each view, using guidelines to align the top, bottom, and sides of the three views. Use a straightedge to keep things aligned. Start with the front view and extend guidelines into the other views.

4. Lay out details and features, such as the drawers and top-edge overhang seen in Figure 2-28. Reinforce straight lines with a straightedge. Add freehand detail as required.

5. Darken the lines of the object. Do not erase the guidelines, as those will aid in discussion of how the final product was constructed (Figure 2-29).

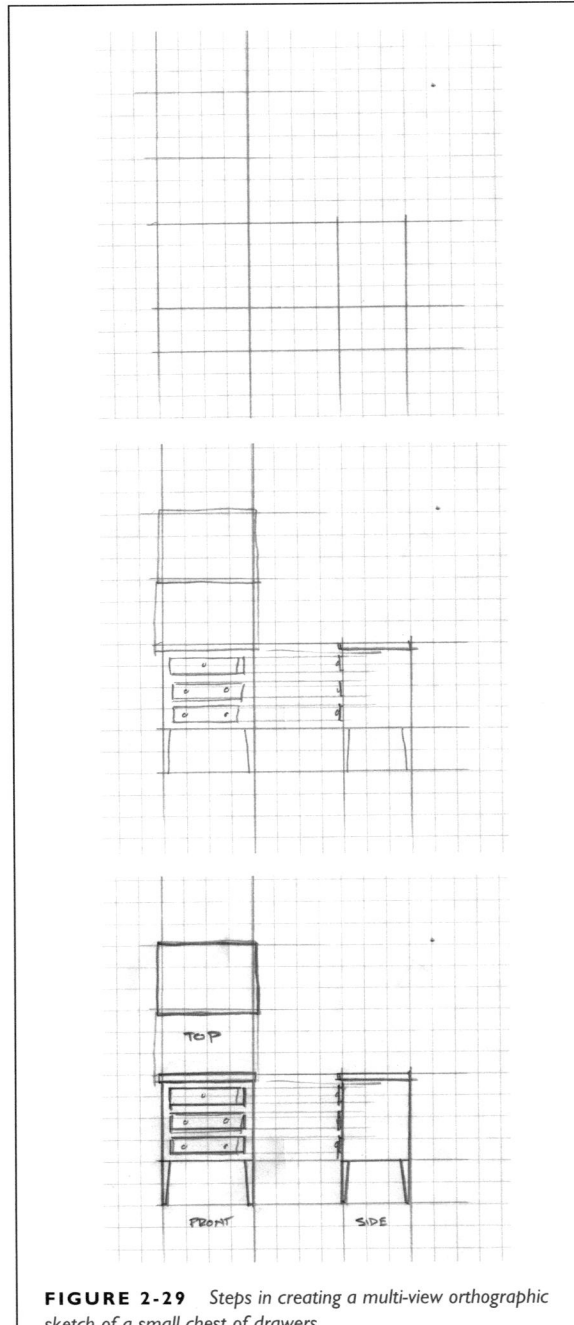

FIGURE 2-29   *Steps in creating a multi-view orthographic sketch of a small chest of drawers*

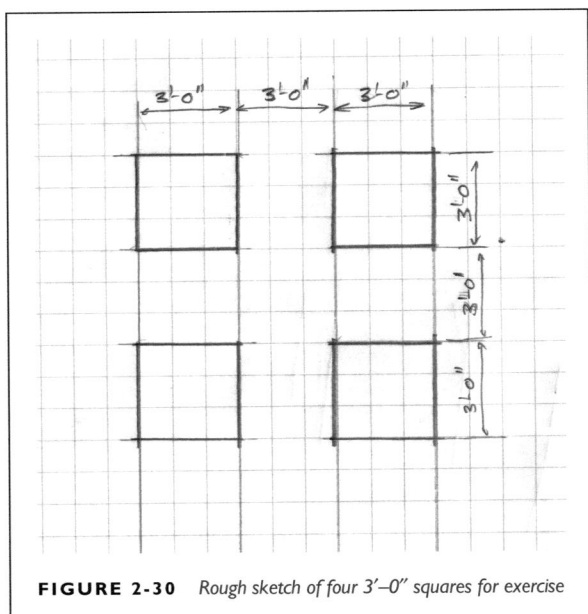

FIGURE 2-30   *Rough sketch of four 3'–0" squares for exercise*

MODULE 3: DERIVING THE MEANING OF LINE WEIGHTS AND TYPES

Visualizing an object goes hand in hand with selecting appropriate line style. The following exercise provides an introduction to SketchUp and also provides practice visualizing 3D objects with their 2D representations.

Exercise: Building Four Cubes in SketchUp

Whereas Vectorworks works primarily with lines and endpoints, SketchUp works with volumes enclosed by planes. The program's designers want the user to begin drawing as quickly as possible, so beyond selecting a template at start up, there is little setup aside from selecting desired toolbars and palettes beyond the basic tool set.

The initial steps of this exercise involve laying out four equal squares. Each square's sides will be 3'–0". The four squares will be laid out in two rows and two columns on a horizontal plane (Figure 2-30). There will be a space of 3'–0" between the rows and between the columns.

1. On the start-up screen, a variety of templates are offered for use. Select the "Simple/Inches" template.

2. A drawing space will appear (Figure 2-31). Three colored axes run through the space. Note their colors, as the colors appear whenever you draw a line that aligns with one of the three axial directions: height, width, and depth. A human figure is present to provide a sense of proportion. The space is divided into sky and ground, with a horizon line far off in the distance. This is to aid viewer orientation while zooming, panning, and orbiting. The ground is infinitely below the intersection of the three axes located near the human figure.

3. Along the top of the drawing space, there should be a menu of tool icons. If this is not present, go to View on the main toolbar and select Toolbars from the drop-down menu. Check the Boxes for Getting Started and Large Tool Set. When you Close and return to the drawing space, it should look like Figure 2-31.

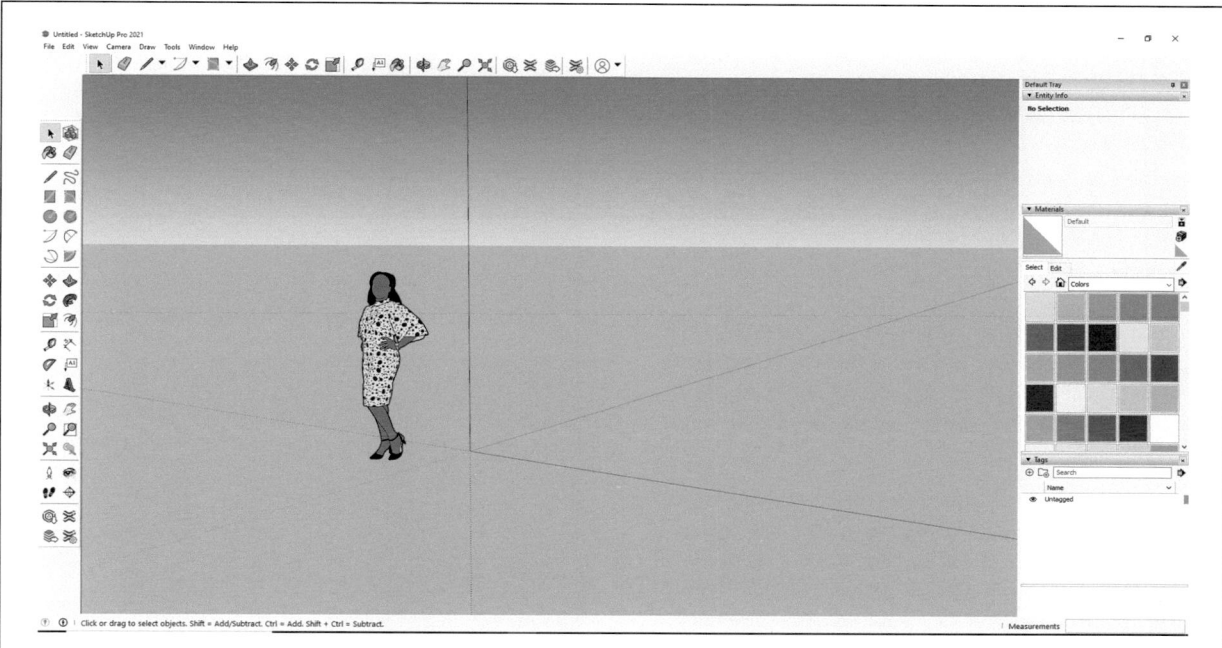

FIGURE 2-31   *The SketchUp drawing space, with Getting Started and Large Tool Set toolbars, as well as Default Tray on right side of screen*

A number of tools repeat between these two toolbars. On the right side of the screen, there is strip that reads Default Tray at the top. Click on the corner X to close this; it will reappear as needed.

4. Hover the cursor over the various icons in the Getting Started toolbar. The names of each icon will appear. Locate the following tools: Zoom, Pan, Tape Measure, Rectangle, Line, Erase, Offset, and Push/Pull (Figure 2-32).

5. Experiment with the Zoom and Pan tools. As with Vectorworks, it is necessary to zoom in and out, moving the point from which the object is viewed. Because you're working in a 3D space, orbiting allows you to swivel your viewpoint in order to observe the object from any direction. If you get lost in your model, the Zoom Extents tool snaps to a default point in front and above the model. The wheel on your mouse can also be used for zooming.

6. Select the Tape Measure tool (the tiny tape measure) to lay out guidelines. This tool also reads the measurement

of a drawn line by clicking on the line's endpoints. There are two ways to draw guidelines:

a. Click on a point along an already drawn line or axis to drag out a guideline parallel to that line. Click to affix the guideline at the desired distance. This produces a guideline that extends to infinity in either direction. You can type the desired separation distance into the keyboard, and hit Enter.

b. Click on an intersection to drag out a guideline of a desired length. A guideline will not appear unless there is an existing point or line to anchor the start point.

On the screen, the red axis is the default horizontal guideline and will be used to denote the forward edge of the first row of squares. The squares will be drawn in a horizontal plane extending back from the red axis, laying out their depth and width. The first set of guidelines drawn will mark the depth of the squares and the space between rows.

   With the Tape Measure tool selected, click on the red axis (horizontal: width) and drag the cursor

FIGURE 2-32   *The Getting Started toolbar*

back from the axis line toward the horizon. A dashed guideline will appear, parallel to the red axis. There will also be an arrow connecting the axis to the guideline; when this arrow is the same color as the green depth axis, the guideline is in the same horizontal plane as the origin line. You can also drag the cursor along the green axis to ensure that the guideline is in the desired plane. It may take a bit of micro-movement with the mouse to get the guideline into the correct alignment.

While this tool is still active, type 3′. This measurement appears in the text box in the lower right-hand corner of the screen. Hit Enter. The guideline will snap to a distance 3′–0″ away from the red axis. This guideline marks the rear edge of the first row of squares.

SketchUp does not require typed measurements to include a dash/hyphen between feet and inches. As long as the foot measurement has a single prime mark and the inch measurement has a double prime mark, it will understand your intention. Measurements with both feet *and* inches should be entered without separation: 3′6″.

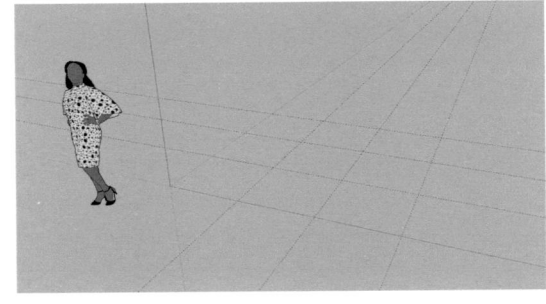

FIGURE 2-33 *Laying out guidelines for the array of squares*

The human figure may be standing right where you are working, so it's time to move it aside. Click on the Move tool icon. Left click on the human figure. A blue box will appear around the figure's perimeter, showing that it is active. Holding down the mouse button, drag the figure away from your working space. Release the mouse button to affix the figure in its new location.

7. To place the forward edge of the second row of squares, select the Tape Measure tool. Click on the guideline, and hold the mouse button down to drag out a second guideline, parallel to the first. Type 3′. Hit Enter. The new guideline will snap into place. Be sure that it intersects the green axis and is therefore in the same plane as the other guidelines.

8. Repeat this process to lay out guidelines for the squares' width. Use the green depth axis as the left-most edge of the array and drag out guidelines to the right of and parallel to it. When complete, the set of guidelines should look like a tic-tac-toe grid made of dashed lines (Figure 2-33).

One of the issues many students struggle with in their early use of SketchUp is creating lines and surfaces that are aligned with other lines and surfaces. Because of the viewer's floating vantage point in a 3D drawing environment, it's easy to draw lines slightly askew or parallel to their intended location. It's worth taking a moment to rotate the view and double-check line placement to prevent errors later on. The axis colors are there to help you, and the appropriate color (red, blue,

green) will show up to keep a line oriented in the desired axial direction. Undo is the first choice on the Edit drop-down menu, and will back you up step-by-step.

9. Use the Line tool to outline each of the four squares. Click on a corner and drag the line to another corner of the square. Click to snap the line into place. Intersections will be indicated with a small red box or a small box with a red x in it. Don't outline the space between the squares. Use the Erase tool (or Undo under Edit in the upper toolbar) to delete erroneous lines. There is a Rectangle tool, but as in the Vectorworks exercises, shortcuts will be saved for later.

As you complete each square, SketchUp automatically fills in the area bounded by the lines to create a surface. Surfaces can be deleted by using the Erase tool or Selecting the surface and right clicking to choose Erase from the menu that appears. Erasing the surface leaves the boundary lines in place. If you erase a boundary line, the surface also vanishes. For this exercise, the surfaces will be used to Push/Pull the squares into cubes.

You've just drawn four 2D polygons on a plane, much as you would on a sheet of paper (Figure 2-34). Each square is a surface bounded by lines. Each square has width and depth but no height.

10. The three axes shown in the drawing environment also provide references for standard orthographic view picture planes. SketchUp considers the side of the

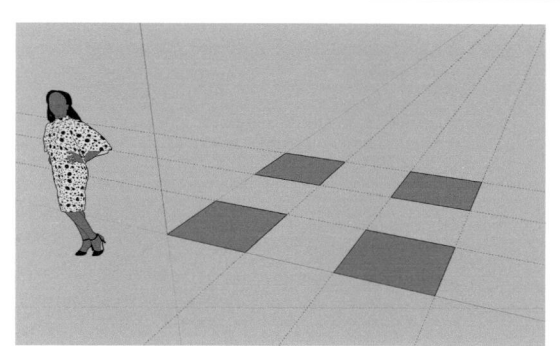

FIGURE 2-34   *The four squares, outlined, with filled surfaces*

object facing the red axis as the front. To demonstrate this, go to Camera on the upper toolbar and open its drop-down menu (Figure 2-35).

At the top of the menu is Standard View, with a small arrow to the right. Slide the cursor to the right, and a number of choices become available. Select Front. The screen will reorient to position the viewer directly in front of the object. Since the default setting is Perspective view, lines will recede toward the distant horizon but only a sliver of the surface will be visible.

Select Top. Now the image shows all four squares in their square array, as though you were looking down at a piece of paper.

11. Use the Orbit and Pan to rotate the view to look at the squares from a more angled viewpoint. To make the squares cubes, choose Push/Pull and click on the surface of one of the squares. A grid of tiny dots appears on the surface to show that it is active. Left click the mouse. Drag the cursor up slightly. The surface of the square follows the cursor and vertical planes will automatically rise from the outline of the square to create sides. While the surface is active, type 3′; hit Enter. The surface will snap to a height of 3′, creating a cube that is 3′ on all faces. Repeat this process for the three remaining squares (Figure 2-36).

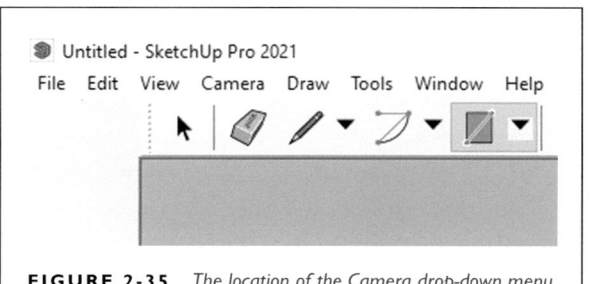

FIGURE 2-35   *The location of the Camera drop-down menu on the main toolbar*

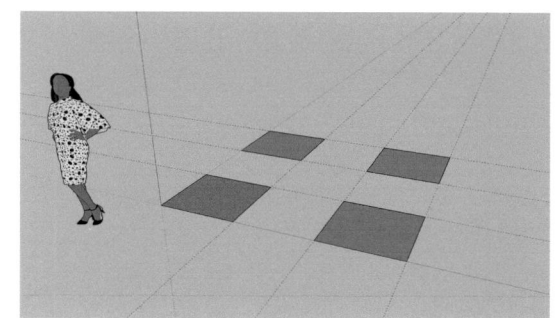

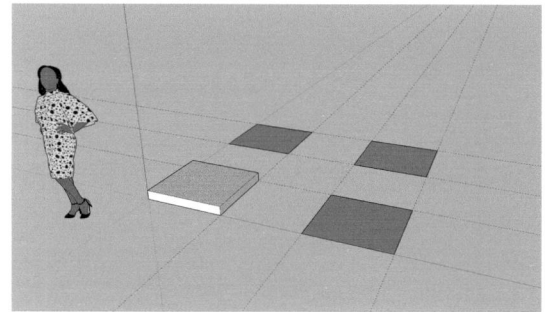

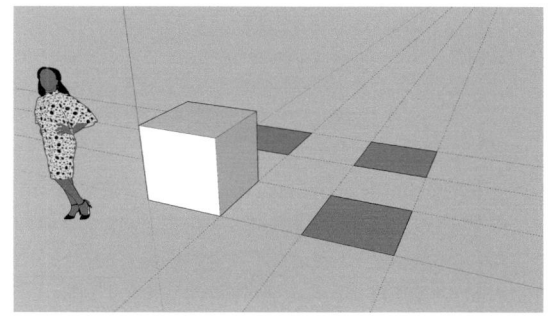

FIGURE 2.36   *Using Push/Pull to create a volume from a flat surface*

12. Look at the cubes with the standard camera views. Switch to **parallel projection** via the Camera drop-down menu. Projection choices are listed.

13. Go to the standard views and select Top. This view of the cubes appears identical to the top view of the squares; however, there is a shift in the meaning of the lines. When this was a drawing of 2D squares on a 2D plane, the lines represented only the outlines of the squares. Now the lines represent:

  – The boundary of the cube's top surface
  – The intersections of the cube's top and side surfaces
  – The edge view of the side surfaces

Note that in the top view, only width and depth are shown. If a reader was presented with this image, they

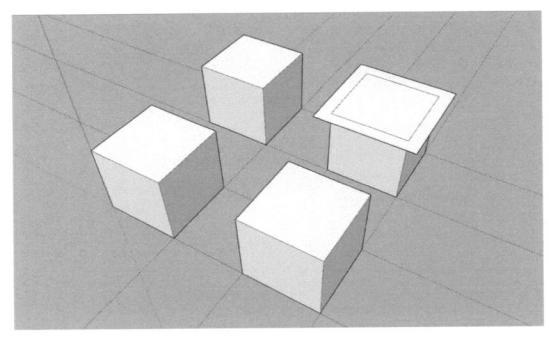

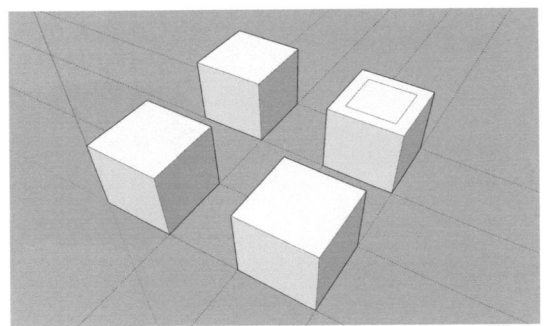

FIGURE 2.37   *An offset toward the exterior of the rear right cube's top surface; offset into the interior of the cube's top surface*

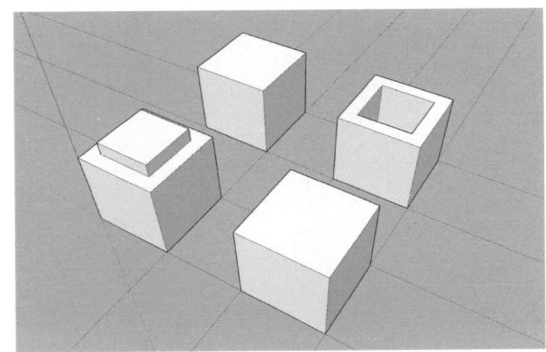

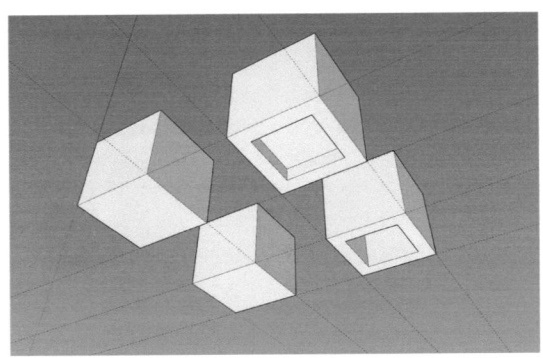

FIGURE 2.38   *Rear right cube with a hole all the way through it; front left cube with a raised area; orbiting the view to see cubes from beneath*

would be unable to tell whether the shapes were squares or the top surfaces of cubes.

14. Switch to Front view. Only two cubes will be visible, as they are positioned directly in front of the other two. Only height and width are visible. Change the view to Right. Again, only two cubes are visible. Only height and depth are shown.

15. Return the Camera to Perspective. Three cubes will now be altered to create contexts for different line styles. Find the Offset Tool (on the Large Tool Set). When the cursor is clicked on a line, this tool creates a second line that follows the perimeter of the first line at a selected distance. Click the cursor on the outline of the top surface of the back right cube. When the cursor is dragged outside the square, a line will surround the top surface in the same plane as the surface. Drag the cursor into the interior of the top surface. Type 6″ and press Enter. A square will snap into place 6″ within the outline of the top surface (Figure 2-37).

16. Use Push/Pull to push the inner square's surface all the way to the bottom of the cube (3′). You can also type 3′, Enter while the interior square is active. When push/pulling, if you click the cursor on an edge parallel to the desired depth, the surface will snap to that height or depth. SketchUp will then help out by automatically

erasing the surface within the boundaries of the push/pull action (in most cases). A square hole now passes through the cube from top to bottom. This hole cannot be seen from the front or side standard views.

17. Use Offset on the top surface of the front left cube. Place the inset line 6″ within the top's outline. Use Push/Pull to pull the interior square up 6″. This feature can be seen in the front and side views, but from the top, there is no indication whether this is a raised area or depressed area (Figure 2-38)

18. Orbit so that your viewpoint is beneath the array of cubes. On the bottom of the front right cube, use Offset to draw an interior square 6″ away from the outline. Push this interior square up into the volume of the square 6″. The bottom of the cube now has a cavity that cannot be seen from the top, side, or front views (Figure 2-38).

19. Change the camera view to parallel projection. Select the standard top view. Only two cubes have features visible in the top view. The hole running vertically through the upper right cube is discernable as a hole because SketchUp shades the ground below. Without the shaded field, the top views of the upper right and

lower left cubes would appear identical. The lower right cube shows no extra features at all. The inset area on the bottom of the cube is totally hidden. A viewer seeing only the top view would never know this feature exists.

20. Return to perspective projection and orbit the image so that all four cubes are visible from above. Open File on the main toolbar to find Export. Choose Export 2D Image. Export as a PDF to your desired file location. Orbit the view so that the undersides of all four cubes are visible. Export this view as another PDF. Be sure to give it a name different from the first PDF. Print out both PDFs, as they will be used as reference images when these cubes are drafted in the following exercises.

    You can also select Print from the File menu to print whatever image is currently on the screen.

21. Save the SketchUp file with a useful name.

Exercise: Choosing Appropriate Line Styles

Instructor

The students will draft the top views of the four cubes constructed in the previous SketchUp exercise. This is an exercise to lay out related objects and practice drawing the various line styles. Line styles will be selected to describe the features of each cube. This exercise should be done both by hand and with Vectorworks. Decide which version to do first, and whether one or the other is done in class or as homework. Borders should be included.

Students

Draft the top views of the four cubes constructed in the SketchUp exercise using appropriate line styles for the features. Be sure to have the printed (or otherwise accessible) images of the four cubes at hand for easy reference. Draft in ½″ = 1′–0″. Your instructor will let you know whether to draft in Vectorworks first, or by hand first. Figure 2-39 presents a rough sketch of the layout of the drawing to be produced.

Hand Drafting Version

1. Use 8½″ × 11″ paper, landscape orientation.

2. Lay out the guidelines for the border. Since this drawing will not be bound and a little extra drawing space will be useful, use ½″ spacing around the full perimeter. Do not darken the border until the drawing is complete.

3. Do layout math to situate the center of the array in the center of the drawing space.

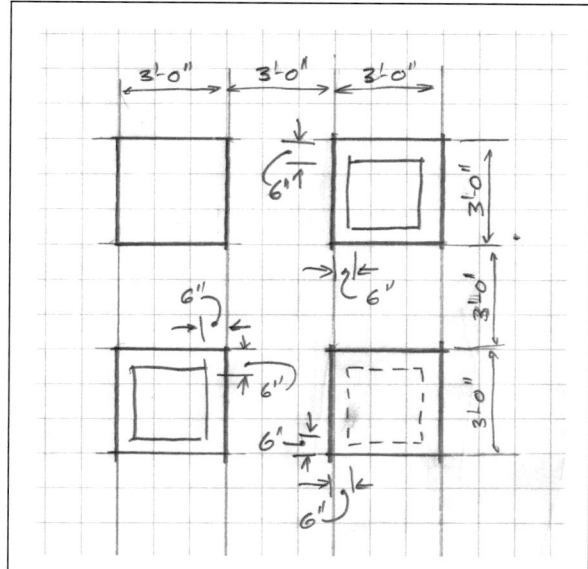

FIGURE 2-39 *Rough sketch of four cubes, top views with interior features and measurements*

4. Use guidelines to lay out the outlines of the array. Lay out the full perimeter of the array, and then lay out the interior guidelines (think globally) rather than first drawing the perimeter of each square individually. Once the full perimeter is laid out, measurements only need to be placed along one horizontal axis and one vertical axis. Use your T-square and triangles to transfer measurements across the page (Figure 2-40).

5. Lay out guidelines to place the interior features. Again, since all of the cubes are aligned in their array layout, measurements only need to be marked out along one vertical line and one horizontal line.

6. When darkening lines, it is typical to work horizontally from the top of the sheet to the bottom, and then vertically from left to right. When hand drafting, this approach limits smudging. This is another reason to think globally when laying out the drawing, as darkening outer perimeters and interior detail typically occurs together as you work across the page. Since all of the lines of the drawing have been established with guidelines, work your way from top to bottom drawing the appropriate line style over the guidelines. Keep the following in mind:

    a. Since the upper left cube has no interior features, no interior lines will be added.

    b. The hole running through the upper right cube is a major feature and is best described by using the same weight as the object outline.

    c. The raised portion of the lower left square is also an object outline, and should be darkened with the same weight as the object outline.

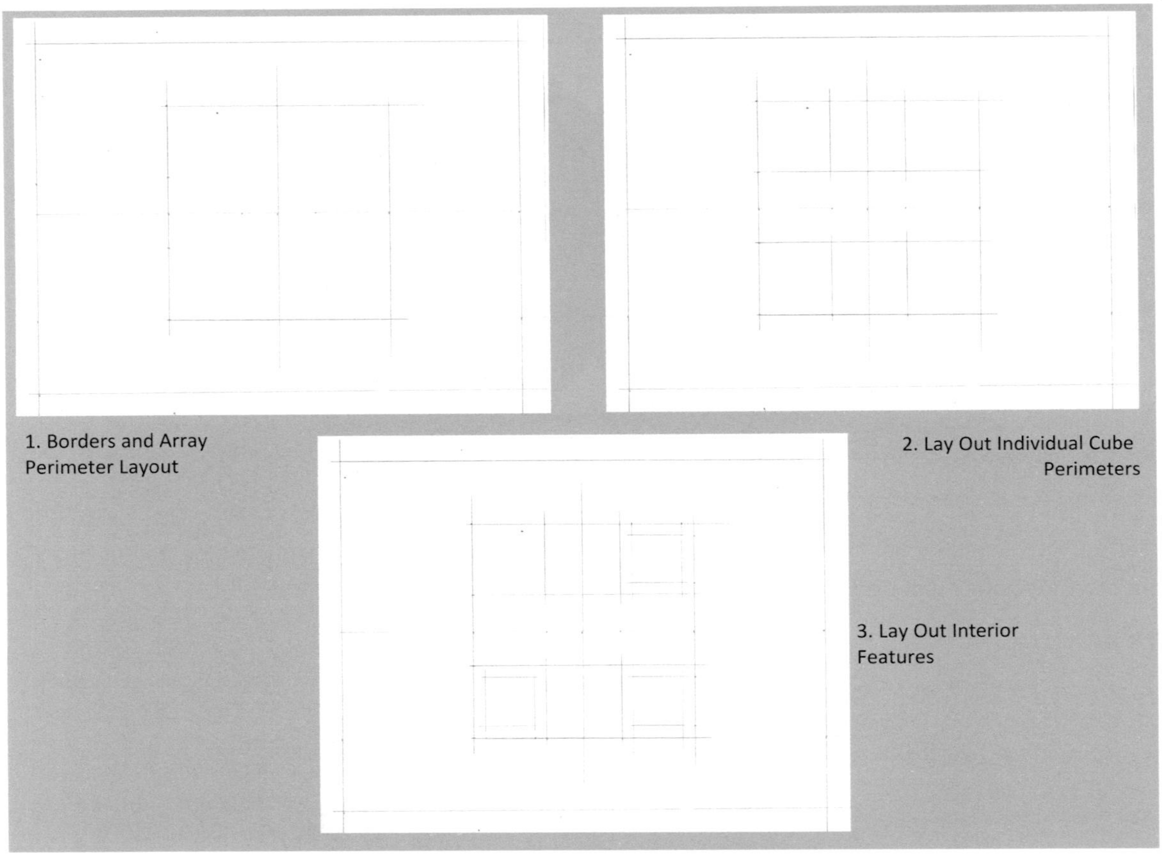

1. Borders and Array
Perimeter Layout

2. Lay Out Individual Cube
Perimeters

3. Lay Out Interior
Features

FIGURE 2-40 *Laying out the cube top views: (1) centering the array, (2) laying out the cube perimeters, (3) laying out interior detail*

d. The bottom inset of the lower right square cannot be seen in the top view. Use hidden lines to represent the outline of this feature.

7. After the features are darkened in with the appropriate line type, darken borders with extra-thick visible lines.

8. Letter your name, class, date, scale (full-size) and project title ("Top Views of Four Cubes") in the bottom right corner of the sheet. Use guidelines for all lettering.

Rules for Hidden Lines (Figure 2-41)

1. At its beginning and end, a hidden line (a full dash) intersects a visible line *except* when the hidden line's dash will appear to be an extension of the visible line.

2. Corners and intersections that are wholly described by hidden lines should be shown with dashes that meet at the intersection (rather than leaving a gap at the intersection).

3. Whenever a hidden line crosses a visible line, the visible line should fall into a gap between dashes. This reinforces the sense that the hidden line is behind the visible line.

4. When a hidden line intersects a corner made by visible lines, a gap is left between the corner and the hidden line.

5. When a hidden line meets a place where a visible line meets and terminates at another visible line, such as a

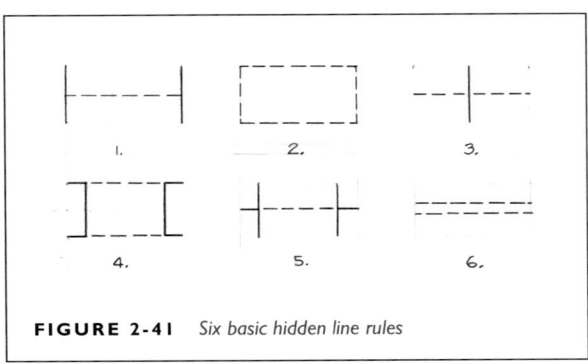

FIGURE 2-41 *Six basic hidden line rules*

T-intersection, a gap is left between the intersection and the hidden line.

6. When drawing parallel hidden lines close to each other, stagger their dashes.

7. If the view of an object has multiple layers of hidden lines, the hidden lines closest to the viewer take precedence. The hidden lines further back interact with the nearer hidden lines as if they were visible lines, following the rules listed above.

Vectorworks Version

1. Open your Vectorworks base file.

2. This drawing will be in landscape orientation. To rotate the drawing area (reference grid), open the main toolbar's File drop-down menu. Select Page Setup. In the dialog box, click the button for Printer Setup. The Printer Setup dialog box will appear. Select Landscape.

    Note that there is a text box titled Scaling and that the option reads "100%." This means that if a line is 1″ long on the drawing, then that line will be 1″ long on the printed page. Later, when you create or receive files, this selection in Page Setup can be useful to reduce or enlarge a drawing by a known proportion, rather than allowing Shrink to Fit to choose something for you (Figure 2-42).

    OK back to the drawing. The reference grid will be in landscape orientation.

    Save As a new file: [Your Name] Vectorworks Square Tutorial. You may also wish to save this file as a second base file so that you do not need to rotate the reference grid to begin drawing in landscape orientation.

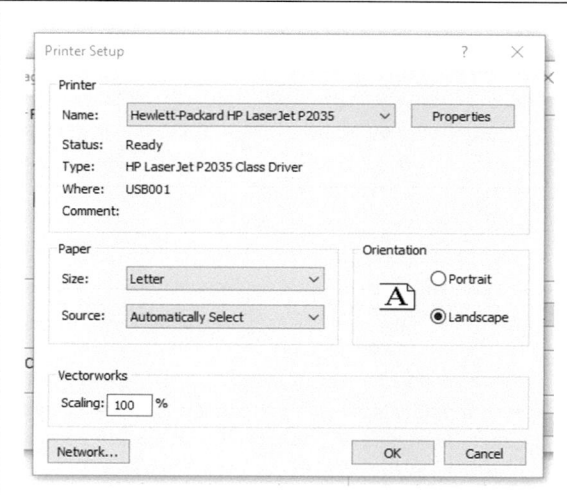

FIGURE 2-42 *In Page/Printer Setup: reorienting the page to Landscape; making note of the scaling selection*

If you do so, be sure that your file names indicate which is Portrait and which is Landscape.

3. As drawings become more complex, more planning is required to facilitate the placement and layering of information. Take a moment to consider which design layers will best organize the drawing process.

    For example, a layer for border guidelines and another for the drawing's guidelines will be useful. There should be a layer for the drawing's final lines—what will be darkened. Eventually, there may be separate layers for dimensions and notes. If multiple objects are to be included on a single printed sheet, each object may have its own array of design layers. New design layers can be created at any point along the drawing process, but planning before drawing helps to control the workflow. When creating and labeling new design layers, give them descriptive and easily understood names.

    In this exercise three design layers will be of use:

    – Border guidelines
    – Drawing guidelines
    – Finished cubes

    Click on the arrow that opens the Design Layer drop-down menu.

    Select New Design Layer to open its dialog window. Since there's already a default Design Layer-1, the New Design Layer text box will give the new layer the next available number (Figure 2-43). Type the new name of the layer into the Name text box and click OK. For this exercise, the other default settings can remain as they are. When you click OK, the new layer will be automatically activated in the drawing space. You can also add new layers through the Layers Organization window by selecting New.

    The other selection offered by the drop-down menu is the creation of a Sheet Layer. These layers function a bit differently than Design Layers and are used to create the polished printable version of the drawing. Sheets Layers will be discussed in Chapter 3.

    Create the three new design layers listed above. Go to the Layer Organization window to select the scale of the layers. Click on any of the layers listed and then click on Scale in the dialog box that appears. Select ½″ = 1′–0″ from the Imperial side of the menu. Remember to checkmark All Layers, so that all of the Design Layers are working in the same scale. When done, open the Design Layer drop-down menu and click on the Border Guideline layer to activate it.

4. In the Navigation-Classes window, select the Guidelines class. Using the same approach used in the Module 1 tutorial, lay out guidelines ½″ from the edge of the drawing space at the top, bottom, and sides. At the end of this exercise, the Border tool will be used to place a

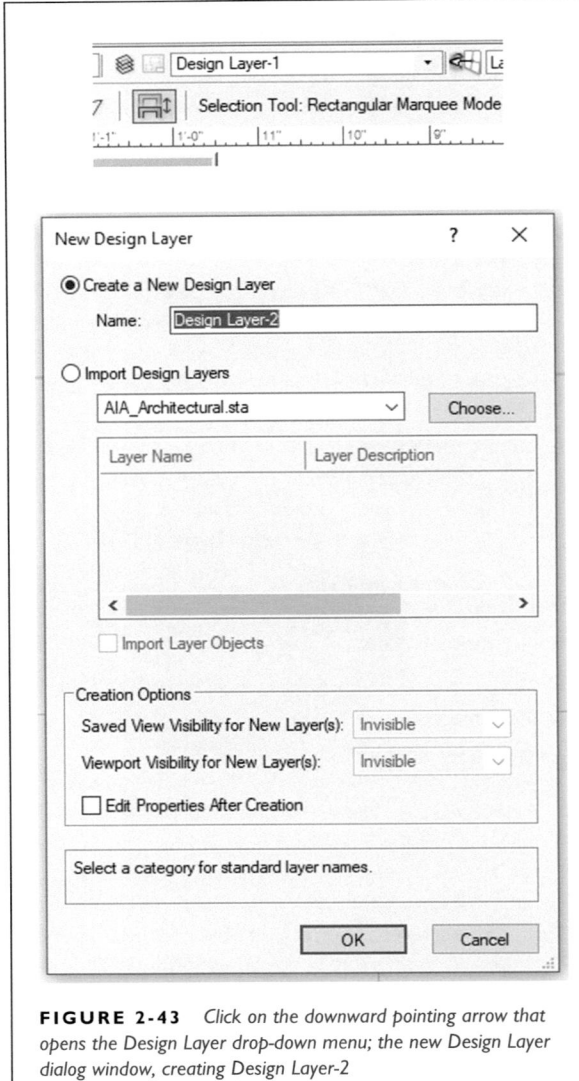

FIGURE 2-43 *Click on the downward pointing arrow that opens the Design Layer drop-down menu; the new Design Layer dialog window, creating Design Layer-2*

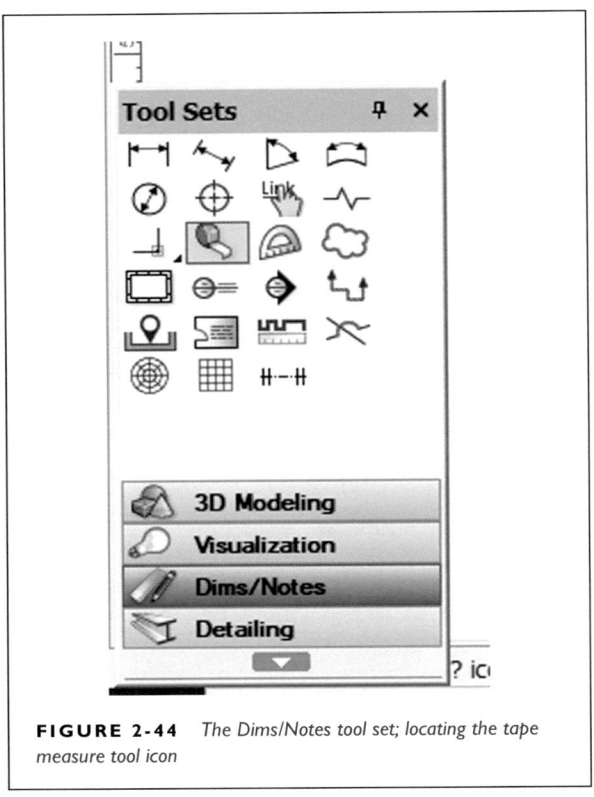

FIGURE 2-44 *The Dims/Notes tool set; locating the tape measure tool icon*

chose Palettes, and then checkmark Tool Sets. In the Tool Sets window that appears (usually lower left corner of the screen), there will be a bar that reads Dims/Notes with an icon of a small pencil and ruler on it. When the cursor hovers over this bar, it opens a text box that reads Dims/Notes. Click to open the tool set. One of the icons is the Tape Measure tool (Figure 2-44). Select it. Move the cursor to one top corner of the border outline, click and drag it down to the bottom corner in order to measure the height of the reference grid. When placed on the bottom intersection, a text box will note Endpoint and a measurement will appear. Measure the drawing space within the border guidelines, and calculate the center of the drawing space for your layout needs.

Vectorworks also indicates a line's midpoint as the cursor is dragged over it. However, this midpoint is noted between the nearest intersections, not for what might be perceived as the full length of the line. If a line is constructed with two segments end to end, Vectorworks considers it to be two independent lines even though it appears to be a single line. You can also draw a fresh guideline through the drawing space, from border to border; this single line will have a midpoint in the desired mid-space location.

border atop the guidelines. Since you're working in ½″ = 1′–0″, translate ½″ in reality into what it would be in scale to lay out the border guidelines.

5. Change the active layer to Drawing Guidelines. Note that the guidelines in the Border Guidelines layer become slightly muted. They cannot be altered while in other layers, but new lines can still snap to them.

    If the hand drafting version of this exercise was already completed, pull out your layout measurements. If not, do layout math to center the array in the center of the reference grid.

    To measure the space within the border guidelines, use the Tape Measure tool. On the left side of the screen, the Tool Sets menu should be open. If there is no tool menu that looks like the one in Figure 2-44, open it by going to the Windows drop down menu,

6. Use the Line tool to lay out guidelines for the array's perimeter (Figure 2-45). Draw lines over the border

1. Lay out perimeter of array

2. Lay out outlines of squares

3. Trim no longer needed guidelines

FIGURE 2-45 *Laying out the array perimeter; laying out the individual cube tops; trimming extraneous guidelines. Reference grid not shown to aid clarity of example*

guidelines to measure and mark their placement. Since you're in a different layer, the border guidelines will not be affected.

7. Lay out the outlines of each cube's top.

8. The interior features are 6″ within each outline. Draw guidelines to place the interior features. Since the array is aligned, it's only necessary to measure along one vertical line to place the horizontal guidelines, and along one horizontal line to place the vertical guidelines. Drag guidelines from the measurement all the way across both squares in the row or column.

9. Activate the Finished Cubes design layer. Since smudging is not an issue in Vectorworks, there's more leeway in the order in which objects are darkened. Use the Line tool to darken the outlines of the squares with the medium visible line class (Figure 2-46).

10. The upper left cube has no interior features, so no interior lines are added.

11. The hole running through the upper right cube is a major feature and therefore the same line weight as the object outline.

12. The raised portion of the lower left square is also a major feature, and therefore the same line weight as the object outline.

13. The bottom inset of the lower right square cannot be seen in the top view. Use the hidden lines class to outline this feature.

14. Turn off the visibility of the Guidelines class (via the Navigation-Classes window) to see the final drawing more clearly.

15. In this exercise, the border will be drawn in the Finished Cubes design layer using the Border tool. Turn on the visibility of the Guidelines class to help with alignment. In the Dims/Notes tool set menu, find the Title Block Border tool. It looks like a double-lined rectangle (Figure 2-44).

When this tool is selected, a dashed border outline appears in the drawing space. The outer edge of this outline represents the edge of the page; the inner rectangle represents the border.

Drag this rectangle so the outer line matches up with the outer edge of the reference grid. Click the mouse to affix it to the drawing space. The rectangle will highlight orange to show that it is active, and the Title Block Border navigation box will appear on the right-hand side of the screen. A Title Block Border Preferences window will open. Click OK to accept the default settings (Figure 2-47).

You may need to Escape or choose the Select tool to prevent extra borders from appearing and floating around the drawing space.

16. The border may contain numbered sections, or **zones** (Figure 2-47). Zones are used on drawings for large projects: for example, directing the reader to "see sheet 6, zone 4." If present, turn them off. In the Title Block Border navigation window on the right of the screen, click the button mid-way down that reads Title Block Border Settings.

A dialog window with a number of choices in a left sidebar menu will appear. To turn off the zones,

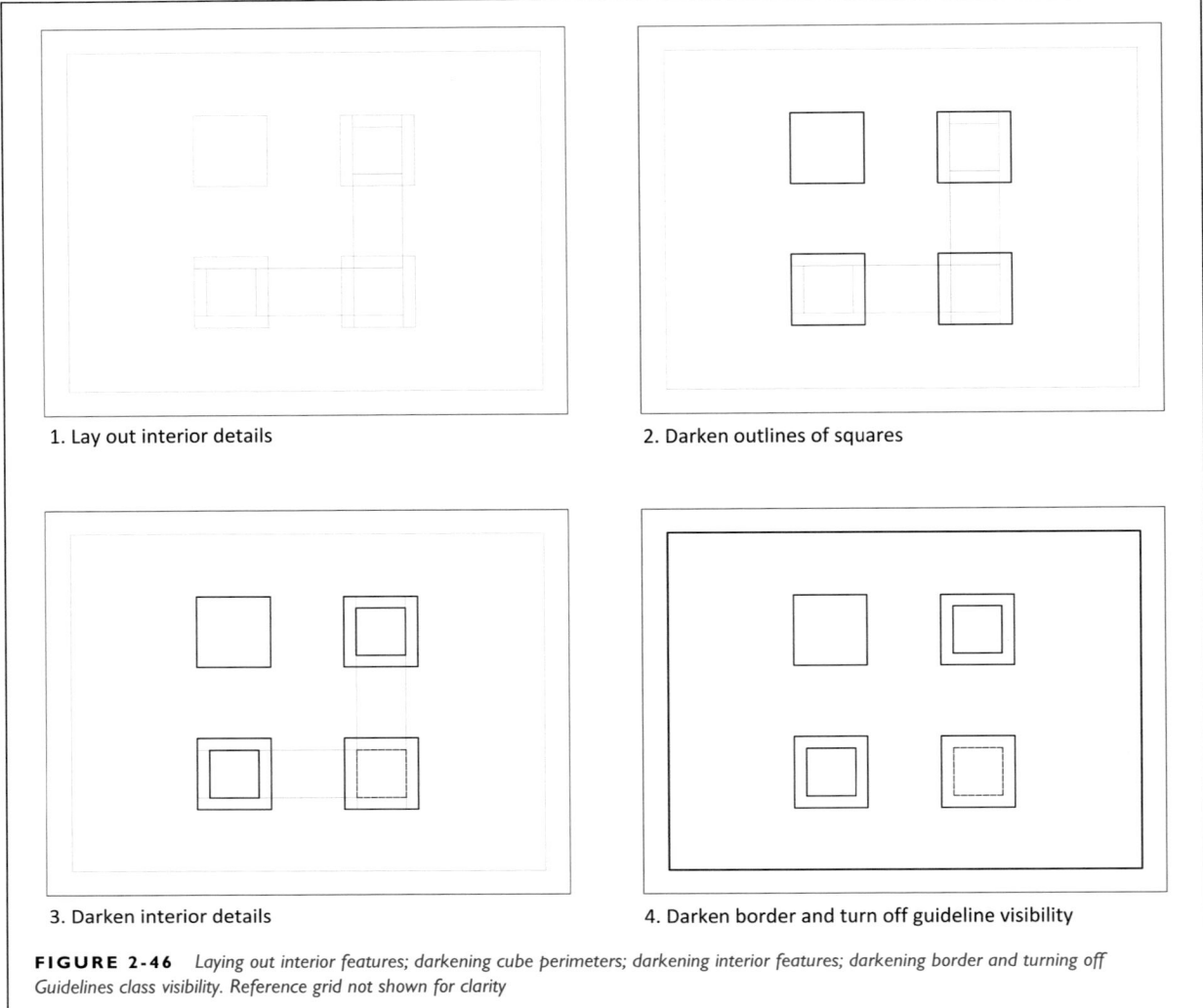

1. Lay out interior details

2. Darken outlines of squares

3. Darken interior details

4. Darken border and turn off guideline visibility

FIGURE 2-46 *Laying out interior features; darkening cube perimeters; darkening interior features; darkening border and turning off Guidelines class visibility. Reference grid not shown for clarity*

select Sheet Zones/Grids. The information to the right of the menu will change, and there will be selections for Vertical Zones and Horizontal Zones. Change these values to 0. OK back to the drawing. The zones will have vanished (Figure 2-48).

17. The border is drawn in the weight of the currently active line class. The border should be the thickest line on the sheet to emphasize the boundaries of the drawing. Since an extra-thick line class was not created for the base file, this weight will be set manually.

Select the border so that it is highlighted and active. In the Attributes window, select Solid next to the pencil (line style), and a thickness of 0.70 mm from its drop-down menu. The border will shift to the new line weight (Figure 2-49).

Add a Border (or Extra-Thick Visible) line class to your classes next time you open your base file.

18. For this exercise, instead of creating a title block (an option in the border navigation box), use the Text tool to add your name, project title, scale, date, and class as a block of text in the lower right corner. Turn off the visibility of the Guideline class before saving as a PDF.

19. Save the file.

20. Export as a PDF. When viewed or printed, there will be two borders—the inner being the actual border, the outer being the size of the reference grid. This is because the current Vectorworks page size is not 8½″ × 11″, but slightly smaller.

A Reminder

In this exercise, only the top views of the cubes were drawn. The line styles selected provide *some* information about the

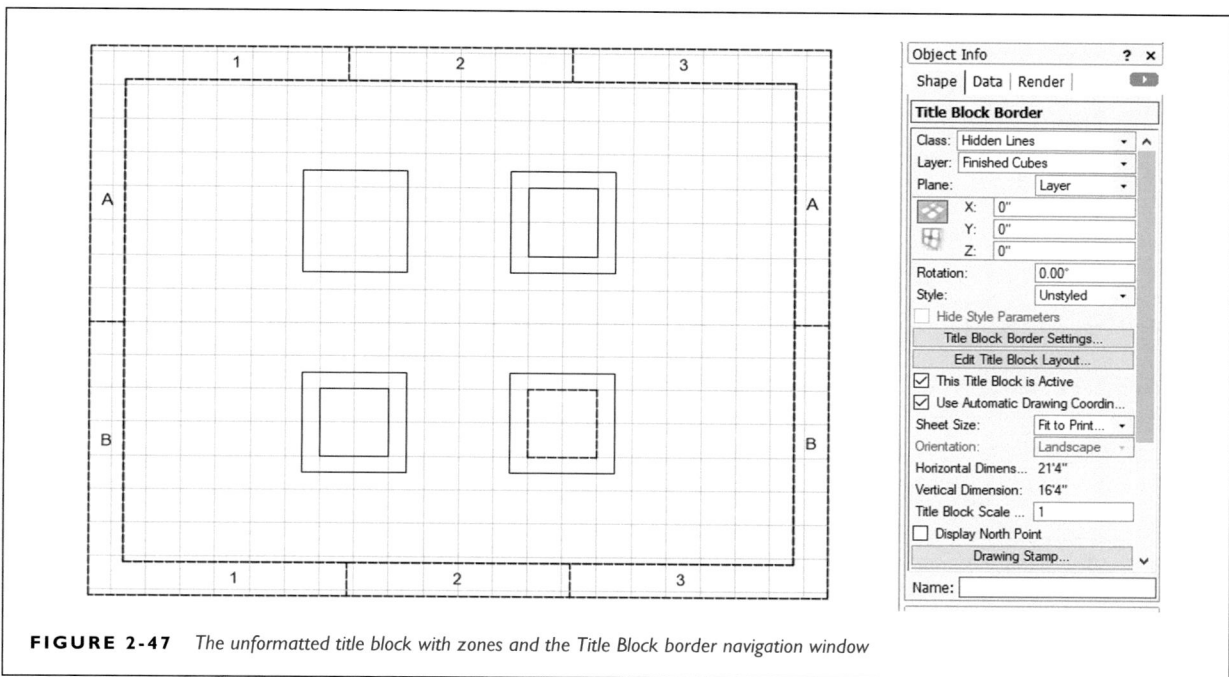

FIGURE 2-47   *The unformatted title block with zones and the Title Block border navigation window*

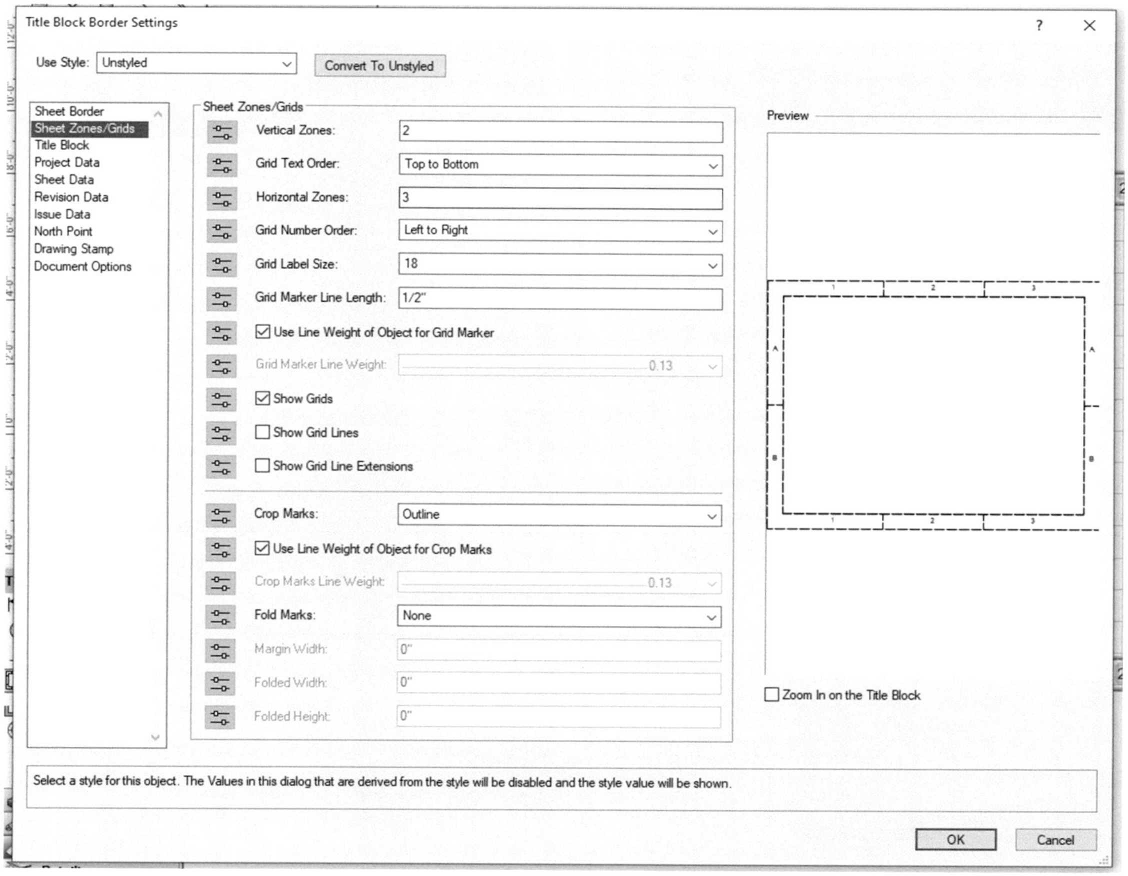

FIGURE 2-48   *The Title Block Border Settings window*

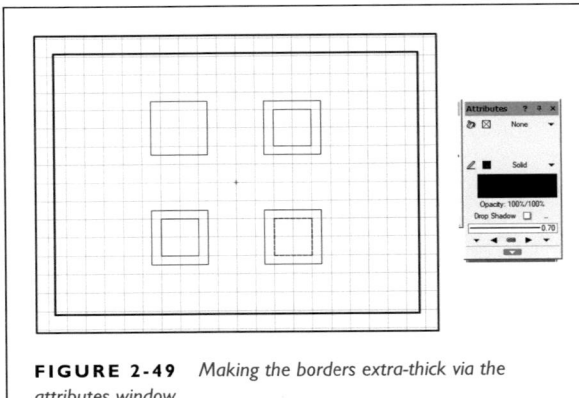

FIGURE 2-49 *Making the borders extra-thick via the attributes window*

interior features, but not all. From the top view alone, it is impossible to know whether the thick line representing the hole in the upper right cube extends all the way through the object, nor is it possible to know the height of the features indicated in the bottom two cubes. Furthermore, the hidden lines drawn in the lower right square could indicate either an inset or an extrusion. Remember that at least three aligned orthographic views are required to fully describe an object. Drawing full orthographic descriptions of these cubes will be the subject of a future exercise.

Homework: Using the Rectangle Tool

The Rectangle tool allows you to drag out a rectangle as a single object rather than building it line by line, which results in four separate objects (Figure 2-50).

A shape that Vectorworks treats as a single, unified object is said to be **composed**. If you attempt to trim/cut one portion of a composed object, all of it will be deleted. There is a selection in the Modify drop-down menu that allows you to **decompose**, but if you expect to trim away segments of a polygon later on, it's sometimes better to stick with the Line tool.

The Rectangle tool creates shapes that are composed until they are intersected by other lines. After drawing a rectangle with this tool, clicking on any side with the Trim tool deletes

the whole shape. If you draw a line bisecting the rectangle from top to bottom and then use the Trim tool on one side of the rectangle, that side of the shape will be cut. The remaining side of the rectangle does not devolve into independent lines, however, and will continue to act as a cohesive unit.

1. Repeat the previous project until arriving at the point the cube outlines are to be darkened.

2. Use the Rectangle tool to drag out a rectangle for each square. Set the Rectangle tool in Corner to Corner mode (upper toolbar). Click the mouse at one corner of the intended rectangle and drag the mouse to the opposite corner, clicking to snap the rectangle into place.

3. Use this tool to draw all of the outlines, being sure to choose the appropriate line style for each. Since no Fill was assigned to any line classes, the interior of the squares should remain un-Filled. If a Fill appears, use the Attributes window to turn it off then double check the class's settings via the Class Organization window.

Homework: Geometric Construction in Vectorworks

Instructor

The students will construct four drawings in Vectorworks, laying out and darkening arrays of simple geometric objects. They will all be in 1:1, with ½″ borders. This is a math and layout exercise. See Figure 2-51. The steps below are outlined for Vectorworks, but the steps for hand drafting remain similar.

Students

Draw each shape array (Figure 2-51) on a separate page, for a total of four drawings. Center each array on its page.

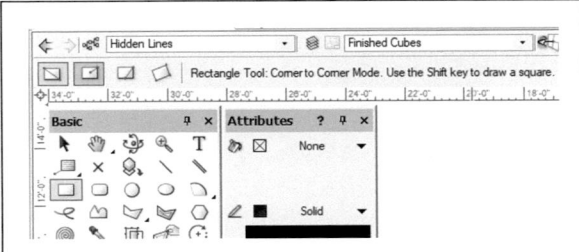

FIGURE 2-50 *Location of the Rectangle tool icon and its modes*

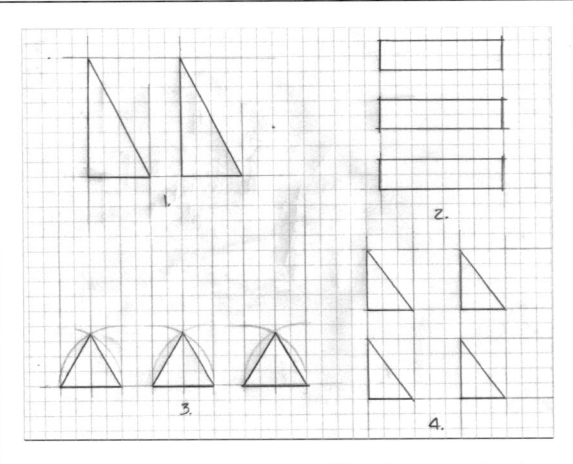

FIGURE 2.51 *Rough sketches of four shape array drawings*

This is a problem-solving exercise, so you may have to do some experimenting and layout math to achieve each graphic solution. Draw in 1:1. Include Borders, and name and class information as a block of text.

1. For each drawing, determine how many layers you will need.

2. Lay out ½″ borders with guidelines in a Border Guidelines design layer.

3. Use the measurements listed below to do your layout math, then use guidelines to lay out each array.

   a. Two right triangles, standing side by side; 2″ wide, 4″ high, 3″ between them.

   b. Three rectangles, lying horizontally, in a stack; 4″ wide, 1″ high, 1″ between them.

   c. Three equilateral triangles; all sides 2″ long, in a row, 1″ between each.

   d. Four right triangles; base 1½″, height 2″; place in square array (two rows of two).

4. After the array is laid out and deemed correct, darken the shapes.

5. Use the Border tool to add a border in the finished shape design layer.

6. Use the Text tool to add name, class, scale, assignment title and date in the lower right-hand corner of the drawing space.

7. Save, Export as PDF, Print.

MODULE 4: CIRCLES AND ARCS

Drawing a circle is relatively straightforward. If you are using a compass, the point of the compass is stuck into the paper and the compass is twirled about this center point. When using the Circle tool in Vectorworks or SketchUp, the cursor is clicked on the point to be the center and the circle is dragged out to its desired diameter. In all cases, the center point of the circle must already have been determined.

In hand drafting, compasses have largely been abandoned in favor of plastic circle templates. They're far less fiddly and it's easier to draw a consistent line weight. However, circle templates offer circles in a limited selection of predetermined sizes; the larger the circle, the fewer that fit onto a single template. Beyond diameters of 4″ or so, circle templates stop being quite so convenient. To draw very large circles and arcs, a beam attachment can be added to a bow compass to extend its reach.

When looking at a circle template, note that there are crosshair markings around each hole at top, bottom, right, and left (Figure 2-52). These marks align the circle to be drawn with already drawn horizontal and vertical axes. Avoid

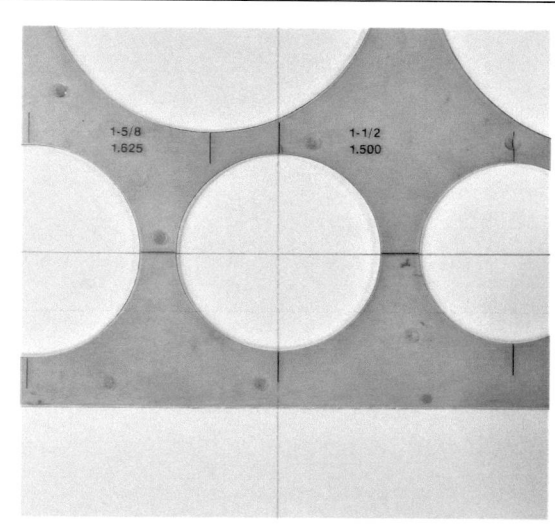

FIGURE 2-52 *Aligning the circle template with already drawn axes*

the temptation to draw a circle before drawing its axes. With no pinprick left by a compass point to mark center, finding a center after the fact requires excavation via geometric construction. SketchUp and Vectorworks remember where the center was placed, but since the axes are the manner in which the circle relates to the rest of the object, it's best to lay out this information before drawing the circle.

If you are using a compass, draw a guideline and mark the radius along it. Place the point of the compass on the start point and open it until the point of the lead reaches the endpoint. This is more accurate than opening the compass along a scale ruler and prevents nicking the ruler with the compass's point. In hand drafting, it is not typical to outline the circle with guidelines and then darken it. Once the center lines are placed, the circle is not drawn until the overall darkening phase.

Exercise: Drawing Circles with a Template

Instructor

The students will need circle templates that go up to a 2″ diameter circle. This is a layout and line control exercise. Scale is 1:1. No borders are required. It can be drawn on an 8½″ × 11″ sheet. This exercise will be repeated in Vectorworks and SketchUp. See Figure 2-53 for the exercise layout information.

Students

For this exercise, axes will be laid out and used to place a series of circles using a template, as per the sketch in Figure 2-53. Use 8 ½″ × 11″ paper, 1:1 scale, no border is required.

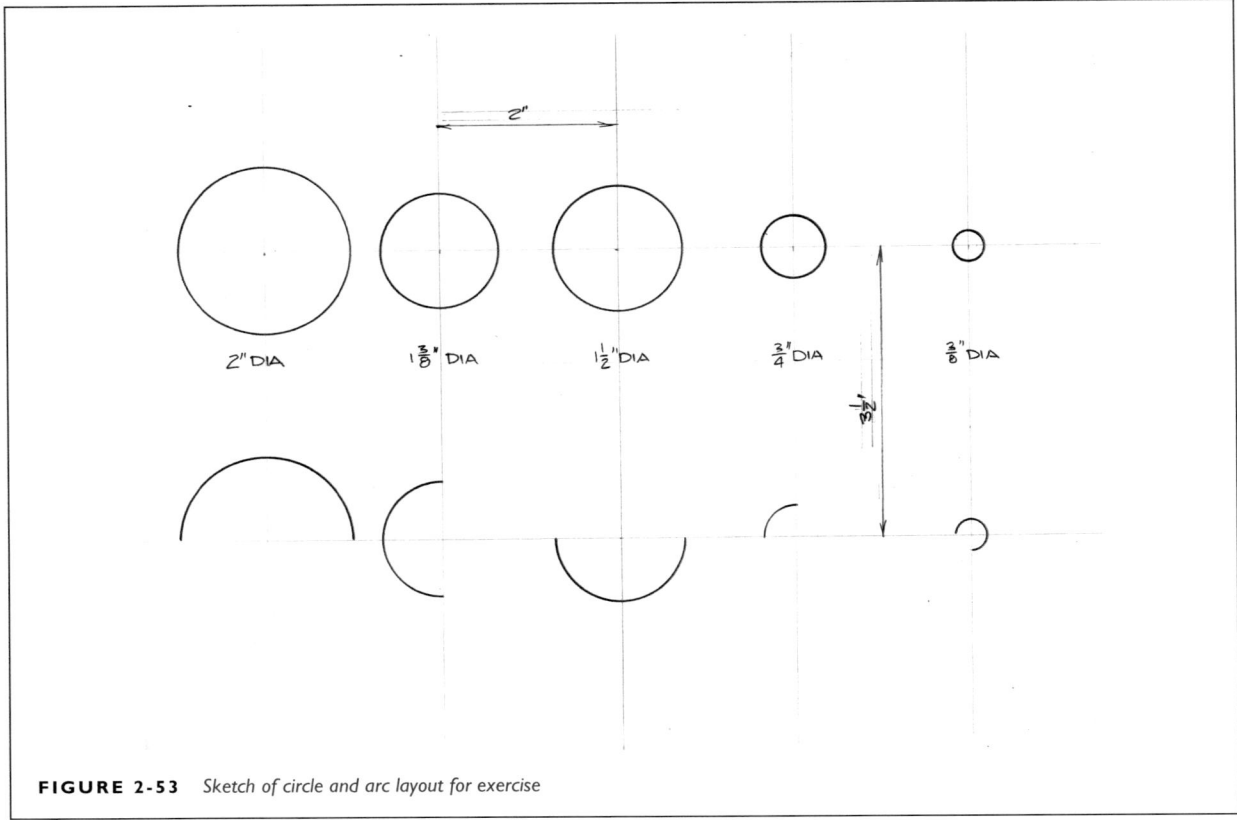

FIGURE 2-53 *Sketch of circle and arc layout for exercise*

1. Lay out the axes, following the distance measurements in Figure 2-53. Do your best to center the array on the page.

2. Along the top horizontal guideline, draw five circles with the diameters indicated. Align the crosshairs of the template with the guidelines. Use one hand to keep the template firmly in place. Keep the pencil vertical so that the point does not dig under the edge of the template nor drift away from it—this will result in a lopsided circle. Aim for a single, firm, slow stroke that goes all the way around the circle. The more times the pencil stops and starts, the choppier the circle will be. Try different pencil hardnesses to see which gives the best result.

3. Along the lower horizontal guideline, draw arcs rather than complete circles. Use the same diameters as in Step 5, but vary the length and placement of the arc as seen on the sketch.

How easy or difficult was it to draw a full circle, as compared to an arc? Is pencil control any different from drawing a straight line? How easy or difficult was it to keep the template aligned on its axes? What pencil hardness worked best for you? Was it the same hardness that you used for optimal straight visible lines?

Exercise: Drawing Circles in Vectorworks

Students

Repeat the circle template exercise in Vectorworks. Use the same layout and diameters. When using the Circle tool, Vectorworks asks you to type in the radius, rather than the diameter (Figure 2-54).

1. Open the base drawing. Apply landscape orientation. No borders are needed. Set up design layers for the guidelines and for the finished circles.

2. Lay out the axes in the guideline layer.

3. Activate the finished circle design layer. Change the class to medium visible line. Select the Circle tool. When selected, a series of mode options appears in the upper toolbar. Choose Radius Mode: Set Circle Center (Figure 2-54)

    To use the Circle tool, click the cursor at the intersection of the axes. Drag the cursor slightly outward and type in the radius of the circle. Hit Enter. Left click the mouse. The circle will snap to the desired diameter.

4. The Circle tool creates complete circles. There is an Arc tool, but that will be saved for later. For this exercise,

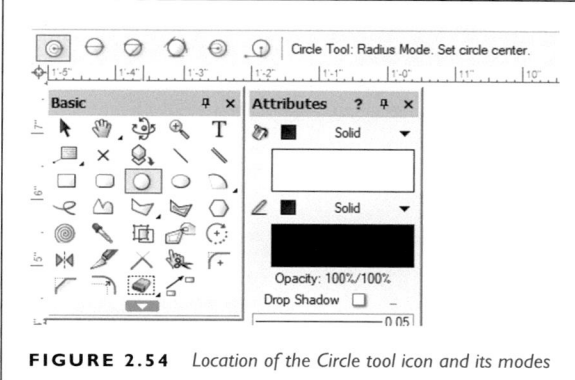

FIGURE 2.54 *Location of the Circle tool icon and its modes*

use the Trim tool to cut the bottom row's circles down to the desired arcs. The guidelines are in a different layer and do not create viable intersections for trimming. Reestablish the axes as needed in the current layer, then delete after the arc is drawn (Figure 2-55).

5. Save. Export as PDF. Print.

Exercise: Drawing Circles in SketchUp

Students

Repeat the exercise in SketchUp. On the Getting Started toolbar, the Circle tool is found as an option within the Rectangle tool icon. It can also be found as its own icon on

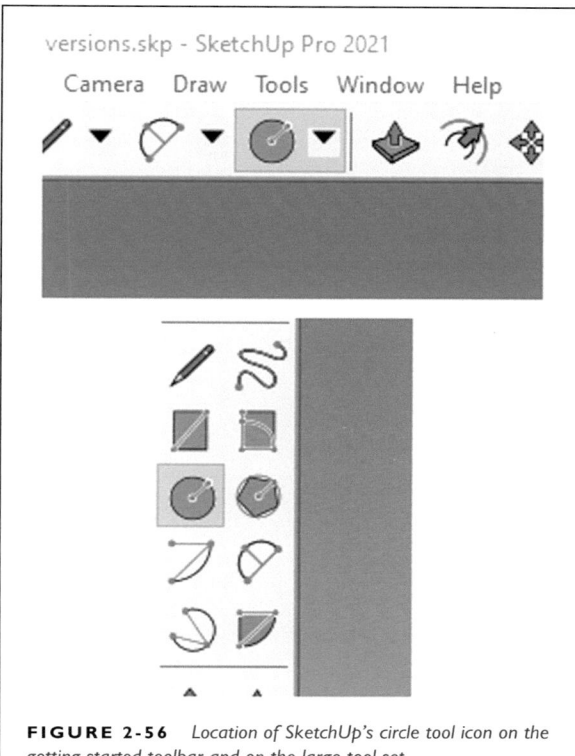

FIGURE 2-56 *Location of SketchUp's circle tool icon on the getting started toolbar and on the large tool set*

the Large Tool Set. On the Getting Started toolset, click on the downward pointing arrow next to the Rectangle tool icon and select the Circle option (Figure 2-56).

To draw a circle, left click to set the center and drag the cursor outward to the desired diameter. Left click again to affix. Radii can also be entered on the keyboard; while the circle is highlighted, enter the radius and hit Enter to affix the circle.

1. Open SketchUp. Select the Simple/Inches template.

2. Convert the circle and arc measurements listed above to feet rather than inches so that it's not necessary to zoom in quite so far. Use the Rectangle tool to drag out an 8'–6" by 11'–0" horizontal rectangle in landscape orientation to the red axis. Move the human figure aside. This rectangle will serve as the "paper."

3. Use the Tape Measure tool to lay out guidelines for the axes (Figure 2-57). The diameters are given in inches; since we're converting inches into feet, convert imperial to decimal notation by dividing numerator by denominator. 1 3/8" = 1.375". SketchUp prefers that you enter radii; divide the diameter in half to get the radius. 1.375 ÷ 2 = .6875. Do the math before drawing the circles.

4. Select the Circle tool. Click the cursor at the intersection of the axes and drag out a circle. Type in the radius

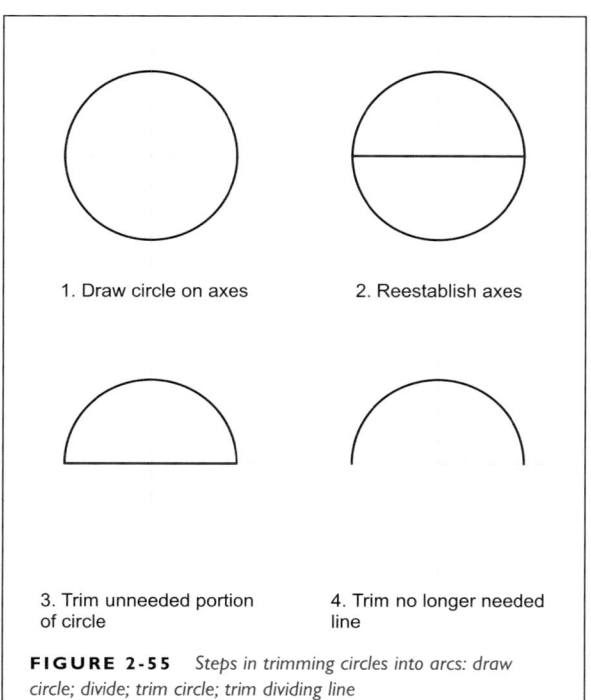

1. Draw circle on axes

2. Reestablish axes

3. Trim unneeded portion of circle

4. Trim no longer needed line

FIGURE 2-55 *Steps in trimming circles into arcs: draw circle; divide; trim circle; trim dividing line*

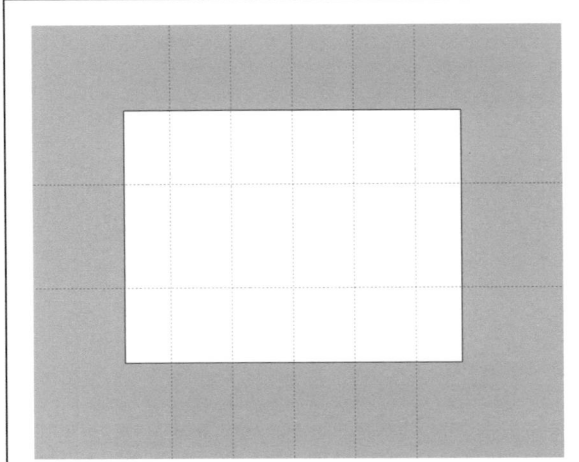

FIGURE 2-57 *Laying out the drawing Space and guidelines for circle exercise*

and hit Enter. In SketchUp, any circle or arc drawn will actually be a polygon: a series of short line segments creating the illusion of a true circle. The direction in which the circle is dragged out and affixed matters, as it is possible to draw two circles of the same size with the same center that don't quite line up with each other. This can cause problems later if the circle is part of a larger construction (Figure 2-58). It's good practice to align the cursor with one of the axes so that concentric circles or lines tangent to the circle will line up.

5. To cut circles into arcs, lines need to be drawn since the guidelines do not create viable intersections for erasure. Use the Line tool to draw vertical and horizontal

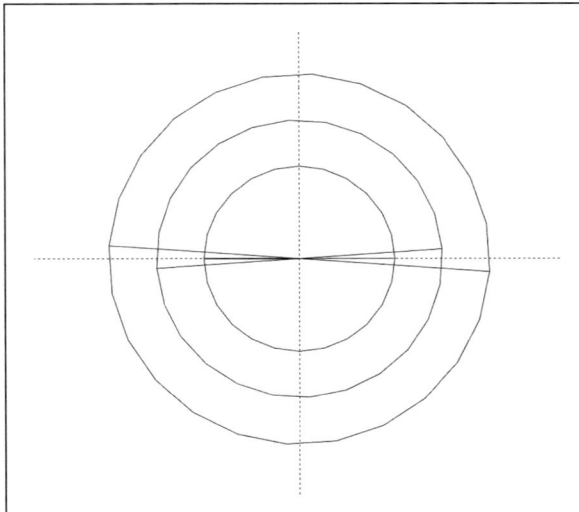

FIGURE 2-58 *SketchUp circles as polygons; the importance of aligning the Circle tool with axes when drawing*

axes over the guidelines that intersect the circle's circumference (you don't need to go beyond the circumference). Use the Eraser tool to delete the undesired circle segments. Erase the straight lines to leave only the arc. Since the lines sit directly atop the guidelines, this erasing action may also delete the guidelines. You can also choose Delete Guides from the Edit drop-down menu.

6. Save the file. Change the camera view to parallel projection and select the top standard view. Zoom until the image substantially fills the screen. Export as a 2D graphic (JPEG). Print the JPEG if a hardcopy is required.

When Arcs Meet Lines

When drafting an object in which a circle or circular arc meets a straight line, it's typical to use guidelines to lay out the circle's center axes at the same time the rest of the object's straight lines are laid out. When the drawing is darkened, the circles and arcs are darkened first. Since hand drafting always carries a degree of inaccuracy, there are times the compass or template doesn't quite perfectly meet the intersecting line. By darkening the circle or arc first, the intersecting line can be minutely adjusted to create a smooth connection. If this adjustment results in visibly slanted lines, it's a sign that something else is seriously in error; measurements should then be double-checked and guidelines redrawn.

Even when using a CAD program in which alignment is pretty much assured, it's still a good habit to follow this same layout and darkening process.

Exercise: Drawing Arcs Meeting Lines

Instructor

The students will hand draft three shapes on separate 8½″ × 11″ sheets, landscape orientation, for a total of three drawings. Include borders. The focus is on creating smooth joins between arcs and their intersecting straight lines. The same steps apply if this project is to be drafted in Vectorworks or laid out in SketchUp, which may be a follow-up project. See Figure 2-59.

Students

Three shapes will be hand drafted as three drawings. Use 8½″ × 11″ paper, landscape, 1:1 scale. Figure 2-59 presents a sketch with measurements.

1. Do layout math to determine placement of each object on its sheet. Rather than drawing the object's outer perimeter, lay out axes first.

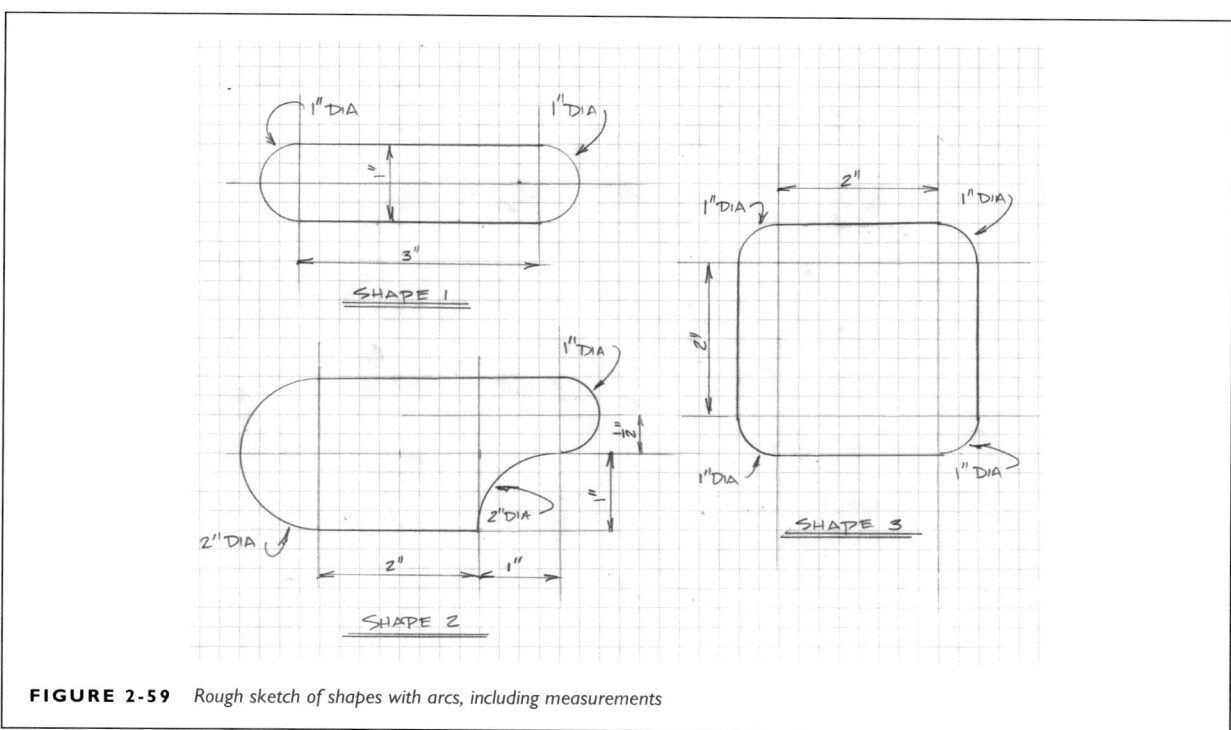

FIGURE 2-59 *Rough sketch of shapes with arcs, including measurements*

2. For shape one, lay out and mark the length of the horizontal axis. Draw vertical guidelines through the endpoint marks to complete the set of axes for each arc. Be sure the guidelines extend beyond the expected radius of the circle so there is ample room for template alignment.

3. For the second shape, three centers need to be placed. Lay out the full-length horizontal axis of the shape. Draw a vertical line at the left endpoint mark to complete that set of axes. On the right side, draw a vertical line at the right endpoint in to find the semi- and quarter-circle vertical axes. Measure up and down to find the horizontal axes of the quarter- and semi-circles.

4. For the third object, lay out the inner square formed by the intersections of the axes. Be sure the guidelines extend far enough out to accommodate placement of the circle template.

5. For all objects, darken the arcs first, then darken the lines between the arcs, being careful to make the connections smooth and seamless.

NOTES

1. Frederick E. Giesecke, Alva Mitchell, and Henry Cecil Spencer, *Technical Drawing*, 4th ed. (New York, NY: The MacMillan Company, 1958), p. 9.
2. Rendow Yee, *Architectural Drawing: A Visual Compendium of Types and Methods* (New York, NY: John Wiley & Sons, 1997), p. 67.
3. Giesecke, Mitchell, and Spencer, p. 482.
4. Thomas E. French and Carl. Svensen, *Mechanical Drawing*, 7th ed. (New York, NY: McGraw-Hill, 1968), p. 66.
5. Dennis Dorn and Mark Shanda, *Drafting for the Theatre* (Carbondale, IL: Southern Illinois University Press, 1992), p. 58.
6. Dorn and Shanda, p. 58.
7. French and Svensen, pp. 89–90.

CHAPTER 3

ORTHOGRAPHIC VISUALIZATION AND AUXILIARY VIEWS

The first modules of this chapter focus on visualizing objects via multi-view orthographic projection and continue to link specific line styles to the features they describe. The miter line is introduced as a method to transfer information between non-adjacent orthographic views. Auxiliary views show the true shape of inclined and oblique planes. At times, a section view may be required to understand an inclined plane and will be generated in order to create the auxiliary view. The chapter concludes with a checklist review of Chapters 1, 2, and 3.

TOPICS AND GOALS

- Orthographic visualization with SketchUp

- Transferring information between diagonal views with a miter line

- Auxiliary views

- Sections and cutting plane lines

MODULE 1: VISUALIZING ORTHOGRAPHIC PROJECTION

Exercise: Using SketchUp to Visualize Orthographic Views

Students

This exercise features the construction of an L-shaped 3D object. This object will be copied and rotated. The copies will be oriented to the front picture plane in a simulation of orthographic projection views. See Figure 3-1 for the appearance of the finished object.

DOI: 10.4324/9781003154921-4

1. Open SketchUp. Choose the Simple/Inches template. When the file opens, Move the human aside.

2. Using the Rectangle tool, set the cursor at the intersection of the three axes and drag out a rectangle. Before clicking it into place, type in 6′,3′. The comma is important in SketchUp as it separates width and depth (or depth and height, etc.) measurements. When you left-click the mouse, a rectangle will snap into place.

    Drag another rectangle out elsewhere along the red axis but this time type 3′,6′ before clicking to place it. Because the measurements are typed in a different order, the rectangle has a different orientation. Erase or Undo this additional rectangle.

3. Use the Push/Pull tool to raise the surface of the rectangle to 2′.

4. Orbit and Pan to reorient the view if the top needs to be seen a little better. Additional lines will be laid out on the top surface.

    Use the Tape Measure tool to drag a guideline parallel to the green axis (front to back) to bisect the top surface. If you drag the cursor along the back or front edge of the object, a blue dot will appear when you reach that line's midpoint. A textbox reading 3′ will also appear when you've reached the midpoint. You can also begin the drag and type in 3′, Enter.

5. Use the Line tool to draw a line over the guideline on the top surface. Click the cursor where the guideline intersects the front edge. A small box with a red X will appear, with a text box saying Intersection. Drag the line to the back edge. Click to secure. Another red X will appear when the cursor hits that intersection (Figure 3-1).

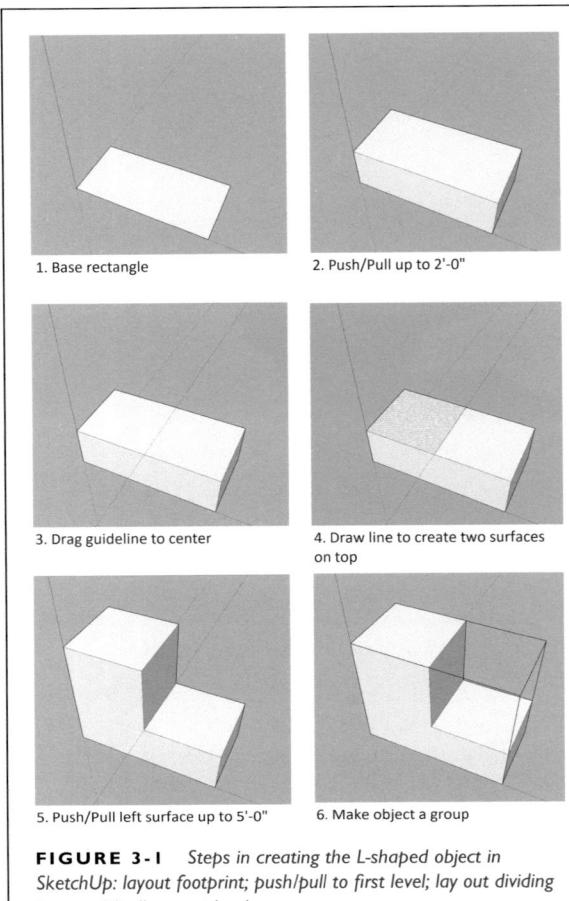

1. Base rectangle

2. Push/Pull up to 2'-0"

3. Drag guideline to center

4. Draw line to create two surfaces on top

5. Push/Pull left surface up to 5'-0"

6. Make object a group

FIGURE 3-1 *Steps in creating the L-shaped object in SketchUp: layout footprint; push/pull to first level; lay out dividing line; push/pull to next level*

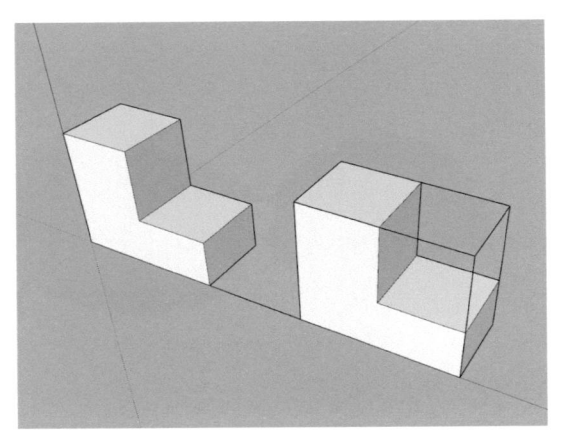

FIGURE 3-2 *Copying an object by selecting Move and holding down the Ctrl key*

6. This line separates the top plane into two independent surfaces. Use Push/Pull to raise the left-hand portion of the top surface to 3' (total of 5' high).

7. The object is now complete. To prevent the object from being accidentally distorted or altered, make this object a **group** so that all the lines and surfaces of the object bind together as a single unit. Use the Select tool to drag a box around the whole of the object by left-clicking the cursor to one side of the object and holding down the mouse button as you drag the cursor to the other side of the object. Avoid including the human figure. All of the edges of the object should turn blue.

   With object selected, right click the mouse. On the menu that appears, choose Make Group. The whole of the object will be enclosed in a blue frame box. Click the cursor anywhere else on the drawing space and the blue box will disappear, releasing the object from control. Escape will also deactivate the Group.

   In SketchUp, objects can be easily duplicated. This comes in handy when arranging stock flats or platforms, or building a table with four identical legs. Simply build

one leg and make three copies of the original. The Make Component command (also in the right click menu) links copies together so that changes made to one are made to all. This object will be copied twice—once for the side view, and once for the top view. The copies will be rotated and placed so that when viewed from the front, three views of the object in correct orthographic relationship will be seen.

8. The copy representing the object's right view will be made first. Select the object. A blue box should appear around it. While it is active, select the Move tool. Hold down the Ctrl (Control) key on the keyboard and left click the mouse on the object. Drag the cursor to the right, along the red axis. A copy of the object, framed in blue, appears. Placing the copy is easier if you click on a corner of the object, rather than in the middle of a surface or line. A red dotted line will appear to help keep your drag aligned with the red axis. Left click to release the object (Figure 3-2).

9. Next, rotate the copy so that the right-hand side of it faces the front picture. Select the Rotate tool (Figure 3-3).

   This tool is somewhat fiddly. Depending on the placement of the cursor and minute shifting of the mouse, the protractor that appears can not only be

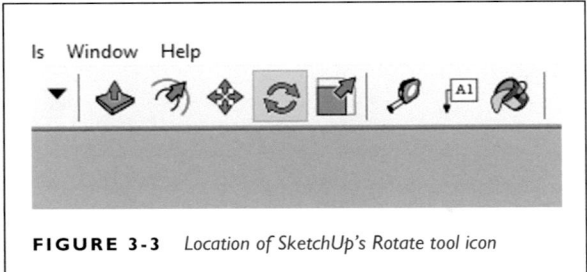

FIGURE 3-3 *Location of SketchUp's Rotate tool icon*

aligned with one of the three axes (red, green, blue) but also can be angled obliquely (black).

There are three steps required to rotate an object:

a. Placing the center of rotation.
b. Telling the protractor where a baseline of 0° is located.
c. Rotating the object around the pivot point the desired number of degrees from the baseline's zero.

Place the cursor at the bottom left corner of the copy. The object should be boxed in blue to indicate that it is selected. Adjust the mouse until the protractor is blue, and lies flat along the bottom of the object. The blue protractor means that you will rotate the object around a vertical axis. You may need to pan and orbit to get the protractor to align correctly.

Once the blue protractor appears, click to secure its center to the object's corner. Drag the cursor to the other end of the front of the copy (the right-hand corner) and click to create the 0° baseline along the front edge of the object. When the cursor is moved, the object will rotate around the point at the center of the protractor (Figure 3-4).

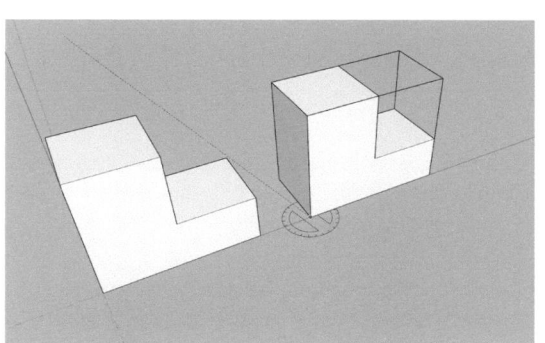

1. Setting the Rotate tool protractor at the pivot point

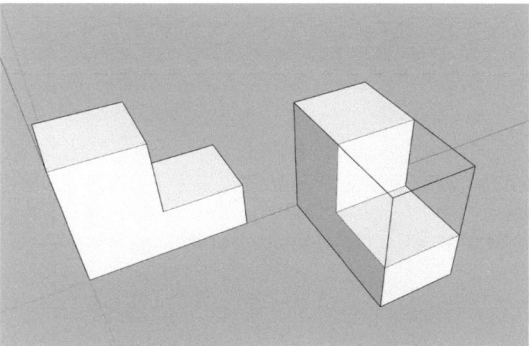

2. Turning the object 90°

FIGURE 3-4 *Rotating the object: placing the protractor, swinging the object 90°*

Drag the cursor toward the front and type 90, then Enter. The object will snap to a new orientation 90° from the original placement. Select or Escape so that the object will not be accidentally further rotated.

10. There should always be enough space between orthographic views to permit the uncrowded placement of notes and dimensions (even if you're not going to add any). Use the Line tool to drag out a 4'–0" line from the lower right corner of the original object along the red axis to provide a snapping point. You can also use a guideline, but it may be difficult to see on top of the red axis.

Using the Move tool, click the cursor on the lower left corner of the copy and drag it back to the endpoint of the 4'–0" line. A green circle with the textbox Endpoint should appear when the object is there. Click the mouse to secure the object to the line (Figure 3-5).

Erase the line that you drew for alignment. If you used a guideline, wait until all the views are placed and then use the Delete Guidelines selection in the Edit drop-down menu to remove them all at once.

11. Make another copy of the original. Drag it straight up so it hovers above the original. A blue dotted line will appear to help you stay aligned.

12. To rotate the new copy so that the top surface faces the front picture plane, secure the Rotate protractor to a lower left corner of the copy and adjust the mouse until a **red** protractor appears. Rotate the object forward 90°.

13. Orbit and Pan so you can see the left side of the original object better. From the upper front corner of the original object, draw a 4'–0" spacing line upward (along the blue axis). Move the newest copy to the endpoint of that line. Erase the line after the object is placed. Erase extraneous lines. This is a good time to delete guidelines, as well (Figure 3-5).

14. Under Camera, pick Standard View, Front. The image will reorient. With the camera in perspective projection, parts of the sides and tops of the objects are visible. Switch to parallel projection to create an orthographic view appearance. Under Camera, select Parallel Projection. The image will flatten out (Figure 3-6).

The copies are now arranged in a 3D environment to create the appearance of an orthographic drawing. If the standard views of Top or Side are selected in the Camera menu, the view will shift to show the top or side of this array of copies.

While SketchUp's standard camera views provide the modeler with orthographic images of the modeled object, making copies and arranging them as done above can help the draftsperson understand the views' relationships in their glass box configuration. When doing layout math and thinking about how to arrange views on a page, creating this view array is a useful reference tool.

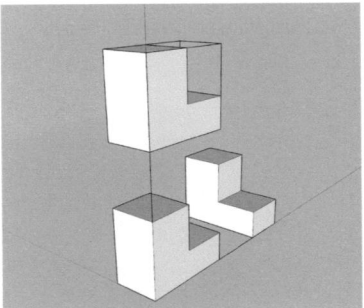

1. Copy to right spaced at 4'-0";
making a second copy for "top view"

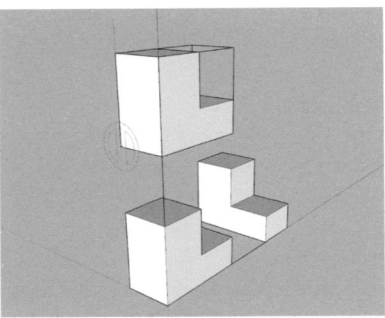

2. Position the Rotate tool protractor
(red) at lower left corner of top copy

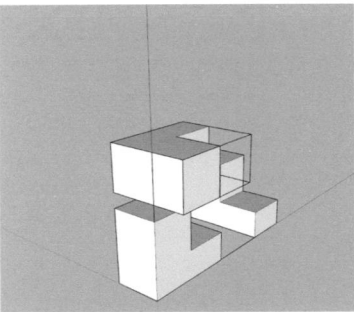

3. Rotating the top copy forward 90°

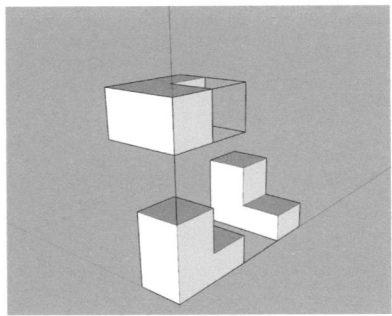

4. Moving top copy into alignment 4'-0"
above original

FIGURE 3-5 *Positioning the first copy 4'–0" away from the original; copying, moving, and rotating a second copy*

Exercise: Additional SketchUp Visualization

Instructor

This exercise provides additional practice with SketchUp and provides more practice translating a real object into a virtual object. If the students did not retain their sketches of the physical objects used to first visualize orthographic views, redistribute the objects. Have the students make quick, rough orthographic sketches before moving to modeling so that measurement information is at hand. The questions below can help guide discussion.

Students

Pull out the orthographic projection sketches of the small physical 3D objects drawn in Chapter 2, Module 2. Exchange yours with another student's.

Create SketchUp models of each object. After completing each object, switch the camera to parallel projection and look at the object in each of the various standard views. Compare what you see in each view to each view in the sketch.

SketchUp offers a **wireframe** view, found in the View drop-down menu, under Face Style. Wireframe hides surfaces, leaving only the lines that mark the edges and intersections of planes. Switch to the wireframe view and look at each view of the object in turn (Figure 3-7). In wireframe view, is it easier or more difficult to tell what lines are behind other lines? Can you tell which would be drawn with hidden lines? Does looking at the wireframe give you more, less, or different information than looking at the object with all the surfaces visible? In what instances would a wireframe view be useful?

Switch back to Shaded with Textures (under Face Style) after exploring the wireframe mode. If time allows, copy the object and arrange the copies to create an array of aligned top, front, and side views as seen from the Camera's standard front view.

Homework: Drafting the SketchUp Object

Students

Draft the L-shaped object constructed in the earlier SketchUp exercise both by hand and in Vectorworks. Use 8½" × 11" paper, landscape, with ½" borders on all four sides. Scale is ½" = 1'–0". Draw three aligned orthographic views: front, top, and right side (Figure 3-8).

Array viewed from front in
perspective projection

Array viewed from front in
parallel projection

FIGURE 3-6 *Front view of the object array in perspective and parallel projections*

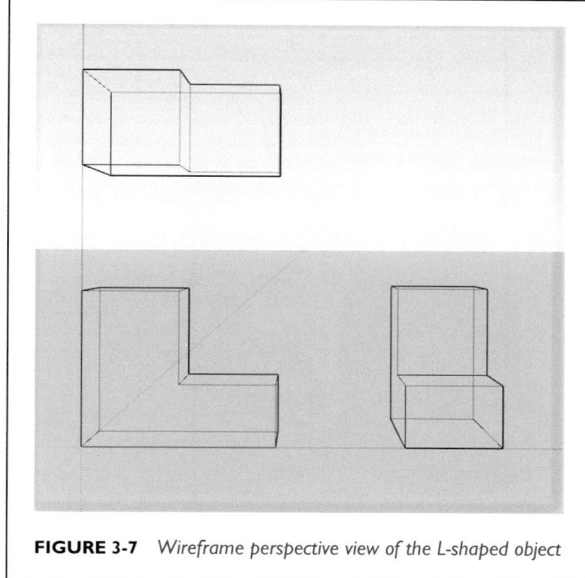

FIGURE 3-7 *Wireframe perspective view of the L-shaped object*

top and side views will be discussed in the next module (the miter line).

4. Once the drawing is laid out, darken with appropriate line styles.

5. Darken the border. Add your name, project title, class name, date, and scale in a block of text in the lower right corner of the drawing.

Homework: Adjacent Views with Missing Information

Instructor

A traditional orthographic drawing exercise is to complete unfinished views of objects. This promotes the use of logic and layout, and provides further practice visualizing a 3D object from the given 2D information. This exercise may be distributed to students as a handout, or as a Vectorworks file.

For the first exercises, ensure missing information can be deduced from information in an adjacent view, rather than from views situated diagonally from one another. The first exercises should feature only one view with missing information, and only one or two missing lines.

See Figure 3-9 for examples. Compare and discuss when complete; the questions in Step 4 can help guide discussion.

The range of objects traditionally used for this type of exercise are generic "machine parts" of no specific function—they exist simply as geometric objects. However, this means that students are presented with unfamiliar objects for which they have no real-world reference. Be sure that the first series of exercises also include a perspective or isometric image of the object to help with visual navigation.

1. Do layout math to center the array of views in the drawing space. Don't cram views against the borders or against each other; always leave enough white space for notes and dimensions. For Vectorworks, determine how many design layers you'll need. Create and name them.

2. Lay out the border and perimeter of the view array.

3. Lay out the front view first. Extend guidelines up into the area where the top view will be constructed. Extend horizontal lines into the area where the chosen side view will be placed. Width information can be moved vertically between the top and front views. Height information can be moved horizontally between the front and side views.

    Depth information needs to be measured and marked separately for the top and right views. A method to transfer depth information diagonally between the

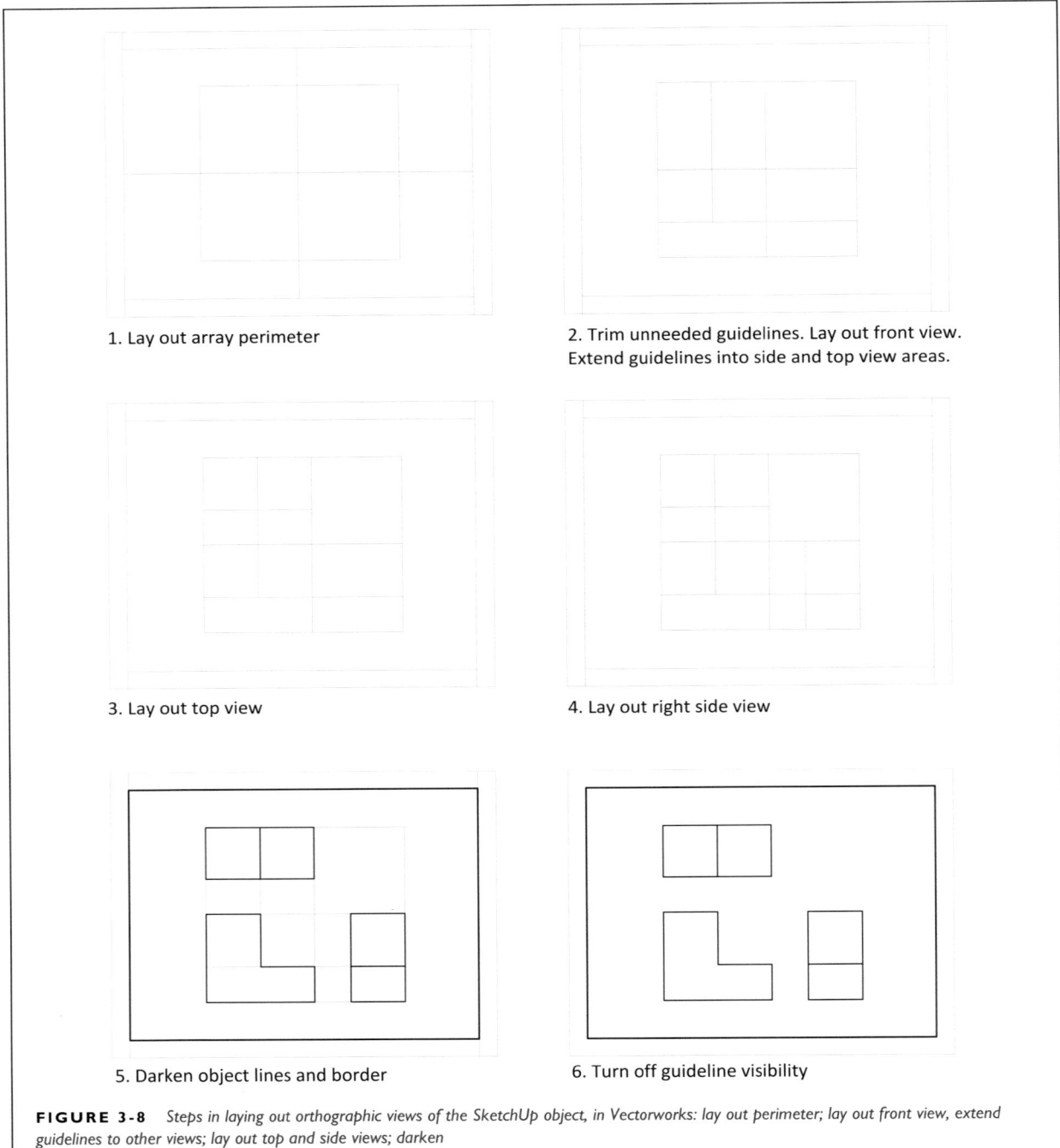

1. Lay out array perimeter

2. Trim unneeded guidelines. Lay out front view. Extend guidelines into side and top view areas.

3. Lay out top view

4. Lay out right side view

5. Darken object lines and border

6. Turn off guideline visibility

FIGURE 3-8 *Steps in laying out orthographic views of the SketchUp object, in Vectorworks: lay out perimeter; lay out front view, extend guidelines to other views; lay out top and side views; darken*

Students

On the handout or file distributed by your instructor:

1. Examine the views and isometric image of the object to determine what information is missing from which view.

2. Use a straightedge to draw guidelines that transfer information laterally and/or vertically between views.

3. Darken the required line with the appropriate line style (visible or hidden).

4. Discuss your choices in class. What led you to decide what was missing? How did you decide if lines were visible or hidden lines? If you missed something, what do you think caused you to miss it? If you added an unnecessary line, what do you think led you to think it was needed?

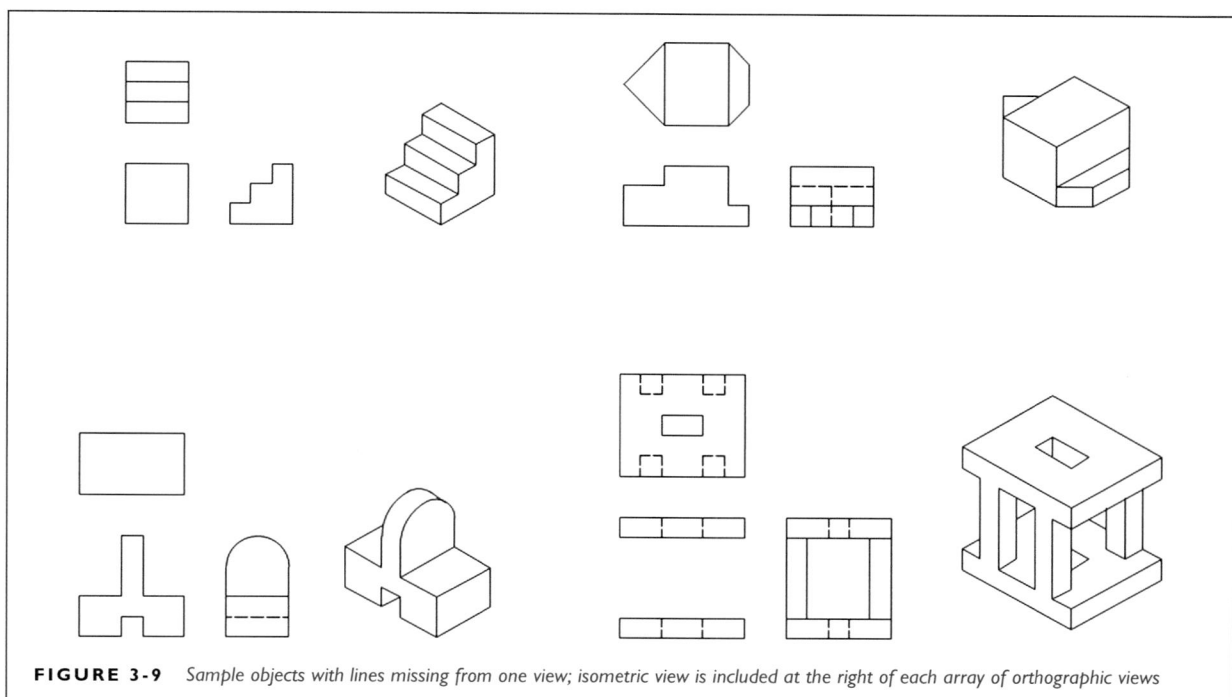

FIGURE 3-9 *Sample objects with lines missing from one view; isometric view is included at the right of each array of orthographic views*

MODULE 2: MITER LINES

It is relatively easy to move information between adjacent orthographic views by extending guidelines directly between adjacent views.

Often though, information needs to be transferred from views that are diagonal to one another, such as from the top to a side view. In these cases, the draftsperson can use a **miter,** or **reflection line**. Use of this method is formally called the "miter line transfer method."[1] If a high degree of accuracy is required in hand drafting, using dividers or a scale ruler to move measurements is preferred, but the miter line method is accurate enough for most purposes and can be quicker (Figure 3-10).[2] In computer-aided design (CAD), using a miter line can be a speedy way to transfer information without loss of accuracy.

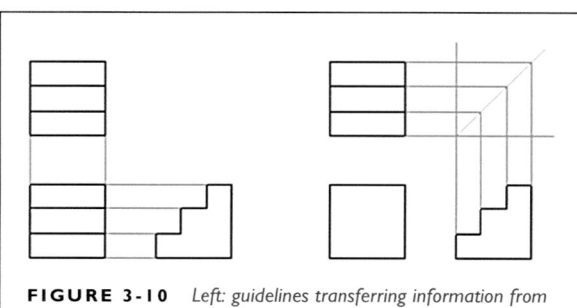

FIGURE 3-10 *Left: guidelines transferring information from adjacent views; right: miter line used to transfer information from diagonal views*

The miter line is a guideline angled at 45° off to the side of the top view and above the side view. Horizontal lines in the top view are extended laterally to intersect this angled line. At the point of intersection, a vertical line is dropped down into the side view, thus transferring depth information from the top view to the side view. Similarly, guidelines can be extended up from the side view, intersect the miter line, and then extended horizontally into the top view (Figure 3-10).

To determine the placement of a miter line starting from the front view (Figure 3-11):

1. Extend vertical guidelines from the extreme right and left points of the front view to establish lateral placement of the top view.

2. Consider how much notation and dimensioning might be needed in your drawing. Accommodating that space, draw a horizontal line that intersects both verticals to establish the front (lower) edge of the top view. Extend the horizontal guideline (for the front surface of the object) into the area above where the side view will be placed.

3. Extend horizontal guidelines from the extreme top and bottom points of the front view laterally into the area of the desired side view (right or left).

4. To place the side view, repeat the white space accommodation laterally along the bottom horizontal guideline (right or left). Draw a vertical line from that tick mark to intersect the horizontal line drawn in Step 2.

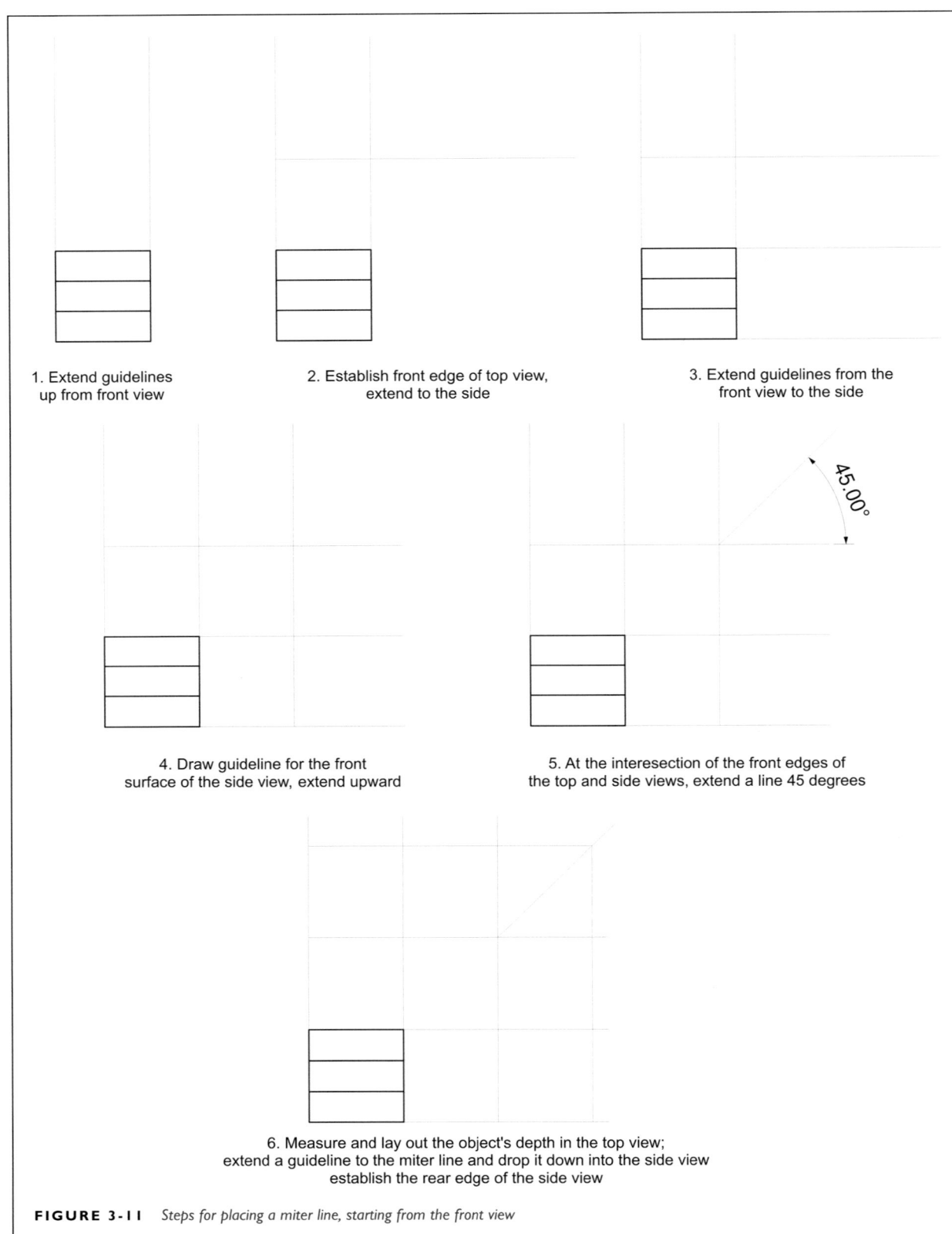

1. Extend guidelines up from front view

2. Establish front edge of top view, extend to the side

3. Extend guidelines from the front view to the side

4. Draw guideline for the front surface of the side view, extend upward

5. At the interesection of the front edges of the top and side views, extend a line 45 degrees

6. Measure and lay out the object's depth in the top view; extend a guideline to the miter line and drop it down into the side view establish the rear edge of the side view

FIGURE 3-11 *Steps for placing a miter line, starting from the front view*

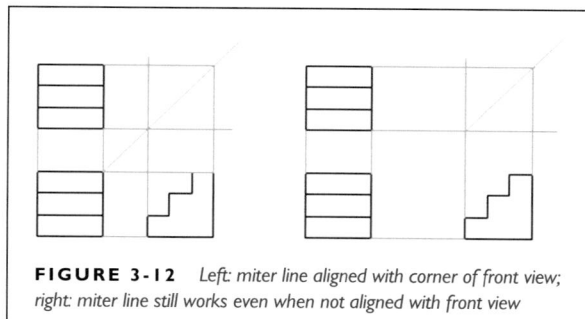

FIGURE 3-12 *Left: miter line aligned with corner of front view; right: miter line still works even when not aligned with front view*

5. At the intersection of the two guidelines, draw a 45° guideline that extends away from the front view. This is the miter line.

6. Information can now be transferred from top view to side view, and vice versa. Mark the depth measurement along the leftmost vertical guideline in the top view. At this tick mark, draw a guideline spanning the full width of the object and extending all the way to the miter line. To place the rear of the object in the side view, drop a vertical line from this intersection with the miter line to the bottom horizontal guideline of the side view.

   The perimeters of all three views are now laid out and aligned, with a miter line in place for further information transfer.

If the perimeters of the top and side views were laid out before determining a miter line, extend the front edges of both views until they intersect. Draw a 45° line from that intersection. As long as the miter line aligns with the views between which information will be exchanged, it will work (Figure 3-12). Use this approach to add miter lines to solve the homework exercises below.

If you've laid out the whole view array perimeter, you can simply draw a 45° through the upper right corner of that rectangle. The perimeter already did the work of aligning the rear of the top view with the rear of the side view.

Homework: Diagonal Views with Missing Information

Instructor

In the previous homework assignment, students were asked to fill in missing lines by moving information either vertically or horizontally between adjacent views. In this exercise, they will fill in missing lines by moving information from a side view to the top view (or vice versa) using miter lines. Exercises should include a perspective or isometric drawing of the object to aid visualization of the missing parts. For the first exercises, be sure that missing information can

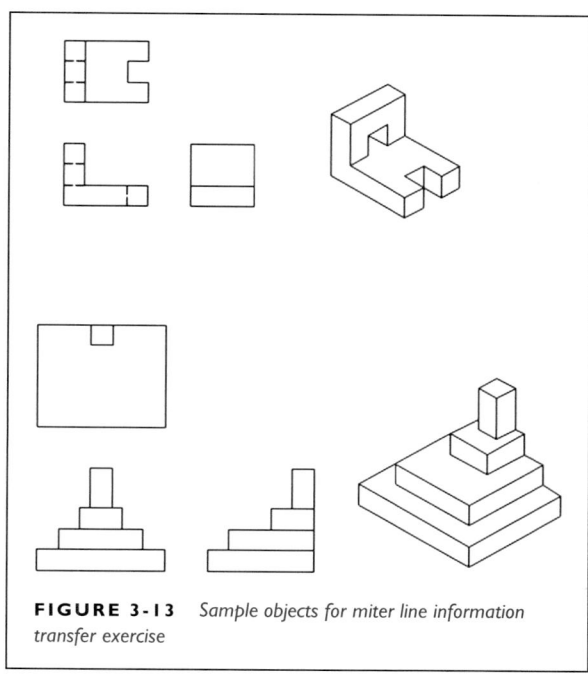

FIGURE 3-13 *Sample objects for miter line information transfer exercise*

be deduced from information in the other views. The first exercises should feature only one view with missing information. See Figure 3-13 for examples.

As the students become more confident in reading orthographic views, this exercise can be modified to wholly omit one view, asking students to provide it from information provided in the remaining views and 3D representation. The objects themselves can also become increasingly complex, with more hidden features and arcs.

Students

1. Examine the views and image of the object (such as the ones in Figure 3-13) to determine what information is missing from which view.

2. Draw a miter line between the top and side view.

3. Use a straightedge to draw guidelines that transfer information to the reflection line and then into the appropriate view.

4. Darken lines with the appropriate line style.

5. Don't erase the reflection line when you're done so that your instructor can see how the work was accomplished.

MODULE 3: AUXILIARY VIEWS

There are times an angled plane or edge cannot be accurately represented in a standard orthographic view. An **auxiliary** view is required to show its true size and shape.

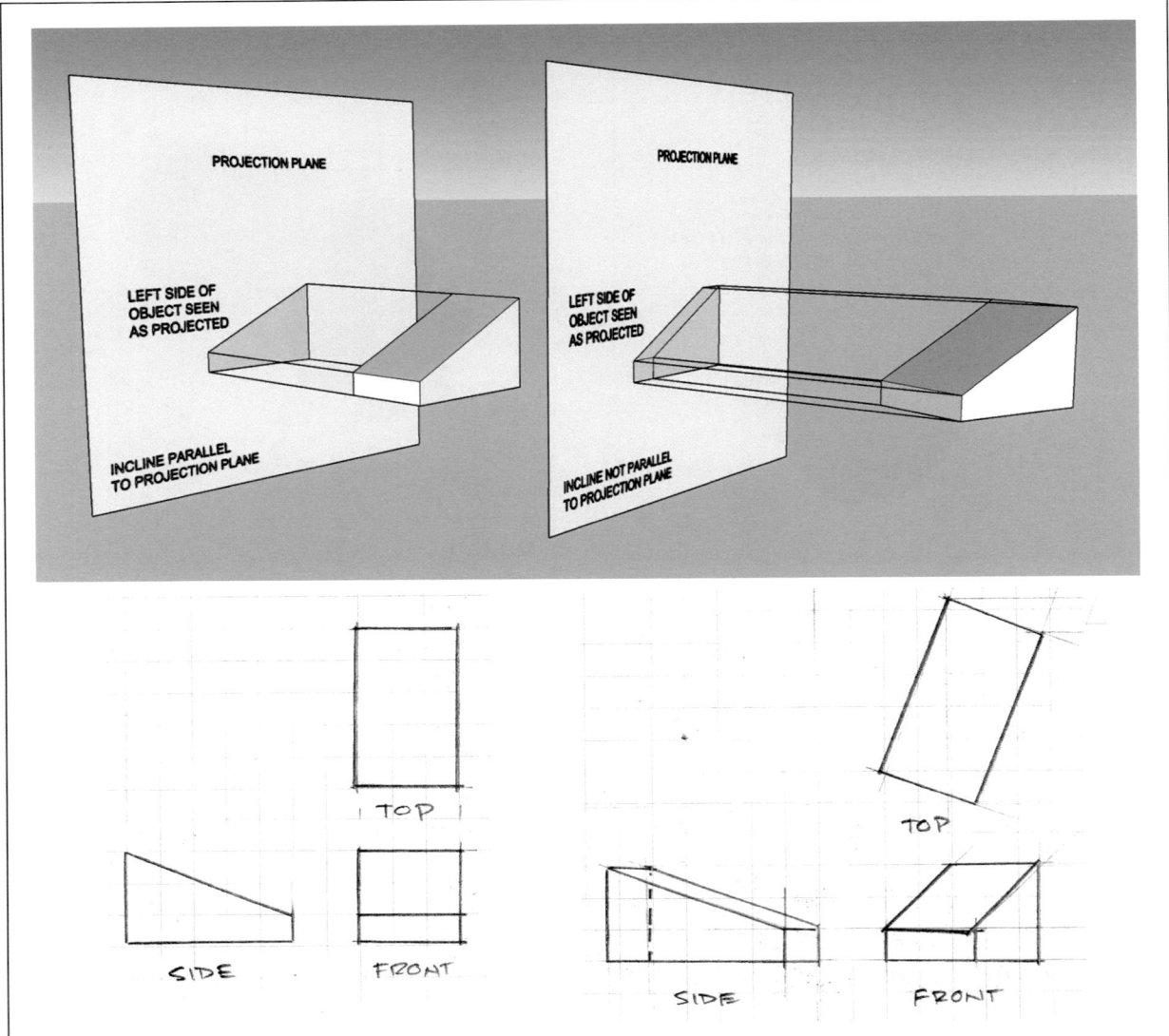

FIGURE 3-14   *Left: object with incline having an edge parallel to a projection plane; right: object rotated so that incline has no edges parallel to the projection plane; below: orthographic sketches of object in original and rotated positions*

There are two categories of angled planes and edges (Figure 3-14):

- **Inclined**: one of the plane's edges is parallel to a picture plane but the surface itself is inclined toward an adjacent picture plane. An edge appears in true length on the picture plane to which it is parallel; it will appear foreshortened in other views.

- **Oblique**: the plane and its edges are not parallel to any picture plane, and therefore appear foreshortened in all views. No edges appear in true length in any standard view. An oblique surface is also called "skewed."[3]

An auxiliary view does not replace any of the standard orthographic views, nor does it typically show the whole of the object since that would present a distorted representation of the area around the incline. For oblique planes, it is usually necessary to derive the auxiliary from a section of one of the standard views.

Layout planning should ensure that placement of the auxiliary view does not crowd other views. The auxiliary view should be near to the view it references for context, which means it may be necessary to push nearby standard views a little farther out (Figure 3-15).

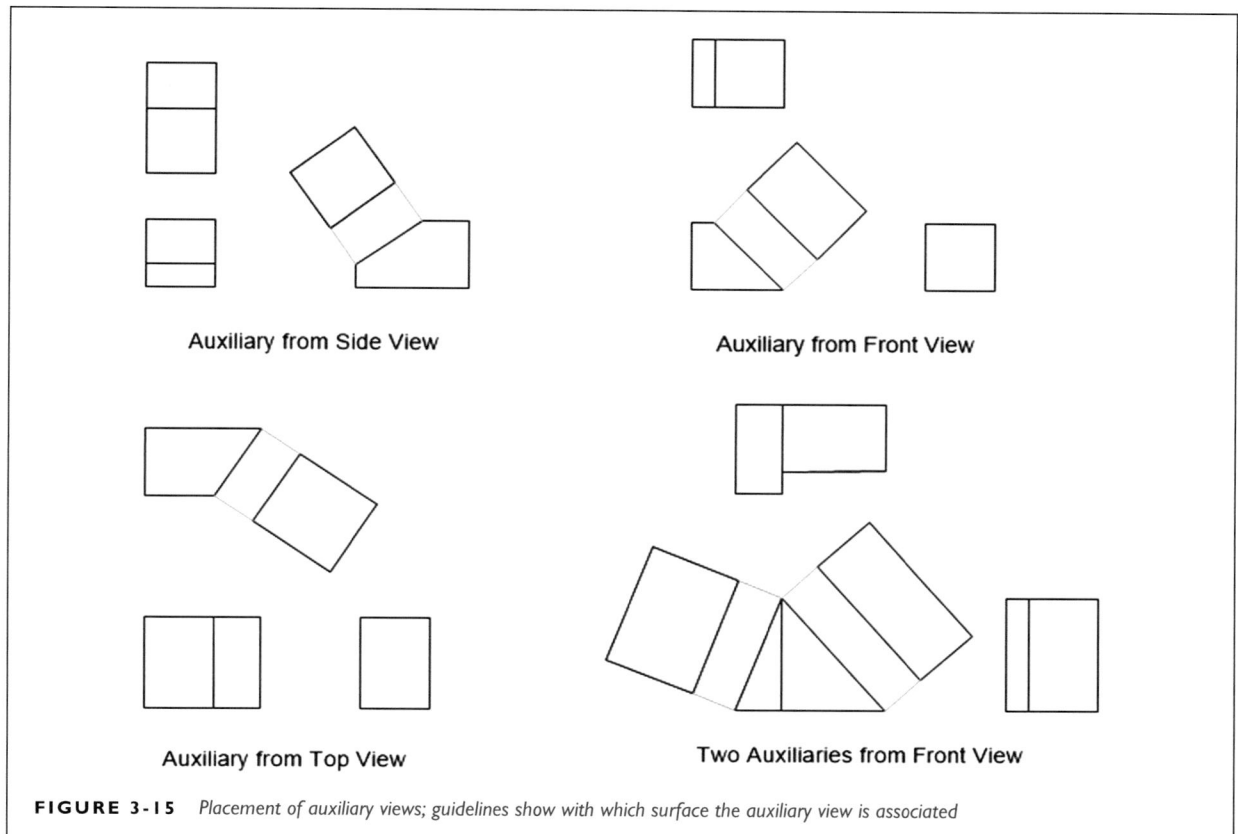

Auxiliary from Side View

Auxiliary from Front View

Auxiliary from Top View

Two Auxiliaries from Front View

FIGURE 3-15 *Placement of auxiliary views; guidelines show with which surface the auxiliary view is associated*

The Two-Triangle Technique

When you need to draw a rectangle whose lines do not fall into the range of standard triangle degrees (15°, 30°, 45°, etc.), the two-triangle technique is useful.

One triangle is used as a reference edge. A T-Square may be used, but they're a bit cumbersome for this purpose, and the transparency of triangles allows the draftsperson to better see what they're doing. Since both 45° and 30°/60° triangles have a 90° corner, either triangle can serve as the reference edge.

First, draw one of the sides of the angled rectangle. Align one of the 90° edges of one triangle with the angled line. Be sure the hypotenuse edge is a short distance away from the initial object edge or you will be unable to draw perpendicular lines that fully pass through your measurement tick marks.

Holding this first triangle firmly in place, butt the hypotenuse of the second triangle (the reference triangle) against the hypotenuse of the first (Figure 3-16).

With the reference triangle firmly held down, the first triangle can now slide back and forth along its hypotenuse to draw lines perpendicular and/or parallel to the original angled line.

Exercise: Using the Two-Triangle Technique to Draw Angled Shapes

Instructor

This exercise provides practice using the two-triangle technique in preparation for hand drafting auxiliary views. The steps below outline a student choice version of the exercise. Depending on time and the needs of the class, worksheets can be created featuring arrays of angled lines and a list of shapes to be drawn.

Students

Practice using the two-triangle technique to draw lines perpendicular and parallel to an angled line. If your instructor does not distribute a handout with an array of angled lines, draw your own (see Step 2).

1. Tape a sheet of 8½" × 11" copy paper to the drawing table, landscape orientation. No border is needed.

2. Use a triangle to draw an angled guideline anywhere on the page. Be random in placement. Make the line about 3" long.

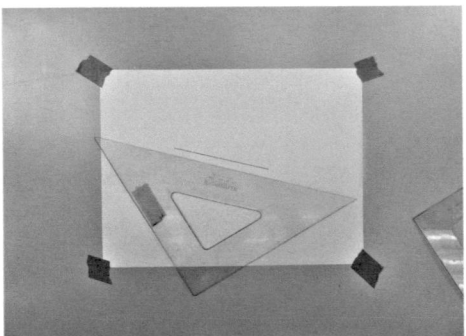

1. Draw an angled line

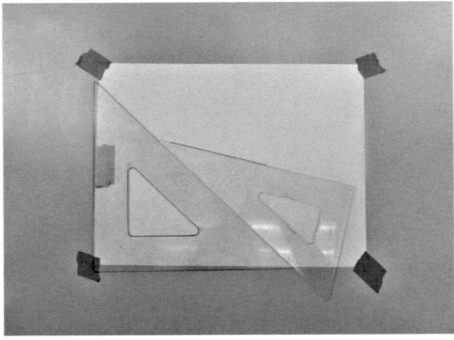

2. Line up triangle with angled line; butt second triangle against hypotenuse of first triangle

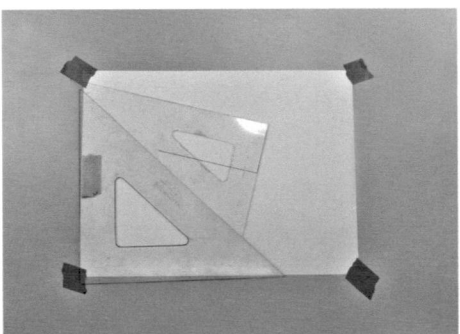

3. Slide first triangle along second triangle to draw a line perpendicular to original line

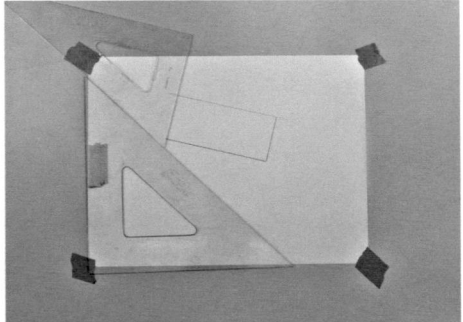

4. Slide first triangle to draw remaining sides of the angled rectangle

FIGURE 3-16   *Two-triangle technique: (1) draw an angled line; (2) align triangle edge; (3) slide triangle along edge of other triangle to draw perpendicular lines; (4) parallel lines*

3. Place tick marks along the line to mark the endpoints of a measured line segment (your choice as to measurement).

4. Using the two-triangle technique, create an angled square using the chosen measurement. Lay out two perpendicular lines. Measure and place a tick mark along one of the perpendiculars to set the rear edge of the shape. Draw the far side (parallel to the first line). Remember to lay out the whole of the shape with guidelines first, and when complete, darken with visible lines.

5. Repeat. Try drawing L- and H-shapes. For shapes more complicated than a rectangle, lay out the overall perimeter and then measure and mark the interior features along the perimeter guidelines. Connecting the dots is permitted, but be as precise as possible.

Exercise: Modeling an Incline in SketchUp

Before drafting auxiliary views, it will be useful to model an incline and take a look at it via the standard views.

1. Open SketchUp. Select the Simple/Inches template. Move the human figure aside.

2. From the axes' intersection, drag out a rectangle that is 6′ wide (red) and 2′ deep (green).

3. Use Push/Pull to drag the rectangle's surface up to a height of 2′ (blue).

4. Draw a line angling from corner to corner across the front, splitting the front surface into two independent surfaces (Figure 3-17).

5. Push/Pull the upper triangular area to the rear of the object so that it vanishes, leaving behind an inclined surface.

1. Lay out base rectangle

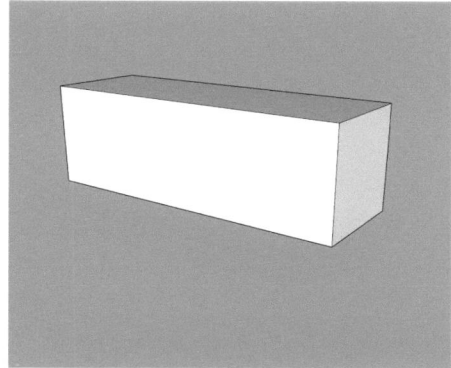

2. Push/Pull surface to height

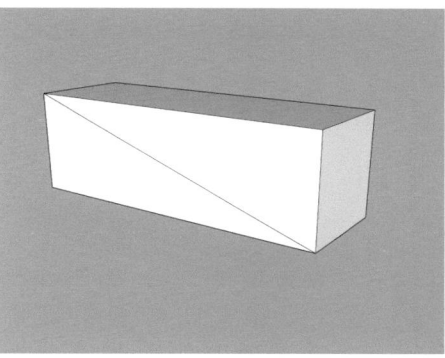

3. Draw a line across the front of the box, from upper corner to lower corner, to divide the front into two surfaces

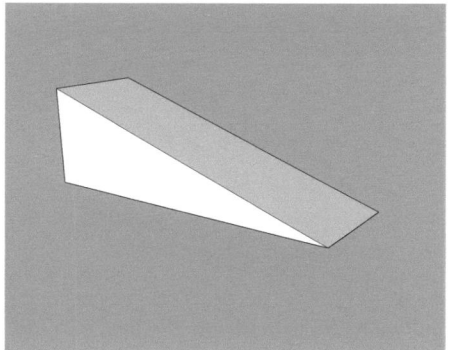

4. Push/Pull the upper triangular surface back the full depth of the box (it will disappear)

FIGURE 3-17   *Modeling the incline: (1) lay out rectangle; (2) push/pull up; (3) lay out incline; (4) push/pull back*

Switch the camera to parallel projection and look at the ramp in the standard views. In which views is the ramp foreshortened? Which views feature an edge of the ramp in true length? Which view shows the angle of the ramp in true length? Use Orbit and Pan to look at the incline without foreshortening. What happens to the rest of the object when you find that viewpoint?

Drawing Auxiliary Views

At its simplest, drawing an auxiliary view takes the two-triangle technique and applies it to the inclined surface of a drawn object. The draftsperson must determine which picture plane shows an edge that is the true length of the incline. For a ramp such as the one modeled in Figure 3-17, the incline's true length is seen in the front view; therefore, the auxiliary view is drawn in reference to the front view. Since the auxiliary view should be visually associated with its reference view, the top and side views are placed a little farther away to provide space for it.

To draw an auxiliary view of a simple incline (Figure 3-18):

1. Select the view of the object in which the true length of the incline can be seen. Extend perpendicular guidelines from the endpoints of the incline's edge.

2. Leaving enough space above the referenced view for dimensions and notes, draw a line parallel to the incline that intersects the perpendicular guidelines.

3. Along one of the perpendicular guidelines, measure and mark the depth of the incline. If the incline is a rectangle, use the two-triangle technique to draw the rear edge of the incline.

Exercise: Hand Drafting a Simple Auxiliary View

Students

This exercise uses the two-triangle technique to draw the auxiliary view of a ramp.

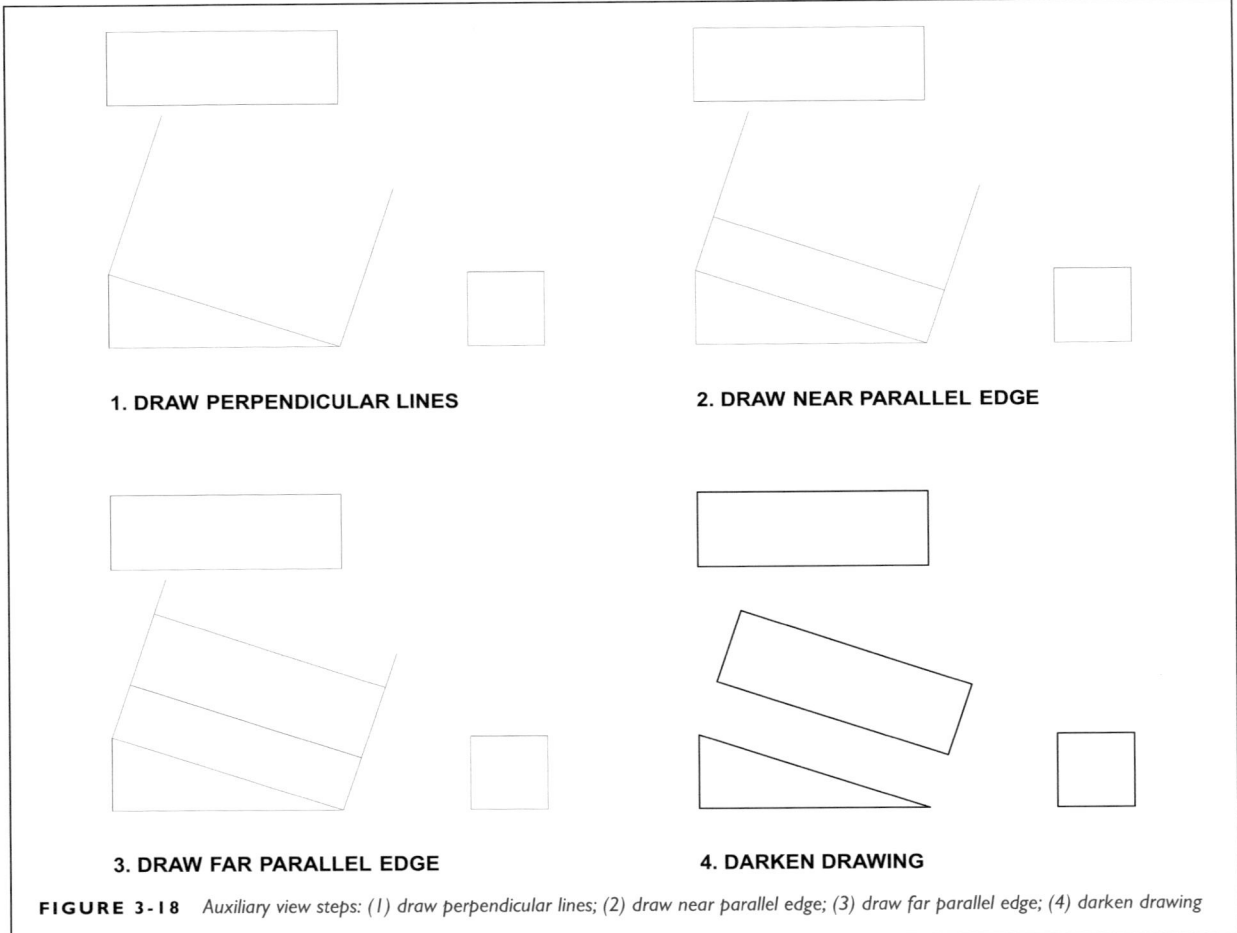

1. DRAW PERPENDICULAR LINES

2. DRAW NEAR PARALLEL EDGE

3. DRAW FAR PARALLEL EDGE

4. DARKEN DRAWING

FIGURE 3-18   *Auxiliary view steps: (1) draw perpendicular lines; (2) draw near parallel edge; (3) draw far parallel edge; (4) darken drawing*

1. Use 8½″ × 11″ paper, landscape orientation. Use the measurements of the SketchUp model in the previous exercise (6′–0″ wide, 2′–0″ deep, 2′–0″ high). Draft the object in ½″ = 1′–0″. Include borders that are ½″ from the page edge on all four sides of the sheet.

2. Do layout math to center the drawing on the page. Remember to leave space for the auxiliary view.

> As drawings become more complex, initial layout math becomes more approximate. Sheets that contain multiple drawings of disparate objects may require other layout approaches, such as placing drawings in rows and/or columns. This is one of the reasons guidelines are drawn light enough to be easily erased and darkening is held off until the end—you may lay out a view and discover the placement is not optimal. Erase it and draw it again. As you become more proficient in judging distance and mentally planning the work, a quick rough sketch may be all that you need to plan a sheet's layout.

1. Lay out guidelines for the perimeters of the front, top, and side views.

2. Lay out the front view.

3. Use the two-triangle technique to draw two guidelines perpendicular to the incline. Lay out the front and back edges (depth) of the incline in the auxiliary view.

4. Take a look at the overall layout. Did you leave enough space for the top and side views? If views look crowded, erase and redraw.

5. Once you have laid out all elements of the drawing and deemed them correct, darken.

6. Letter your name, assignment title, class, and date in the lower right corner. Justify the block of text to the left. Darken the border.

Homework: Drafting a Simple Incline in Vectorworks

Having drafted the ramp and its auxiliary view by hand, repeat the drawing in Vectorworks.

Landscape orientation. Scale is ½″ = 1′–0″. The border will be at ½″ on all sides.

1. Create the desired number of design layers. Lay out the border in a Border Guideline layer.

2. Rather than laying out the perimeter of the whole array, lay out each view's perimeters with guidelines in the guideline layer. If the array looks off-center or cramped, use Select to drag a box around a view (or all of the views) and drag it to a better location.

3. Lay out the incline in the front view. You may need to switch the Line tool to unconstrained mode. The constrained mode will allow you to draw lines perpendicular to angled lines, but the start point must be on the angled line.

4. When drawing the perpendicular guidelines to place the auxiliary view, the Line tool will tell you the angle at which the intersecting line is drawn. Click to set the perpendicular line's start point, drag the line out, and look at the blue text box that appears near the cursor. **L:** tells you the length of the line, in decimal notation. **A:** tells you the angle. The angle may be negative, depending on the direction in which you've dragged the line. The keyboard's Tab key allows you to shift between L: and A:. Hit the Tab key once, and L: will be highlighted. Hit the Tab key a second time, and A: will be highlighted, allowing you to enter the desired angle. Hit Enter, then click to affix the line.

5. Lay out the perimeter of the auxiliary view.

6. Double-check the placement of the other views. Move them if crowded.

7. Switch to the finished drawing design layer and darken.

8. Use the Border tool to place the border (switch off zones). Use the Text tool to write your name, assignment title, class, and date in the lower right corner. Justify text to the left.

Irregular Inclines and Section Views

Not all inclines are rectangular. In Figure 3-19, the true angle of the incline is seen in the front view, even though the front edge of that incline is broken into segments.

When the plane is irregular polygon, start the auxiliary view with a center line down its long axis rather than by drawing the forward or rear edge. Use the center line to measure and mark the various corners of the object, then connect the dots. Label points to keep track of them.

In some cases, the incline may appear to offer no true-length edges that are parallel to a picture plane. In

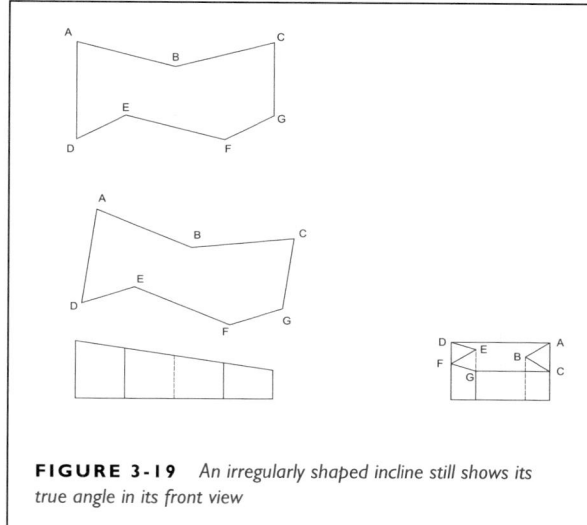

FIGURE 3-19 *An irregularly shaped incline still shows its true angle in its front view*

Figure 3-20, the forward and rear edges of the incline are shown true length in the front and top views, but the incline itself and its side edges are foreshortened. In the side view, the line denoting the edge of the inclined surface does not present as true length as it is angled in relationship to the picture plane. Even though the edge of the inclined surface is not shown true length in the side view, the true angle of the incline is shown.

If you cut through the object along its center axis from front to back, this sliced view, or **section**, shows a true-length edge of the incline that is parallel to the right view picture plane. The sectioned side view becomes the reference for the auxiliary view (Figure 3-21).

FIGURE 3-20 *An incline whose true angle is seen in the side view, even though the true length of the edge is not evident*

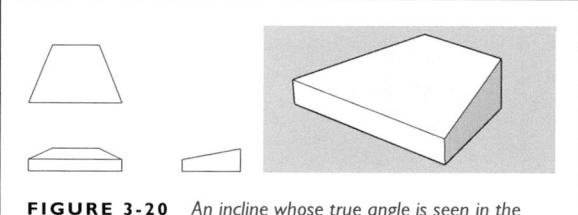

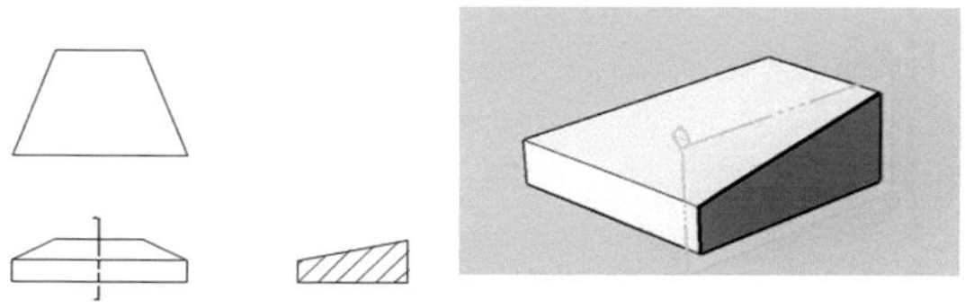

FIGURE 3-21 *Cutting through the incline to create a section view, revealing an angled line with true length*

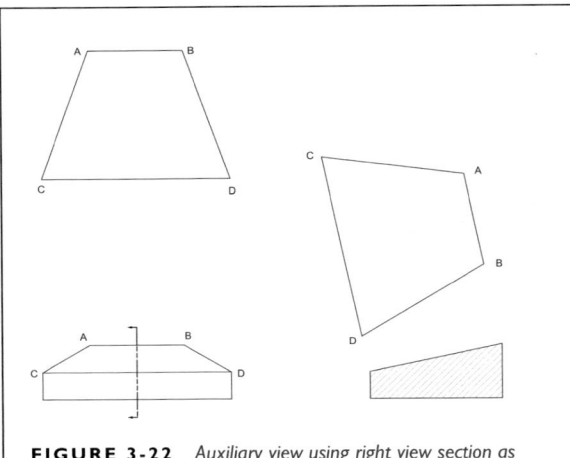

FIGURE 3-22 *Auxiliary view using right view section as reference view*

In the auxiliary view, the front and rear edges are perpendicular to the incline as seen in the section. The front and rear edges are seen in true length in both the front and top views, despite foreshortening of the incline itself. Their measurements can be transferred to the auxiliary view by placing a centerline through the short axis and then measuring outward from it (Figure 3-22).

A more complex arrangement of inclined planes can be seen in Figure 3-23. In this roof-like object, none of the upper surfaces are parallel to any picture planes—their bottom edges are, however.

In order to demonstrate the creation of section views for this object, it will be modeled first.

Exercise: Modeling a Tent-Shaped Object in SketchUp

Students

In this exercise, you will construct an object with compound angles. The Push/Pull tool pushes and pulls volumes perpendicular to the selected surface, so using this tool on angled surfaces can produce undesired results. Instead, the object will be constructed plane by plane (Figure 3-24).

1. Open SketchUp. Select the Simple/Inches template. Move the human figure aside.

2. From the axes' intersection, drag out a rectangle 8'–0" wide (red) by 4'–0" deep (green).

3. Draw a guideline through the long axis of the rectangle (2'–0" from front edge).

4. Draw a guideline 2'–0" in from either end (parallel to the sides).

5. Draw a 3'–0" vertical line up from the two intersections of the guidelines within the rectangle.

6. Connect the top endpoints of the two vertical lines.

7. Connect the top endpoints of the two vertical lines with the corners on their side of the rectangle. The angled surfaces will snap into place as each outline is completed.

   If time allows, construct a box of the same overall dimensions. Lay out the angles of the right- and left-angled planes on the front surface. Push/Pull them to create angled surfaces at either end. Lay out the angles of the front and rear planes on the right surface.

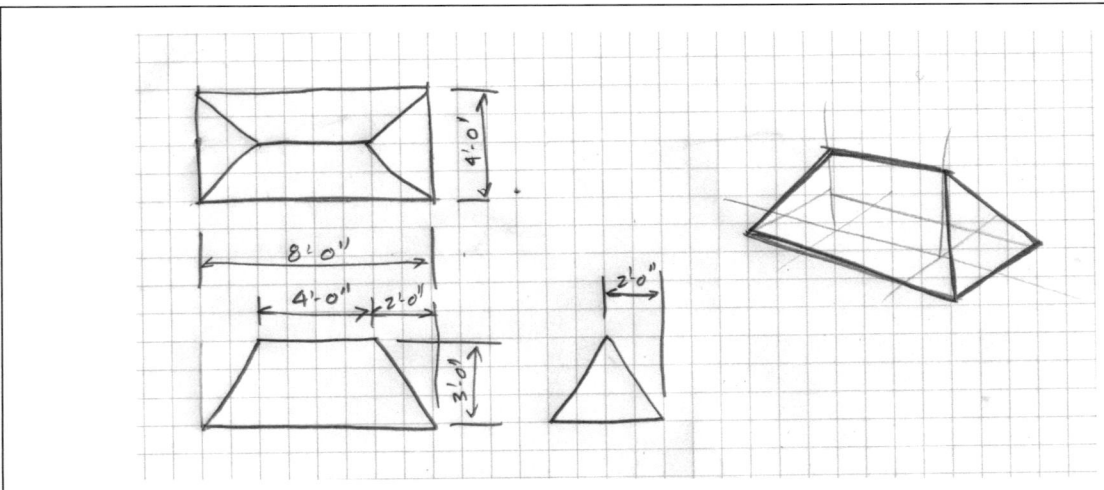

FIGURE 3-23 *Rough sketch of a tent-like object with four inclined planes*

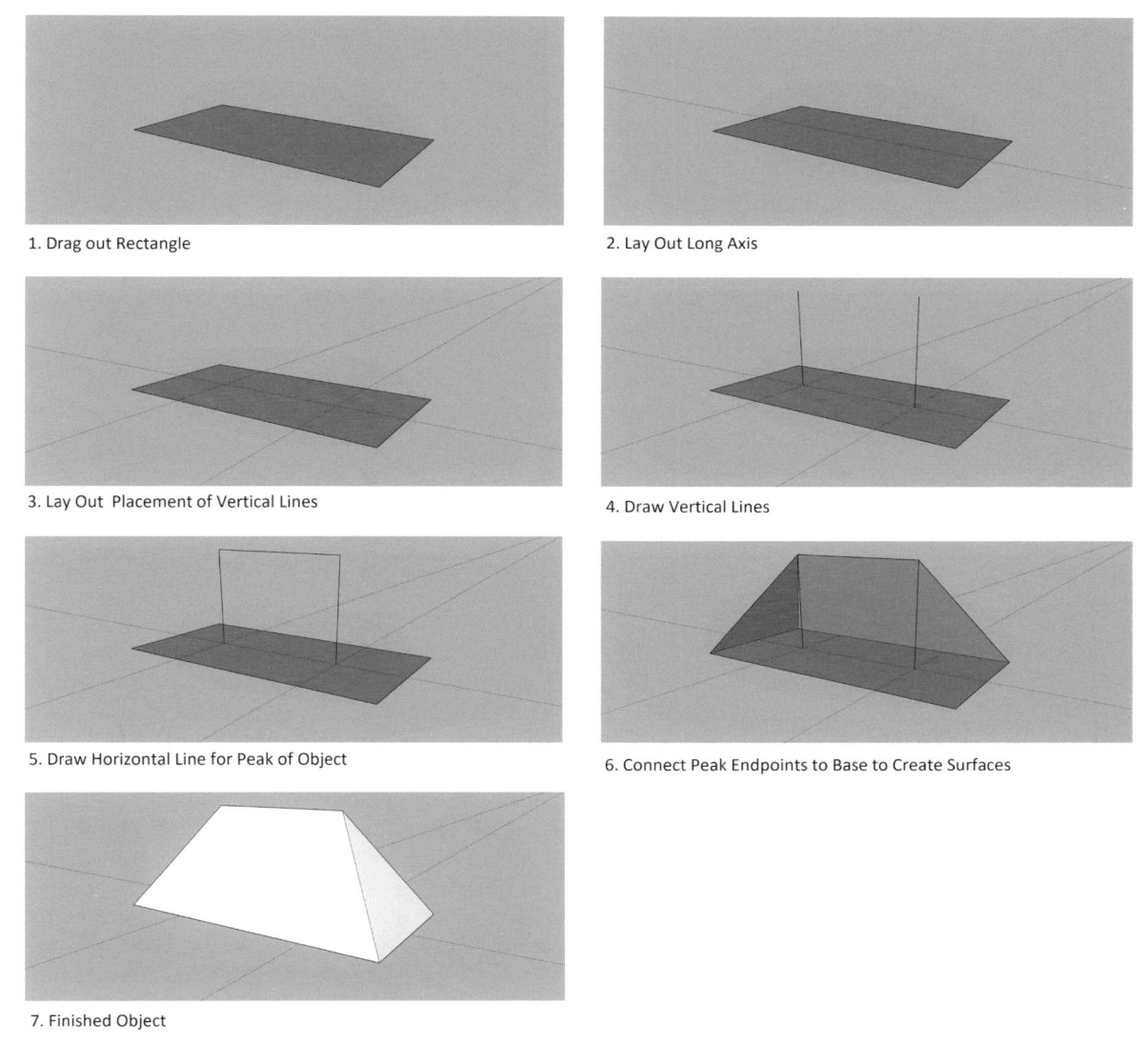

1. Drag out Rectangle

2. Lay Out Long Axis

3. Lay Out Placement of Vertical Lines

4. Draw Vertical Lines

5. Draw Horizontal Line for Peak of Object

6. Connect Peak Endpoints to Base to Create Surfaces

7. Finished Object

FIGURE 3-24   *Steps in modeling the tent-like object: lay out footprint; locate axis and draw guidelines; draw vertical lines; outline planes*

See what happens when you use Push/Pull in an attempt to create the front and rear angled planes from an already angled plane.

8. Look at the object via the standard views. In which views can you see the true angle of which incline? Return to perspective projection. There is a Section Plane tool in SketchUp that allows you to slice through an object. At the bottom of the Large Tool Set, find the icon that looks like a circle on top of a triangle. This is the standard symbol used to label a section view's **cutting plane** (Figure 3-25).

9. When the Section Plane tool is selected, an orange outlined window-like area appears. This window can be aligned with a surface by shifting it with the mouse. In this exercise, a vertical cutting plane is desired, but since the object has no vertical planes, the Section Plane tool insists on angling the cutting plane parallel to one of the angled surfaces. Draw a vertical rectangle at one end of the object to give the Section Plane tool a vertical surface to reference, then reselect the tool. A Name Section Plane window will appear. For this exercise, accept the default number.

10. Click on the section plane window to make it active; it will turn blue. Click on the Move tool and drag the cutting plane through the object until you reach its middle. As the plane passes through the object,

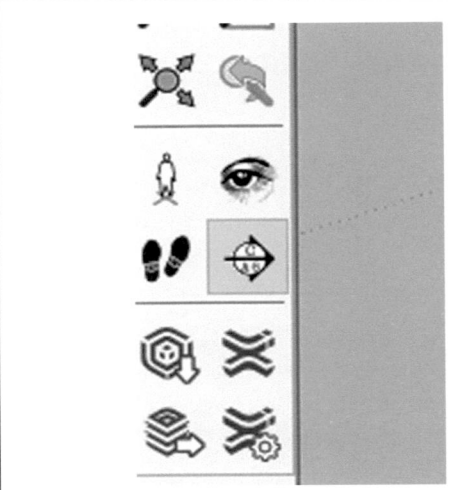

FIGURE 3-25 *The location of SketchUp's Section Plane tool icon*

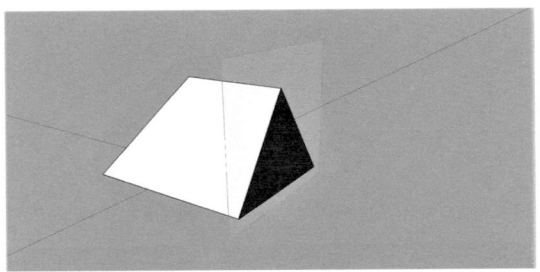

1. Section from End

2. Section from Front

FIGURE 3-26 *Using the Section Plane tool to create vertical sections along the object's long and short axes*

everything it passes through will vanish from view. Look at the object again in parallel projection with the standard views. Has your perception of any of the lines or planes changed (Figure 3-26)?

11. Erase the section plane and the rectangle that was drawn to help place it. Draw a vertical rectangle at the front of the object and open a new section plane. Drag the plane back until it meets the object's peak. Again, look at the object in the standard views and note whether the appearance of any lines or planes have changed.

In the section views, not only is an incline's true angle shown but also the line created by the section is true length. For complex objects, multiple cutting planes can be placed to create multiple section views within a drawing. For an object like this one, two auxiliary views are drawn, one describing the front and rear inclines, and one describing the right and left inclines. For each auxiliary view, a section is required.

Cutting Plane Lines

A cutting plane is indicated by drawing a **cutting plane line** in the view to which the proposed section is perpendicular. When an object is entirely cut through, a **full section** is created. Imagine opening the object along this cut and swinging the halves back to reveal their interiors. The half swung to the left becomes a left-view section; the half swung to the right becomes a right-view section (Figure 3-27).

At its endpoints, the cutting plane line turns to indicate the direction toward which the section will be viewed, terminating with arrowheads and a label. A bubble half-surrounded by a triangle is a typical label format. Sections

are usually labeled with letters rather than numbers to note that they are subset drawings within a larger array (e.g., "See Drawing 4, Section B-B").

The cutting plane line style is thick (USITT recommends 0.50 mm), with long dashes alternating with two short dashes (in Vectorworks, line Style ISO-09: Long Dashed Double Short Dashed). The cutting plane line may be gapped rather than running through the object, particularly for large items with a fair amount of interior detail. Create a section line class in your base file; for this class, assign it a black Fill so that the arrow around the bubble is filled in. Simple arrowheads can also be used, instead of a bubble (Figure 3-28).

In Vectorworks, the Section-Elevation Marker tool is found in the Dims/Notes toolbox. The icon looks like a stepped line with arrowheads at either end, pointing upward. The step in the line is reflective of how a cutting plane line may be drawn for half, partial, or offset sections. The Reference Marker tool looks like SketchUp's Section Plane tool: a bubble set within a triangle. It too can be used when drawing section lines, but indexes information differently (Figure 3-29).

Before selecting the Section-Elevation Marker tool, draw a guideline to place the intended section line. The endpoints should lie beyond the object's perimeter to allow enough space for the labels/bubbles. Cutting plane lines should not cross dimension lines, so any dimensioning on that side of the object should be moved farther out. A gapped section line

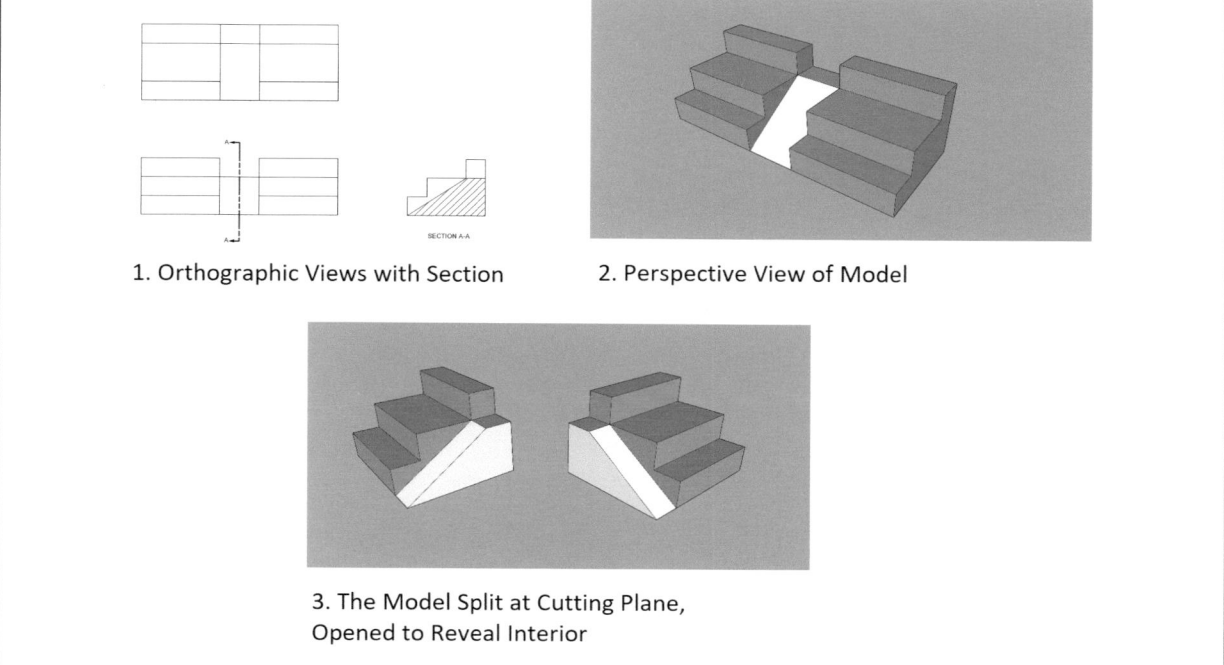

1. Orthographic Views with Section   2. Perspective View of Model

3. The Model Split at Cutting Plane,
Opened to Reveal Interior

FIGURE 3-27   *(1) An object shown with cutting plane line and vertical section view; (2) a model of the object; (3) the model swung open at the cutting plane*

could also be used, placed beyond the tiers of dimensioning, but this runs the risk of losing association with what is being sectioned.

Activate the Cutting Plane class. Select the Section-Elevation Marker tool (Figure 3-30). Options will appear on the upper toolbar. Click on the wrench-and-pencil icon to open the object properties window. Marker Size can vary,

depending on the scale of the printed sheet; 3/8″ is a good bubble size for this exercise. Uncheck Use Gapped Line; upcoming exercises don't feature interior detail that might be obscured by a cutting plane line running through the object. This feature can be turned back on when drawing complicated objects. Uncheck Text Auto Rotate; this turns the notation within the bubble to the direction in which the

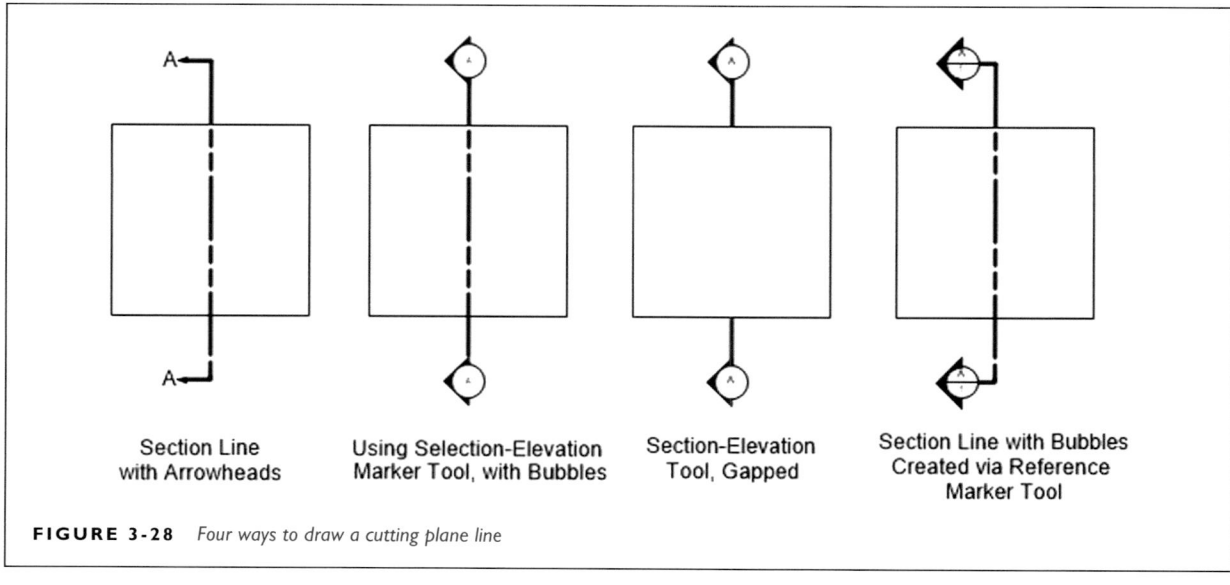

Section Line
with Arrowheads

Using Selection-Elevation
Marker Tool, with Bubbles

Section-Elevation
Tool, Gapped

Section Line with Bubbles
Created via Reference
Marker Tool

FIGURE 3-28   *Four ways to draw a cutting plane line*

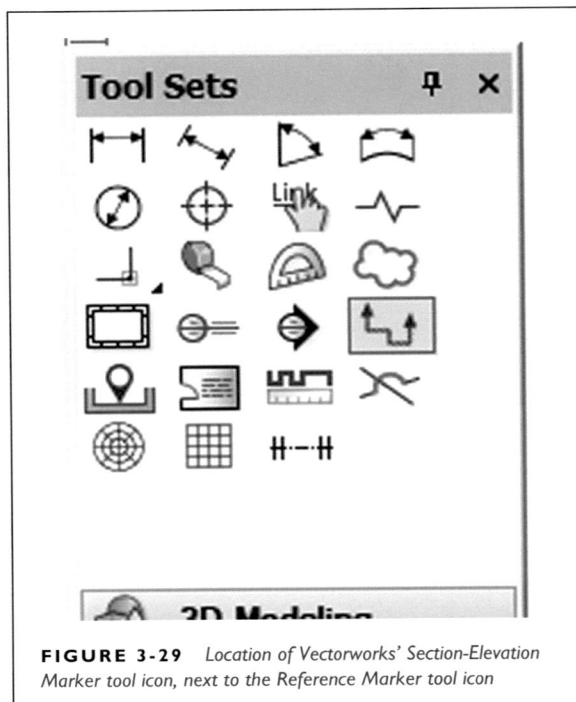

FIGURE 3-29 *Location of Vectorworks' Section-Elevation Marker tool icon, next to the Reference Marker tool icon*

arrow points. Unchecking this box orients the bubble text to be read from the bottom of the page. Checkmark Text Both Ends so the section name appears in the bubbles at both endpoints. Drawing and sheet numbers can be changed via the navigation window that appears when the cutting plane line is drawn. Click OK to return to the drawing.

Click the cursor at the cutting plane line's start point (one end of the guideline), then drag the cursor to the other endpoint and double-click. The cutting plane line, with bubbles, will snap into place. The Object Info navigation window will appear on the right side of the screen (Figure 3-31). In this window, select a text style. There are three buttons near the top of the window. The middle one, Reverse Direction, allows you to flip the direction of the bubble/arrows. The bottom button allows you to change the style of the bubble/arrow, depending on project and industry preferences. The top button creates section viewports when working in 3D.

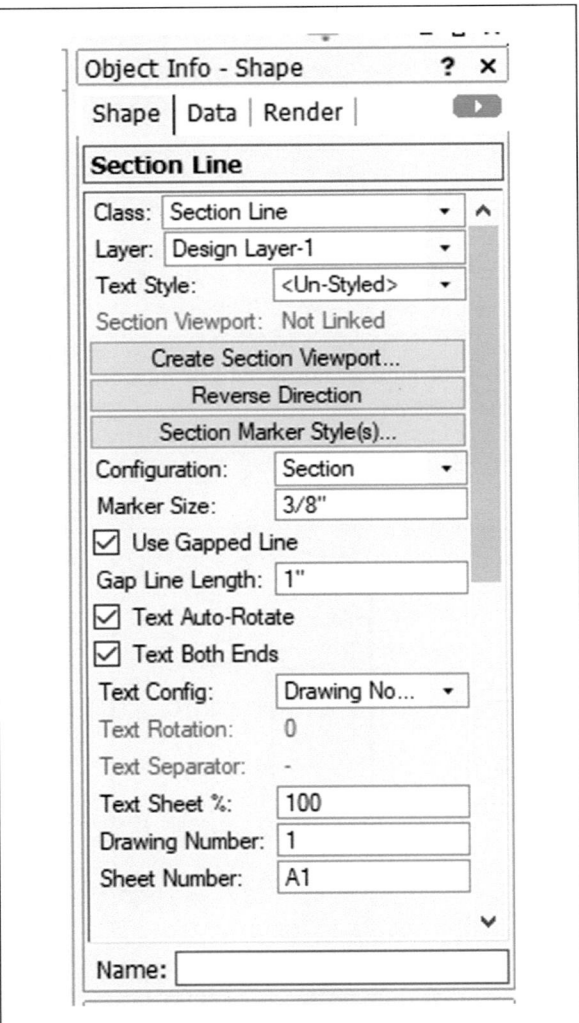

FIGURE 3-31 *The Object Info window that appears when a cutting plane line is drawn*

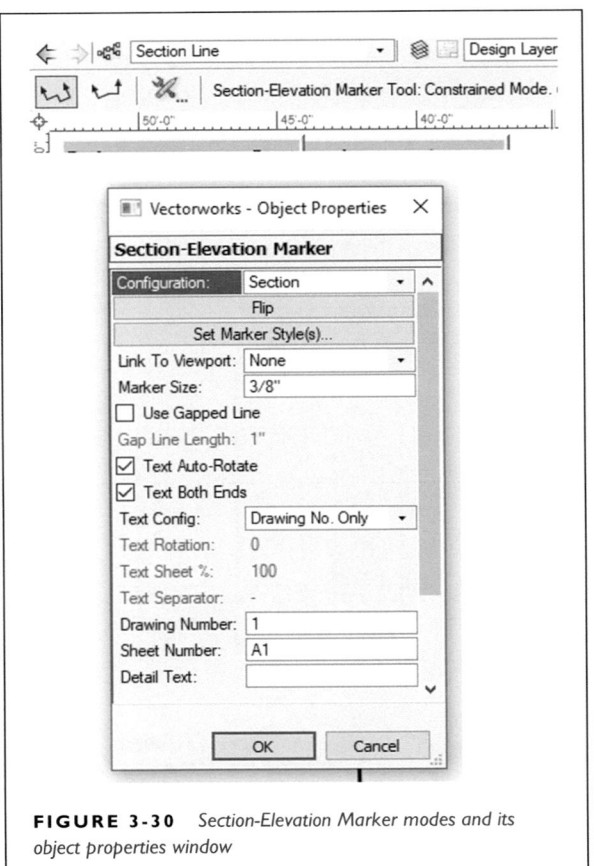

FIGURE 3-30 *Section-Elevation Marker modes and its object properties window*

Crosshatching the Section View

Solid portions of an object that have been cut through need to be indicated as being solid. This is done with **crosshatching.** Elements of the object visible beyond the cut-through portion are drawn as per a normal side view. Since the purpose of the section view is to provide information about the object at the cutting plane, elements of the object still hidden beyond the cutting plane are not typically included in the sectioned view.

The basic crosshatch pattern is a series of thin black parallel lines (USITT recommendation: 0.13 mm), angled at 45° and spaced 1/8″ apart, regardless of the drawing's scale. When crosshatching by hand, a piece of tape affixed 1/8″ from the edge of triangle makes this process a little less tedious (Figure 3-32).

In Vectorworks, Fills add crosshatching and other patterns to the interior of a shape. A Fill can be selected when a class is created; when the outline of an object is completed, the interior is automatically filled. An outline can also be composed with a Fill selected from the Attributes window. Crosshatching in Vectorworks can be done manually; the Double Line tool makes the task slightly less tedious (Figure 3-33).

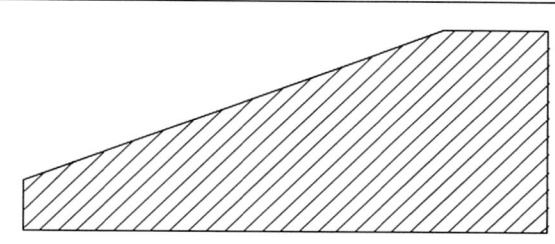

FIGURE 3-33 *In Vectorworks, crosshatching a cut-through mass with a hatch from the fill menu*

Depending on the material being cut through, different fill patterns can be selected to indicate the type of material: concrete, steel, glass, wood, etc. (Figure 3-34). These variations are found more frequently in architectural drawings; for most theatrical drawings, a basic crosshatch usually suffices. In construction drawings where components meet, the angle of crosshatching can be adjusted to ensure that components are read as different objects.

Vectorworks also features a library of Patterns, which are smaller fills than the Hatch fills. If a Hatch seems to vanish when applied to a shape, try a Pattern instead. This library is also accessed via the Attributes window. The following exercise includes instructions on adding Fills to shapes.

Exercise: Adding Crosshatching and Fills to Shapes

Instructor

There are two parts to this exercise. The first features crosshatching by hand, using the tape-on-a-triangle technique. Students should aim for line weight consistency across the whole crosshatched field. The second asks the student to draw two columns of squares in Vectorworks and then apply a variety of Fills.

Students

Hand Drafting Version

1. On 8½″ × 11″ paper, landscape orientation, lay out two 3″ × 3″ squares.

2. Elsewhere on the paper, draw a 45° guideline. Reorient your triangle to draw a second guideline intersecting and perpendicular to the first. Measure and mark the second line 1/8″ away from the first. Draw a line parallel to the first line through this mark.

3. Line up the edge of your 45° triangle with the third line (the one parallel to the first). You should be able to see the first line through the triangle's transparent plastic.

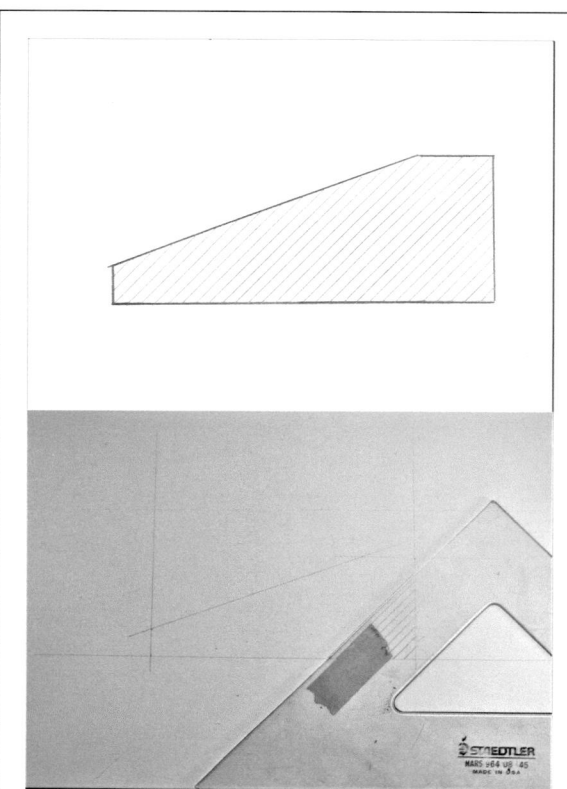

FIGURE 3-32 *Crosshatching a cut-through mass by hand; using a piece of tape affixed along the edge of a triangle*

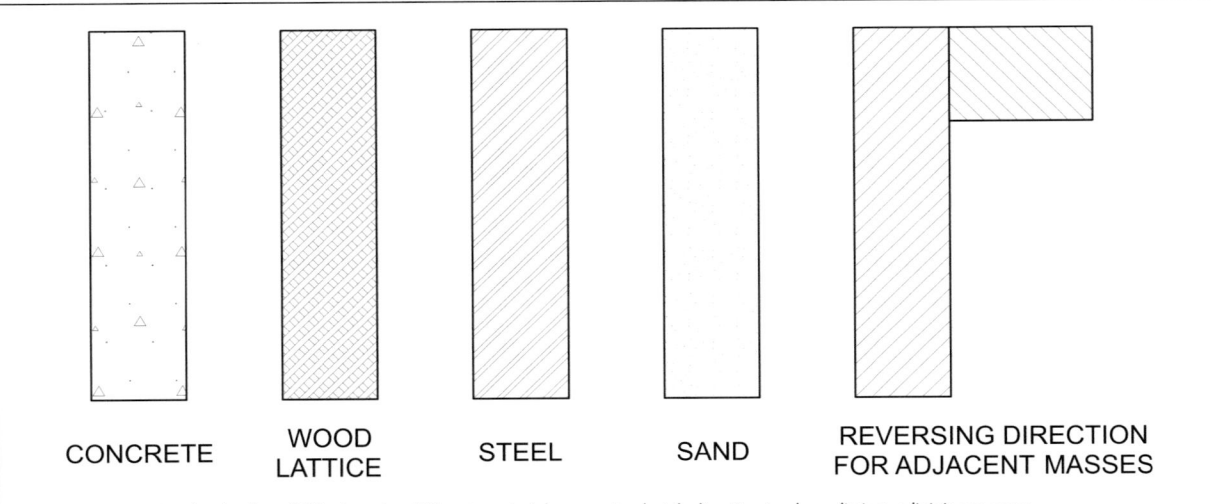

CONCRETE WOOD LATTICE STEEL SAND REVERSING DIRECTION FOR ADJACENT MASSES

FIGURE 3-34 *A selection of Fills denoting different materials; reversing hatch direction to show distinct adjoining masses*

Affix a piece of drafting tape to the triangle, aligning its edge with the first line; there should be 1/8″ between the edge of the tape and the edge of the triangle.

4. Pick one of the squares. Because crosshatching is among the final steps and is a laborious process, they are typically drawn dark rather than laid out with guidelines first. Set the triangle edges just within one corner of the shape, angled at 45 degrees to the square's outline. Draw a thin black line across the square's interior. Slide the triangle so the tape lines up with the line just drawn. Draw another line. Slide the triangle. Draw a line. Repeat until the square is filled with parallel 45° lines.

5. Reverse the triangle and fill the second square with crosshatching perpendicular to the crosshatching in the first square.

Vectorworks Version

1. Portrait orientation, 1:1 scale. Draw guidelines in a guidelines in a guideline layer. Lay out two columns of 2″ × 2″ squares, centered down the page, six squares total (Figure 3-35).

2. In the final drawing layer, darken each square's outline using the Rectangle tool.

3. Select one of the squares in the top row. While it is highlighted, open the Fill options in the Attributes window (the paint can icon). From the drop-down menu, select Hatch. Click on the drop-down menu that appears next to the wrench-and-pencil icon to open it and access the hatch libraries (Figure 3-36). Double-click on any of the hatches; that hatch will appear within the square. If no hatch seems to have been applied, it may be because the hatch pattern is too big for the size of

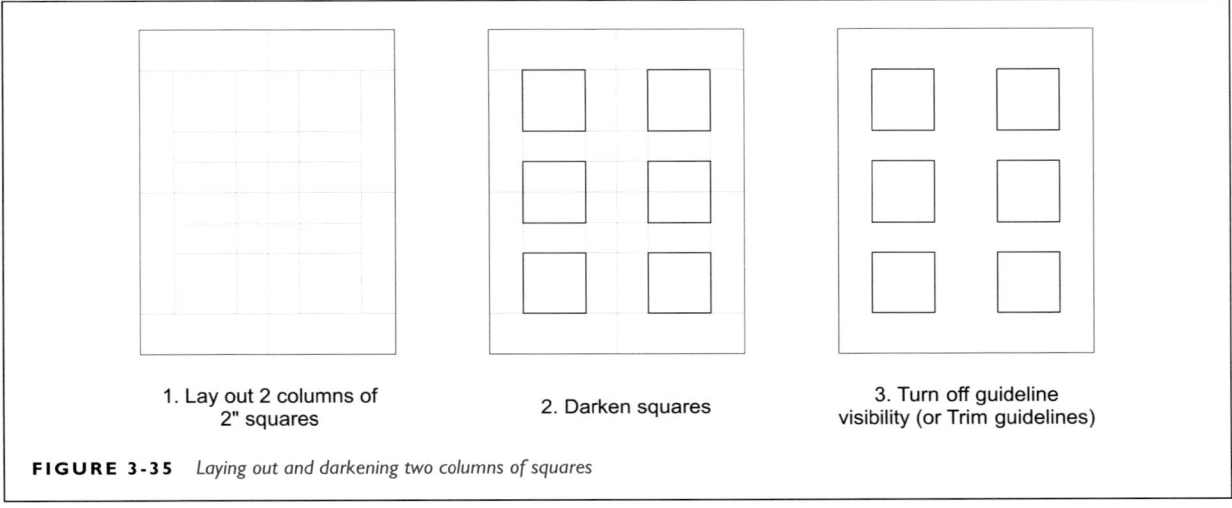

1. Lay out 2 columns of 2″ squares

2. Darken squares

3. Turn off guideline visibility (or Trim guidelines)

FIGURE 3-35 *Laying out and darkening two columns of squares*

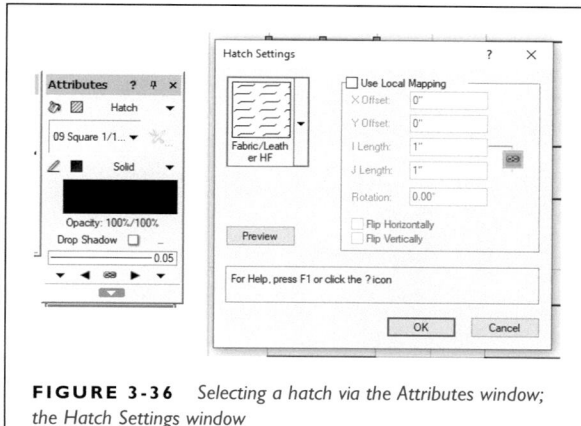

FIGURE 3-36 *Selecting a hatch via the Attributes window; the Hatch Settings window*

the selected shape. It may reappear when the shape is imported to a sheet layer in a different scale.

4. As mentioned previously, it's sometimes necessary to reverse a hatch's direction. Clicking on the wrench-and-pencil icon opens the Hatch Settings window. Checkmark the Use Local Mapping box, so the selections below it become available. Try them out on some of your squares.

5. The Fill drop-down menu also contains options for Patterns, Tiles, Gradients, as well as importing images. If time allows, lay out a fresh array of squares and try out some of these options.

Exercise: Drawing Auxiliary Views

Instructor

Having modeled the tent-like object in a previous exercise, the students will now draft the object's orthographic and auxiliary views. Using the measurements from the model, it should be drafted in $\frac{1}{2}'' = 1'-0''$ on appropriately sized paper. Spacing of the standard views should accommodate both auxiliary views. This exercise can be done either by hand or via Vectorworks.

Students

Draft the orthographic and auxiliary views of the tent-like object depicted in Figures 3-23, 3-24 and 3-26. Use the measurements from measurements from Figure 3-23. Draft in $\frac{1}{2}'' = 1'-0''$ on an appropriate size sheet of paper (larger than $8\frac{1}{2}'' \times 11''$ is suggested), landscape orientation. Sections and auxiliary views will describe the left/right and front/rear inclines.

1. Lay out the array of standard views. Allow enough space between them to place the two auxiliary views.

2. The first auxiliary view will be that of the right/left incline. While the front view shows that incline's true angle, neither it nor the side view provides the true lengths of any of its lines except for the bottom edge. However, since the height and depth of the object are known, the auxiliary view can be mapped out in a dot-to-dot fashion. Lightly label the points of the right-hand incline to keep track of them (Figure 3-37).

3. This auxiliary view will be extended from the sectioned front view, using the angle of the right-hand incline for reference. Since the angle is true, perpendicular lines can be extended from its endpoints. Draw a guideline through the side view to indicate placement of the cutting plane. Lightly label it A-A.

4. The incline's base is seen in true length in both the top and side views. From the overall dimensions of the object, it's known that the triangle is symmetrical along a center axis. In Figure 3-37, points B and C are equidistant from the base's midpoint.

5. The base of the auxiliary view will be placed along the line extended from the front view's points C and B. Measure along the line to mark the center of the incline's base; be sure the center's mark is far enough along the line so that the auxiliary view does not overlap or crowd the front view. Draw a line through the center mark perpendicular to the base guideline. It should intersect the line that was extended for the top of the auxiliary view (Figure 3-38).

6. In the side view, measure the base's length in either direction from point D; the distance to points C and B. Transfer that measurement to the auxiliary view. Draw lines connecting the auxiliary view's points C and B to the apex of the triangle (point A). The auxiliary view now shows the true shape of the right/left incline (Figure 3-38).

7. Using this same process, construct the auxiliary view for the front/rear incline. Draw a guideline in the front view to indicate the cutting plane line. The true angle of the incline is seen in the sectioned side view, so the side view will be the reference for the auxiliary view. Figures 3-37 and 3-38 show the exercise drawn on $8\frac{1}{2}'' \times 11''$ paper; this is unlikely to be big enough to accommodate the second auxiliary view. In Vectorworks, you can draw beyond the edge of the reference grid or simply change the size of the reference grid by selecting a different paper size.

8. Darken the drawing. Crosshatch the sections. Add borders. Use the Text tool to add name, class, etc.

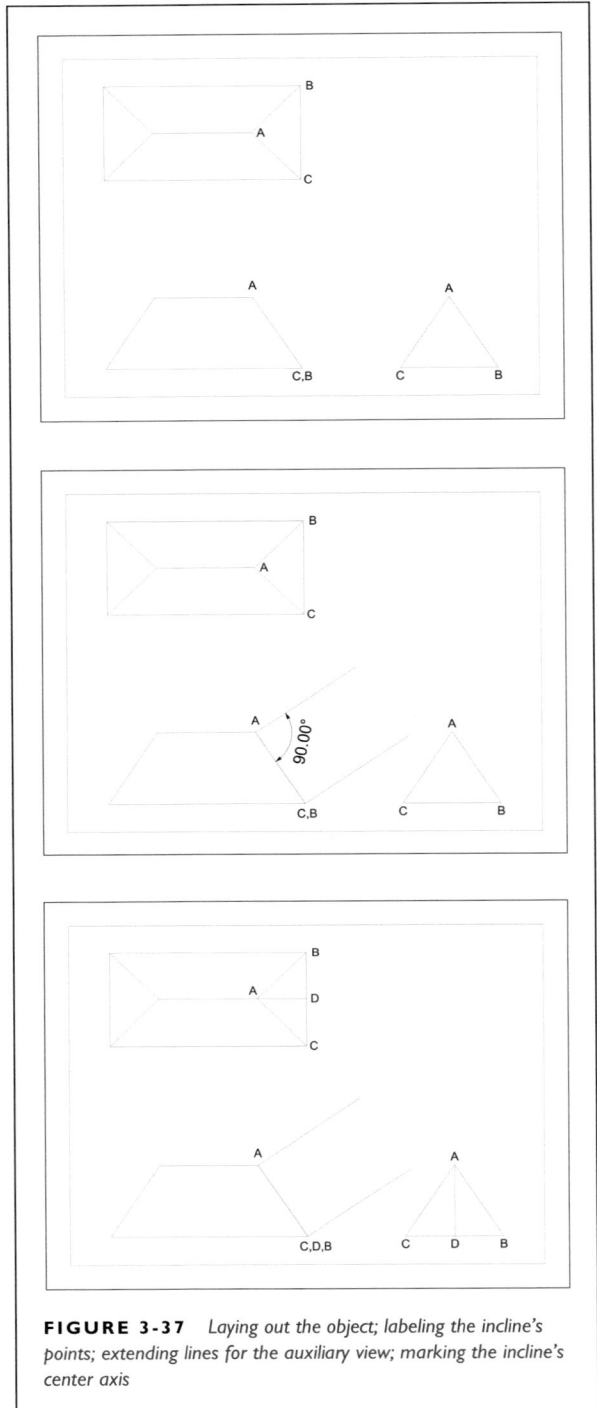

FIGURE 3-37   *Laying out the object; labeling the incline's points; extending lines for the auxiliary view; marking the incline's center axis*

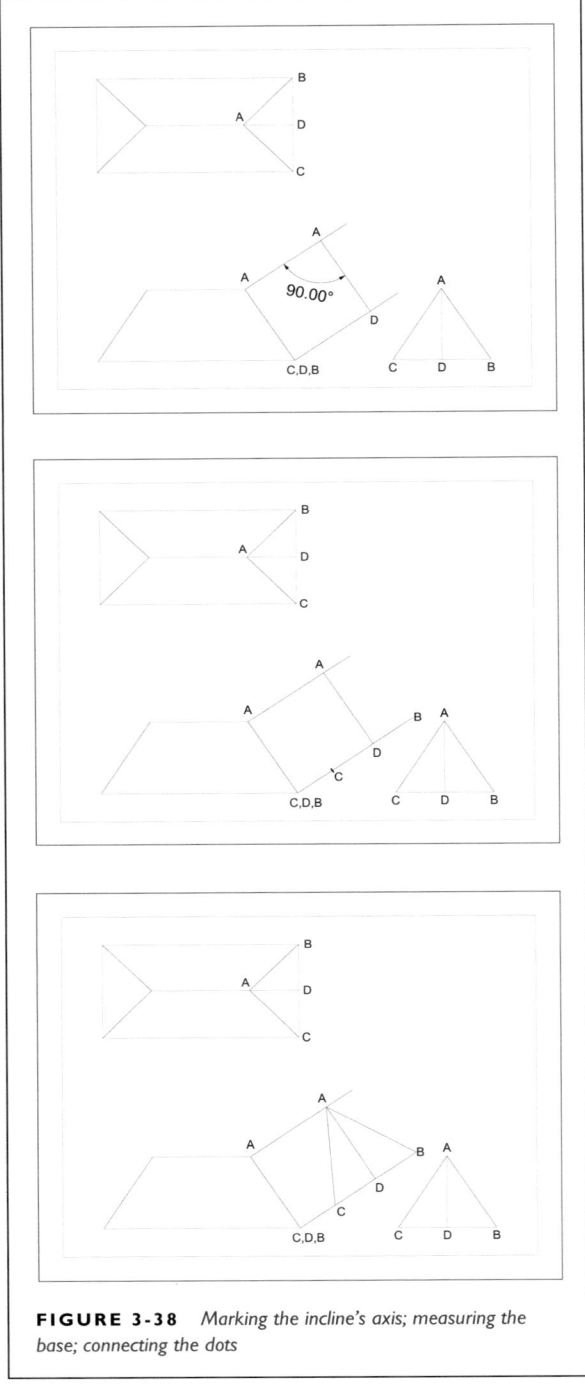

FIGURE 3-38   *Marking the incline's axis; measuring the base; connecting the dots*

Homework: Constructing More Auxiliary Views

Instructor

Distribute a handout or file with a number of objects drawn in orthographic projection. Each object should have at least one inclined plane. Using the views provided, the students should construct the appropriate auxiliary view from the correct reference view. Figure 3-39 provides an example of handout/file. Include an isometric or perspective view of the object (see Chapter 2, Module 2 for information on isometric views). If modeling is to be included as a follow-up, provide the students with measurements for each object.

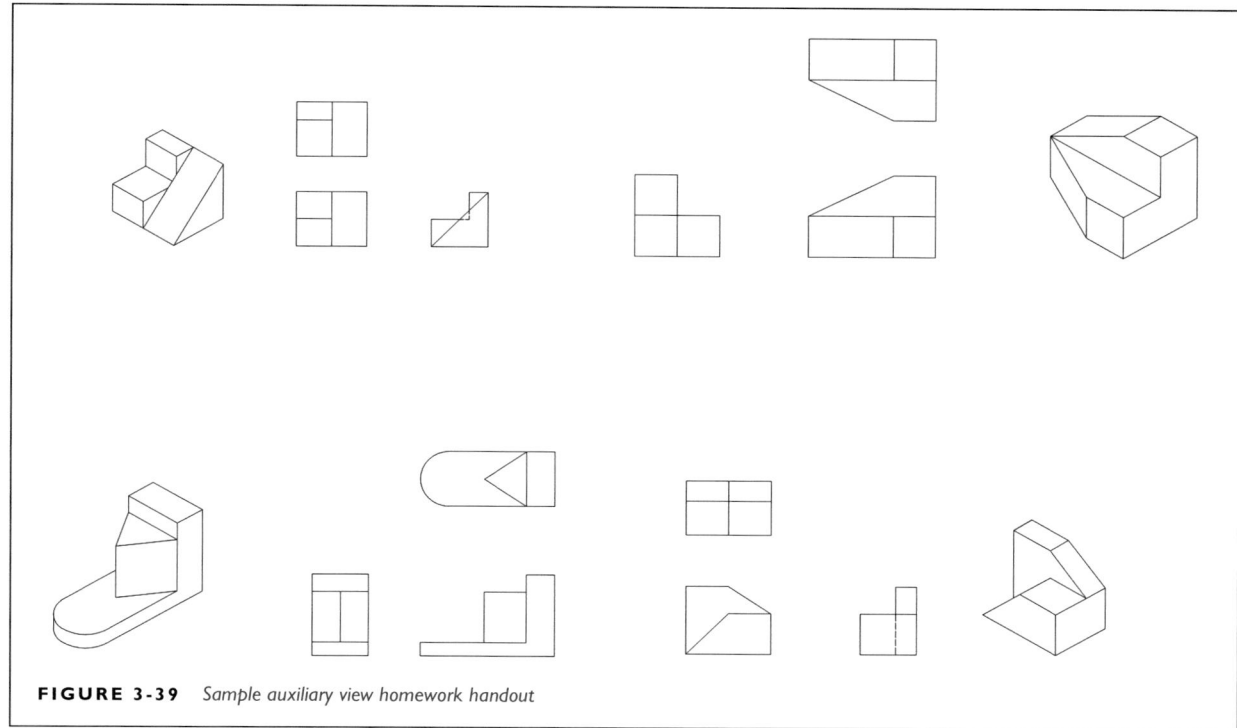

FIGURE 3-39   *Sample auxiliary view homework handout*

Students

Using the handout/file distributed by your instructor, construct the required auxiliary views for each object (see Figure 3-39 for examples). If hand drafting, tape the worksheet to your work surface and use straightedges.

For each object, determine the number of inclined planes and which view shows their true angle. Use that view as the reference for that plane's auxiliary view. Construct or indicate sections as required.

If modeling is included as a follow-up, use measurements provided by your instructor to construct the objects in SketchUp.

CHAPTERS 1–3 DRAFTING REVIEW CHECKLIST

- All drawings are for the end user. Clarity promotes understanding; understanding facilitates fabrication.

- Measuring tools are for measuring. Drawing tools are for drawing.

- Borders tell the reader that they have received the whole drawing.

- Be familiar with what you are about to draft. Keep sketches and research images near at hand for easy reference.

- Adhere as much as possible to the graphic standards and practices recommended by USITT and other mainstream industry organizations. Standards ensure that the reader can understand the drawing.

- Plan the drawing before drafting. Determine what will go on each sheet. Sketch the layout. Do the layout math.

- Think globally. Construct the drawing as a unified whole. When hand drafting, darken the drawing only after it is fully laid out with guidelines and the information is verified to be correct.

- Select line styles according to the information that line is meant to provide. Line styles require context to express their full meaning. In Vectorworks, select the intended line style from the classes that have been set up.

- Describe objects through the conventions of multi-view orthographic projection, as this projection system permits undistorted scale measurement of drawn objects. Use other projection systems such as isometric or cabinet to augment orthographic description as needed.

- Correct placement of standard front, top, and side orthographic views creates the context by which the meanings of lines can be fully understood.

- Choose the standard orthographic views that provide the most and the clearest information. Front, top, and one side are the most common required views.

- Use auxiliary views to describe angled surfaces. Accommodate their placement in your layout plans.

- Use section views to reveal features that cannot be adequately described via outer surfaces alone. Sections views are also used to construct auxiliary views when an inclined surface does not present a true length edge to any of the standard picture planes.

NOTES

1. Dennis Dorn and Mark Shanda, *Drafting for the Theatre* (Carbondale, IL: Southern Illinois University Press, 1992), p. 60.
2. Frederick E. Giesecke, Alva Mitchell, and Henry Cecil Spencer, *Technical Drawing*, 4th ed. (New York, NY: The MacMillan Company, 1958), p. 165.
3. Patricia Woodbridge and Hal Tine, *Designer Drafting and Visualizing for the Entertainment World*, 2nd ed. (New York, NY and London: Focal Press, 2013), p. 64.

CHAPTER 4

DIMENSIONING AND NOTATION

An object is not fully described until it has also been dimensioned. The first module presents dimension and notation standards, recommendations, and practices, including leader lines and labels. Module 2 revisits the four cubes drawn in earlier exercises, drafting their orthographic views (including sections) and dimensioning them. Title blocks are discussed. For Vectorworks, viewports and sheet layers are introduced.

TOPICS AND GOALS

- Parts of dimensioning

- Dimensioning hierarchy

- Dimensioning systems

- Leader lines

- Dimensioning angles

- Sections

- Title blocks

- Vectorworks sheet layers and viewports

MODULE 1: DIMENSIONING

Dimensioning is the systematic placement of measurements and notation so that the reader does not have to personally measure each line. Historically, dimensioning was also the backstop for accuracy. A line drawn with a pencil and straightedge always has some degree of inaccuracy. For example, if the drawing is used to fabricate machine components with narrow tolerances, it's best to tell the reader that the length of a certain line is 2.357″ as it's impossible to draw a line with that level of accuracy by hand.

Even a computer-aided design (CAD) drawing, when printed, shares these inaccuracies. Lines represent edges and while the edge might not have inherent thickness, a drawn line does. Does the reader measure the outside of an intersection, the inside of the intersection, or the middle of the drawn line? Depending on the scale of the drawing, the thickness of the line itself can inadvertently add inches to a measurement lifted with a scale ruler. To get a true measurement from the drawing, the reader can consult the original CAD file, but the complete file is not usually part of the package delivered to the shop. The shop may not have the same CAD program as the draftsperson, and small companies may not even have a CAD program available on their office systems—if they have an office system at all.

There are three aspects to dimensioning:[1]

1. Choosing the format system

2. Knowing standard layout and arrangement

3. Determining which features will be dimensioned

Dimensioning Format Systems

Dimensions and notes consist of a number of parts. Standard systems of appearance direct the placement of these parts.

A typical dimension has three parts (Figure 4-1):

1. **Extension Lines**: thin straight lines indicating to which point on the object the measurement refers. They are also called **witness** lines. USITT recommends them to be 0.25 mm.

2. **Dimension Lines**: thin lines spanning extension lines with a standard terminator, such as an arrowhead. They may be arcs when dimensioning angles. They indicate

DOI: 10.4324/9781003154921-5

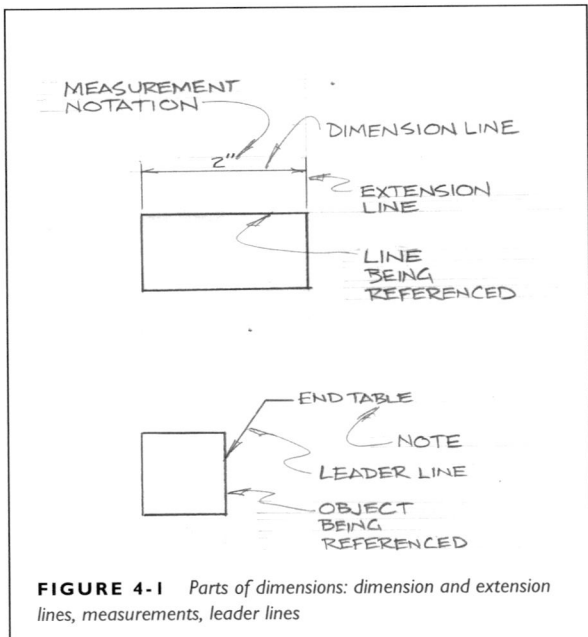

FIGURE 4-1 *Parts of dimensions: dimension and extension lines, measurements, leader lines*

the direction and extent of the measurement. USITT recommends them to be 0.25 mm.

3. **Measurement Notation**: the numeric expression of the measurement. Measurement notation is typically located along and aligned with the dimension line.

When it is necessary to point at a feature and place information a short distance away, a **leader line** is used to indicate that the note is connected to that feature of the object. Leader lines may also be used to associate measurement notation with a space between extension lines that is too narrow to permit legible lettering. In Vectorworks, Callout Boxes are a leader line variant.

Regardless of the scale in which a drawing is drafted or printed, the size of lettering remains constant to ensure readability. Lettering for dimensions and notes is typically 1/8″ (about 12pt font size) in height. While this is relatively easy to accomplish when hand drafting, in Vectorworks, adding dimensions to very large objects drawn in 1:1 can result in what appears to be very, very tiny measurement notation; the font size must then be adjusted accordingly.

When a layer scale is chosen via the Layer Organization Window, the scale selection window includes a Scale Text check box. This selection keeps text proportional to the drawing when changing that design layer's scale. If you dimension an object in ½″ = 1′–0″ and then change the drawing's scale to 1:1, the text enlarges along with the drawing.

If you wait to dimension the drawing after it is imported onto a sheet layer, the dimensioning reflects the measurements of the lines on the sheet, not the lines of the

object in scale; e.g., a line 10′–0″ long in ½″ = 1′–0″ will be dimensioned on the sheet layer as being 5″ long.

In Vectorworks, dimensions may be added via viewport annotation[2]—a halfway space between the drawing layer and the sheet layer (sheet layers and viewports will be discussed later in this chapter). In upcoming Vectorworks exercises, dimensions will be added in their own design layer before viewports are created. This is to reinforce thinking about dimensioning not only as it applies across the array of orthographic views but also as an essential element of the drawing as a whole, rather than as a last step.

Since Vectorworks' drawing space is infinite, it may seem like a good idea to draft in 1:1 all the time. On the other hand, since you can also zoom in exceptionally close, there's not a tremendous difference between drafting in 1:1 and ½″ = 1′–0″ except for the drawing's relationship to the reference grid. As your Vectorworks knowledge increases, you'll begin to decide for yourself what scale best suits drafting a given project and in which layer or viewport dimensioning is best placed.

Extension Lines

Extension lines have two primary functions:

1. They indicate what line/edge is referred to by the measurement

2. They move notation away from the object to prevent visual clutter that may obscure portions of the object.

Extension lines are drawn perpendicular to the line being measured, with the dimension line parallel to the line being measured. Each line and distance requires extension lines projecting from its endpoints. A small gap (1/16″) is left between the endpoint of the extension line and the object; otherwise, the extension line might appear to be part of the object itself (Figure 4-2).

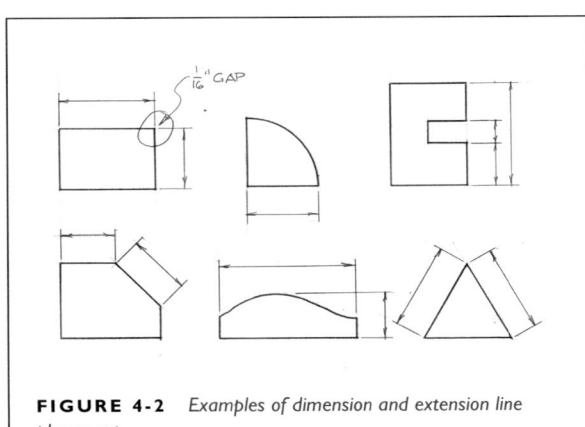

FIGURE 4-2 *Examples of dimension and extension line placement*

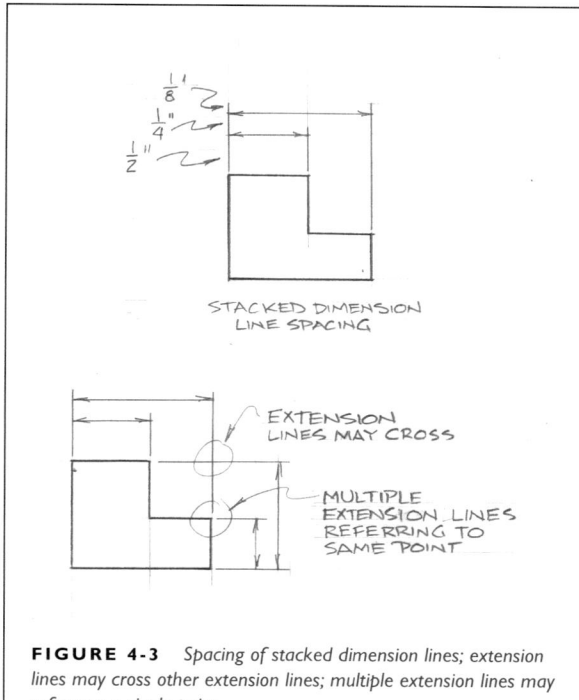

FIGURE 4-3   *Spacing of stacked dimension lines; extension lines may cross other extension lines; multiple extension lines may reference a single point*

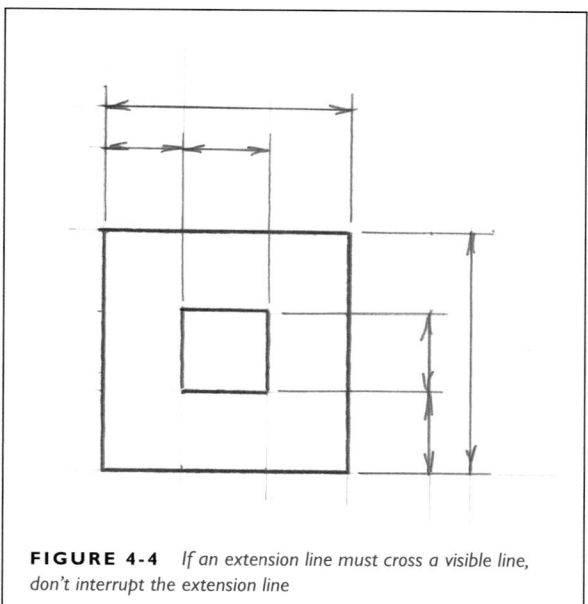

FIGURE 4-4   *If an extension line must cross a visible line, don't interrupt the extension line*

Extension lines extend slightly past their intersection with the dimension line (about 1/8″) so that the dimension line is visibly bracketed within its pair of extension lines. A clean corner at the intersection of an extension and dimension line looks too much like a corner of an object.

Depending on the complexity of the object, dimensions may be "stacked" (Figure 4-3). A single extension line can serve multiple dimension lines, along its length and to either side of it. The stacked dimensions lines should be between ¼″ to 3/8″ apart; enough space to allow un-cramped lettering (wider for fractions). The extension line extends about 1/8″ beyond its intersection with outermost dimension line in the stack. Extension lines may cross other extension lines when dimensioning a corner feature. Multiple extension lines may reference a single point.

Generally, extension lines should not cross an object's visible lines for the same reason a gap is left between them and the point being referenced. However, there are plenty of times a detail or feature within the object requires breaking this rule. Line weight shows its importance here, as the thin black extension line should not be mistaken for a thick or medium visible line (Figure 4-4).

Center Lines

Center lines are part of the dimensioning family. The center line also has a specific line style: thin (0.25 mm) with alternate long and short dashes (in Vectorworks: ISO-08 Long Dashed

Short Dashed) (Figure 4-5). Center lines denote axes of objects and locate the centers of circles and arcs. The next time you open your base file, add a center line class.

When used to locate the center of a circle or arc, the intersection of the horizontal and vertical axes is marked with the crossing of two short dashes. In smaller drawings or objects, the center line typically continues outward to meet dimension lines without additional dashes or gaps. When the center line travels farther, the long/short dash style is continuous. While extension lines avoid intersecting the lines of the object, centerlines often must and do cross them without interruption (Figure 4-6).

Vectorworks does not automatically create crosshairs denoting a circle's center. Whether using the center line type found in the library or using the Line tool while in the Dimension line class, it may be necessary to zoom in and draw the crosshairs as separate line segments.

Center lines are the primary reference lines used to locate items in a venue's space or scenic environment. Scenic groundplans feature a center line running along the depth axis. Light plots also include a center line; working from

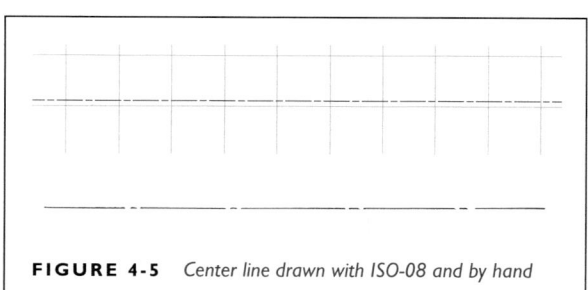

FIGURE 4-5   *Center line drawn with ISO-08 and by hand*

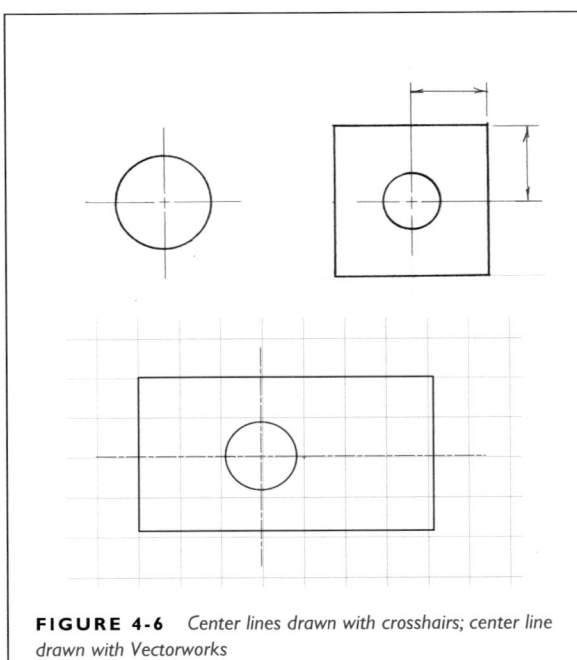

FIGURE 4-6   *Center lines drawn with crosshairs; center line drawn with Vectorworks*

center out is the typical way to place fixture symbols when drafting as well as when hanging the actual fixtures.

On large drawings, a label (**CL**) is added at either end of the center line to ensure that this important reference line is not missed or misinterpreted. On groundplans and light plots, the center line is drawn through/over everything it crosses. To avoid visual confusion, in CAD, the center line may be drawn in color.

There are a few basic looks to the CL label; most of them involve crossing the upward line of the L with the bottom arc of the C (Figure 4-7). CAD programs push you toward discreet letters, though you can design something to import and paste. Scenic designers of a more architectural bent sometimes use the CL as an added moment of artistic

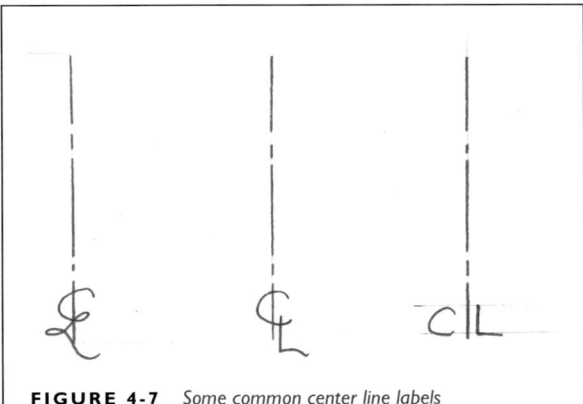

FIGURE 4-7   *Some common center line labels*

expression; if the CL becomes too fancy, though, it risks being misinterpreted as a scenic element.

Dimension Lines and Terminators

Dimension lines carry measurement notation. They run parallel to the line being referenced and never cross an object's lines. Dimension lines should be no closer than ½″ to the feature being dimensioned or, if possible, the overall perimeter of the object.

There are multiple style systems for terminating dimension lines where they meet extension lines, as well as for placing measurement notation along them. Terminators visually clarify the distance referred to by the measurement. They indicate that *this* set of lines is indeed notation and not part of the object. Whichever style of terminator is chosen, use them consistently across the drawing. There are three basic choices (Figure 4-8):

1. Arrowhead

2. Slash

3. Dot

Arrowheads are the older, more traditional terminator. When hand drafting, they should be slender, about 1/8″ long from tip to flare, and may be filled in. They are instantly identifiable as to function, as in most cases, no other drawing feature looks like a dimensioning arrowhead.

Slashes are about 1/16″ long, just enough to emphasize the intersection of extension and dimension lines. They should all slant (45°) in the same direction, across the entirety of the drawing. Dots should be just large enough to emphasis the intersection. An array of large dots (or fat arrowheads) scattered across the drawing diminishes readability by pulling focus.

A general rule is that dimension lines should not terminate at any of an object's lines. Placing the dimension beyond the perimeter of the object is always the first choice. However, for objects containing a fair amount of interior detail or a number of interior features, terminating the dimension line at a visible line may be the only option. Visually, when a dimension line must meet a visible line, slashes and dots physically join the dimension line to the object's line; arrowheads indicate the object's line (Figure 4-9).

Measurement Notation Placement

Measurement notation uses 1/8″ (12pt) lettering, centered along the dimension line. If the space between extension lines is too narrow to fit the numerals, the notation is placed a short distance away, connected to its dimension line or gap between extension lines with a leader line. In very

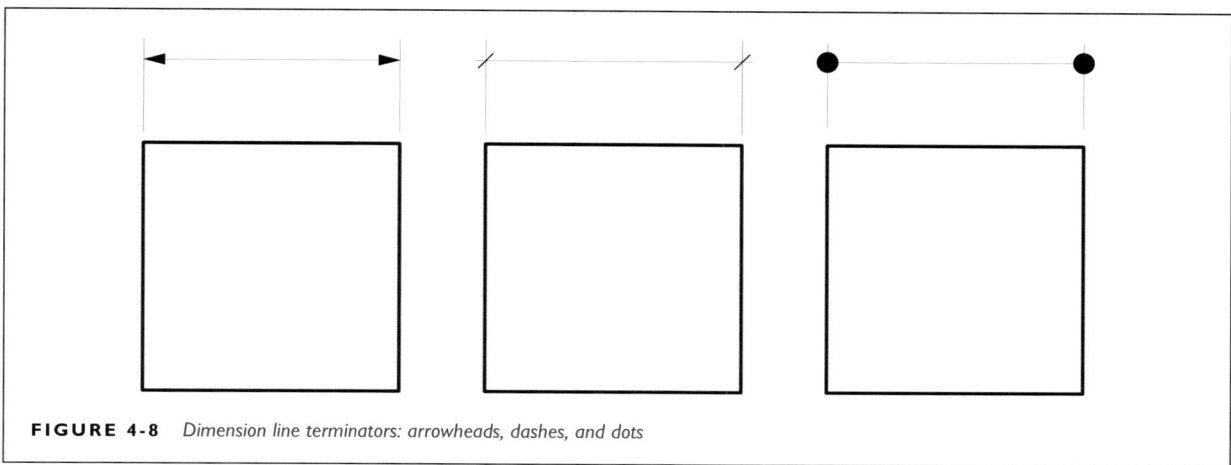

FIGURE 4-8   *Dimension line terminators: arrowheads, dashes, and dots*

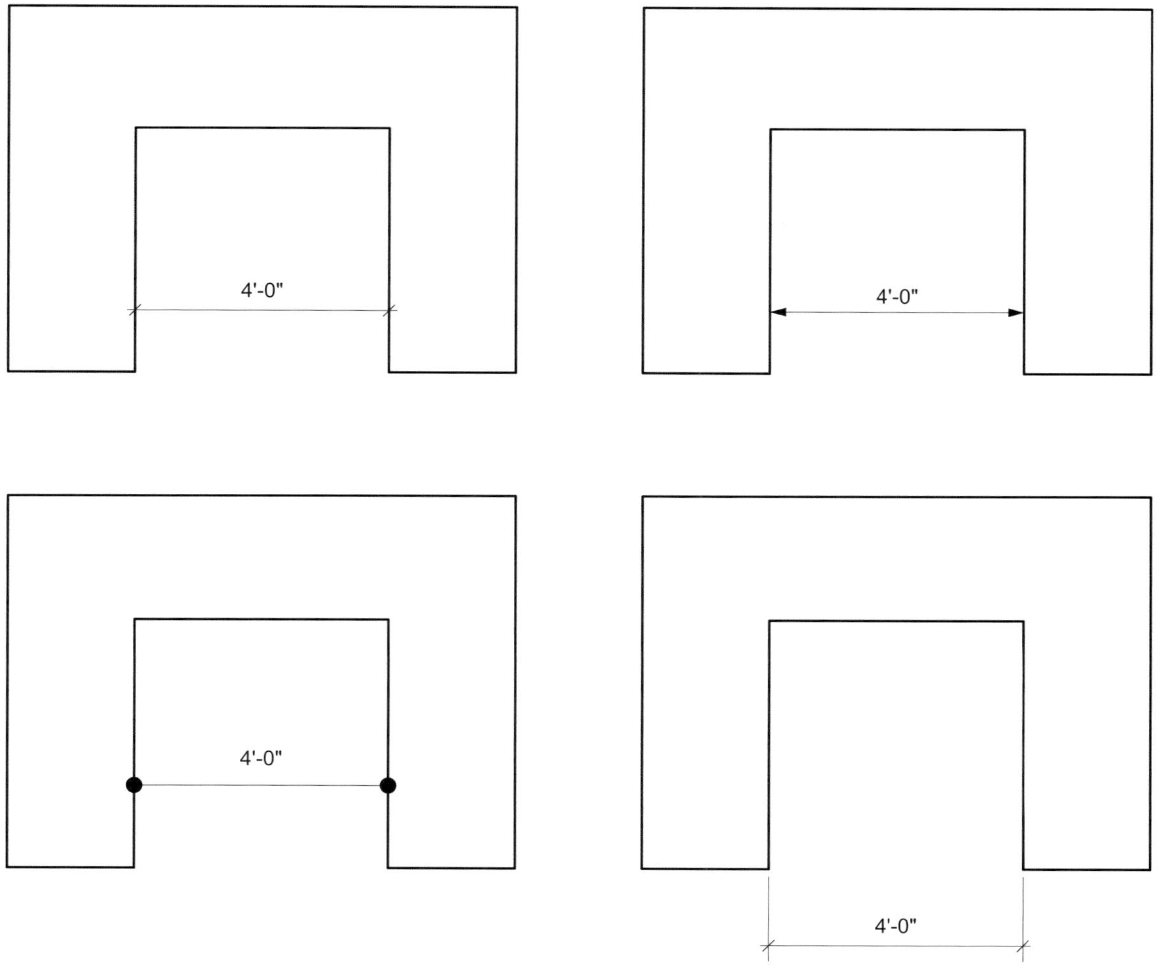

4'-0"

4'-0"

4'-0"

4'-0"

FIGURE 4-9   *Arrowheads versus slashes and dots when dimensioning within the feature cannot be avoided; preferable to locate dimension outside of object perimeter if possible*

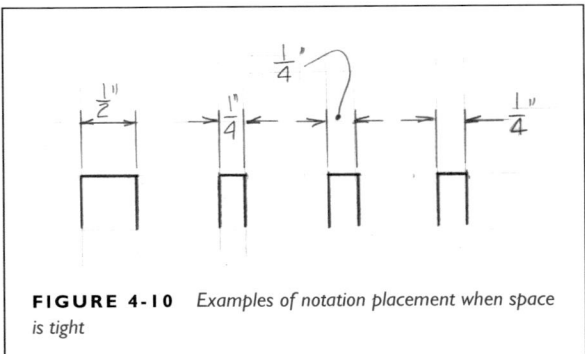

FIGURE 4-10   *Examples of notation placement when space is tight*

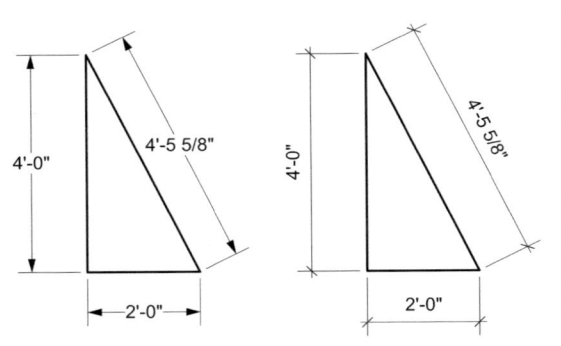

FIGURE 4-12   *Left: the unidirectional system with interrupted dimension lines and arrowhead terminators; in Vectorworks, this is the "ASME" option; right: the aligned system with slash terminators; in Vectorworks, this is the "Arch" option*

narrow spaces, the dimension terminators may be placed outside of the extension lines (Figure 4-10). Don't shrink or otherwise crowd notation.

There are overall two style systems approved by ANSI: unidirectional and aligned.[3] Both systems minimize the need to turn the page to read notation. The selected system must be used consistently across the drawing and series of drawings (Figure 4-11).

In the **unidirectional** system, the notation is aligned horizontally so it can be read from the same direction: the reader located at the bottom of the sheet.

In the **aligned** system, the measurement notation is oriented along the length of the dimension line. When the dimension line is horizontal, the measurement is written horizontally; when the dimension line is angled, the measurement is angled; when vertical, vertical. Placement must be consistent and readable either from the bottom or the right-hand side of the page.

Within the systems, there are two further choices: to interrupt the dimension line or to not interrupt the dimension line. Figure 4-12 presents the unidirectional system with interrupted dimension lines, and the aligned without interruption.

In the uninterrupted system, the notation floats just above the dimension line. When hand drafting, don't use the dimension line as a bottom guideline for lettering. If the measurement looks cramped between the dimension line and the object (as can happen for fractions), reposition the dimension line a little farther away from the object.

The author prefers to use arrowheads, the aligned system, and uninterrupted dimension lines, primarily because these choices allow a bit more speed in hand drafting. Subsequent hand drafted illustrations and examples will feature these choices. In Vectorworks examples, the Arch dimensioning selection will be used, which features slashes rather than arrowheads.

Applying Dimensions in Vectorworks

In Vectorworks, the dimensioning tools can be found in the Dims/Notes Tool Set (Figure 4-13). The first icon is the Constrained Linear Dimension Tool. It places dimensions horizontally and vertically. The icon next to it is the Unconstrained Linear Dimension tool and places dimensions at angles (parallel to the line being referenced). The third icon is the Angular Dimension tool, for dimensioning angles. Selecting one of these icons and then clicking on the point or intersection to be dimensioned will open the dimension navigation window.

When an icon is chosen, options appear in the upper main toolbar. Dim Std (which also appears in the Navigation window) is where the overall system of dimensioning is chosen. For the following exercises, select Arch. This gives you the aligned system with slashes as terminators, and notation that does not interrupt the dimension lines (Figure 4-14).

To dimension an object, select a tool. Click on one endpoint of the line to be dimensioned. Click the cursor on the other endpoint and drag the cursor away from the object. A blue outlined textbox appears; **L:** gives you the distance

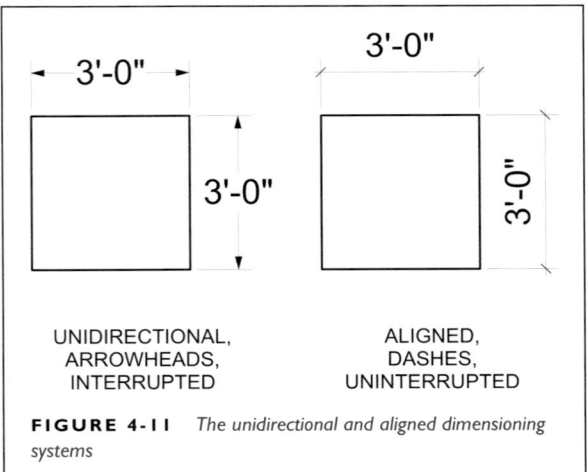

FIGURE 4-11   *The unidirectional and aligned dimensioning systems*

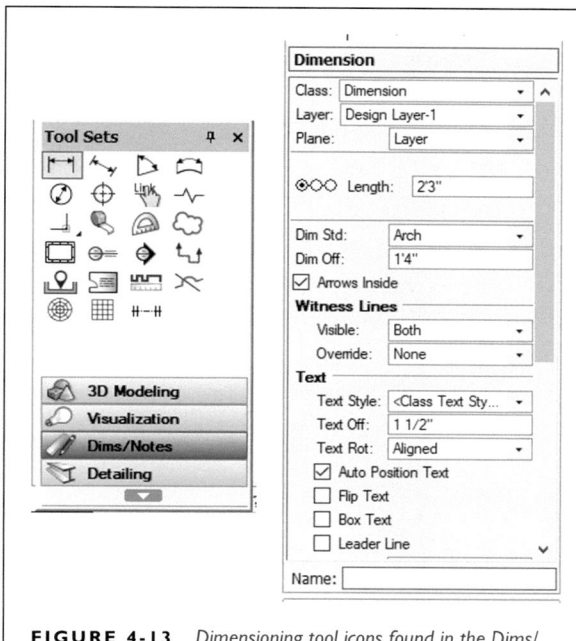

FIGURE 4-13 *Dimensioning tool icons found in the Dims/ Notes tool set; the Dimension Navigation window*

from the measured line to the dimension line. By hitting Tab on the keyboard, this value will be highlighted and you can enter a distance, such as ½″. Click the mouse a third time and the dimension line will snap into place.

Leader Lines

Leader lines are thin black lines (0.25 mm) that connect notes to the feature being referenced. There are two styles of leader lines:

1. Straight angled line with straight shoulder (shoulder somewhat optional)

2. Freehand curve

The reference end of the leader line terminates with an arrowhead or dot, depending on the nature of the item being referenced. Arrowheads point at something; dots indicate spaces or surfaces (Figure 4-15).

The notation end of the leader line meets text midway up the letter at the beginning or at the end of the note, never in the middle of a word or coming up from below or down from above the text. Leader lines may cross visible lines and extension lines but *not* dimension lines. As much as possible, they should not cross other leader lines.

When hand drafting, freehand leader lines are acceptable. They have the advantage of being instantly recognizable as a leader line, though they are more difficult to execute well. The curve should be smooth and elegant. A switchback prevents it from being mistaken for an arc in the drawing. Elegant freehand leader lines are difficult to construct in CAD.

Leader lines are one way to provide dimensioning for circles and arcs. If the circle is large enough, a dimension line can bisect the circle through the center; the notation is placed along this line. Arrowheads are oriented toward the *inner* edge of the circumference (Figure 4-16).

If there is not enough space within a circle, the leader line begins at the center and continues outward past the circumference, angling at the shoulder to lead the eye to the notation. The arrowhead remains within the circle, oriented toward the circumference. This longer leader line is an option even if the circle is large enough to contain notation. For very small circles, a leader line terminating at the outside edge of the circumference is acceptable, since there's no room for an arrowhead within the circle (Figure 4-16).

Circles are typically dimensioned by their diameter (DIA) and arcs by their radius (RAD), though offering the radius of a circle means less math for the reader fabricating the object. Always include the DIA or RAD abbreviation with the measurement, e.g., 3′–0″ DIA.

In Vectorworks, leader lines can be constructed by using the Line tool and manually placing an arrowhead or dot at the termination, or by using either the Callout Box tool or the Leader Line Simple tool (Figure 4-17). A Callout Box surrounds the note with a box; the Leader Line Simple tool provides only the leader line. Be sure that you are in the dimension line class when using these tools so their line weights are correct.

The Leader Line Simple tool is a multi-click tool. Place the text before adding the leader line. Click at the first word of text to start the shoulder of the leader. Drag the cursor horizontally to affix the other end of the shoulder, click again, and then drag to the feature being referenced. Click again, and the arrowhead will appear.

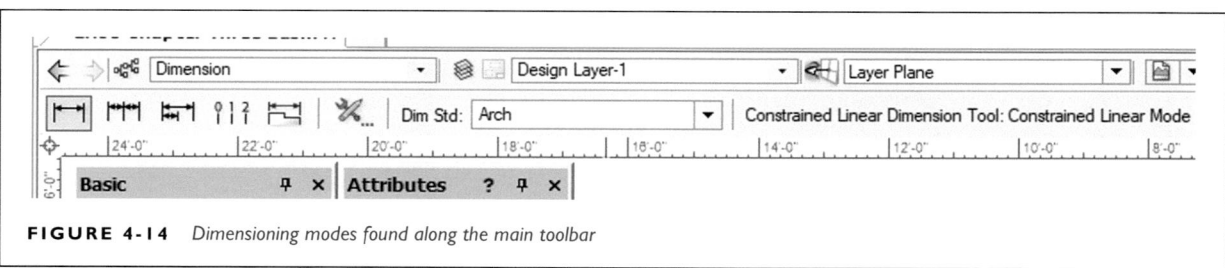

FIGURE 4-14 *Dimensioning modes found along the main toolbar*

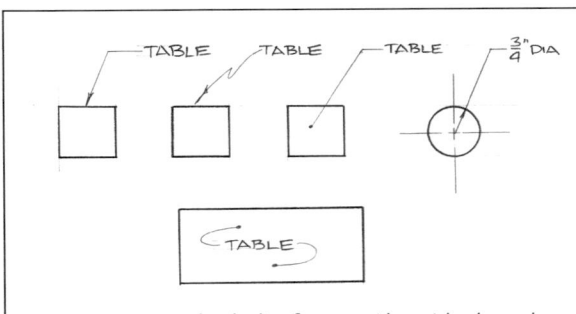

FIGURE 4-15 *Leader line formats: with straightedge and freehand; indicating via outline or via interior; use in dimensioning circles*

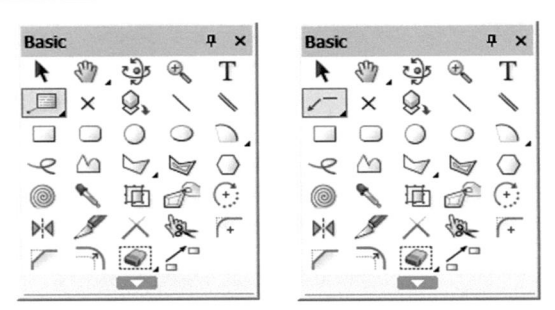

FIGURE 4-17 *The Callout tool; left-clicking on the icon and holding the button opens a menu to find the Leader Line Simple tool*

The Callout tool also starts at the text placement, but the text is added after the leader line is situated. Click to affix the start point and then drag the cursor to the object. When you click to affix the arrowhead, a window will appear asking you to enter the information you wish to place in the callout box. A callout navigation window also opens on the right side of the screen. This window allows you select the style of container that surrounds the box (the bubble style), text style, and other features (Figure 4-18).

When drawing leader lines manually, line termination selections are found at the bottom of the Attributes window (Figure 4-19). The downward pointing arrow on the right, Line Endpoint Style, opens a menu. The downward pointing arrow on the left opens the Line Startpoint Style. Choose one of the menu selections and it will appear at the line's endpoint when you click on the Line End Toggle Marker (the arrow pointing to the right in the Attributes window).

To select an arrowhead from the Line Endpoint Style menu, click on the selection featuring a small black arrowhead pointing to the left with the note (0.12 ″ × 0.000″ 15° 0.05 mm). When the Line End Marker Toggle is highlighted, an arrowhead

will be placed at the endpoint of the line being drawn. Be sure to de-select it by clicking it a second time, or every subsequent line drawn will have an arrowhead.

Dimensioning Angles

If the opening of the angle is wide enough, notation is placed within it along an arced dimension line, with arrowhead terminations. The center of the arc is the hinge point of the angle. This is an exception to the rules, as the dimension line may contact the object being dimensioned. If the angle is too narrow to place the degree note within it, a leader line can be used. If necessary, use extension lines to extend the angle beyond the perimeter of the object. (Figures 4-20 and 4-21).

In Vectorworks, select the Angular Dimension tool. Click the cursor on one line of the angle, then click again on the other line of the angle. Drag the cursor out toward the mouth of the angle, as far as you wish to set the dimension. Extension lines will appear if you go past the angle's endpoints. Click to affix the dimension.

Exercise: Hand Drafting Dimensioning Practice

Instructor

Prepare a handout or file featuring a number of simple, regular polygons. Some shapes should include angles and circles. Examples can be seen in Figure 4-22. Announce the scale and the preferred alignment/style system. The students should measure and dimension the objects as fully as possible. Stacking and creating chains of dimensions will be discussed further; for now, see what solutions students come up with on their own.

Students

On the handout or file provided by your instructor, dimension the objects as fully as possible. Use the alignment system and scale announced by your instructor.

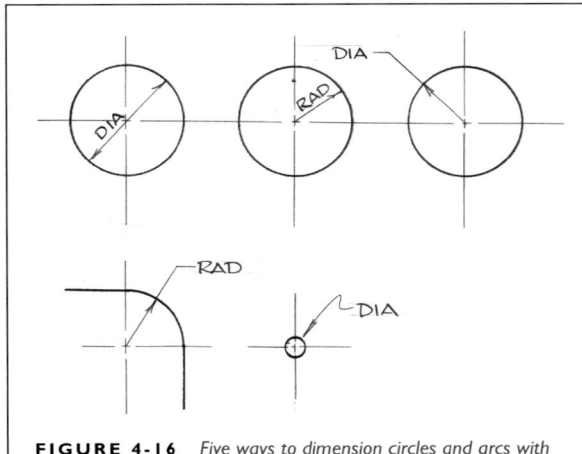

FIGURE 4-16 *Five ways to dimension circles and arcs with leader lines*

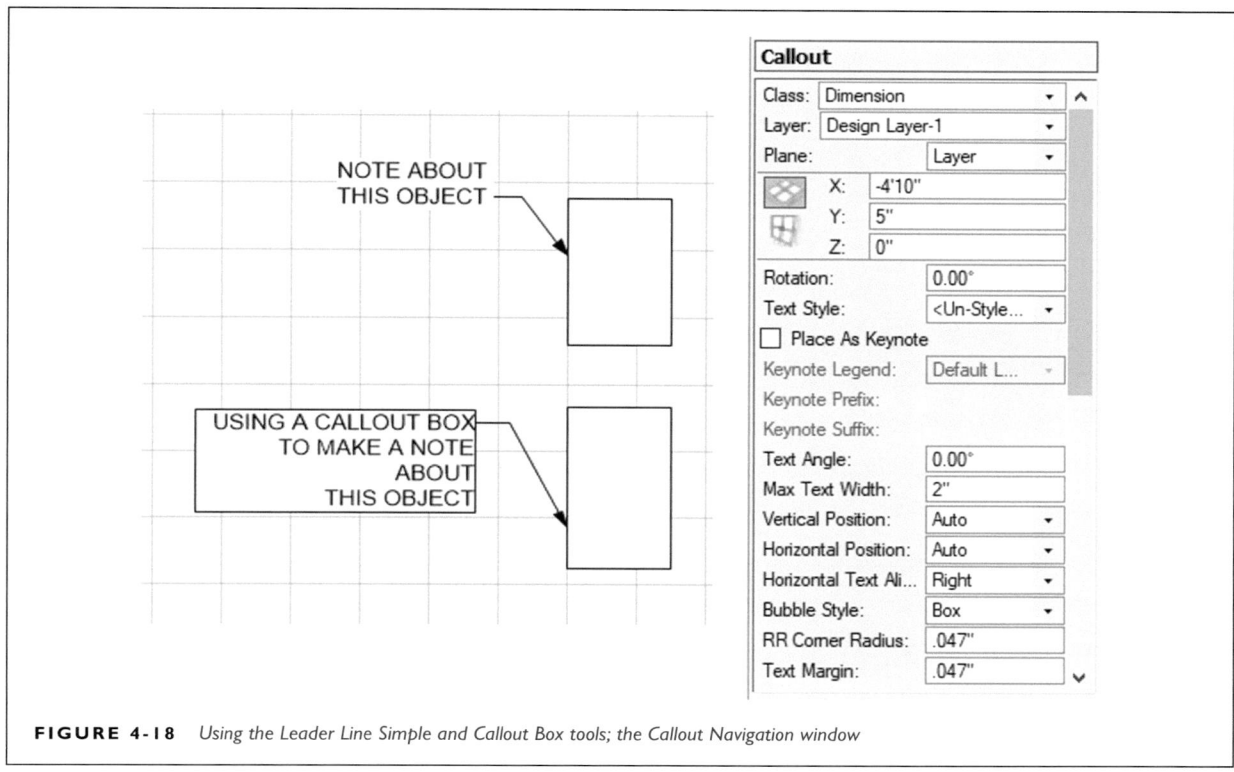

FIGURE 4-18 *Using the Leader Line Simple and Callout Box tools; the Callout Navigation window*

Select a feature of the object, measure it, draw extension lines, dimension lines, select terminators and place notation. If drafting by hand, use the two-triangle technique to draw extension and dimension lines that reference angled lines. Circles and arcs include dimensions that place

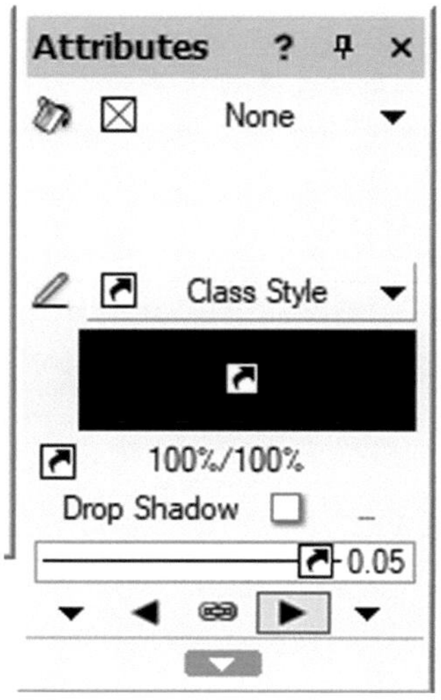

FIGURE 4-19 *Locating the Line End Toggle Marker and Line Endpoint Style buttons in the Attributes window*

their centers as well as note their diameter or radius. If this exercise is done in Vectorworks, follow these setup steps before you begin dimensioning:

1. Open your base file. If you haven't already created a Center Line class, do so.

2. A dimension class of lines is already set up, but the text style is unstyled because a text style has not yet been chosen. In the organization window, double-click on the Dimension class to open the editing window (or click on Edit). Open the Text Style drop-down menu and choose Dimensions 12pt. Now, whenever you use a Dimensioning tool while in the dimension line class, the measurement will automatically appear in 12pt font. OK back to the drawing.

   If using the interrupted style of dimensioning, select a white Fill for this class. When notation appears, it is placed within a borderless white box that obscures lines beneath it. If a Fill is not selected, the dimension line will run through the measurement text.

3. In the Dims/Notes tool set, choose the Constrained Linear Dimension tool. In the upper main toolbar, select Arch as the Dim Std (unless otherwise instructed).

4. OK back to the drawing. Re-save your base file. Dimension the shapes distributed by your instructor.

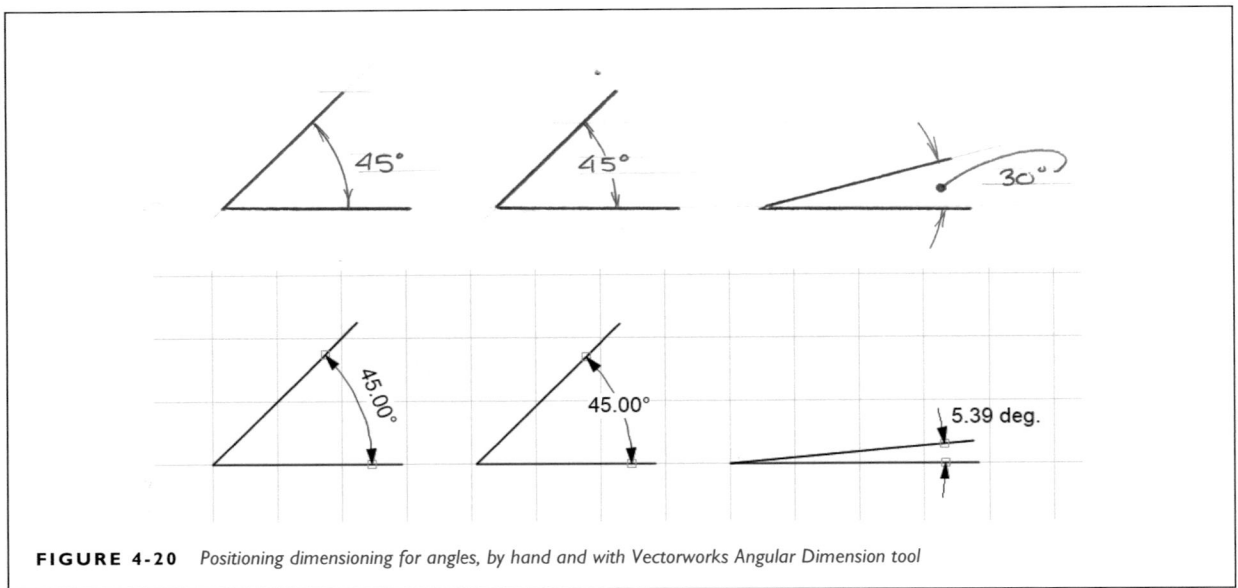

FIGURE 4-20   *Positioning dimensioning for angles, by hand and with Vectorworks Angular Dimension tool*

Determining the Choice of Dimensions

Dimensioning requires planning. It is a holistic process that goes hand-in-hand with how you think about the object and what information you wish to communicate about the object.

Deciding what features to dimension determines the placement of extension and dimension lines. If possible, plan dimensions as you plan the drawing's layout. Since the entertainment industry crafts many unique and irregular objects, the draftsperson must exercise considerable discretion in determining what features to dimension and where to best place the notation. In all cases, however, the draftsperson must remember that dimensioning must serve the reader.

When stacked, dimensioning follows a hierarchy. The outermost tier references the overall width, height, or depth of the object, depending on which orthographic view is being dimensioned. The middle tier references larger features and the placement of smaller details. The innermost tier references smaller details (Figure 4-23). Two tiers are usually sufficient.

The middle tier should always include a measurement to the object's perimeter so that an interior feature's distance from the edge of the object is automatically known. Dimensions sharing a tier should be grouped together as much as possible so that they can be read as a continuous string (Figure 4-23).

Discussion and Exercise: Determining Placement of Dimensions

To demonstrate the decision-making process, this discussion walks through the placement of dimensions on orthographic views of a simple object. The instructor should prepare a handout featuring the object pictured in Figure 4-24.

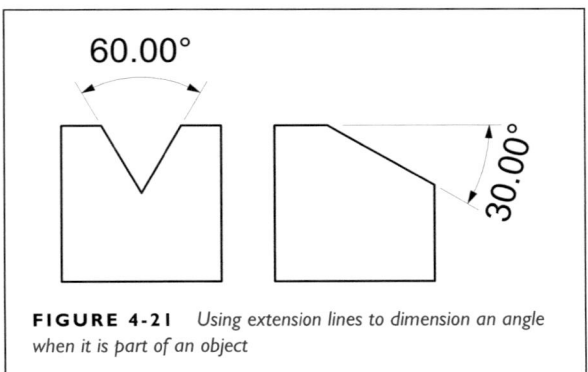

FIGURE 4-21   *Using extension lines to dimension an angle when it is part of an object*

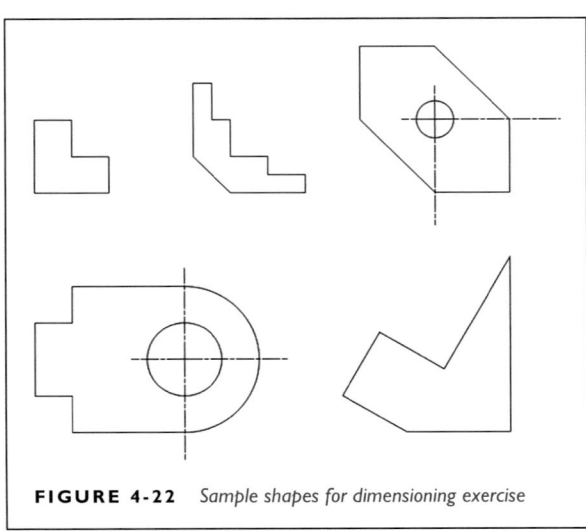

FIGURE 4-22   *Sample shapes for dimensioning exercise*

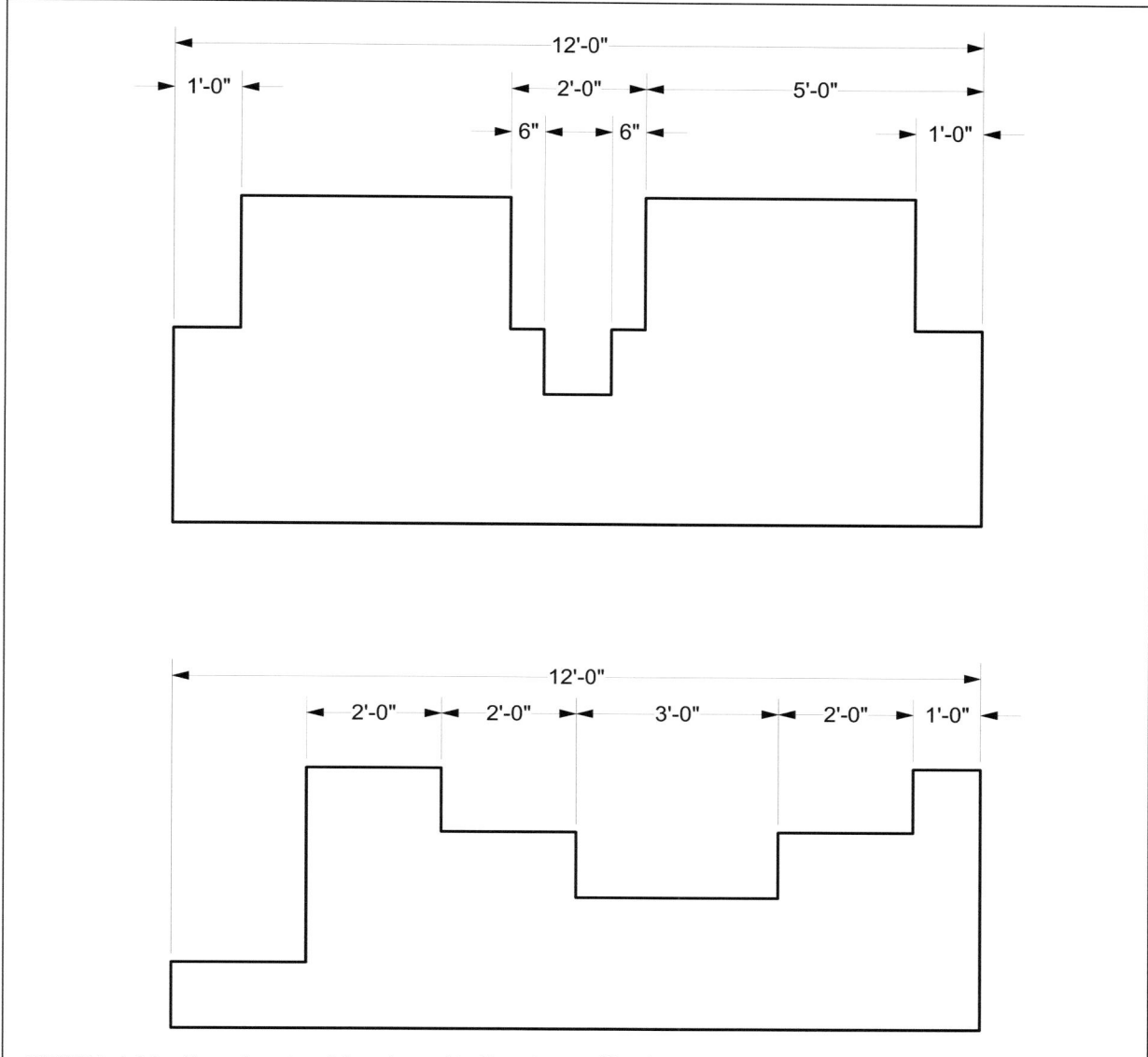

FIGURE 4-23   *Upper: three tiers of dimension stacking hierarchy: overall length in outer tier; larger features and placement of details in middle tier; smaller Details in inner tier; lower: a tier of dimensions grouped as a continuous string*

Distribute two copies: one for planning, one for drafting. It does not need to be in any particular scale, as for this exercise, the placement of dimensions does not require their measurements.

1. The front, top, and right side views are provided. The first consideration is which views provide what information. Height and width are found in the front view. Depth and height are found in the side view. Width and depth are seen in the top view. Are there any features that can only be described in one particular view?

2. It is usually best to begin with the front view. Whatever cannot be dimensioned in the front view is then dimensioned in the top view. Anything remaining is then dimensioned in the side view. It is not uncommon for a side or top view to feature no dimensioning at all—provided, of course, all information is provided via the other views. This does not mean that the view itself is not required; inclusion of the view clarifies the information seen in the other views.

In the front view, the overall height and width of the object comprise the outermost tier. The question then is where the dimensions will be placed—preferably along the side of the view where the feature is most clearly seen (Figure 4-25).

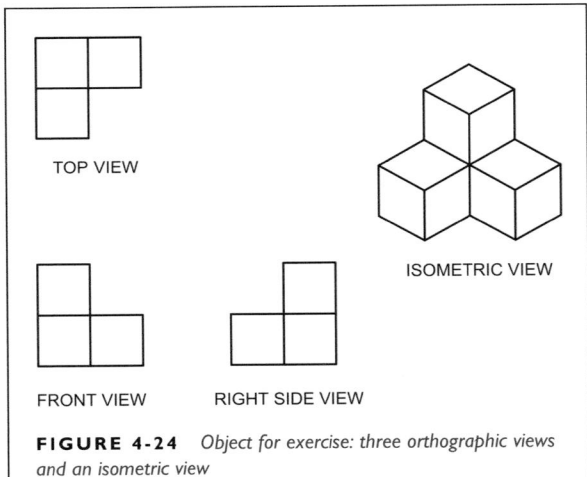

FIGURE 4-24 *Object for exercise: three orthographic views and an isometric view*

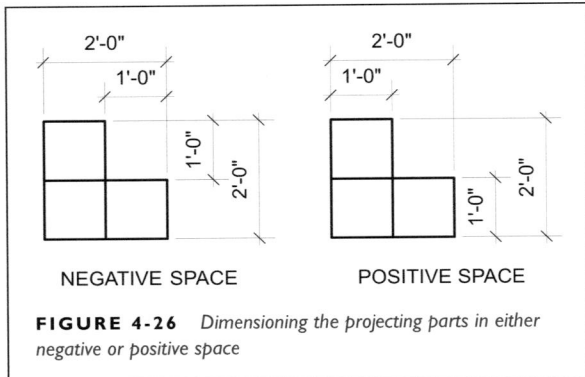

FIGURE 4-26 *Dimensioning the projecting parts in either negative or positive space*

projection itself. The negative space choice dimensions the space the projection creates (Figure 4-26).

For this object, either choice is reasonably clear to the reader. A more complex object might drive the decision toward one or the other. It is not necessary to dimension both positive and negative spaces. When laying out the object, whether on paper or plywood, either measurement is enough to place the projection's inner edge.

4. Consider whether there are any other features best dimensioned in the front view. If not, it's time to turn to the top view. Since the front view features all of the width and height measurements, the overall depth of the object and the depth of the projections can be placed in the top view (Figure 4-27).

In Figure 4-27, depth dimensions for the positive space choice are placed to the right of the view, as this location offers the most distinctive view of the features.

If the dimensions are placed below and to the left of the view, they reference the projecting parts where they are least obviously project. Since the projections themselves require dimensioning, placing the dimensions above and to the right of the view allows them to be grouped together.

Lines running through the whole view should be dimensioned at their most distinctive end. It's not necessary to dimension both ends of a clearly continuous horizontal or vertical line.

3. You now have a choice. The projecting parts can be dimensioned either in positive or negative space. The positive space choice places the dimension over the

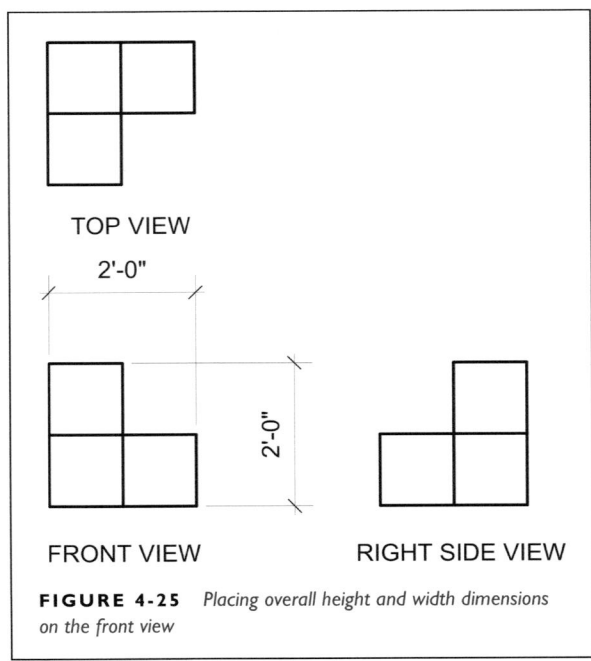

FIGURE 4-25 *Placing overall height and width dimensions on the front view*

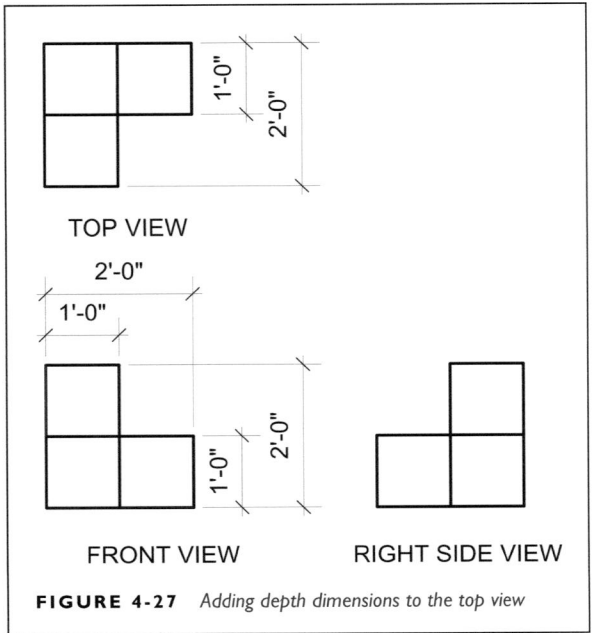

FIGURE 4-27 *Adding depth dimensions to the top view*

If there are no other features to be addressed in this view, it's time to move to the side view.

5. Width and height are addressed in the front view. Depth is addressed in the top view. If there are any features that are more clearly described in the side view, their dimensions can be shifted to that view. If not, dimension placement is completed.

6. Draft the dimensions on the second handout using correct line styles and selected dimensioning system.

If you're unsure whether you've included all the required dimensioning, one way to be sure is to redraft the object using only the dimensions determined during your planning. If you can't draft a feature, you have likely missed a useful dimension.

Exercise: Dimensioning in SketchUp

Adding dimensions in SketchUp can be useful when sharing preliminary information with fellow production team members, giving rise to questions sooner rather than later ("It says here the overall height of the booth signage is twenty-four feet; you do know the convention hall is only twenty-two feet high, right?"). Adding dimensions to a 3D object in a 3D drawing environment can be tricky. It's easy to connect dimensioning to points that *appear* to be aligned in one camera view but are seen to be misaligned when the view is orbited. Furthermore, dimensions themselves are objects within the overall 3D environment, shifting as the viewpoint is changed (Figure 4-28).

1. Open SketchUp. Use the Feet and Inches template. Move the human figure aside.

2. Drag out a horizontal rectangle 8'–0" wide and 4'–0" deep.

3. Push/Pull the rectangle's surface up to a height of 3'–0".

4. On the Large Tool Set, the Dimension tool icon looks like a set of dimension lines with the number 3 above them (Figure 4-29).

5. Dimension the height of the object along the front right-hand edge. Click on the Dimension tool. Click the cursor on the top corner. A measurement will appear as you drag the cursor to the bottom corner. Click the mouse to affix the extension line to the bottom corner. Drag the cursor to the side a short distance to draw out the extension lines. Click the mouse to affix the dimension to an adequate distance from the object's line (Figure 4-30).

6. Add a width dimension above and along the front face of the object.

7. Add a depth dimension along and above the right-hand side of the object.

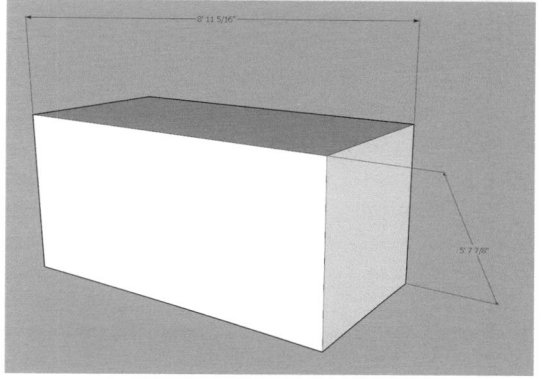

FIGURE 4-28 *Dimensions in front view seen to be misaligned when viewed in perspective*

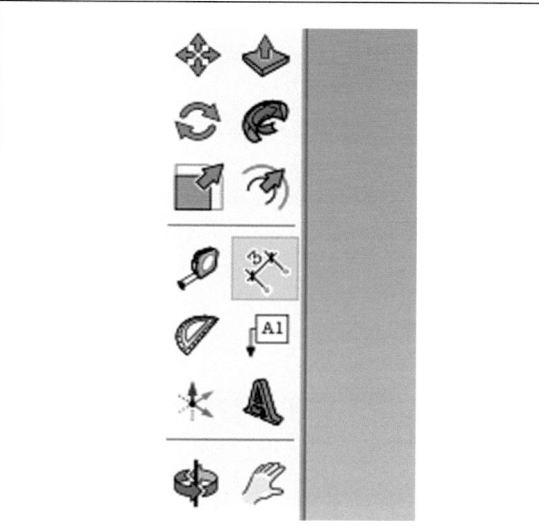

FIGURE 4-29 *Location of the Dimension tool icon in the large tool set*

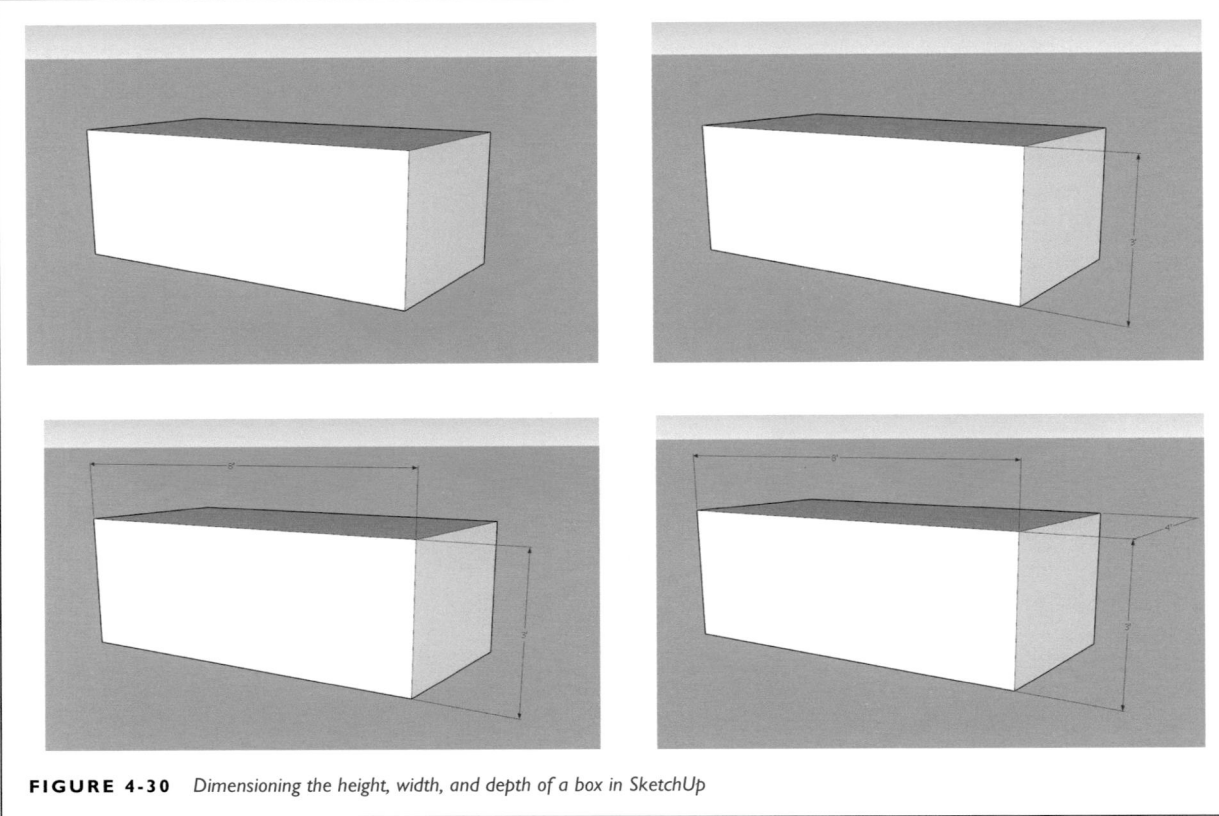

FIGURE 4-30 *Dimensioning the height, width, and depth of a box in SketchUp*

8. Pan and Orbit to see how the dimensions shift with the viewpoint. In Camera, switch to parallel projection and look at the dimensions in the standard views. How easy are they to read? Are there better placements for them? Do they change size as you zoom in and out? If you were to dimension a large, complex object, how readable would you expect the dimensions to be if you printed out a view of the whole thing?

9. If you have saved previous SketchUp exercises, select one, open it, and place dimensions around the object. Look at them in the standard views. Is there any way to dimension these objects so that the dimensions are more readable in the standard views? Is it better to try to dimension while in a standard view?

Discussion and Exercise: Dimensioning with Auxiliary Views and Hidden Lines

Not every shape is rectilinear, with easy-to-see-features. Nor is it always possible to neatly group together the majority of a view's dimensions. The object in Figure 4-31 includes a ramp as well as features that must be described with hidden lines in either side view.

Instructor

Distribute the image in Figure 4-31, using it to walk through the process in class. After dimensioning decisions have been made, the object should be drafted with its dimensions either by hand or via Vectorworks. As the example images were drafted in Vectorworks, measurement notation is added to the figures as the exercise progresses. The final images contain all of the information required for students to draft their own versions.

Students

In Figure 4-31, the front, top, and both right and left views of an object are presented. An isometric drawing provides added description. In the right view, part of the ramp is hidden behind the right-hand platform area and the left-hand platform's step is hidden behind the ramp. In the left view, the base of the ramp is hidden, as is the upstage portion of the right-hand platform. The ramp requires an auxiliary view to describe its true size. You'll need to decide which view, (right or left) provides the most and the clearest information for a section view, or whether multiple sections are necessary.

1. Consider the front view. Which features are most clearly described? The heights of all levels are present, but are they apparent as levels? Overall height and

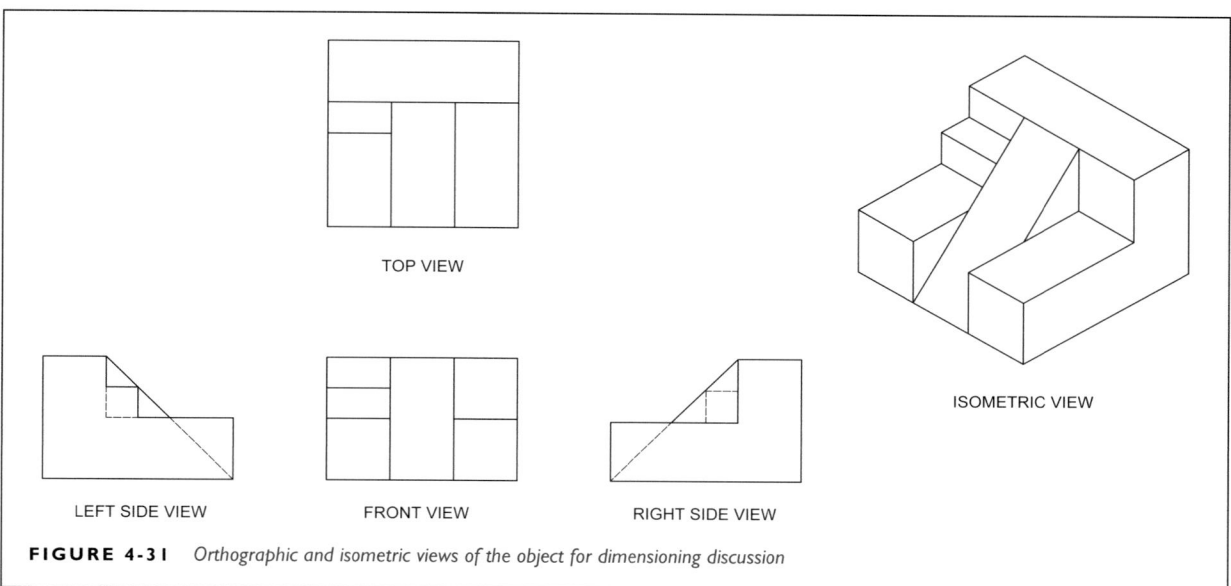

TOP VIEW

ISOMETRIC VIEW

LEFT SIDE VIEW   FRONT VIEW   RIGHT SIDE VIEW

FIGURE 4-31   *Orthographic and isometric views of the object for dimensioning discussion*

width are evident in this view, so perhaps at least the outermost tier of dimensions can be placed. In Figure 4-32's first example, they're placed to the right and below the front view just to get the ball rolling. The vertical details of the front view (the parallel edges of the ramp) are continuous from top to bottom, so there are no measurements at the top of the view that can't also be placed below the view. This is the planning stage; you can change your mind later.

2. The top view presents width and depth. Overall width is described in the front view. The forward edge of the upper platform area runs unbroken from side to side, so it can be dimensioned to the right or the left. The left side of the top view also includes the step, the measurements for which can be added as a middle tier.

    Having the front view's height dimensions on the right, while the grouping in the top view is on the left, makes the drawing look unbalanced. Furthermore, the height of the step is seen on the left side of the front view and has not yet been dimensioned. Grouping the height dimensions on the front view's left side may be preferable.

3. The widths of the ramp and right and left platforms are seen in both the front and top views. If the width measurements are moved from below the front view to above the front view, they gain visual proximity to the top view (Figure 4-32).

4. Reconsider the step on the left platform. It is best seen as a step in the right and left views. In the right view, however, it's described by hidden lines. In the left view, its lip meets the ramp. Dimensioning to this corner is okay, but not preferable. Since dimensioning to the upper or lower edge of the object is necessary to

place the step within the object's overall height, height dimensions can be moved from the front view to the side view to avoid redundancy with the front view dimensioning. Figure 4-33 shows height dimensions moved to the left view. It's not bad, but it does feel a little awkward.

5. The height of the right-hand platform has not yet been addressed. It looks as though the top of the right-hand platform is at the same height as the top of the left-hand platform, but since they are not joined by a continuous horizontal line, that's an assumption rather than a fact. The height of the right-hand platform needs its own dimension.

    If you place the dimension in the view where the platform is most evident, the right view might be the best place for it. However, the ramp still has to be addressed. Since the incline is seen in its true angle in both the right and left view, one of those views will serve as the reference view for the auxiliary view.

    A major question now arises: do you need both right and left views of this object? The step is most evident in the left view, but its height and width can be adequately described in the top and front views. If you include an isometric or perspective sketch of the object on the drawing, there's even less possibility of misinterpretation. Figure 4-34 shows how it looks if the left view is eliminated and height dimensions are transferred back to the front view.

    The height of the right-hand platform has been added to the front view. The left side view can be retained even if the right side view is sectioned and becomes reference for the auxiliary view, but it's not particularly necessary.

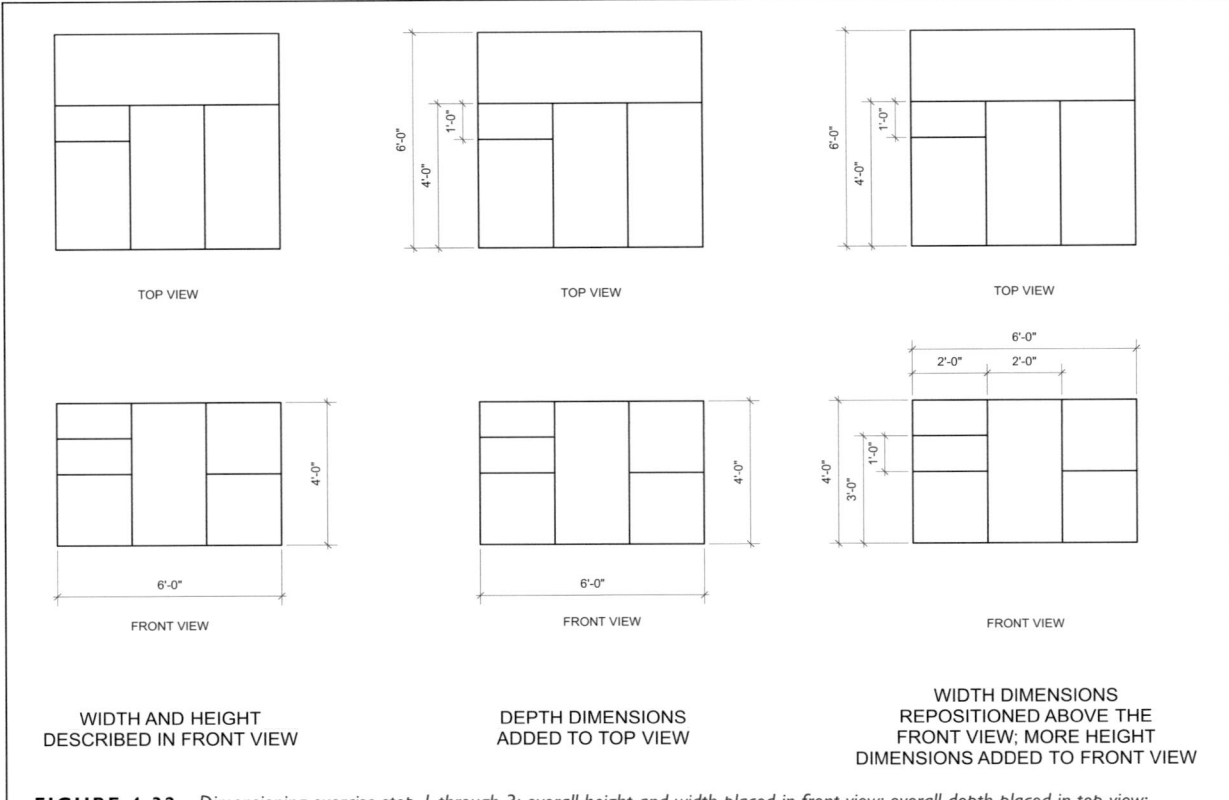

FIGURE 4-32 *Dimensioning exercise step 1 through 3; overall height and width placed in front view; overall depth placed in top view; repositioning of dimensions in front view*

6. Since all of the height, depth, and width information has been placed around the front and top views, the right side view's primary function is now to serve as reference for the ramp's auxiliary view and to place the measurement of the ramp's angle.

    Draw a guideline to locate a cutting plane slicing vertically through the center axis of the front view, dividing the ramp equally. Lightly label the cutting plane line with an A at either end, to denote the section as Section A-A.

7. Since the right view is now Section A-A, it will be redrawn as a section view (Figure 4-35). This should be placed a little farther away from the front view to provide space for the auxiliary view. Section views do not typically include the hidden lines of features beyond the cutting plane.

8. Draw the auxiliary view parallel to the ramp's incline. Place the auxiliary view far enough away from Section A-A so the ramp's angle dimension won't be crowded by it.

9. Lay out the auxiliary view's dimensions. While the true width of the ramp is found in both front and top views, it's still useful to include it in the auxiliary view so that all

of the measurements of the true shape are available in a single location (Figure 4-36). In this case, redundancy is useful.

10. Finally, add angle notation for the ramp (Figure 4-37). Avoid placing notation within the crosshatching of a section. Choose whether to place it vertically at the bottom or horizontally at the top.

11. Review the drawing. Are there any features not dimensioned? Draft the object with dimensions, either by hand or in Vectorworks. Crosshatch the section with 45° thin black lines.

MODULE 2: DRAWING FULLY DIMENSIONED ORTHOGRAPHIC VIEWS OF FOUR CUBES IN VECTORWORKS

Instructor

The following tutorial returns to the four cubes whose top views were drawn in previous exercises. Dimensioning will be applied to the three standard orthographic views: top, front, and side. Since all four cubes are vertically symmetrical, right and left views provide the same information; for this exercise,

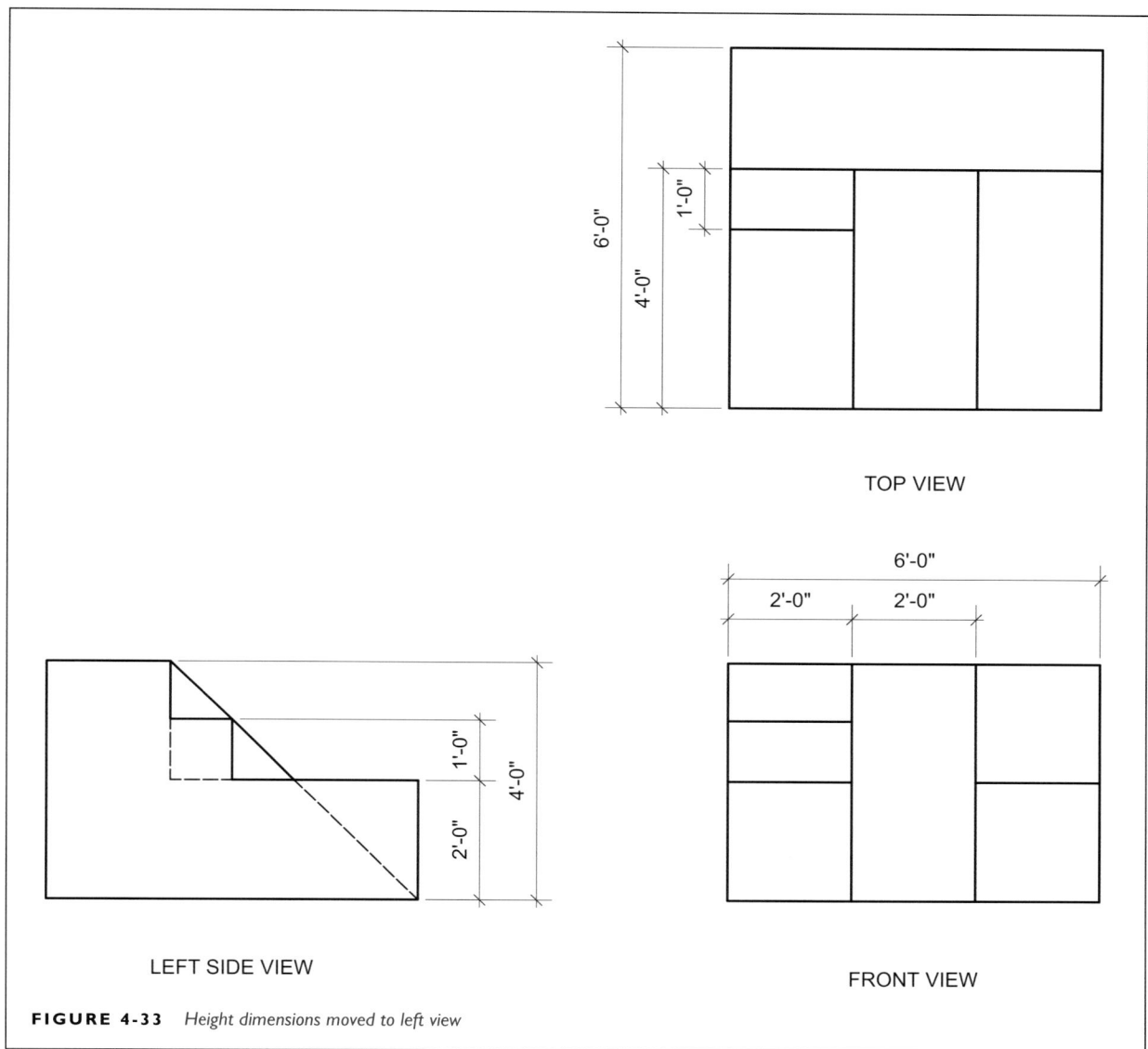

FIGURE 4-33   *Height dimensions moved to left view*

the left side view will be used. Miter lines are used to transfer information between diagonal views. This tutorial introduces viewports, sheet layers, and title blocks. Once complete, you may choose to have students repeat it as a hand drafting project on 24″ × 36″ paper.

Students

The cubes whose top views were previously drafted will now be drawn in full orthographic description. See Figure 4-38 for a rough sketch of the project with measurements. Draft all four cubes and their views on a single Arch D-sized sheet in landscape orientation. Draft in ½″ = 1′– 0″. Some cubes have interior features that will be hidden in some of the views; section views will be constructed in order to

dimension those features. Each cube has a base measurement of 3′–0″ for height, width, and depth. Interior features are inset 6″ from the perimeter.

During layout planning, determine what dimensions will be placed in which view of each cube.

Remember that these are four different objects, so each dimension applies only to the cube that is directly referenced.

1. Open your base file. Double check that a font style has been selected for the Dimension class (Dimension 12pt). If you have not already done so, add classes for Cutting Plane/Section Lines (50 mm thick, with line type ISO-09), and Extra-Thick Visible Lines (0.70 mm). Re-save your base file. Save As with a new name for this tutorial.

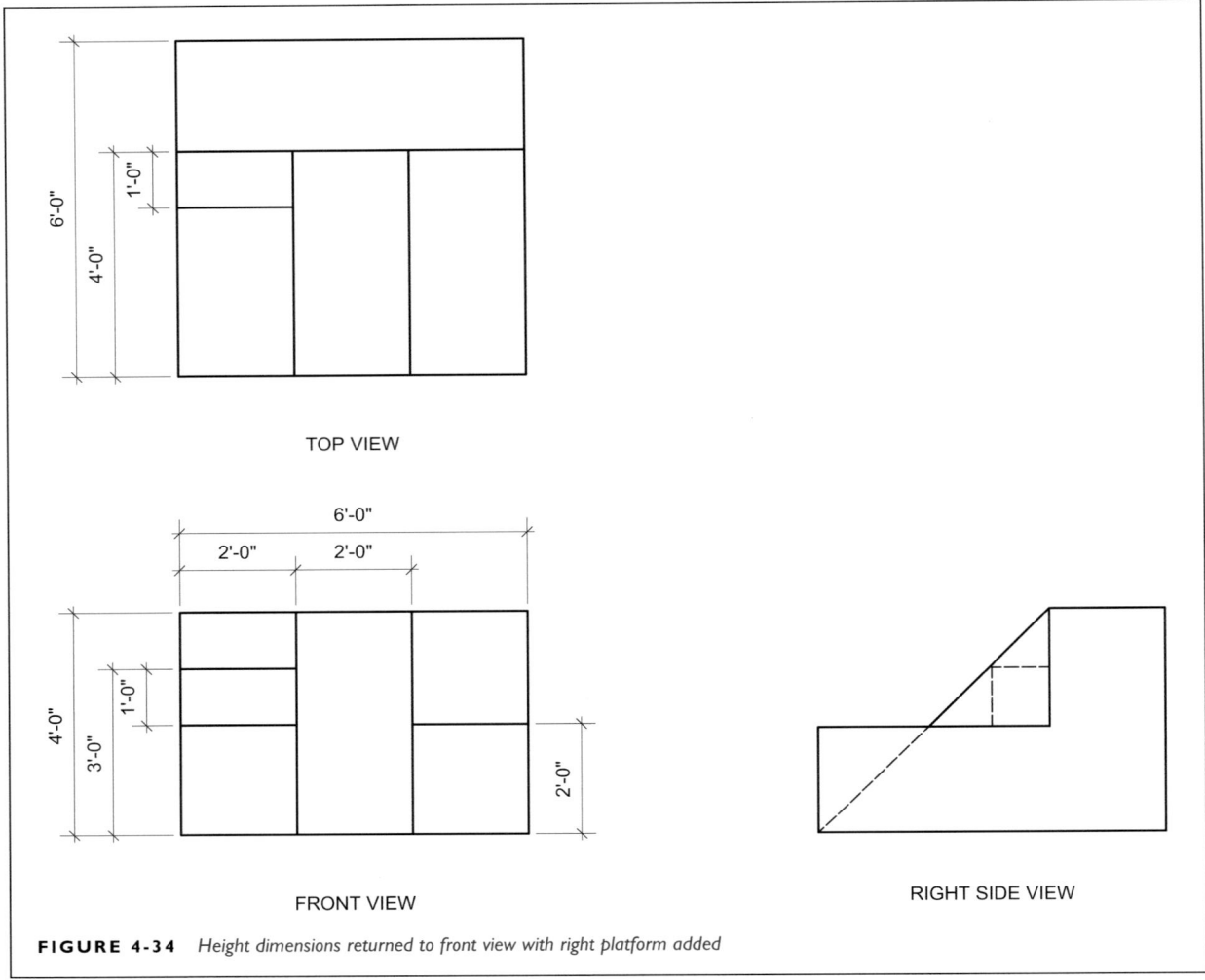

TOP VIEW

FRONT VIEW

RIGHT SIDE VIEW

FIGURE 4-34 *Height dimensions returned to front view with right platform added*

2. Change the paper/reference grid size to Arch D. Don't worry if you don't have a printer or plotter that can't print this size sheet. The project will be saved as a PDF which can be resized to fit on the largest size sheet of paper your printer can handle. From the File menu, select Page Setup.

   If your printer can handle only letter-sized paper, check the box that says "Choose size unavailable in printer setup." The drop-down menu should allow you to choose Arch D. If the computer is connected to a plotter or large format printer, it should be possible to choose Arch D without checkmarking the "unavailable" box.

3. Consider how many design layers are required. It is suggested that each cube should have its own final design layer to provide practice moving between design layers. More layers can be added later (as well as deleted). You might create a layer for all of the dimensions, or have separate dimension layers for each object. You might include the dimensions in the final

drawing layer of each cube. Each choice has advantages and disadvantages. For this tutorial, dimensions will be placed in the final design layer of each individual cube.

   Suggested layers include:

   – Border Guidelines
   – Guidelines
   – Cube One
   – Cube Two
   – Cube Three
   – Cube Four

4. In previous projects, all borders were placed ½" from each page edge. For this exercise, place the left border the standard 1" from the left edge of the paper. Use the Guideline class in the Border Guideline design layer to lay out the borders. Remember that you will have to translate 1" in real size into ½" = 1'–0" increments to lay out the border. The final border will be placed in a sheet layer, which will be discussed later.

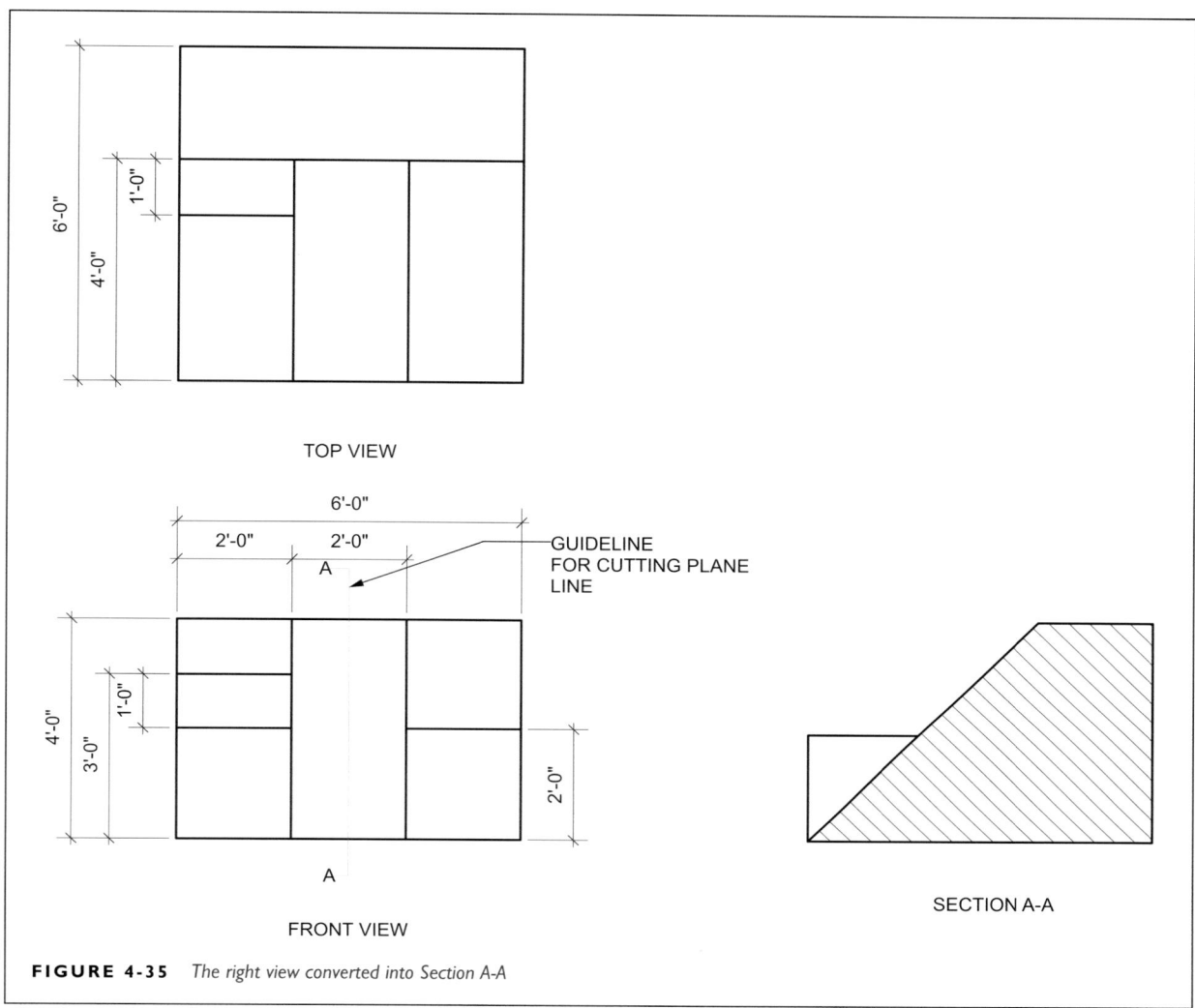

TOP VIEW

GUIDELINE
FOR CUTTING PLANE
LINE

FRONT VIEW

SECTION A-A

FIGURE 4-35 *The right view converted into Section A-A*

5. Do layout math. Each cube is 3'–0" on each side. Some have features that extend beyond that measurement, but not so far it needs to be taken into account during this layout phase. Generous white space ensures that all features and dimensions have adequate room. Calculate distances between views and between each array of views. Since the drawing will be laid out as a whole, you'll quickly discover whether or not there is adequate spacing. Don't hesitate to delete and redraw if necessary.

In the Guideline layer, use your calculations to draw a rectangle that represents the outer perimeter of all four sets of views. While each array of cube views needs to be orthographically aligned, aligning all four arrays of the overall drawing can provide further visual clarity (Figure 4-39).

Is your rectangle centered on the page? Does it seem balanced? Does the rectangle feel like it will crowd everything too tightly in the middle of the page or against one edge of the reference grid?

6. There are a couple of ways to proceed:

a. All the views for all the objects can be laid out in one go. The drawback is that there will be a lot of guidelines, making later steps a bit more confusing than they need to be.

b. Lay out each cube's views separately. After darkening a cube's array, extend selected guidelines into the next cube array. Since the arrays will be aligned and share many measurements, guidelines will transfer information not just between individual views of a single object but also between the arrays of different cubes.

For this tutorial, the second option will be employed. Begin in the upper left corner with the simplest cube.

7. In Figure 4-39 the perimeters of all the cube arrays are shown laid out. Figure 4-40 shows a placement process for the views of Cube One. Knowing that the views are

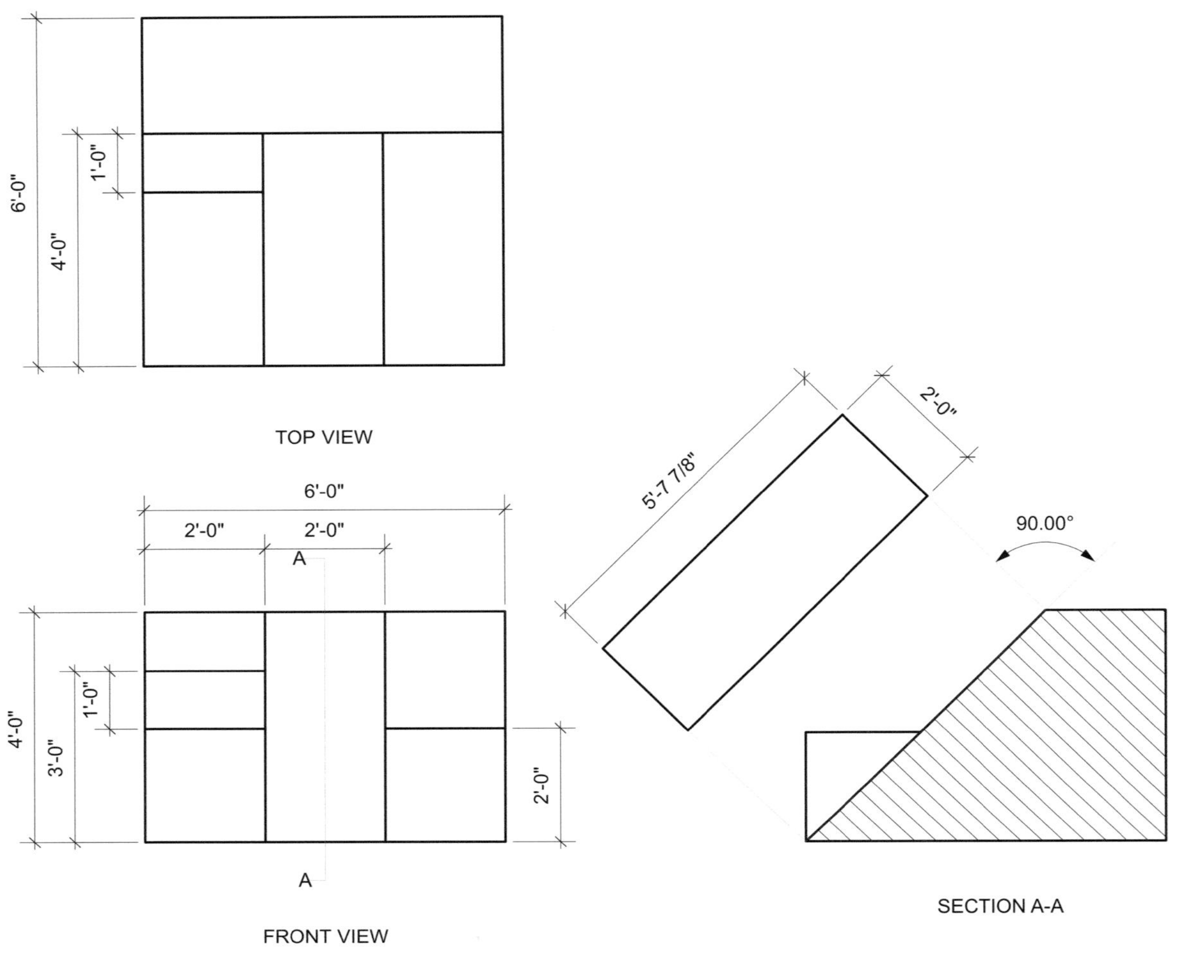

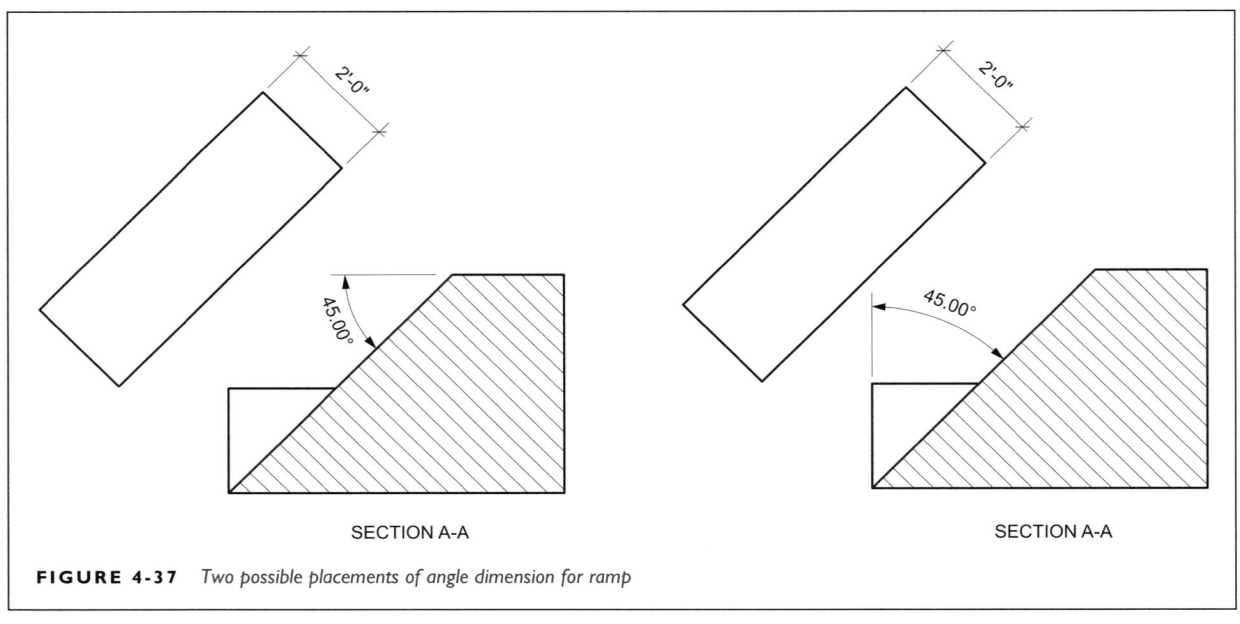

TOP VIEW

6'-0"

4'-0"

1'-0"

FRONT VIEW

6'-0"

2'-0"   2'-0"

A

4'-0"

3'-0"

1'-0"

2'-0"

A

5'-7 7/8"

2'-0"

90.00°

SECTION A-A

FIGURE 4-36 *Auxiliary view drawn in reference to Section A-A; auxiliary view dimensioned*

2'-0"

45.00°

SECTION A-A

2'-0"

45.00°

SECTION A-A

FIGURE 4-37 *Two possible placements of angle dimension for ramp*

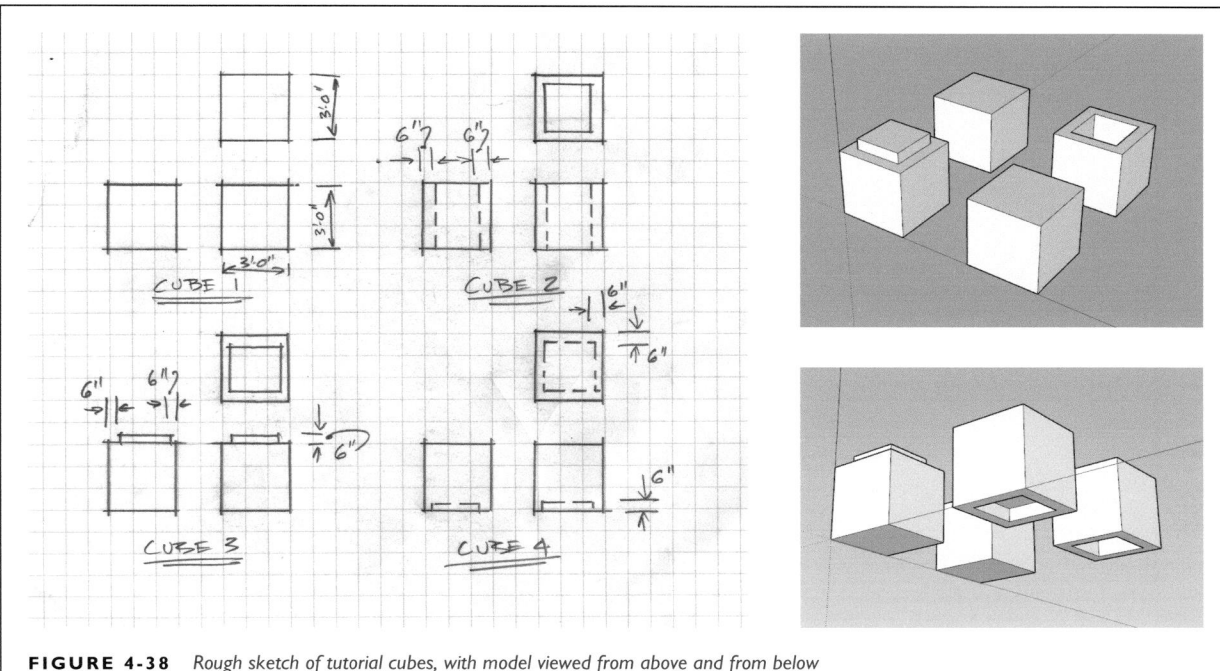

FIGURE 4-38 *Rough sketch of tutorial cubes, with model viewed from above and from below*

aligned, the edges of the views share guidelines. For example, a single guideline is drawn to place the top edges of both the top and left side views.

8. Rather than darkening everything at the very end, darken each cube when its layout is complete to keep steps clear. If each finished cube is drawn in a different layer, it's more difficult to accidently alter completed portions of the drawing. Activate the design layer for the finished drawing of Cube One. Use medium visible lines to darken the front, top, and side views of Cube One. As it has no additional features, all three views appear identical.

9. Return to the Guidelines layer and the Guidelines class. Lay out a grid for Cube Two in the upper right-hand area of the overall perimeter. Extend horizontal guidelines from Cube One to situate Cube Two's views. Add required vertical guidelines. (Figure 4-41).

10. Cube Two's hole is inset 6″ from the perimeter. In the top view, lay out the perimeter of the hole. Drop vertical guidelines into and through the front view. The hole cannot be seen in the front view, but will be indicated with hidden lines drawn atop these guidelines. There are two ways to place guidelines for the hole in the left view. Either measure the hole directly into the view, or use a miter line to move information from the top view to the side view (Figure 4-41). As in the front view, the hole is a hidden feature in the side view.

11. Activate the design layer for the finished Cube Two. Darken object lines. Use hidden lines to draw the sides of the hole in the front and side views. Because it's a vertically symmetrical object, the side and front views are identical.

12. Return to the Guidelines layer and the Guidelines class. Lay out a grid for Cube Three in the lower left corner of the drawing, starting by dropping vertical guidelines from Cube One. As with Cube Two, lay out the interior square in the top view and drop guidelines down to the front view. The raised area is inset 6″ from the outer perimeter of the cube. The surface of the raised area is 6″ above the top plane of the cube. Measure the height of the raised area in either the left or front view and extend a single guideline across both views (Figure 4-42).

13. Activate the design layer for the finished Cube Three. Darken the object lines. Since the raised portion is an exterior feature that contributes to the overall outline of the cube, darken it with medium visible lines.

14. Return to the Guideline layer and Guidelines class. Lay out a grid for Cube Four in the lower right corner of the page. The cube's bottom recess is inset 6″ from the outer perimeter of the cube, and ascends 6″ into the body of the cube.

    When Cube Four is laid out, activate its finished cube design layer and darken. Use hidden lines for the recess, which cannot be seen in any of the views. For an

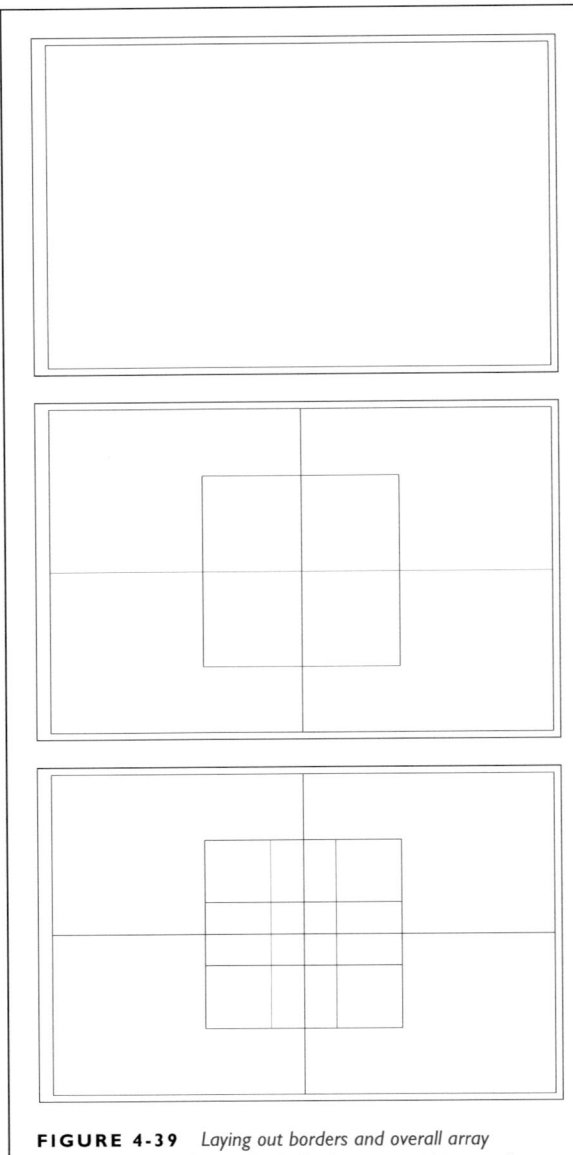

FIGURE 4-39 *Laying out borders and overall array perimeter; layout of the individual cube arrays within overall perimeter; reference grid not shown*

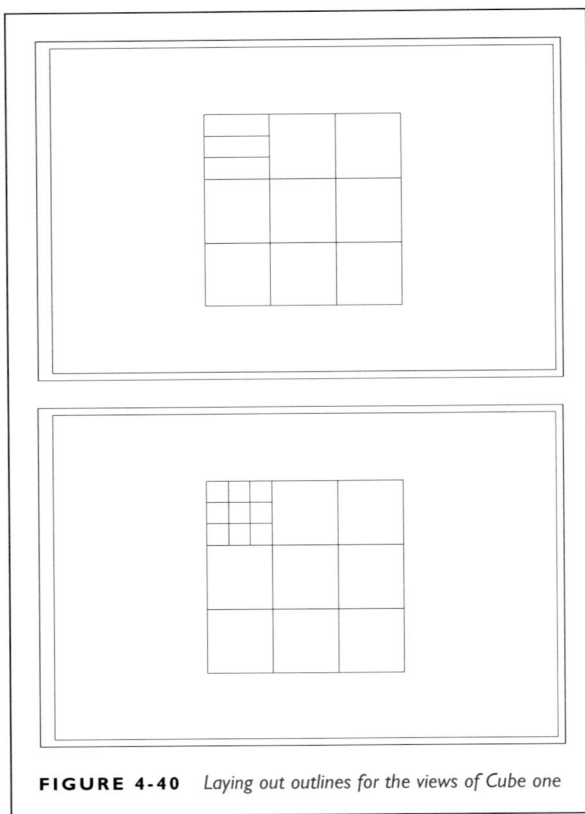

FIGURE 4-40 *Laying out outlines for the views of Cube one*

object with an important feature that can only be seen from beneath, a bottom view may be desired.

15. Dimensions should not reference hidden lines, so it's necessary to convert the side views of Cubes Two and Four into section views. Start with Cube Four since its layer is currently active. Draw a guideline along the center vertical axis of the front view to mark the cutting plane line for Cube Four's Section A-A.

   Use the Select tool to drag a box around the side view of Cube Four and Cut (delete) it. The guidelines should still be visible, provided the Guidelines class visibility is turned on. Reestablish the outline of the

cut-through portion of the cube. The line representing the far bottom edge of the recess will be added later; first, this shape will be composed so that a Fill can be applied (Figure 4-43).

16. Use the Select tool to drag a box around the outline. While the shape is active, open the main tool bar's Modify drop-down menu. Choose Compose. When you return to the drawing, small blue boxes appear at the shape's corners and midpoints to indicate that this polygon is now considered to be a single object, rather than eight separate lines.

17. In the Attributes window, click on the text window next to the small paint can. It should currently read as None since no Fills were assigned to any classes in the base file. Use the drop-down menu to select Hatch. A default hatch pattern will fill the shape.

18. For this project, thin diagonal lines are preferred as they are the basic, universal crosshatch pattern. Click on the downward arrow in the text box below the paint can icon to open the Hatch library. On the menu tree to the left, find the folder that says ANSI.vwx. Choose this library. Look for Cast Iron HF. It's a good, well-spaced diagonal pattern, so even though you're not cutting through a cast iron object, it will serve the purpose. When you double-click on Cast Iron HF

FIGURE 4-41 *Laying out the grid for Cube two by extending guidelines from Cube one; laying out the interior hole using a miter line to move information from the top view to the left view*

and return to the drawing, the Fill appears within the composed object.

19. Darken the line representing the far edge of the recess. Section A-A of Cube Four is complete (Figure 4-43).

20. Darken the front view cutting plane line with the cutting plane class. Choose the Section-Elevation Marker tool from the Dims/Notes tool box. Click on one end of the vertical axis guideline. Drag the cursor to the other end. Double click. While the cutting plane line is highlighted, its Object Info window will be open. If the bubbles are pointing to right, click on Reverse Direction to orient them to the left. Change the Marker Size (the bubble size) to ¼″ so it's not quite so big on the drawing. Checkmark Auto-Rotate and Text Both Ends if not already checked. Since Cube Four is Drawing 4 on Sheet 1, change the Drawing Number to A (as the section) and the Sheet Number to 4 (as the drawing) so that this information appears in the marker (Figure 4-44).

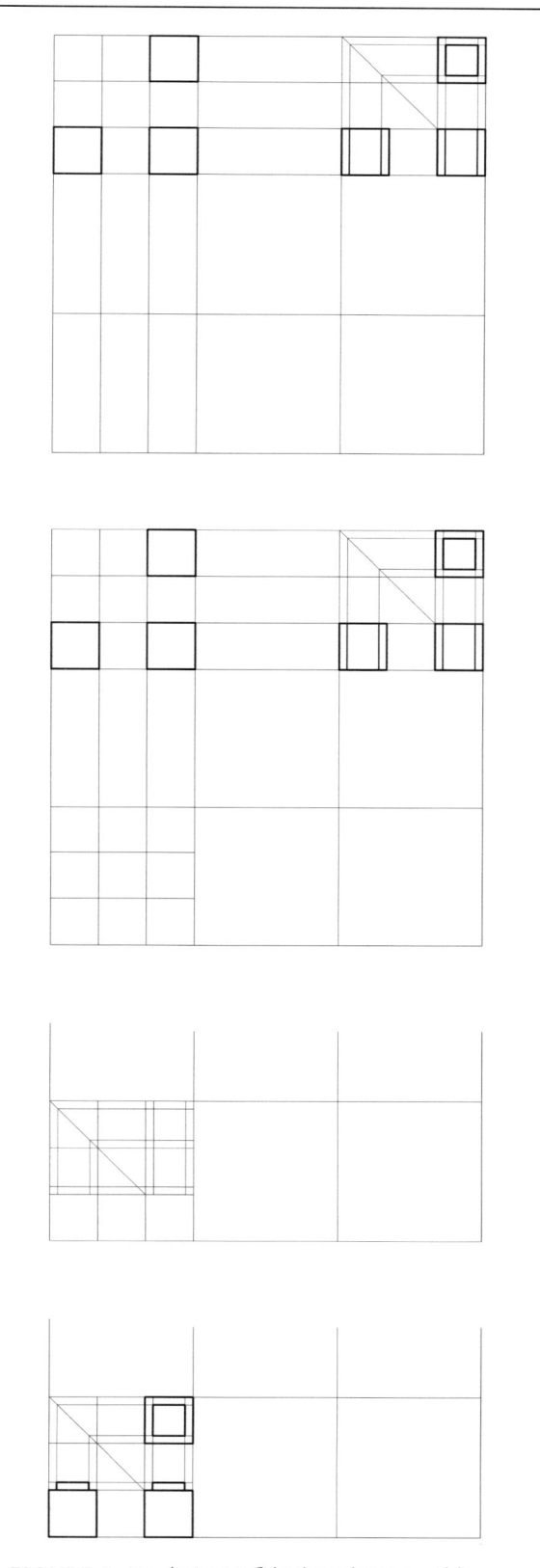

FIGURE 4-42 *Laying out Cube three; dropping guidelines from Cube one; adding horizontals; adding interior detail; darkening*

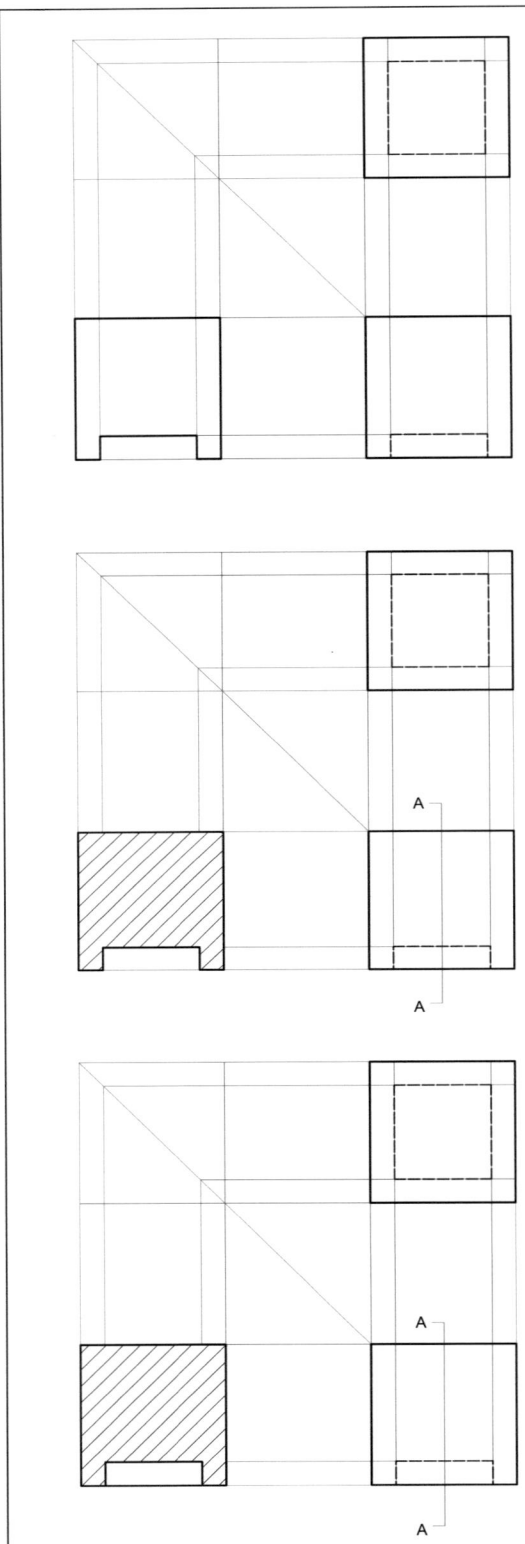

FIGURE 4-43 *Drafting cube four Section A-A; composing the cut-through mass; selecting Cast Iron HF fill; reestablishing the bottom edge*

21. Switch to the Cube Two layer. Delete the side view. For this cube, the hole runs all the way through the object, so the section creates two narrow rectangular areas which require crosshatching. Add a cutting plane line to the front view: label it Drawing 2, Section A-A. Change the settings in the Object Info window to match those used for Cube Four.

    Use the Rectangle tool to outline one cut-through area. The rectangle is automatically composed, as evidenced by the array of blue squares at corners and midpoints. While active, select the Cast Iron Fill from the Attributes window. Repeat for the other side of the cube. Darken the top and bottom edges of the hole (Figure 4-45).

> For another approach, create a class in your base file named Filled Section. For Fill, choose the cast iron crosshatch style. Don't forget to check Use at Creation. When drawing a shape with a polygon or circle tool, the interior will fill with the hatch upon the final click completing the closed shape.

22. For this exercise, dimensions will be placed in each cube's finished design layer. Activate Cube One's design layer. Notice that the crosshatching just applied to Cubes Two and Four appears to have vanished. Don't worry about this. Settings will be adjusted later to ensure that all Fills are visible when a sheet layer is created.

    Consult your dimension layout plan. Add dimensions to Cube One. Use the Constrained Linear Dimension tool. Be sure Dim Stds is set to Arch. If the arrays for each cube are in relatively close proximity, consider placing dimensions, if possible, toward the interior of each set of views (Figure 4-46).

23. Switch to Cube Two's finished design layer and add dimensions. There is a choice in the dimensioning tool bar that helps create stacked dimensions. Try it. Stacked dimensions can also be created by overlaying extension lines while using the basic Dimensioning tool.

    For short distances, the notation may overlap extension lines. Use a leader line to place the measurement. Do this by selecting the dimension line to activate it. In the navigation window, uncheck Show Dim Value; you may have to scroll down to find this option. The measurement disappears, leaving the dimension line empty. Use the Text tool to write the measurement nearby and draw a leader line to connect the notation to the dimension line (Figure 4-47).

    Dimension the hole's side walls in the section view where they are seen without hidden lines. The height of

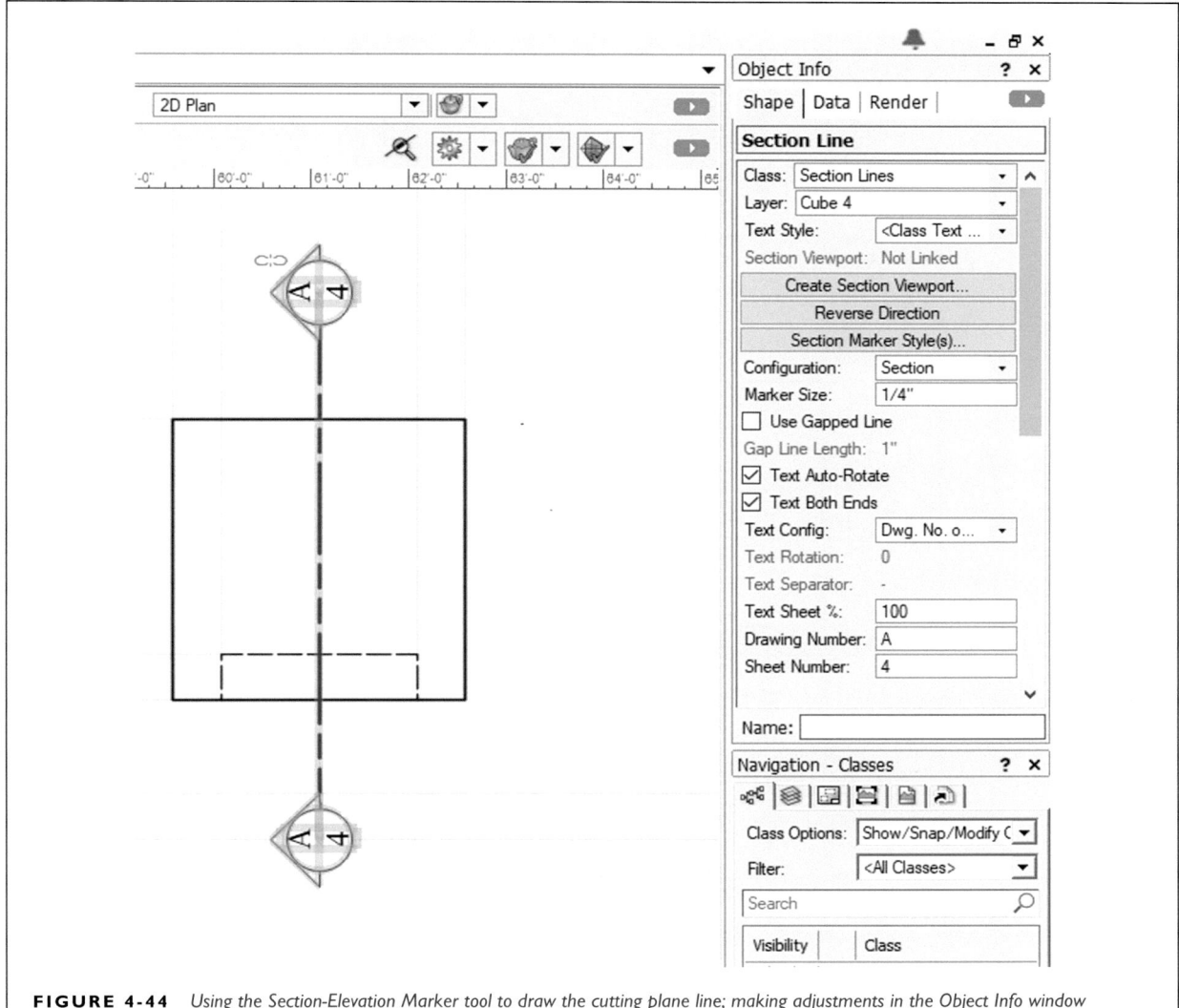

FIGURE 4-44　*Using the Section-Elevation Marker tool to draw the cutting plane line; making adjustments in the Object Info window*

the hole is shown in both the side and front views. The width of the hole can only be seen in the front and top views. The views have not been sectioned, so where can this dimension be placed?

Multiple sections are permissible. A cutting plane line could be added to the top view, pointing toward the front. In this case, the front view would also be redrawn as a section (Section B-B). This creates two sections that appear identical because of the object's symmetry. To fully express the hole as a feature, it might be best to leave the front view and its hidden lines as is, and place the hole's width dimensions in the top view.

While it's not preferable to have extension lines cross visible lines, conventions can be bent if clarity is maintained. It is permissible for extension lines that refer to an interior feature

to terminate beyond the perimeter of the object, provided it is clear what feature is referenced by the extension line (such as a window located in the center of a large wall).

In Vectorworks, the first click of the dimensioning tool should be on a corner of the interior object. The second click should be on a corner of the object's outline. An unbroken extension line is created that extends from the interior feature, crosses the object outline, and places the notation beyond the object's perimeter.

If a continuous extension line crossing visible lines is not desired, extend a guideline from the corner of the interior feature to the object's outline. This creates a viable intersection for the Dimension tool to grab on to. Dimension from the guideline intersection to a corner of the object's outline.

Some designers will allow extension lines to "jump" the visible line—leaving a small gap on either side of the line

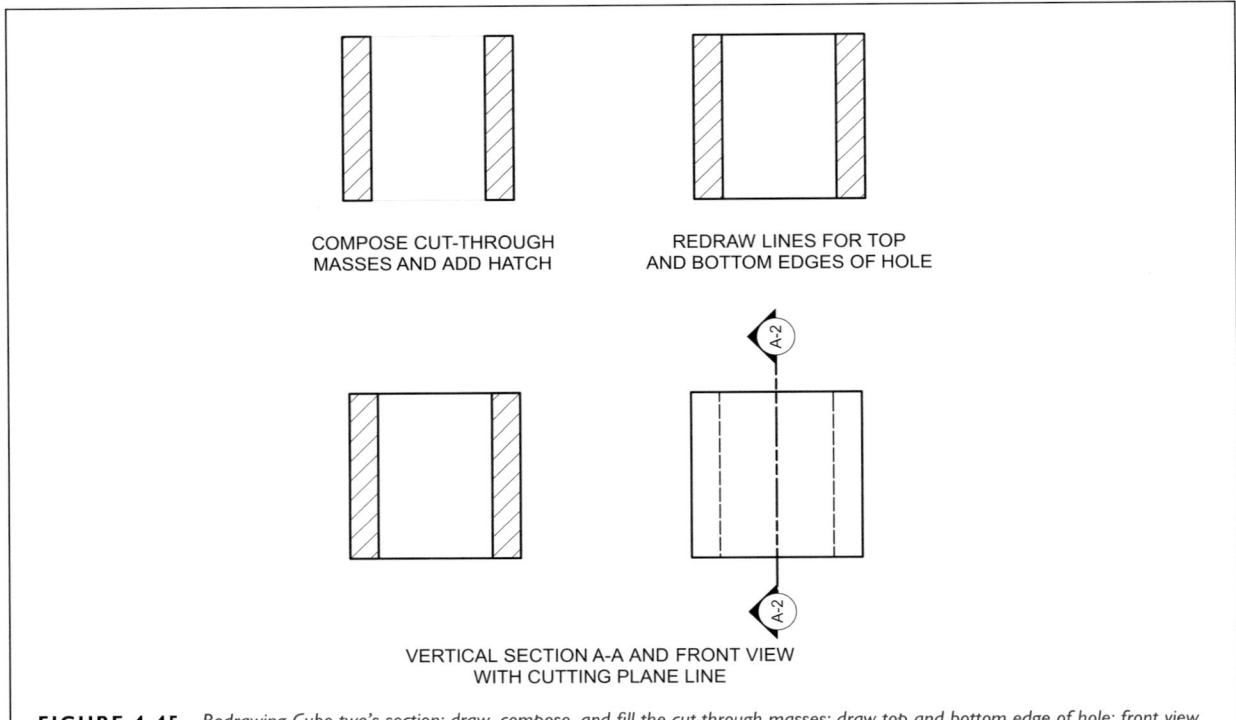

FIGURE 4-45 *Redrawing Cube two's section: draw, compose, and fill the cut-through masses; draw top and bottom edge of hole; front view with cutting plane line*

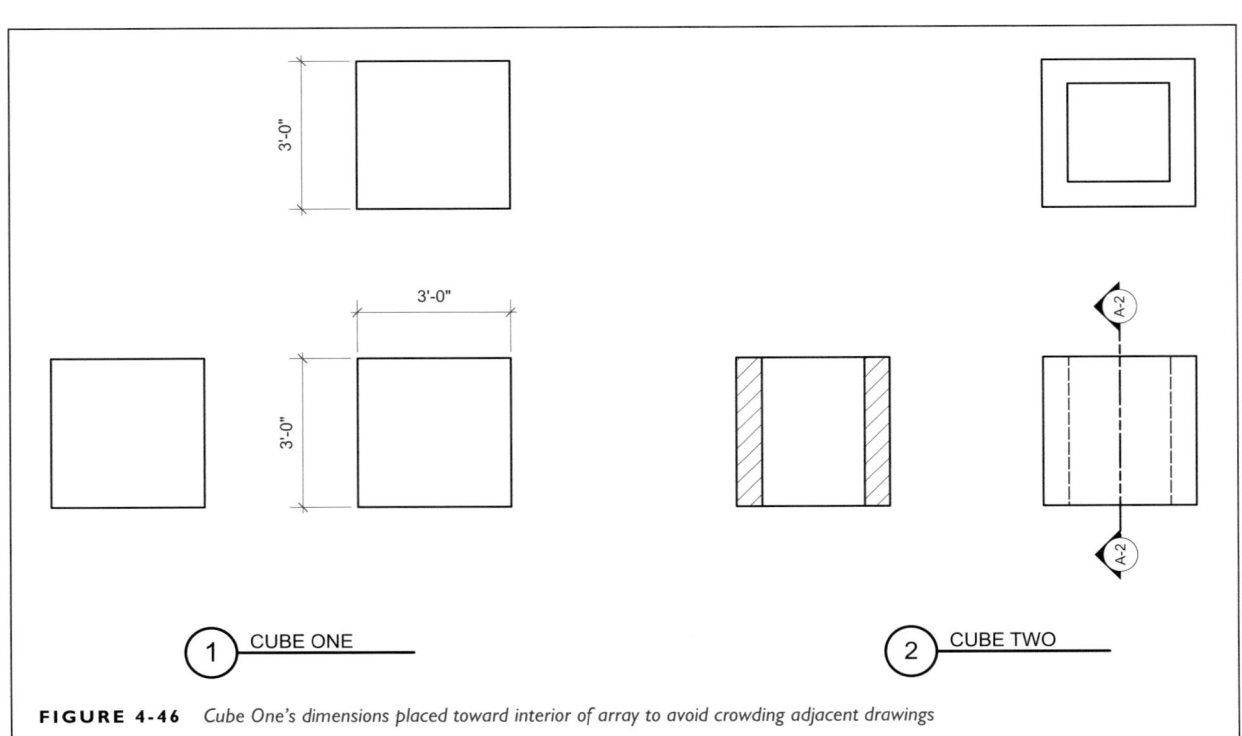

FIGURE 4-46 *Cube One's dimensions placed toward interior of array to avoid crowding adjacent drawings*

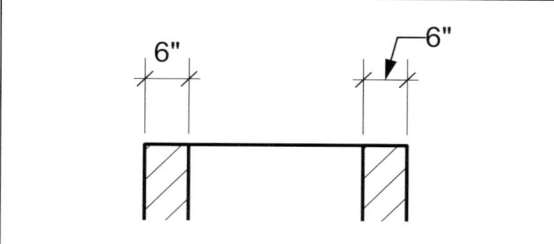

FIGURE 4-47 *If the space between extension lines is too narrow for notation, use a leader line*

being jumped. This, however, breaks the continuity of the extension line and is not as optimal as simply crossing the visible line (Figure 4-48).

24. For Cube Three, the raised area is clearly shown in the front and side views. The depth of its inset can be grouped in the side view. Its width can be grouped in the front view.

25. Cube Four presents the same questions as Cube Two. The depth of the recess is shown without hidden lines in the section. Its height is seen in the section, but extension lines would need to cross not just visible lines but crosshatching as well. A sectioned front view provides width but does not solve the height problem.

    Including a bottom view still requires extension lines crossing visible lines in order to dimension the recess, but at least they won't have to cross crosshatching. Figure 4-49 shows this option.

Planning Cube Four's dimensions demonstrates that graphic recommendations sometimes only take a draftsperson so far; discretion based on experience then comes into play. Discuss dimensioning options for Cube Four in class. Which option is most preferred? Why? Is there another solution that

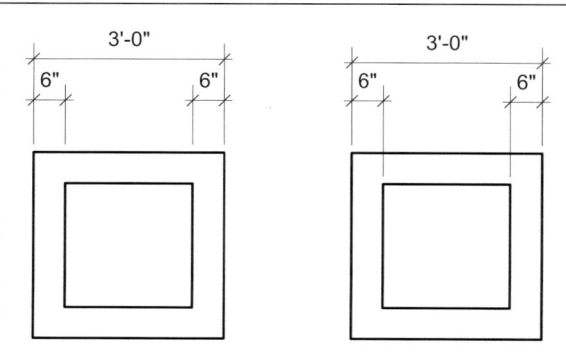

FIGURE 4-48 *Two ways to dimension Cube two's hole: (1) extension lines beyond the object's perimeter; (2) extension lines crossing the perimeter to reference the hole directly*

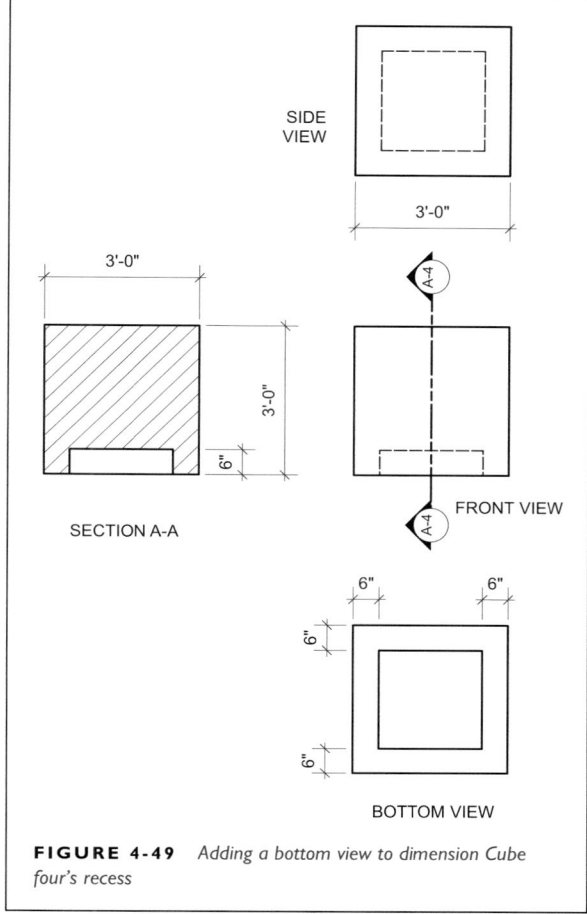

FIGURE 4-49 *Adding a bottom view to dimension Cube four's recess*

hasn't been considered? If you hand this drawing to someone outside of the class, could they construct the object from the information given?

Titles, Labels, and Notes

Labels and titles are included so that the reader instantly knows what they're seeing, as well as how the drawing/object relates to the larger project. There is a certain amount of overlap in the purposes and use of the terms Titles, Labels, and Notes. **Titles** name the view or drawing as a whole. Titles are also called **identity labels**.[4] Titles are typically found beyond the dimensioning, below a view or drawing. **Labels** typically refer to a single view or a collection of elements within a drawing: a "general label."[5] Labels are placed in proximity to the object referred to, sometimes connected with a leader line.

Notes refer to text that provides additional information about a drawing/object. **Local notes,** or, "specific labels,"[6] refer to elements and pieces within a drawing/object and are connected to that item with a leader line. **General notes** are typically found above or to the left of the title block, or sometimes placed below the drawing title. USITT recommends using local notes rather than general notes whenever possible.

The following is a list of labels and titles found on a typical drawing:

- Descriptive name: hallway, border, table, etc. When a sheet describes a single object, the name of the object is typically placed in the title block rather than as a title below the array of views.

- Orthographic view name: front elevation, top view, Section A-A, etc.

- Indexing numbers and letters for a coded labeling system; e.g., a groundplan mapping out a series of flats may label individual flats by letters. Detail drawings of the individual flats later in the package are then labeled with the corresponding letter (Flat K). The more complex the project, the more important a coded labeling system becomes.[7]

Titles for hand drafting use ¼″ high lettering. Labels for views are typically ¼″, while labels for features within a drawing are 1/8″. USITT recommends 3/16″ (20pt font) for CAD labels, and ¼″ (26.5pt font) for titles. Notes are typically 1/8″ or 10–13.5pt font. In hand drafting, a double underline can help indicate that the text is a label. Vectorworks uses a bubble-and-line combination as standard label format. Coded labeling systems use a bubble enclosed by an arrow to reference objects (Figure 4-50).

Titles/labels are typically centered beneath a drawing, as that is usually the first place a reader looks. If that location is not available, the title may be placed elsewhere but it should be in such proximity that it clearly references the view/drawing.

In this tutorial, drawings of four separate cubes were constructed: four objects share the sheet and are titled as separate objects (Cube One, Cube Two, etc.). There are two sections amongst the arrays; those should also be labeled. Since the rest of the views have been organized in standard orthographic relationships, the draftsperson has discretion whether to label each and every view. For this exercise, it's pretty apparent what each view represents—for larger objects and more complex projects, labeling the views becomes more imperative. When in doubt, provide a label.

26. Find the Drawing Label tool in the Dims/Notes tool set. It looks like a bubble with lines extending toward the right (Figure 4-51).

    In the Guideline layer, draw a guideline across the page below the Cubes One and Two; align their labels. Switch to the Cube One layer. Select the Drawing Label tool and click on the guideline. A bubble and line appear, as does a navigation window. Drag the cursor along the guideline and click to affix the label below Cube One. If it's off-center, it can be selected and moved to a more desirable location after information is entered. The label's line style will be that of whatever class is currently active.

    In the navigation box, name the drawing Cube One. This text appears above the label's mid-line.

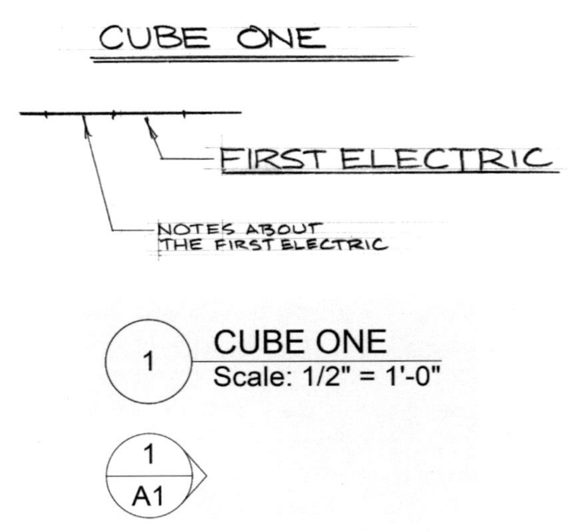

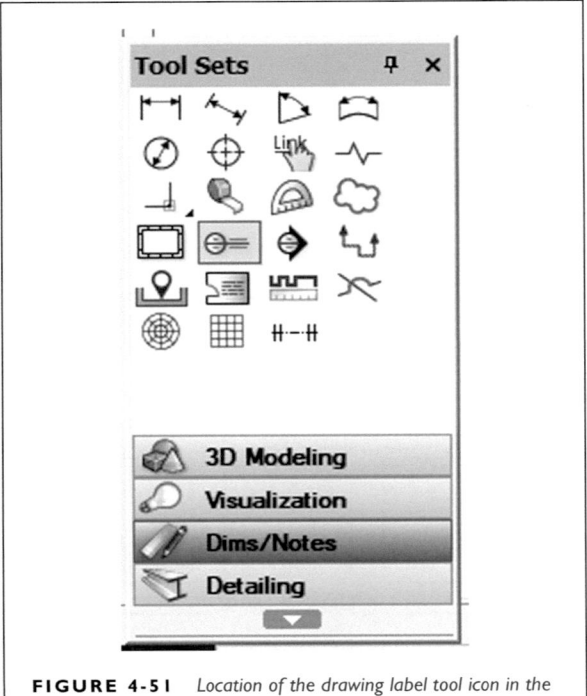

FIGURE 4-51 *Location of the drawing label tool icon in the Dims/Notes tool set*

FIGURE 4-50 *Examples of label and note formatting*

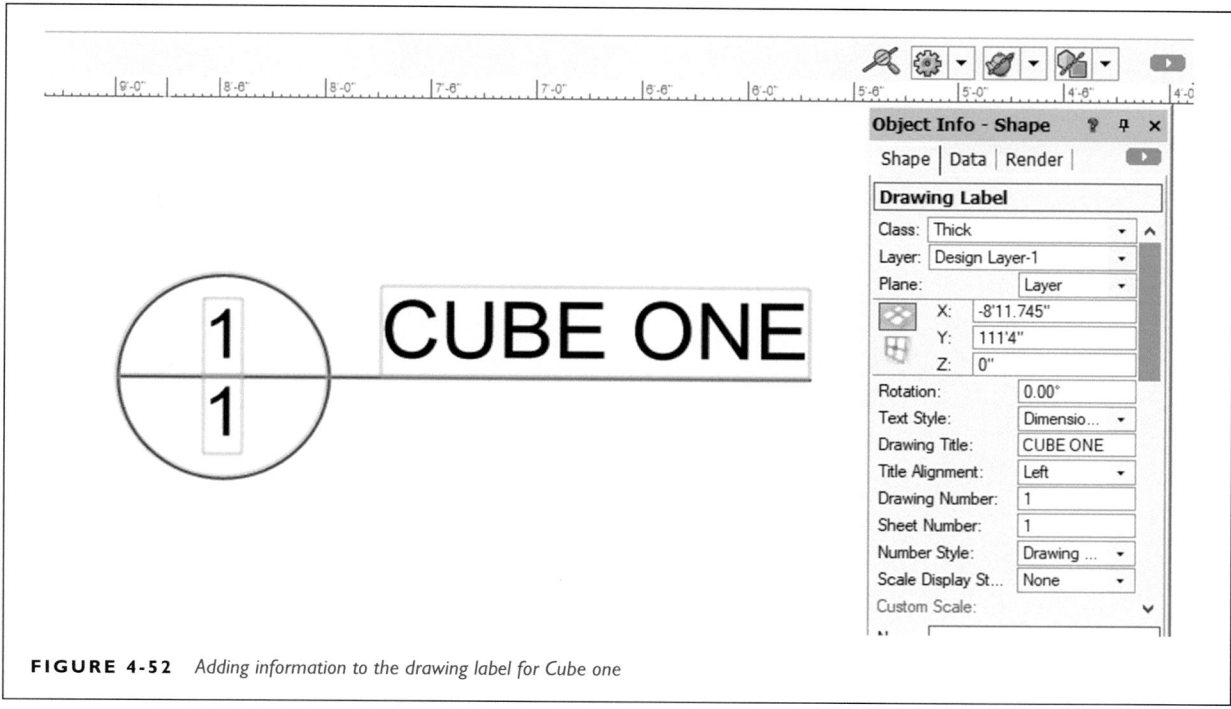

FIGURE 4-52 *Adding information to the drawing label for Cube one*

Change the Number Style to Drawing and Sheet so the upper half of the bubble features the drawing number and the bottom half the sheet's number. Cube One is Drawing 1 on Sheet 1. Cube Two is Drawing 2 on Sheet 1, etc. Change the Scale Display Style to None since the whole sheet is drawn in ½″ = 1′–0″. The overall sheet scale will be noted in the title block. If time allows, explore the other label settings to see what they do (Figure 4-52).

Place labels and enter information for all cubes. Label the section views. Label the standard orthographic views, if desired.

Viewports and Sheet Layers

In previous Vectorworks exercises, printing was done directly from a design layer. To have more control over what appears on a printed sheet, portions of design layers are imported to sheet layers via **viewports**. When the drawing spans Vectorworks' infinite drawing space, the draftsperson can select various items and gather them onto a single sheet. Viewports can also break a drawing apart across a number of sheets. For example, each of the four cubes can be exported via four separate viewports to four different sheets. The scale of the printed sheet can be selected or changed during viewport creation.

For this exercise, all four cubes will be imported onto single sheet; only one viewport will be created.

27. Turn off the visibility of the Guideline class; otherwise they'll appear in the final drawing. Use the Rectangle tool to drag a rectangle around everything you want to place on the finished drawing (Figure 4-53).

28. While the rectangle is active, go to the View menu on the main toolbar. Open the drop-down menu. Choose Create Viewport. Create Viewport. A window appears, asking if you wish to use the selected area as the crop. Click Yes. The Create Viewport window will appear (Figure 4-54). Only one sheet is to be generated and all four cubes have already been labeled. Uncheck the Create Drawing Label option. If you want the viewport to label the drawing, this box would remain checkmarked. The Scale selection box opens a drop down menu; if it does not already read ½″ = 1′—0″, open the menu and select that scale. This will be the scale of the printed drawing.

29. There is a button that reads Layers. Click on this to open the Viewport Layer Properties window (Figure 4-55). This allows you to choose which layers appear in the sheet layer, allowing the Fills in various layers to be visible. Turn on the visibility for all the Cube layers and any Dimension layers by clicking that row's first column to make the eye icon appear. Turn off the Guidelines layer. OK back to the Create Viewport window.

30. Keep the other default settings for this exercise. Click OK at the bottom. The New Sheet Layer window will

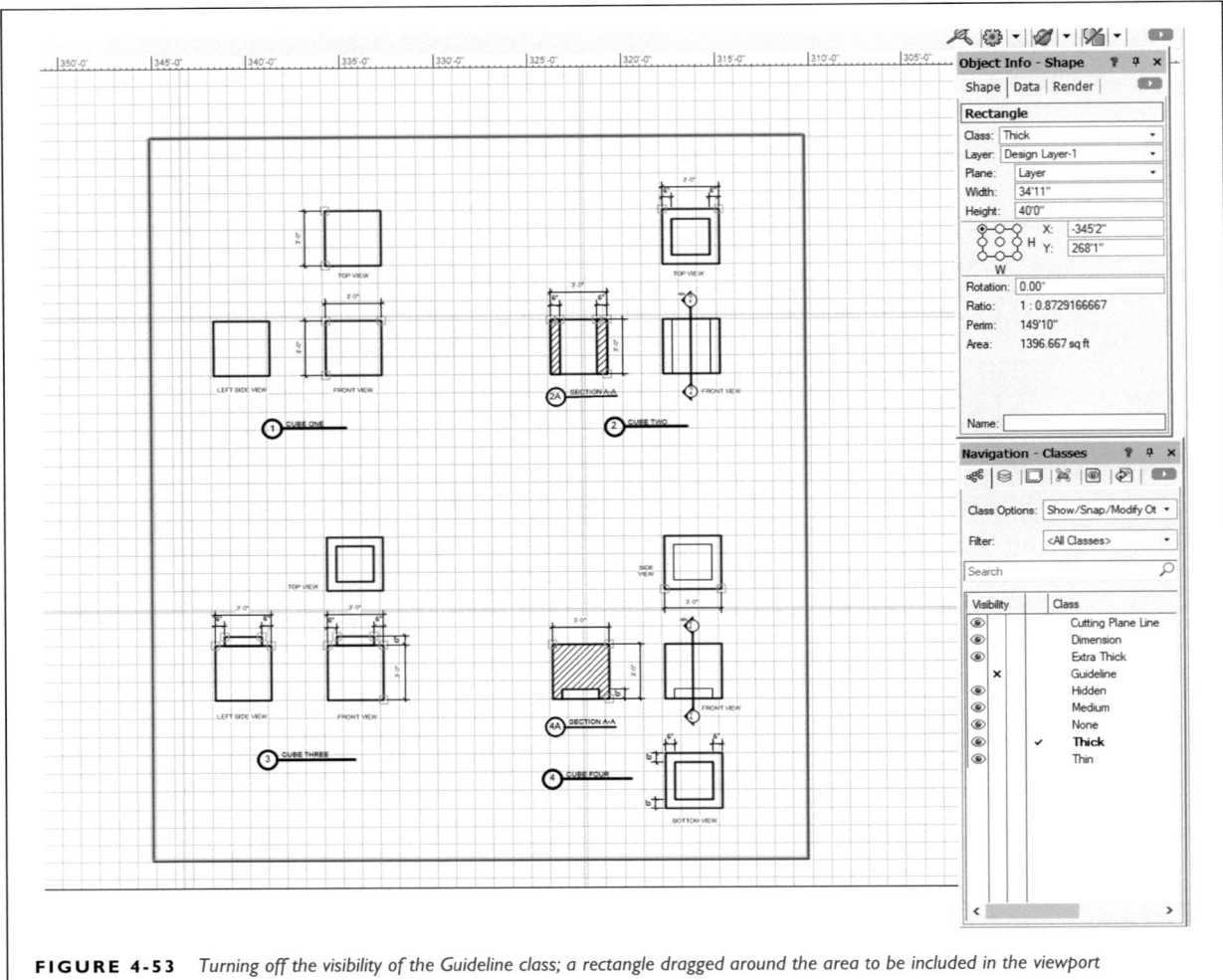

FIGURE 4-53   *Turning off the visibility of the Guideline class; a rectangle dragged around the area to be included in the viewport*

appear. Sheet Number can remain Sht-1 as there is only one sheet generated in this tutorial. Title the sheet Four Cubes. OK back to the drawing. It should look something like Figure 4-56.

The orange rectangle is the area enclosed by the viewport's crop. The reference grid remains the size of the selected page size (Arch D). If part of the drawing is missing, delete the sheet layer via the Design Layer organization box on the upper toolbar and redo the Viewport. There are editing options for Viewports, but deleting is the swiftest action for the time being.

When you Escape or click on the Select tool icon, the crop outline will disappear. The viewport can be reactivated by moving the cursor over the area until the crop outline appears. You can then click and drag the viewport to a more desirable location on the page.

Title Blocks

Title Blocks are mandatory for any drawing prepared for a job. They provide important organizational information for the project as well as reference notation to place each drawing within a series.

USITT recommendations state that title blocks shall contain:[8]

- Producing organization name (and contact information as space allows)

- Show/Project title

- Plate/Sheet title

- File name (if CAD)

- Director name

- Designer name (and contact information as space allows)

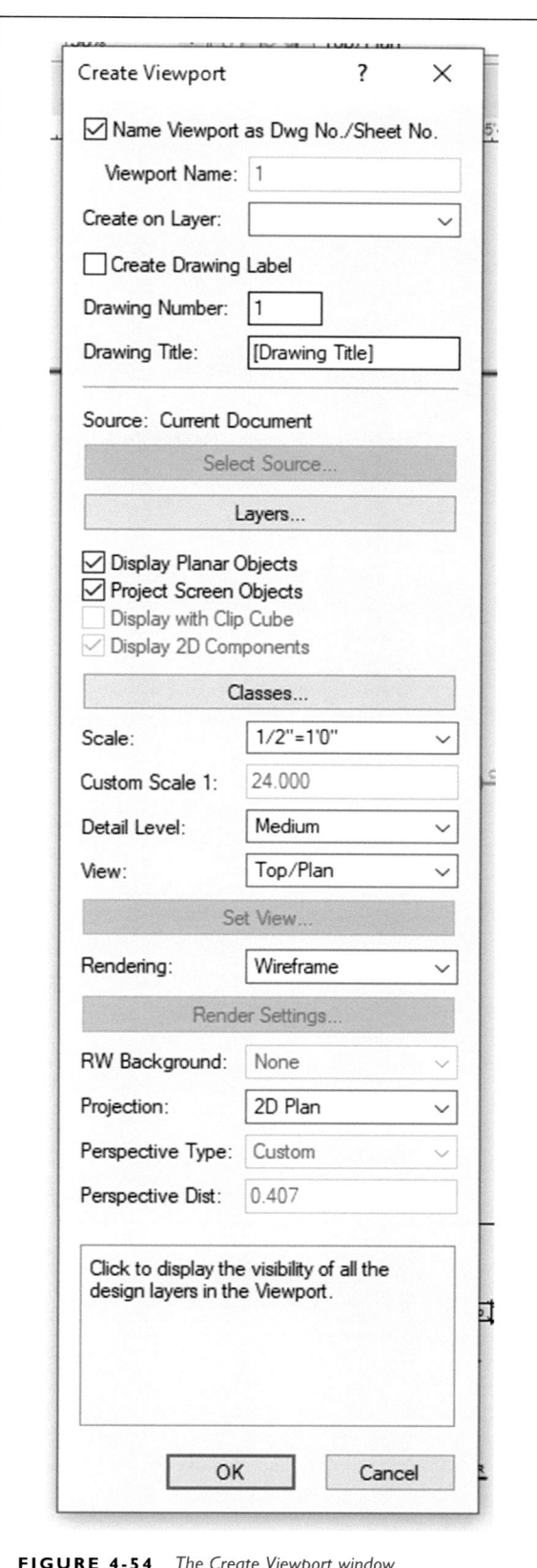

FIGURE 4-54 *The Create Viewport window*

- Technical director name

- Draftsperson name (if different than designer or TD)

- Checker/Approval name; if not required, list as Not Applicable (NA)

- Plate/Sheet number (X of Z)

- Drawing date (date submitted, not date drawn)

- Scale of plate/sheet (if varies, scale is designated "As Noted")

- Full file string same (if CAD)

They may also contain:

- Union stamp, if the designer is a member of United Scenic Artists (USA).

- Other production team members, such as the master electrician, production manager, etc. (who is to receive the drawing?).

- Disclaimers. This informs the reader as to which party bears responsibility for the safe execution of the design. Typically, a designer will state that this responsibility falls with the production company as the drawing represents only a suggestion for technical execution. Disclaimers mean that technicians should be empowered to speak up in unsafe situations, and that designers are duty-bound to make problem-solving compromises.

> United Scenic Artists (USA) is a labor union and professional association of designers, artists, craftspeople, and department coordinators, organized to protect craft standards, working conditions, and wages for the entertainment and decorative arts industries. Local USA 829 is a member organization of International Alliance of Theatrical Stage Employees (IATSE).[9] Website: www.usa829.org.

> The International Alliance of Theatrical Stage Employees (IATSE) is a labor organization that represents behind-the-scene workers in theater, film and television production, trade shows and exhibitions, television broadcasting, and concerts as well as the equipment and construction shops that support all these areas of the entertainment industry.[10] Website: www.iatse.net.

There are three standard locations for title blocks (Figure 5-57):

- as a block in the lower right corner

- as a vertical strip along the right border

- as a horizontal strip along the bottom border

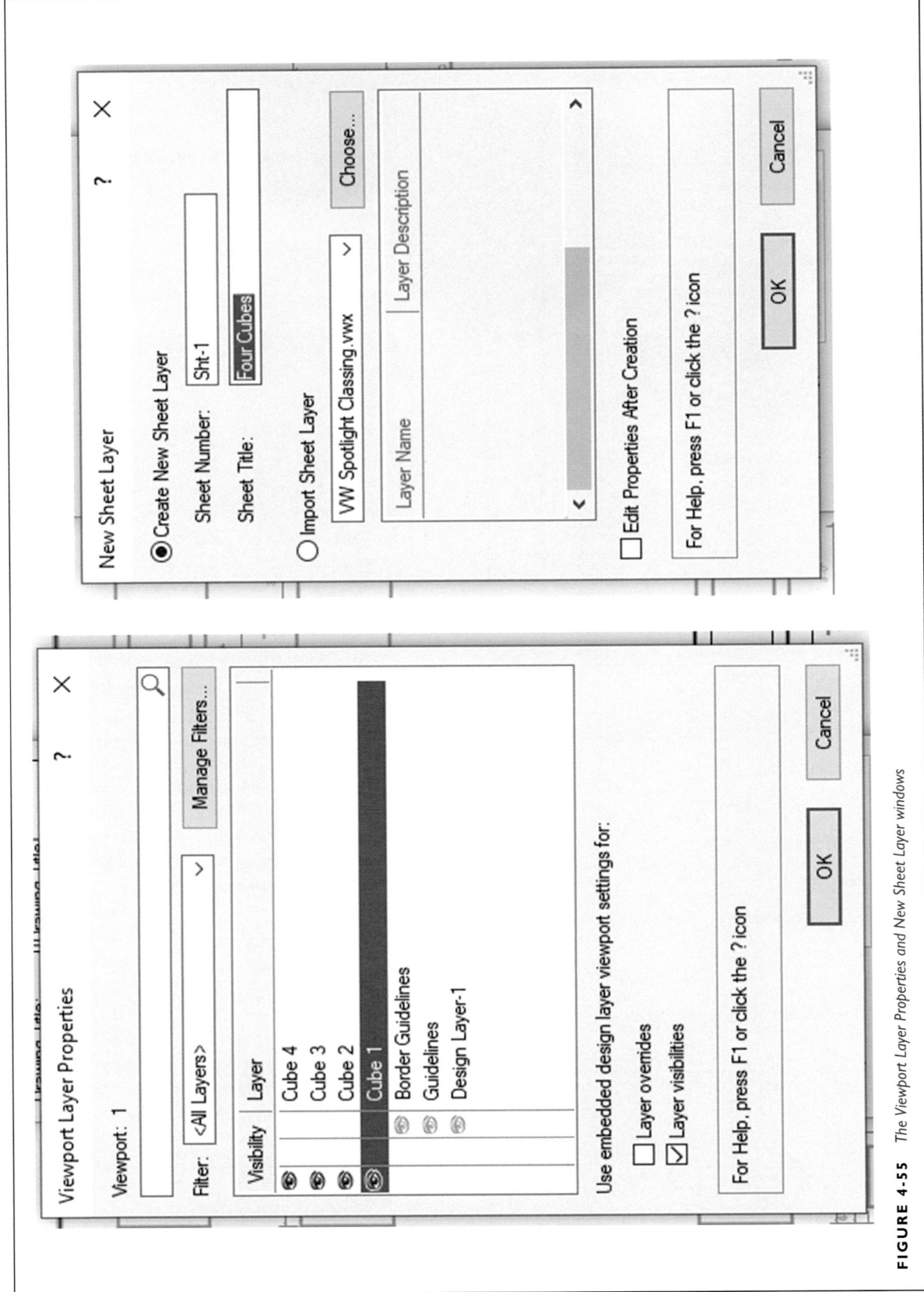

FIGURE 4-55 *The Viewport Layer Properties and New Sheet Layer windows*

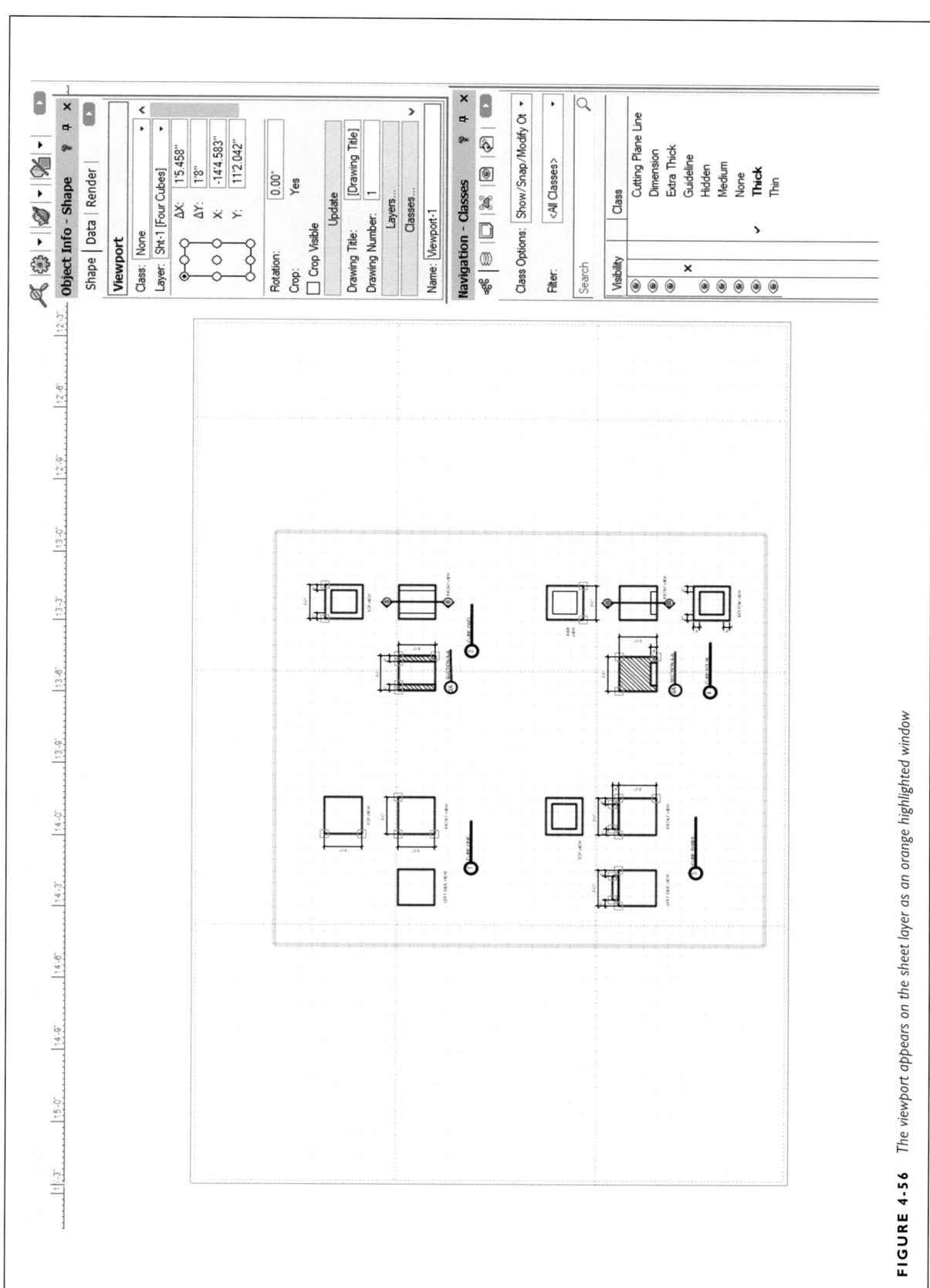

FIGURE 4-56 *The viewport appears on the sheet layer as an orange highlighted window*

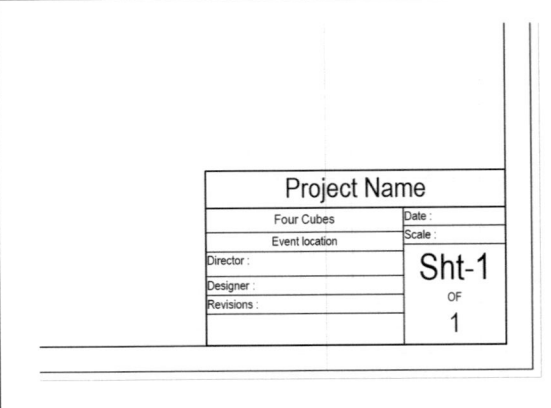

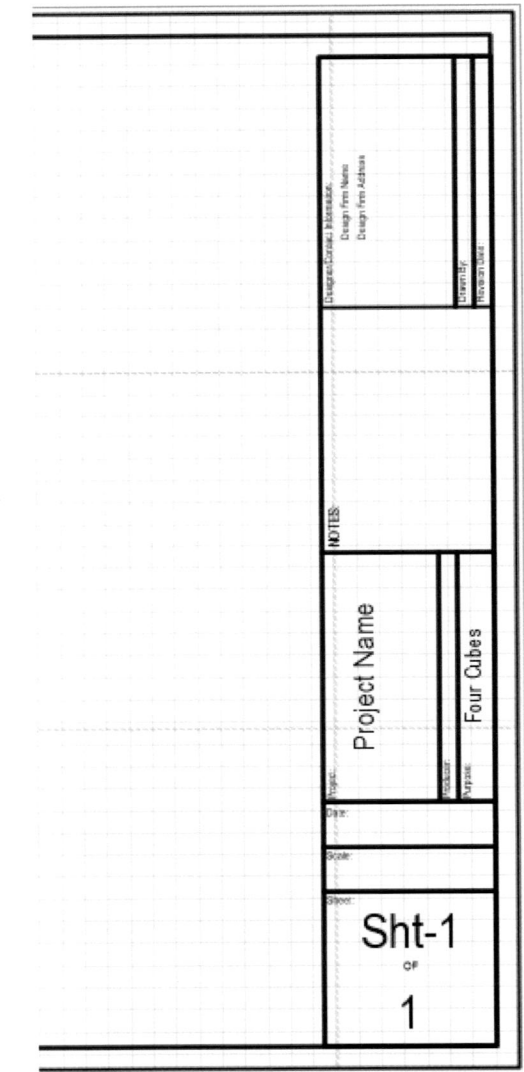

FIGURE 4-57   *Two USITT recommended title blocks in the Vectorworks title block library: corner and right border strip*

Title blocks are placed in the same location on every sheet of a series. When rolled or folded, the title block should end up on the outside so that a drawing can be instantly identified without unrolling or unfolding it.

CAD makes it easier for a draftsperson to create personalized title blocks, which can be useful when branding oneself or one's business. It has also become common practice to import a project's logo into the title block. Always remember that clarity takes precedence over aesthetics.

31. In the sheet layer, use the Border tool to place a border using the extra-thick visible line class (.70 mm).

32. With the border highlighted and active, click on the Title Block Border Settings button found about halfway down the navigation window. The settings window will appear. If necessary, in left-hand menu, choose Sheet/Zones Grid and reset the number of Zones to 0. Change the left border margin to 1″ (Sheet Border). Open the Title Block option (Figure 4-58).

    Click on the button that reads Import Title Block Graphics. If USITT Title Block Corner is not immediately offered as a selection, open the drop-down menu for Use Style at the top of the window. This opens the libraries. Open the Spotlight Title Blocks USITT selection. Double-click on the USITT corner title block. It appears on the Preview side of the window. It appears on the drawing when you OK your way back to it (Figure 4-59).

33. The Title Block Border Settings menu offers a number of choices for placing information into the title block. For this tutorial, most of the needed information can be added via Project Data. Fill in information for Project Date, Project Name, Project Title, Venue, and Designed By. The scale for the sheet is entered via Sheet Data. OK back to the drawing.

34. **Save**. The sheet layer is now part of the overall drawing file and can be accessed, along with all the design layers, via the layer navigation option on the upper toolbar.

35. While in the sheet layer, export the drawing as a PDF. Once saved, the PDF can be printed full-sized on a plotter that accepts Arch D paper. If a plotter is not available, it can be printed as full-sized Letter tiles on a standard office printer and taped together, though information can be lost at the edges and seams. Use Shrink to Fit to reduce the drawing onto a letter-sized sheet. If shrunk to fit, make a note somewhere on the page that it is Not Printed to Scale.

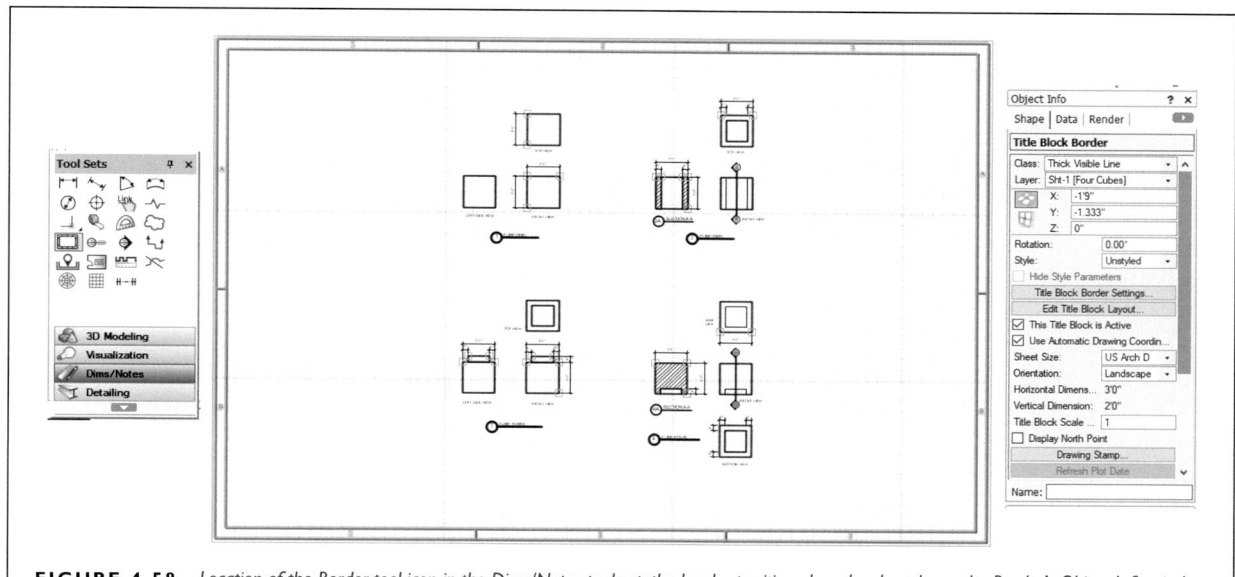

FIGURE 4-58 *Location of the Border tool icon in the Dims/Notes tool set; the border positioned on the sheet layer; the Border's Object Info window*

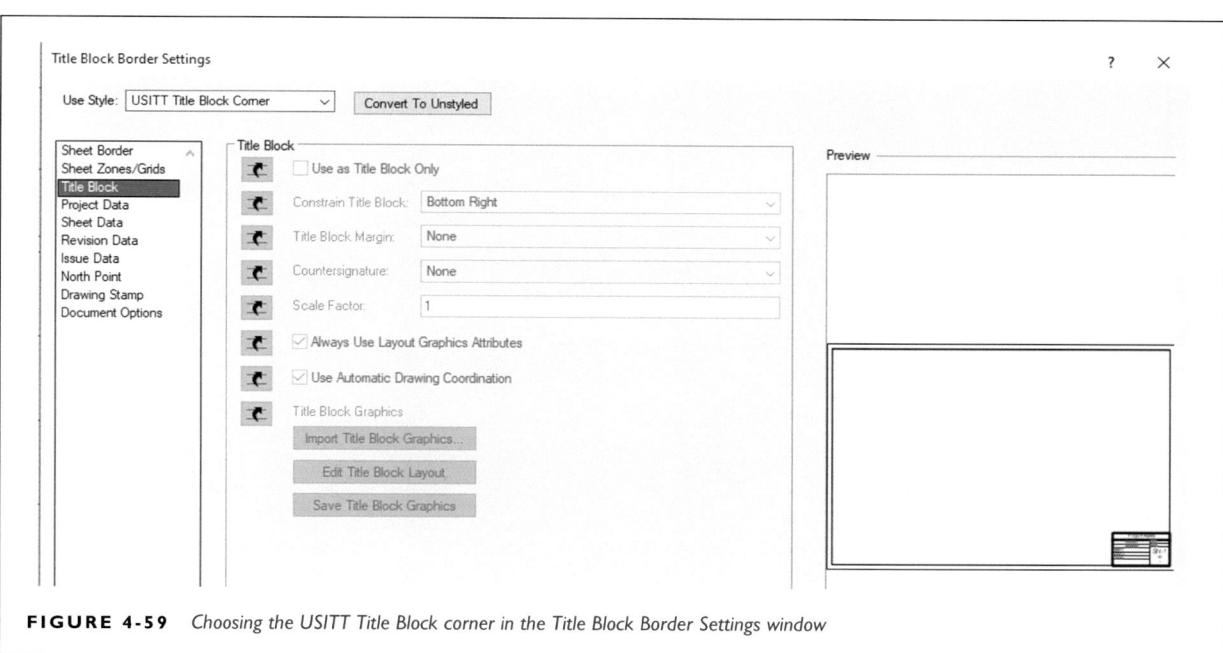

FIGURE 4-59 *Choosing the USITT Title Block corner in the Title Block Border Settings window*

NOTES

1. Frederick E. Giesecke, Alva Mitchell, and Henry Cecil Spencer, *Technical Drawing*, 4th ed. (New York, NY: The MacMillan Company, 1958), pp. 299–300.
2. Kevin Lee Allen, *Vectorworks for Entertainment Design: Using Vectorworks to Design and Document Scenery, Lighting, and Sound* (New York, NY and London: Focal Press, 2015), pp. 87–88.
3. Dennis Dorn and Mark Shanda, *Drafting for the Theatre* (Carbondale, IL: Southern Illinois University Press, 1992), p. 69.
4. Patricia Woodbridge and Hal Tine, *Designer Drafting and Visualizing for the Entertainment World*, 2nd ed. (New York, NY and London: Focal Press, 2013), p. 50.
5. Lynn Pecktal, *Designing and Painting for the Theatre*, (New York, NY: Holt, Rinehart, and Winston, 1975), p. 141.
6. Pecktal, p. 141.
7. Woodbridge and Tine, p. 123.
8. USITT Education Commission, *Graphic Recommended Best Practices*, (Syracuse, NY: United States Institution for Theatre Technology, 2020), pp. 10–11.
9. www.usa829.org, accessed 12/15/20
10. www.iatse.net/about-iatse, accessed 12/15/20

CHAPTER 5
THE SYSTEM OF WORKING

INDIVIDUAL DRAWINGS

When hand drafting there is a traditional order to a drawing's phase. The exercises in Chapters 1, 2, 3, and 4 have largely adhered to these steps, but it's time to explicitly review this system of working. Thomas E. French and Carl L. Svensen, in *Mechanical Drawing*, offer the following steps as a process framework that can be applied while drafting either by hand or by computer-aided design (CAD): "A systematic method of working should be followed to ensure accuracy and understanding. All views should be carried along together. Do not attempt to finish one view before starting the others… The following order is suggested:

1. Consider the characteristic view

2. Determine the number of views

3. Locate the views

4. Block in the views with light, thin lines

5. Lay [out] the principal measurements

6. Draw the principal lines

7. Lay [out] the measurements for details

8. Draw the circles and arcs

9. Draw any additional lines needed to complete the views

10. Brighten the lines where necessary to make them sharp and black of the proper thickness."[1]

For students new to drafting and design, do your best to be as familiar with the object(s) to be drafted beforehand by doing research and generating sketches. Plan the overall drawing layout before committing to drafting. Rough placement of dimensions, labels, and notes is a part of this preliminary work. For lighting design, a rough plot should be laid out in pencil before the final plot is drafted. For sound and video, signal paths can be mapped out in preliminary pencil sketches. Active design work should be kept to a minimum so that drafting can focus on graphic description and its problems, rather than design problem-solving. As you become more proficient in drafting and visualization, as well as more assured as a designer and/or technician, more design decisions will happen as you draft. Accomplished designers may sit down at the computer and jump right into design work; they're not detouring around process and planning, but they have become experienced enough that design and drafting can indeed happen simultaneously.

Keep an array of pertinent doodles, images, and sketches in your workspace. Hiding these visuals in computer files and tabs or across various devices reduces their accessibility. Print things. Pin them up. Import them directly into the Vectorworks drawing currently in process. If you're committed to a totally digital process, multiple monitors may be useful in keeping sketches and research images ready at hand.

As you draft, new things *will* be discovered. Details will be refined and measurements finalized. But without a plan in the first place, no revision of that plan can occur.

The planning phase includes these considerations:

• Which objects will be included on a printed page?

• Which views of the object(s) will be necessary?

• What scale is to be used for drafting or for printing?

• What size paper is needed to accommodate the drawing in the selected scale?

• Which dimensions, labels, and notes will be required to provide enough information for the reader?

DOI: 10.4324/9781003154921-6

Consider any drawing as a whole rather than as a series of fragments that are attacked and completed one by one. CAD, with its infinite drawing space, can encourage fragmentary thinking since disparate bits and pieces can be stitched together at a later phase to create the appearance of a unified drawing. Don't draw orthographic views independently and then use viewports to arrange them on a sheet layer. Drawing unified orthographic views helps the draftsperson avoid error by considering the entire object rather than just one view of it.

On the other hand, CAD's infinite drawing space means that you can draw very large items, such as a scenic groundplan, and then use it to draw the set's front elevation and center-line section in their correct orthographic placements. Vectorworks' viewports allow you to then transfer each view to its own sheet layer. Doing this by hand requires the translucency of vellum and an extra-large drawing table; it's not nearly as convenient.

After the planning phase, drafting begins by using guidelines to lay out the elements of a drawing as an integrated whole. Errors can be more easily detected and corrected in this early stage. In Vectorworks, keeping guidelines in their own design layer allows easy deletion and clean-up without affecting other portions of the drawing.

When drafting by hand or using the Vectorworks reference grid to aid page layout, lay out borders to establish the boundaries of the available drawing space. Remember, the drawing space is not the entire sheet of paper. Lay out the title block, as this is a significant area of real estate.

Layout begins with the big picture and works toward the details:

1. Overall array perimeters and shared baselines
2. Object view array perimeters
3. Object view outlines
4. Center lines for circles and arcs
5. Object outline features and details
6. Object interior details and features
7. Dimensioning layout
8. Placement of titles, labels, and notes
9. Review for omissions, mistakes, and revisions

Once the drawing is laid out, erase or delete extraneous guidelines that may cause confusion regarding what is to be darkened. The following is a series of steps for hand drafting, but still applicable to CAD work.

1. Darken object arcs and circles.
2. Darken freehand object lines.

3. Darken object straight lines.
4. Darken crosshatching.
5. Darken center lines (unless it's a full-drawing center line, as found on a groundplan).
6. Darken dimensioning and cutting plane lines.
7. Add notes, labels, and titles.
8. Darken the border and title block.
9. Add title block information
10. Review for omissions, mistakes, and late revisions.

While hand drafting, visible lines are darkened at the end of the process. To minimize smudging, begin by darkening horizontal lines at the top of the page. Once the bottom of the page is reached, begin to darken vertical lines from left to right. With CAD, smudging is not an issue and it may be useful to darken each object's array of views as it is completed.

The Drawing Package

Each craft field requires a different package of drawings. Film and television requirements are a bit different from theatrical ones. Trade shows, corporate events, museums, and photo shoots all have their own requirements. Regardless of the field, there is a commonality of packages moving from big to small, from overview to detail. It's not that a television studio requires a unique style of light plot; it's that certain types of lighting instruments standard in television are different from the ones found in most theaters. If you know what lighting unit is represented by what symbol, you can still read the plot and hang the instruments.

While the system outlined above pertains to a single drawing, most projects are described through a series of drawings. Knowing what drawings are to be part of the **package** is also part of a systematized work process.

The first drawings of a package are typically overviews of the whole object/project and include the groundplan and center line section and might include an illustration of the project, scenic sketch or rendering, or photos of a model. Images of virtual models largely replaced flat graphic representations such as painted renderings.

A typical package for a theatrical scenic design contains, in order (Figure 5-1):[2]

1. Perspective rendering/illustration of the project
2. Groundplans/staging plans
3. Center line vertical section
4. Platforming
5. Elevations and details of individual elements

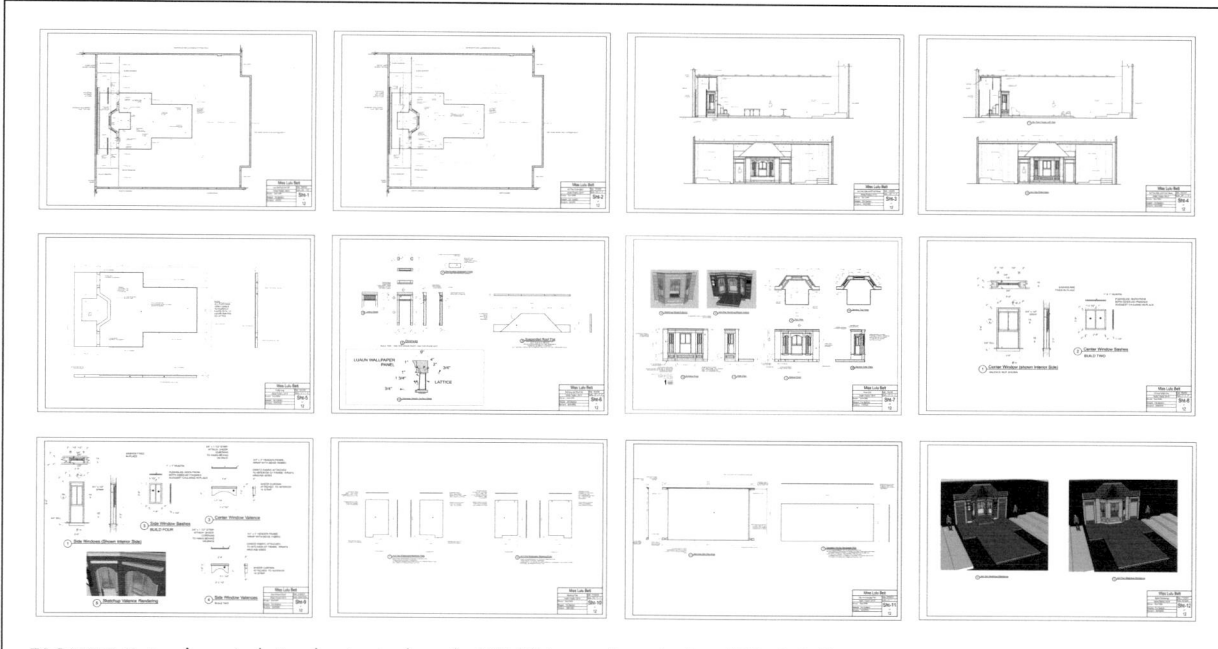

FIGURE 5-1   *A scenic design drawing package for UW-Whitewater's production of* Miss Lulu Bett

6. Other references and examples such as models, technical specifications, and research/reference images; may include drawings of props and set dressing

7. Paint elevations/treatments

Thomas Umfrid, a professor of scenic design at the Cincinnati Conservatory of Music, gives this overview of how he arranges his typical design package: "Plate 1 is always the groundplan. Use consistent labeling system. Plate 2 is the center line section. For multi-scene/set shows I may do multiple plates; it depends on the level of complexity. Then the rest of the drawings are the scenic units. On each plate of a specific scenic piece I sometimes do a ¼″ scale composed front view showing all of the parts assembled, with labels, and then the balance of the drawings are the individual scenic parts laid out in a usually larger scale. I will add detail reference clips, web links, etc. as needed. Old school 'set renderings' are out of fashion now. Fancy 3D digital renderings are in, but rarely show up in designer drawings. Sometimes I may insert a SketchUp 3D study or an isometric to explain 3D."[3]

Sheet, **page**, and **plate** are all terms used to refer to a single sheet of paper and are generally interchangeable. Draftspersons use one or the other depending on their craft field, industry antecedents, and era of initial training. If you're hired into a shop or design office, be sure to review the company's preferred nomenclature.

Patricia Woodbridge and Hal Tine, in *Designer Drafting and Visualization*, include this list of drawings for a film production's package:[4]

1. The studio plan (placement of all sets within the studio)

2. A director plan, showing each set with dressing/décor, reduced for portability by the director

3. A technical plan (master schematic for a set)

4. A reflected ceiling plan, showing ceiling architecture (useful for camera placement)

5. Flooring plan, showing floor treatments

6. Wall elevations

7. Built units such as doors, windows, and stairs

8. Details, explaining construction of important or typical areas of the set

9. Details for cornice and other molding

10. Built backings (things seen outside windows and doors)

A technical director, working from the scenic design package, often creates their construction drawings in the order by which they intend to build, starting with foundational structures such as platforming, stair units, and deck elements. The order may also be influenced by the build schedule and how much time and labor are required to build various elements, as well as the shop labor's level of expertise. The technical director will likely need to complete all construction drawings before a total materials budget can be estimated.

When the shop requires the package to be submitted as a whole, the production schedule must be arranged to allow for this requirement. Directors must be prepared to begin production discussions many months in advance and understand that there is a point where design must freeze so that construction can begin. If time or other factors necessitate a progressive submission of drawings, the designer and the shop must be in constant communication, with the drawings arriving in an order that facilitates a smooth construction schedule.

The package of lighting design drawings begins with the light plot, followed by a center line section of the venue, and then any required detail plates for ladders, booms, and other lighting positions that the main plot sheet(s) could not accommodate. A designer new to a venue should always request a set of base drawings, which should include a floorplan, section, locations of all hanging positions, and circuiting information. Sound design drawings include plans and sections for loudspeaker placement, as well as the location of any sound equipment integrated into the scenic environment. Signal flow diagrams map analog and digital information routed through the equipment. A rack drawing details the equipment to be placed in the system's rack, as well as the equipment's connections. Projection design requires a similar set of drawings.

There are many reference sources regarding materials, construction techniques, and standards. ASME, ANSI, ISO, and USITT all have websites. Remember that the Internet itself is not a reputable source, nor are search engines. If the website offering information is not an authoritative mainstream organization, the information may be suspect. A printed reference mainstay is the *Backstage Handbook*, by Paul Carter. It's chock-full of images, standards, formulas, and terms that can help you visualize and understand what you are about to draw.

Technicians receiving a package of drawings should remember that the draftsperson has done their best to explain the material as completely as possible on the page. When asked to explain a drawing, the draftsperson is likely to verbally review the information they believe to be there without knowing what elements the reader does not understand and therefore cannot address. Examine the drawings and prepare a list of specific questions before reviewing the package with the designer or draftsperson.

Discussion: The Drawing Package

Instructor

Bring in a package of drawings from a production or project and talk through the series with the students. If you do not have drawings on hand, reach out to peers through USITT or the Kennedy Center American College Theater Festival (KCACTF.org) as well as area colleges and universities housing theater, television, and/or film production departments. Drawings are intellectual property, so designers and venues may be wary of lending copies. Don't post drawings into online repositories without permission; digital documents wander easily. The following list of questions can help guide discussion:

- How easy is it to maneuver through the drawings?

- In what order are the drawings placed? Does it seem logical?

- Does information on the first few drawings serve as introduction to details in later drawings?

- Is information indexed/coded? What coding method was used?

- What information is found in the title blocks?

- Is labeling clear and consistent throughout the package?

- Considering what has been discussed in the previous chapters, what aspects of the drawings remain unfamiliar or confusing?

- If this package was delivered to you, what might be your first questions at the next production meeting?

NOTES

1. Thomas E. French and Carl L. Svensen, *Mechanical Drawing*, 7th ed. (New York, NY: McGraw-Hill, 1968), p. 76.
2. John Blurton, *Scenery: Drafting and Construction for Theatres, Museums, Exhibitions, and Trade Shows* (New York, NY: Theatre Arts, 2001), pp. 23–24; Rich Rose, *Drafting Scenery for Theatre, Film, and Television* (Cincinnati, OH: Betterway Books, 1990), p. 117.
3. Thomas Umfrid, Email Correspondence, 2020.
4. Patricia Woodbridge and Hal Tine, *Designer Drafting and Visualizing for the Entertainment World*, 2nd ed. (New York, NY and London: Focal Press, 2013), pp. 337–350.

CHAPTER 6

DRAFTING FLATS AND PLATFORMS

This chapter puts into practice principles and techniques discussed in the previous chapters by drafting traditional scenic elements such as flats and platforms. In keeping with the idea that a draftsperson should be familiar with what is to be drafted, modules begin with a review of components and common construction techniques. Flats and platforms are introduced before more global drawings such as groundplans so that the students can better understand the makeup of these smaller, contributory elements before their incorporation into more complex structures.

TOPICS AND GOALS

- Construction techniques for basic scenic elements
- Translation of construction goals into graphic language
- Construction vocabulary
- Drawing flats
- Drawing flats with doors and molding
- Drawing platformed areas

MODULE 1: FLATS

Learning the name of a thing is the first step toward being able to communicate information about it. It's the difference between "please put the thing on the thing" and "please put the toggle on that Hollywood style flat."

Whether you are a designer, technician, director, or producer, having a grasp of construction vocabulary and a familiarity with common shop processes enables you to more easily interpret the lines of a drawing—even if you already have a firm grasp of drafting's graphic language. The label identifies a

shape, for instance, as a flat. Knowing how a flat is constructed leads to a more instant understanding of which lines represent surfaces and objects and which lines represent openings or enclose white space. Knowing how to merge construction knowledge of common objects with their representation on the page means that the reader is better situated to correctly interpret representations of unusual and unique objects.

American scenic designer Howard Bay in his text *Stage Design* insists that designers have a grasp of how their visions are to be constructed if they wish the end product to match their intention: "Before the layout drawings (which break down the scenery into manageable segments and specify all construction details) are distributed you must check to avoid obvious cracks. You discuss butt joints or lap joints, the spacing of visible cables and how to disguise visible braces. Certain members may need beefing up for support or rigidity—the shop knows best, so do a little redrafting… The standard flatwork and the platform bulk need to not take up your valuable time, but you should hover over the trim, both wood and metal, and all sculptured pieces. Metalwork, which plays a leading role these days both structurally and as skeletal tracery, must be watched; bent tubing can appear particularly graceless if it isn't the right gauge or if a full-size pattern has not been provided. If you have left the choice of special hardware until now, you may settle for second best; but if you keep an up-to-date file of molding, hardware, and decorative trim catalogs the right numbers would be included in your drawings."[1]

What is a Flat?

A **flat** is a frame covered by a lightweight material such as fabric or thin plywood. They are used to create vertical surfaces, such as walls, in a scenic environment. The typical

DOI: 10.4324/9781003154921-7

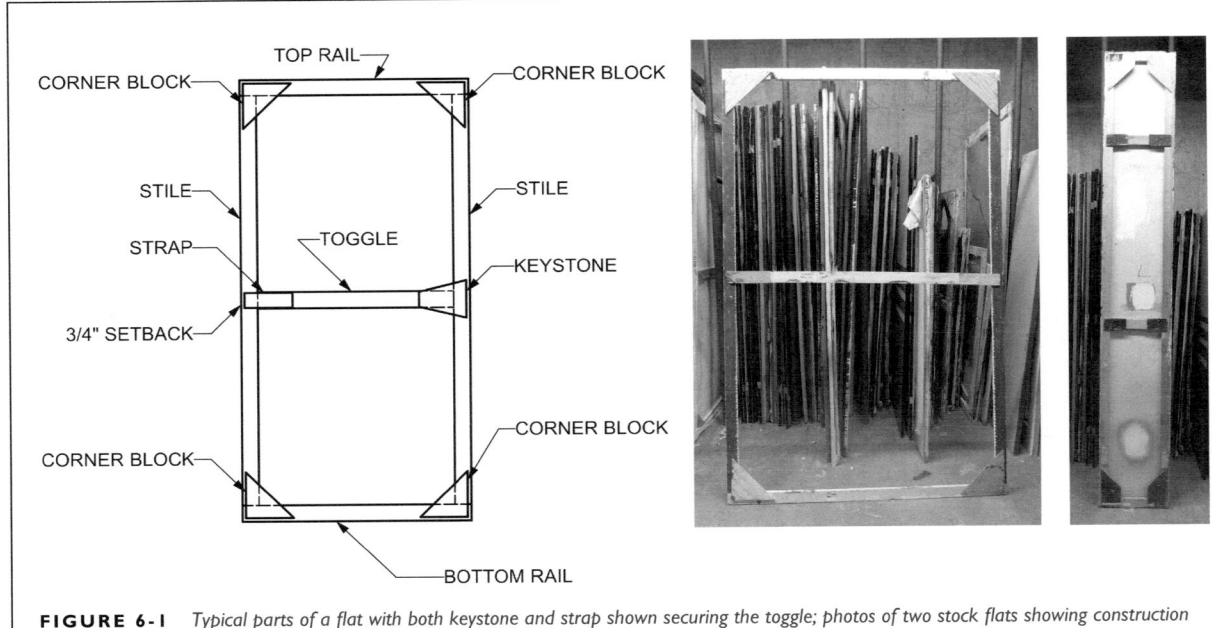

FIGURE 6-1 *Typical parts of a flat with both keystone and strap shown securing the toggle; photos of two stock flats showing construction*

fabric used is **muslin**, a cotton cloth. The thin plywood most often used is **lauan**, named for a family of hardwoods from which its veneer originates. One-quarter-inch thick lauan is usually preferred. While a flat can be covered with any material from silk to plastic to drywall, muslin and lauan are lightweight, cost-effective, paint-friendly surfaces. While many larger shops have moved to hollow square steel tubing as the framing material, wood remains the basic go-to material.

Parts of a Flat

Flats have standard named parts, regardless of the framing material (Figure 6-1). The horizontal top and bottom pieces are **rails**. The vertical side pieces are **stiles**. Stiles sit between the rails so that when the flat is dragged along the floor, stress is taken by a single piece of wood (the rail) rather than the joint between rail and stile. Horizontal members within the flat are **toggles**. Toggles contribute to structural stability and keep stiles parallel to each other. Together with interior stiles, they frame out features like windows and doors. Toggles are placed three to four feet apart.

The drawing in Figure 6-1 depicting the parts of a flat is a **rear elevation.** The framing as well as the straps and corner blocks cannot be clearly described in a front elevation, which makes the rear elevation the standard construction drawing view for flats.

Other framing members include 45° diagonal **corner braces** that prevent the flat from parallelogramming. Braces should be included on flats wider than 4′–0″.[2] **Sweeps** create curved openings and profiles (Figure 6-2).

Two Common Styles of Flat Construction

There are two basic styles of wooden flat construction. The older style, in which framing lumber is laid flat, originates in theatrical practice. It goes by several names,

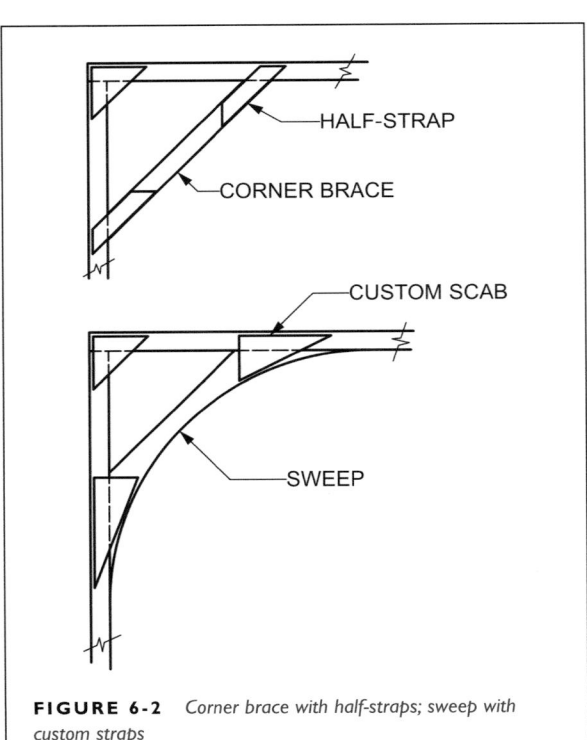

FIGURE 6-2 *Corner brace with half-straps; sweep with custom straps*

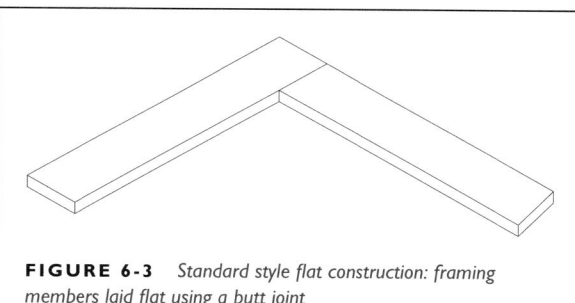

FIGURE 6-3 *Standard style flat construction: framing members laid flat using a butt joint*

including **traditional**, **standard**, **stage**, and **Broadway** (Figure 6-3). They are typically soft-covered with muslin.

Standard style flats use **butt joints**: one piece of lumber simply butting against another. Because driving a fastener (nail, staple, or screw) through the width of the plank is difficult, butt joints are secured with **scabs**.

Scabs for flats are typically pieces of ¼″ plywood (not lauan) laid atop a joint and then glued and fastened in place. Scabs have various names depending on their shape and placement. Triangular **corner blocks** secure the corners of flats. Rectangular **straps** secure toggles to other framing members. Trapezoidal **keystones** do the same job as straps. Narrow trapezoidal **half-straps** secure angled braces to vertical and horizontal members (Figure 6-4).

Scabs are set back from the edge of the frame, allowing another frame to butt tightly against it with no gaps. The setback is the thickness of the framing lumber, typically ¾″. Scabs are placed on the rear of the flat, otherwise they interfere with the material covering the front.

The other flat construction style came into being when film and television required walls with increased stability and less need for portability. This style places its lumber on edge and is variously called a **studio**, **film**,

Hollywood, or **television** flat. It is typically hard-covered with lauan. Because the lumber is on edge, fasteners can be driven through the thickness of the wood; no scabs are required (Figure 6-5). Some sources call any hard-covered flat Hollywood style and any soft-covered flat a stage flat, regardless of orientation of the framing members.

When flats exceed the size of a standard sheet of lauan (4′–0″ × 8′–0″), there will necessarily be joints between multiple cover sheets. To support sheet edges and create a clean joint, additional pieces of wood oriented flat against the face of the flat may be added in either the stile or toggle orientation (Figure 6-6).

Lumber Size: Nominal and Actual

Lumber sizes are broken into two categories: **nominal** and **actual.** Nominal refers to the rough cut of the plank before it is milled smooth. The familiar two by four (2 × 4) is a nominal size name, as its *actual* measurements are 1½″ × 3½″.

The overall dimensions of a flat are whatever the designer requests—it's when the carpenter measures out framing members that nominal versus actual comes into play. 1 × 3 is a nominal name; it is not written with the prime marks that denote inches. The actual depth and height of a 1 × 3 is ¾″ × 2½″. This is where the ¾″ setback for corner blocks and straps is derived. 1 × 3 lumber is different from 1″ × 3″ lumber.

If the overall height of a flat is 8′–0″ and the drawing requests that 1 × 3 be used, the results will be different if the carpenter does construction math with the nominal size instead of the actual size. The stiles should be 7′–7″ long; if the carpenter does not convert to actual sizes, they'll end up 7′–6″ long, resulting in flats that are too short.

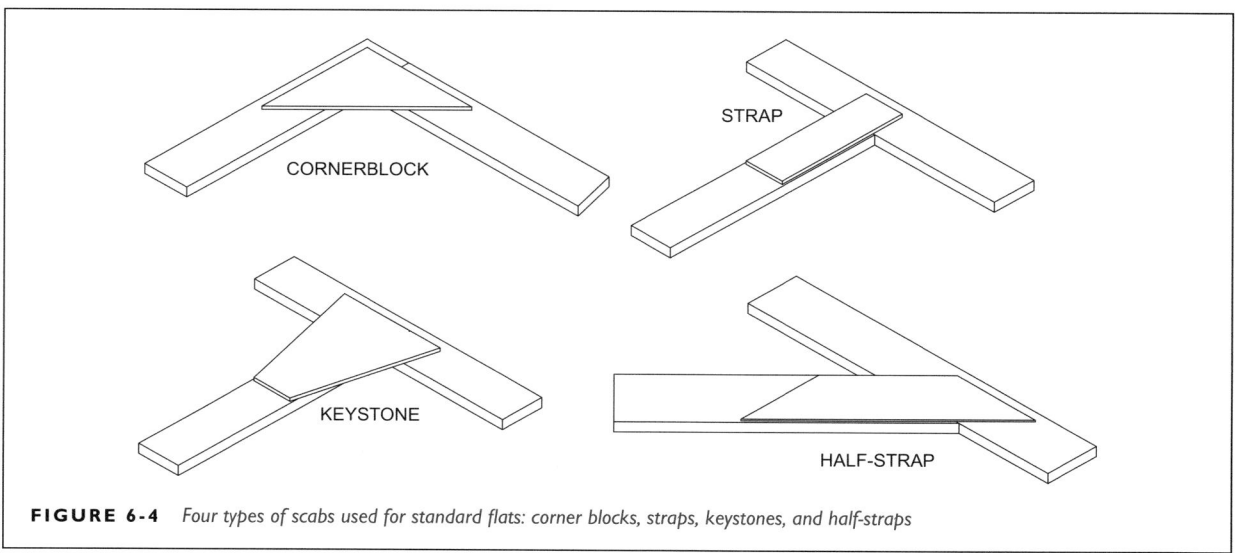

FIGURE 6-4 *Four types of scabs used for standard flats: corner blocks, straps, keystones, and half-straps*

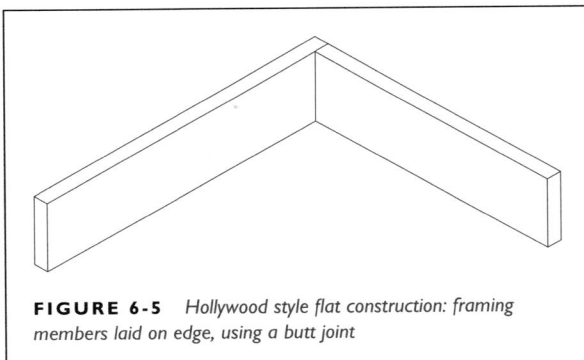

FIGURE 6-5 *Hollywood style flat construction: framing members laid on edge, using a butt joint*

Exercise: Drafting a Standard Style Rectangular Flat

Instructor

The front and rear elevations of a Standard flat will be drafted. The first iteration is hand drafted. The second is drafted in Vectorworks, with instructions following the hand drafting steps. The two drawings will be substantially identical, and use the same layout steps. The students should draft in $1'' = 1'-0''$.

Students

Draft the front elevation of a standard style flat, both by hand and with Vectorworks (Figure 6-7). Draft in $1'' = 1'-0''$, landscape orientation; use layout math to determine the best paper size. The larger scale makes it easier to draft the rear elevation's corner blocks and straps. Include borders and title block. The first set of steps leads you through layout by hand; the second set of steps via Vectorworks.

The flat's overall dimensions are $4'-0'' \times 8'-0''$, using 1 × 3 for its framing members. It will be soft-covered with muslin. A side view is not included as the flat contains no detail features; noting construction method and covering material is sufficient. The sketch shows straps used to secure the toggle; keystones may be substituted.

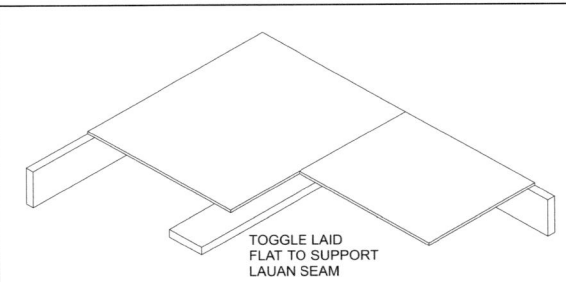

FIGURE 6-6 *Using a flat toggle to support a lauan seam on a Hollywood style flat*

TOGGLE LAID
FLAT TO SUPPORT
LAUAN SEAM

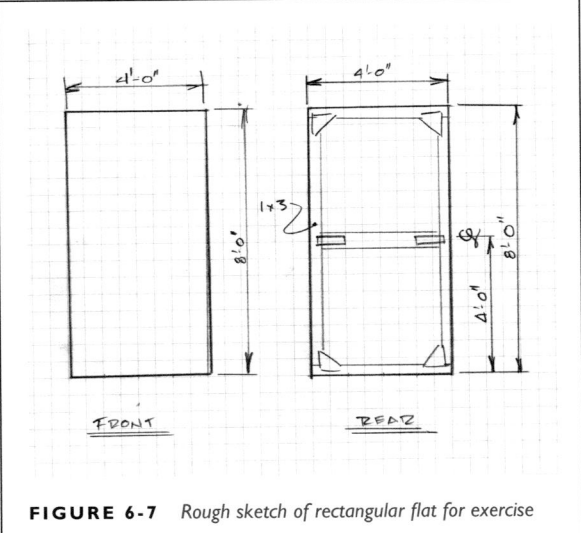

FRONT   REAR

FIGURE 6-7 *Rough sketch of rectangular flat for exercise*

When drafting flats, it's usual to break a few drafting conventions. Corner blocks and other scabs are not typically shown so that framing members can be seen more clearly. A note regarding use of scabs is usually sufficient. If scabs are drawn, hidden lines are not typically used to show framing edges beneath the scabs, as the extra lines in a tight space can obscure the scab itself.[3] In this exercise, draw the scabs without hidden lines.

If this exercise were solely a design drawing, the front elevation alone might suffice as long as there are no depth details such as windows, doors, or cornice that require description. A designer would not typically add a rear elevation detailing the flat's construction; that is generated by the technical director. The designer is more concerned with the appearance of the front of the flat and has not worked out the framing details; therefore, the typical front view does not include hidden lines that refer to the framing.

The **rear elevation** is not a recreation of the front view with the surface lifted away. Recall the orthographic view glass box concept to remember that the rear view, when unfolded and flattened into the drawing plane, is in reverse orientation to the front view. On a rear elevation, features like doors and windows are found on the opposite end of where they appear on the front elevation.

By Hand

1. Do layout math to determine the optimal paper size. Consider dimension placement; measurements are included on the sketch, but are they in their optimal positions (Figure 6-7)?

2. Affix the paper to the drawing table. Lay out the borders and title block.

3. Remember that these two elevations are still related orthographic views. The tops and bottoms share layout guidelines. Lay out the outer perimeters of each elevation.

4. The front elevation is complete, except for dimensions and notes. A note about the covering will be added in the dimensioning phase of the drawing.

5. The framing lumber is 1 × 3 (nominal). Look up the actual size of the lumber, then lay out guidelines within the rear elevation to indicate the interior edges of the rails and stiles.

6. The center line of the toggle is centered along the flat's horizontal axis. Lay out the flat's horizontal axis. From it, measure up one half the width of the lumber to find the upper edge of the toggle. Measure down one half the width of the lumber to find the bottom edge (Figure 6-8).

7. The scabs are set back ¾″ within the flat's outer perimeter. Lay out guidelines to mark the outer edges of the scabs.

8. A corner block is typically 10″ along its perpendicular sides. A strap's width is ¼″ narrower than that of the framing lumber to which it is attached (e.g., if the framing stock is 2½″ wide, the straps are 2¼″ wide).[4] Choose one corner of the rear elevation. Along the scab setback guideline, measure 10″ from the corner in one direction. Use a 45° triangle to add the hypotenuse. Repeat for the other three corners (Figure 6-9).

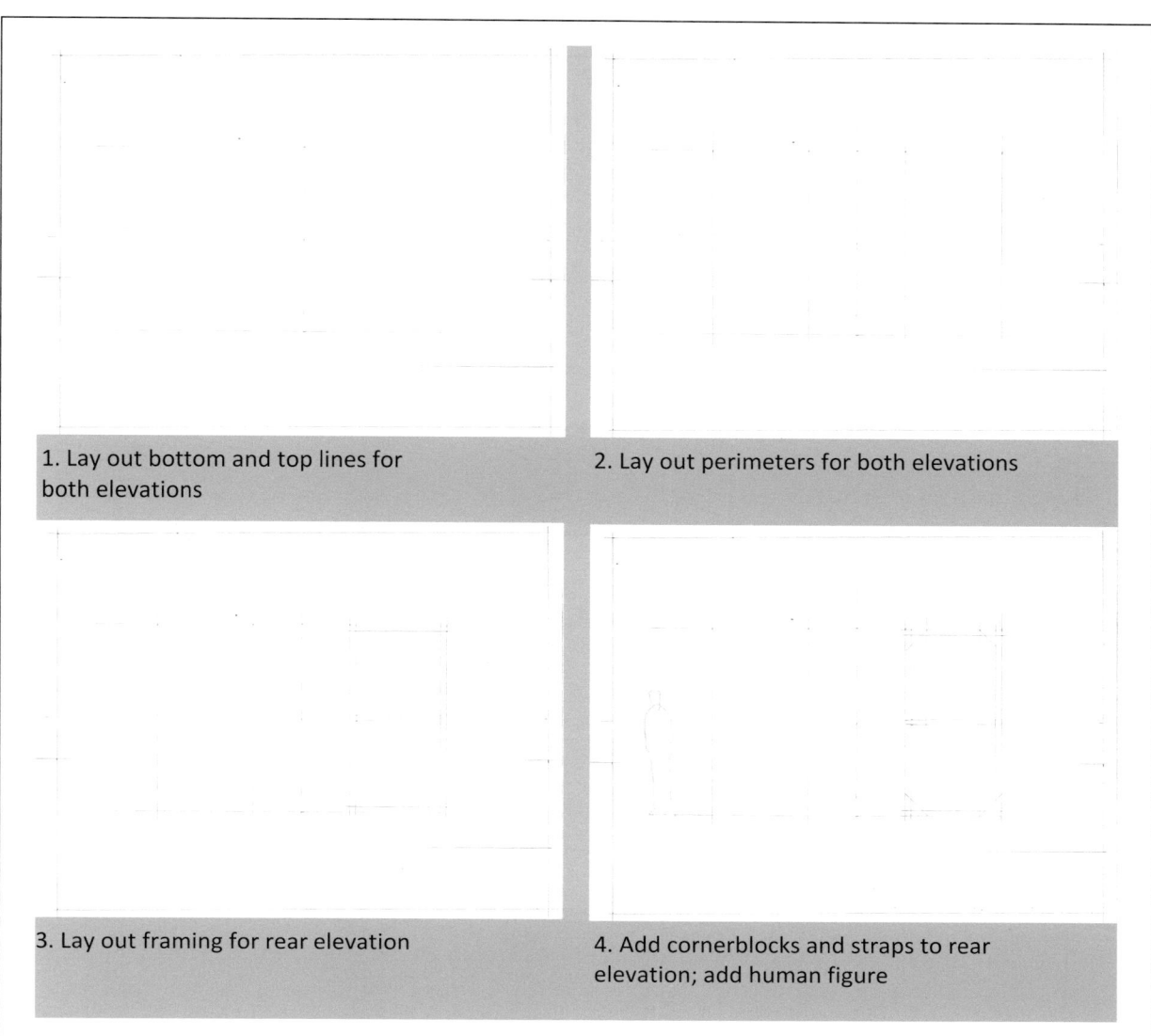

1. Lay out bottom and top lines for both elevations

2. Lay out perimeters for both elevations

3. Lay out framing for rear elevation

4. Add cornerblocks and straps to rear elevation; add human figure

FIGURE 6-8 *Layout steps 1 through 9 (shown on smaller paper): lay out borders, title block, upper and lower baselines; front and rear view perimeters; rear elevation framing members; human figure, corner blocks and straps*

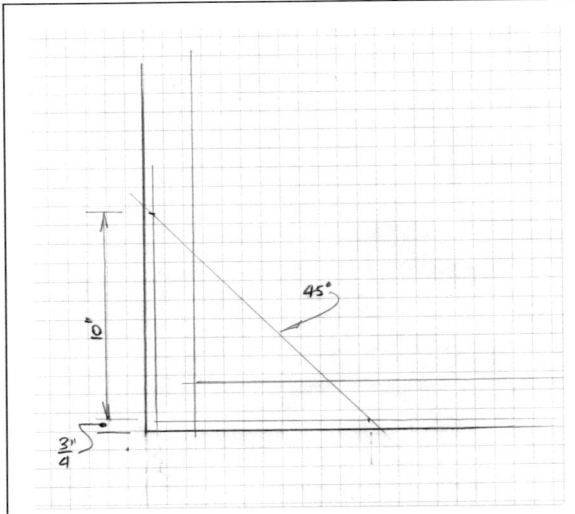

FIGURE 6-9 *Laying out a corner block: edges first; measure length of side; use triangle to draw hypotenuse*

should be included in a notes section. You can also use the abbreviation **TYP**, short for **typical**. When there are multiple identical items, dimension one and include TYP after the measurement to indicate that it applies to all of the copies.

The toggle is centered in the frame and dimensioning to the edges of the toggle may generate a measurement with fractions. Dimensioning to the toggle's center line is, therefore, generally preferred. Either run a center line through the toggle's long axis or leave the usual 1/16″ gap between the extension line's endpoint and the frame's perimeter (Figure 6-10). Both choices are labeled with a CL.

Adding a Human Figure for Proportion

In design drawings (less so for construction drawings), a scaled **human figure** is included to provide instant understanding of the scenic element's size. Be sure that the figure is of an average human height and in the same scale as the drawing. An outline or partial outline is sufficient. If there are completed costume renderings, adding a figure connected to the production can be a nice touch (Figure 6-11).

If you are not confident in your freehand sketching ability, the following techniques are useful:

9. The top and bottom of each strap is inset 1/8″ from the top and bottom of the toggle. Even in 1″ = 1′–0″ this will be difficult to measure accurately; a narrow double line may suffice. Lay out guidelines across the length of the toggle, and then measure in 8″ from the scab setback guidelines. The front and rear elevations are now laid out.

10. Lay out dimensions. Scabs are stock items and typically do not need to be dimensioned, though a size note

a. When hand drafting, photocopy or scan an image to size it to the desired scale. Print, then slide the figure beneath the paper. Align the bottom of its feet with the flat's base line. Lightly trace the figure's outline. Darken with the rest of the drawing.

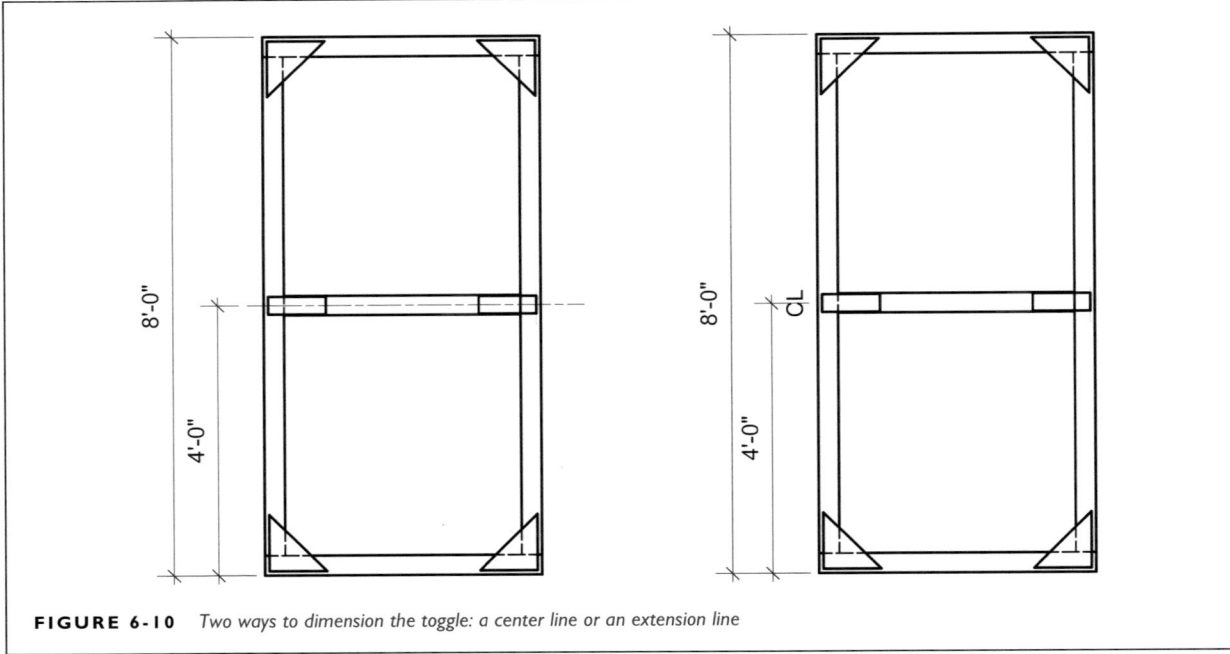

FIGURE 6-10   *Two ways to dimension the toggle: a center line or an extension line*

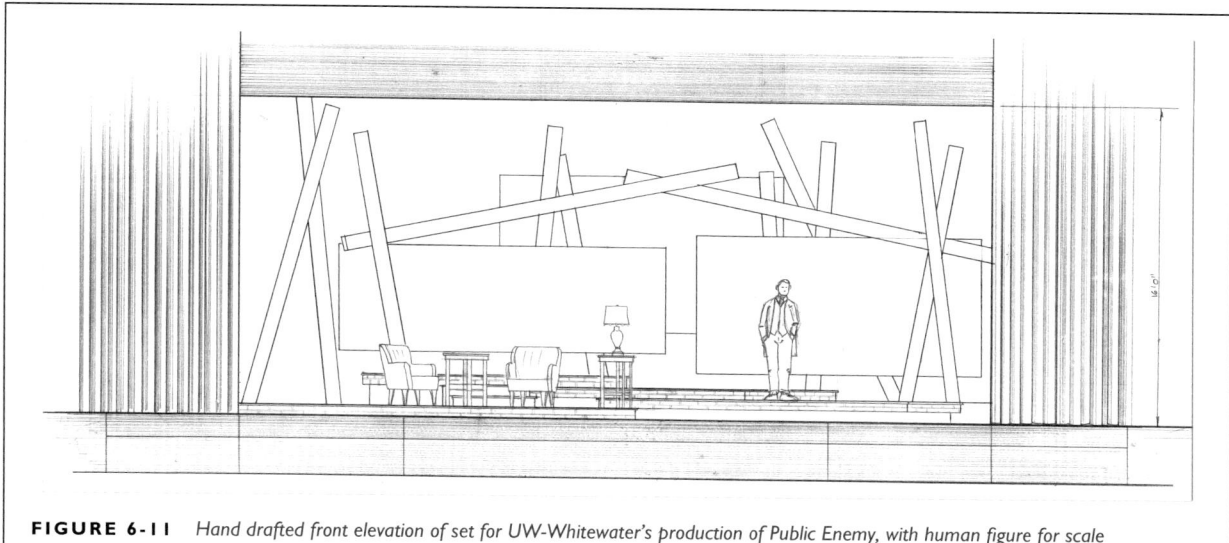

FIGURE 6-11   *Hand drafted front elevation of set for UW-Whitewater's production of Public Enemy, with human figure for scale*

b. For paper with limited transparency, rub pencil graphite over the back of the image's paper. Lay the figure's paper, graphite side down, on the drawing. Trace with a firm hand to transfer the outline to the drawing. Lightly trace over the transfer outline to clean it up. Darken along with the rest of the drawing.

Once a stock human figure is created, scan and save it as a JPEG file. Import it into Vectorworks drawings as needed.

11. Add a human figure to one side of the front elevation. Don't crowd dimensions or other views.

12. Darken the drawing. Remember that hidden lines will not be included in either view.

13. Use a leader line to indicate the flat's soft-covering material: in this case, muslin (Figure 6-12).

14. Label each view. Determine what other, if any, information a reader requires to build this flat. A block

of notes (such as scab information and construction style) may be placed to the side of the drawing above the title block.

15. Fill in the title block information. Use the class name as the project, and Flat A as the drawing title. Darken the title block and borders.

In Vectorworks

1. Open your base file. Use Page Setup to select the desired paper size.

2. Consider design layers and create the desired number. Name them appropriately. It may be useful to import the human figure into its own layer. Set the scale for All Layers.

3. In a border guideline layer, lay out the border and the title block's outline.

4. In a guideline layer, lay out the perimeters of the views.

5. On the rear elevation, lay out guidelines for the interior edges of the framing members, including the toggle. Include a horizontal axis guideline to help with dimensioning.

6. Lay out the scab setback distance.

7. Lay out the corner blocks. Lay out the straps.

8. In the chosen design layers, darken the drawing.

9. Add dimensioning and notes. Add labels.

10. Place the human figure. In Vectorworks, a JPEG or PNG file of the figure can be imported. Be sure the image is already trimmed down as closely as possible to the figure's outline. In the File drop-down menu, open

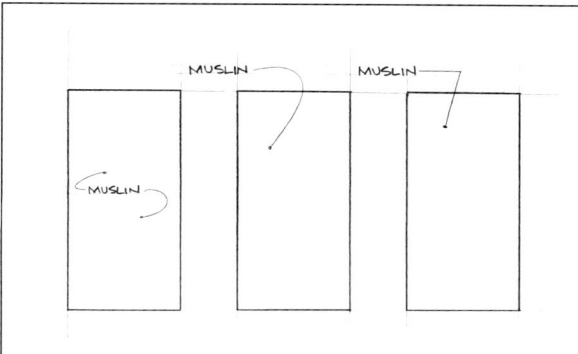

FIGURE 6-12   *Three different leader line styles for labeling a surface's material*

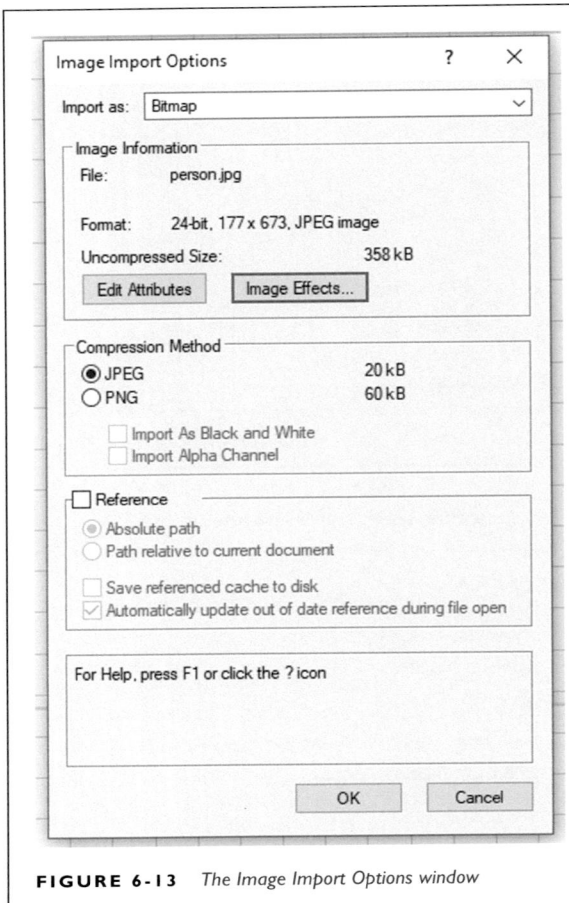

FIGURE 6-13   *The Image Import Options window*

Import, and then Import Image File. Find the image in your files, and double-click. The Image Import Options window appears, featuring buttons that allow you to edit the image before pasting it into the drawing (Figure 6-13).

Edit Attributes button allows you to select color and change the dimensions of the object. Image Effects button offers you a range of settings, including Contrast, Saturation, and Soft Edges. Import an image and play with the settings to see what they do. Click on Apply to save the changes, then OK back to the drawing. If you are unhappy with the look of the figure, trace the figure's outline in a new layer and delete the imported image (Figure 6-14).

16. Once all dimensions, labels, and notes are complete, turn off the guideline class's visibility. Drag a rectangle around the drawing and create a viewport. Be sure to click on the Layers button in the Create Viewport window to select which layers will be visible on the sheet layer. Select 1″ = 1′–0″ for the viewport scale. Leave the Viewport Drawing Label unchecked to prevent an additional label from being generated.

17. Center the drawing on the sheet layer. Add a border and title block.

18. Edit the border as needed (1″ left margin, zero number of zones, etc.). Fill in the title block information.

19. Review for completion. If something is missing, return to the design layer to make revisions. Changes will automatically appear in the viewport as long as they are done within the viewport's crop area.

20. Save the file. Export and save as a PDF. Print the PDF.

Exercise: Modeling a Flat in SketchUp

There are two main rationales behind modeling: experimenting and presenting. Experimental models generate information used for later drafting and presentational modeling phases. Presentational models are a representation of the finished product using finalized measurements. Traditional white models are sometimes built by cutting up copies of finished production drawings, pasting the parts onto stiff paper, and then assembling them.

When modeling a flat, the construction approach reflects the modeler's intentions. Is it a quick stand-in wall, or meant to provide a true pictorial representation of the final product? Is the flat to be part of a more extensive scenic environment? Will it fit together with other flats? Building a scenic model where all the walls are paper-thin planes may be fast but not useful as a reference for measurements, nor will it look like the intended final product unless all the walls are meant to be sheets of paper or fabric.

In this exercise, three different modeling approaches are explored.

Approach #1

This approach produces a ¾″ thick slab, with no features on the front or rear (Figure 6-15).

1. Open SketchUp. Select the Feet and Inches template. Move the human figure aside, if necessary.

2. Use the Rectangle tool to drag out a vertical rectangle (blue axis) 8′–0″ tall and 4′–0″ wide. Run the base along the red axis.

3. Push/Pull the surface of the rectangle forward ¾″. This is a soft-covered flat; the thickness of the muslin is negligible and will remain a plane without thickness for this exercise..

4. Orbit the object. A fully surfaced object is not always created when a face is extruded; a five-sided open box may be created, rather than a six-sided closed box. If the rear of the box is missing, draw a line along any rear edge. A surface will snap into place when the second endpoint is affixed.

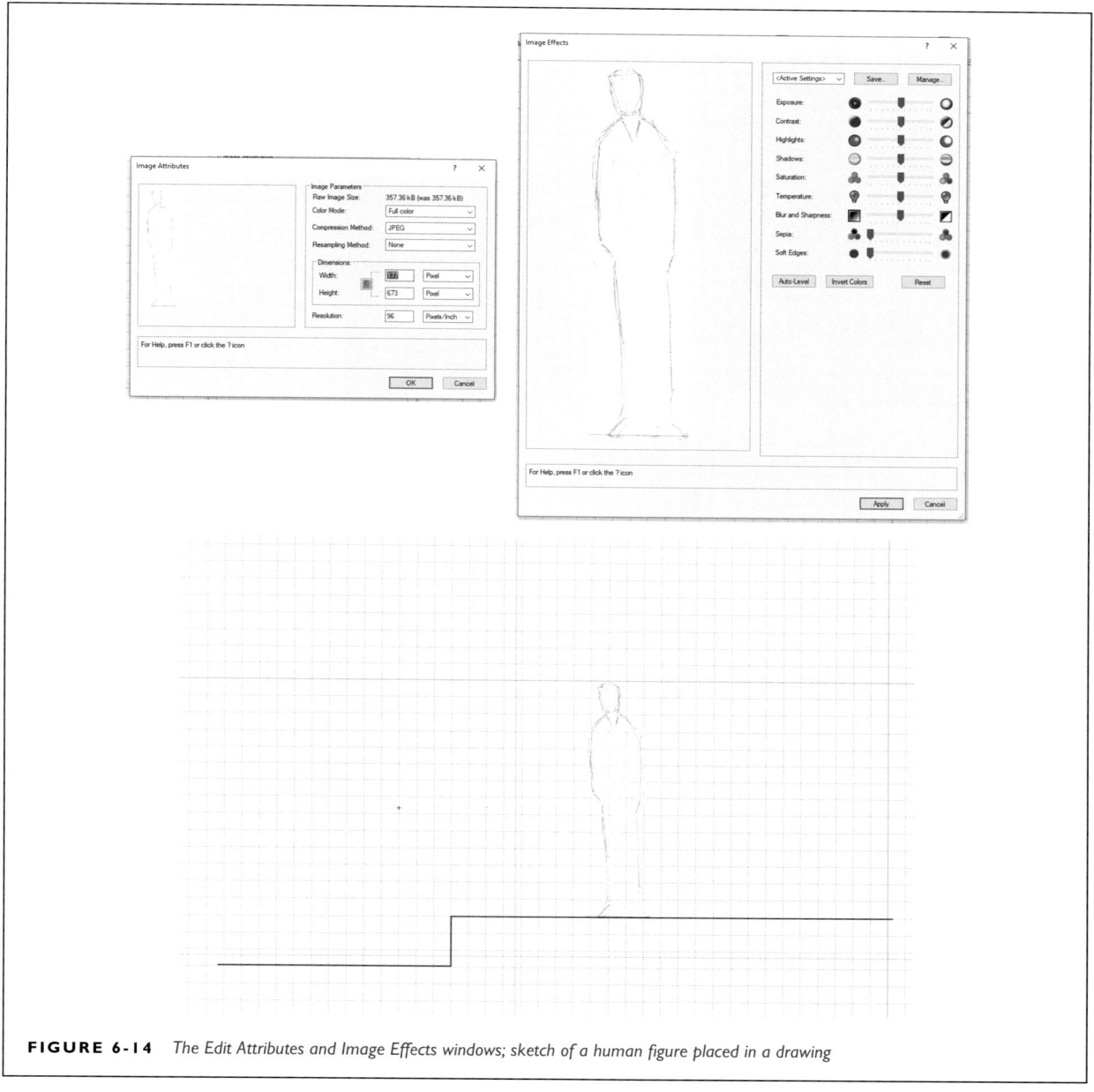

FIGURE 6-14   *The Edit Attributes and Image Effects windows; sketch of a human figure placed in a drawing*

5. In the toolbar, find the icon that looks like a small paint bucket: the Paint Bucket tool (Figure 6-16). When selected, a navigation box, or Tray, with menus appears. This allows you to import and edit colors and textures, as well as to access a default library. Explore the library until you find an appealing color or texture. Choose it and then click on the flat's front surface; the texture will appear.

6. Use the Select tool to drag a box around the whole object. Once the object is highlighted, right click the mouse to open its menu. Choose Group. To alter the flat later, choose Edit Group from the same menu.

When painting, if the object is already a group, all of the surfaces will be painted with that texture; to paint selected surfaces within a group, open the group via Edit Group and then paint.

7. Make a copy of the flat. Select the flat so that it is highlighted. Select the Move tool and hold down Ctrl on the keyboard. While holding down the Ctrl key, click the cursor on a bottom-front corner of the flat and drag it to the side. A copy of the flat will emerge. The copy will carry the highlighting, rather than the original.

8. Use Move and Rotate to butt the new copy against the original to make a corner.

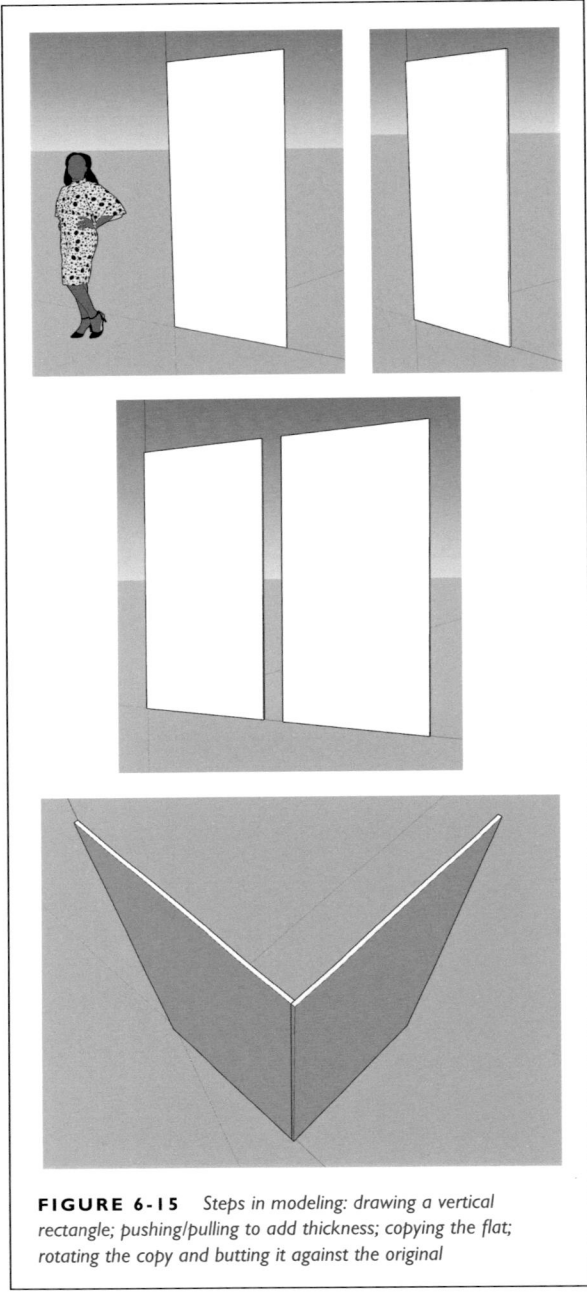

FIGURE 6-15 *Steps in modeling: drawing a vertical rectangle; pushing/pulling to add thickness; copying the flat; rotating the copy and butting it against the original*

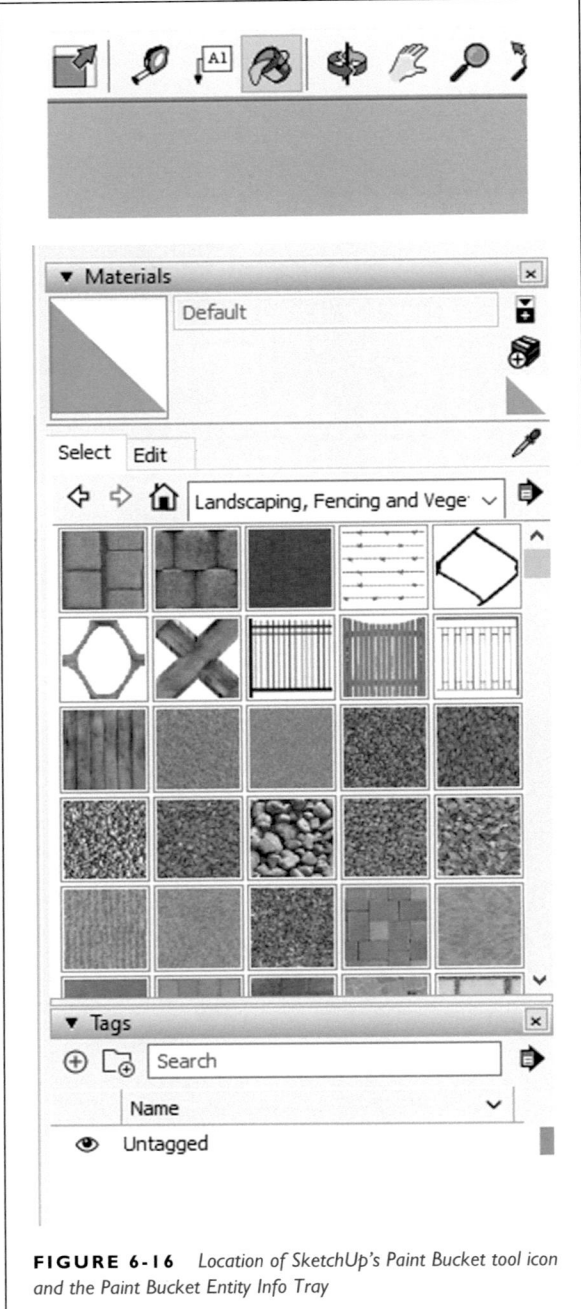

FIGURE 6-16 *Location of SketchUp's Paint Bucket tool icon and the Paint Bucket Entity Info Tray*

If the two flats are joined only by their corners as though they're hinged on the rear, ¾″ (the thickness of the flat) is added to the total length of each section of the wall with a crease running down the corner. To create a clean corner, butt the side of one flat butt against the rear of the other. This means that ¾″ is added to one flat's length (Figure 6-17). Three-quarters of an inch may not seem like much, but with multiple corners, there's accumulation. For thicker Hollywood flats, a single corner can add three or four inches to the overall length of a wall. Construction drawings must take this into consideration when working up measurements.

9. Save. Use this file for the next exercise.

Approach #2

This approach is useful if framing is to be indicated but not fully modeled. The framing is constructed as a continuous object rather than as individual pieces of lumber (Figure 6-18).

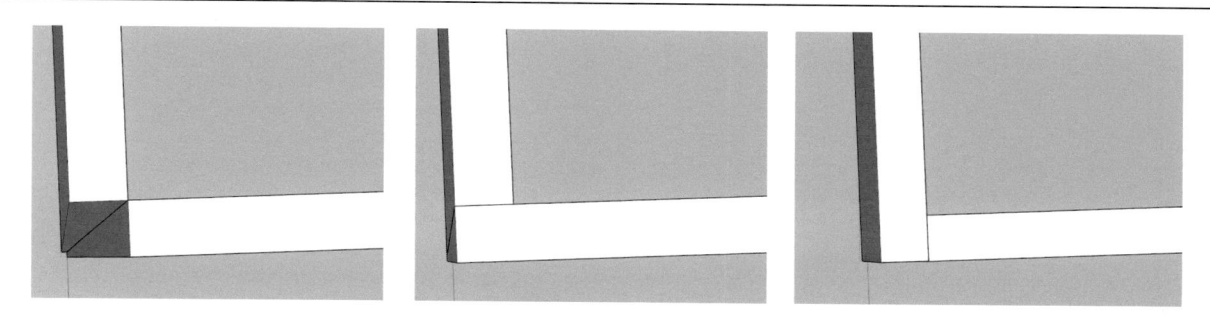

FIGURE 6-17   *Three joints adding length to at least one surface: open corner (hinged on rear); front overlapping side; side overlapping front*

Construct this version further along the red axis from the previous flat. The framing lumber is 1 × 3. Scabs are ¼″ thick. Overall dimensions are the same as for Approach #1.

1. Drag out a vertical rectangle seated on the red axis.

2. Use the Offset tool to drag an outline 2½″ within the perimeter of the front surface to mark the inner edges of the rails and stiles.

3. Place a guideline through the horizontal axis of the rectangle (4′–0″ up from the bottom).

4. Draw a horizontal guideline 1¼″ above the horizontal axis to place the top edge of the toggle. Place the bottom edge of the toggle an equal distance below the axis.

5. Use the Line tool to draw the top and bottom edges of the toggle, intersecting the stiles. Four independent surfaces are created: the rails and stiles together, the toggle, and two interior rectangles.

6. Erase the surfaces of the upper and lower interior rectangles, leaving just the framing members.

7. Erase the lines separating the toggle from the stiles to create a single, unified surface.

8. Push/Pull the surface forward ¾″. Orbit to check whether the rear is filled; if not, draw lines along edges until a rear surface appears.

9. Use the Select tool to drag a box around the frame. Make it a group.

10. To one side of the frame, drag out a 4′–0″ × 8′–0″ rectangle to be the muslin cover. Push/Pull the rectangle to give it a thickness of 1/16″. This is less to replicate the thickness of muslin as to allow the plane of the rectangle to float slightly above the surface of the flat, preventing the two surfaces from trying to inhabit the same plane.

11. Use the Paint tool to apply a texture/color to the front surface of the muslin cover. Make the cover a group.

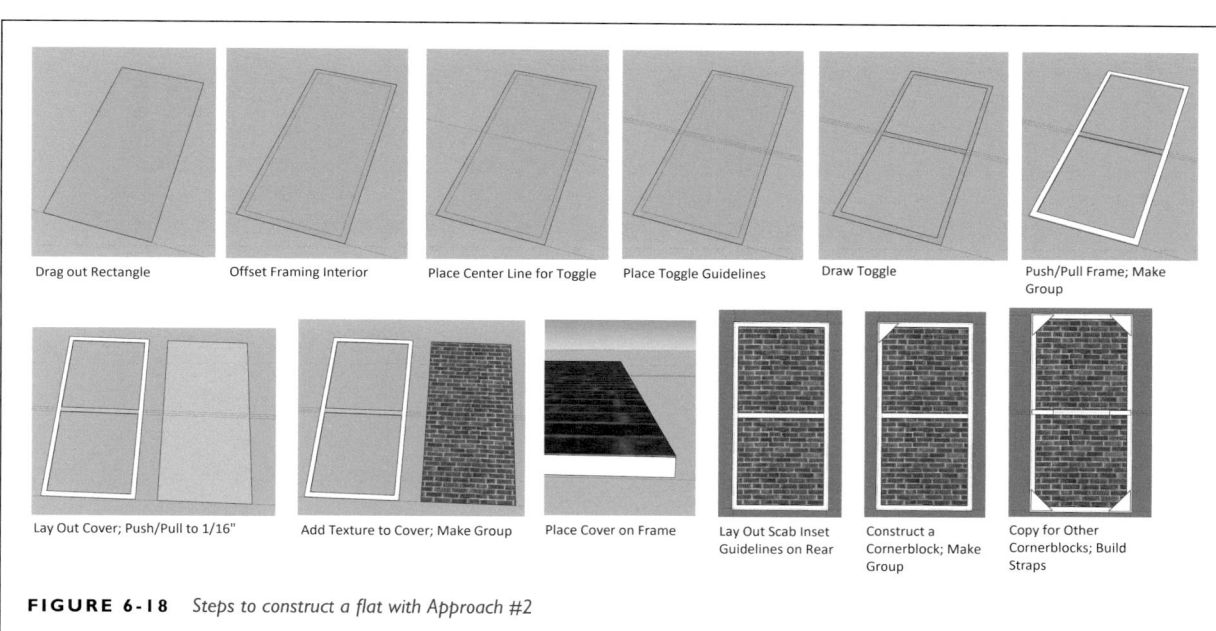

Drag out Rectangle   Offset Framing Interior   Place Center Line for Toggle   Place Toggle Guidelines   Draw Toggle   Push/Pull Frame; Make Group

Lay Out Cover; Push/Pull to 1/16″   Add Texture to Cover; Make Group   Place Cover on Frame   Lay Out Scab Inset Guidelines on Rear   Construct a Cornerblock; Make Group   Copy for Other Cornerblocks; Build Straps

FIGURE 6-18   *Steps to construct a flat with Approach #2*

Move the cover and affix it to the front of the frame. Grabbing the cover by one lower corner will make its placement easier.

12. Orbit the flat to add scabs to the rear. Drag guidelines in from the perimeter to mark the scabs' ¾″ setback. Since the flat is a group, the Offset tool won't grab the edges.

13. Pick one corner to construct the first corner block. From the setback's corner, draw 10″ lines outward along the perpendicular guidelines. Connect the two endpoints to make a triangle. A surface will snap into place. Push/Pull the surface up ¼″.

14. Make the corner block a group. Since the frame and cover are already groups, none of their points or lines will be incorporated into the corner block group.

15. Four corner blocks are required. Use Move and Ctrl to duplicate the first corner block. Move and Rotate the copy into position at another corner of the flat, aligned with the setback guidelines. Repeat two more times.

16. Draw a guideline through the center axis of the toggle, unless the original axis guideline is still in place. The straps are 2¼″ wide and 10″ long. From the axis, drag a guideline up 1⅛″ and another one down 1⅛″.

Add a guideline 10″ from and parallel to the scab setback guideline. Draw a rectangle to outline the scab. Push/Pull up ¼″. Make it a group. Use Move and Ctrl to copy it. Move the copy into position on the other side of the toggle.

17. Make the whole flat a group. Copy it, and butt the copy against the original. Look at the rear to see how the scabs relate to the butt joint. Does the texture/color wrap around the corner? What does this mean if it does not?

18. Save.

Approach #3

This approach is useful when explicit construction illustrations are necessary. All components will be constructed separately, made into groups, and then assembled (Figure 6-19). A parts list, or **cut list**, needs to be generated first. A cut list provides all of the lumber lengths and sizes required so that all can be cut in a single session at the saw, rather than on a piecemeal, as-you-build basis.

2   4′–0″ long pieces of 1 × 3 (rails)
2   7′–7″ long pieces of 1 × 3 (stiles)
1   3′–7″ long piece of 1 × 3 (toggle)

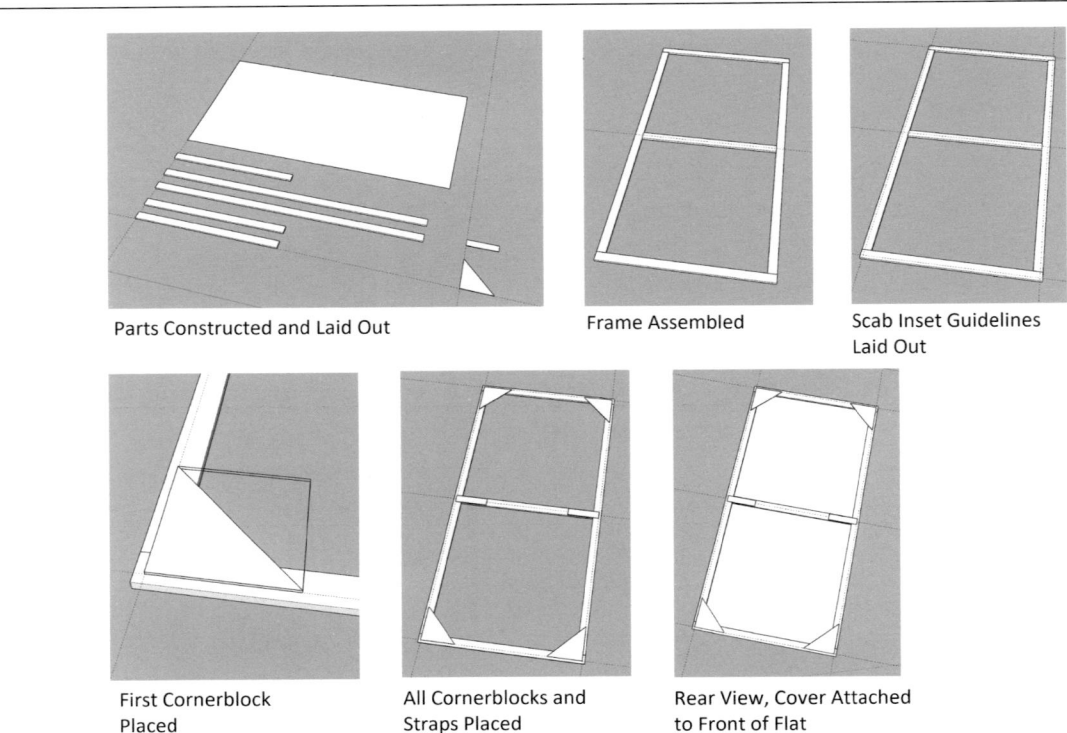

Parts Constructed and Laid Out   Frame Assembled   Scab Inset Guidelines Laid Out

First Cornerblock Placed   All Cornerblocks and Straps Placed   Rear View, Cover Attached to Front of Flat

FIGURE 6-19 *Steps in assembling the flat from a cut list (Approach #3)*

4 10″ corner blocks

2 2″ × 10″ straps

1 4′–0″ × 8′–0″ cover (muslin used for an actual flat would be larger, as it typically folds over the flat's edges); add 1/16″ thickness for modeling purposes.

1. Construct the rails. Drag out a rectangle that is 4′–0″ long and 2½″ wide. Push/Pull the surface up ¾″. Check to see if the reverse side is closed. Make it a group. Copy and Move the group to create two rails.

2. Construct the stiles. Drag out a rectangle that is 7′–7″ long and 2½″ wide. Push/Pull. Make it a group. Copy and Move to make a second stile.

3. Construct the toggle. Drag out a rectangle that is 3′–7″ long and 2½″ wide. Push/Pull. Make it a group.

4. Construct a corner block. Drag out a 10″ × 10″ square. Draw a line connecting opposite corners. Delete one side of the square to leave a triangle. Push/Pull up ¼″. Make it a group. You can build four now, or Copy and Move once the first corner block is placed.

5. Construct one strap. Drag out a 10″ × 2″ rectangle. Push/Pull up ¼″. Make it a group. You can build two now, or Copy and Move once the first strap is placed.

6. Construct the muslin cover. Drag out a 8′–0″ × 4′–0″ rectangle. Push/Pull 1/16″. Apply a texture/color. Make it a group. Without the 1/16″ thickness, the plane of the cover will co-inhabit the same plane as the surface of the flat framing and when a texture is applied, it will be only partially visible.

7. You now have all the parts. Use Move and Rotate to move the framing pieces into their positions. If desired, make the frame a group.

8. Once the frame is assembled, draw scab setback guidelines on the rear of the frame. Move and Rotate the corner block into position. Copy and Move the copy to the next corner. Repeat for the final two corner blocks.

9. Move the first strap into position. Copy. Move the copy to the other end of the toggle.

10. Move and Rotate the cover into position on the front of the frame.

11. Select the whole object and make it a group.

12. Copy the flat and butt the copy to the original to examine joints and unpainted areas.

13. Save.

Homework: Modeling a Hollywood Style Flat

Use the same overall dimensions as the previous exercises to construct a Hollywood flat using each of the three modeling

FIGURE 6-20   *Rear view of the model of a Hollywood flat*

approaches. Since the framing members are on edge, some of the cut-list measurements will be different. The instructor may distribute a cut list or ask students to generate their own. No corner blocks or straps will be needed. To hardcover the flat, Push/Pull the cover to ¼″ thickness to represent a sheet of lauan (Figure 6-20).

Homework: Drafting a Series of Flats

Instructor

Figure 6-21 features rough sketches of the front elevations of four flats. Decide which construction style (Standard or Hollywood) the students will use to draft them. If you don't offer a framing sketch, discuss framing options as this is a layout exercise rather than an engineering one. Discuss layout options as a class: how many drawings will fit on a single sheet?

Students

Figure 6-21 features rough sketches of the front elevations of four flats. Draft each flat's front and rear elevations, either by hand or via Vectorworks, with either Standard or Hollywood construction as determined by your instructor. Be sure each flat's label/notes indicate the construction style. If your instructor chooses not to have scabs drawn on Standard flats, be sure to include notes about them. Your instructor may provide a sketch of framing suggestions. The printed

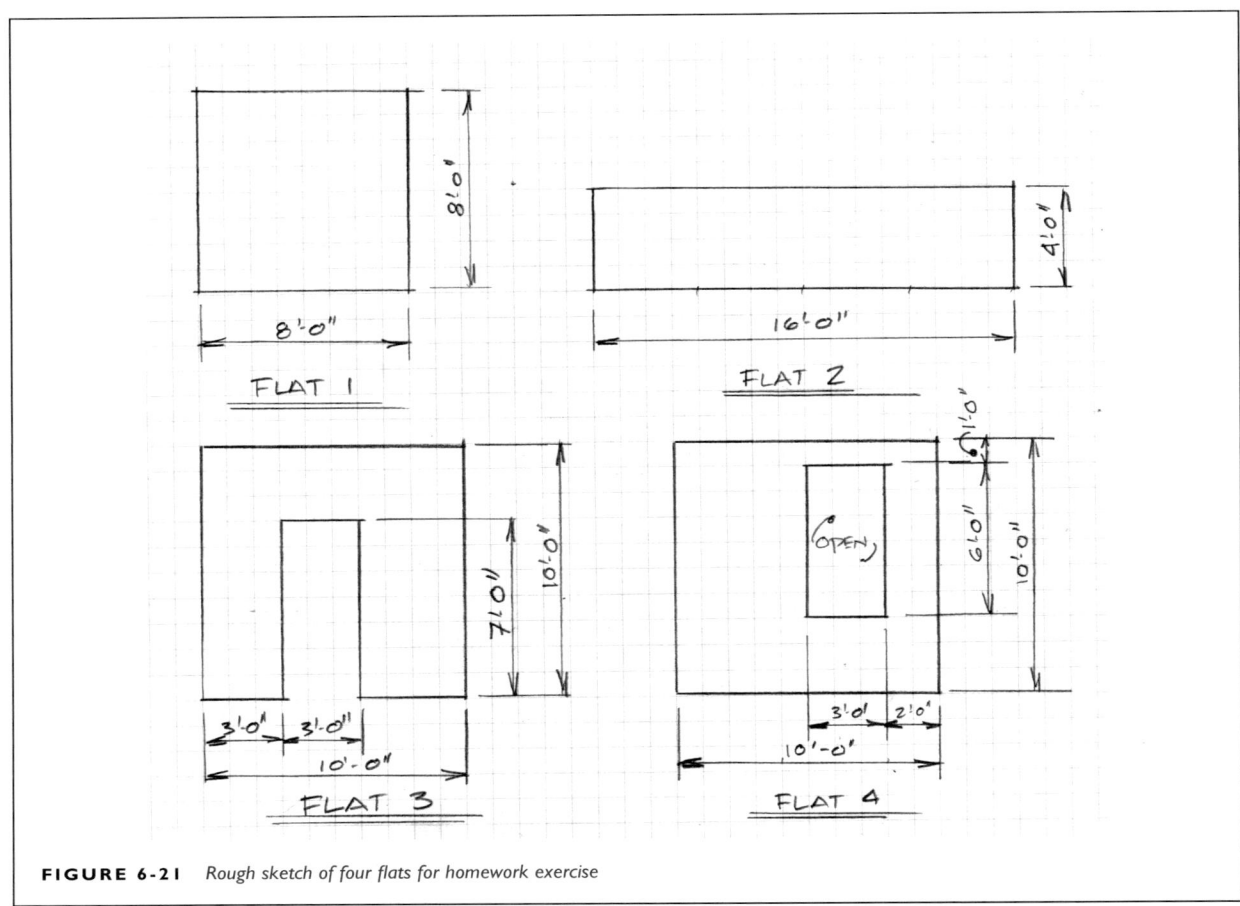

FIGURE 6-21   *Rough sketch of four flats for homework exercise*

scale may be ½″ = 1′–0″; if scabs are drawn, use 1″ = 1′–0″. Choice of paper size will be driven by scale choice.

In your planning phase, determine how many and which flats to place on each sheet. Each row of flats on a sheet should share a baseline for their bottom edges. Include a human figure on each sheet. Dimension both views for each flat. Label each flat as well as the individual views.

Include borders, title blocks, and notes.

When constructing these drawings in Vectorworks, two different approaches may serve.

1. After determining which flats will appear on which sheet, create each drawing as a separate file. For a series of drawings, this approach means the information that auto-populates the title block will not be generated and must be manually entered.

2. Use the infinite drawing space to draft each elevation seated along a common baseline. When complete, use viewports to transfer each elevation to its assigned sheet layer. In this case, checkmark the label option when creating the viewport to include the title of the flat.

MODULE 2: REVEALS, DOORS, AND MOLDING

Flats often feature portals such as doors, archways, and windows. Unless nothing more than an opening is desired, drawings must detail the parts of these portals. Flats also often feature **molding** (also commonly called "trim"): cornices, baseboards, chair rails, casing, etc. Sections and side views (**profiles**) are used to depict molding detail and construction. This module discusses simple openings with depth, basic doors, and the description of molding.

Simple Openings with Depth

Because a Standard flat is only ¾″ thick, a **reveal** can be added to create the illusion of wall thickness at an opening. Reveals add stability, as well as contribute toward scenic realism. In an architectural doorway, the piece of wood that finishes the sides of the opening is called a **jamb** (a door sits within a door jamb). Reveals are typically strips of wood attached to the inside edge of a flat's opening. While a Hollywood flat could be built with lumber wide enough to suggest the full thickness of a wall, it is

FIGURE 6-22   *Photograph of an apartment's simple archway*

it's too wide for most doors and is assigned to the category of archway. If it were narrow enough to accommodate a door, it would more likely be called a **doorway**. The walls in the photo are about 6″ thick. Grab a ruler and measure the wall-depth within a few nearby doorways and arches to get a sense of typical portal depths.

Figure 6-23 describes a Hollywood flat with an archway, based on Figure 6-22. The perspective sketch shows two sections of the reveal, which continues around to the right-hand side of the opening. The orthographic views include a vertical section (B-B) and a horizontal section (A-A). Since the top of the archway is above the A-A cutting plane, the flat appears as two pieces. The side and top views show the back sides of the reveals. The section shows the interior surface of the reveals (Figure 6-24).

The designer determines how wide the opening will be. The technician must take into account the thickness of the reveal's lumber as well as its placement when determining measurements for the construction drawing. If a 3′–0″ wide opening is requested, framing a 3′–0″ opening and then installing ¾″ reveals results in a 2′–10½″ opening. A horizontal section through the opening clarifies reveal placement (Figure 6-25).

If no door is to be installed, ¼″ lauan or plywood is often sufficient. Doors require sturdier structure for hinge installation and to bear the door's weight. Hollywood construction is preferred for operational doors due to its greater structural stability.

usually more cost effective to construct the flat with narrower stock and add reveals as required.

Figure 6-22 presents a simple opening or **archway**. Even though the opening does not feature a curve at the top,

Exercise: Draft a Flat with Opening and Reveal

The exercise can be done either by hand or in Vectorworks. Use Arch D sized paper, landscape orientation. The printed

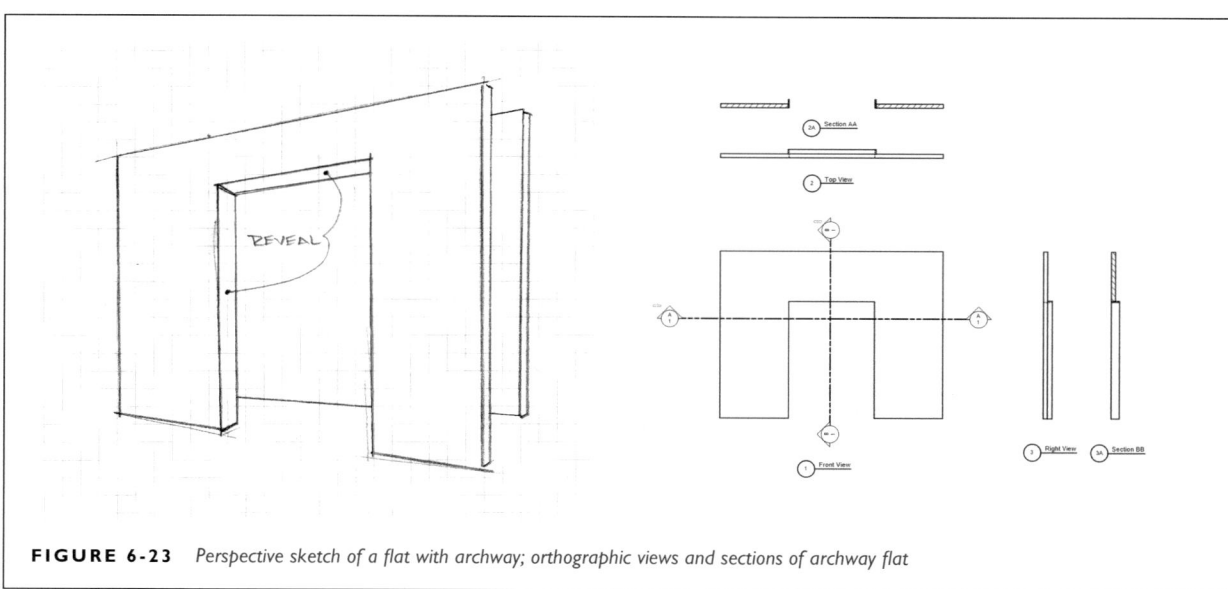

FIGURE 6-23   *Perspective sketch of a flat with archway; orthographic views and sections of archway flat*

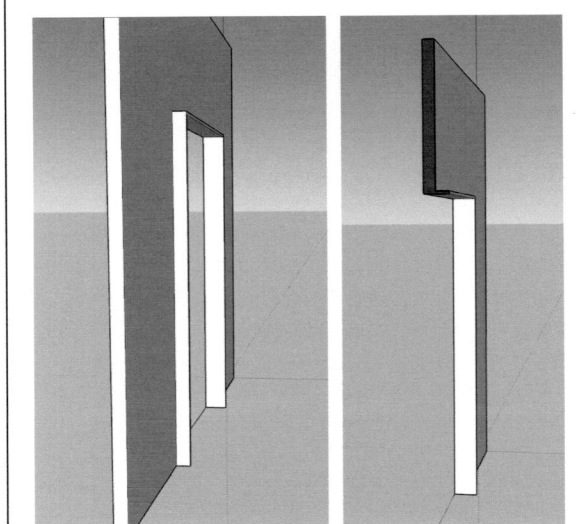

FIGURE 6-24 *The side view of the archway flat, showing the rear of the reveals; vertical section of the flat showing interior face of reveals*

scale is 1'–0". A design drawing and a construction drawing will be drafted. The steps below treat the exercise as a two-sheet project, though if scale allows it may be drafted/printed on a single sheet. Depending on how many sheets are used, discuss titles: which one is Drawing 1 on Sheet 1, etc.

Figure 6-26 provides sketches for a design drawing and a construction drawing. The flat is Standard style with a doorway and 6" reveal. Framing lumber is 1 × 3. The covering is muslin. The reveal is attached to the back side of the flat, using 1 × lumber. Dimension fully. Include necessary notes. Include borders and title block. Measurements are given in the design sketch; they may not be in their optimal locations. Extrapolate those dimensions into the construction drawing. The rear elevation sketch provides

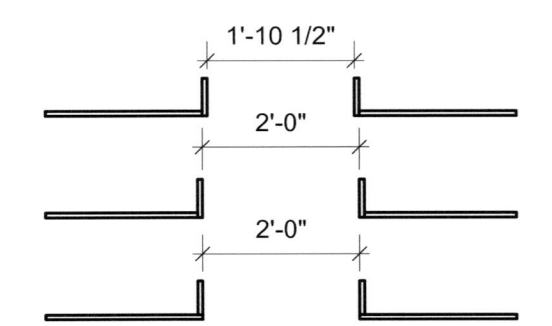

FIGURE 6-25 *A 2'–0" opening with reveals within opening, reducing it to 1'–10½"; a 2'–0" opening with reveal attached to back of flat, preserving full opening width*

suggested framing; when toggles or stiles frame an opening, dimensioning should reference the edge of the lumber. If the toggles and stile lie within the flat, dimensions should reference center lines. Replace the top or side view with sections, if useful for full description.

Do layout math. Include adequate white space between views for dimensions, notes, and possible sections. Create rough layout sketches for each drawing that include dimension and note placement. For Vectorworks, determine how many layers you will need; create and name them.

Design Drawing

1. Lay out guidelines for borders and title block.

2. Lay out the perimeter of the front elevation; extend guidelines to help place the other views. Front and side views should share a baseline.

3. Lay out the doorway in the front elevation. Extend guidelines into the side and top views.

4. If sections are desired, lay out and lightly label cutting plane lines. In the top view, lay out flat thickness and reveal as seen from above. If a horizontal section has been chosen, add required detail as seen at the cutting plane.

5. In the side view, lay out flat thickness and reveals as seen from the side. If a vertical section has been chosen, add required detail.

6. Darken the drawing, adding final dimensions, labels, and notes.

7. If hand drafting, darken the borders and title block. Add title block information. If in Vectorworks, create a viewport and import into a sheet layer. Add border and title block. Add title block information. After the sheet layer is complete, return to the design layers to lay out the construction layers.

Construction Drawing

1. If you use the same file as the design drawing, use Move Page to slide the reference grid to a new area of the drawing space. Lay out guidelines for the borders and title block.

2. Lay out the perimeter and doorway for the rear elevation. Extend guidelines for the top and side views. Lay out the perimeter of the lauan placement front elevation.

3. On the rear elevation, lay out the reveal around the opening, then lay out rails, stiles, and toggles.

4. Lay out the lauan placement front elevation.

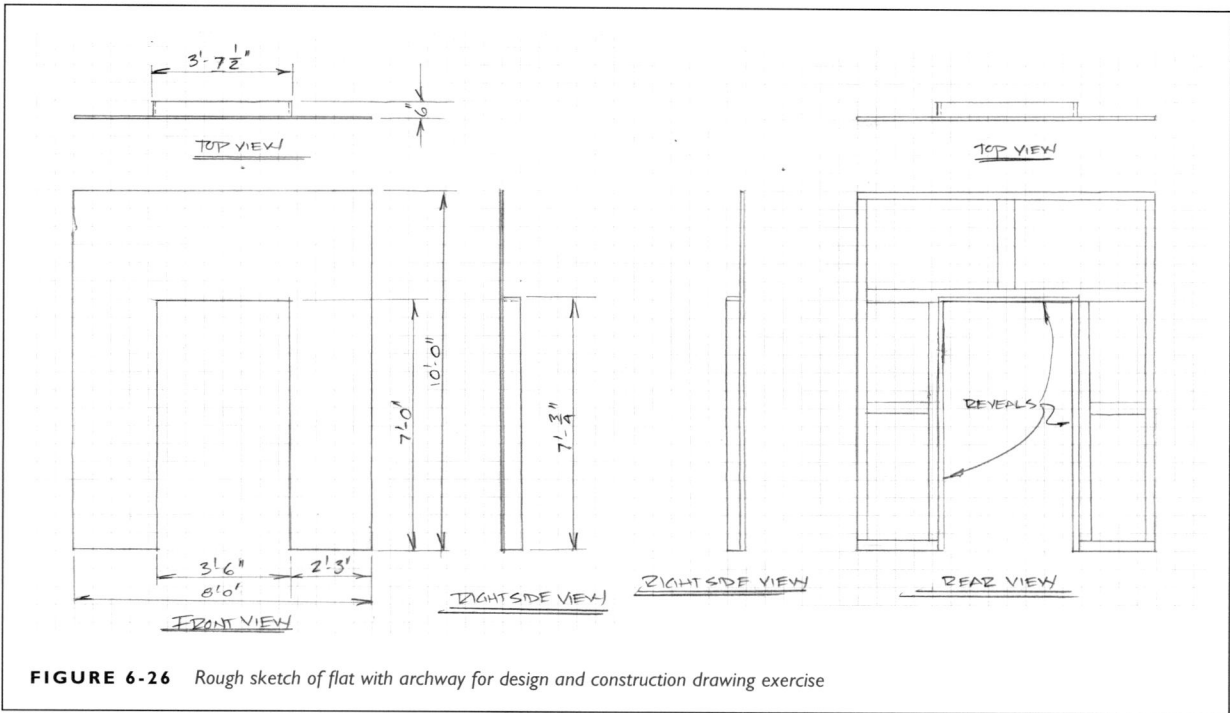

FIGURE 6-26   *Rough sketch of flat with archway for design and construction drawing exercise*

5. If sections have been chosen, lay out and lightly label their cutting plane lines. Lay out details in the top view/ horizontal section. Move information up from the rear elevation with guidelines as needed.

6. Lay out details in the side view/vertical section. Move information laterally from the rear elevation with guidelines as needed.

7. Darken the drawing, adding final dimensions, labels, and notes.

8. If hand drafting, darken borders and title block. Add title block information. If in Vectorworks, create a viewport and import into a new sheet layer. Add border, title block, and title block information.

9. If in Vectorworks, Save the file. Export all sheet layers as PDFs. Print the PDFs.

Homework: Modeling a Flat with Opening and Reveal

Using the first and third approaches from the SketchUp exercise in the previous module, model the Standard flat just drawn.

Approach #1

1. Drag out a 8′–0″ wide by 10′–0″ tall rectangle.

2. Use guidelines to lay out the opening. Use the Line tool to outline the opening.

3. Erase the surface within the opening. Erase the line at the bottom of the opening.

4. Since this is a blunt representation of the flat, Push/Pull the surface forward 6″. Orbit to check that the rear of the flat is closed; if not, add a line along a rear edge to snap a rear surface into place.

5. Use the Paint tool to select a color/texture for the front surface and reveals.

6. Make it a group.

While this is the fast and easy construction method, the fact that the whole flat is 6″ thick can pose problems when linking these extra-thick walls together to build a set. You can also Push/Pull the flat to ¾″, then outline the reveal on the rear of the flat. Push/Pull the reveal to the desired depth (Figure 6-27).

Approach #3

Use the cut list below to model the components. Make each part its own group. All lumber is 1 × 3, except for the reveal. The reveal is attached to the rear of the frame; its pieces are 5¼″ wide (actual).

1 top rail	8′–0″ long
2 bottom rails	2′–6″ long
2 stiles	9′–7″ long
1 toggle	7′–7″ long (top of doorway)

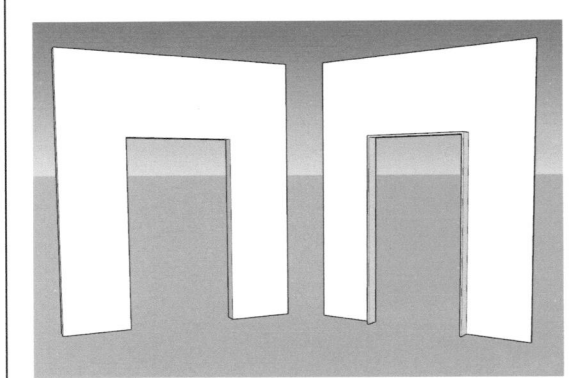

FIGURE 6-27   *Rear views of two blunt approach flat models; one pulled to 6″, one with just the reveal pulled to 6″*

2 stiles	6′–8½″ long (sides of doorway)
1 stile	2′–11″ long (centered above doorway)
2 toggles	2′– 1″ (either side of doorway)
2 reveals	7′–0″ long, 5¼″ wide (sides of doorway)
1 reveal	3′–1½″ long, 5¼″ wide (top of doorway)
1 covering	8′–0″ × 10′–0″ overall, with 3′–0 ″ × 7′–0″ opening, 1/16″ thickness for modeling purposes 6 corner blocks (triangle 10″ × 10″, ¼″ thick)
8 straps	(2¼″ × 8″, ¼″ thick)

Construct all components. Move into place. Copy scabs as required. Choose a color/texture for the cover before making it a group.

Doors and Molding

Before drawing a door, the draftsperson needs to know what it looks like by researching the relevant architectural style. Follow up the style research by visiting door supply websites or catalogs for measurement information. Doors function; therefore, the flat in which they are installed requires particular elements. For the following discussion and exercises, a plain, flush hollow core door such as the one in Figure 6-28 will be used. **Hollow core** refers to the door's construction: a frame covered by thin plywood. **Solid core** means that the interior of the door is filled with wood block. A **panel** door consists of multiple wooden panels held in place by a frame of thicker lumber. Don't forget that your residence is also a research tool: go measure a door.

The door jamb includes a narrow strip of wood along all three sides called a **door stop** (or "slam strip") to keep the door from swinging past the point at which it latches into the jamb. This strip also prevents drafts and light leaks. A **threshold** is a wooden or metal plate at the base of the

FIGURE 6-28   *Photo of an interior hollow core door with casing*

doorway that hides the join between the flooring of one space and another.

The **doorknob** (or door handle) turns a mechanism which extends and retracts a bolt that aligns with a hole in the jamb. In most cases, this latch is spring-loaded so that it automatically extends into the jamb. Knobs and handles without latching mechanisms can be installed, but alternative hardware such as a cabinet catch is required to keep the door shut. Doorknobs are typically centered 3′–2″ above the floor. The distance of the knob from the edge of the door is determined by the latching mechanism and length of its bolt. A **strike plate** is attached to the jamb to reinforce the bolt hole and guide the bolt home.

Hinges connect the door to the jamb, allowing the door to swing. Hinge placement determines in which direction the door will swing. The weight and style of the door will determine whether two or three hinges are required. The center of the bottom hinge is typically placed 10″ above the bottom of the door. The center of the top hinge is typically placed 9″ below the top of the door. The center hinge is centered between the upper and lower hinges (Figure 6-29).

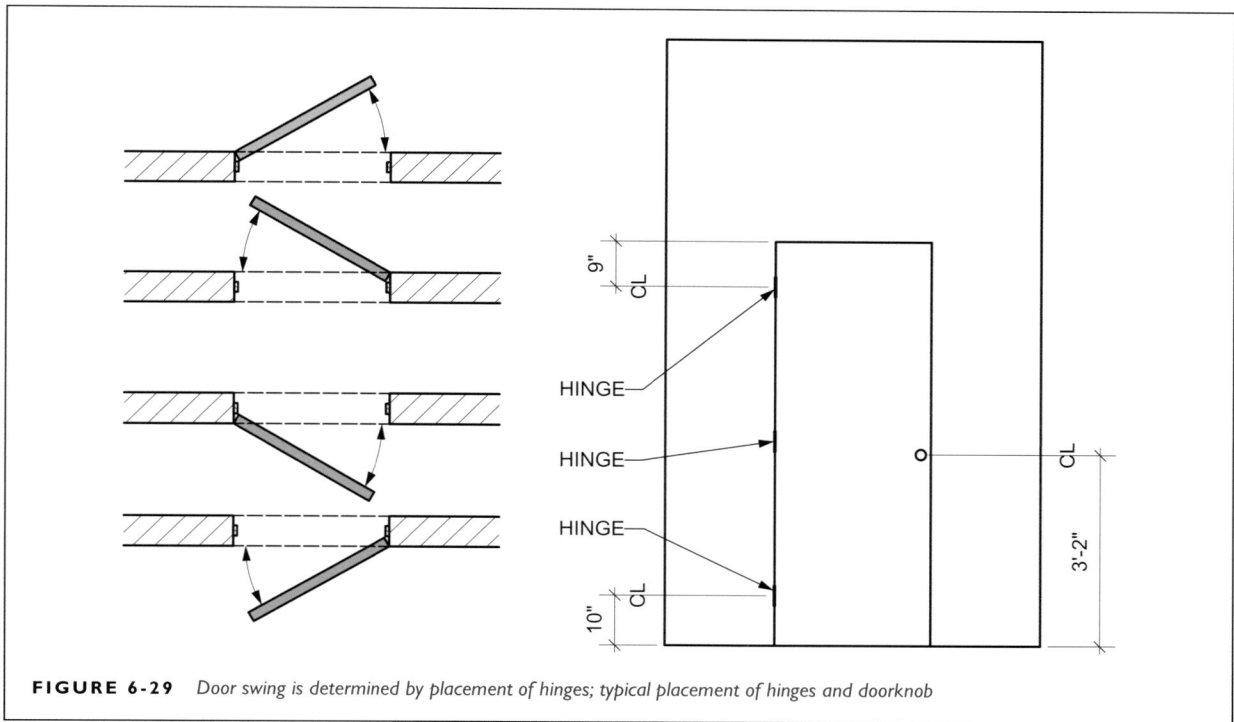

FIGURE 6-29   *Door swing is determined by placement of hinges; typical placement of hinges and doorknob*

Molding Nomenclature

The primary function of molding is to hide joints and seams. Decoration is secondary, though molding can certainly be installed for purely decorative purposes. In a scenic environment, molding can have both the aesthetic goal of reinforcing the period décor of a setting and gracefully covering up seams between flats.

Casing is the term for molding that outlines a door or window opening. Molding running along the top of the wall where it meets the ceiling is called **crown molding** or **cornice**. The strip of wood at the base of the wall where it meets the floor is **baseboard.** A **plinth** may be placed at the bottom of a door's side casing to emphasize its intersection with the baseboard (Figure 6-30).

Two further decorative pieces of molding are the **chair rail** and the **picture rail.** Chair rails divide the upper and lower sections of a wall, often separating different wall treatments such as wainscoting below and paint above. The chair rail gets its name from being placed at about the height of the top of chair's back. Picture rails are rails from which pictures are hung so that the wall itself is not damaged by nails and hooks. Both tend to be more of an old-fashioned or period feature.

There are many molding styles. Research directs the choice of style, but the choice must then be described on the drawing via notes and a **profile**, or end view. Again, lumber supply stores are a good place to find measurements and profiles.

Depending on the complexity of the door and designer preference, the door may be drawn as a separate element on the sheet. The flat must then feature a note indicating that a particular door is to be installed: e.g., "Install Door A in Flat A."

Exercise: Drawing a Door Flat with Details

Instructor

Two drawings will be constructed for this exercise. They may be drawn by hand or with Vectorworks. Students should use Arch D paper. The first drawing is a design drawing incorporating molding detail profiles and imported doorknob images. The second is a construction drawing including lauan layout on a front elevation. If desired, provide molding and doorknob catalog information for inclusion on the drawings. Framing is suggested in the rough sketch; if an alternative framing solution is preferred, provide a sketch.

Students

The door photo in Figure 6-28 will be used as the research image for a Hollywood flat featuring a hollow core door and molding. Two drawings will be constructed. Use Arch D paper; select the scale that best suits the drawing: either ½″ = 1′–0 or 1″ = 1′–0″. Use 1 × 3 stock for framing, with a 6″ reveal. Hard-cover with lauan. The door will have three

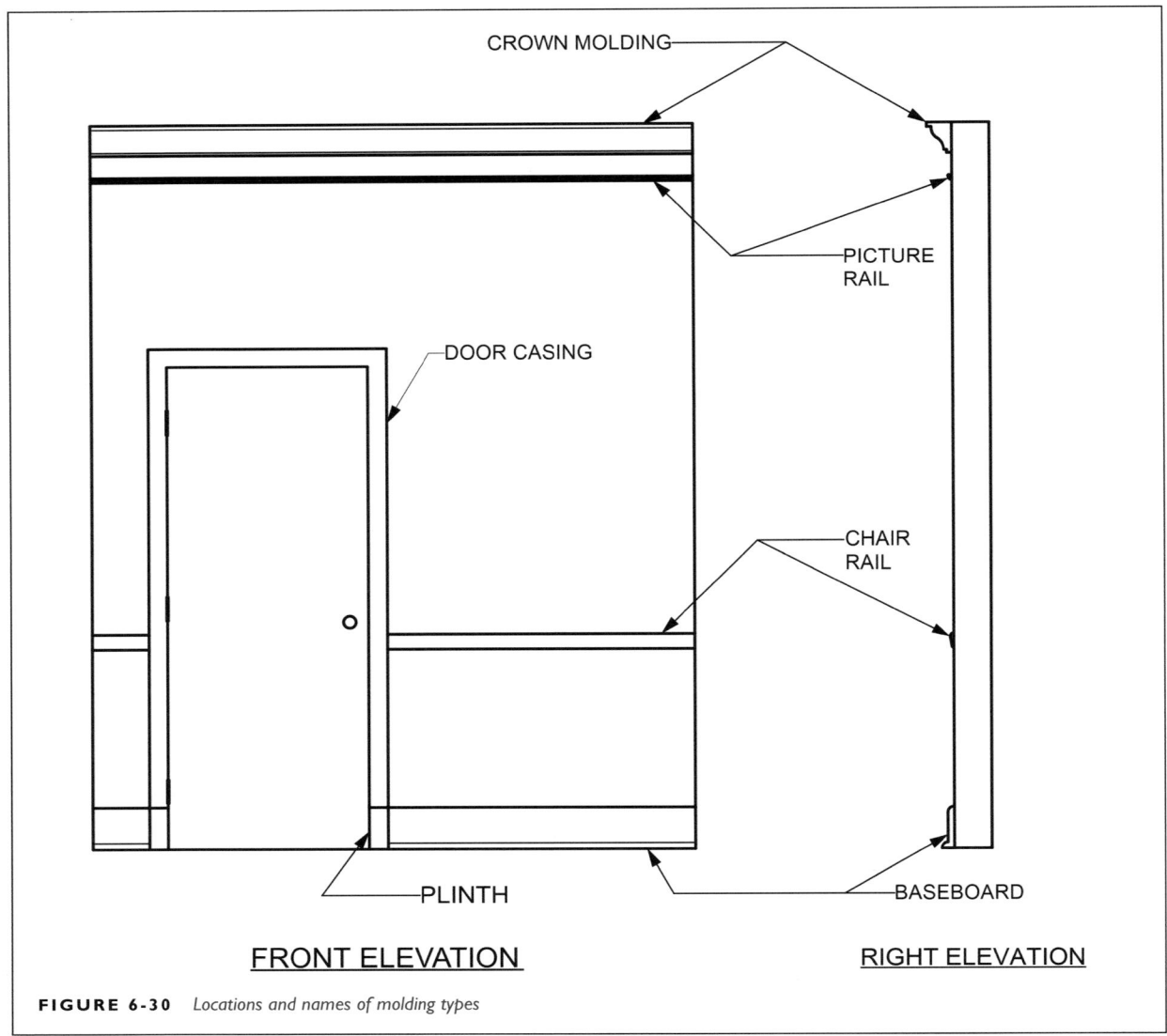

FIGURE 6-30 *Locations and names of molding types*

hinges and working doorknob/latch. It is your choice whether to draw the door installed or off to one side.

Figure 6-31 presents rough sketches of both drawings, with measurements. The sketch does not include construction drawings top and side views; construct them from the given information.

Decide whether sections will help describe the flat; include them as needed. The reveal pieces will be set *within* the door opening, which means that the doorway framing will need to be slightly larger to accommodate the pieces.

Figure 6-31 provides a framing suggestion for the rear elevation. See the additional front elevation, showing suggested lauan sheet placement. You may lay framing members flat against the face of the flat to support lauan seams.

Figure 6-31 also includes rough molding profiles. Before beginning the project, consult a lumber supply website or catalog to seek matches. Along with the profiles, include the names and details of each style as notes to help the shop purchase the exact molding required, or find an affordable near-equivalent. When hand drafting, print out profiles and trace them onto the drawing (a French curve helps with drawing irregular curves). The more brutal approach is to affix a printed photo to the paper with transparent tape. In Vectorworks, downloaded profile images can be imported into the drawing. Depending on research findings, the dimensions shown in the rough sketch may change slightly.

Collect research images and measurements of a hollow core door and a simple doorknob. Include images and product information on the drawing. Since a working knob is desired, notes should indicate its expected functionality.

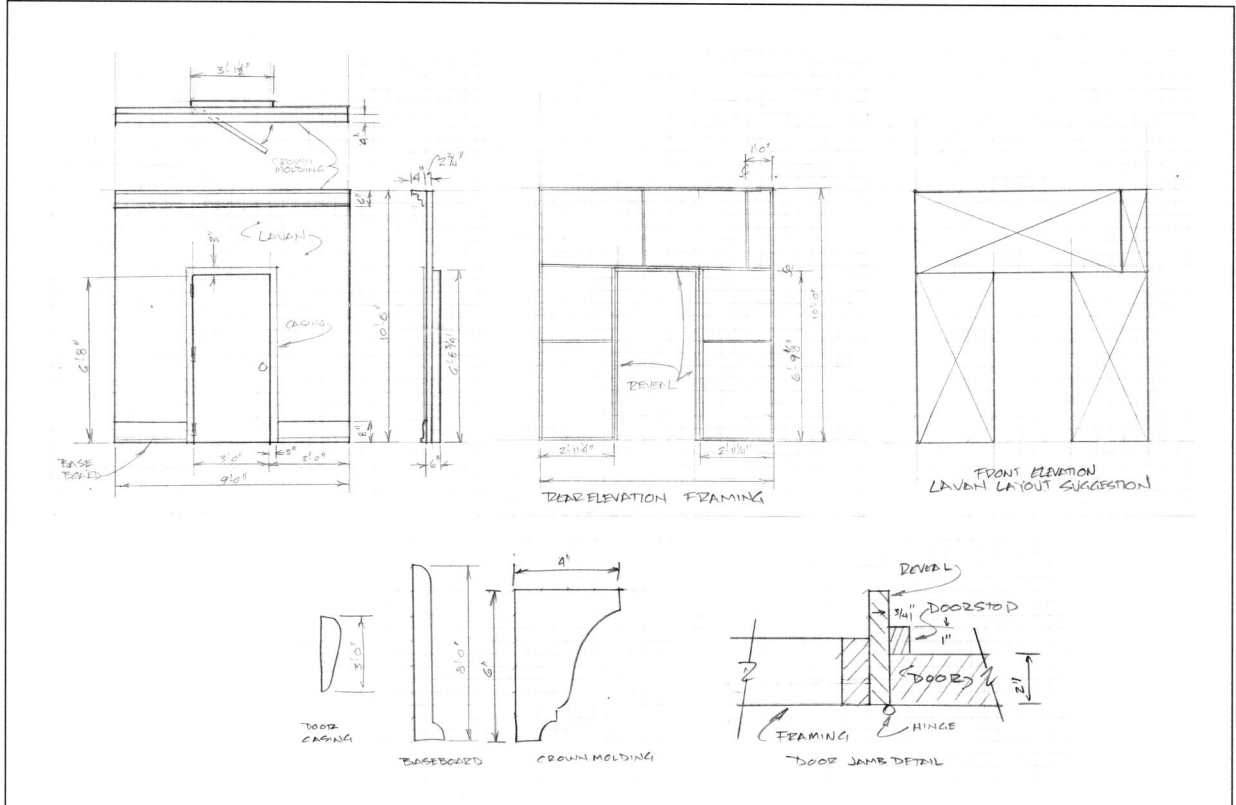

FIGURE 6-31   *Rough sketch of exercise drawings: design drawing with molding profiles and imported images; construction drawing with lauan placement*

Drawing 1 of 2 (Design)

If working in Vectorworks, determine how many design layers are required. Create them. Do preliminary layout math.

1. Lay out the perimeter of the front elevation. Extend guidelines to the right and above to place the top and side views/sections, as well as the door. Reserve and lay out space for molding profiles, knob photo, and notes. If you choose to draw the door separately, reserve space for that as well.

2. In the front elevation, lay out the doorway. Extend guidelines into the other views. Lay out guidelines locating the centers of hinges and the doorknob. If drawing the door separately, lay it out to one side of the front elevation.

3. On the front elevation, lay out door casing, baseboard, and crown molding. Consult the measurements in the rough sketch, as well as your research. Note that the baseboard and cornice have interior edges/planes. The door casing is curved; its only drawn lines are its two outer edges. The baseboard is curved at the top, with a piece of quarter-round protruding at the bottom; the

top of the quarter-round creates a line. To represent the baseboard, three lines are required: the top, the bottom, and the top edge of the quarter-round. The crown molding has three interior edges/planes; five lines are needed (Figure 6-32).

When darkening, the outer edges of the trim are medium thick. Interior lines are thin. If an outer edge coincides with the perimeter or outline of the object itself, that line remains the original object's line weight.

4. Depending whether and which section views were chosen, lay out their cutting plane lines and lightly label them.

5. Lay out top view/section and side view/section details. Framing is not a concern of the design drawing, so the wall is shown as a sliced-through solid mass. Consult Figure 6-31 for information about the door stop and jamb. Is a detail drawing required to adequately describe the door stop and jamb? Does not including the door in the front elevation make drawing the section more or less clear?

If a vertical section is selected, the cornice and above-door casing are cut through. The interior of the reveal, the door stop, and edge view of the door casing are seen. The quarter-round of the baseboard may be seen jutting out beyond the side door casing. If the

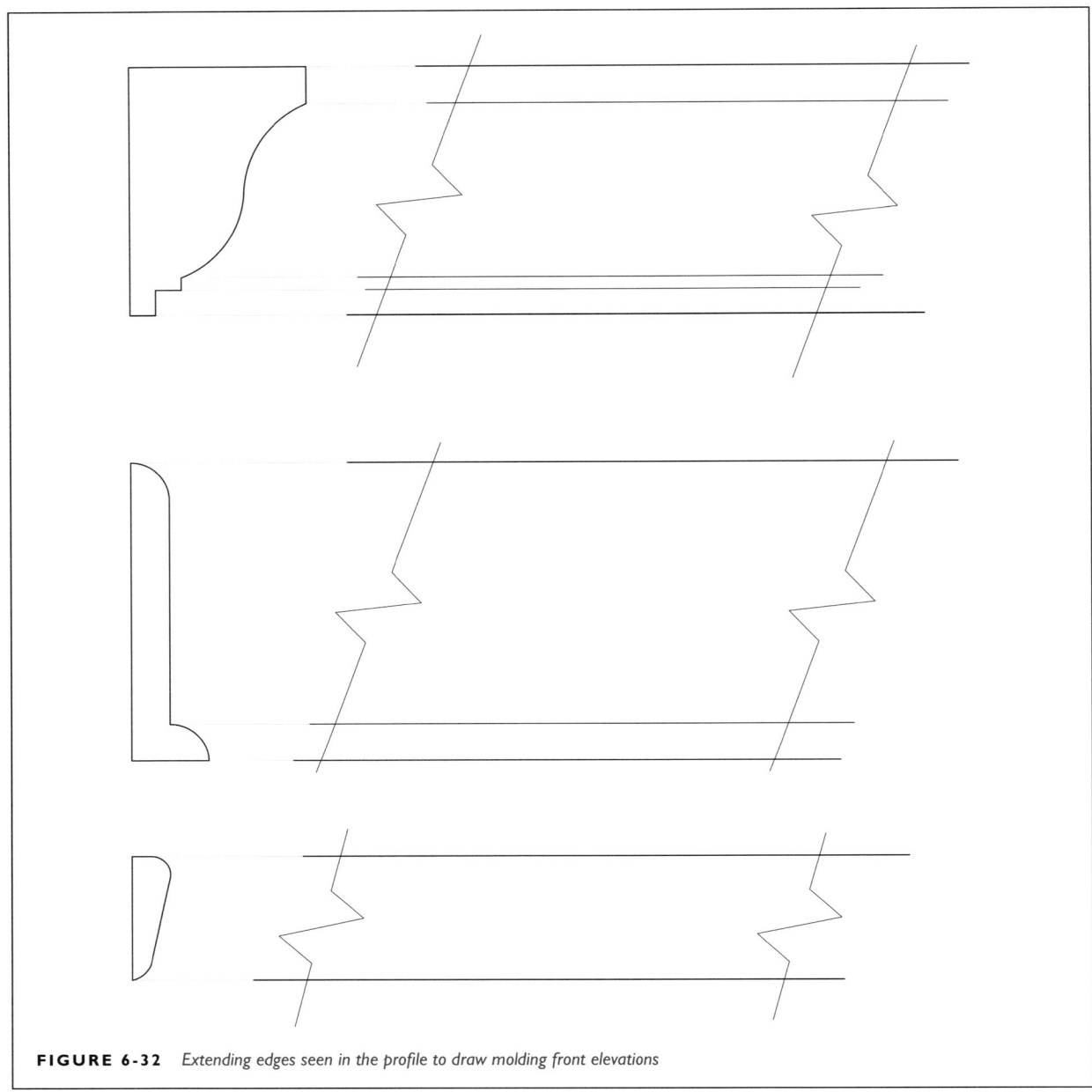

FIGURE 6-32   *Extending edges seen in the profile to draw molding front elevations*

side view is not a section, the ends of the cornice and baseboard are seen, as are the edge of the casing and rear of the reveal (Figure 6-33).

6a. In ½″ scale, the side views of the molding are very small. When hand drafting, it is acceptable to freehand the round bits; attempting to use circle templates or compasses is virtually impossible. The profile details elsewhere on the drawing will be drawn in a larger scale; a French curve will be useful.

6b. In Vectorworks, you can zoom in close enough to draft the molding profile regardless of the scale

you're working. Each profile is a collection of straight lines and arcs. The baseboard is simple. There is a quarter-circle at the top, and a quarter-circle for the protruding quarter-round at the bottom. The cornice features a quarter circle at the bottom, but an **ogee** (S-shaped) curve in the main section. The Arc tool has an option that does not require finding a center first (Figure 6-34).

Determine the endpoints of the desired arc. Select the Arc tool; choose the Point on Arc Mode from the upper tool bar. Click the cursor on the top endpoint of the upper arc, then click on the bottom point. An

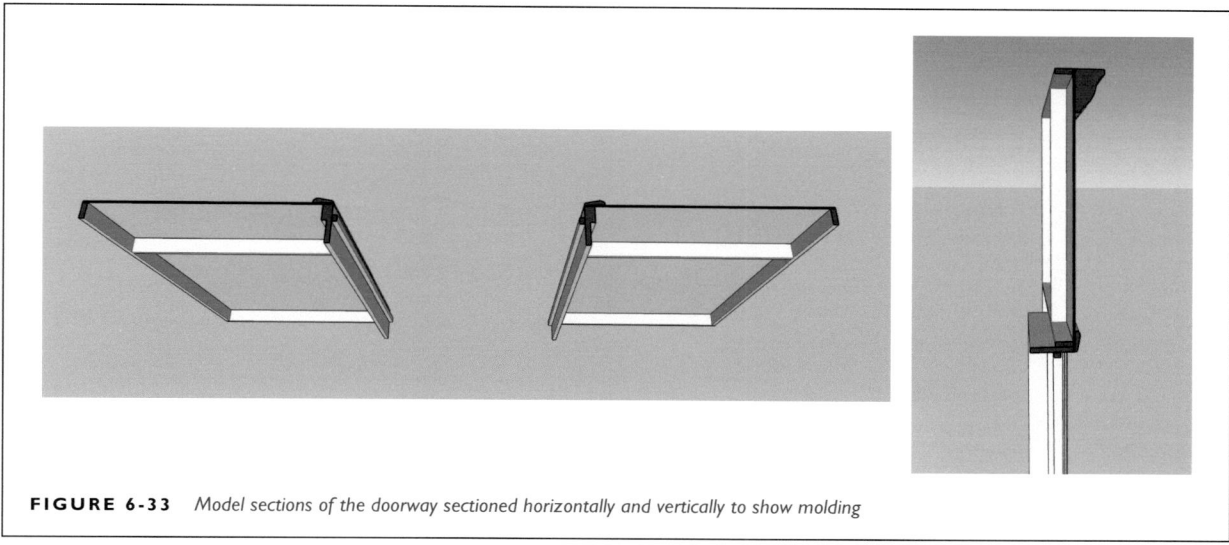

FIGURE 6-33   *Model sections of the doorway sectioned horizontally and vertically to show molding*

arc will appear. You can drag the belly of the arc out to create the desired curve. Click to affix the arc.

In Vectorworks, individual layers can be assigned different scales. Try creating a new design layer with a larger scale. Uncheckmark the All Layers box. What happens when you return to the drawing space? Redraw the molding profiles in the new layer. Is there any visual disconnect between this layer and the other layers? What benefits might there be in drawing different items in differently scaled layers?

7. Darken the drawing. Lay out dimensions. Use leader lines to label molding, coverings, etc. For very small features like the door stop, leader lines with notes may be the best way to include their dimensions. A note with leader line to the doorknob should direct the reader to the detail image elsewhere on the sheet for knob details. Be sure to dimension to the center of the doorknob, as well as to the height of the hinge centers. Include overall molding dimensions on the front elevation as well as on the profiles.

Since scale varies across the sheet, each drawing has scale included as part of their title label. The flat's views do not need to include individual scale notes. Number each drawing on the sheet (e.g., Dwg 1: Door Flat; Dwg 1A: Front Elevation; Dwg 1B Top View; Drawing 2: Trim Profiles; etc.).

In Vectorworks, drawing numbers can be added via the Create Label feature of the Create Viewport window. If you are placing both drawings on one sheet, allow Vectorworks to create labels for each Viewport rather than placing individual labels in the design layer.

8. Add doorknob information in its designated space. When hand drafting, taping a photo to the paper is a

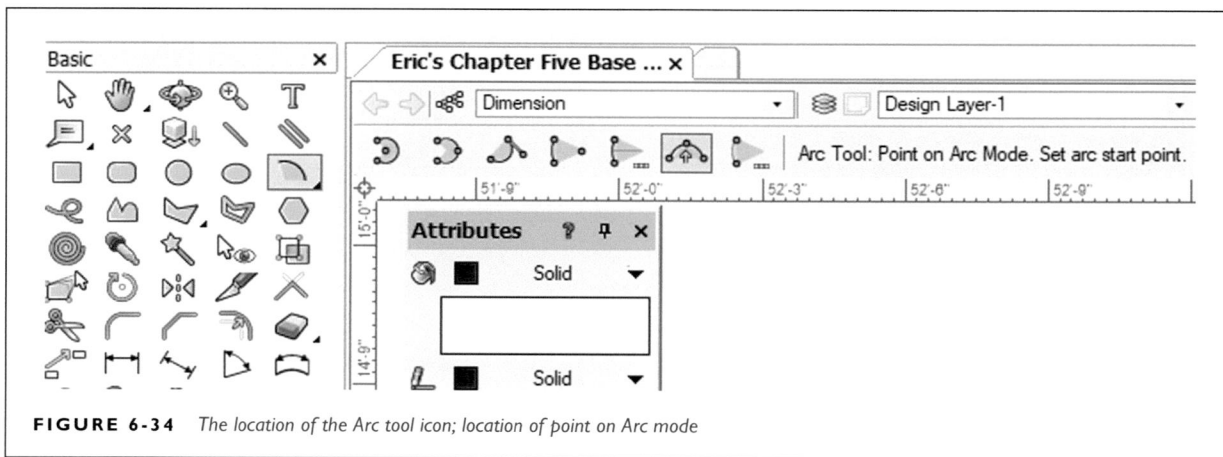

FIGURE 6-34   *The location of the Arc tool icon; location of point on Arc mode*

final step as the tape's edges will collect graphite from continued tool use. In Vectorworks, import the image into its own layer (be sure to turn on the layer when creating the viewport).

9. Add a human figure, standing on the flat's baseline.

10. If working in Vectorworks, create viewports and import the drawings to the sheet layer.

11. Darken/add the border and title block. Enter information into the title block. This drawing is 1 of 2. Scale for the sheet is As Noted. Review for completeness.

Drawing 2 of 2 (Construction)

1. Lay out the perimeters of the front (lauan) and rear (framing) elevations, sharing the same baseline. Extend guidelines to place the top view/section and side view/section. Lay out the door opening in both elevations. Extend guidelines into the other views. This drawing focuses on flat construction; the door is not included, though placement of hardware is. Include a note referring the reader to the location of door detail drawings.

2. Review framing layout. Use the framing suggestion in the rough sketch (or your instructor's preferences) to lay out the framing members. Reveals are attached within the doorway opening. To ensure that the opening has enough clearance for the door's swing, consider adding 1/8″ to ¼″ to the opening's width and height. Extend guidelines into the other views as necessary.

3. Lay out desired cutting plane lines through the rear elevation. Lay out any remaining guidelines in the top and side views/sections.

4. Lay out lauan sheets on front elevation. A full sheet is 4′–0″ × 8′–0″. When laying out sheet goods on construction drawings, it is typical to draw thin lines corner-to-corner to emphasize individual sheets. Label sheets to facilitate creation of the parts list.

5. Lay out dimensions, labels, and notes. Dimensions may repeat between front and rear elevations, as the drawings have two separate purposes.

6. Darken the drawing. On this sheet, the front elevation featuring lauan placement can be considered as one drawing (Dwg 1: Front Elevation Lauan Placement), and the rear elevation and its associated views/sections as another, with the top and side views/sections as its subsets (Dwg 2A: Top View).

7. If in Vectorworks, create viewports and import to the sheet layer (Sht-2). Since all of the drawings on this sheet are in the same scale, no scale is required on individual drawing labels. The scale for the entire sheet is placed in the title block. Add borders and title block.

8. Darken the border, title block. Add title block information. This is Sheet 2 of 2.

9. In the Vectorworks, save the file. Export each sheet layer as a PDF. Print both PDFs.

MODULE 3: PLATFORMS

What is a Platform?

A platform is a structure that provides a weight-bearing performance area raised above a permanent floor or deck. A typical stock platform consists of a frame and a covering, and is usually 4′–0″ wide by 8′–0″ long, as that's the standard size of a sheet of plywood. Individual stock platforms can be joined together to create a larger raised area. Budget and production needs determine construction materials (such as wood or steel for the frame). Platforms must always be strong and reliable, as performer safety is paramount. Irregularly shaped platforms are frequently constructed, but this module will focus on describing basic rectangular platforms.

Parts of a Platform

The most common basic style of platform is the rigid wooden platform. It uses 2 × 4 lumber for the frame and ¾″ plywood for the top, or lid. Legs, also typically 2 × 4, are usually bolted inside each corner and inside at the midpoint of the long sides. They are spaced no more than 4′–0″ apart for this style of platform. Interior framing members are often called **joists** rather than toggles. The platform size determines the number of joists; they are typically spaced 2′–0″ apart to prevent the plywood top from flexing when weight is added (Figure 6-35).

Joints are glued, and fasteners are driven through the thickness of the framing members to secure them. The lid is fastened to the frame. This style of platform is heavy and awkward to transport, but budget-friendly; it gets the job done. Another common style of frame is built using 5¼″ wide strips of ¾″ plywood and requires more complex joints

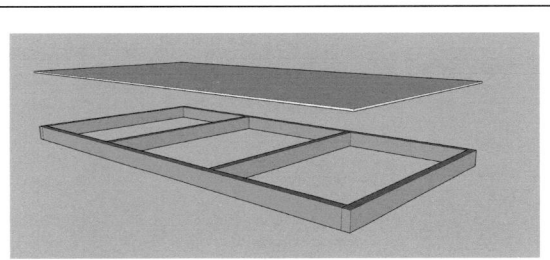

FIGURE 6-35 *Model of a 4′–0″ × 8′–0″ platform frame with ¾″ plywood lid*

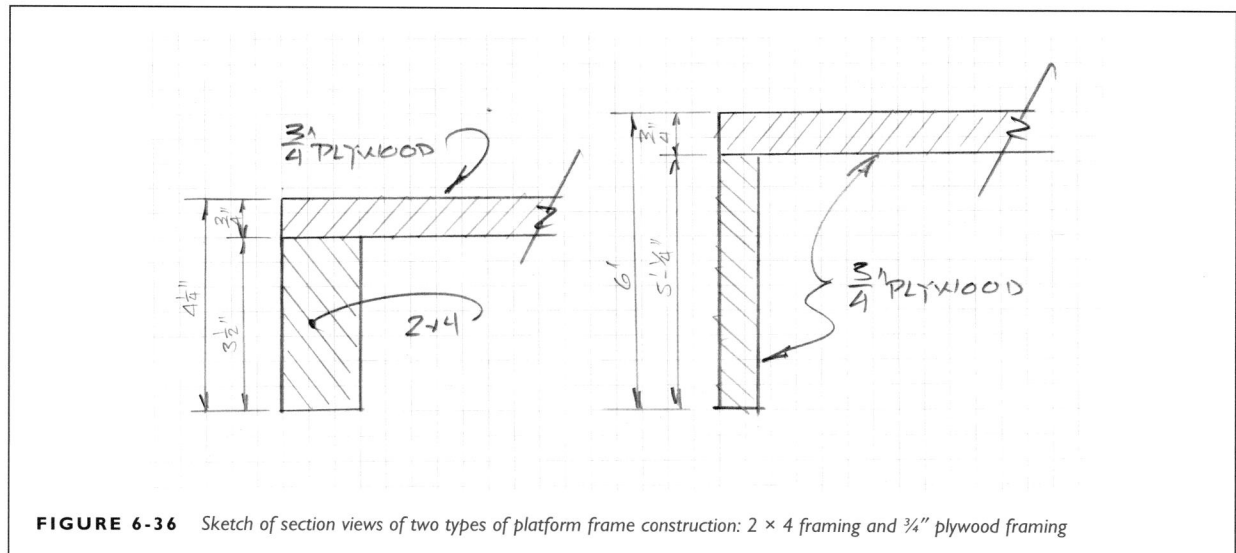

FIGURE 6-36 *Sketch of section views of two types of platform frame construction: 2 × 4 framing and ¾″ plywood framing*

and extra joists; it is lighter and provides an even 6″ platform height rather than the 4¼″ overall height of a 2 × 4 framed platform (Figure 6-36). There are multiple legging schemes, but whichever type of legging or support is used, it must be stable and secure.

Crossbracing and/or **stretchers** are attached to the legs to prevent parallelogramming. A stretcher is a horizontal piece of wood connecting legs near their bottoms; they prevent legs from splaying outward. For a large raised area, platforms may rest upon and be secured to 2 × 4 **stud walls** (also called **knee walls** when they are less than 3′–0″ tall). The vertical members of this frame are called studs, and are typically spaced either 16″ or 2′–0″ apart. The lower horizontal members are called **plates**.

The upper pieces are called **caps.**[5] Diagonal crossbracing is also added to stud walls to prevent paralellogramming (Figure 6-37).

In most cases, the plywood lid will be covered with other sheet materials to create a more finished or paint-friendly surface. Masonite is a sheet material made from compressed wood pulp. It is brittle and does not support weight on its own, but provides a smooth paint surface.

Sound-deadening material such as Homosote or Celotex (neither of which are weight supporting) can also be layered in between Masonite and plywood.

Each addition of sheet material to the top of a platform adds height that affects the height of the legs or stud wall. If a masonite-covered platform is intended to have a total height

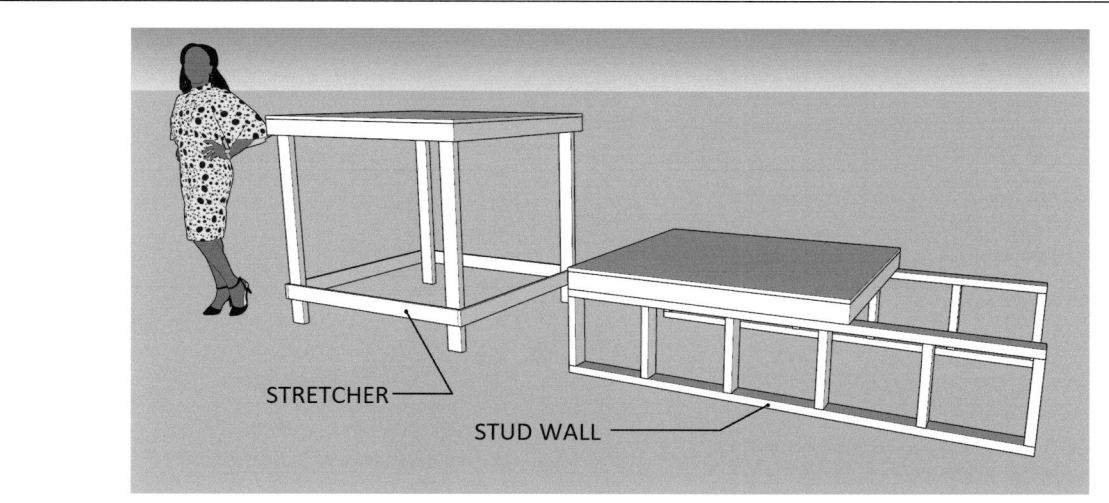

FIGURE 6-37 *Square platform with stretchers around legs; platform sitting atop two stud walls*

of 2′– 0″, the legs must be ¼″ shorter than they would be without the Masonite. It seems like very little, but it's enough to misalign flats and other elements that butt against the platform.

To hide structure beneath platforms, **facing** may be installed on sides facing the audience. Since facing is not a structural element, there is more leeway in choice of materials. Lauan and masonite are typical, especially for straight runs. Bendable lauan is an option for rounding curves. Platform sides can also be covered with fabric, such as pleated velour panels.

Nosing is a strip of wood or decorative trim attached along the top edge of the facing to emphasize the platform edge. Nosing is frequently attached to the front lip of steps in a staircase.

Baseboard may be attached along the bottom of the facing.

As they are with flats, designers are more concerned with the overall area and exterior appearance of platformed areas. The technical director and carpenters will be concerned with the construction details of the platforms. In Figure 6-38, the designer drawing on the left shows the full extent of the platformed area with front and side views. Individual stock platforms are not indicated; rather, the important feature is where there is a change in height.

The construction drawing in the lower half of Figure 6-38 shows a top view of individual platform placement alongside a drawing describing how an individual platform frame is to be built. Since the raised area will consist of eight platforms joined together, it's only necessary to describe one frame and include the note "Build Eight." Leg lengths are dictated by platform height; a note indicates how many of which length are required.

The height of platforms and stairs is indicated by writing the measurement in total inches on the top view. Each change in level is marked. A plus sign indicates that the measurement denotes the surface's distance above the venue deck or **zero deck**. The zero deck is indicated by + 0″. Surfaces below the deck, such as those accessed through trap doors, are indicated with a minus sign (e.g., – 24″). The height measurement may be enclosed by a bubble. Heights may also be included on side and front views of platforms as dimensions (Figure 6-39).

Exercise: Drawing Platforms (Design)

Instructor

This exercise constructs the design drawing of a large platformed area. Figure 6-40 presents a rough sketch with measurements. This is an array of straightforward

orthographic views, plus a detail drawing. It can be drawn either by hand or with Vectorworks.

Students

Figure 6-40 presents a rough sketch of large stepped platform. Use Arch D paper, landscape orientation, ½″ = 1–0″ printed scale. Dimension fully, notate, and label. Include borders and title block. Draw either by hand or with Vectorworks. Lauan facing is affixed to the front of each rise, as well as to the right and left sides. 1″ × 2″ nosing is affixed to the lip of each rise. Include a detail drawing of the nosing profile in a larger scale.

If working in Vectorworks, determine how many design layers are required. Create them.

In layout planning, reserve space for a detail drawing of the nosing profile.

1. Since the platformed area is clearly described in the top view, consider beginning by laying out that view's perimeter. Drop guidelines down into the front view area. Use a miter line to extend guidelines into the side view.

2. In the top view, lay out the front edges of each tier of platforming. Extend guidelines into the side view via the miter line.

3. In the front view, lay out the heights of each tier. Extend guidelines into the side view.

4. The nosing is attached to the front of each step. In the top view, lay out the front edge of the nosing 1″ from and parallel to the edge of each tier.

5. In the front view, lay out the bottom edge of the nosing, beneath and parallel to the edge of each tier. Extend guidelines into the side view. In the side view, complete layout of the nosing end views.

6. Lay out the nosing detail drawing to one side of the page.

7. Lay out dimensioning, notes, and labels. Include notes indicating facing and nosing. In the top view, include height notation on each tier. Decide how to number the drawings and views. Since scale varies, include scale as part of the drawing labels (this can be done as part of viewport creation).

8. Darken the drawing.

9. If in Vectorworks, create viewports and import into a sheet layer. The nosing detail should have its own viewport, set to its own scale.

10. Add the border, title block, and title block information. The title block scale is As Noted.

11. If in Vectorworks, save the file. Save as a PDF. Print.

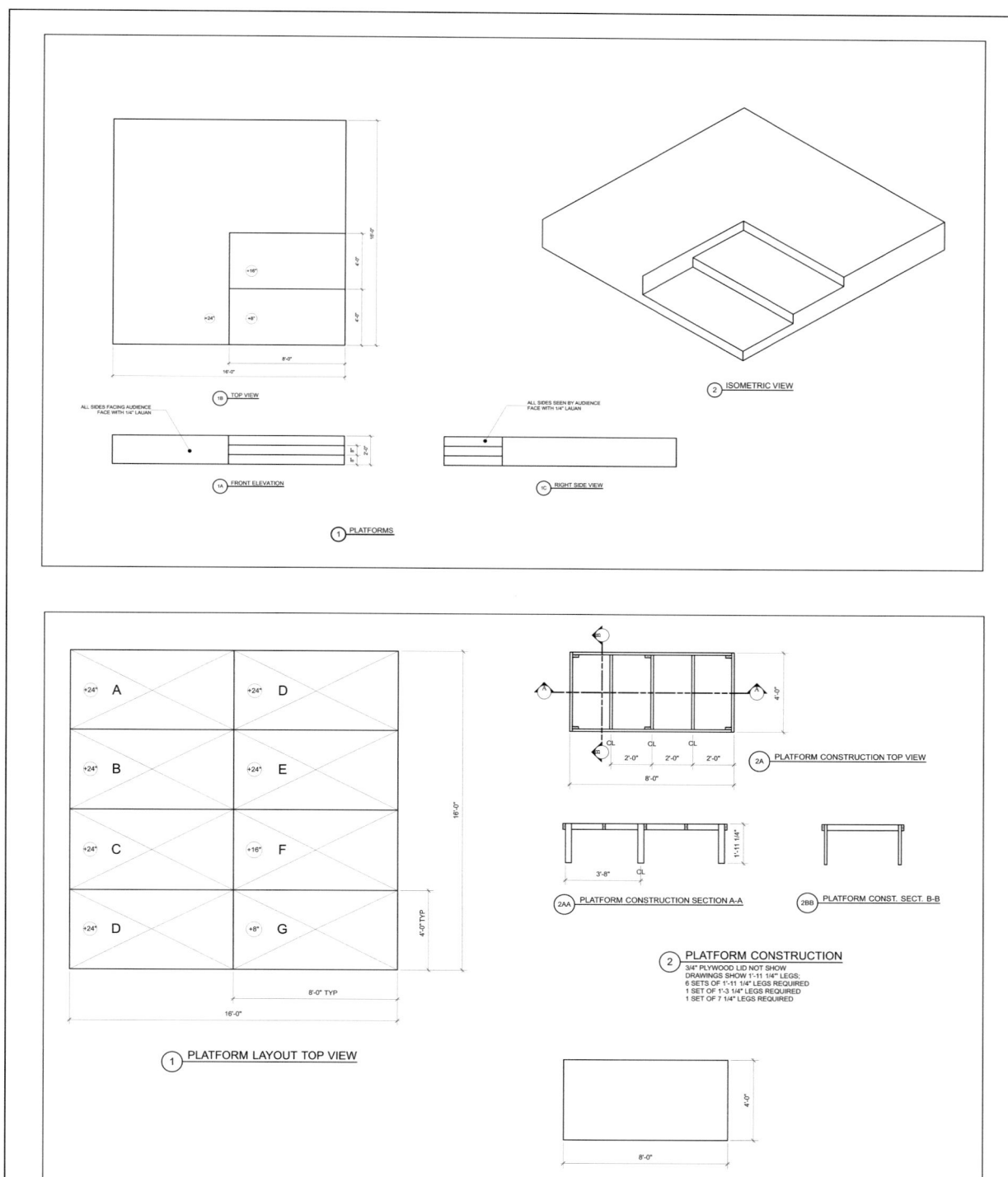

FIGURE 6-38   *Designer and construction drawings for a platformed area*

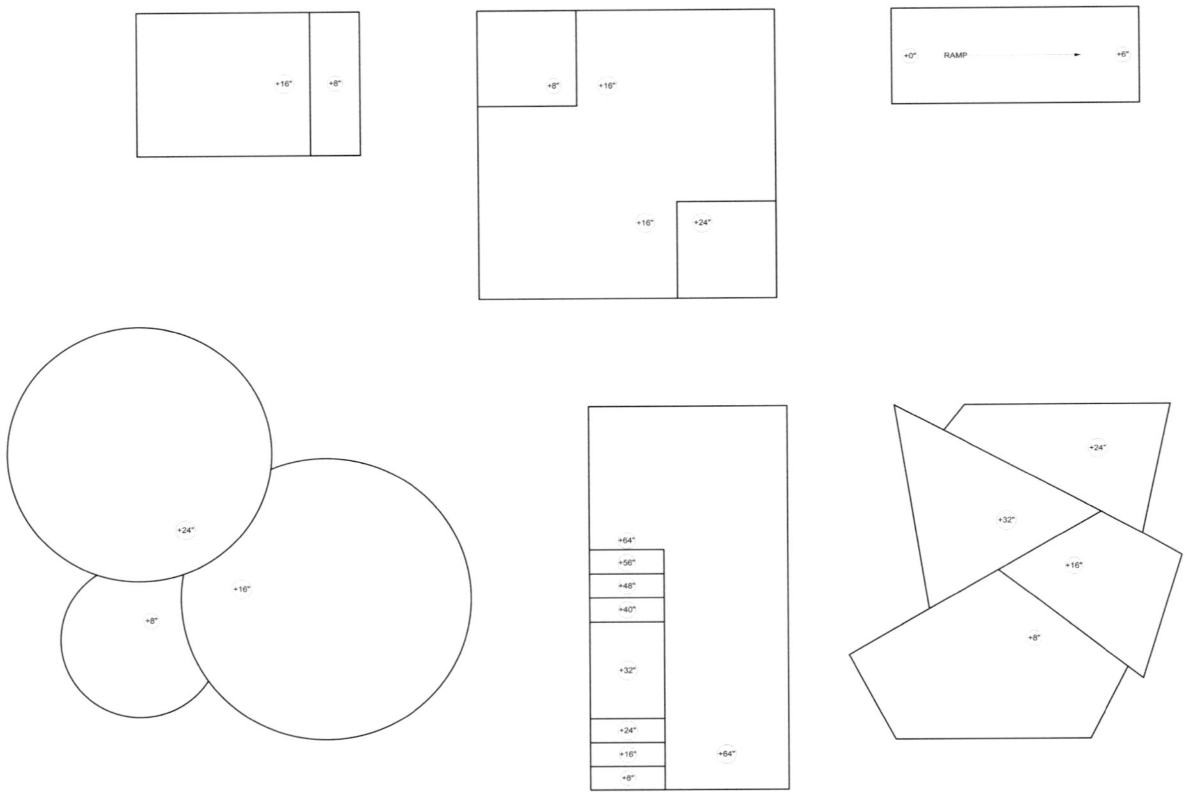

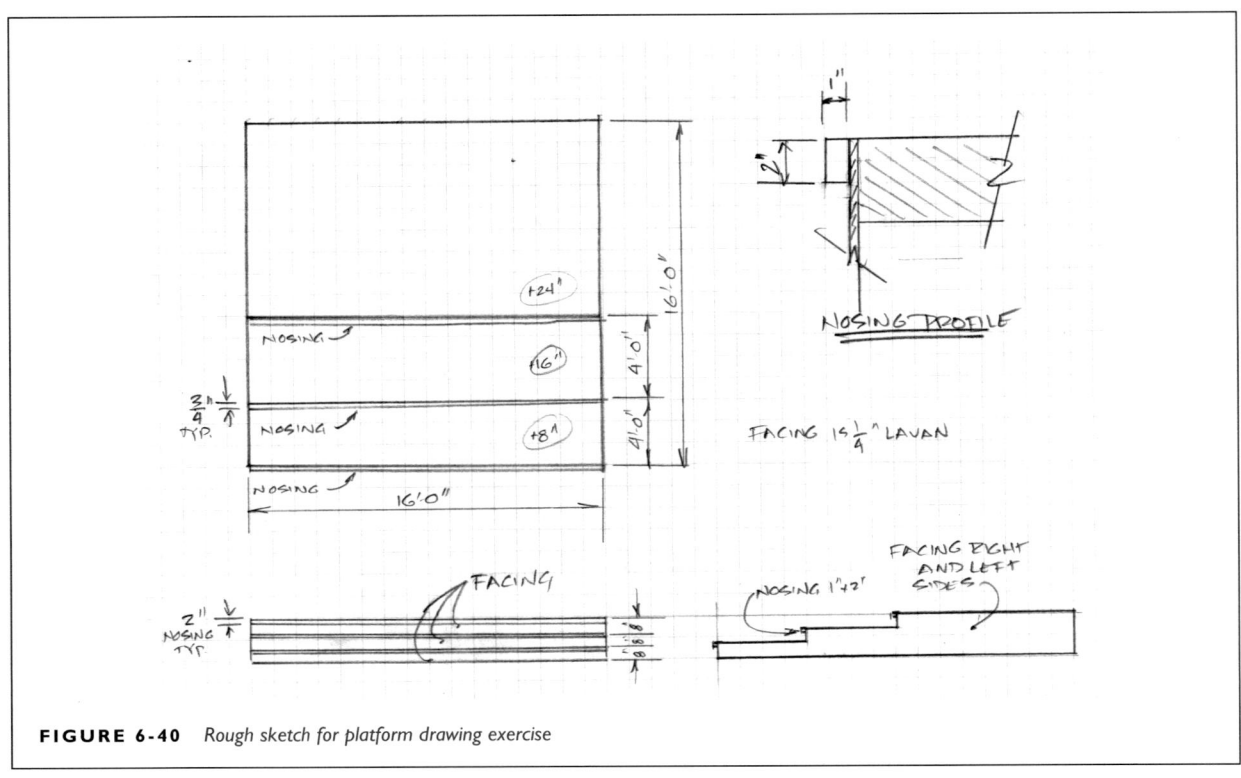

FIGURE 6-39 *Examples of height notation on a variety of platform configurations*

FIGURE 6-40 *Rough sketch for platform drawing exercise*

Homework: Modeling a Platformed Area

In SketchUp, use the drawing generated in the previous exercise to construct a model of the platformed area. Refer to your drawing for dimensions.

1. Open SketchUp. Move the human figure aside. From the axes' intersection, drag out a horizontal square.

2. Raise the surface to the height of the first step. Lay out the edge of the second tier upstage from the front edge. Draw a line over the guideline to create two independent surfaces.

3. Raise the rear surface to the height of the second tier. Lay out and draw a line to mark the edge of the final tier. Raise the rear surface to the height of the third tier. All three levels are now in place. If desired, use the Paint tool to select a texture/color for the tops and front faces of the platform. Make the object a group (Figure 6-41).

4. Zoom in on the right-hand side of the lowest step. Orbit to view the tier from the right side. From the corner of the step, drag out a rectangle for the end plane of a strip of nosing, perpendicular to the step's face (Figure 6-42).

5. The Follow Me tool allows you to select a shape and then extrude it along a series of chosen lines (Figure 6-43). This is useful when, for example, modeling crown molding that goes around a series of corners. Follow Me can also be used to build spheres or lathe-turned furniture legs by drawing a half-profile and then having the tool follow a circle.

    Since the platforms are a group, they will not interact with the rectangle representing the end view of the nosing. Draw a line from one end of the step to the other, over the existing edge. Click on the line running along the edge of the step to highlight it. Click on the Follow Me tool. Click on the surface of the nosing rectangle. A strip of nosing will appear, running the length of the tier's front edge. To follow a series of lines, hold down Shift on the keyboard as you select the lines.

6. Make the nosing a group. Use the Paint Bucket tool to select a texture for the nosing.

7. Copy and Move to place nosing against the edges of the remaining two tiers.

8. Save the file.

Exercise: Drawing a Stepped Platform (Design)

Instructor

This exercise features a platformed area featuring stepped, round areas (Figure 6-44). Students should lay out the centerlines first in order to draw the circles in their correct locations. Only the top view is given, but the height notation provides enough information to construct the front and side views.

Students

Figure 6-44 presents the top view of a group of round platforms. Construct front, top, and right side views.

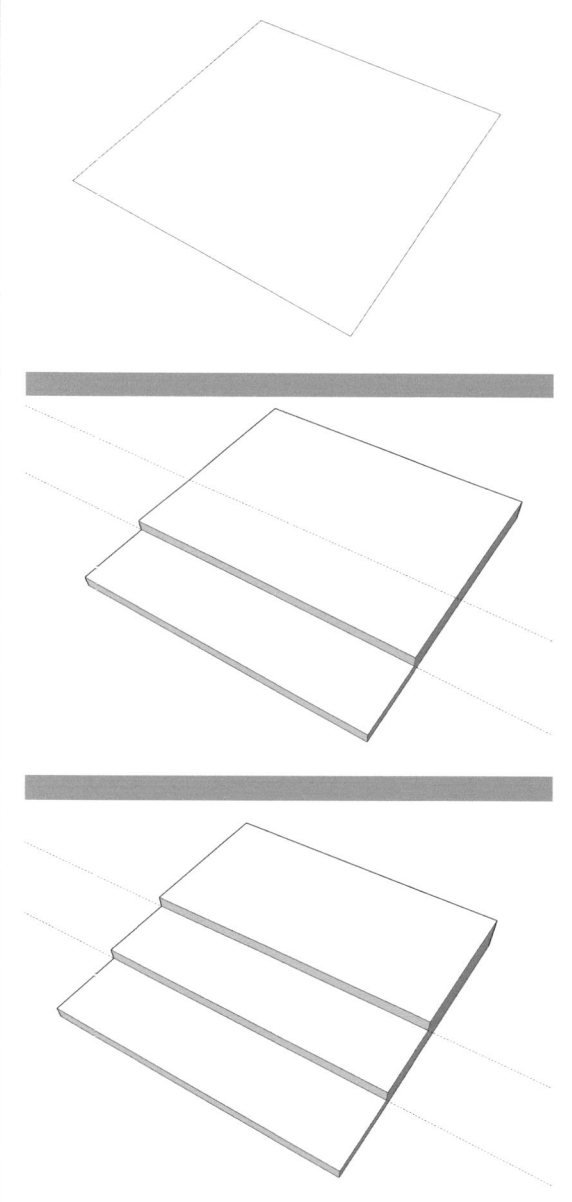

FIGURE 6-41   *Steps in modeling the tiered platform: (1) lay out base, push/pull; (2) lay out edge of second tier, push/pull; (3) lay out edge of third tier, push/pull*

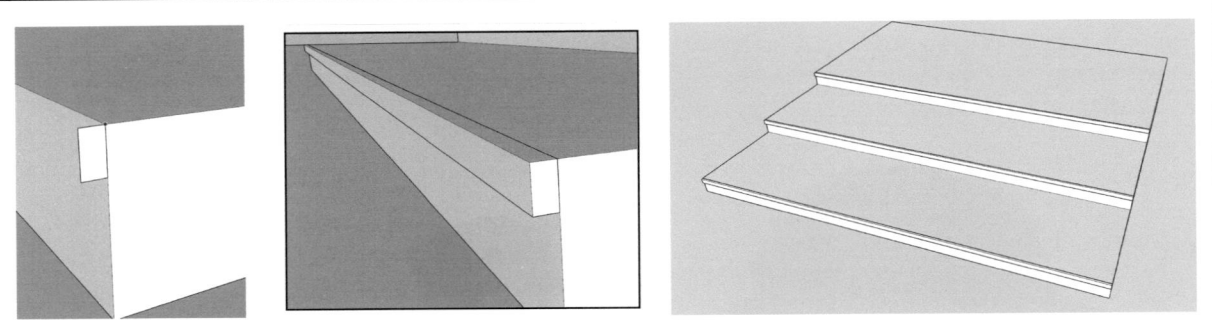

FIGURE 6-42 *Drawing the nosing profile; using Follow Me to extrude along step; copy and move to add nosing to remaining tiers*

The top view's height notation provides enough information to construct the front and side views. Use Arch D paper, landscape orientation, ½″ = 1′–0″ printed scale. Dimension, notate, and label. Include border and title block. The platforms feature lauan facing around their full circumferences.

If hand drafting, do layout math to situate the views on the sheet; the diameters and graph paper increments should help you calculate rough perimeter measurements. In Vectorworks, the reference grid can be shifted once a substantial portion of the drawing is laid out; so layout math is less essential. Since all of the platforms are circular, the first step is to lay out their centers. Since the top view is given, begin by laying it out.

1. The + 8″ and + 24″ platforms share a center line. In the top view area, lay out a vertical guideline for this center line.

2. Determine how far the vertical center lines for the + 32″ and + 16″ platforms lie from the first center line. Lay them out.

3. Draw a guideline to mark the horizontal center line of the + 24″ platform. From this guideline, measure and mark the horizontal center lines for the other platforms.

4. Since these are large circles and you'll need their perimeters to import information into the front and side views, lay out the circles with guidelines. Trim/delete the portions that disappear under an adjacent platform (Figure 6-45).

5. Use guidelines to drop the width of each platform into the front view.

6. In the front view, lay out the height of each platform between their respective width guidelines.

7. Using a miter line, transfer depth information into the side view. Add height information. Note that the + 8″ platform has a portion hidden behind the + 16″ and + 24″ platforms. This will be indicated with hidden lines.

8. Lay out dimensions, notes, and labels. Use leader lines to place diameter/radius information. Use leader lines to indicate faced areas.

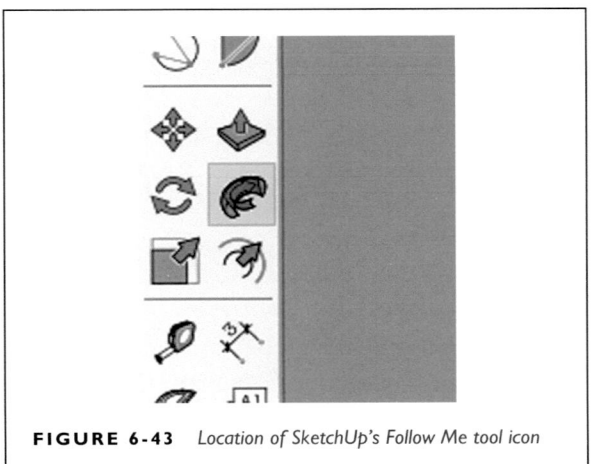

FIGURE 6-43 *Location of SketchUp's Follow Me tool icon*

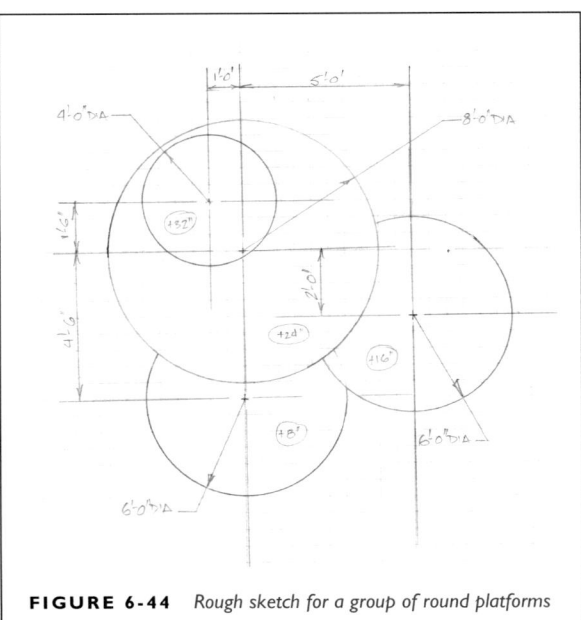

FIGURE 6-44 *Rough sketch for a group of round platforms*

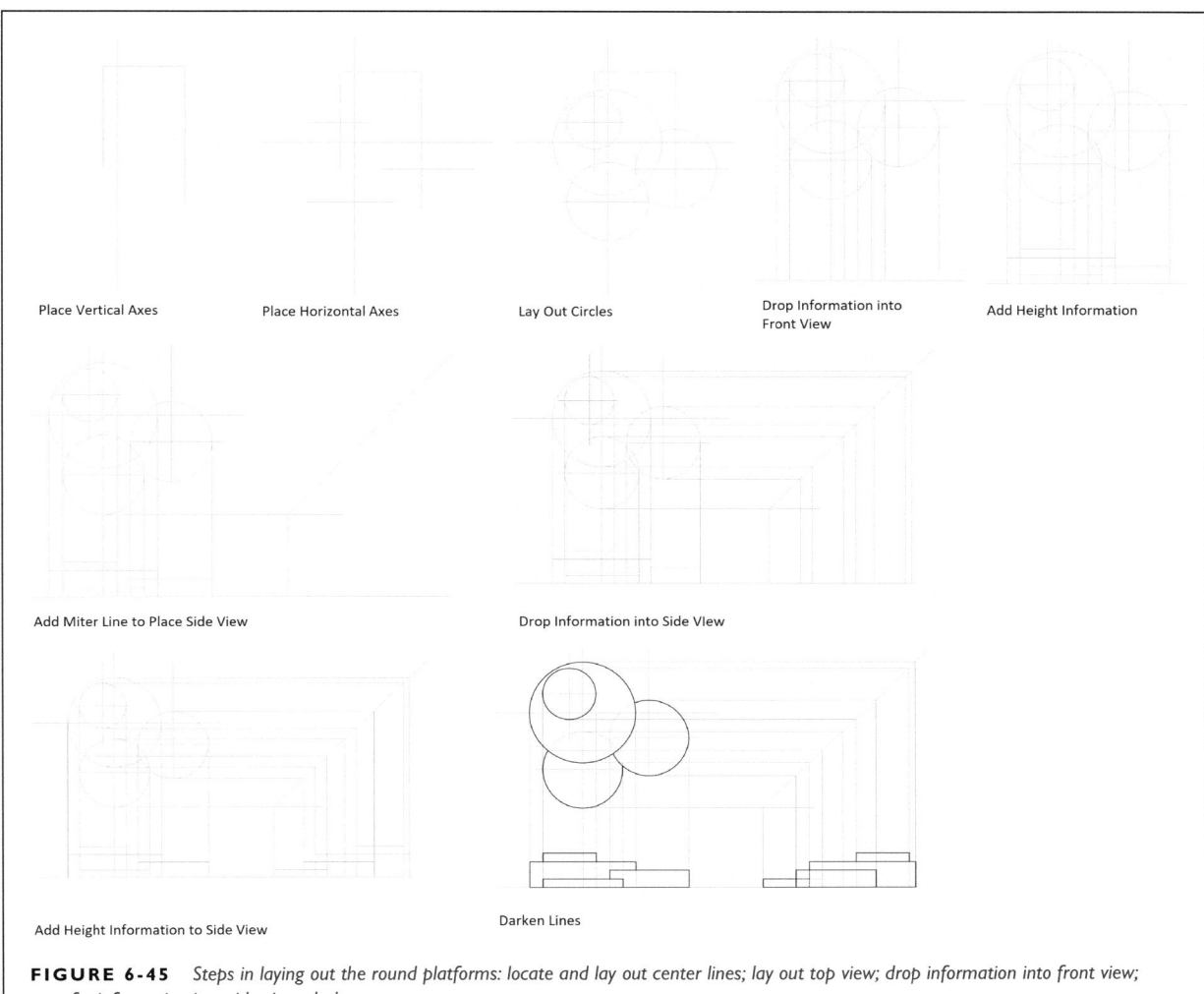

Place Vertical Axes   Place Horizontal Axes   Lay Out Circles   Drop Information into Front View   Add Height Information

Add Miter Line to Place Side View   Drop Information into Side View

Add Height Information to Side View   Darken Lines

FIGURE 6-45   *Steps in laying out the round platforms: locate and lay out center lines; lay out top view; drop information into front view; transfer information into side view; darken*

9. Darken the drawing.

10. If in Vectorworks, create a viewport and import into a sheet layer.

11. Add borders, title block, and title block information.

12. If in Vectorworks, Save the file. Save as a PDF. Print the PDF.

Exercise: Construction Drawing of a Platformed Area

Instructor

Figure 6-46 shows the construction elements of a platformed area supported by stud walls. The platforms are stock 4″–0″ × 8″–0″ platforms. The image does not present platform framing structure; consult drawings earlier in this chapter if you wish the students to include them. If not, a note instructing the reader to use stock platforms will suffice.

Students

Use Figure 6-46 to draft the construction drawing of platforms supported by stud walls. Your instructor may direct you to include a platform framing drawing; if not, include a note indicating that stock platforms are to be used. Include a stud wall detail drawing.

Use Arch D paper, landscape orientation. Printed scale is ½″ = 1′–0″; the stud wall detail drawing may be in a larger scale. Fully dimension, notate, and label. Include borders and title block. In Vectorworks, if desired, darken views as completed.

1. Do layout math. Determine how many design layers are required. Reserve space for stud wall detail and notes.

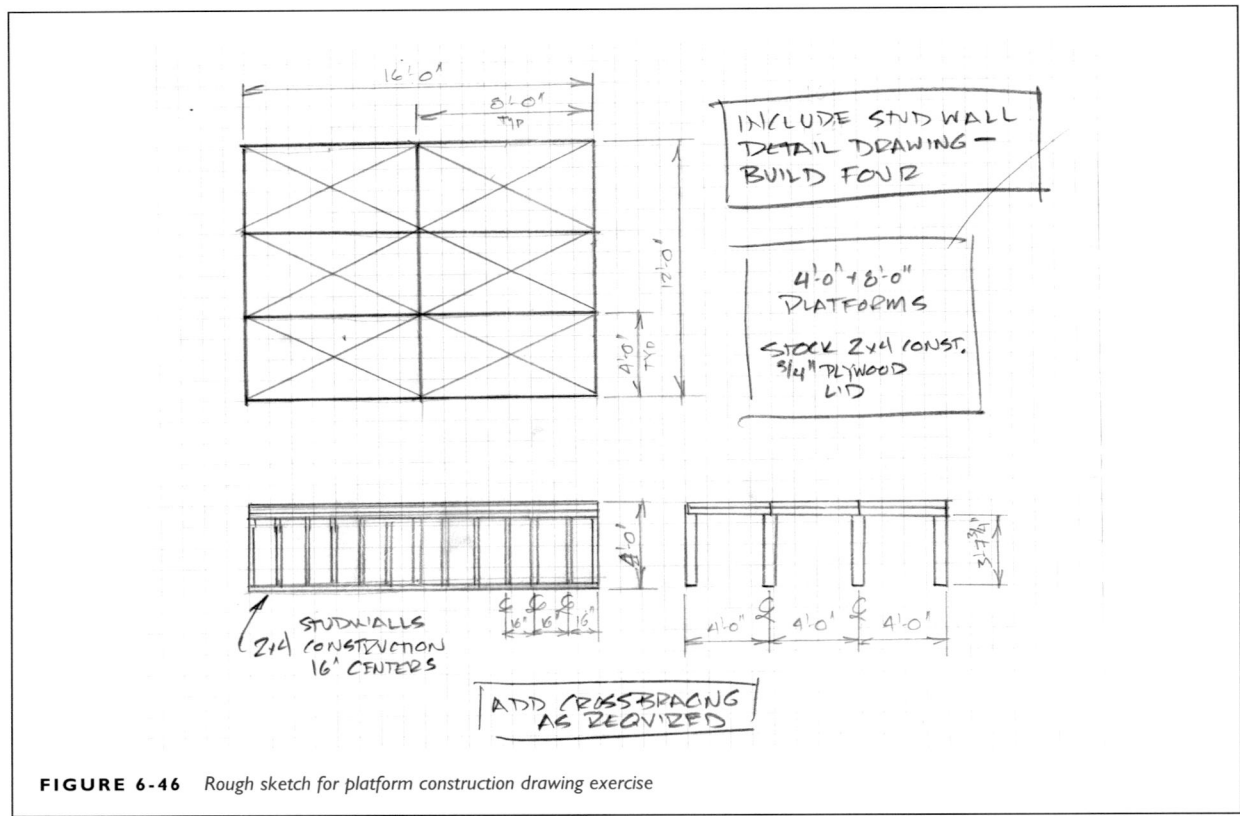

FIGURE 6-46   *Rough sketch for platform construction drawing exercise*

2. Lay out the perimeter of the top view. Drop guidelines into the front view area.

3. Lay out the perimeter of the front view. Extend guidelines into the side view area.

4. Lay out the perimeter of the side view, either by measuring or using a miter line.

5. In the top view, lay out individual platforms, including an X from corner to corner. Lay out the front and rear edges of the stud walls (beneath the platforms). These will become hidden lines. Lightly label the platforms (A, B, C, etc.)

6. In the front view, lay out the elements of the stud wall (cap, plate, studs), and the front edges of the platforms. The studs are on 16″ centers; if the overall length of the stud wall is not divisible by 16, chose one end to have a space less than 16″.

7. In the side view, lay out the ends of the stud walls and edges of platforms. Include the ends of caps and plates.

8. Since the stud wall's front elevation has been drawn as part of the front view, in Vectorworks a copy can form the basis of the detail drawing. Use the Select tool to drag a box around the front view. Right click the mouse, select Copy. Move the cursor to its designated area and Paste into the designated layer. Adjust placement as necessary. Cut/Delete extraneous lines. The stud wall does not require a top and side view, though they can certainly be included. If a top view is drawn, include hidden lines for all studs. Note that the stud wall is 2 × 4 construction.

9. Lay out dimensions. Dimension to the vertical center lines of the studs. Dimension to the centers of the stud walls in the side view. TYP may be used to reduce repetition. Add notes and labels. Decide whether title labels are to be added to design layers or during viewport creation.

10. Darken the drawing.

11. If in Vectorworks, create viewports and import to a sheet layer. Use a separate viewport for the stud wall detail (and platform frame, if included).

12. Add borders, title block, and title block information. Scale is As Noted since scales should appear with each viewport's label.

13. Save. Export as PDF. Print the PDF.

FLATS AND PLATFORMS DRAFTING CHECKLIST

- Lay out from big to small. Start with perimeters, work toward detail.

- Multiple elevations sharing a sheet should share a baseline. Standard orthographic view placement helps keep the sheet tidy and clear.

- Use sections to describe interior features within flat openings, such as doors and windows.

- Add larger scale detail drawings to describe elements that cannot be described adequately in the scale of the orthographic view.

- Insert photos, model images, and other research as needed to further describe details.

- Describe molding by including profile detail drawings.

- Design drawings of flats do not typically include framing; minimize the use of hidden lines.

- Construction drawings for flat framing do not typically include scabs; be sure that notes include information on securing joints.

- Use leader lines and notes to place information that standard dimensioning practice cannot accommodate.

- Design drawings for platforming are most concerned with overall area and appearance.

- Construction drawings describe structure and use of stock items.

- Be sure that scale is included in each drawing's label if it varies across the sheet, as and when detail drawings are included. In the title block, the recorded scale is As Noted.

- Include notes indicating covering materials, such as muslin or Masonite.

NOTES

1. Howard Bay, *Scene Design* (New York, NY: Drama Book Specialists/Publishers, 1974), p. 129.
2. Dennis Dorn and Mark Shanda, *Drafting for the Theatre* (Carbondale, IL: Southern Illinois University Press, 1992), p. 183.
3. John Blurton, *Scenery: Drafting and Construction for Theatres, Museums, Exhibitions, and Trade Shows* (New York, NY: Theatre Arts, 2001), p. 82.
4. Bill Raoul, *Stock Scenery Construction Handbook* (Shelter Island, NY: Broadway Press, 1990), pp. 26–28.
5. Dorn and Shanda, p. 193.

CHAPTER 7

DRAWING AND MODELING THE PLAN OF A PROSCENIUM STYLE VENUE

Chapters 8, 9, and 10 address the three major orthographic views of a scenic environment: the groundplan, the center line section, and the front elevation. To prepare for those discussions, this chapter presents a basic proscenium venue, using it to draft a plan and construct a model. This process will begin to familiarize students with what to expect on a venue's base drawing. Theatrical vocabulary terms will be reviewed.

Whether drafting something small or something large, the drafting process remains consistent. A building may seem like an overwhelmingly huge and complex structure, but it too can be broken down and organized by outlines and shapes that are then plugged into a typical system of working. For a ground or floor plan, the center line becomes the spine of the drawing, giving the draftsperson a starting point from which to build outward.

TOPICS AND GOALS

- What is a base drawing?

- Performance space vocabulary terms

- Drafting a venue plan/base drawing

- The line schedule

- Modeling a venue from a drawing

MODULE 1: DRAFTING THE PLAN OF A VENUE

Base Drawings

A scenic structure is usually placed within an architectural space—whether theatre, convention hall, or park pavilion. Always request a set of **base drawings** when beginning

any project. If no overview architectural drawings exist, you may have to survey and measure the space yourself to create them. A package of base drawings provides such information as how much space is available for the project, the relationship of the audience/viewers to the project, and specifics on equipment and resources.

Depending on a company's resources, base drawings may be distributed in a variety of formats, from photocopied bluelines to PDFs to computer-aided design (CAD) files, whether Vectorworks, SketchUp, or AutoCAD, among others. CAD files may need to be converted into formats for which you have access; the DWG file format can help in this conversion.

When receiving Vectorworks base drawings from different companies, you'll discover that classes and layers may not be organized in the way that you might approach the drawing. Spend time exploring the file to see what will impact your own drafting preferences. You may need to create new layers and trace over existing layers to create a version of the file with which you can more easily work. On the other hand, a company may have things set up in a manner that optimizes their workflow; you might even pick up a few new techniques and approaches. If you are submitting full CAD files rather than PDFs, check with the client regarding preferences. The larger the company, the more structured their workflow is likely to be. Time is money, and they're not going to want to redraft your work.

Classes in Vectorworks can be used to draw objects, not just individual lines. A specific weight, style, and color can be used to draw, for example, all of the auditorium seating. Some draftspersons connect all line styles to specific types of objects throughout a drawing; this means that it can be difficult to add a line to a drawing if there's no basic "pencil box" of line styles included amongst the myriad classes. If

DOI: 10.4324/9781003154921-8

this is the case, create your own pencil box. You might then trace features of the drawing into new layers, turning off the visibility of no longer needed classes and layers. You can also delete classes and layers. Be sure to save alterations as a new file, as you'll want to be able to refer back to the original.

Vocabulary Terms

Knowing these theatrical vocabulary terms will aid in the comprehension of the following exercise's instructions:

Upstage: (toward) the rear of the stage

Downstage: (toward) the front of the stage

House: where the audience is seated (auditorium)

Deck: the permanent floor of the stage

Stage Right: when facing the audience, to the right

Stage Left: when facing the audience, to the left

House Right: when facing the stage, to the right

House Left: when facing the stage, to the left

Proscenium: a permanent wall with a picture frame-like opening through which the audience sees the stage (the opening is called the **proscenium arch**); many theatres are configured to resemble the proscenium format without actually having an architectural arch.

Apron: in proscenium theatres, an extension of the deck into the House, downstage of the proscenium arch

Thrust: a stage configuration in which the deck is surrounded on multiple sides by the audience

Arena or In-The-Round: a stage configuration in which the deck is entirely surrounded by the audience

Plaster Line: in proscenium theatres, the upstage side of the proscenium, labeled as PL on drawings. If there is no proscenium, there is no plaster line.

Wings: the areas to stage right and left of the proscenium arch (wing-space); typically hidden from audience view and considered a backstage area

Off-stage: directionally, moving away from the center of the stage; locationally, another name for backstage areas

Smoke Pocket: in proscenium-style theatres, metal rails that guide the fire curtain as it drops to the deck. The fire curtain is sometimes called an Asbestos, because they used to be made of that flame-retardant material

Masking: collective name for flats, curtains, etc. whose purpose is to hide the backstage and overstage areas from audience view

Sightlines: a line drawn from an extreme seat to an object on stage (such as masking) that determines what the audience can and cannot see

Battens or bars: pipes suspended horizontally above the stage, typically raised and lowered by a lineset as part of a counterweight pulley system. If a batten is suspended at a fixed height, it is said to be **dead-hung**.

The term **pipe** is used interchangeably with batten, even though battens were originally made of wood

Electric: a batten with a functional lighting instrument hung on it becomes an electric

Lineset: a system of pulleys and weights used to raise and lower a single batten, as well as the batten itself. A full counterweight system includes multiple linesets: "Bring in lineset twelve"

Lighting Bridge: a suspended catwalk, typically over the stage. It may be a lighting position or provide access to overstage lighting positions

Line Schedule: a chart or table listing the names, numbers, and distances of battens operated by linesets.

Fly Rail: linesets are operated at the fly rail. This grouping of ropes, weights, and pulleys is usually placed against the far wall of a wing. Also called the **locking rail,** or simply, the rail.

Pit: short for Orchestra Pit. Typically a sunken area between the audience and the deck housing musicians when live music is performed as part of the production

FOH: abbreviation for Front-of-House, referring to items located in the auditorium (such as FOH lighting positions)

Crossover: a route used by performers to get from one side of the stage to the other without being seen by the audience. This is typically upstage behind scenery or the cyclorama, but can also refer to passage through dressing room areas or even routes outside the venue (such as through an alley)

Exercise: Drafting a Venue Plan in Vectorworks

Instructor

Using the information in Figure 7-1, the students will lay out and draft the plan of a generic proscenium venue. This drawing will be used for subsequent exercises. Construction of the drawing is broken up into the following phases, reflective of the previously discussed system of working:

- Laying out reference lines

- Laying out the deck and auditorium footprint

- Adding width to walls and placing wall openings

- Laying out features within the spaces, such as seating tiers and the fly rail

- Darkening the drawing

- Adding the line schedule and battens

Students should draft in ½″ = 1′–0″ on an Arch D reference grid. The drawing will exceed the boundaries of the reference grid; final printed scale and reorientation will be selected during viewport creation.

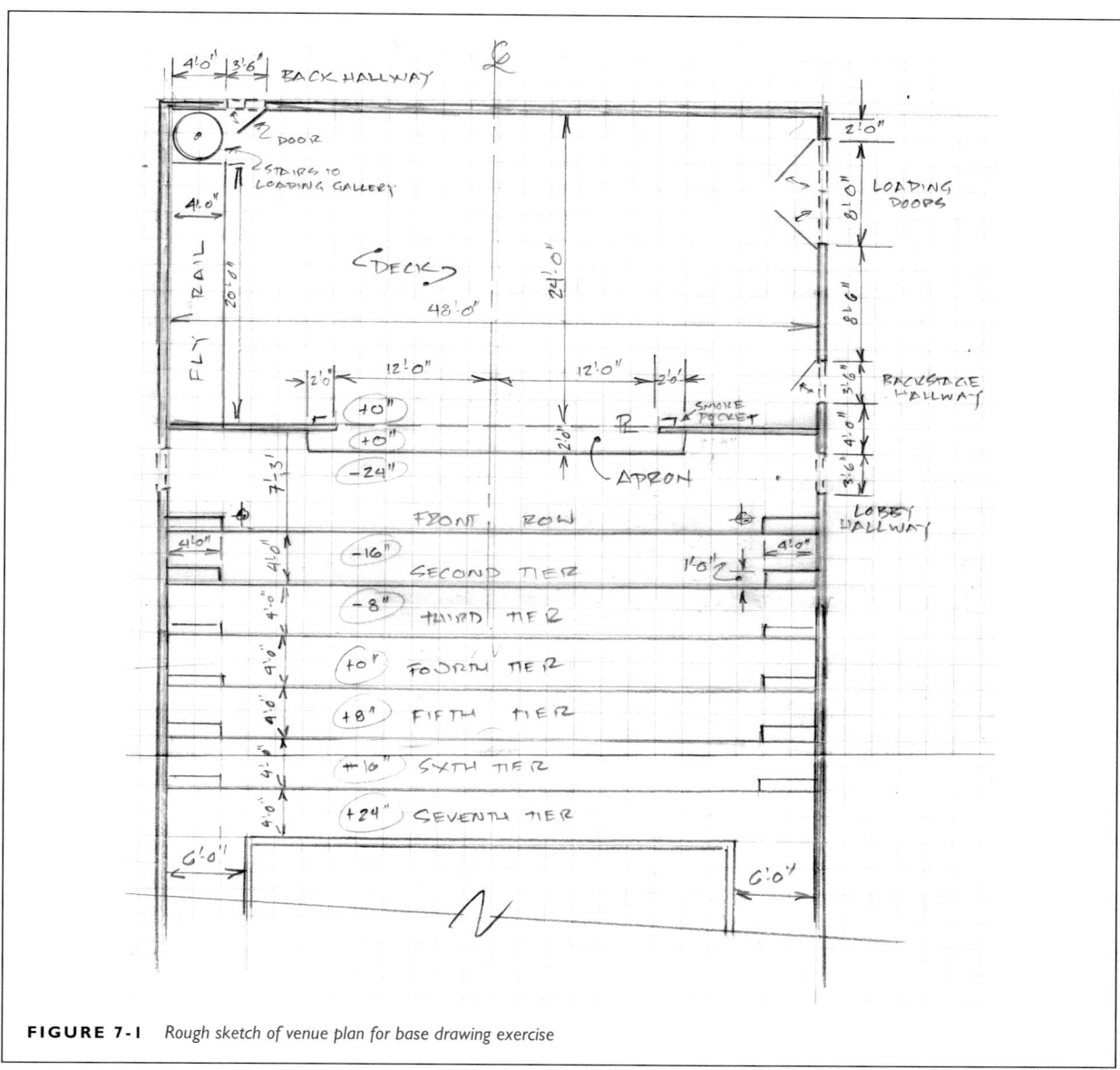

FIGURE 7-1   *Rough sketch of venue plan for base drawing exercise*

Students

Figure 7-1 features the plan of a proscenium style theatre. The drawing constructed in this exercise will become a base drawing for projects in the subsequent chapters. For purposes of clarity, some example figures will not include the reference grid. Figure 7-2 presents images of the first series of layout steps.

Ground and floor plans are section views. The cutting plane slices horizontally through the building/environment at about shoulder height above the deck. There are graphic conventions applied in the depiction of various objects and features—especially those that extend or hang above the cutting-plane line but remain important to the reader. These conventions are discussed in Chapter 8.

Three further line styles are used in this exercise. Create classes for:

*   **Thin Visible lines**: 0.13 mm, solid, no Fill.
*   **Plaster Line**: 0.25 mm, ISO-02, no Fill.
*   **Reference Guidelines**: as thin as possible, draftsperson-preferred color, no Fill.
*   **Phantom Lines**: 0.30 mm, ISO-09 Long Dashed Double-Short Dashed

After creating these classes, save your base file. Then Save As your base file with a new name specific to this project. Work in the new file.

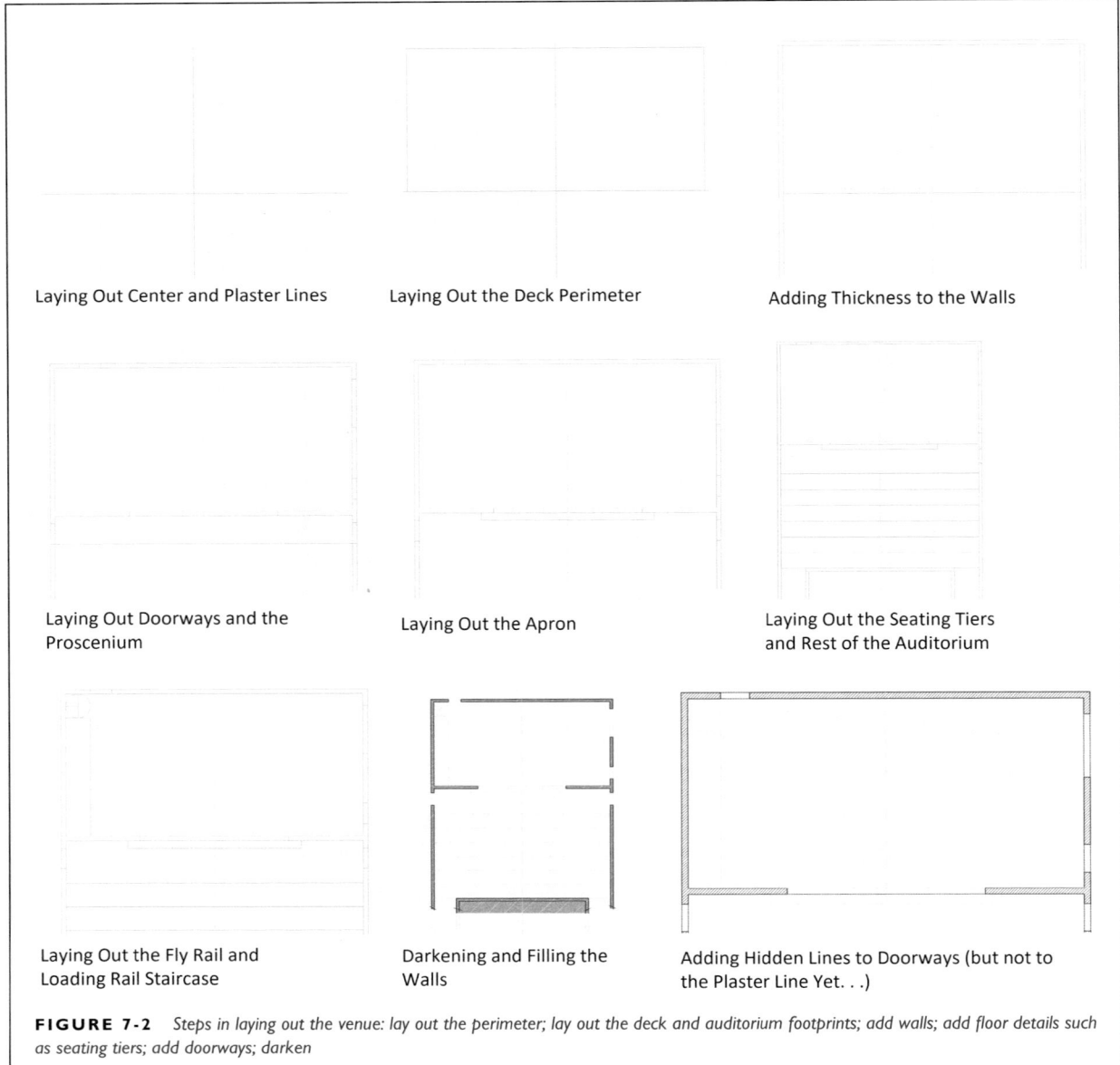

Laying Out Center and Plaster Lines

Laying Out the Deck Perimeter

Adding Thickness to the Walls

Laying Out Doorways and the Proscenium

Laying Out the Apron

Laying Out the Seating Tiers and Rest of the Auditorium

Laying Out the Fly Rail and Loading Rail Staircase

Darkening and Filling the Walls

Adding Hidden Lines to Doorways (but not to the Plaster Line Yet. . .)

FIGURE 7-2 *Steps in laying out the venue: lay out the perimeter; lay out the deck and auditorium footprints; add walls; add floor details such as seating tiers; add doorways; darken*

Only guidelines for the center line and plaster line will be drawn with the Reference Guidelines class. Give them a color distinct from your regular guideline class to keep them visually separate. Should you need to turn off regular guideline visibility, these will remain behind.

How the drawing's layers are set up will affect later work. Placing the venue's architecture in its own set of dedicated design layers allows them to be turned on or off by other users as needed.

The suggested design layers for this drawing are:

- Guidelines

- Venue Architecture

- Venue Architecture Dimensions and Notes

- Line Schedule and Battens

- Reference Lines

1. Work in ½″ = 1′–0″. This drawing will exceed the bounds of the reference grid, and in any case, a viewport will be used to export portions of it into sheet layers. No preliminary layout math is required. Copy and print a copy of Figure 7-1 to have it handy, or scan and import the rough sketch's image into a design layer. Be aware that the image will only appear when its design layer is active.

2. For most plan drawings, it's best to start with reference lines and work out from them. For a proscenium style venue, those lines are the center line and the plaster line. Not all spaces are symmetrical, so it's good practice to always start with the center line and work outward.

   Consult the sketch. The center line runs along the upstage/downstage (depth) axis. The plaster line is located along the upstage side of the proscenium wall. These two lines will be drawn with the Reference Guidelines class.

   Draw a horizontal line to situate the plaster line. This creates the baseline from which the rest of the drawing will be constructed, and its placement on the reference grid can be somewhat arbitrary. The deck is 48'–0" wide. The walls are 9" thick. We're not concerned with architecture beyond the stage and auditorium, so the line representing the plaster line does not have to be longer than 49'–6".

   > In venues where there is no plaster line, the draftsperson chooses some other lateral reference line for beginning layout, such as the downstage edge of the deck. A reference line at the front of the playing space is preferred to one at the rear of the playing space; if there is to be accumulated error, it should accumulate in somewhat less critical areas of the stage. This is the same reason load-ins begin by installing scenery and lighting at the center line and work toward the wings.

   The center line intersects the plaster line at its midpoint. At the midpoint of the plaster line, draw a vertical line upward. The deck is 24'–0" deep. The back wall is 9" thick. Extend the center line 15'–0" below the plaster line to extend it into the auditorium. It will be extended further later on. Switch to the Guideline class.

3. Lay out the deck. From the plaster line's midpoint, draw a line to stage left that is half the deck's width (24'–0"). Do the same toward stage right. From those two endpoints, draw vertical lines upstage (24'–0"). Add a horizontal line for the rear wall.

   To begin layout of the seating area, drop vertical lines from the lower right and left corners of the deck that are at least as long as needed to place the seating tiers. These lines will become the inside walls of the auditorium.

4. Use the Double Line tool to draw the thickness of the walls. The icon can be found next to the Line tool icon in the Basic Tool Box (Figure 7-3). As with the Line tool, options appear in the upper toolbar when it is selected. One mode places the second line to the right of the primary line; another mode will place it to the left, and yet another runs both lines equidistant from a center

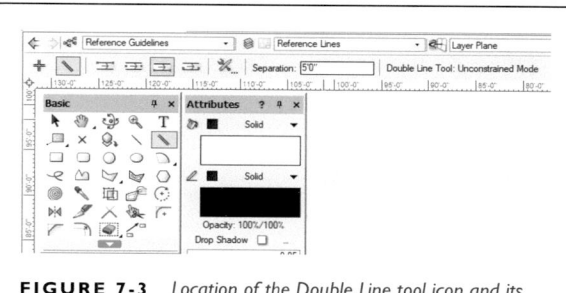

FIGURE 7-3   *Location of the Double Line tool icon and its modes*

axis. The side on which the second line appears also depends on the direction in which you draw the line. A text box allows you to set the distance (separation) between the parallel lines. Set the Separation to 9".

Select an option and draw over one of the already-drawn deck perimeter guidelines. If the second line falls to the outside of the original line, keep going. If the second line falls to the inside, you've just lost 9" of stage space. Undo, change your selection or direction, and try again. Along the plaster line, be sure that the second line appears downstage of (below) the first. Some corners may need to be extended farther to close them.

5. Locate wall openings: the door in the rear of the stage, two doors stage left, the proscenium opening, and doorways on either side of the first seating tier. Use guidelines to mark the placement of these openings within the walls. The two auditorium doorways are in line with each other; use guidelines to transfer the placement of the first doorway to the opposite wall for the second doorway.

6. Lay out the apron. Its front lip extends 2'–0" downstage of the plaster line. It extends house right and left 14'–0" from center. Work from the center line out and from the plaster line down.

7. Lay out the auditorium seating tiers. The first tier's edge is 7'–3" from the proscenium wall. Lay out the steps at either end of the tiers. Extend the walls and center line to lay out the seating tiers all the way to rear of the auditorium. Add the wall and entrances at the rear of the auditorium.

8. Lay out the Fly Rail area and the spiral staircase footprint on stage right. The spiral staircase leads up to the lineset system's weight loading rail; its footprint is 4'–0" DIA.

9. The architectural outlines are now laid out. Switch to the Venue Architecture design layer to darken lines. Since the walls are cut through by the cutting plane, they are shown as filled (Figure 7-4). You can use either

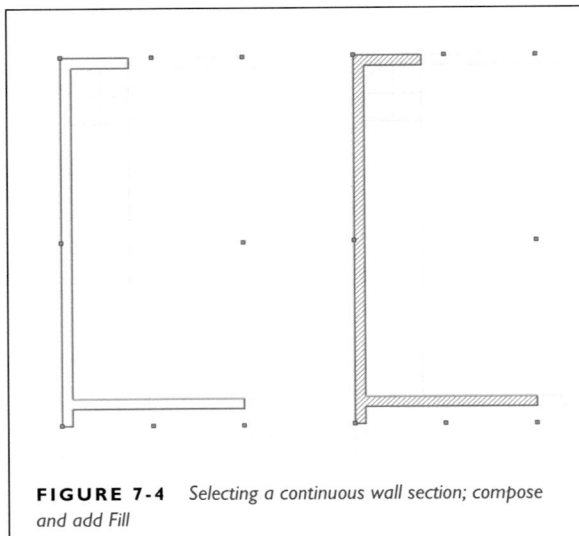

FIGURE 7-4 *Selecting a continuous wall section; compose and add Fill*

diagonal crosshatching (Cast Iron HF), or choose a concrete-style Fill. The open-ended auditorium walls require an extra step, explained in Step 10. There are two good approaches to drawing these filled outlines:

a. Darken a closed section of wall. Use the Select tool to drag a box around it. Compose. While it is active, select a Fill from the Attributes window. Work section by section, from doorway to doorway; this way portions of other walls won't accidently be included when selecting for composing.

b. Create a Filled Architecture class with medium line weight and the desired Fill. With this class, use the 2D Polygon tool in the Basic Tool Set to draw a section of the wall. When the shape is closed, it will automatically be composed and filled.

10. Open-ended objects can be composed; the Fill field will angle from endpoint to endpoint on the open side. Compose the auditorium walls below the first tier doorways and then draw a **break line** across the open ends (over the hatch) to indicate that the walls continue past what is drawn (Figure 7-5). Draw the break lines with the Dimension class.

Break lines are typically angled across the object or space being broken, so slant the break line across the open wall ends just within their endpoints. When the Break Line tool is chosen, a Navigation window appears, allowing you to change the size of the "heartbeat" indicating that the line is a break line.

When composing and filling the rear wall of the auditorium, the hatching may fill the area behind the wall as well, rather than being contained within the wall outline. Don't worry about this for the time being, as that back area is an architectural space that will not affect the overall meaning of this exercise's drawing. Solutions to this hatching overflow issue will be discussed later.

> You can create a line style class with no line. This is useful for closing shapes for fills without adding a line across the open side. Create a new class with no fill and no line. Decompose the back wall of the auditorium and use this class to close the open ends of the wall. Compose again and Fill. The Fill will appear only within the wall's outline.

The seating tiers and steps can be drawn with a line thinner than that used for the walls. The tiers do not require the visual precedence that the walls do or that features like lighting positions will need. The lip of the apron is an important feature and should be drawn with a thicker line.

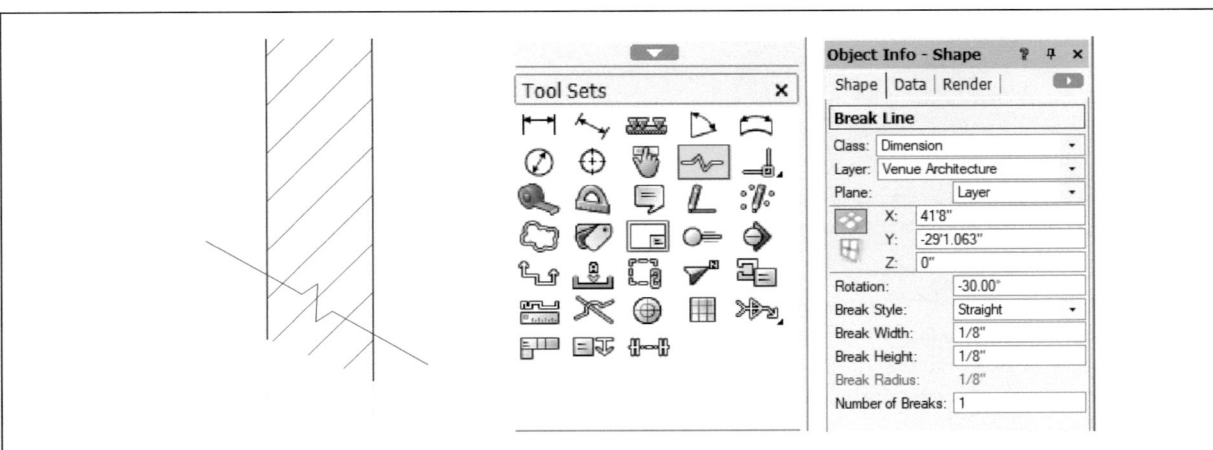

FIGURE 7-5 *A break line indicates that an object continues past what is drawn; location of the Break Line tool icon and its Navigation window*

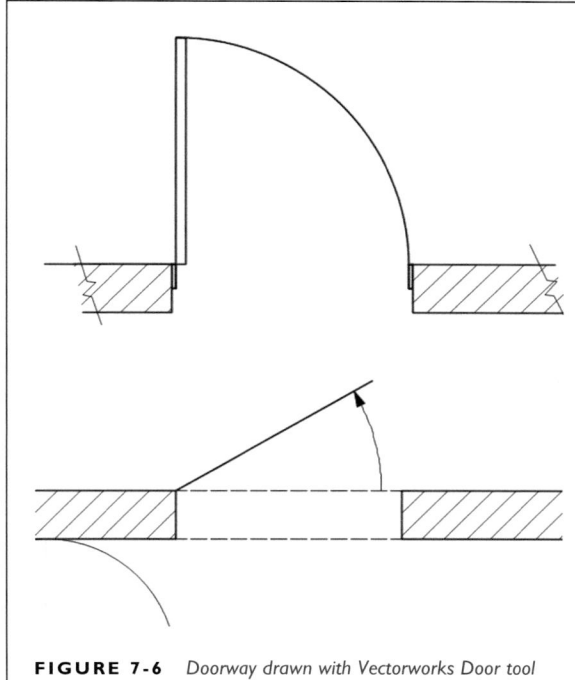

FIGURE 7-6 *Doorway drawn with Vectorworks Door tool (above) and standard groundplan door graphic symbol (below)*

11. Scenic groundplans are generally less concerned with the construction detail of doorways than with the placement of them; therefore, their depiction is more schematic. Dashed lines through the doorway indicate that the wall continues unbroken above the opening; the Hidden Line class works for this. An angled line extending from the hinge point denotes the door itself (Medium Thick), and an arc (Dimension) with an arrowhead indicates direction of swing (Figure 7-6). You may also draw the door as a filled narrow rectangle. Most standard doors are about 2″ thick.

Add dashed lines and doors to the drawing. Lines for doors are the same length as the opening's width. The shop doors are a double-door; each door is half the width of the opening. The auditorium openings do not have doors; as doorways they require only dashed lines.

Vectorworks' Spotlight and Architect workspaces have a tool to place doors on plans, located in the Building Shell tool set. It includes jamb elements, a door with thickness, and an arc without arrowheads that runs from the jamb to the end of the door. On groundplans, especially scenic groundplans, they risk being interpreted as a doorway with a curved step in front of it. In Figure 7-6, the lower image presents the traditional USITT groundplan door graphic recommendation.

---

12. In the Reference Line layer, use the Plaster Line class to draw the plaster line. Label it PL at one end. The "PL" typically lies on top of the line. Enlarge the font and turn off the fill (Attributes) so the line is visible through the label. Use the Center Line class to draw the center line. Label with CL at the top and bottom. Return to the Architecture layer. The center and plaster line will remain visible, but muted. Turn off the visibility for the Reference Guidelines class.

13. Use thin visible lines to darken the Fly Rail outline and spiral staircase footprint. Thin visible lines are also used when marking the footprint of objects placed on the stage, such as chairs and tables.

14. Return to the Guideline layer to lay out the smoke pockets. They are L-shaped and attached to the upstage side of the proscenium wall. Dimensions and placement are noted in Figure 7-7. Lay out one on each side of the proscenium arch. Smoke pockets are important to note when present, as scenery must not block a fire curtain from descending completely to the deck. Return to the Venue Architecture layer to darken the smoke pockets.

15. The **sightline symbol** is a circle divided into quadrants with opposite quadrants filled in with black.[1] When the quadrants are not filled in, the symbol is used to denote a rigging point. The filled version is also used to denote a datum line. Datum lines denote a reference line when the structure does not provide one, such as placing an imaginary zero deck where needed. Vectorworks does not have a sightline symbol tool; it must be manually drawn.

The sightline symbol is placed where the head of the audience member seated in an extreme seat is located. Ideally, if this patron can see everything that's meant to be seen, the rest of the audience will also be able to see everything (Figure 7-8). In the Guideline layer, draw a 2′–0″ × 2′–0″ square to represent a chair 4″ away from the first tier's step. Draw intersecting guidelines to mark the square's center. Use this center

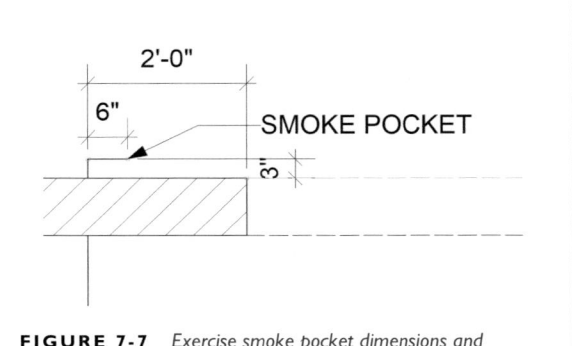

FIGURE 7-7 *Exercise smoke pocket dimensions and placement on upstage side of proscenium*

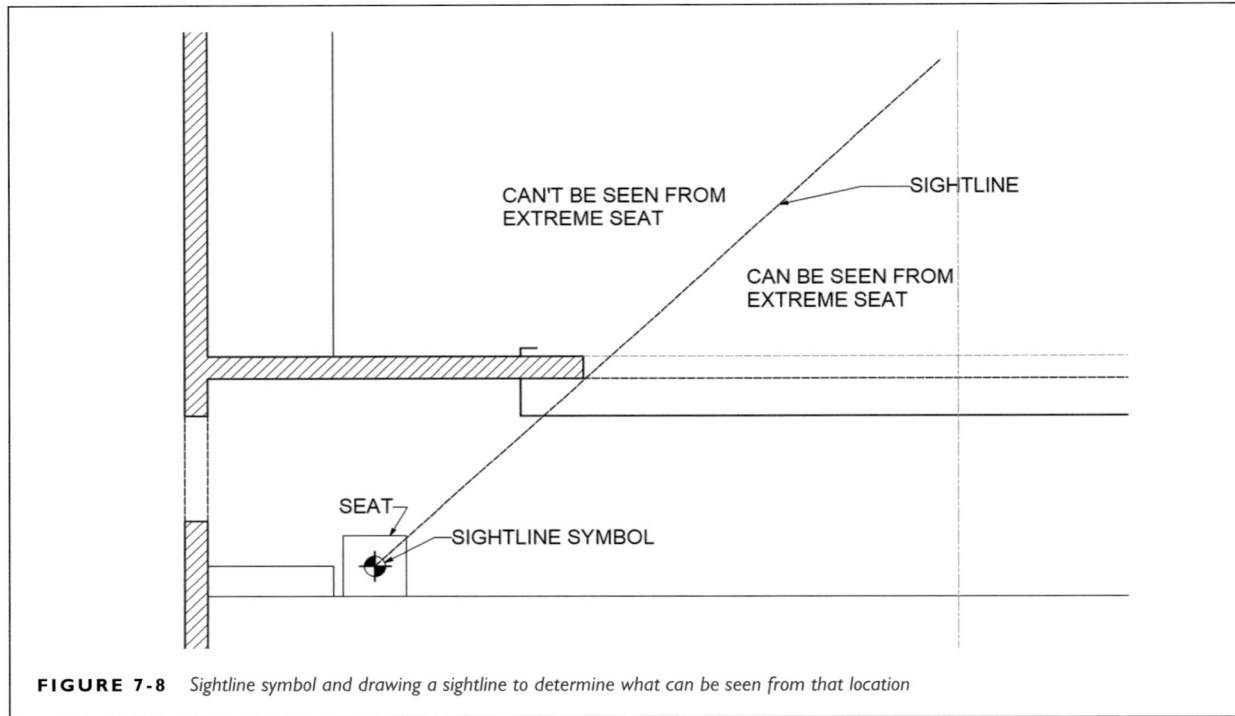

FIGURE 7-8 *Sightline symbol and drawing a sightline to determine what can be seen from that location*

to place the sightline symbol. Lay out sightline symbols at both ends of the first row of seats.

Switch to the Reference Line layer and darken the sightline symbols. The Arc tool's Center Mode allows you to draw quarter circles that can be composed in order to fill opposite quadrants. There is also a Quarter Arc tool accessible via the Arc tool icon's corner tab.

Turn off Guideline and Reference Guideline classes' visibility. Review the drawing and compare to the rough sketch. Are all the features drawn? Is everything darkened with appropriate line styles?

16. Switch to the Venue Architecture Dimensions and Notes design layer. Groundplans are an instance where dimensioning rules are frequently broken. Since the groundplan is a large drawing, placing dimensions beyond the perimeter may not allow all of the information to fit onto a single page unless reduced to a smaller scale. This also places much dimensioning far from the items to which they refer. To utilize the center and plaster lines as reference tools, dimensioning for the deck area is often placed on the deck. The center and plaster lines are used as extension lines.

Add dimensions to the important features of the space. Label smoke pockets and sightline symbols. The spiral staircase does not need to be dimensioned, but label it and the Fly Rail area. Include labels to indicate where doors and doorways lead. Label the seating area (Figure 7-9). In non-traditional spaces with moveable seating, clearly labeling the front row can prevent load-in disasters.

Scale Bars

Sometimes, a **Scale Bar** is added to a drawing. This is a bar broken into one-foot increments, spanning the space being described. Because the scale bar reduces and enlarges along with the rest of the drawing, it provides a consistent visual scale reference regardless of reproduction size. It's particularly useful for non-traditional spaces, where there are few regular features whose size can be readily interpreted. If you include one, place it where it won't obscure other information. Placing it in its own design layer allows it to be turned on or off as needed (Figure 7-10).

In Vectorworks, the Scale Bar tool icon is found in the Dims/Notes toolset (Figure 7-11). When selected, a dashed outline appears in the workspace with one end attached to the cursor. Click to affix the end in the desired location; a Navigation window will open. The Minor Length refers to the first increment of the bar, divided into desired segments (such as inches). The Major Length refers to the length of the bar beyond the Minor Length. You can think of Minor Length as the section on a scale ruler where the foot increment is divided into inches, and the Major Length as the total foot increments along the further edge of the ruler. The Major and Minor Lengths, though, refer to measurements in reality, not in scale. To set a scale bar on the venue drawing, type in the settings seen in Figure 7-11. The major length is 2'–0", which in ½" = 1'–0" equals 48'–0". You can also draft your own scale bar and keep it in a layer of your base drawing file. It can then be edited as required for different projects.

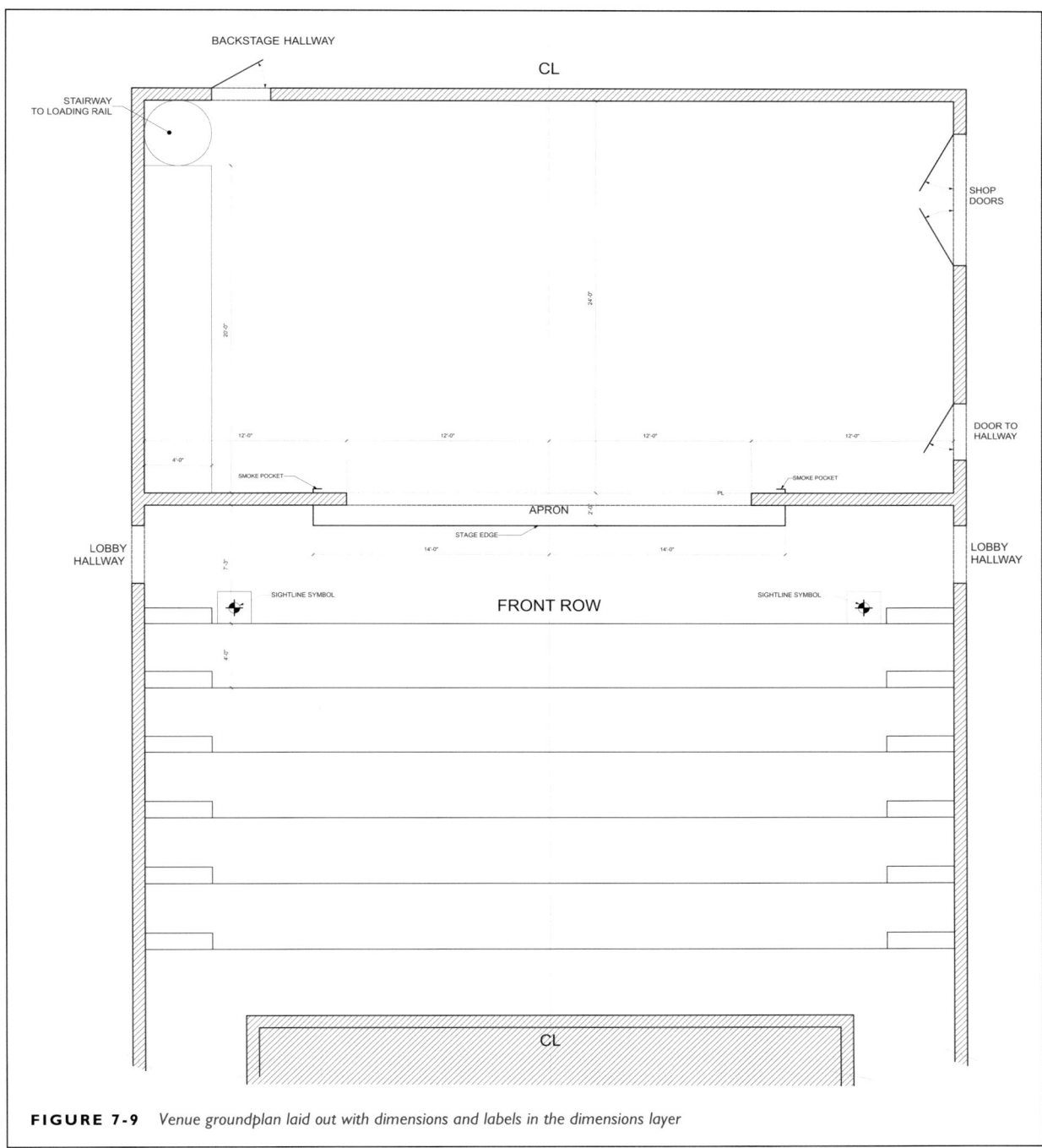

FIGURE 7-9 *Venue groundplan laid out with dimensions and labels in the dimensions layer*

The Line Schedule

The line schedule is a list, divided into rows and columns, of all linesets comprising a venue's counterweight fly system. Battens are numbered starting with 1 nearest to the plaster line. The distance of each batten from the plaster line is included, as is the name of the item placed on the batten. If the batten is empty, the name slot is left empty. Venues with permanently hung items such as band shells, film screens, and dead-hung travelers will include these on the line schedule so that those using the space are aware of everything to be found above the deck.

A column may be added for **trim heights**. A trim height is the distance above the deck at which a batten is to be set during performance. If a batten flies in and out at different times during a performance, the trim height typically specifies the in, or lower trim.

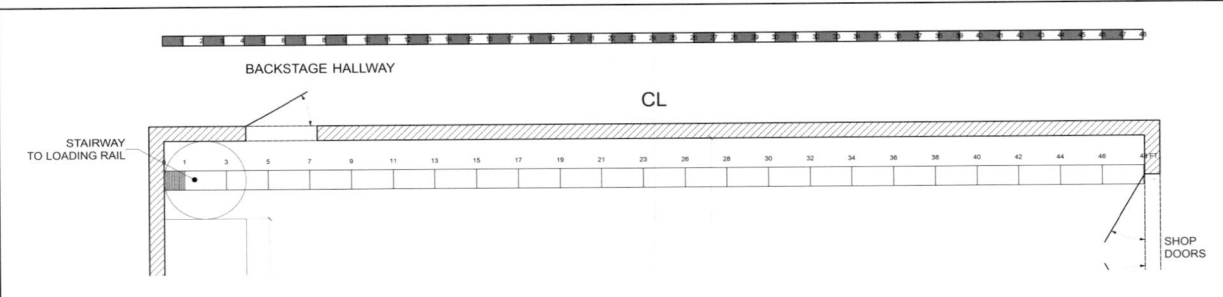

FIGURE 7-10 *Two Scale Bars: above, custom drafted Scale Bar; below, Scale Bar created with the Vectorworks Scale Bar tool, inserted across rear of deck*

Depending on how the venue was engineered, one sometimes finds linesets numbered out of sequence at the fly rail because the pulley system has to avoid ductwork or structural beams. The lineset label at the fly rail should reflect the order of the battens in the air; e.g., batten 3 is always operated by lineset 3, even if lineset 3 is physically operated by the sixth position at the rail.

The **act curtain** (also called the **grand drape**, known as colloquially as the **main rag**, sometimes called the **main act**) sometimes is—and sometimes is not—one of the linesets found on the fly rail. When its control is separate from the fly rail, it is not included in the line schedule. A **show curtain** is a drape specifically constructed and painted for a particular show and may replace the act curtain or be placed within a show portal.

Drapery can fly in and out on battens (in which case the act curtain's movement is said to be "guillotine" style) and also can **travel** along a track (traveler track), opening and closing laterally across the stage. It is usually enough to indicate the line along which the act curtain operates, unless its opening width during performance is required. In this exercise, the act curtain is traveler-style, and plays 9" upstage of the plaster line to allow clearance for the fire curtain. It is not included on the line schedule, but will be drawn in the Line Schedule layer.

Teaser is a term for the first border, especially if it matches the décor of the act or show curtain. It may also be called the **grand valance**.[2] **Tormentor** is a term for the first set of legs, especially when they match the décor of the act curtain. Teasers and tormentors may be hard-covered flats that are semi-permanent to a proscenium space. A **false proscenium,** or **show portal**, is a proscenium arch specifically constructed for a production and placed just upstage of the architectural proscenium opening. It may replace a teaser and tormentors. A show portal provides a visual transition between the auditorium and the world of the performance.

In this exercise, the first set of legs is simple black masking and referred to as legs rather than tormentors.

A vertical lighting position attached to the upstage side of the proscenium wall to either side of the opening is called a **torm position**, as it is located either directly offstage of a tormentor, or where a tormentor would be placed if there was one.

On the groundplan, the line schedule is placed as close as possible to, if not on, the fly rail's footprint. Older styles of line schedules align the batten with the line upon which the text sits. Newer styles place a tick mark for the batten along the onstage vertical line of the schedule, centered vertically for each row. Written distances are included as reinforcement. CAD versions often place each line of text in its own rectangular bubble (Figure 7-12).

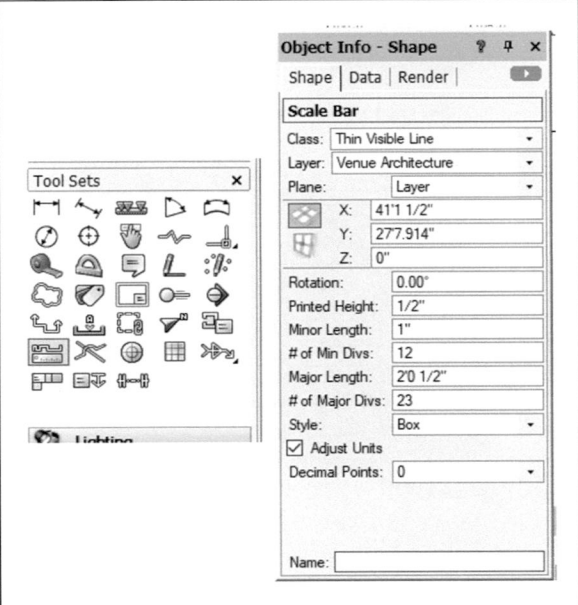

FIGURE 7-11 *Location of the Scale Bar tool icon; the Scale Bar Object Info window*

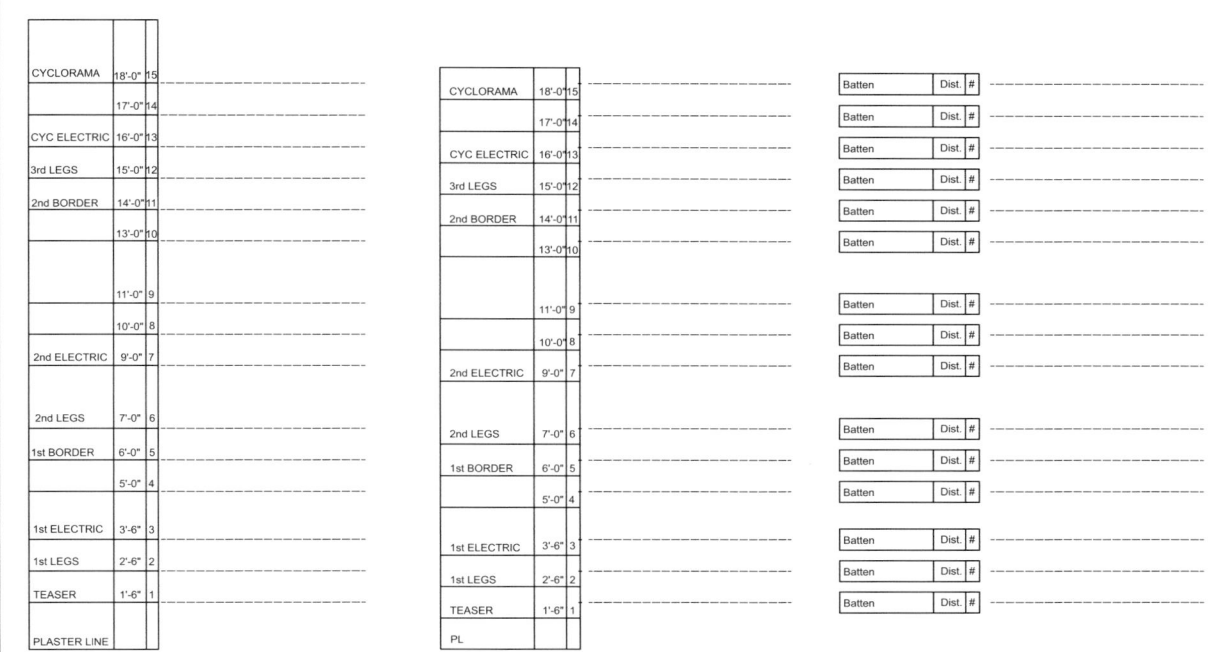

FIGURE 7-12 *Three line schedule formats: (1) battens aligned with line schedule text base lines; (2) battens aligned with side tick marks; (3) battens aligned with horizontal axis of bubbles*

Depending on the drawing size and available page space, the line schedule may be placed just outside the venue's wall, covering the wall, or within the wall in the area reserved for the fly rail's footprint. Placing the line schedule and battens in their own design layer allows those who use the file (such as the lighting designer) to more easily adjust its placement depending on their needs.

For this exercise, the line schedule information can be found in Figure 7-12. If you wish to add a column for trim heights, do so. The names included on this schedule refer to battens with dedicated purposes (a "dedicated lineset"), for which the items hung upon it cannot or should not be moved by productions visiting this venue. The steps outlined below match the baseline of the schedule's text row with the distance of the batten from the plaster line, instead of using a side tick mark (Figure 7-13).

1. Open the Guideline layer and clear out old guidelines to start afresh. In the Guideline layer, use the Rectangle tool to trace the outline of the Fly Rail area. Drag it off to the side to begin editing it.

2. Measure, mark, and lay out horizontal lines spanning the rectangle at their designated distances from the plaster line.

3. Add vertical lines to mark columns. You may want to add sample text in order to space the columns. Use guidelines to help align columns and rows of text.

4. In the Line Schedule layer, trace over the column and row lines with visible lines. Add text and distances from the plaster line. When complete, use the Rectangle tool to drag a rectangle around the line schedule. Add a white Fill. The text will disappear under the fill. Right click on the fill and open Send from the menu. Choose Send to Back. The Fill is "pushed back" and the text will reappear.

    Select the whole line schedule. Under the Modify drop-down menu, choose Group. This will turn all of the line schedule's elements into a single item, making moving and copying it easier. To break a grouped object apart, choose Ungroup from the Modify menu. A group can be edited by choosing Edit Group.

5. Grab a corner of the schedule and drag it into place within the Fly Rail footprint.

6. A base drawing should include the location and length of each batten. In many cases, a venue's battens are a variety of lengths. For this exercise, each batten will be 38'–0" long, with their midpoints on the center line. It is not typical to dimension batten lengths (though this information can be included as a column in the line schedule).

    In the guidelines layer, draw a guideline for lineset 1's row horizontally across the deck until it meets the center line. At the center line, extend the guideline horizontally to another 19'—0". Draw a vertical

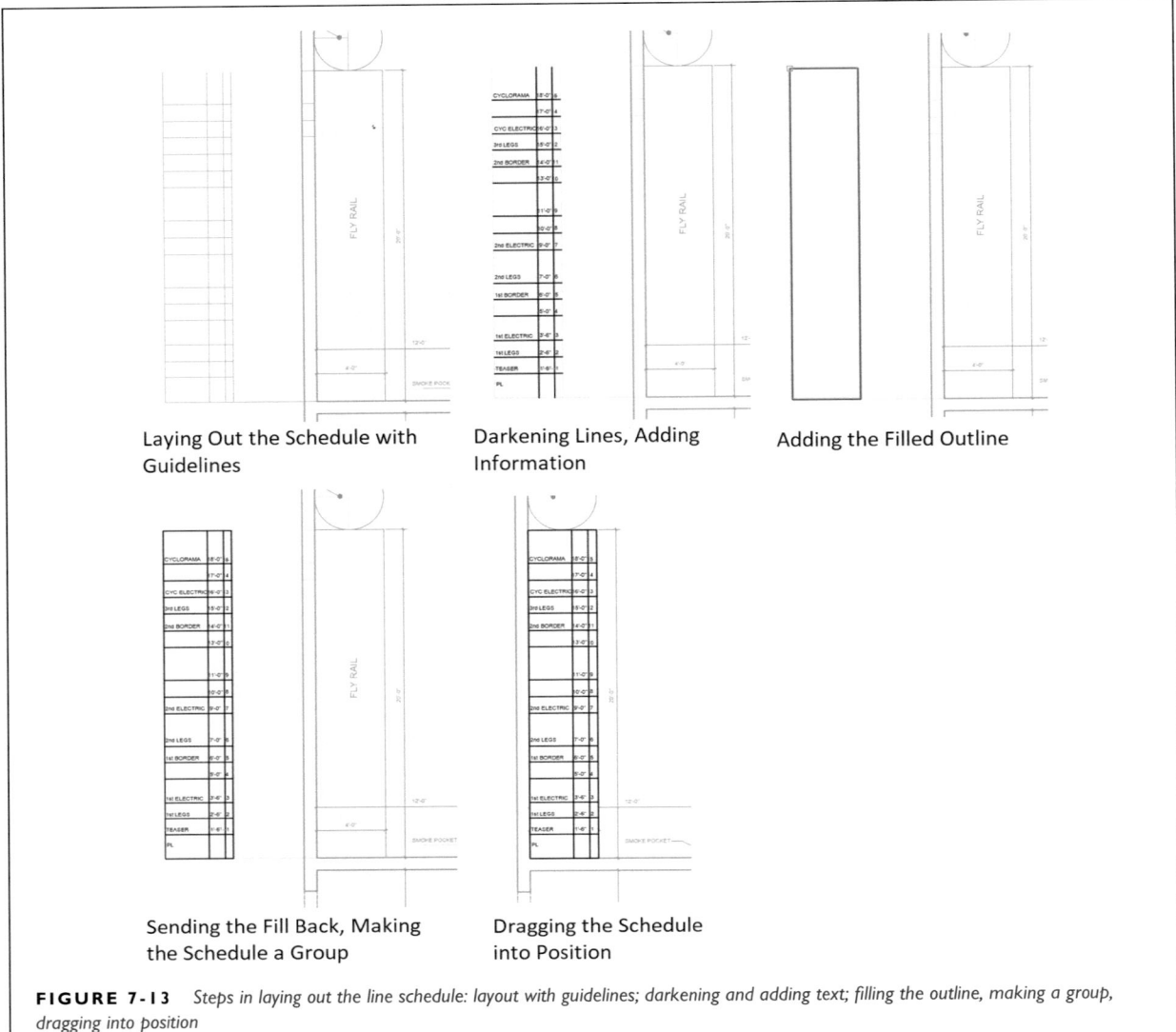

Laying Out the Schedule with
Guidelines

Darkening Lines, Adding
Information

Adding the Filled Outline

Sending the Fill Back, Making
the Schedule a Group

Dragging the Schedule
into Position

FIGURE 7-13   *Steps in laying out the line schedule: layout with guidelines; darkening and adding text; filling the outline, making a group, dragging into position*

guideline from the endpoint so that it extends upward the full depth of the line schedule.

Draw a horizontal guideline for each lineset, extending it to meet the vertical line 19'–0" beyond the center line. When all horizontals are in place, measure back 19'–0" from the center line back toward the schedule. Draw a vertical line. All of the linesets will fit between these two vertical lines. Extend the verticals toward the plaster line as well, as the main act curtain will also be 38'–0" wide.

If the bubble or tick mark style of line schedule is preferred, laying out the battens before drawing the schedule may be preferred. Extend the batten guidelines all the way into the area in which the schedule will be placed, then place the tick marks or bubbles accordingly.

7. Return to the Line Schedule layer and darken each batten. Draw the first batten. To copy the line, while it is selected hold down the Ctrl key on the keyboard and click on the line. Drag the cursor up to the second batten position. When you release the mouse button, a copy of the line will be placed. Use this process to place battens for all of the linesets (Figure 7-14). Use a phantom line for the main act curtain. Label the Main Act Curtain.

8. Turn off the visibility of the Guideline class and Guideline layer. Review the drawing. Are dimensions placed in their optimum positions? Revise as necessary.

9. Create a viewport and sheet layer (Arch D, landscape, ½" = 1'–0"). How much of the drawing fits on the sheet? If reoriented to portrait, will it fit? Does it fit when reduced to ¼" = 1'–0"? Is the drawing still

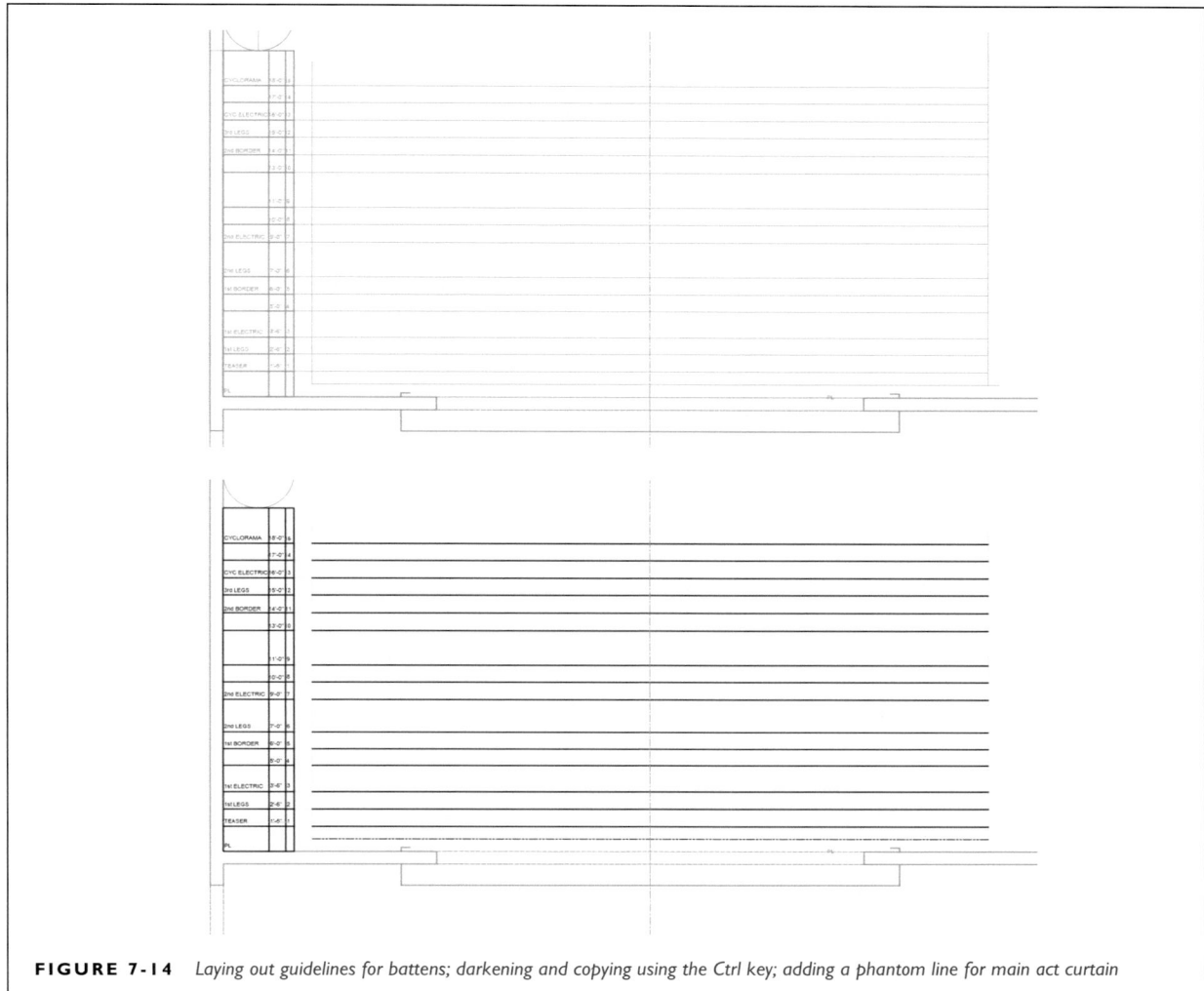

FIGURE 7-14 *Laying out guidelines for battens; darkening and copying using the Ctrl key; adding a phantom line for main act curtain*

readable at a reduced scale? Try creating two viewports, one for the overstage and one for the front of house, and import them onto two separate sheets.

10. Save. Save PDFs of the ¼″ = 1′–0″ sheet as well as the ½″ = 1′–0″ overstage and FOH sheets.

Base Drawing Groundplan Lay Out Checklist

- Begin with the center and plaster lines. Work outward from them.

- Lay out the perimeters of the space. Include as much of the seating area as required.

- Lay out wall thicknesses.

- Lay out boundaries of the playing space, such as deck edges and apron lips.

- Lay out doors, doorways, proscenium arch, etc.

- Place deck features, such as the fly rail and other pertinent equipment locations.

- Lay out auditorium details as required. Include sightline symbols in extreme seats.

- Lay out the line schedule, battens, and act curtain.

- Darken the drawing. Crosshatch cut-through walls. Include dashed lines in doorways and include door swing direction.

- Dimension, note, and label.

Homework: Copy an Existing Plan

Instructor

Provide hard copies of existing base drawing plans. If you have a set of drawings for a venue you have access to, use those. Just as art students visit museums to paint copies of master paintings, copying drawings is a good exercise in analyzing technique and approach, as well as practice in paying attention to detail. If older plans were drawn by hand, this is a chance to turn them into Vectorworks files. Copying a CAD generated plan means the student ends up with a file they know intimately because they've set it up and drawn it.

Students

Use the existing drawing provided by the instructor to draft your own copy in the requested format (hand or CAD). Analyze the drawing to become acquainted with its content and features. Pay attention to the details so that everything is reproduced; if you don't understand a feature, be sure to ask about it. If you are using Vectorworks, how many design layers are required?

Use the layout steps outlined previously to construct the copy. Be logical and global in your approach, rather than drafting detail by detail.

MODULE 2: MODELING THE VENUE

The model constructed in the following exercise will be used to house a simple box set built in the following chapters. Constructing the set as a 3D model and situating it in the venue allows the draftsperson to view the structure from different vantage points to better understand the set's relationship with the overall space. It will also provide a visual comparison between cutting plane sections of the model and drafted plans, sections, and elevations.

Exercise: Modeling the Venue in SketchUp

The model will be constructed to reflect what is seen on the groundplan. At this point, nothing above the cutting plane will be modeled to make it a little easier to insert scenery and view the space and scenery from above. As with the drafted groundplan, the model begins with reference lines. The overall shape of the space is then laid out, walls are extruded, and details added. Planning the model's construction as you would plan a drawing's layout helps keep the process on track. The model will be organized by making elements groups (Figure 7-15).

1. Open SketchUp. Choose Simple/inches template. Move the human figure aside.

2. Consult a hardcopy of your finished groundplan or the rough sketch in Figure 7-1. Use the Tape Measure tool to drag out a center line guideline parallel to and 24'-0" (half the width of the deck) from the green axis

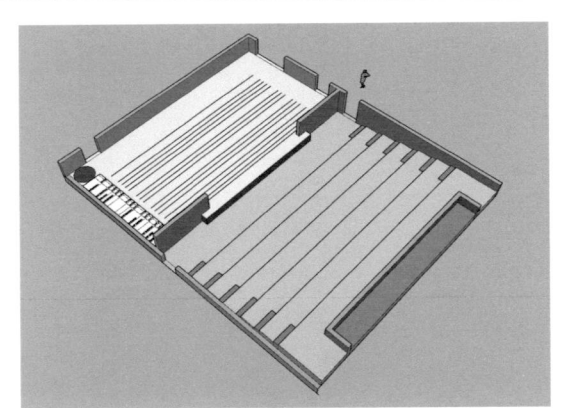

FIGURE 7-15   *The finished SketchUp model of the venue's plan*

(perpendicular to the red axis). This serves as the center line. The red axis itself will serve as the plaster line.

3. Lay out the perimeter of the deck with guidelines. Drag out a guideline to the right and parallel to the center line to set the stage left edge of the deck. Drag out a guideline parallel to the red axis to mark the upstage edge of the deck at the correct distance.

4. Drag out a rectangle from corner to corner of the deck area, creating a surface. Use the Offset tool to give the venue's walls thickness. Click the cursor anywhere on the rectangle's outline and drag the new outline outward 9". Click to affix. There are now two rectangles, the first within the second.

5. Drag out guidelines to lay out the auditorium's perimeter. Drag out a rectangle to turn the seating areas into a surface. Use guidelines to lay out the thickness of the walls, and then go over them with lines. Close the ends of the walls at the rear of the auditorium or they will not become extrudable surfaces.

6. Use guidelines to mark the door and proscenium openings. Draw lines within the wall thickness to mark the edges of each opening. This divides the enclosed wall thicknesses into independent surfaces. Erase any lines remaining within walls that are not associated with an opening (Figure 7-16).

   If proliferating guidelines make layout confusing, use the Delete Guides menu choice under the Edit drop-down menu. You have a choice of deleting the most recent guideline or all of the guidelines. Reestablish the center line as required.

7. Lay out the apron. Draw lines to outline the apron. Erase the proscenium lines that cross the apron so that the apron becomes part of the deck's surface

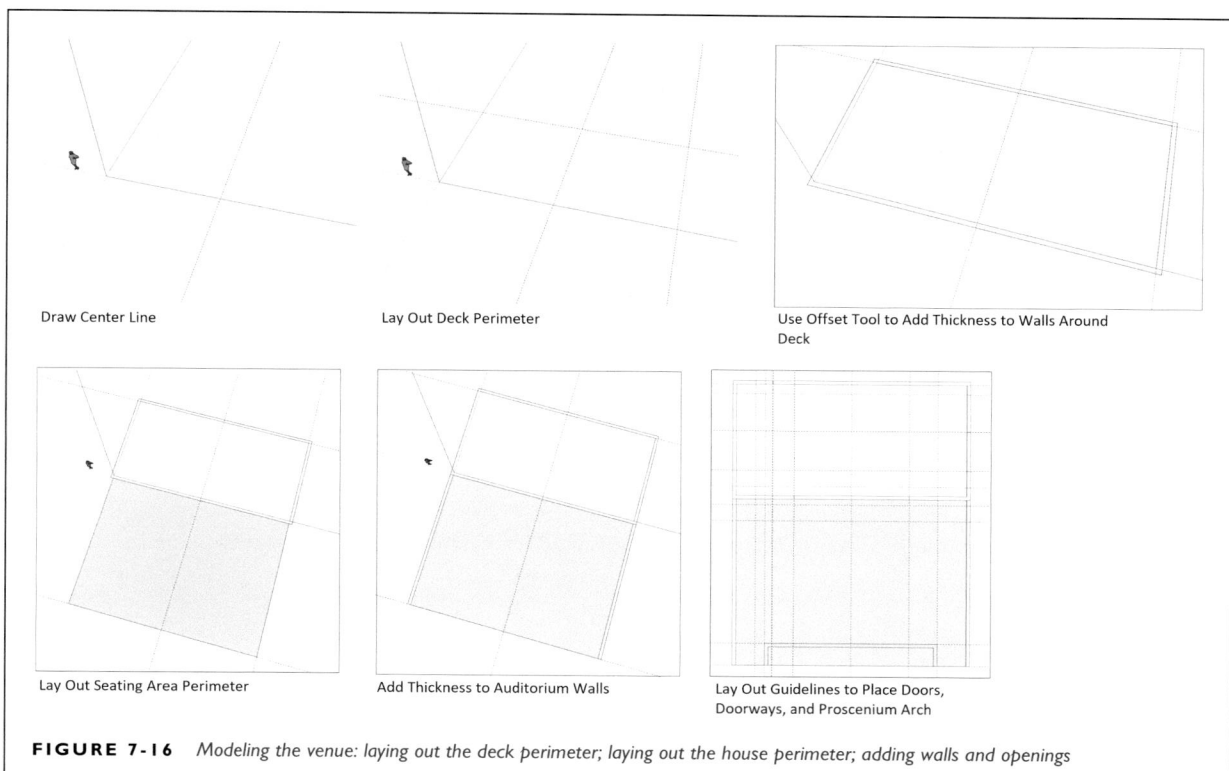

Draw Center Line

Lay Out Deck Perimeter

Use Offset Tool to Add Thickness to Walls Around Deck

Lay Out Seating Area Perimeter

Add Thickness to Auditorium Walls

Lay Out Guidelines to Place Doors, Doorways, and Proscenium Arch

FIGURE 7-16   *Modeling the venue: laying out the deck perimeter; laying out the house perimeter; adding walls and openings*

8. Lay out the seating tiers and steps. Draw lines to create independent surfaces.

9. Lay out the footprint of the fly rail area and the spiral staircase. Draw lines over the fly rail guidelines and use the Circle tool for the staircase. Erase the surface within the fly rail footprint, leaving a hole in the deck into which the line schedule will be inserted. Select the whole plan and make it a group (Figure 7-17).

10. The line schedule will be created as a group, and then inserted into the deck within the fly rail footprint just created.

     In the open space beyond the left side of the floor plan, drag out a rectangle the same size as the fly rail footprint. Consult the line schedule previously drawn to lay out the rows and columns. Make the grid a group to prevent work on the text from accidentally pulling it out of shape.

     Use the 3D Text tool to add the batten numbers, distances, and names of dedicated battens (Figure 7-18).

     Orient text flat on the surface of the line schedule. You can choose the depth of text (in SketchUp, text is a 3D object) as well as various text sizes and styles. Select text with no depth, a clean sans-serif font, and a size that fits easily within the rows of the schedule. Text is 1:1, so letters should be about 4″ to 6″ high for this exercise.

When complete, make the line schedule a group. Move the line schedule and align it with the deck's fly rail hole. It should fit the hole exactly (Figure 7-19).

The line schedule could have been drawn directly on the stage floor, but it's often easier to build objects to one side and then Move them into a model. Making the line schedule its own group has certain advantages that will be seen when working on later steps of the model, such as adding scenery on battens.

Both Vectorworks and SketchUp can export and import DWG format files. You can also import PDF's, JPEG's and PNG's. Image files usually need to be resized after importation to scale them appropriately. Importing a groundplan into SketchUp may seem like a shortcut, but there can be glitches in the transfer of information that lead to troublesome misalignment issues as the model is constructed. Once this project's model is complete, experiment by making a DWG file of the Vectorworks plan and import it into SketchUp to see how it reacts when you use it to build a second copy of the model.

11. Open the venue's group by right-clicking and choosing Edit Group. Since the features of the floor plan are all

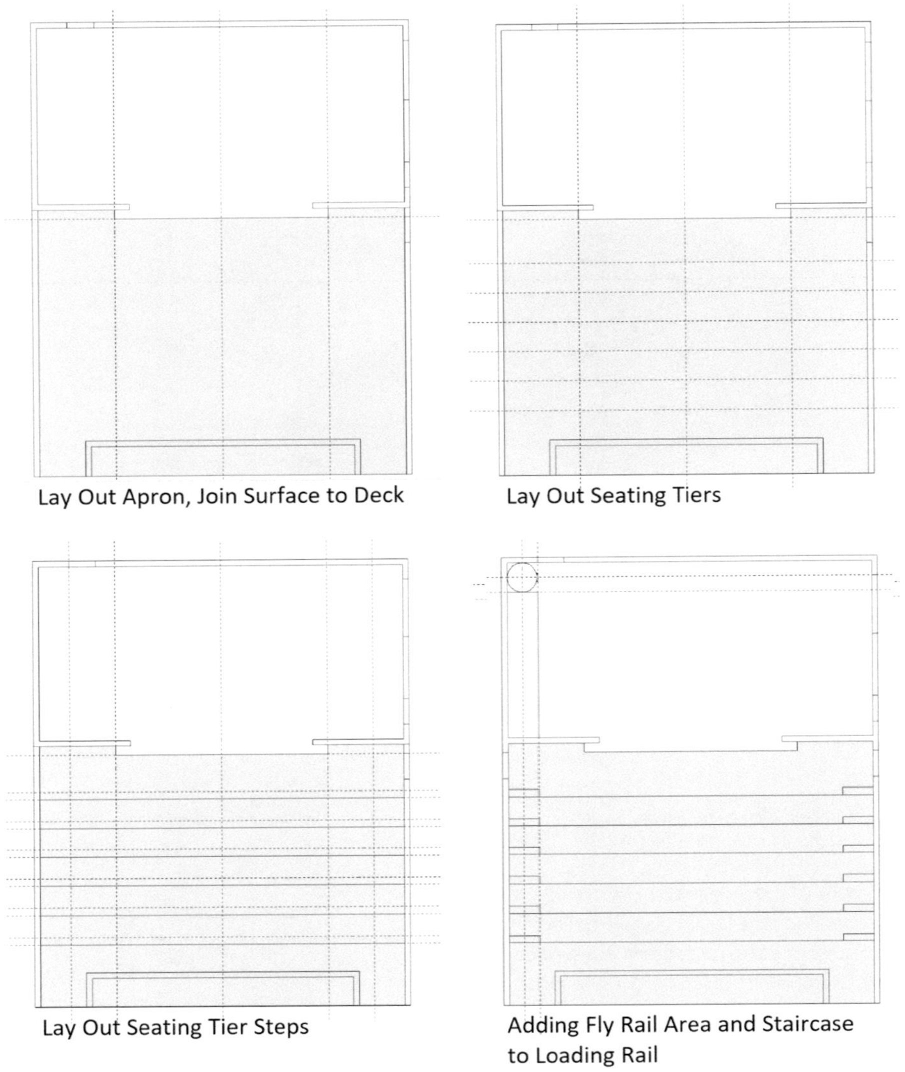

Lay Out Apron, Join Surface to Deck

Lay Out Seating Tiers

Lay Out Seating Tier Steps

Adding Fly Rail Area and Staircase
to Loading Rail

FIGURE 7-17   *Modeling the venue: lay out the apron; lay out seating tiers; lay out steps; lay out fly rail and staircase*

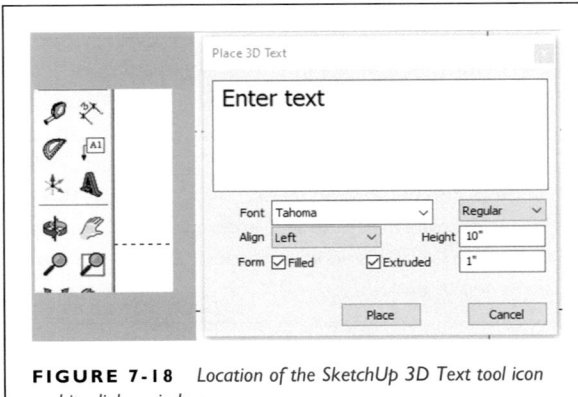

FIGURE 7-18   *Location of the SketchUp 3D Text tool icon and its dialog window*

independent surfaces, use Push/Pull to raise and lower them to their desired heights. The deck remains at + 0″. The floor of the front row of the auditorium is − 24″. Use Push/Pull to drop it below the height of the deck. Begin the Push/Pull action and type 24″ (or, 2′). Hit Enter. As long as the extrusion is going in the desired direction, you do not need to add a plus or minus sign to the measurement.

If a surface adjacent to the one you're working on is already set to a height, the current surface will stop at that level with a message (Offset Limited) that it can't go any farther. Match the level, and then Push/Pull again to go past it. Remember to keep track of how far the surface was already Push/Pulled so you end up at the

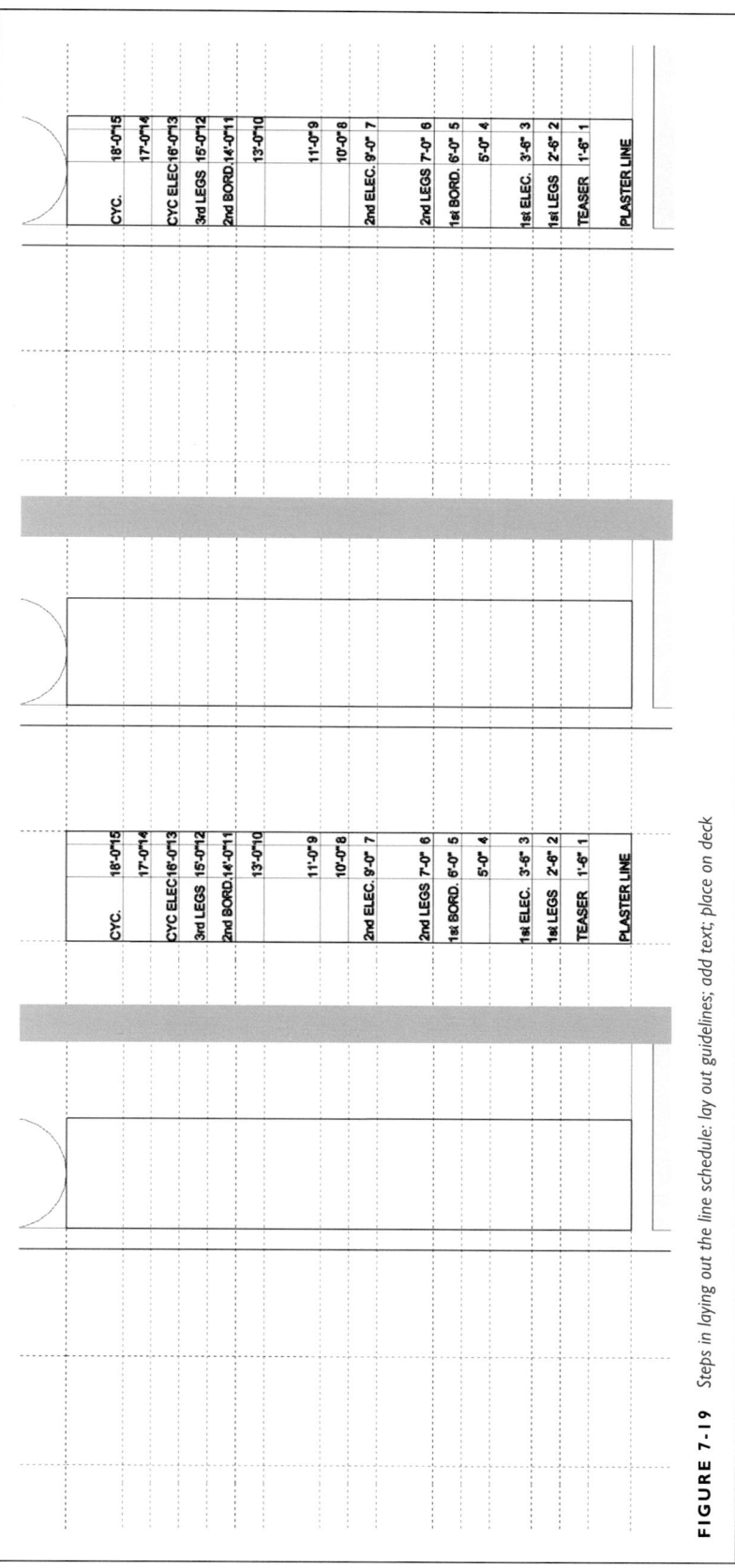

FIGURE 7-19 *Steps in laying out the line schedule: lay out guidelines; add text; place on deck*

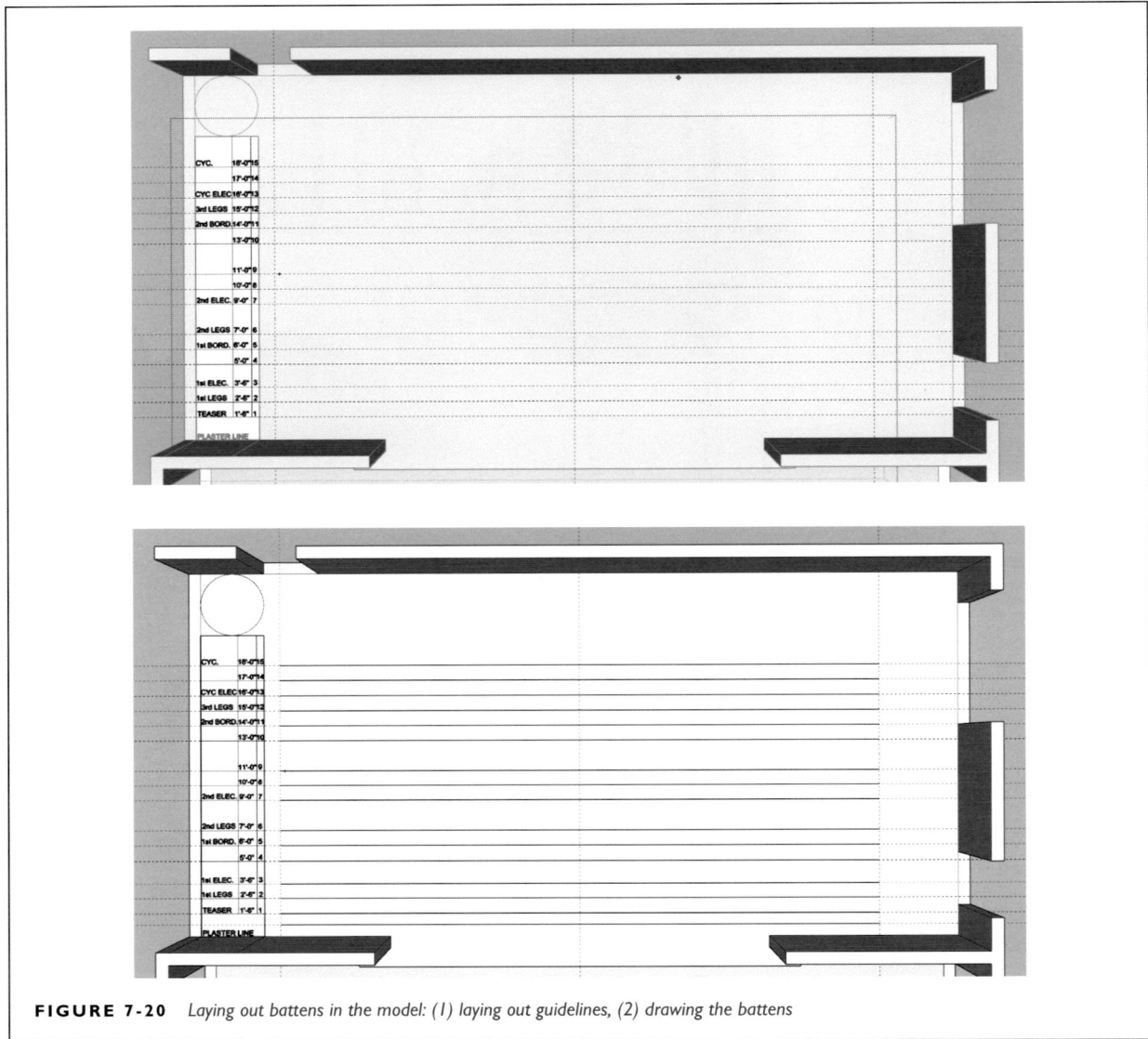

FIGURE 7-20 *Laying out battens in the model: (1) laying out guidelines, (2) drawing the battens*

correct height. Erase any vertical lines that track onto the wall surfaces.

12. Since a groundplan's horizontal cutting plane is at about shoulder height, Push/Pull the stage walls up to + 4'–0". Leave the door openings at + 0". Work around the space until all of the walls are set at that height. Leave the stage right wall behind the line schedule at + 0" to allow easier access to the backstage area.

13. The auditorium's doorways thresholds are at – 24". Push/Pull those door openings down to match the level of first row's floor. Since interior doors and doorways are typically 6'–8" tall, the top of the auditorium doorways remains above the cutting plane (Figure 7-15).

14. Use the Paint Bucket tool to add color/texture to surfaces, as desired.

15. Close Edit Group by hitting Escape on the keyboard. You can also close Edit Group by choosing Select and clicking the cursor elsewhere in the workspace.

16. Select the line schedule group and right click to choose Edit Group. Use the line schedule to lay out batten and act curtain placement on the deck. If necessary, reestablish the center line and plaster line. Drag guidelines across the deck from the plaster line to the depth of each lineset as marked on the line schedule. Drag out guidelines from the center line to mark the ends of the battens (Figure 7-20).

Draw a line over one batten, and then use Copy and Move to draw the rest of them. Once done, close Edit Group. Delete guidelines.

17. Save the file.

SketchUp features an online warehouse with access located via the main toolbar's File drop-down menu. You will be asked to open an account to access the models uploaded to the warehouse. Do so, if you wish. Once the warehouse opens, search for a chair suitable for auditorium seating. Be as specific as possible in your search terms to narrow down results.

When you have found a chair, download it into your model. After a moment of processing, the chair will appear somewhere in your model. Depending on how the chair was built, it may not be scaled to fit into your model. If the chair is too big or too small, use the Scale tool (Figure 7-21). Selecting the object and then the Scale tool causes a box to appear which features an array of small green cubes at its corners, midpoints, and centers. Grabbing a cube allows you to pull or push the object in a direction aligned with that cube. Grabbing and dragging the box's corners enlarges or reduces the object while maintaining its proportions.

Move the chair to the end of the front row. Grab the chair from the bottom edge of a leg to make it easier to place. Copy and Move to start building a row. Once four or five chairs are placed, make them a group. Copy and Move the group to cut down on steps. Add a row of chairs to the second row as well. If the chairs are complex models, the more you add, the harder your computer's processor works. You may notice more frequent pauses.

Since the architecture is already a group, you can Hide it, then Select all the chairs to make them their own group. If desired, search the warehouse to see if you can find a spiral staircase appropriate to a backstage environment. Import, scale, and place.

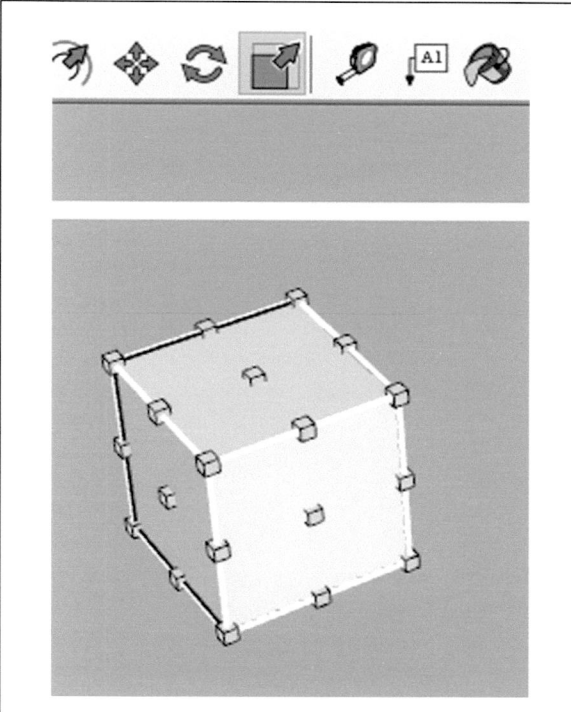

FIGURE 7-21 *Location of the Scale tool icon and a cube ready to be scaled*

NOTES

1. Dennis Dorn and Mark Shanda, *Drafting for the Theatre* (Carbondale, IL: Southern Illinois University Press, 1992), p. 172; USITT Education Commission, *Recommended Practice for Theatrical Lighting Design Graphics* (Syracuse, NY: United States Institution for Theatre Technology, 2006), p. 6.
2. J. Michael Gillette, *Theatrical Design and Production: An Introduction to Scenic Design and Construction, Lighting, Sound, Costume, and Makeup*, 6th ed. (New York, NY: McGraw-Hill, 2008), p. 65.

CHAPTER 8

SCENIC GROUNDPLANS

This chapter enlarges the discussion of groundplans begun in Chapter 7. The first module discusses what is expected on a typical scenic groundplan and how this information is used by various members of a production team. The second module outlines the construction of the model of a simple box set. This model is then placed into the venue model to help visualize the groundplan constructed in Module 3.

TOPICS AND GOALS

- What is a groundplan?

- Essential elements of a groundplan

- Reading a groundplan

- Drafting common groundplan graphics

- Constructing a groundplan

- Constructing a model from a groundplan

MODULE 1: WHAT IS A GROUNDPLAN?

A groundplan is a horizontal section of the performance space with its cutting plane about + 4'–0" above the zero deck (+ 0'–0"). The height of the cutting plane can vary. + 4'–0" to + 10'–0" is the usual range.[1] Since the cutting plane is at a low height, a majority of items on a stage are cut through; however, not all of them are shown as sliced-through masses. The groundplan is more schematic than pictorial and therefore graphic conventions come into play. Figure 8-1 presents the top view of a model in perspective and a top view in parallel projection. Figure 8-2 shows two

drafted versions, by hand and by Vectorworks, presenting the same material using typical graphic conventions.

Research, sketching, and many, many conversations contribute to generating the groundplan; the drafted groundplan firms up decisions and presents the final information in a scaled environment.

The groundplan is also a representation of the space directly above the plan—an indication of what is to come.[2] As a distillation of technical information about the design the groundplan is not meant to be a standalone drawing. To interpret it as correctly as possible, a rendering, model (whether virtual or physical), or elevation(s) must accompany it: "Oh, I see, this wall on this picture of the model is indicated on the groundplan by this line here." When the reader is familiar with the standard repertory of graphic symbols they can still gain a substantial understanding of the environment from a groundplan, even without the additional information imparted by pictorial images.

How Are Groundplans Used?

Scenic groundplans serve a number of different production departments. The director relies upon them to know how and where performers can move within the designed environment. As well as using reduced copies for blocking notation, stage managers use groundplans for setting the stage, managing scene changes, and arranging the backstage. Technical directors use the designer's groundplan to index construction drawings and plan the construction process. The lighting designer (LD) uses them to determine how light will be controlled, what lighting unit shines upon what part of the stage, and where that unit will be hung in the venue. Sound designers use them to determine speaker and microphone placement. The prop manager can use them to gauge

DOI: 10.4324/9781003154921-9

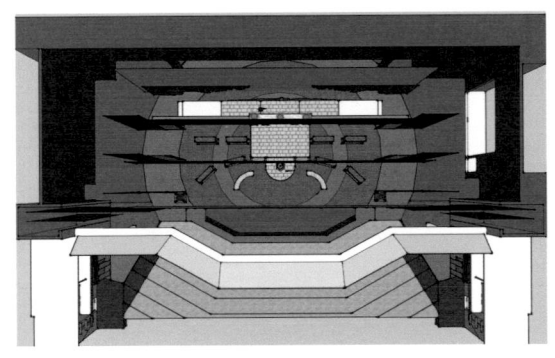

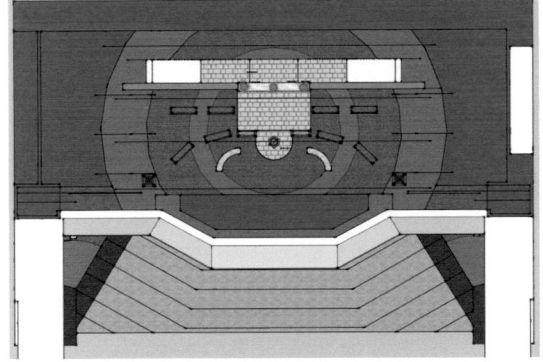

FIGURE 8-1 *SketchUp model of UW-Whitewater's production of* Triumph of Love: *Top view in perspective, top view in parallel projection*

the relative size of furniture and other set dressing to be procured. Even the costume department will use a groundplan to determine placement of quick-change areas as well as identify areas where floor treatments may not be amenable to footwear choices (e.g., stiletto heels on metal grating).

The groundplan presents the scenery within and in relationship to the venue's architecture. If an event or performance takes place in a parking garage, it's important to know where the pillars, ramps, and speed bumps are located. The width of a venue's doors restricts the size of structures that can be brought in.

Groundplans map all areas accessible to performers; where they can and cannot go. Stairs and platforms may rise above the cutting plane but are shown on the groundplan regardless of their height. Scenic features beneath platforming, such as a doorway beneath a balcony, are described with hidden lines. If platformed playing areas are stacked, the highest level is drawn, with notes and hidden lines indicating levels beneath. Additional groundplans may be required for exceptionally complex playing areas.

If the scenic environment includes many scene changes, multiple groundplans are generated, indicating what piece of scenery is on the stage at a given time (Figure 8-4). A **master groundplan** depicts the environment's permanent

architectural and scenic features, while simplified secondary plans depict the movement of individual elements.

The stage manager can take these simplified scene change plans and create **shift plots**, which are diagrams noting the movement of scenery as well who moves it. If a production has a particularly large number of items to be flown on battens, a separate **hanging plot** should be created.

The groundplan should not become so cluttered and cramped that it becomes unreadable. Breakout drawings are permitted; be certain that the master groundplan has notes and labels directing the reader to the sheet containing the information they require.

What's on a Groundplan?

The groundplan is a standardized map of a scenic environment. Every design and every venue is different, but there are commonalities to the graphics and items included on a groundplan.

Center Line

If the venue is a proscenium theatre, the center line runs along the depth axis of the space.

If the venue is a black box or non-traditional space, the center line may run along the center axis of the scenery. Television and Film studios vary. Small spaces may use an architectural center line to place scenery; in large spaces with multiple stages or sets, center lines tend to be placed through individual locales.

Plaster Line

If a venue has an architectural proscenium, the plaster line is indicated.

Venue Architecture

This includes the permanent walls of the space, entrances, at least a portion of the auditorium and seating (for an LD, the whole auditorium is generally required, as lighting positions are found throughout), line schedule and batten placement, smoke pockets (if the space has a fire curtain), stage edge/boundary (apron), and orchestra pit (if present). For black boxes, studios, and environmental spaces, note the edge of the playing space, whether a platform edge or perimeter of painted stage floor. Pertinent equipment and other technical features are noted.

Sightline Symbols

These are placed in extreme seats, typically at either end of the first row. They are also found in balcony and box seats (see Figure 7-8).

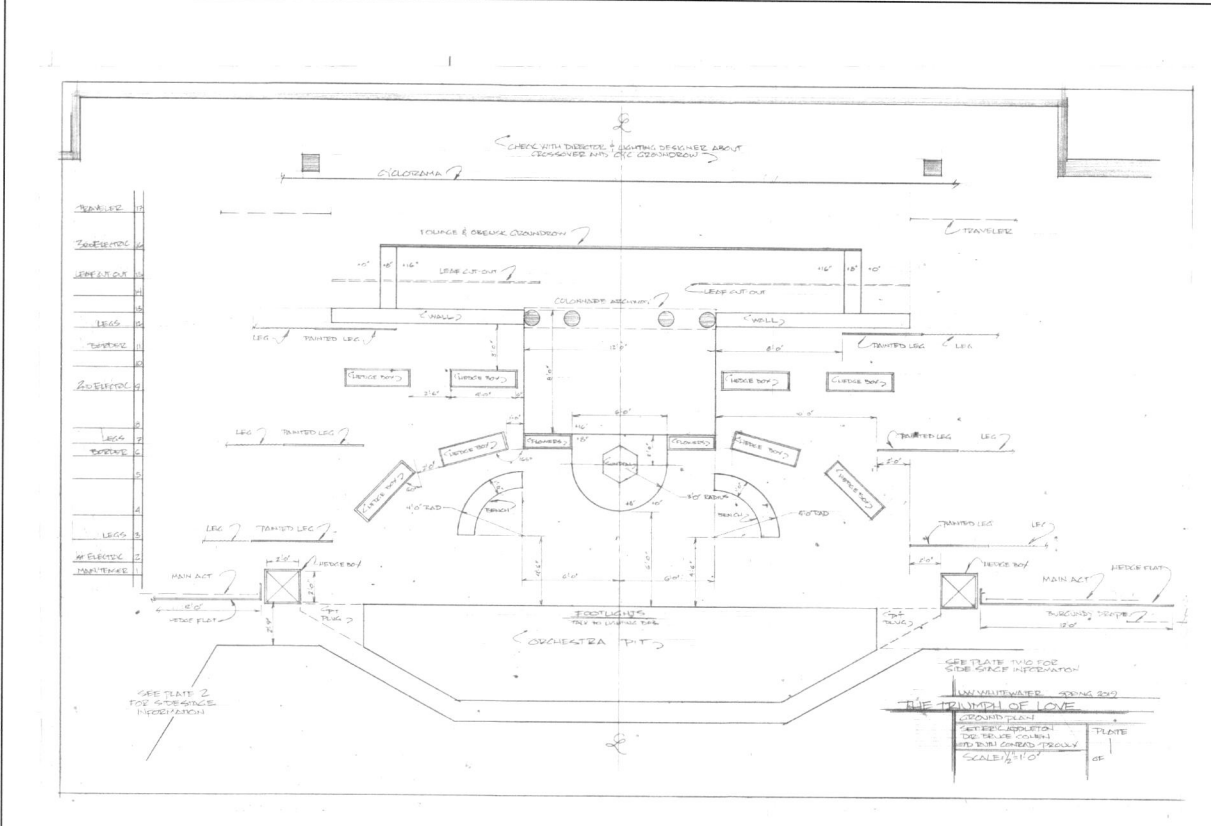

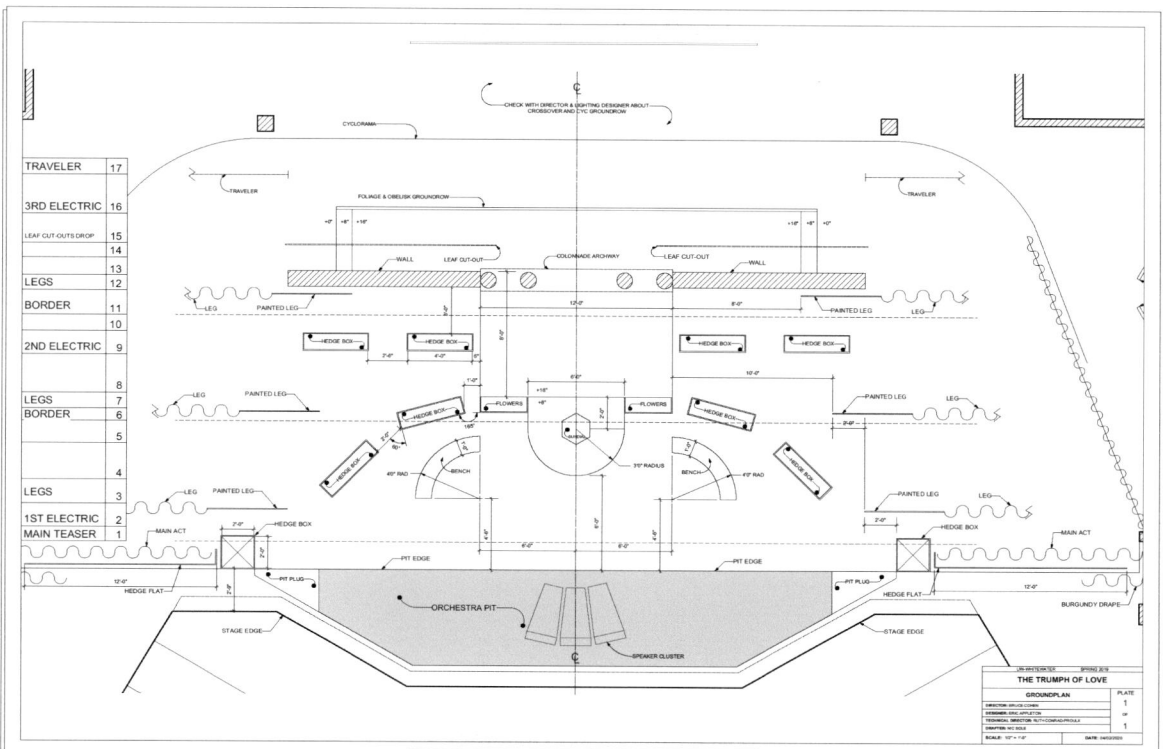

FIGURE 8-2   *Hand drafted groundplan of* Triumph of Love, *and Vectorworks' copy by student. Vectorworks' drawing by Nicolas Sole, used by permission*

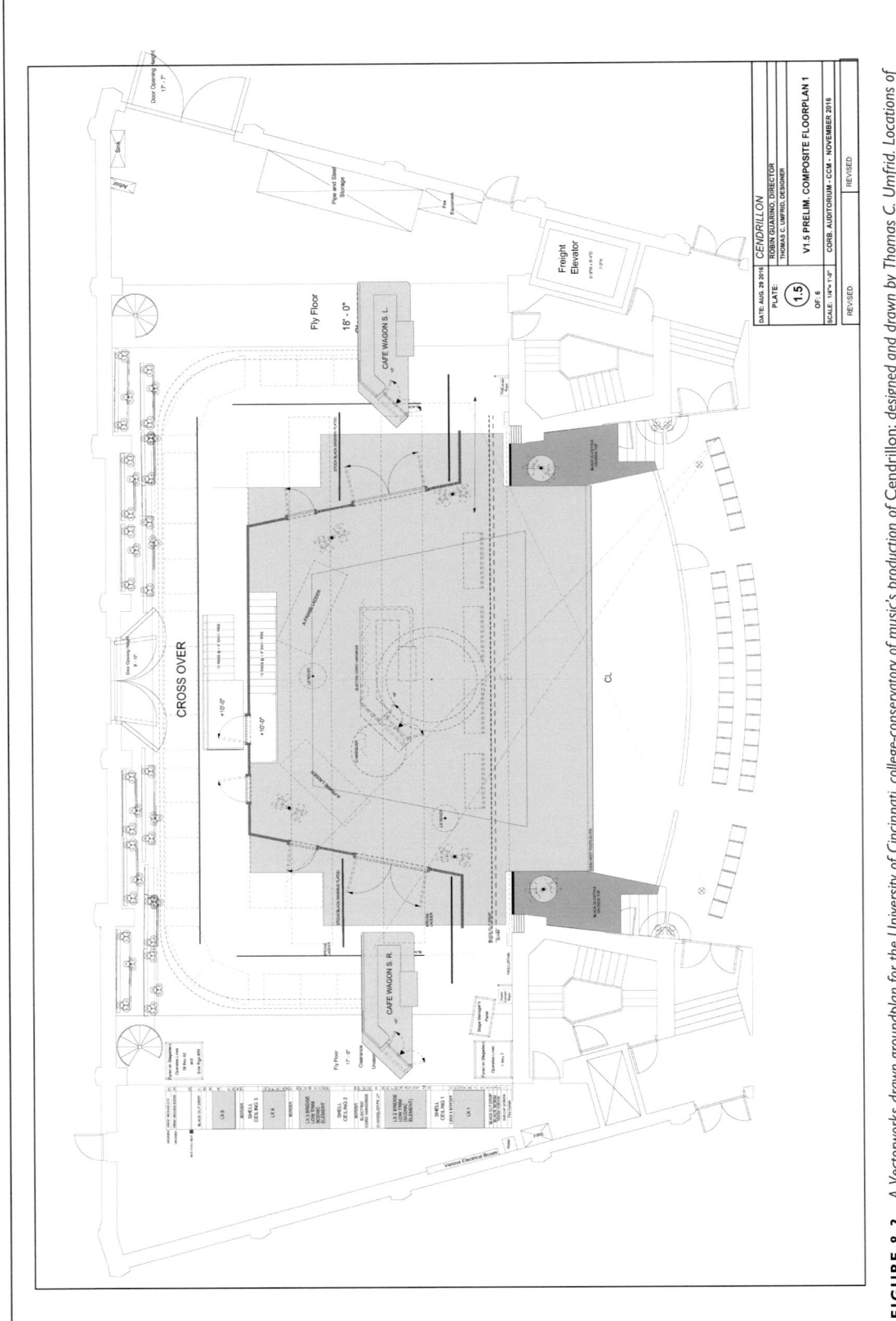

FIGURE 8-3 A Vectorworks drawn groundplan for the University of Cincinnati; college-conservatory of music's production of Cendrillon; designed and drawn by Thomas C. Umfrid. Locations of moveable scenery indicated with dashed line; playing space (including wagons) indicated by shaded areas. Used by permission of Thomas C. Umfrid

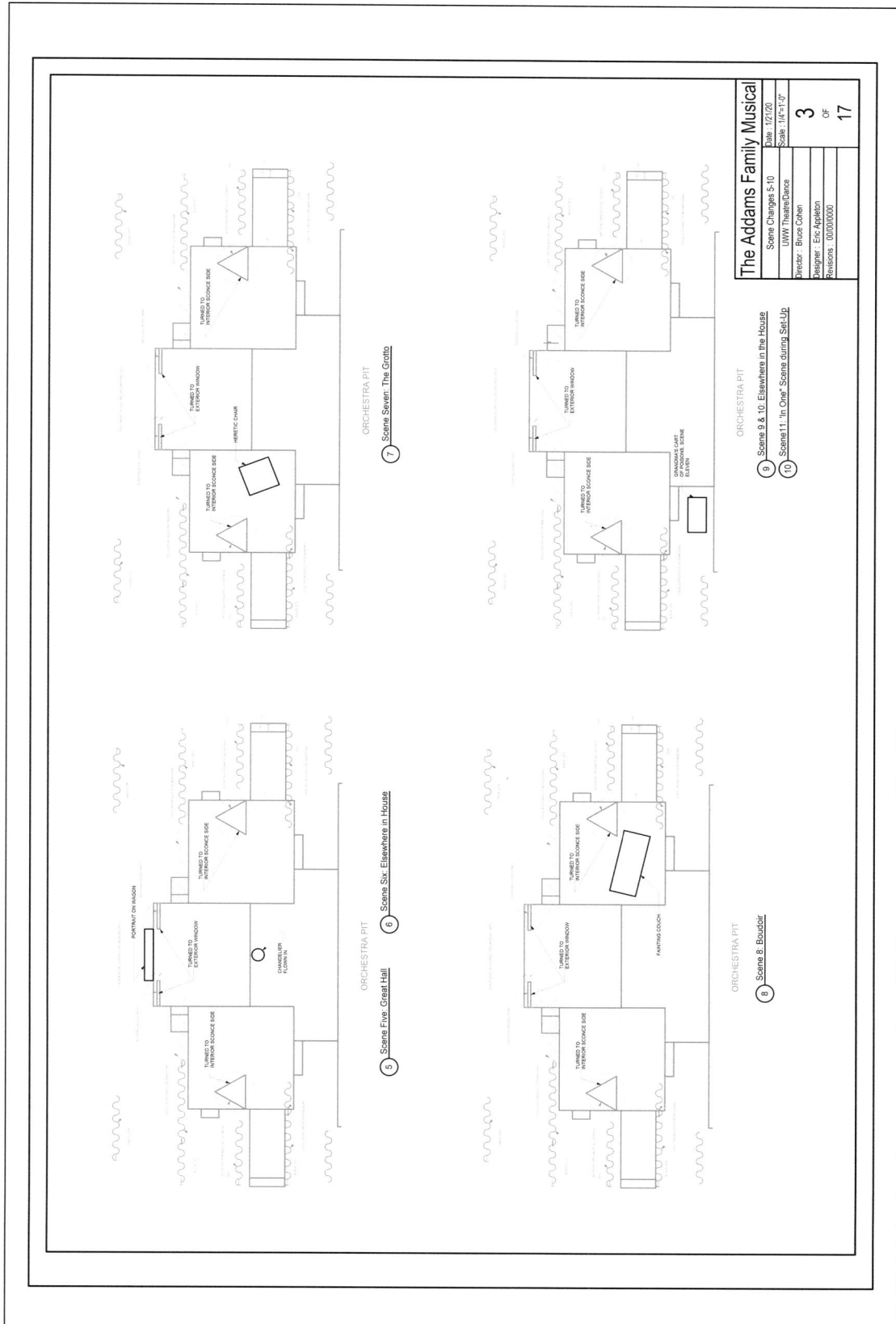

FIGURE 8-4 *Scene change groundplans for the UW-Whitewater production of* The Addams Family Musical

Line Schedule and Hanging Items

The line schedule is drawn in the area of the fly rail so that its labels line up with batten placement. Battens are not typically drawn, though the items suspended from them are. Dead-hung items and pick-line locations are indicated on the plan itself. The Main Act Curtain, if present, is drawn. If the curtain travels, indicate how far onstage it is to be set. All legs, borders, and other masking elements are listed on the line schedule and drawn on the plan. Borders and other items above the cutting plane are shown with dashed lines. Cycloramas, travelers, and other items that cross the cutting plan are shown with solid lines.

Item Placement

Footprints of all permanent scenic elements, including platforms, wagons, stairs, flats, columns, are included. Scenic elements that move are denoted with phantom lines. Stairs and platforms that rise above the cutting plane line are included. Height is noted on all platforms and stairs, including the zero deck. Footprints of furniture and other important set pieces are noted with thin lines. Electrical devices that are part of the scenery are called **practicals**. Depending on their size, they may be indicated with a simplified over-all shape such as a dashed circle representing a chandelier. Equilateral triangles are the standard symbols for practicals on light plots.

Dimensioning and Labels

Only as much dimensioning as needed to place major scenic elements is included. Label everything that is possible to label.

Scale Bar (Optional, but Common)

Place the scale bar in a location where it does not obscure important scenic information.

Floor Treatment

If the floor is painted or otherwise surfaced, the type of surface and edges of treated areas are indicated.

Offstage Storage of Scenery

If major scenic elements are stored offstage when not in use, consider including suggested offstage placement to facilitate actor traffic and scene-change choreography. This may be a separate, secondary groundplan.

Title Block with All Show Information
Other Technical Information as Needed

Since the LD works from the scenic designer's groundplan, it's a courtesy to include permanent lighting positions and lighting bridges found within the area covered by the groundplan, as long as they don't obscure or confuse the representation of scenic elements. This way, at early production meetings, the LD can instantly see the relationship between, for example, the house right box boom and the upright Act Two Bed Unit. Speaker arrays, permanent microphones, and wireless mic receiver box locations are important to the sound designer. If applicable, the technical director will want to know the placement of equipment such as turntable motors or deck tracking shift motors; for turntables, note the direction of rotation.

Discussion: Reading a Groundplan

Instructor

Bring in a number of groundplans as examples: your own, from past students, and from colleagues. If possible, present companion renderings or elevations to enable comparison between a pictorial image and its schematic representation. Figures 8-1 through 8-7 can be used to begin the discussion. Print drawings on easily distributable 8½″ × 11″ handouts so everyone can have a copy while a larger version is projected or laid out on a table. The questions asked in the section below can be used to guide the conversation. You can also use this opportunity to demonstrate how you would talk a collaborator through a freshly presented groundplan.

Students

Look at one of the groundplans presented by your instructor (or in Figures 8-1 through 8-7). Spend a few moments observing and exploring the drawing before trying to describe or explain what you see. Take notes and write a list of questions regarding things that are unclear to you. After the class has had time to examine the drawing, consider and discuss the following questions.

When you first look at the drawing, where does your eye land first? What drew your eye to that point of the drawing? Where did you look next? Why? What do you think is the most important piece of information on the groundplan and where is it located? Is it easy to find? Why or why not? Why do you think that item is the most important piece of information?

Can you find and/or identify all of the elements in the previous list? Is there anything missing? How easy is it to discern between scenic elements and architecture? Why or why not: line weight, line style, labeling, etc.? Are there any graphic symbols or representations that confused you or with which you're not yet familiar? Was it easy to connect the line schedule (if present) with elements on the stage? Are there things on these groundplans drawn differently than this book recommends? Does this other approach provide added clarity? Does the information on the drawing start to reveal itself more readily as you zero in on the

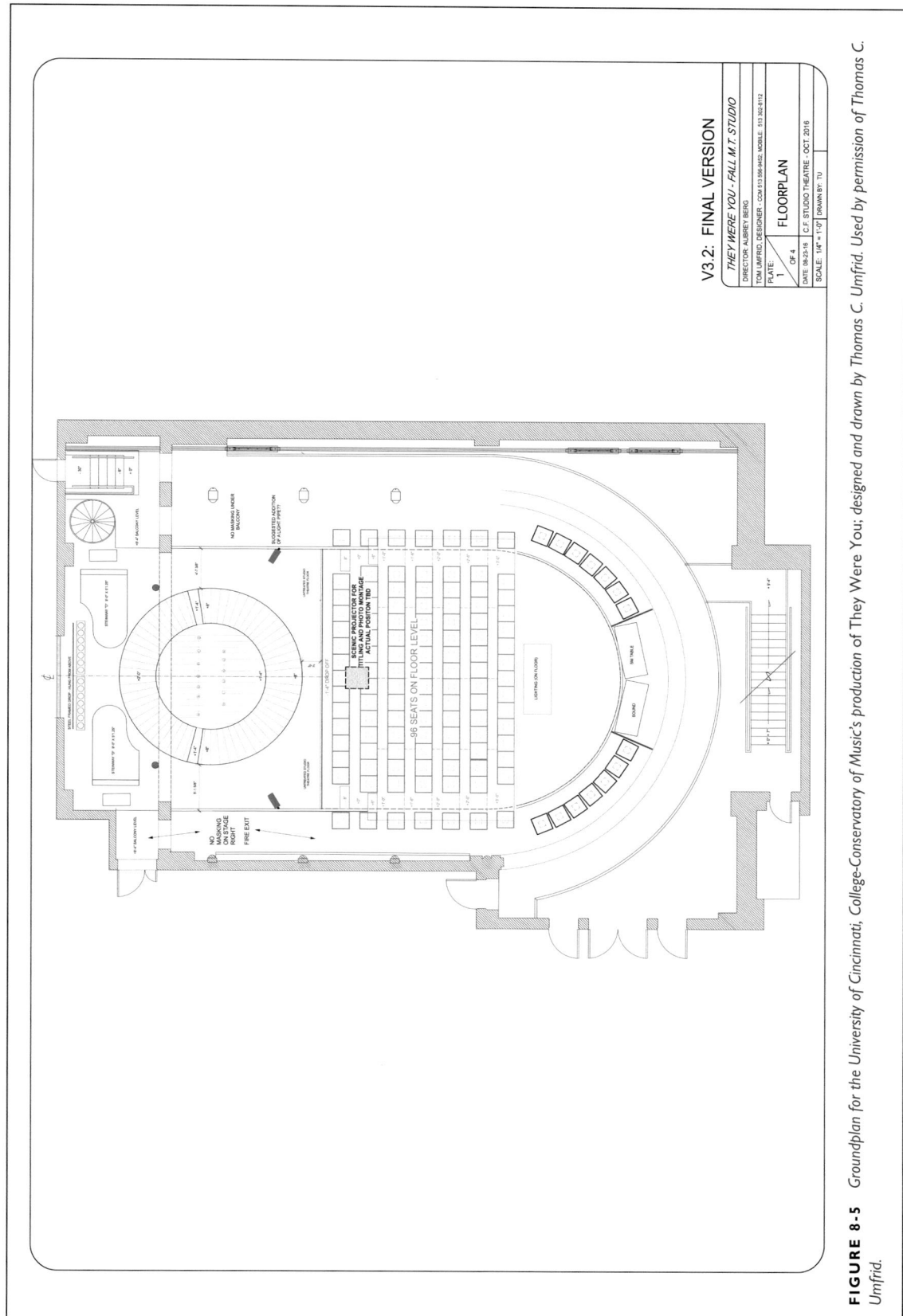

FIGURE 8-5 *Groundplan for the University of Cincinnati, College-Conservatory of Music's production of They Were You; designed and drawn by Thomas C. Umfrid. Used by permission of Thomas C. Umfrid.*

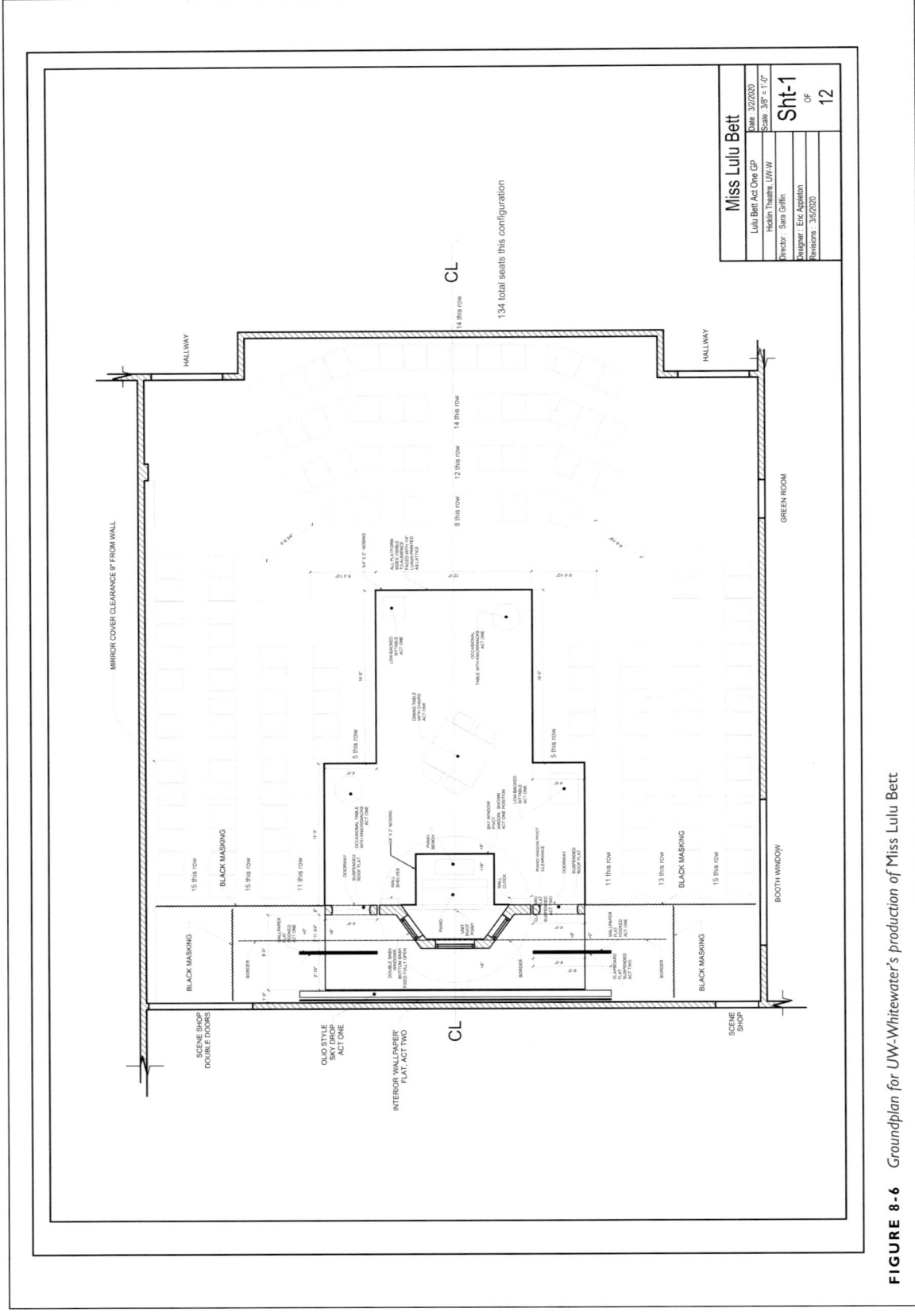

FIGURE 8-6 *Groundplan for UW-Whitewater's production of Miss Lulu Bett*

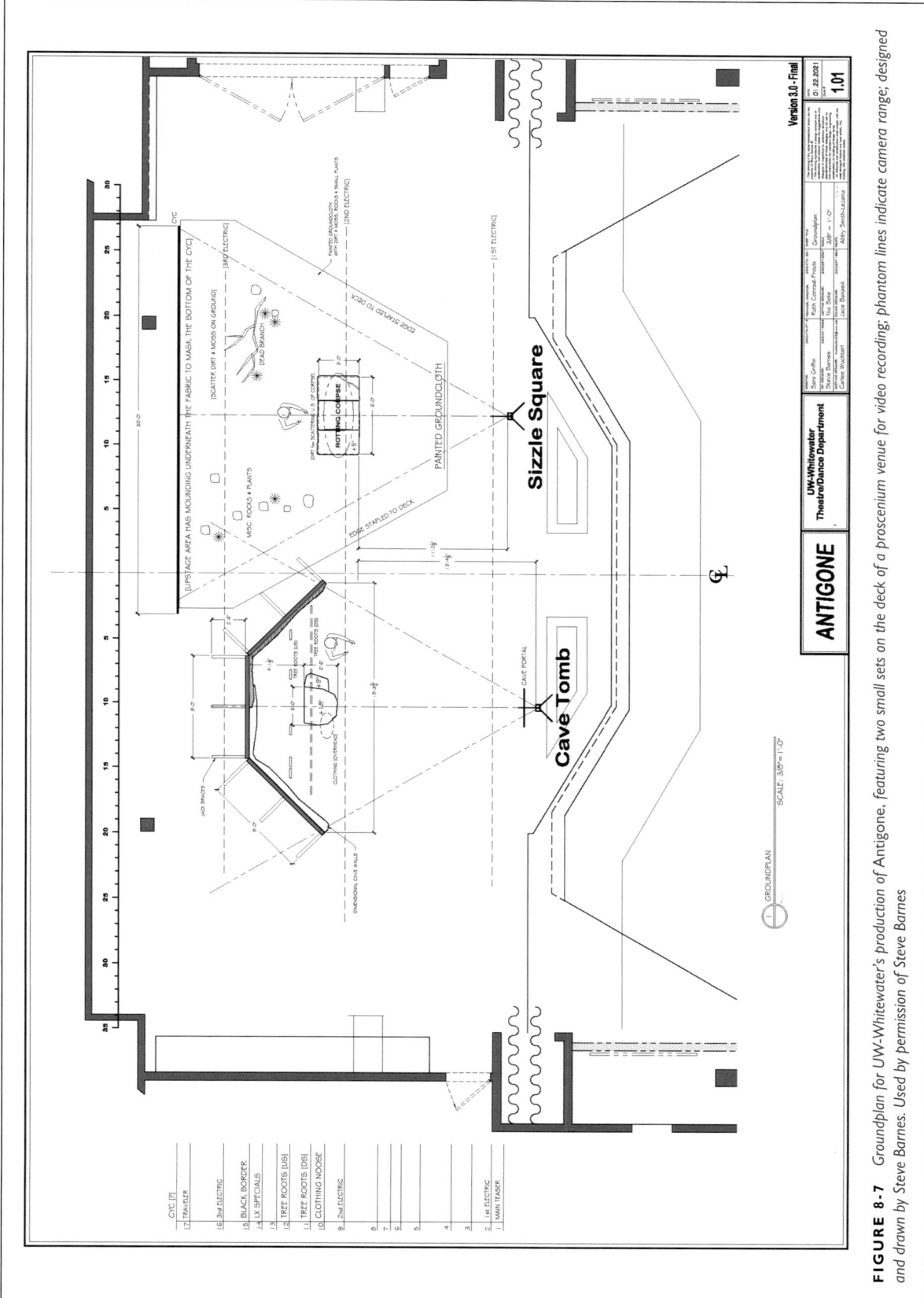

FIGURE 8-7 Groundplan for UW-Whitewater's production of Antigone, featuring two small sets on the deck of a proscenium venue for video recording; phantom lines indicate camera range; designed and drawn by Steve Barnes. Used by permission of Steve Barnes

details? Is there enough information presented that you can begin to visualize the environment without the aid of a rendering or elevation?

In general, the permanent scenic elements should be the most prominent visual features of the groundplan; this is usually accomplished through line weight. For secondary groundplans (and shift plots), the moving scenic piece(s) should be most prominent, as that is the most important feature of that drawing. On light plots, the lighting unit symbols are the most important feature and therefore the most prominent.

The first impulse when encountering a groundplan is often an attempt to take in the whole drawing in one glance. Instead of trying to read the whole plan at once, consider encountering the drawing in a manner that is a little more information-segregated. About which specific item do you want information first?

Perhaps you first want to consider the scenery. Find the center line. Find the plaster line. Determine where permanent scenic elements cross the center line and work your way along the scenery in one direction, identifying various features, like doors, windows, archways, steps. When you encounter a door or window, move backstage to see what is behind the opening. Choose a particular feature and see if there are dimensions that help you locate it on the stage: what is it measured from? The center line, the plaster line, some other piece of scenery? Are objects labeled clearly? Where are furniture and smaller scenic items situated on the deck? Can you determine the nature of those items from the given information (a chair, a table)?

Perhaps you want to first examine the line schedule. How many linesets are there? How many have assignments, and what are those assignments? Are there any dedicated battens (for travelers or electrics?). Pick an assigned lineset and follow it onstage to locate the item suspended from it. Legs? Border? Chandelier? Other named flying scenic element?

Take a look at the venue's architecture. How much is shown? Depending on the size of the space, the scale of the drawing, and the size of the sheet, architecture is often only partially depicted. Is there enough architecture shown to help you understand how the scenery fits in the space? Is there a proscenium? If not, how is the performance space arranged and oriented within the architectural space? What is the audience's relationship to the performance space? Can you find the sightline symbols? Are individual chairs drawn, or is the seating area labeled as a general area?

Use your finger to follow lines and see how things connect. Groundplans are working drawings: feel free to take a pencil and make notes on the drawing itself. Making notes with colored pencils can help ensure you don't lose or forget things you wanted to ask about.

Compile a list of questions you still have after reviewing the drawing. What requires clarification? What are you unsure of? Be specific when framing questions. A comment like "I don't get this" is not useful to the draftsperson since they intended the drawing to be reasonably self-explanatory and therefore may be at a loss as to why you don't understand that element. If you aren't sure how to frame a specific question, something like "Please tell me more about how this part over here works" is one way to start the conversation without accusing either the reader or draftsperson of willful ignorance.

Exercise: Drawing Standard Groundplan Graphics

Instructor

This exercise introduces some of the USITT groundplan graphics recommendations: flats, soft goods, platforms, and various entryways.[3] These drawings present only top views of these items; layout is concerned with overall alignment and grouping rather than presenting orthographically related arrays. See Figure 8-8 for a rough sketch of the project, fully laid out. The steps use Vectorworks; if hand drafting, the steps remain similar, though students should wait to darken the totality of the drawing after it has been completely laid out. This is also an exercise in laying out disparate elements on a shared sheet to ensure that each separate drawing area is seen as an independent entity.

Students

Figure 8-8 presents a selection of USITT recommended groundplan graphics. While the graphics presented are by no means all of them, these are among the most common and will aid you in drafting your first groundplans. As these are groundplan graphics, only schematized top views are depicted. Elements cut through by the cutting plane are shown as sections. Elements above the cutting plane are drawn with dashed lines.

The dimensions provided are to aid in overall layout of the drawing. Additional dimensions are provided in the steps below. In Vectorworks, each portion of the drawing will be darkened as it is completed. Draft in $\frac{1}{2}'' = 1'-0''$, Arch D paper, with borders and title block. Dimension the graphics, not the spaces between them; include labels.

1. Open your base file.

    Create a new a new class in the base file: **Phantom Lines**: 0.010 mm, with line style ISO-9 (Long Dashed Double-Short Dashed), un-Filled. Phantom lines will be used to represent movement in the sliding door graphic. Save the change into your base file.

    Save As with a new file name. Use Page Setup to select Landscape orientation and Arch D paper.

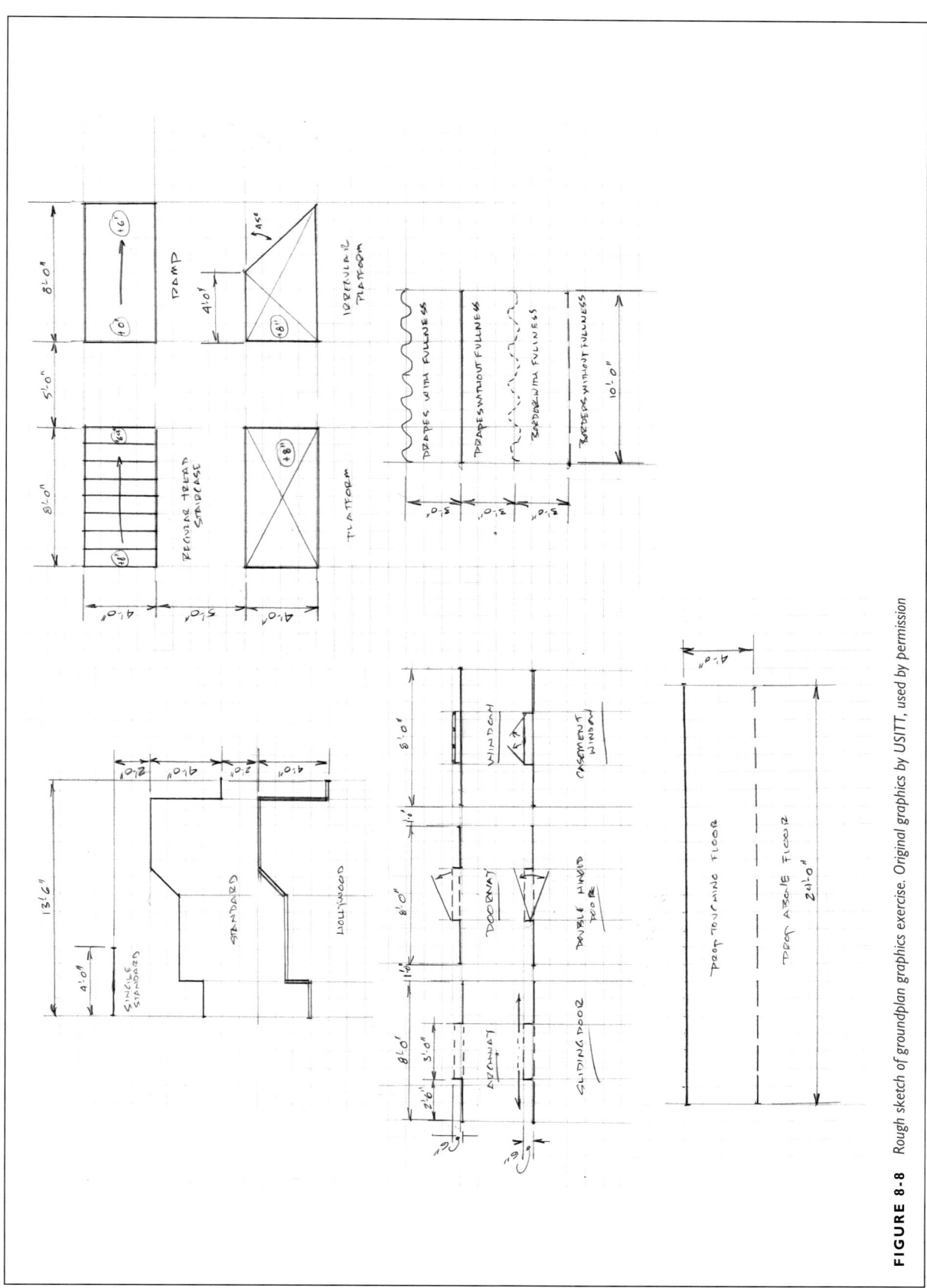

FIGURE 8-8 Rough sketch of groundplan graphics exercise. Original graphics by USITT, used by permission

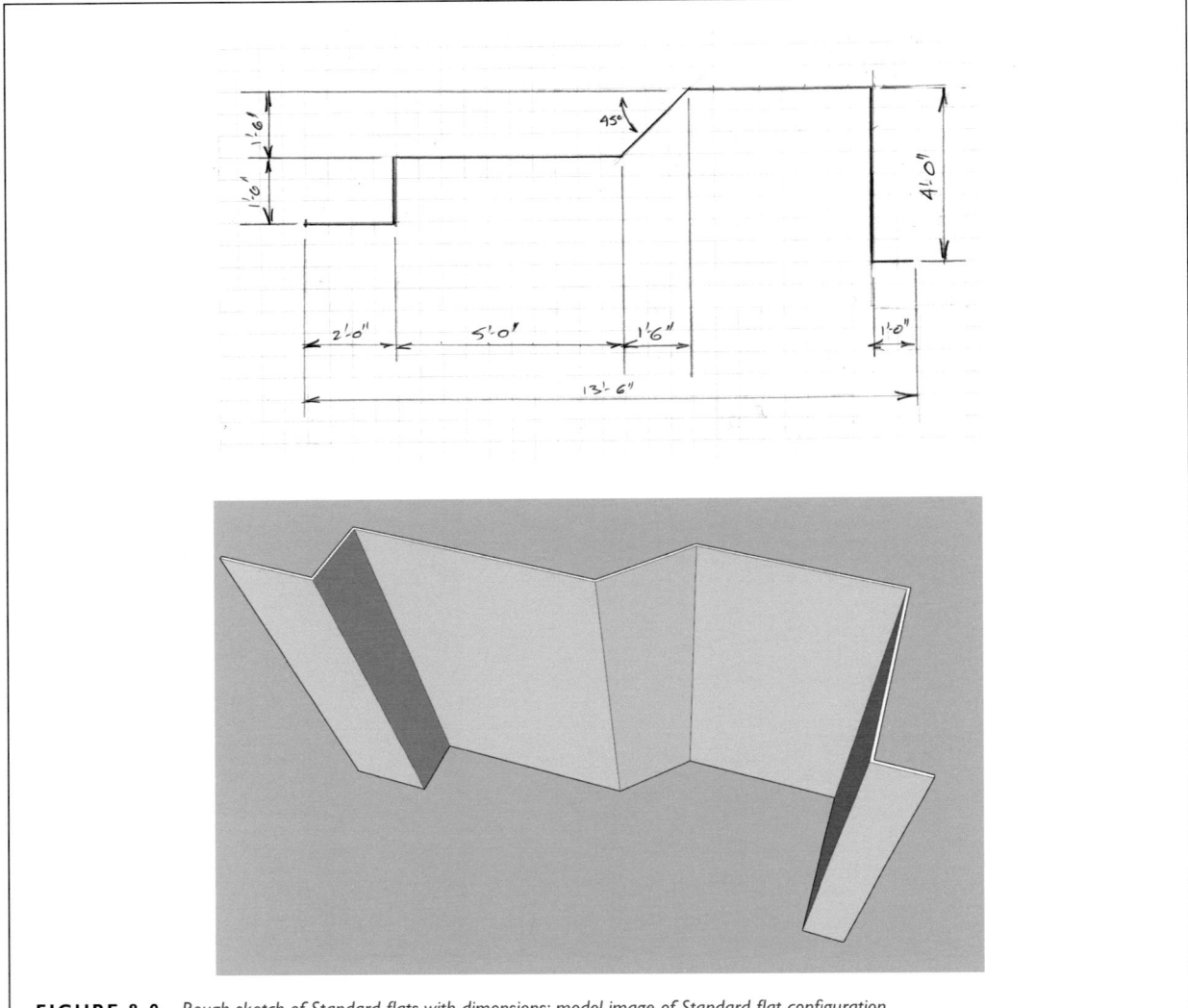

FIGURE 8-9   *Rough sketch of Standard flats with dimensions; model image of Standard flat configuration*

2. Suggested design layers for this project:
   - Border and Title Block Guidelines
   - Guidelines
   - Flats
   - Doors, Archway, and Windows
   - Drops
   - Platforms, Ramps, Stairs
   - Drapery
   - Dimensions and Notes

3. In the Border/Title Block Guidelines layer, lay out the borders and title block.

4. Switch to the Guidelines layer. The upper left quadrant of the sheet features the top view of a 4'–0" wide Standard flat and the top views of two continuous series of flats, drawn as both Standard and

Hollywood flats (Figure 8-9). To refresh yourself on Standard and Hollywood flat construction, refer back to Chapter 6, Module 1 (or Paul Carter's *Backstage Handbook*).

Standard flats are ¾" thick, so it's difficult to show their thickness in small scales. A thick solid line suffices. In Vectorworks, a double line with ¾" separation could be used, but depending on the scale the sheet is printed it may just read as a single thick line.

Hollywood flats have more depth and are represented by a double line indicating the flat's suggested thickness (3", 4", etc.).

Consulting your layout math, lay out the perimeter for the array of flats. Lay out the perimeters of the two continuous walls within the array perimeter. Consider the bottom of the sheet to be downstage.

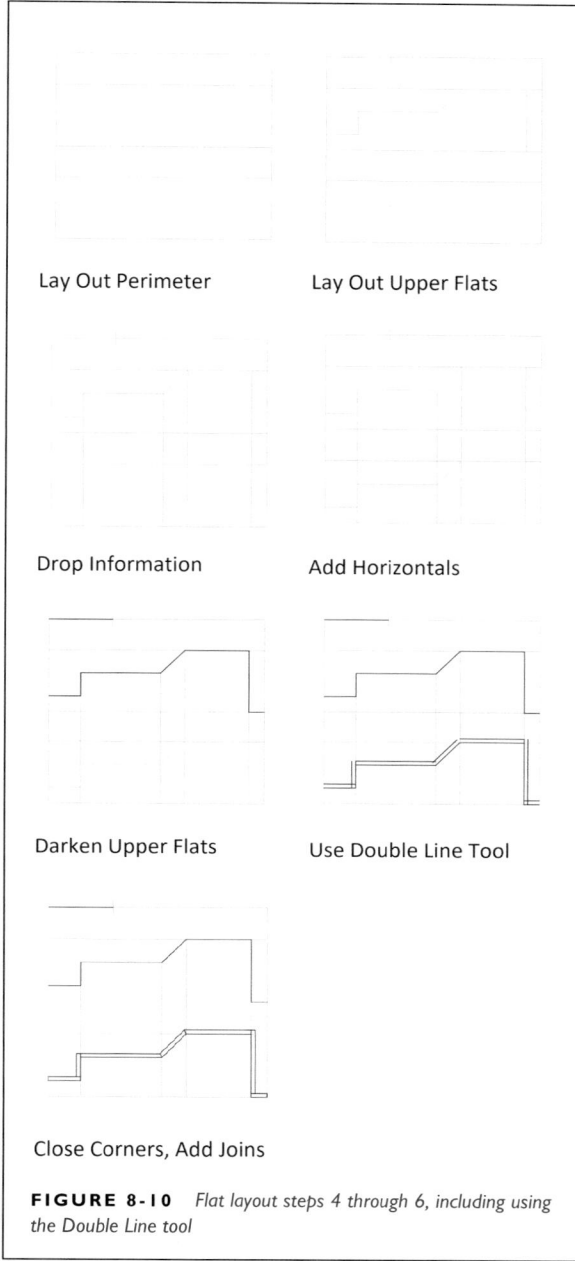

Lay Out Perimeter

Lay Out Upper Flats

Drop Information

Add Horizontals

Darken Upper Flats

Use Double Line Tool

Close Corners, Add Joins

FIGURE 8-10 *Flat layout steps 4 through 6, including using the Double Line tool*

5. The single flat can align with the top edge of the array perimeter. Lay out the continuous walls within their perimeters; see Figure 8-8. Since the dimensions for both flats are the same, lateral measurements can be made in the top drawing and dropped into the lower drawing. Even though the Hollywood flat has thickness and will eventually be drawn with a double line, first lay it out as a single line since the downstage (audience-facing) measurements are to be the same as that of the Standard flat (Figure 8-10).

6. Switch to the Flats layer and darken the Standard flats. For the Hollywood flats, use the Double Line tool. Set

the separation to 3″. Choose the option that you think will place the second line upstage of the first. If the second line falls downstage of the guideline, inches will be trimmed off the audience side of the flat, resulting in flats that are not the intended size or in the correct location. Close up open corners and ends with the Line tool and Trim any excess lines within and beyond the flats' outline. Since a wall like this is seldom constructed as a single unit, indicate how individual flats butt against each other at their joints (Figure 8-8). Try to minimize how many seams directly face the audience (downstage). Compose and crosshatch.

Except for labels and dimensions, the top views of the walls are now complete.

7. Switch to the Guideline layer and Guideline class to lay out the arch, doorway, and window graphics. See Figure 8-8 for dimensions and layout. Figure 8-11 presents the graphics as presented in USITT recommendations. Figure 8-12 presents models depicting each graphic. Figure 8-13 presents layout steps.

8. Lay out the perimeter of this drawing array. Leave enough space below it for the drawing of drops (see Figure 8-8). All six graphics have the same base measurements. Lay out the widths of the openings in the top row and drop guidelines down to the bottom row. All of the graphics have 6″ reveals, so guidelines can be drawn across all three items in the same row. The archway will be automatically laid out once the array includes reveal depth and width of openings (Figure 8-11).

9. The Doorway is identical to the Archway except for the addition of the angled door and indication of the door's swing.

10. The final drawing for the top row is the multi-paned window. The panes of glass, or lights, are separated and supported by strips called **muntins.** The groundplan's cutting plane slices through the window opening, so the muntins are seen as cut-through features. If this drawing were in a larger scale, the glass panes would also be shown as cut-through features. The sill and bottom of the window frame, being below the cutting plane, are visible. See Figure 8-14 for muntin dimensions. Fill the muntins during the darkening phase.

11. Switch to the Doors and Windows layer to darken the top row. Use single visible lines to represent the Standard flats' ¾″ depth as well as the reveals. Since the front and rear of the door openings are above the cutting plane, they are drawn with dashes (as was the case for the doorways in Chapter 7's venue plan).

To represent the swing of the door, use the Dimension class. Draw an arc with the Radius Mode option, then choose the Line Startpoint Style from the Attributes window to open its menu. To put an arrowhead

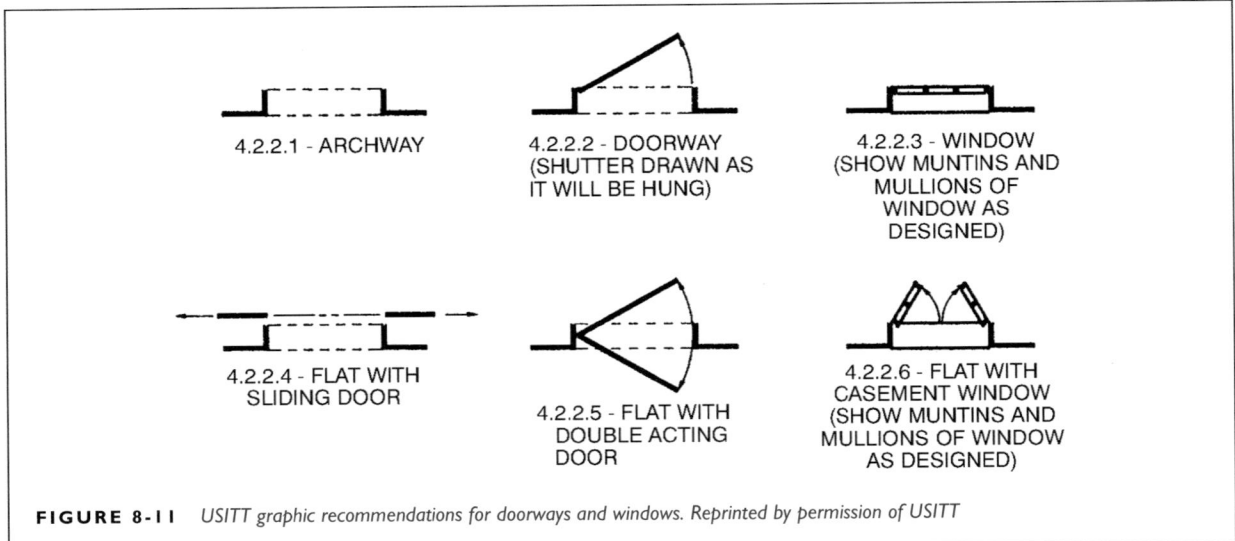

FIGURE 8-11   *USITT graphic recommendations for doorways and windows. Reprinted by permission of USITT*

on the end of the line contacting the door, checkmark the arrowhead pointing to the left (0.125″ × 0.000″ 15° 0.05 mm). Unselect the Line Starter Marker Toggle when finished, otherwise the next line drawn will also have an arrowhead. You'll use these arrowhead toggle choices again for all three graphics in the bottom row.

12. For the multi-paned window, use the Rectangle tool to draw horizontal sections of the individual vertical muntins. Choose a solid black Fill from the Attributes window; the area is too small for crosshatching. Darken the muntins first, then connect them with lines representing the front and rear of the bottom of the window frame (see Figure 8-14). When alternating between design layers,

the Fill may vanish. It's still there, and will reappear on the printed drawing when the layers are checkmarked for visibility in the Create Viewport window.

13. Return to the Guidelines layer to lay out the bottom row's three graphics. The sliding door features two

1. Lay out perimeter of full array

2. Lay out outlines of top and bottom rows

3. Lay out openings

4. Darkening basic archway graphic

FIGURE 8-13   *Steps in laying out the Arch, Door, and Window Graphics: (1) Lay out perimeters; (2) Add depth; (3) Lay out opening widths; (4) Lay out additional details and darken completed graphic*

FIGURE 8-12   *SketchUp models of the six archway, door, and window graphics*

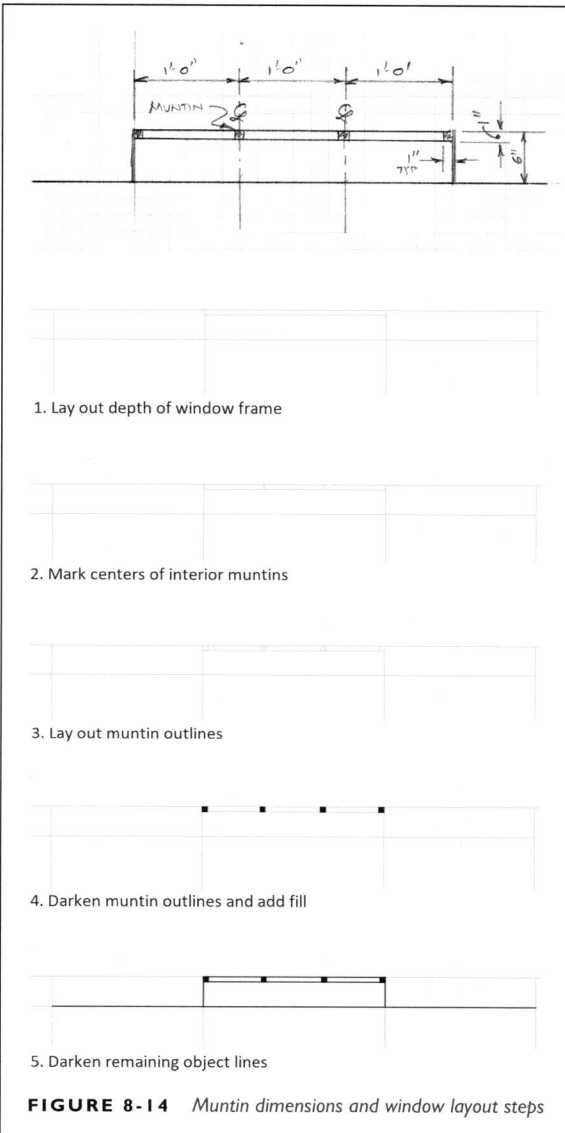

1. Lay out depth of window frame

2. Mark centers of interior muntins

3. Lay out muntin outlines

4. Darken muntin outlines and add fill

5. Darken remaining object lines

FIGURE 8-14 *Muntin dimensions and window layout steps*

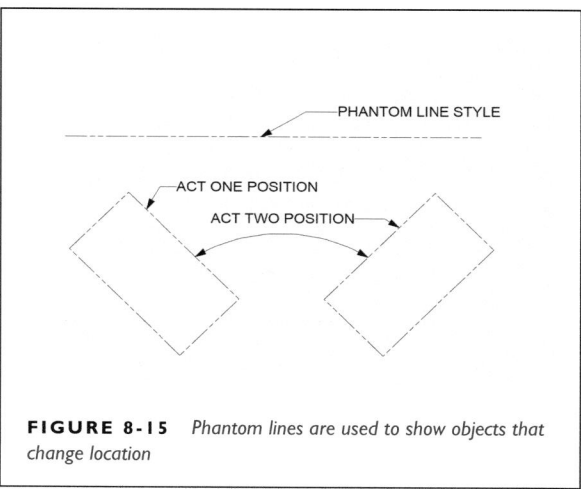

FIGURE 8-15 *Phantom lines are used to show objects that change location*

panels hung upstage of an archway. The panels are accompanied by lines and arrows indicating their movement. A **phantom line** is used to show their movement. Phantom lines are used to outline or indicate a location of an object that moves; e.g., if a rolling platform, or wagon, has a position onstage for Act One and moves to a different position for Act Two, both locations are outlined with phantom lines, then labeled (Figure 8-15). Figure 8-16 presents the bottom row's finished layout.

When laying out the parts of the sliding door, leave small gaps (1/16″) between the endpoints of the guidelines for the phantom lines, panels, and movement arrows. This will prevent the darkened lines from becoming a continuous object.

The center graphic depicts a door that swings in both directions, such as one might find at the entrance to a restaurant's kitchen. The hinge is located in the center of the reveal. The door is shown in two positions, spanned by an arc with arrowheads at both ends.

The third graphic is a **casement** window. This window is comprised of two frames, hinged on their sides, with two columns of panes within each frame. The cutting plane slices through the stiles of each frame as well as through a center muntin. Arcs with arrowheads indicate swing. Note that hinge of each frame is at the rear of the reveal, as well as on the rear of the frame. When closed, the rear of the window frame is flush with the rear of the reveal.

14. Return to the Guideline layer to lay out the perimeter of the drops array below the arch, door, and window array (Figure 8-8). If it seems that there's not enough space, go into the necessary design layer and Select the array to be moved. With the array highlighted, drag it to a new location.

15. A **drop** is a large piece of painted fabric (typically muslin) that spans the full width and height of the stage space, often flown in and out on a lineset. After laying out the two drop graphics, switch to the Drops layer and darken. Use a thin visible line for the Drops Touching Floor, and a thin long-dashed line for the Drops Over Head (consult the line styles library via the Attributes window).

16. Return to the Guidelines layer to lay out the platform, ramp, and stair unit array (Figure 8-8). The stairs on the Regular Tread Staircase have treads 1′–0″ deep. An arrow pointing up the staircase is a permissible way to indicate stairs as long as each step is uniform. Otherwise, indicate overall height on each tread.

The two platforms are shown as they would appear on a construction drawing indicating the placement of

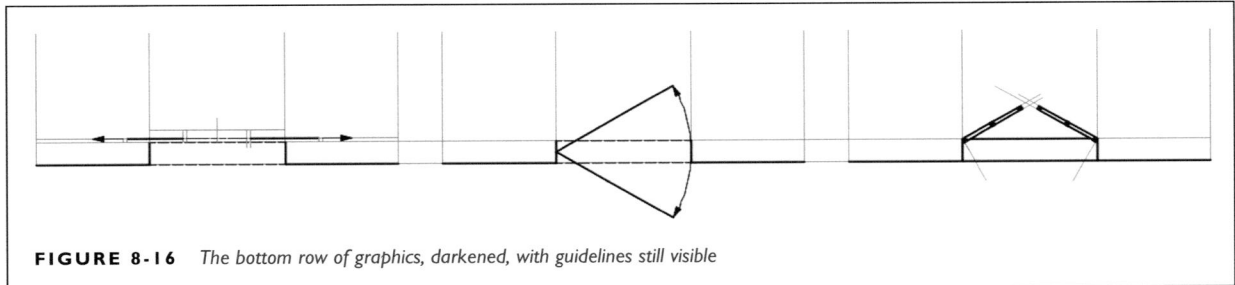

FIGURE 8-16   *The bottom row of graphics, darkened, with guidelines still visible*

stock platforms. Heights and directional arrows will be added during the dimensioning phase.

17. Switch to the Platforms, Ramp, Stair layer. Darken the outlines of the platforms, ramp, and stair unit. The lines crossing the platforms from corner-to-corner are thin lines.

18. In the Guideline layer, lay out the array for the Drapes, or **soft goods** (Figure 8-8).

    Soft goods are any unframed suspended fabric panels and include drops, legs, borders, scrims, and cycloramas. There are typically two ways they are hung: with **fullness** or flat. Fullness is the term used when the fabric is gathered and pleated so that it hangs with vertical undulations across the panel (Figure 8-17).

19. While soft goods without fullness are drawn with straight unbroken lines, the ones with fullness are represented by an undulating line requiring layout in the Guidelines layer.

    It is difficult to draw a consistent undulating line freehand. For hand drafting, templates with wavy lines can be purchased. There are wavy line styles available in the Vectorworks library. In the default libraries, the closest one is Sine Wave, but Sine Wave's peaks and valleys are fairly sharp—not like gently undulating fabric.

    In this exercise, the drapery with fullness has a footprint that is 6″ deep. In the guideline layer, draw a 10′–0″ guideline for a horizontal axis. Use the Double Line tool with a separation of 3″ to draw lines parallel to and above and below the axis. With the Double Line tool still set to a 3″ separation, draw vertical lines dividing these two horizontal rectangles into 3″ wide vertical increments. The intersections will serve as endpoints and centers for the arcs you're about to draw (Figure 8-18).

20. Switch to the Drapery layer to darken the drapes. The first graphic represents drapery that touches the deck and is therefore a solid line. To construct the undulations, choose the Arc tool's Point on Arc mode. Click on the leftmost intersection along the horizontal axis of the strip of squares. Hop over two vertical lines and click the mouse on that intersection. An arc appears, expanding and contracting as you move the mouse. Click on the upper guideline midway the two arc endpoints. A semi-circle will snap into place. Work along

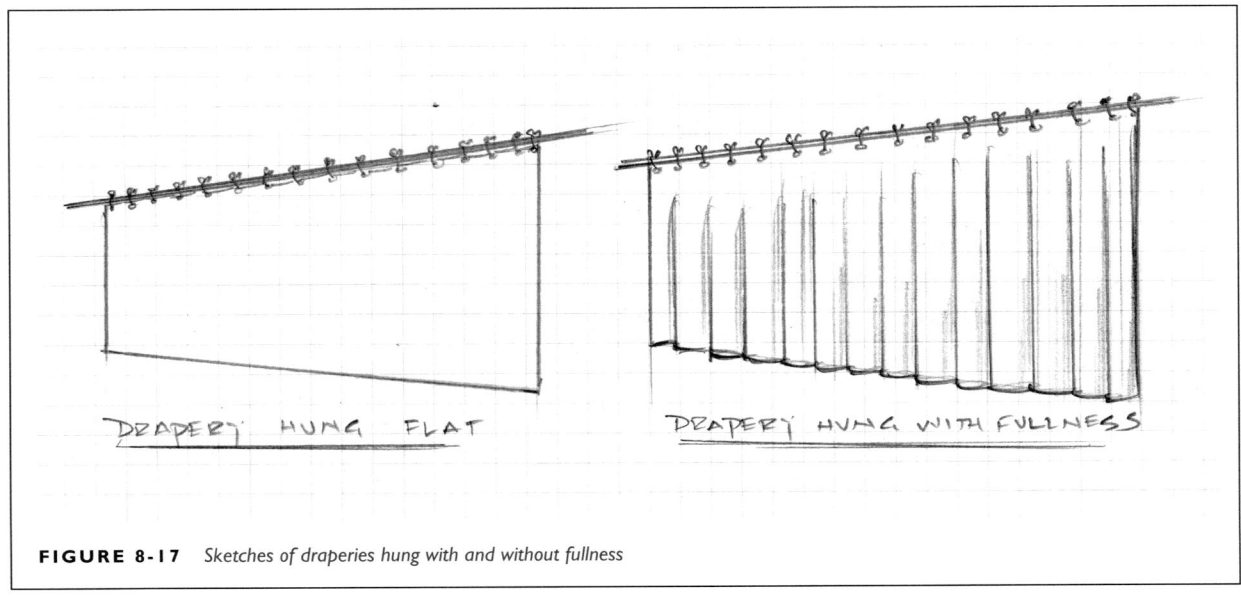

FIGURE 8-17   *Sketches of draperies hung with and without fullness*

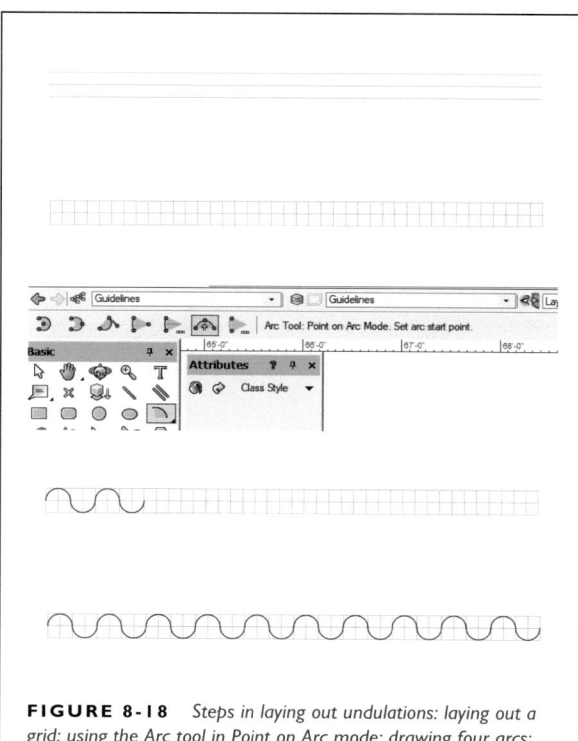

FIGURE 8-18 *Steps in laying out undulations: laying out a grid; using the Arc tool in Point on Arc mode; drawing four arcs; copying and dragging to complete the drapery*

the horizontal axis, alternating upward-bending arcs with downward-bending arcs until you have four arcs (Figure 8-18). Select the four arcs and hold down the Ctrl key as you grab the startpoint. Drag the series over, affixing the startpoint of the copy to the endpoint of the original. Repeat until the drapery is complete. Select the whole drapery and make it a group (via the Modify drop-down menu). This will make it easier to copy, as well as to change the copy's line style to dashed lines.

21. Consider the drawing as a whole. Does it look as though there's enough space for dimensions and notes? Is there space for the title block? If you need to shift an array to create adequate white space, go into that layer, select and move as needed.

    In the Dimensions and Notes layer, dimension all the elements (consult Figure 8-8). Do not include the layout dimensions. For labels, use the labels seen in the USITT graphics sample (Figure 8-11) as well as those in the rough sketches.

22. Create a Viewport for the whole of the drawing. Be sure to checkmark the desired layers so that muntin Fill and crosshatching is visible. The scale remains ½″ = 1′–0″.

23. In the Sheet layer, add the border, title block, and title block information.

24. Save the file. Export and Save as a PDF. Print the PDF.

MODULE 2: CONSTRUCTING THE MODEL OF A BOX SET

Different designers employ different processes to develop the content of the groundplan. Some begin with groundplan doodles and then merge these with their research: in effect, building *up* from the plan. Others begin with more illustrative sketches and extrapolate that information into a floorplan: building *down* from the image. Experimental models are excellent tools by which to engage a director during design development; SketchUp models allow fast revision during meetings so that discussed changes can be shared almost immediately. Regardless of the content generation approach, the goal is to create a rough sketch placing all of the elements of the scenic environment within the venue's plan before drafting begins.

Drafting often begins with certain factors yet to be discovered; dimensions may not be finalized, placement of certain objects may need to be adjusted. These unknowns should be worked out with guidelines. If something goes terribly awry, don't throw out or delete the drawing. Keep it on hand to serve as reference so that you don't repeat the same errors. In Vectorworks, create a new layer and trace what information you need from the erroneous drawing. Turn off the old layer's visibility until you feel comfortable deleting it.

Most theatrical environments can be described as belonging to one of three broad categories: (1) a **box set**, (2) a **unit set**, and (3) a **wing-and-drop set**. A box set is typically presents a space in which the audience sees into a space bounded by walls— for example, looking through a removed wall into a realistic room (the "fourth wall"). A unit set presents an arrangement of permanent structures that support all of the locales of the performance; smaller scenic pieces come and go as required. A wing-and-drop set uses flat, illustrative scenic elements, creating a space in which action is performed in front of scenery rather than within a scenic environment.

Exercise: Creating a Rough Groundplan Sketch

Instructor

Figure 8-19 presents two rough sketches of the box set used for this exercise. The first shows the scenery standing alone; the second shows the scenery placed within the proscenium opening. A human figure can be found in the stage right archway to provide a sense of scale. The students should do their best to situate, in scale, the footprint of the set on a hardcopy of the venue's base drawing floorplan. This is an exercise in thinking in scale,

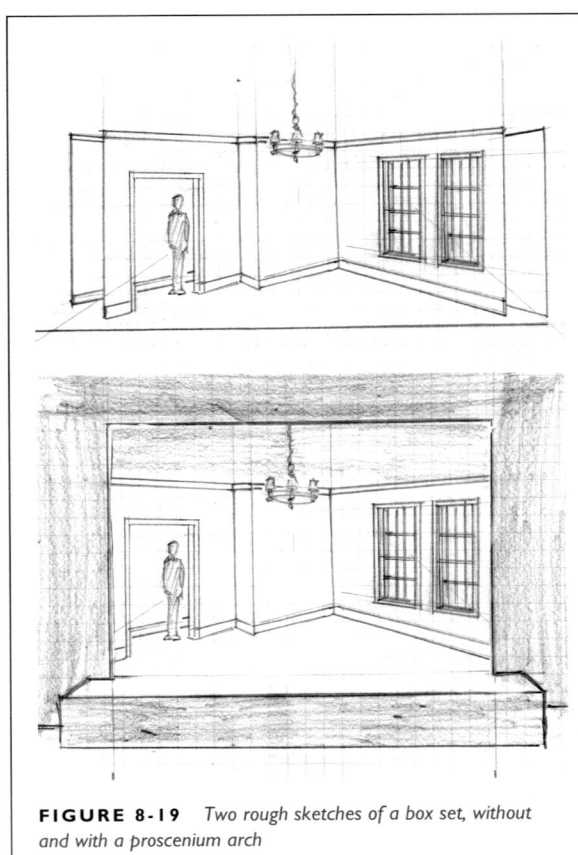

FIGURE 8-19 *Two rough sketches of a box set, without and with a proscenium arch*

as well as interpreting a rendering. Questions found below can guide discussion.

Students

Print a copy of the venue plan from Chapter 7. Looking at the sketches in Figure 8-19, do your best to lay out the footprint of the set's walls on the venue's deck. How many walls are there? How do they relate to one another? Are walls parallel to, or angled with respect to the proscenium? What part of the set falls on the center axis? How far upstage from the proscenium does it appear? Are there clues as to what is seen through the openings? When comparing the standalone sketch with the proscenium-added sketch, what elements appear to be hidden behind the proscenium?

As a class, compare rough groundplans. What did everyone find easily? Was there anything discovered by only one or two members of the class? Did everyone place their footprints in about the same place on the venue deck or was there a wide variety of placements? Why? Does your footprint sketch appear to match the scale/proportions of the venue plan? What information is not offered by the sketches? How might the missing information be obtained or determined?

Exercise: Modeling the Box Set in SketchUp

Building an experimental model from rough sketches is a good way to firm up dimensions and placement and to see if preliminary ideas work. A model, whether physical or virtual, is not a replacement for research and sketching, but rather the next stage: integration of planning and research into a preliminary scale representation of the scenic environment. Having visual goals means that the experimental model building process becomes one of 3D problem solving rather than immersion in a sandbox of bewilderingly infinite possibilities.

Instructor

Have the students compare the rough sketches generated in the previous exercise with Figure 8-20. Discuss similarities and differences, as well as new information. The walls will be constructed as separate groups to facilitate their use as detail elevations in Chapters 10 and 11. Review the rough groundplan in class; height measurements are noted as the exercise progresses.

Model construction will be broken up into a series of phases:

- Laying out the footprints of the flats
- Push/Pulling the footprints into 3D shapes
- Adding molding
- Building and inserting windows
- Adding further decorative items
- Exporting it into the venue model

Students

Figure 8-20 presents dimensions and further detail for the box set presented in Figure 8-19. The plaster line and center line are included on the drawing; these align with the plaster and center lines in the venue drawing and model. An additional line, the **set line**, indicates the farthest downstage points of the set. Set lines are most commonly included on the groundplans of box sets as they indicate the downstage-most points of the set's furthest right and left walls.

There are bubbles containing letters located in close proximity to each flat: these are the index names of the flats (as opposed to their descriptive names). The downstage flats (A, B, C, D) are hard-covered Hollywood flats with 6″ reveals, providing support and realism for the archway and windows. The Hollywood flats are 3″ thick (frame and lauan cover). Flats E and F are soft-covered Standard flats, as they are flat-painted surfaces. They are ¾″ thick. The flats will not be individually framed out in this modeling exercise, but

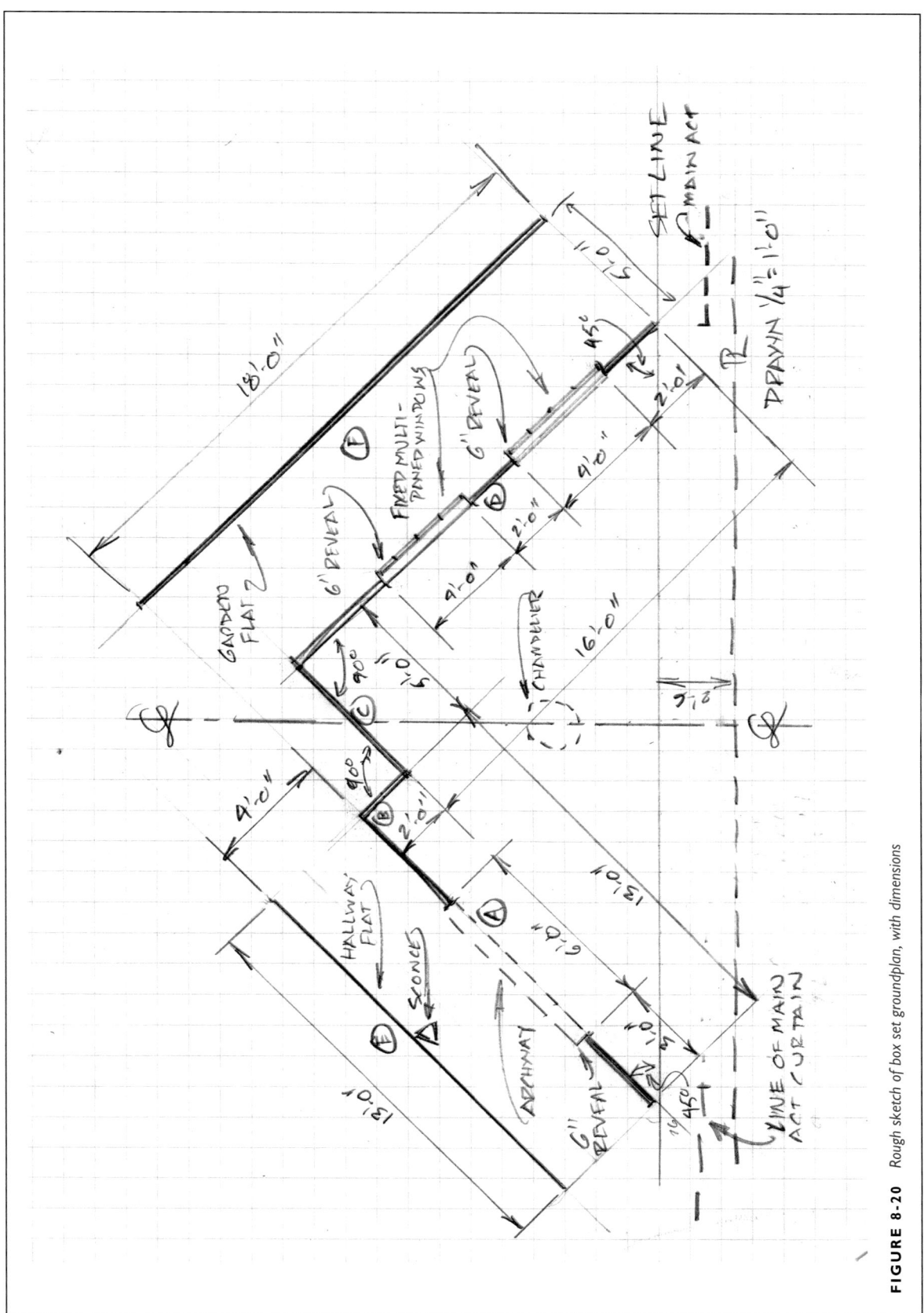

FIGURE 8-20 *Rough sketch of box set groundplan, with dimensions*

simply represented by extruded slabs of the correct depth. All flats are 10'–0" tall. The window bottoms (top of sill) are + 3'–0" above the deck; the window openings are 5'–0" high. The archway is 7'–6" high.

In this exercise, placement dimensions reference either the downstage or the upstage edge of a flat; e.g., Flat F is placed 5'–0" behind and parallel to the upstage edge of Flat D. All corners are 90°. All walls are 45° from the center/plaster line. The downstage ends of walls A and D are an equal distance from the center line. Once the footprint of the walls are drawn, it may be necessary to laterally shift the whole array into alignment with the centerline.

When drafting the groundplan, the center line and plaster line are the first lines to be established. In this case, since we're using a rough sketch to create an experimental model, Flats A, B, C, and D will be laid out first and then shifted as a group into position along the reference lines.

1. Open SketchUp. Choose the Simple/Inches template. Move the human figure aside.

2. At the intersection of the three colored axes, drag out a large horizontal rectangle to use as a drawing surface. Make it at least 36'–0" wide by 20'–0" deep; this is roughly the maximum width and depth of the groundplan, as indicated on the sketch. Make this a group so it can be easily edited or deleted later on. The guidelines can be drafted "in space," but it is easier to ensure that all of the footprint lines inhabit the same plane when there is a surface upon which to draw.

3. At the intersection of the three axes, use the Protractor tool to draw a guideline that extends back and to the right (stage left) at 45°. This marks the downstage edge of Flat A.

4. Use the Tape Measure tool to measure the downstage length of Flat A along this guideline. At the tick mark, use the Protractor tool to place a guideline perpendicular to Flat A. This marks the downstage side of Flat B.

5. Measure the length of Flat B. Draw a perpendicular guideline to situate Flat C.

6. Measure Flat C. Place a perpendicular guideline for Flat D.

7. Use the Line tool to draw lines over the length of each wall.

8. Drag out guidelines parallel to and 3" upstage of each wall's line to denote the thickness of the Hollywood flats. This sets the upstage edges of Flats A, B, C, and D.

9. Consider how the flats butt against each other, and add lines to create each flat's footprint as an independent surface. Remember to close the ends of Flats A and D, making the end perpendicular to the face of the flat (rather than angled along the edge of the drawing surface). Delete guidelines that are no longer needed (Figure 8-21).

10. Draw a line between the downstage points of Flat A and D (the set line). This gives you a midpoint to grab onto when repositioning the group. It will be erased later. Make the flats' footprint a group. Drag a guideline out from the green axis to the midpoint of the original drawing surface to represent the center line of this temporary deck. Select the group, grab it by the blue frame's downstage midpoint. Align it with the downstage midpoint of the deck rectangle. Click to affix it in place.

11. Place the footprints of Flats E and F. Make each of them a group. Since the downstage flats are a group spanning the stage, their group will not be accidentally included in either of these new groups unless your selection of E and F is exceptionally expansive.

12. Right-click the downstage flat group and **Explode** the group (select the group, right-click the mouse, choose Explode; the group returns to its original separate elements). Each flat will now be Push/Pulled separately and made into its own group. Push/Pull Flat A's footprint up to + 10'–0". To make it a group without including adjacent lines, Orbit to a top view and Pan until the top of the flat is horizontal across the screen. As long as the full length of Flat B's lines are not within the selection area, they will not be included.

13. The line marking the joint between Flats A and B became part of A's group. Redraw this line, and then Push/Pull Flat B to match A's height. Make it a group. Repeat for Walls C and D.

14. To add the archway to Flat A, select Flat A and right-click to choose Edit Group. Place guidelines along the flat's face to mark the sides of the archway. Drag a guideline up from the bottom + 7'–6" to mark its top. Draw lines to turn the interior of the archway into an independent surface. Push/Pull the interior of the archway upstage 3" (or type 3" and hit Enter after starting the extrusion). When it reaches the rear surface of the flat, the interior's surface should disappear, leaving an opening.

    Orbit to the rear of the flat and lay out guidelines for the reveal, ¾" outside of the archway opening. If you use the Offset tool for this, a line running around the entire perimeter of the flat will appear. Go over the guidelines with lines, and then Push/Pull the reveal upstage 3" (total 6" depth). Close Edit Group (Figure 8-22).

15. Use this same process to create the window openings in Flat D. The bottom of the window openings are + 3'–0" from the deck; the window openings are 5'–0" high. Remember to add their reveals on the upstage side.

16. Open Flats E and F with Edit Group and Push/Pull them to the same height as the other flats. Construction of

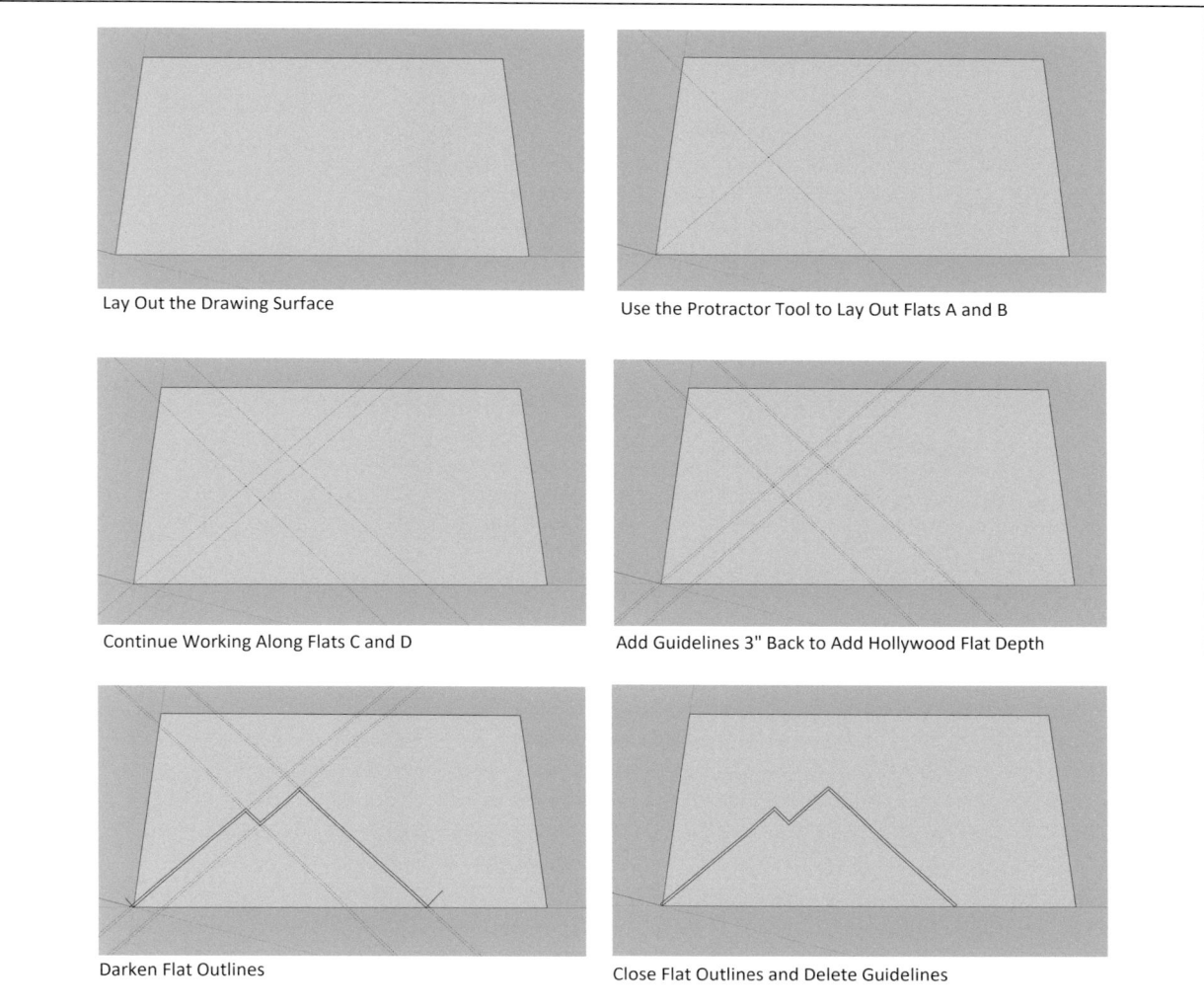

Lay Out the Drawing Surface

Use the Protractor Tool to Lay Out Flats A and B

Continue Working Along Flats C and D

Add Guidelines 3" Back to Add Hollywood Flat Depth

Darken Flat Outlines

Close Flat Outlines and Delete Guidelines

FIGURE 8-21 *Model construction steps: laying out a drawing surface; using the Protractor tool to lay out guidelines; darkening flat outlines*

the walls is now complete. If you wish to add color or texture, for each flat open Edit Group and use the Paint Bucket tool.

17. For this exercise, the casing, baseboard, and crown molding will be kept simple: different widths of 1× stock. The baseboard is 8″, the casing is 3″, and the crown molding is 5″.

18. The crown molding will be run along selected lines using the Follow Me tool. Since the flats are now groups, draw new lines along the top downstage edges of flats A, B, C, and D to give the tool something to follow. The Follow Me tool won't recognize lines within a group unless you're using the tool within Edit Group. Lay out the profile of the crown molding perpendicular to the end of Flat A or D. Activate the lines to follow by holding down the Shift key as you select each line. Click on the Follow Me tool, then on the surface of the

molding profile. The molding should snap into place along the tops of all four flats. Make it a group. Since the flats are their own groups, if you Orbit to a front view and keep your selection box narrow, nothing else should be accidentally included (Figure 8-23).

Creating and grouping the molding as a continuous piece has advantages and disadvantages. This molding group is not associated with individual flats, so when a single flat is extracted for use in an elevation detail, its molding will need to be reconstructed. When using more complex molding, this approach allows SketchUp to take care of the corners and intersections for you.

19. The baseboard will be constructed with a different approach, so that each section is part of each flat's group. Select Flat D and open Edit Group. Select the flat to make it a group within the group; this will designate

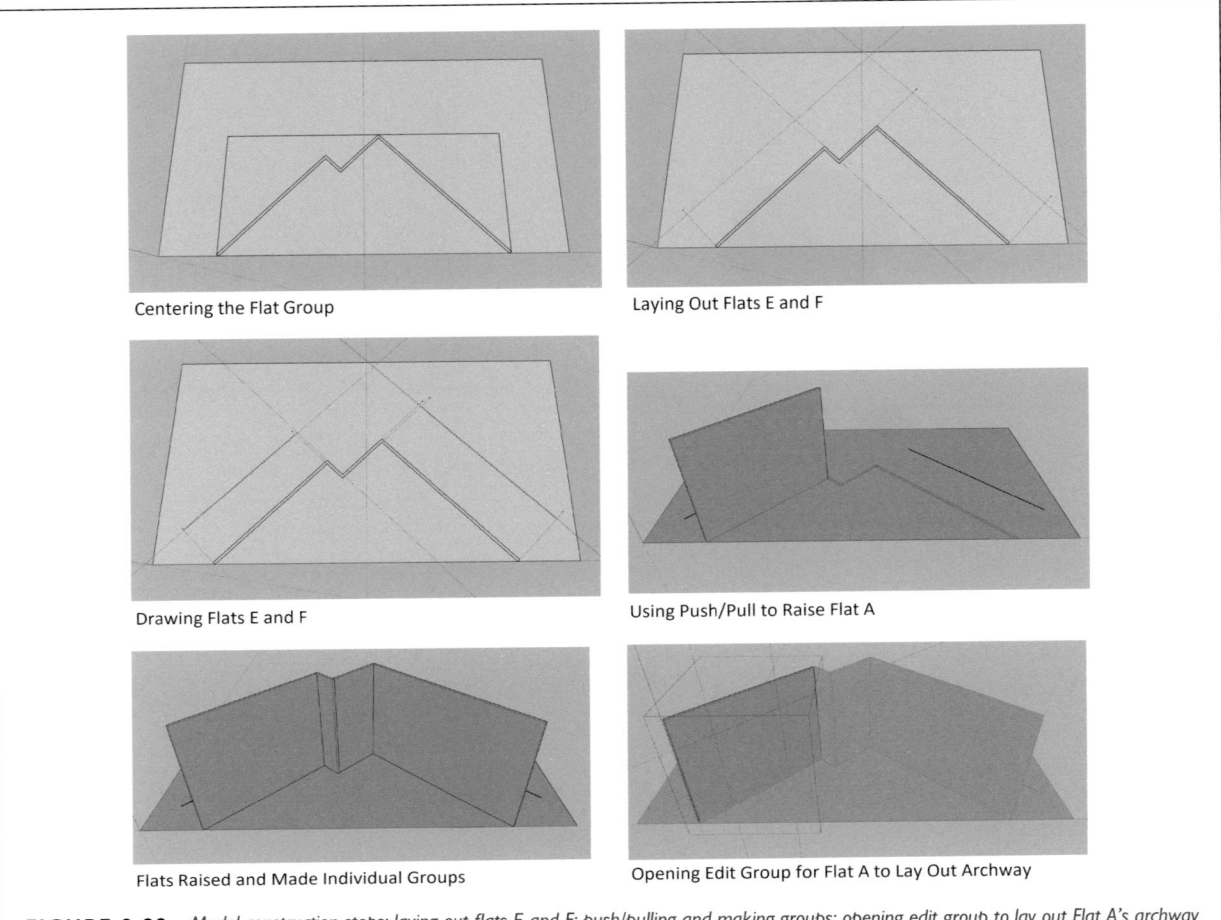

Centering the Flat Group

Laying Out Flats E and F

Drawing Flats E and F

Using Push/Pull to Raise Flat A

Flats Raised and Made Individual Groups

Opening Edit Group for Flat A to Lay Out Archway

FIGURE 8-22 *Model construction steps: laying out flats E and F; push/pulling and making groups; opening edit group to lay out Flat A's archway*

the flat and the baseboard as separate entities contained within the same group.

At the bottom of the flat, lay out and draw a rectangle 8″ high, spanning the width of the flat. Push/Pull the surface downstage ¾″. Make the baseboard a group. Close Edit Group.

Note that the baseboard's onstage end is flush with the surface of Flat C. When adding baseboard to Flat C, start ¾″ in from the end to accommodate Flat D's baseboard.

20. Use this same process to add baseboard to Flats A, B, C, and E. Remember to accommodate the ends of molding on adjacent flats. Leave 3″ on either side of the archway so that the casing can meet the deck. Add crown molding to the top of Flat E. Flat F is a painted representation of a garden and does not feature any molding.

21. To add casing around Flat D's windows, select the flat and open Edit Group. Since the flat is a group within this group, reestablish the windows' outlines with either

the Line tool or the Rectangle tool. Once the rectangle is complete, its surface will be automatically filled in; erase the surface. Use the Offset tool to drag an outline 3″ out from the window outline. Push/Pull this surface downstage ¾″. Drag a box around the pair of windows and make the pair of casings a group. Close Edit Group (Figure 8-24).

22. Use the same process to add casing around the archway in Flat A. The Offset tool will create a line running around the full perimeter of the flat, so use guidelines to lay out the 3″ offset instead.

23. The window frames will now be constructed and inserted into the openings. See Figure 8-25 for their dimensions. The frames are built of 1″ × 1″ strips. Elsewhere in the model space, drag out a rectangle 4′–0″ wide by 5′–0″ high. This can be done either horizontally or vertically, depending on your preference. Regardless of the orientation in which the frames are constructed, they will need to be Moved and Rotated into place. You only need to build one; the second will

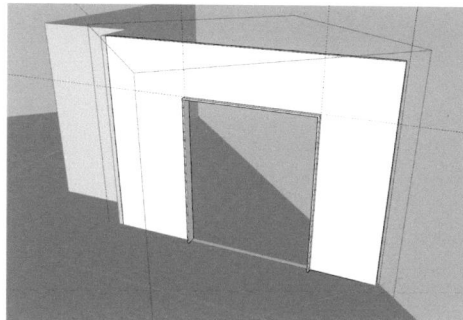

Push/Pulling the Archway Reveal

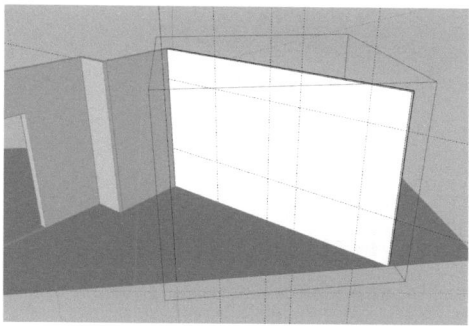

Laying Out the Window Openings

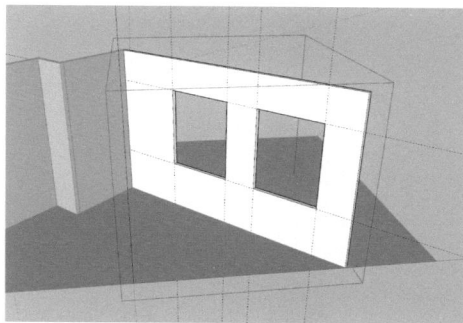

Push/Pulling the Window Openings

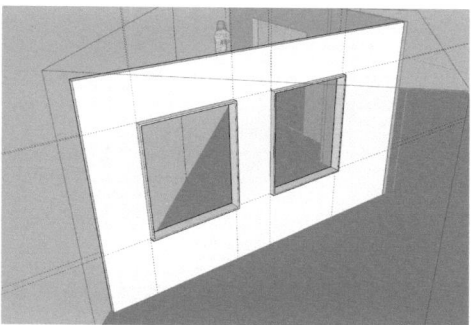

Laying Out and Push/Pulling the Window Reveals

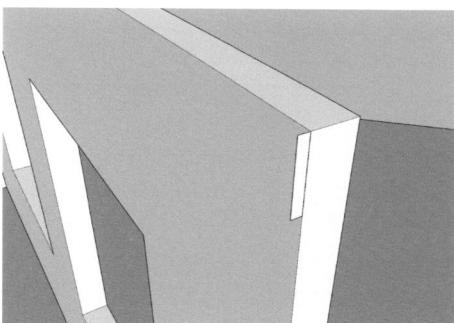

Drawing the Crown Molding Profile

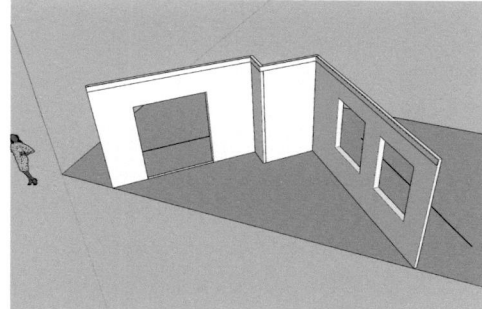

Using the Follow Me Tool to Extrude the Crown Molding

FIGURE 8-23 *Model construction steps: adding the archway reveal; laying out and creating the window openings; adding window reveals; using Follow Me to add crown molding*

be Copied and Moved once the first frame is in place (Figure 8-25).

24. Use the Offset tool to set the inner edge of the frame, 1″ within the perimeter. Use guidelines to lay out the muntins' center lines, both horizontal and vertical. Choose one of the vertical muntin center lines. Drag a guideline out ½″ to its right, and another ½″ to its left. Go over the two lines frame-edge to frame-edge with the Line tool. Repeat for the remaining two vertical muntins.

25. Choose one of the horizontal muntins. Again, drag a guideline ½″ above and another ½″ below it. Draw horizontal lines between the vertical muntins. Repeat for remaining two horizontal muntins. An alternate approach is to lay out all of the muntin guidelines and then use the Rectangle tool to outline each opening.

26. Erase the lines within the intersections of the muntins so that the frame is one surface and the muntins are another. Each rectangular opening will be an independent surface.

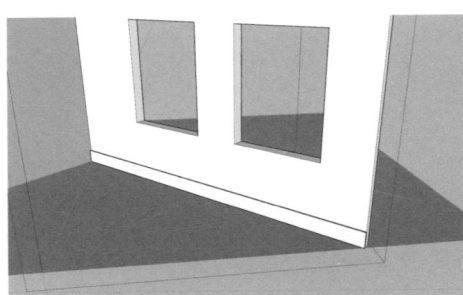

Adding Baseboard Within Flat D's Group

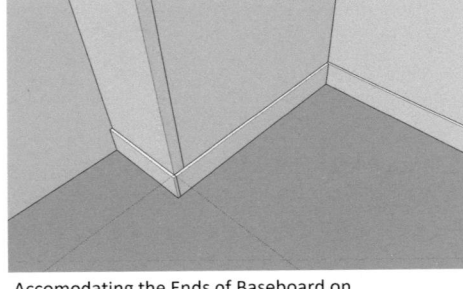

Accomodating the Ends of Baseboard on Adjacent Flats

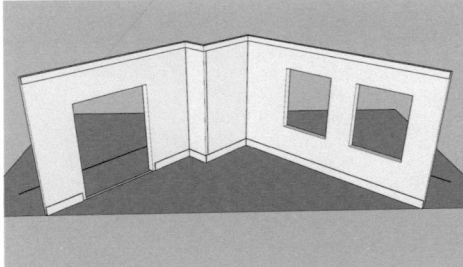

Baseboard Added to Flats A, B, C, and D

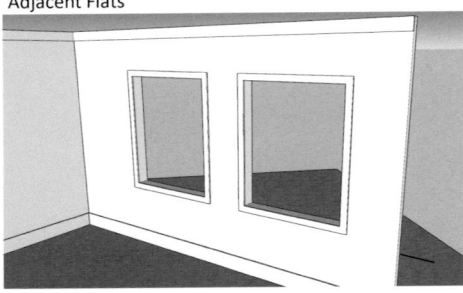

Casing Added to Windows

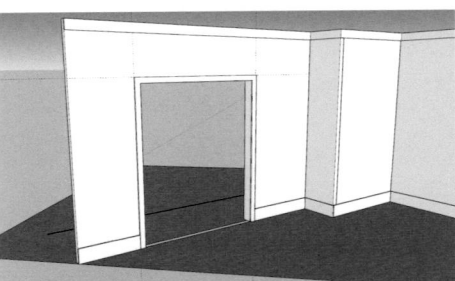

Casing Added to Archway

Flats E and F Push/Pulled to Height; Molding Added to Flat E

FIGURE 8-24 *Model construction steps: adding baseboard and casing; push/pulling Flats E and F to final heights; adding molding to Flat E; human figure returned to model*

27. A real window would have the glass situated in the middle of the muntin's depth; for this exercise, we're going to treat the window as though a sheet of plastic was attached to the upstage side of the frame. Select the Paint Bucket tool. In the materials library, choose the folder for Glass and Mirrors and then the selection for Translucent Glass Sky Reflection. Use this texture to turn each pane into a piece of transparent glass. SketchUp considers all planes to have a front and rear surface so apply this texture to the front of the window, then Orbit and apply it to the rear of the window (Figure 8-25).

28. If you wish, add a color or texture to the muntins and frame. When these shapes are extruded, the color will also appear on the sides. Use Push/Pull to extrude the frame and muntins 1". Make the window a group.

29. If the window is not already vertical, Rotate it. Move it to the upstage side of the wall. Orbit to work from the upstage side of the wall. Select Flat F and Hide it if it is in your way. The Hide command is found in the Edit drop-down menu.

   Rotate the window frame 45° so that it aligns with the wall. Grab the window from a rear upper corner (the glass side) to position it within the opening. Make sure the rear is flush with the rear surface of the reveal. Copy and Move to place a second frame in the other window opening. Unhide Flat F when done.

   If you wish to add the windows to the Flat D group, copy the frames and then erase them. Open Edit Group for Flat D and Paste in Place; the frames should drop back into their places. If they don't, adjust as needed. Close Edit Group.

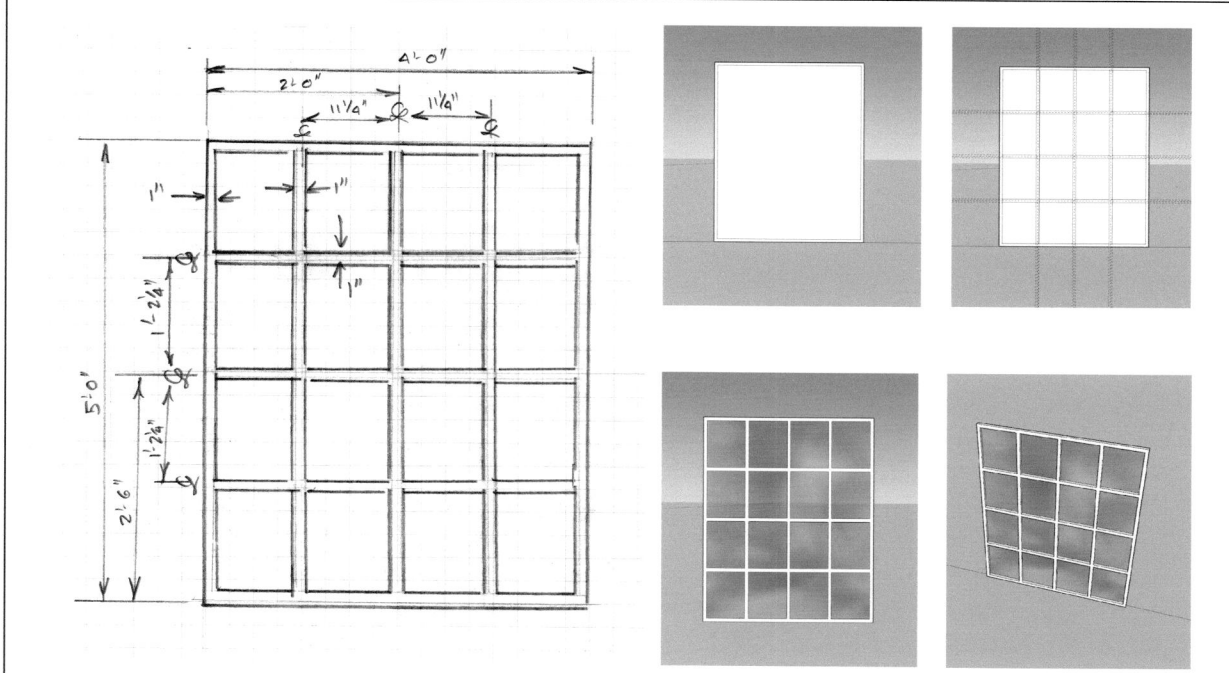

FIGURE 8-25 *Window measurements and steps: lay out frame and muntins; draw lines to turn muntins and frame into a unified surface; add texture to panes; push/pull frame and muntins*

30. Hide the deck group. Select all of the flats and molding. Make them a single group. This will make the walls easier to Hide while editing the deck. Unhide the deck.

31. The deck will now be trimmed to represent the limits of the floor's paint treatment.

    Select the deck. Open Edit Group. Draw a guideline connecting the downstage-most points of Flats A and D (the location of the set line). Drag out a parallel line 2'–6" downstage to mark the plaster line, and then another line 2'–0" further downstage to mark the lip of the apron. Draw lines from the corner of Flats A and B directly downstage to connect with the lip of the apron. Draw a line along the apron guideline between those two lines. Connect the apron surface with the deck surface by erasing any lines that divide them. If the deck surface vanishes, it may be because the depth lines from set to apron lip connect not to the flat, but to the corner of the baseboard, leaving a narrow gap between the apron and the deck. Zoom in to see if this is the case. If it is, Orbit to the underside of the model to see the connections better and redraw as necessary (Figure 8-26).

32. Draw a line connecting the downstage corner of Flat A with the upstage corner of Flat E. Draw a line along the upstage base edge of all of the flats, crossing openings and working your way around the set until you return to the set line at the stage left end of Flat D. Close Edit

Group. Hide the set; an unbroken outline matching the footprint of the scenery should remain on the deck. If there are breaks, return to Edit Group and connect as needed.

33. While in Edit Group, delete the surface and lines of the rectangle beyond the floor treatment outline. For this exercise, the floor treatment will be painted on a ¼" Masonite surface laid atop the venue deck. Orbit the deck and use Push/Pull to pull the deck down ¼". If you pull the deck up, ¼" of the set will be lost. Adding this thickness to the deck also means that when the model is pasted into the venue, the surface of the Masonite and the surface of the venue deck will not try to inhabit the same plane. Orbit back to a top view and double check that Push/Pull did not leave the upper area of the floor treatment un-surfaced. If that's the case, reestablish the floor treatment perimeter until the surface reappears.

34. If desired, go into the 3D Warehouse and search for a chandelier and a sconce. The chandelier will be placed once the set is moved into the venue, as batten placement still needs to be determined; for now, set it near the model.

    The sconce is placed on Flat E in alignment with the center of archway, its wall plate centered + 5'–0" off the deck. Use guidelines to mark its location on the wall and then Move the sconce into position.

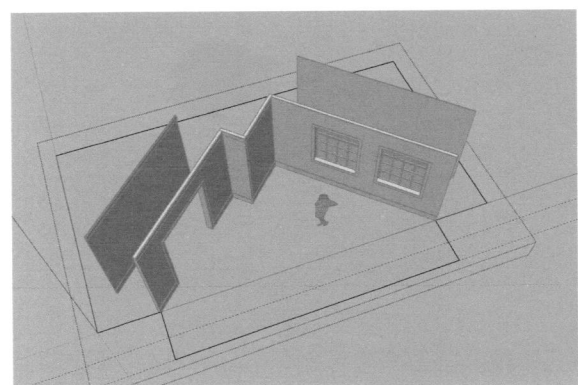

In Edit Group, Adding the Apron to the Deck

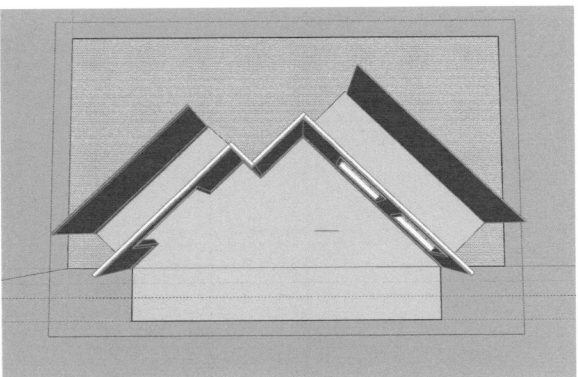

Outlining the Set to Separate the Deck into Two Surfaces

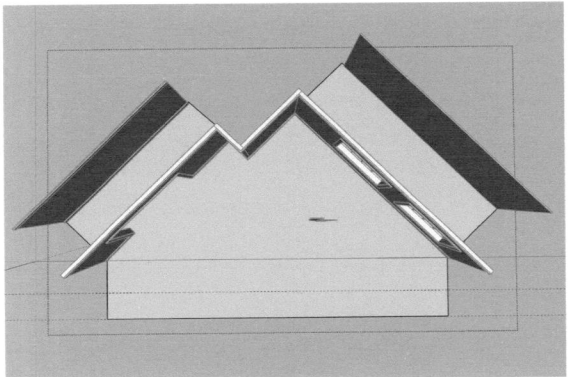

Deleting Extraneous Deck, Leaving the Floor Treatment Area

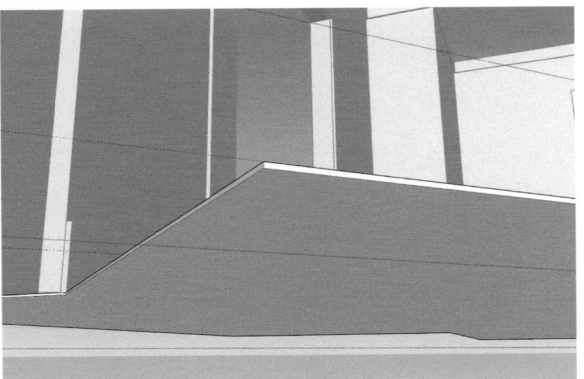

Push/Pulling the Deck 1/4" From Below

FIGURE 8-26 *Model construction steps: adding the apron; outlining the floor treatment area; deleting extraneous deck; push/pulling the deck down ¼"*

35. Save the file.

36. Select the entire model, make it a group, and Copy it via the toolbar Edit menu, rather than via the Ctrl/Move combination. Open your venue model file. Under the Edit menu, select Paste. The set model will appear, though perhaps not in an expected location. While it is still highlighted, zoom in to grab it by the lower midpoint of the Masonite deck's front edge. Align this midpoint with the midpoint of the venue's apron lip (Figure 8-27).

The model is not yet complete; masking still needs to be added, for one. This will be placed after the groundplan is drafted, then added to the model. For now, use Camera and parallel projection to look at the set from the top view. This view is not a true groundplan, as no cutting plane reveals details below the tops of walls.

Return to perspective projection and use the Cutting Plane tool to situate a horizontal cutting plane over the top of the model. Select the cutting plane and use Move to drag it downward until it is about 4'–0" above the deck. It will automatically enlarge as it surrounds a larger area of the model. Leaving the cutting plane in place, return to parallel projection and the top view. What can be seen, and what does it look like? How is this different from an unsectioned top view?

A horizontal section view of a virtual model can serve in a pinch when a plan needs to be shared, but because it does not employ graphic language it can confusing to a reader, especially if textures and colors are included.

Remember to save this file with a new name to indicate that that it contains both set and venue. Export 2D graphics of the model from various angles; save and print these images to use as reference tools in the following exercise.

MODULE 3: LAYING OUT A GROUNDPLAN

Now that the set can be visualized via the model, it's time to return to Vectorworks and construct a full groundplan. The layout process is similar to that used for the venue plan

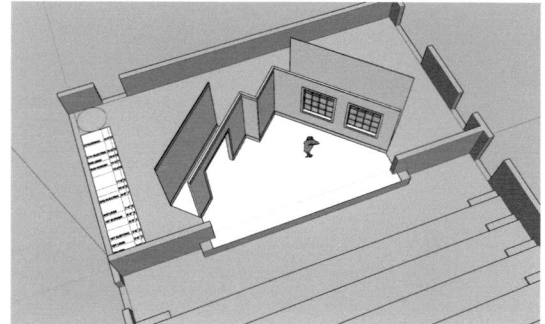

FIGURE 8-27 *The basic box set model, and the set placed in the venue model*

and model layout: begin with the reference lines, establish perimeters and major shapes, and then add detail. Once the set has been laid out, sightlines and masking will be considered and placed.

In this exercise, the set will be drawn in new layers directly atop the venue drawings.

A set can be drawn as an independent drawing and then imported, pasted, or traced into a base drawing; however, this means that the architecture's impact on scenic choices may not be known until the set drawing is incorporated. If there have been miscalculations, they won't be discovered until much work has already been done. Constructing an experimental model and placing it into the venue model is one way to check for error before drafting, whether working in SketchUp or with physical models.

Students

Open the venue plan file. Save As with a new name to indicate that this is a scenic groundplan file and no longer the base drawing of the venue. It will be useful to have copies of the set's rough groundplan at hand. Printing out the front elevation and horizontal section of the model is also useful. If you can, keep the SketchUp model open to obtain any necessary additional measurements. As the venue drawing scale is ½″ = 1′–0″, all layers for this remain in that scale.

Borders and title block will be added after the viewport/sheet layer phase. Turn off the visibility of the Architectural

Dimensions and Line Schedule layers until their use is required. Turn on the visibility of the Guidelines layer.

Suggested new design layers:

- Flats
- Soft goods and masking
- Practicals
- Floor treatment
- Scenic dimensions and notes

See Figure 8-20 for the box set's rough sketch. Have a copy accessible as you work through the following steps.

1. Clean up the Guideline layer. For a box set, placement of the set line is a usual first step. Measure along the venue's center line to place the set line the correct distance from the plaster line. Draw a generous horizontal line across the stage.

2. The rough sketch does not contain a measurement from the center line to the corners of Flats A and D. However, these measurements have been generated in the model. Go into the model and measure the distance from the center line's intersection with the set line to the downstage corner of Flat A (the flat itself, not the front of the baseboard). This should be the same distance from the center line to Flat D.

3. Starting with the Flat A endpoint, draw guidelines of the correct lengths for the downstage edges of Flats A, B, C, and D. The Line tool in its constrained mode will permit 45° lines. If the set line measurements are correct, the guideline for Flat D should terminate at the endpoint already marked on the set line.

    If Flat D does not terminate in the expected place, double check the wall measurements and corner angles. Double check the measurements lifted from the model. If they are all correct, the error is likely the set line distances. Measure the distance between the Flat A and Flat D endpoints; divide it in half and check those distances from the center line. If everything checks out and the walls are still off for some reason, it may be time to make an executive decision and change the length of a wall or two.

4. Use the Double Line tool to add thickness to the flats (3″ separation). Extend lines as needed to close corners. Add perpendicular lines to close the downstage ends of Flats A and D (Figure 8-28).

5. Place perpendicular guidelines within each wall to locate the archway and windows (Figure 8-28). Add guidelines for the reveals in Flats A and D. Consult the graphics standards exercise for each opening's graphic

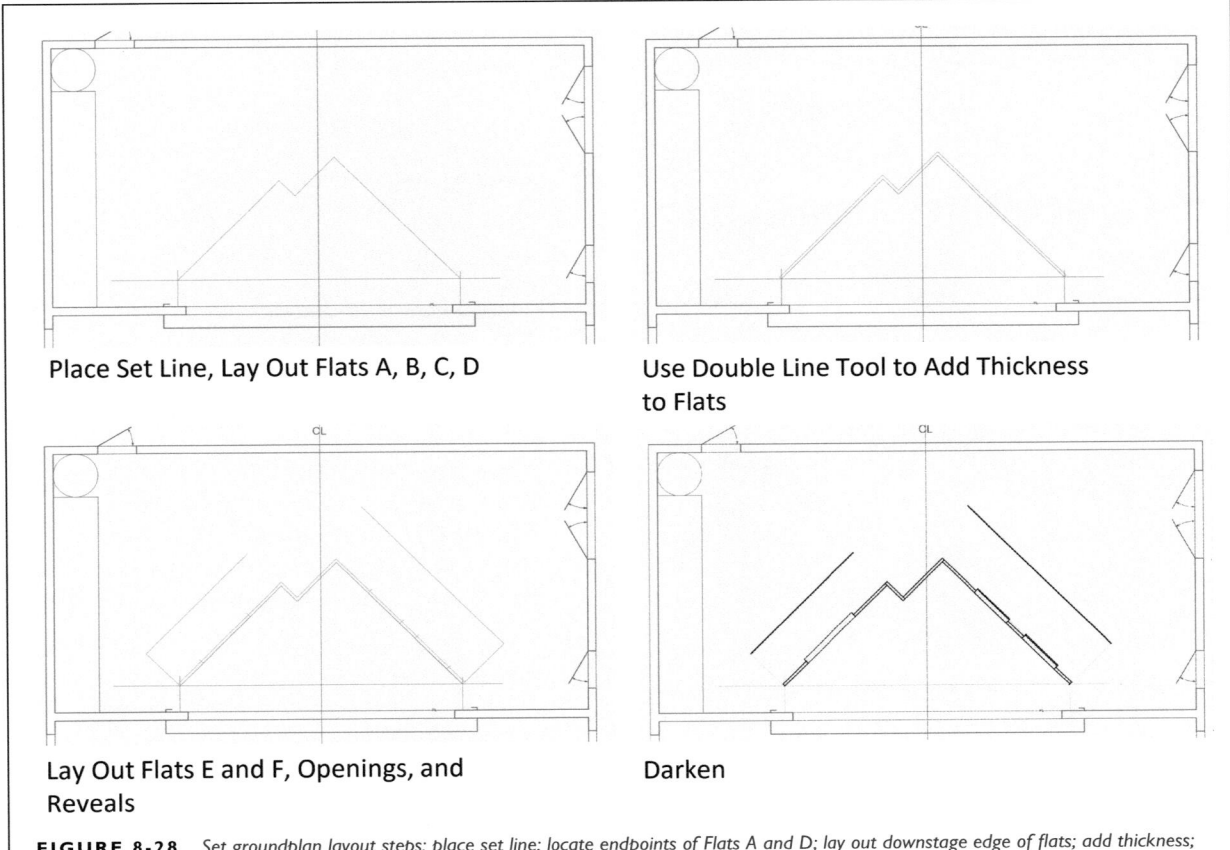

Place Set Line, Lay Out Flats A, B, C, D

Use Double Line Tool to Add Thickness to Flats

Lay Out Flats E and F, Openings, and Reveals

Darken

FIGURE 8-28 *Set groundplan layout steps: place set line; locate endpoints of Flats A and D; lay out downstage edge of flats; add thickness; locate openings; lay out Flats E and F; darken*

recommendation. Remember that the windows have three muntins, as well as the stiles of the frame.

6. To place guidelines for Flat E, draw a 4'–0" guideline perpendicular to Flat A from one of its offstage endpoints (the offstage side of the flat). From its endpoint, draw a line parallel to Flat A that is the length of Flat E. Place Flat F 5'–0" away from Flat D. Flats E and F are Standard flats and require only thick single lines when darkened.

   All of the flats are now laid out. Molding is not often included on groundplans, as the scale is typically too small to depict it well and the focus of the drawing is the footprint of scenic elements.

7. Switch to the Flats layer. Darken the flats and their openings with appropriate line styles.

8. Switch to the Guidelines layer to place the practicals. The chandelier is hung from Lineset 4, and centered on the center line. Turn on the visibility of the Line Schedule layer so you can see Lineset 4. If you imported a chandelier into the model, return to that file and measure the chandelier's diameter. If you did not import

one, the chandelier's diameter is 2'–6". Mark the center of the chandelier with guidelines, then draw the circle. Turn off the visibility of the Line Schedule layer.

9. The sconce is attached to Flat E and centered with the archway. Find the midpoint of the archway opening and project it back to Flat E. The sconce's base is to be secured to the flat 5'–0" above the deck, which puts it just above the cutting plane. However, items attached to the set, such as sconces, are typically *not* depicted with a dashed line.

   If a sconce was imported into the model, determine its overall width and depth measurements if you wish to draw a simplified top view. Otherwise, an equilateral triangle about 9" per side will suffice (an equilateral triangle is the standard symbol for a practical). Set the triangle along the wall, just downstage of it, not quite touching the flat. There is a Triangle tool that can be accessed via the tab on the 2D Polygon tool icon.

10. Switch to the Practicals layer to darken the sconce and chandelier. Since the chandelier is above the cutting plane, its outline is drawn with a long-dashed line. If you have a long-dashed line style set up as a class, use that.

Otherwise, draw the circle and while highlighted use the Attributes window to change the line's style.

11. Before laying out the masking, clean up the Guidelines layer. For this set, the flats themselves provide most of the masking that prevents the audience from seeing into backstage areas. However, exactly what can and cannot be seen through the archway and windows still needs to be determined.

Check the sightlines of the archway first. Draw guidelines from the center of the house right sightline symbol through and past the upstage and downstage edges of the archway.

The interior of this angle is what someone sitting in this seat will be able to see through the archway: basically, a portion of Flat E.

Draw guidelines from the house left sightline symbol to the archway. This will be a much steeper and narrower angle. A first question is whether the sightlines clear the edge of the proscenium. If they don't, that means a number of patrons are going to be unable to see much of the stage right portion of the playing space; the designer will have to think about this and come up with a solution. If the proscenium is imagined to not be an obstruction and the sightlines can project through the archway, do they meet Flat E or extend into the backstage beyond the end of Flat E?

Working with the sightlines reveals that the set is a little far upstage, and should be moved downstage to allow more patrons to see the maximum amount of the stage. For this exercise, change the set line's placement to be 1'–3" upstage of the plaster line. Draw a new set line at the new measurement, then select all of the flats and move them as a group downstage to the new set line. Recheck the sightlines to see if this helped (Figure 8-29).

The chandelier is on Lineset 4; does it need to move downstage as well? If so, are there any available linesets further downstage to which it could be transferred?

Remember to return to the SketchUp model to shift the set downstage.

12. Clean up guidelines. Turn on the visibility for the Lineset layer. Most venues and rental houses have an assortment of stock soft goods. For this project, the venue will have an adequate number (whatever that number might turn out to be) of 8'–0" wide by 24'–0" tall legs and 8'–0" tall by 38'–0" wide borders. The

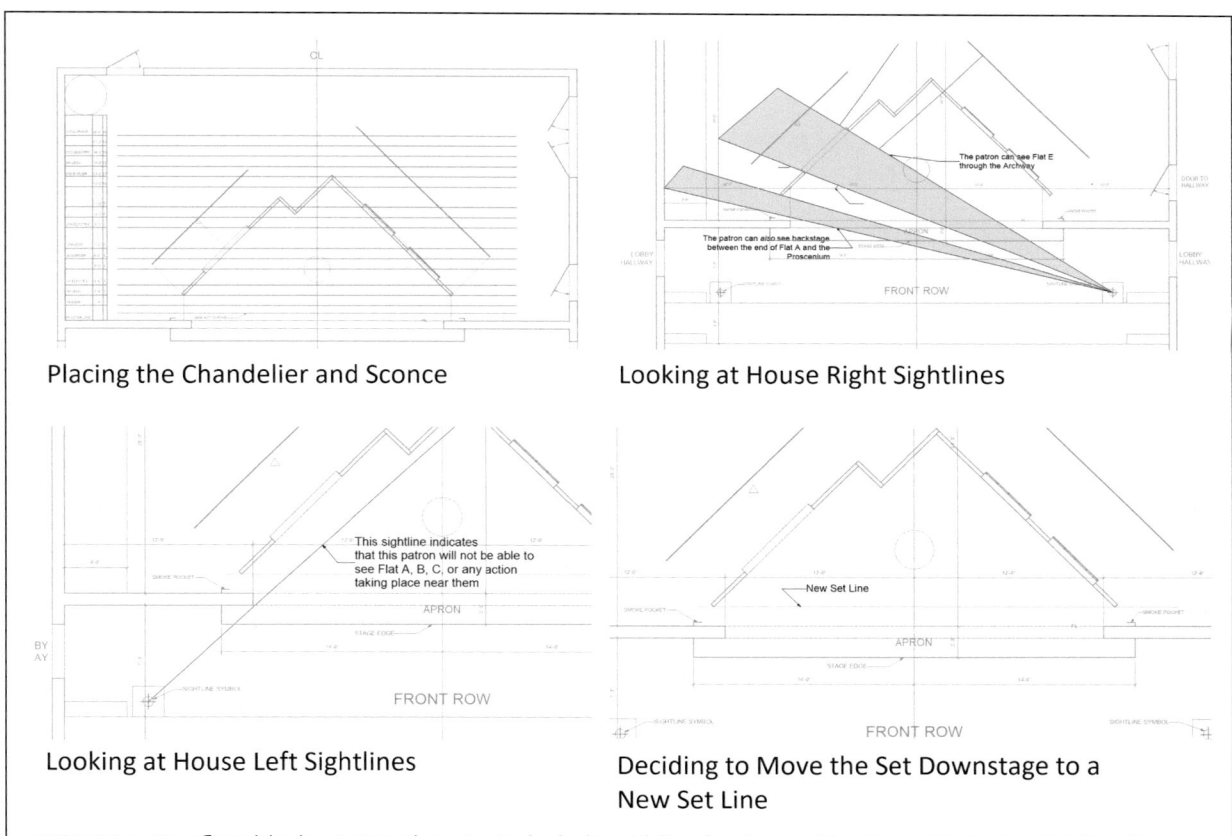

Placing the Chandelier and Sconce

Looking at House Right Sightlines

Looking at House Left Sightlines

Deciding to Move the Set Downstage to a New Set Line

FIGURE 8-29 *Groundplan layout steps: placing practicals; checking sightlines from house right and house left; deciding this information requires moving the whole set downstage*

venue prefers that the cyclorama not be moved from its pipe. Legs can be folded back on themselves to create narrower widths as needed, as well as overlapped to create legs wider than 8'–0". The legs have fullness; the borders are flat.

The first set of legs will be placed on Lineset 2, as suggested on the line schedule. The Main Act Curtain, when open, will also help mask any part of the backstage that might be visible through the gap between the proscenium and Flats A and D, but the act curtain should seldom be considered as a piece of masking. The set is symmetrical where Lineset 2 crosses it, so the same width of leg can be used stage right and stage left.

As a short cut, open the graphics exercise file from earlier in this chapter. Select the graphic for Drapery with Fullness and copy it. Return to the set groundplan file. Switch to the Soft Goods and Drapery layer. Paste. Drag the pasted leg into position so that the midpoint of its right-hand end aligns along Lineset 2 just offstage of Flat A. The graphic drawn in Module 1 is 10'–0" long—wider than the venue's stock legs. Ungroup the legs via Modify's drop-down menu. Trim extra length that extends beyond the offstage end of the batten. Remake the leg into a group (Figure 8-30).

Select the leg. Hold down Ctrl on the keyboard and drag the leg's copy to the stage left end of the batten. Align the onstage edge just beyond Flat D along Lineset 2.

13. For this set in this venue, the upstage legs' primary purpose is to prevent the audience from seeing the backstage through the gap between the top of the set and the borders. Legs are planned to hang on Linesets 6 and 12.

   Drag a copy of one Lineset 2 leg and align its onstage end on Lineset 6 just offstage of Flat E, leaving a small gap between the leg and the flat. Trim excess that extends beyond the end of the batten (or add more leg if necessary). If Lineset 6 extends stage right beyond Flat F, add a leg and Trim it to the end of the batten (Figure 8-33).

14. The onstage edges of the legs on Lineset 12 are in line with the proscenium opening. Extend proscenium opening guidelines to intersect Lineset 12. Copy and drag legs into place. Trim as needed. It is not typical to draw the batten between the legs hanging on it.

15. The teaser and borders run the full length of the battens. Since they are above the cutting plane, they are represented by long-dashed lines. Borders are hung on Linesets 5 and 11; the teaser on Lineset 1. They do not have fullness. The battens as drawn in the venue's Lineset layer are adequate guidelines so there's no need to draw new guidelines. Use a thin

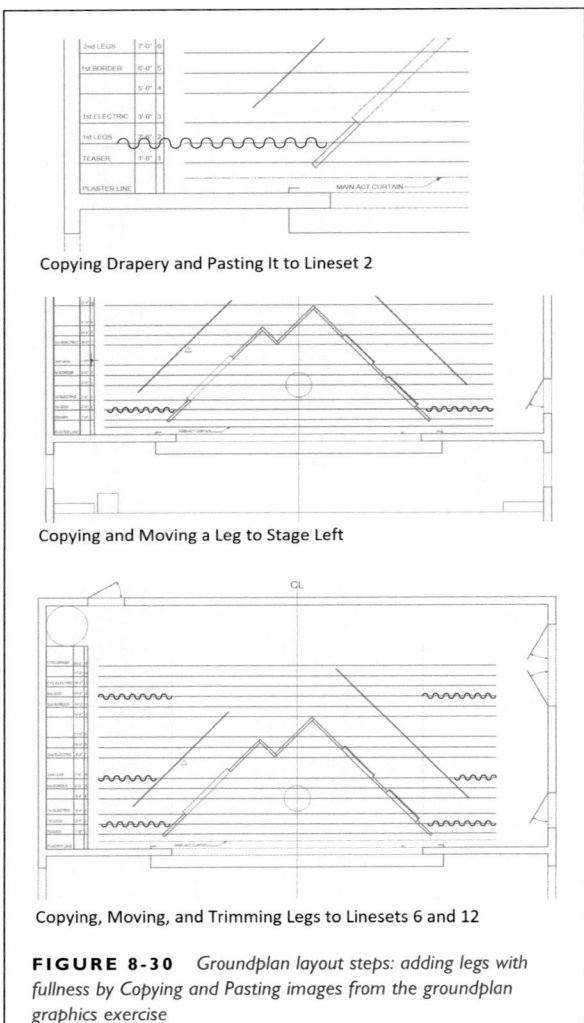

Copying Drapery and Pasting It to Lineset 2

Copying and Moving a Leg to Stage Left

Copying, Moving, and Trimming Legs to Linesets 6 and 12

FIGURE 8-30 *Groundplan layout steps: adding legs with fullness by Copying and Pasting images from the groundplan graphics exercise*

long-dashed line style to trace over the line of each batten (Figure 8-31).

16. In the venue's Lineset layer, the line of the Main Act Curtain was indicated with a phantom line. For our production, it will open until its onstage edges lines up with the edges of the proscenium opening. Since the legs on Lineset 2 are already the correct length (edge of proscenium to end of batten), copy the downstage legs and drag to align the copies atop the phantom line.

   Use the sightline symbols to see how well the legs mask the offstage areas. Should any of them be moved to other linesets? Do you need more sets of legs to adequately mask the space? If so, make these changes. This exercise will continue with the current placement of the legs.

17. Since the venue's Lineset layer will be turned off so that the battens don't run through the set, the line schedule

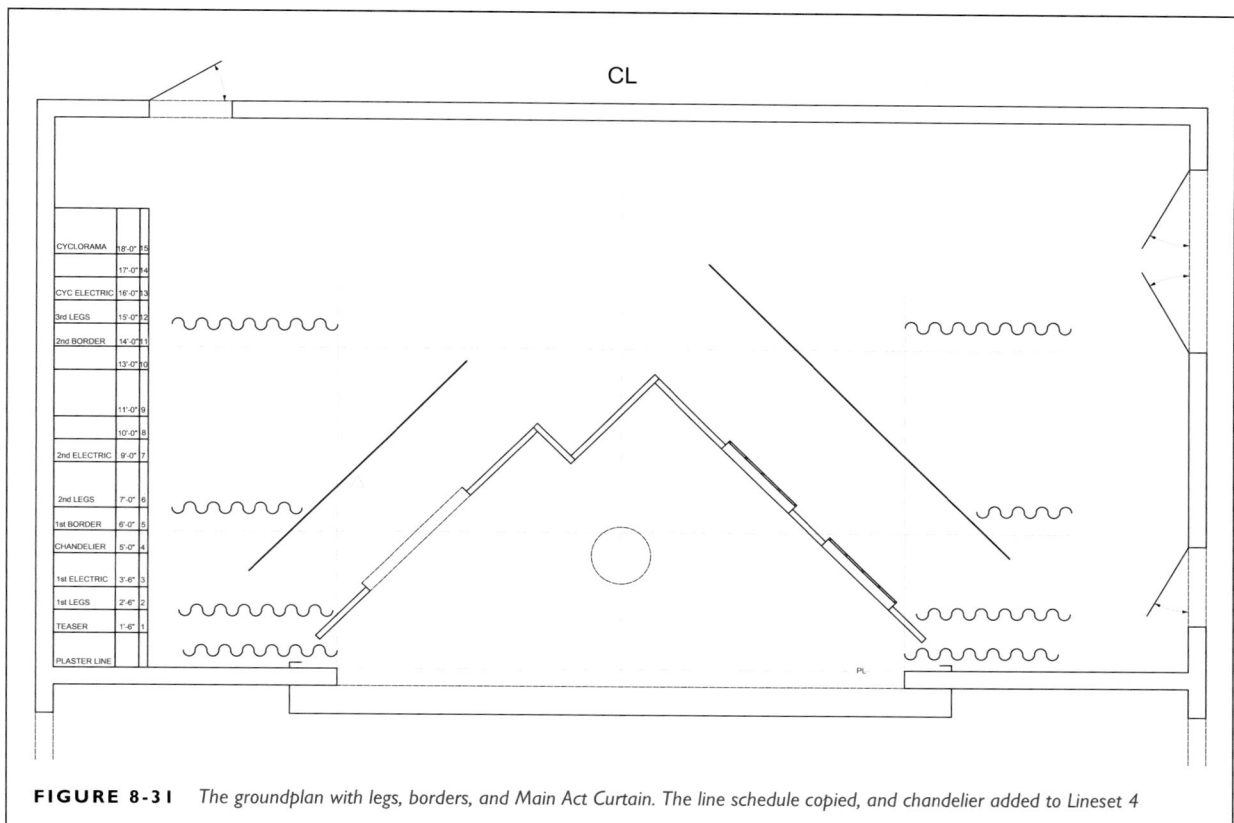

CYCLORAMA	18'-0"	15
	17'-0"	14
CYC ELECTRIC	16'-0"	13
3rd LEGS	15'-0"	12
2nd BORDER	14'-0"	11
	13'-0"	10
	11'-0"	9
	10'-0"	8
2nd ELECTRIC	9'-0"	7
2nd LEGS	7'-0"	6
1st BORDER	6'-0"	5
CHANDELIER	5'-0"	4
1st ELECTRIC	3'-6"	3
1st LEGS	2'-6"	2
TEASER	1'-6"	1
PLASTER LINE		

FIGURE 8-31   *The groundplan with legs, borders, and Main Act Curtain. The line schedule copied, and chandelier added to Lineset 4*

needs to be copied into the set's Soft Goods layer. Go into the Lineset layer to copy the line schedule. Return to the Soft Goods layer and Paste in Place. Drag the line schedule into place, if necessary. Turn off the visibility of the venue's Lineset layer.

Use the Text tool to add additional or revised information to the Line Schedule in the Soft Goods layer (if the schedule is a group, do it via Edit Group in the Modify menu). Trim heights of the soft goods and chandelier will be noted on the scenic section. If you included a trim column on the line schedule, the bottom edges of the borders currently play 2'–0" above the top of the set's walls (+ 12'–0" above the deck). Since the borders are 8'–0" tall, their battens trim at 20'–0". The legs are 24'–0" tall so their battens are set at that trim. Ideally, the borders should mask the battens upon which legs are hung. This will be double-checked in the scenic section. Consult the chandelier chosen for your model to estimate the trim of its batten.

18. Clean up guidelines. As in the model, the Masonite laid to create the floor treatment needs to be outlined. Turn off the visibility of the Soft Goods and Practicals layers.

Lay out the floor treatment perimeter with guidelines. Since the downstage corners of Flats A and D are offstage of the proscenium opening, the designer has decided that the Masonite will extend downstage from A and D's endpoints to the upstage edge of the smoke pockets, jog over to the proscenium opening, and then continue in a straight line downstage to the apron lip,

Switch to the Floor Treatment layer. Use a Thin Visible line to outline the treatment's perimeter. Later construction detail drawings will delineate individual sheets of Masonite and their arrangement.

19. If you placed furniture and set dressing into the model, determine the size and shape of their footprints and add them to the groundplan. Use a thin line; on future projects, they can be color coded with their own class and placed in their own layer. Lay out the footprints with guidelines in the guideline layer. Items that are above or on top of other items may interrupt the outline of the lower item; e.g., a chair placed across the edge of a rug may interrupts the outline of the rug. If an object's outline crosses the outline of the chandelier, the chandelier takes precedence. Label all objects (Figure 8-32). Switch to the Dimension and Notes layer.

20. Many of rules regarding dimensioning are gently broken when it comes to groundplans. Because of the number of objects and their sizes, it's often difficult to place extension and dimension lines without crossing objects

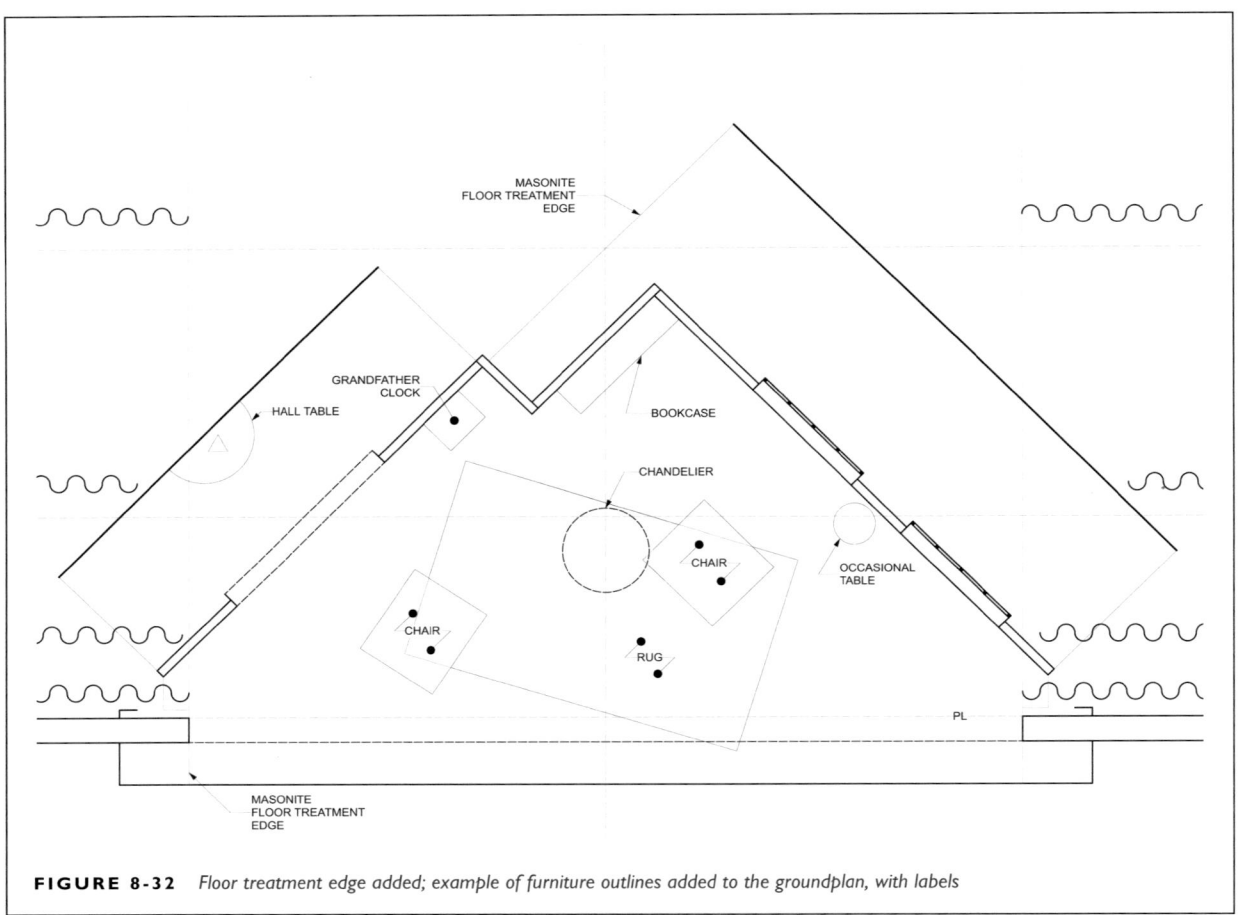

FIGURE 8-32 *Floor treatment edge added; example of furniture outlines added to the groundplan, with labels*

or other extension lines. Dimension lines may need to refer directly to objects without intermediary extension lines. Continue to avoid crossing dimension lines. The center and plaster lines are used as extension lines.

On the other hand, groundplans are generally only lightly dimensioned. They're concerned with the size and placement of major scenic objects upon the deck: lengths of walls and platforms, distances from center line, platform and stair heights, angles of corners, width of doors, etc. The line schedule provides information about flown elements so those items typically do not need to be dimensioned on the groundplan, though they are labeled.

Position labels and notes in open space remaining after dimensioning. Label flats and other scenic elements so they can be indexed to their detail drawings later in the package. Include the measurements a stage manager would require in order to lay out the groundplan in the rehearsal space.

Turn on the visibility of the other scenic layers. Turn off the visibility of guideline, architectural dimensions, and the venue's lineset layers.

Reestablish the new set line in the Dimension layer. Use a thin long-dashed line and label it. The Unconstrained

Dimensioning tool will be the best choice for placing many of this drawing's dimensions. Use the list below to guide dimension layout:

— Lengths of the flats, with the dimensioning placed downstage of them; lengths of openings in the flats, including the distance of one side of the opening from the end of the flat
— Distance of Flat E from Flat A and Flat F from Flat D
— Distance between the center line and the downstage termination of either Flat A or D; angle of the flats in relation to the center or plaster line
— Distance between the set line and the plaster line
— Distance between the upstage terminations of Flats E and F and the center line
— Distance of the sconce's center from one end of Flat E

Review the dimensions you just placed against the rough sketch. Is there anything missing? Anything that it would be useful to add? Do any dimensions need to be placed differently to make them more clear or to

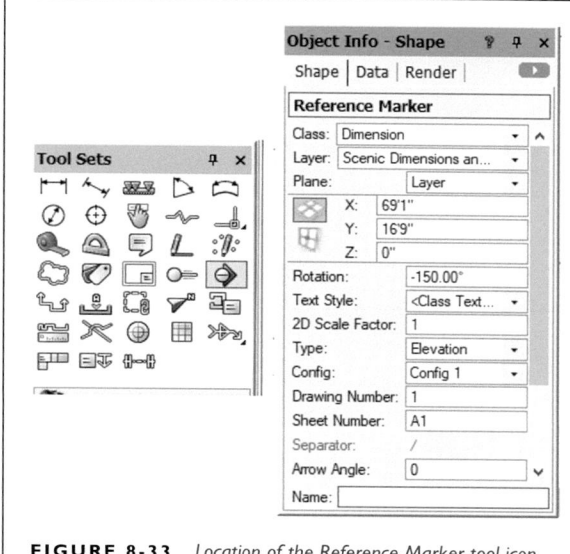

FIGURE 8-33 *Location of the Reference Marker tool icon and its object info window*

interfere less with the set's objects? Would adding a scale bar be useful?

21. Use Leader Lines and the Text tool to label most items on the drawing. Use a bubble to label the walls. The Reference Marker tool can be found in the Dims/ Notes tool set (Figure 8-33). If you have planned the package of drawings, the Reference Marker can contain not just a letter as a label, but direct the reader to the sheet and drawing number on which the detail drawing is to be found.

　　Label:

- — Walls
- — Legs and borders
- — Chandelier and sconce
- — Set line
- — Plaster and center lines if they do not already have their PL and CL labels.
- — Act curtain
- — Sightline symbols, if not already labeled in their layer
- — Edge of floor treatment (treatment may also be indicated: e.g., "Paint as Wooden Planks")
- — Furniture and set dressing

Since descriptions are useful to directors and stage managers, it is a good idea to label the areas of the stage as a performer would use them. The area beyond the archway is the Hallway. The area outside the windows is the Garden. Is there anything else on the drawing that a descriptive label would clarify?

22. Save your file. Slide the reference grid (the Move Page tool accessed via the tab on the Pan tool icon) so that the onstage portion of the venue is reasonably centered within it. Use the Rectangle tool to select the viewport crop, remembering to allow space for the borders and title block. In Create Viewport, checkmark all of the layers intended to be visible on the sheet so that all Fills appear.

　　When OK is clicked and the sheet layer appears, drag the crop area to its best position on the sheet. If it doesn't seem correct, delete the sheet layer and try a different sized viewport.

　　Add borders, title block, and title block information. The total number of sheets is not yet known, but groundplans are typically Sheet 1 of the series. If all other package drawings are created in this drawing space, the sheet layers will be automatically numbered as part of a series as they are created.

23. Save the file again. Export the drawing as a PDF. Print the PDF.

Homework: Extrapolating a Groundplan from an Image

Instructor

Provide each student with a still photo or screenshot of a television program. The image should be of a realistic interior (a box set) with people included to provide a sense of scale. The students will sketch a rough scale groundplan from the set's photo. The assignment can be expanded by asking the students to draft the groundplan.

　　A variation of this exercise is to use a realistic interior from a more complex, more cinematically filmed program. It will likely be necessary to watch a scene from the program, as well as refer to still photos to understand camera movement and how the set might have to change to permit various camera angles—do walls need to come and go? Are ceilings seen?

Students

Using the images provided by your instructor, construct a rough scale groundplan of the set depicted. Identify the major architectural features and do some research on typical measurements of items seen. What furniture is to be found, and where is it placed? Where are the windows, doors, and archways located? Is there a staircase, and where does it lead? Where might escape stairs be located? What is seen through doors, windows, and other openings? Does the human figure help indicate the size of objects?

If instructed to do so, draft a groundplan of the set. Be sure to credit the set's designer (production designer) in the title block information.

REVIEW OF GROUNDPLAN SYSTEM OF WORKING

- Sketches, research, and experimental model construction can establish the information to be drafted.

- Draft on top of the architectural base drawing. If hand drafting, trace the required architectural features onto a fresh sheet of paper and then remove the underdrawing.

- Work outward from the center and plaster lines. If there's no Plaster Line, select a permanent venue feature to provide reference, such as the apron lip or stage edge. If working with a box set, lay out the set line,

- The full playing space should be included; outlying bits of architecture may be cropped. Try to include at least the area of the first row of seating. If necessary, use a smaller scale to fit the entire stage onto the sheet.

- Lay out the perimeters of major elements: walls, platforms, ramps, and stairs.

- If there are tall platformed areas, be sure offstage escape stairs have been considered.

- Lay out features found along and within the major elements, such as doors, windows, bookcases. Staircases may require banisters and newel posts. Include the footprints of items such as columns and balcony railings.

Include all scenic structures and features with which a performer might interact.

- Use the major scenic elements as references to place secondary scenic elements, such as a flat seen through doorways and windows.

- Use the line schedule to place flown scenic elements such as cycloramas, drops, chandeliers, and borders.

- Use the sightline symbols to place masking.

- For venues with counterweight systems, include the line schedule.

- Furniture and other portable scenic pieces should be indicated with thin visible lines.

- Use phantom lines to indicate items with multiple positions.

- Lightly dimension; how much or how little information would a stage manager require to tape the plan out (especially if they are handy with a scale ruler?)

- Label everything.

NOTES

1. Dennis Dorn and Mark Shanda, *Drafting for the Theatre* (Carbondale, IL: Southern Illinois University Press, 1992) p. 165.
2. Robert C. McHugh, *Working Drawing Handbook: A Guide for Architects and Builders*. 2nd ed. (New York: Van Nostrand Reinhold Company, 1982), p. 23.
3. USITT Education Commission, *USITT Scenic Design and Technical Production Graphic Standards* (Syracuse, NY: United States Institution for Theatre Technology, 1992), pp. 9–10.

CHAPTER 9

THE CENTER LINE SECTION

This chapter explores the scenic **center line section**, typically the second plate/sheet included in a design package. The first module discusses what is expected on this drawing and how this information is used by members of a production team. The second module continues the construction of the venue and set models, adding height to the stage house and linesets on which to place hanging items such as legs and borders. A center line section will then be drafted in the third module.

TOPICS AND GOALS

• What is a center line section?

• Essential elements of a section

• Reading a section

• Drafting common section graphics

• Modeling the section view of a set

• Drafting a center line section

MODULE 1: WHAT IS A CENTER LINE SECTION?

A scenic center line section places a cutting plane along the central depth axis of the full set and/or venue to generate a section view that looks to either stage right or stage left. The choice of view depends on which direction provides the most and the clearest information not just about the set but about the set's relationship to the venue's architecture and mechanics (such as the counterweight system).

In theatrical projects, typically only one center line section is required unless the production has a particularly complex scene shift agenda. The cutting plane line is seldom indicated on the groundplan unless it takes a particularly unusual path around scenic elements.

Environments for film and television are often individual structures within a studio that must also accommodate camera movement. Multiple sections are drawn to show scenery along different axes. In this case, cutting plane lines are included on plan views. Labels index the section views found on the following sheets.

While section views of various scenic elements may be found elsewhere in the drawing package, the full-environment center line section is often referred to simply as "The Section." It is also called a "hanging section," as an important feature of this drawing is to describe the placement and trim heights of all flown items. It may also be called a "cross-section,"[1] or a "scenic sectional."[2] The section is not just a slice through the stage; while it is more pictorial than the groundplan, certain graphic conventions and schematic representations are still employed.

The draftsperson has discretion in describing features that lie across the center line. Strictly speaking, everything intersected by the cutting plane should be cut through, with its interior crosshatched. In many cases, though, the cutting plane may be offset slightly, or jog around something that would be made less clear if cut through. For example, if the lower steps of an angled staircase meet the cutting plane, will cutting through the stairs provide a useful representation of the staircase? (Figure 9-1).

In some cases, a side view of the structure placed within the venue may be sufficient. In a black box style venue, when scenery is oriented largely perpendicular to its

DOI: 10.4324/9781003154921-10

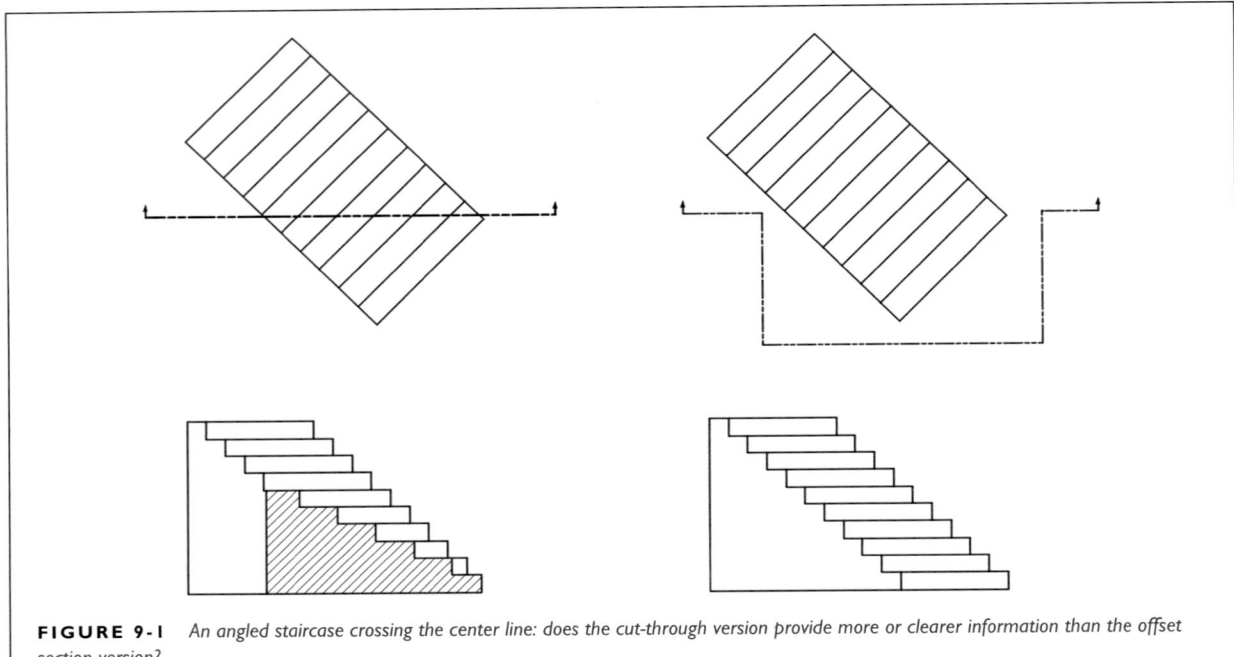

FIGURE 9-1 *An angled staircase crossing the center line: does the cut-through version provide more or clearer information than the offset section version?*

center line, a section is likely to be preferred. If the structure is mostly at an angle to the center line and parallel(ish) to the venue's walls or seating areas, elevations from those directions may be better choice. Showing the relationship of flown pieces to the set is one of the primary purposes of this drawing, so clear representation of these elements should drive this choice.

Important Elements of a Center Line Section

As with the groundplan, designs and venues vary, but there remain commonalities as to what features are found on a typical center line section.

- Architectural structures crossing the center line are shown cut through and crosshatched. This includes the back wall of the venue, the deck, proscenium, apron, and orchestra pit. It is usually not necessary to show the true thickness of the deck unless scenery is delivered up through the stage floor. Discretion is exercised regarding which architectural features are included, depending on sheet size and how features relate to the scenic environment. For example, in a black box theatre where the lighting pipe grid is several feet below the space's actual ceiling, it is usually more important to depict the pipe grid than the ceiling.

- If there is a proscenium, the edge of the proscenium seen from center is included. Include the smoke

pocket on the upstage side of the proscenium wall, if present.

- If there is a main act curtain, include its profile (or as much as can be seen around/above the set).

- If a counterweight system is present, the line schedule is placed horizontally at the top of the drawing, aligned with the placement of the linesets. The sheet may not be large enough to accommodate the theatre's grid at its true height above the zero deck; a break line is used to compress and indicate space not shown. Battens are represented end-on, as small circles. If nothing is suspended from a batten, that batten is often not drawn.

- For black box theatres and studio spaces, as well as environmental and nontraditional venues, the lighting grid and/or ceiling of the space is included, drawn at its true height above the deck if possible.

- The side views of items suspended from battens or the venue's grid. This includes legs, borders, drops, cycloramas, chandeliers, signage, etc. Items are typically drawn at their low trim: the position they play at during the event. Electrics are indicated by drawing a sample lighting symbol on the batten. If lighting ladders, booms, or tail downs are placed offstage, these too are included (at least, what can be seen above/around the scenery).

- A portion of the auditorium/seating area, with sightlines. Include sightlines in balconies and box seats, if present. These sightlines are used to determine the trim heights of flown items. The lighting designer will appreciate the inclusion of front of house lighting positions, if possible.

- Scenery on the deck, cut through at center. Sectioned masses, such as platforming and flats, are crosshatched. All scenery visible when looking in the selected direction from center line is drawn as seen, with detail. Items or portions of items hidden behind other items are typically not included, unless necessary. The unbroken full height of scenery is shown, as this is one of the major descriptive purposes of the drawing. If the rear of a scenic element is seen in the section, the designer has discretion how to describe upstage-facing construction features.

- Selected set pieces, such as furniture. The more that are included, the more cluttered the drawing may become. Which is more important: the flat with the archway, or the dining room set in front of it?

- A human figure. If there are a number of platform levels, include multiple figures.

- Some designers include scale bars for both the depth of the venue (best located below the deck) and the height of the scenic environment (best placed at the front lip of the playing area or behind the venue's back wall).

- Minimal dimensioning. Generally limited to the overall height of the set or major scenic elements. Use the deck as the base line.

- Labels and notes. Label all hanging items and other major elements. Trim heights may be included, as a note at the batten and/or in a column on the line schedule.

- Borders and title block.

Discussion: Looking at Center Line Sections

Instructor

Bring in a selection of center line sections from various productions. The sections seen in Figures 9-2–9-4 can get discussion started. A projected image or individual reduced handouts work well. The questions below can be used to guide the discussion.

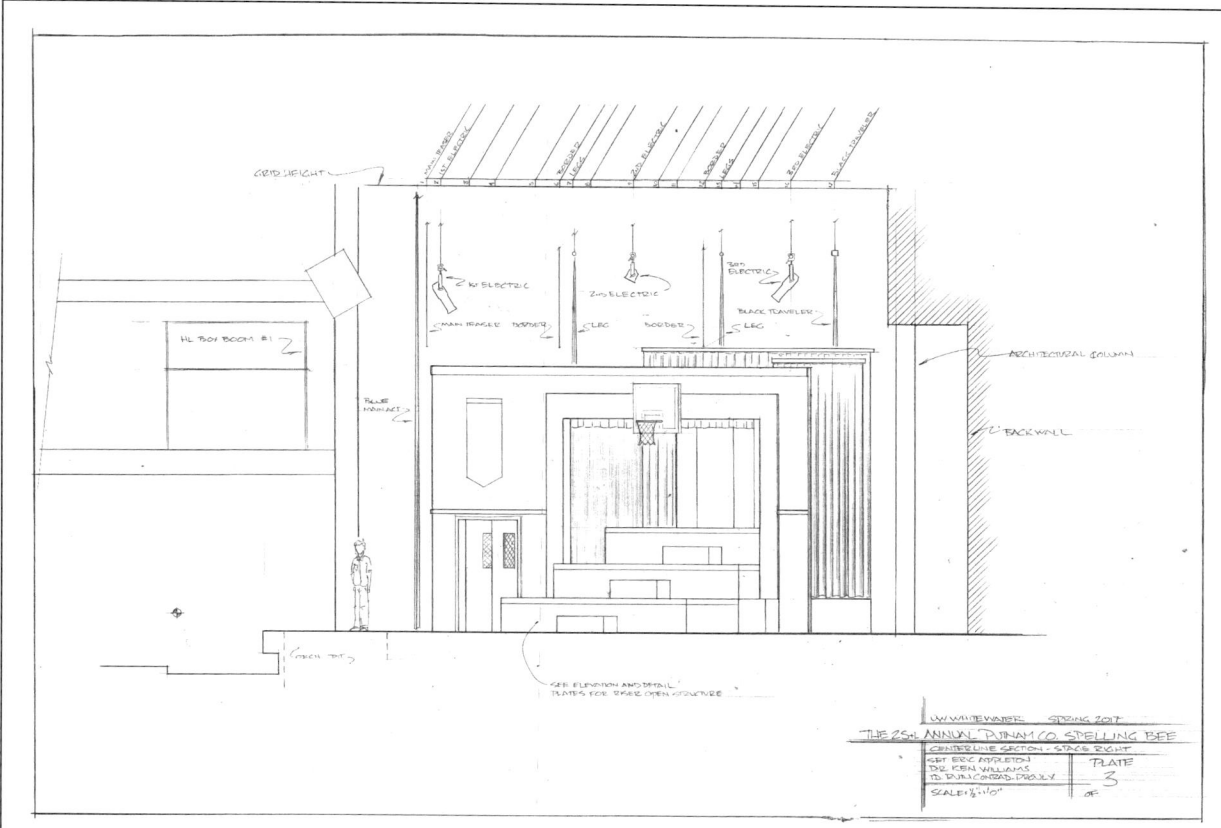

FIGURE 9-2   *Hand drafted center line section of UW-Whitewater's production of* The 25th Annual Putnam County Spelling Bee

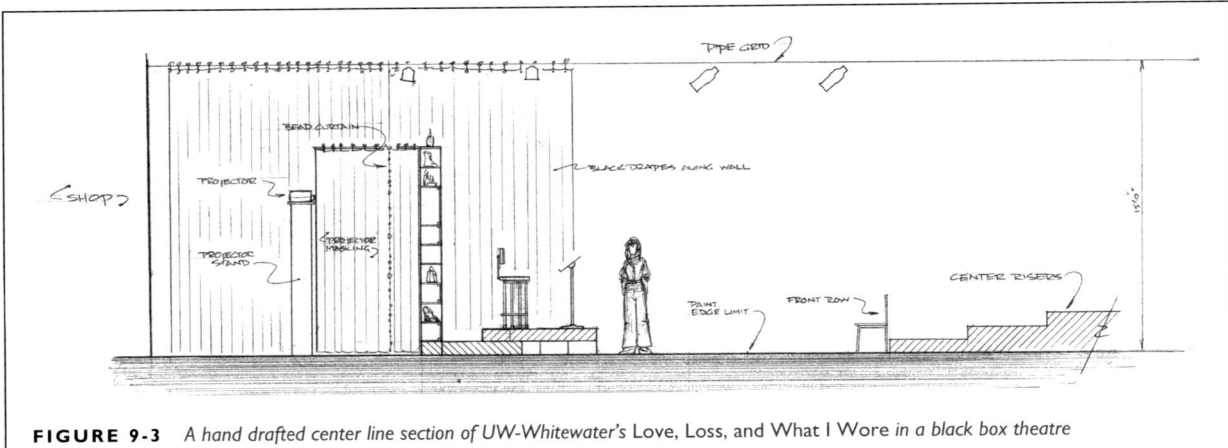

FIGURE 9-3 *A hand drafted center line section of UW-Whitewater's* Love, Loss, and What I Wore *in a black box theatre*

If a venue is accessible to the class, visit it with copies of a section drawn for a production in the space. Stand at center and locate the architectural features included on the drawing. If the set is still in the venue, compare it to the section as well. Does the drawing line up well with reality? Are there features on the drawing that only become evident when in the space, looking at them? How could they be better represented on the drawing?

Students

Examine the sections provided. What important feature(s) draws the eye first? Where should the reader look for information that will put features of the drawing into clearer context?

How is line weight used to describe objects? Because this drawing leans more to the pictorial,

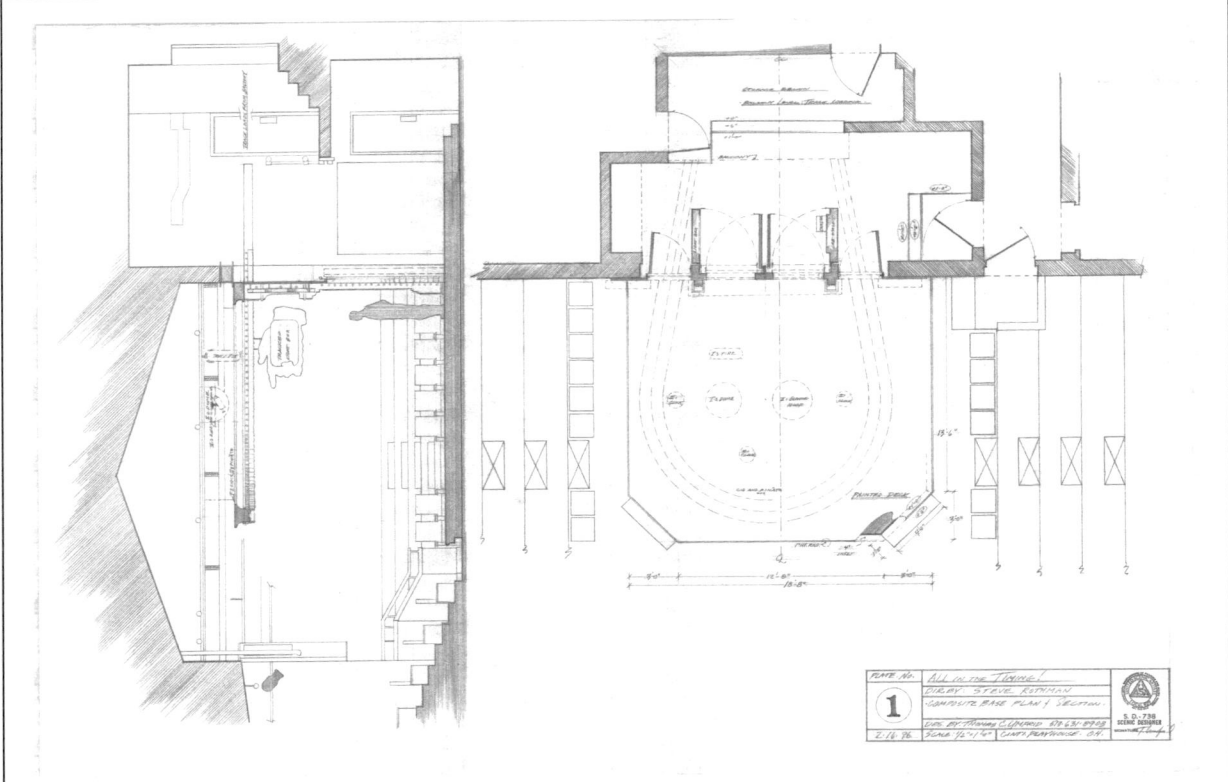

FIGURE 9-4 *A hand drafted plate for Cincinnati Playhouse in the Park's production of* All in the Timing, *thrust configuration, containing both groundplan and center line section; designed and drawn by Thomas C. Umfrid. Used by permission of Thomas C. Umfrid*

how did the draftsperson use line weight to describe scenic objects? Are there elements that are unclear? How might they be better represented? Are there enough labels and are they descriptive enough? Are sightlines indicated? Are there any new or unfamiliar items on the drawings?

Is there a line schedule? How easy is it to determine which flown piece hangs on which lineset? Are trim heights noted? Are trims noted at the batten, or the lower edge of the scenic piece? Are electrics noted? If the stage house is taller than the page can accommodate, what did the draftsperson do to indicate this? Is the drawing printed at a smaller scale to fit the whole space on the page? Are there break lines between the battens and the line schedule? Some other solution?

Is there a human figure? Where is it placed? Does its placement help give a sense of the drawing's scale and the scenery's proportions? Is there a place on the drawing where the figure would be more effective? Would multiple figures in various locations help?

Are there any front-of-house lighting positions included on the drawing? Is there any speaker or microphone placement indicated? Can you find the sightline symbols?

If there are sections drawn by different draftspersons/ designers, what stylistic commonalities do they share? What are their differences? Do you have a preference regarding which draftsperson's technique is clearer?

MODULE 2: ADDING HEIGHT TO THE SKETCHUP VENUE MODEL

The groundplan depicts width and depth. The section depicts depth and height. The venue model constructed in Chapter 7 and used in Chapter 8 currently features walls that rise only to the height of the groundplan's cutting plane. It's time to Push/Pull the venue's walls to their full heights and enclose doorways, as well as add hanging soft goods. Venue heights can be found in Figure 9-5.

1. Open the SketchUp file containing the combined set and venue models. Turn off the visibility of the set by Hiding the group.

    The proscenium opening is 18'–0" high. All standard doors and doorways are 6'–8" high. The shop door is 10'–0" high. The grid is 50'–0" above the deck (the ceiling of the stage house is even farther above the grid). Since the model's line schedule has been placed stage left, that venue wall will be left + 0" to provide a view into the space from the side. The clouds and catwalks, as well as other lighting positions, will be addressed in the next module. They will not be added to the model.

2. Right click on the venue model and select Edit Group. The walls were previously set to a height of + 4'–0". Move the cursor to the top surface of the upstage wall. Push/Pull it up 46'–0" (to a total height + 50'–0").

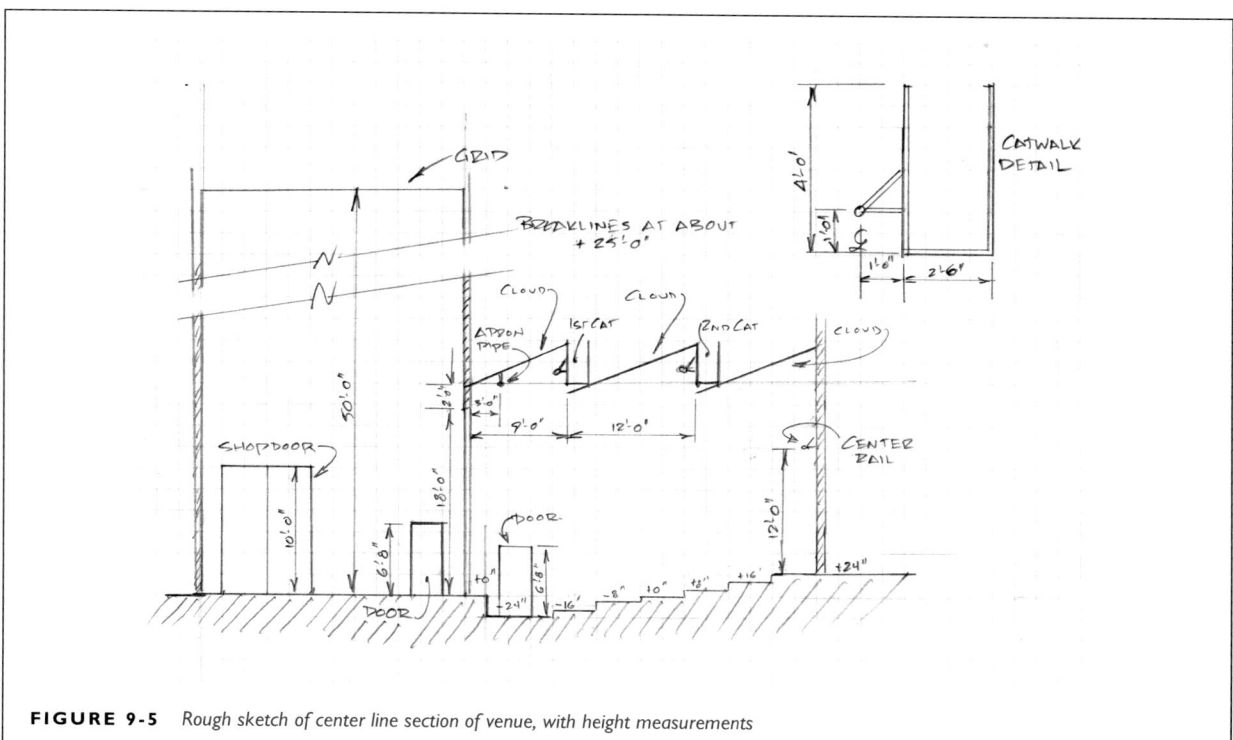

FIGURE 9-5 *Rough sketch of center line section of venue, with height measurements*

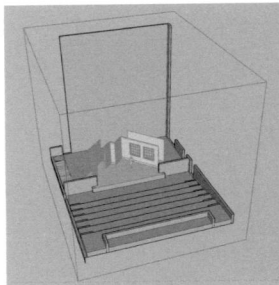

Raising the Upstage Wall

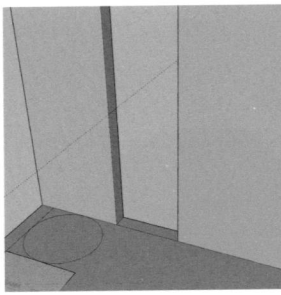

Raising the Wall SR of the Door; Marking Its Height

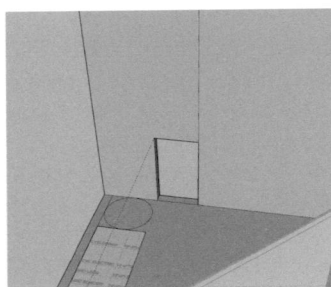

Closing the Space Above the Doorway

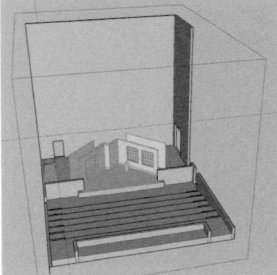

Raising the Stage Left Wall; Closing the Area Above the Shop Door

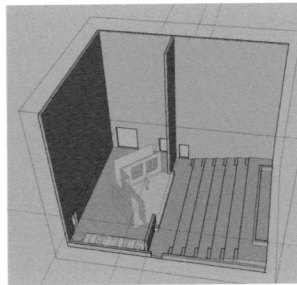

Raising the House Right Wall; Closing the Area Above the Auditorium Doorway

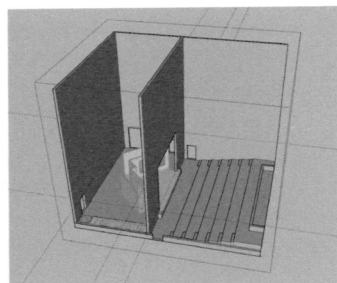

Raising the Proscenium Wall; Closing the Area Above the Proscenium Arch

FIGURE 9-6 *Steps in raising the model's walls: raise the upstage wall; close area above the upstage door; work around the space, leaving house left walls at zero deck level*

Push/Pull the short wall section right of the upstage door to + 50′–0″.

3. The doorways are 6′–8″ tall. Draw a line at this height on one of the jamb surfaces of the upstage door opening to break it into two surfaces: the door jamb and area above the door (Figure 9-6). Push/Pull the upper surface across the gap to meet the other wall. Erase the vertical lines marking the join so that the rear wall is a continuous surface.

4. Work your way around the stage left walls, raising them to + 50′–0″ and closing the tops of the shop doors and the stage left door. Add a line to separate the stage right wall from the proscenium wall so that the walls on the stage right (house left) side of the model remain at + 0′–0″. Close the area above the house right doorway. When complete, close Edit Group.

5. On the drafted groundplan, legs are hung on linesets 2, 6, and 12. Borders are hung on linesets 5 and 11, with the teaser on lineset 1. The stock borders are 38′–0″ wide by 8′–0″ tall. The stock legs are 24′–0″ tall; consult your groundplan for leg widths. All soft goods are black.

First, construct the legs and a single border off to the side of the model. Flat panels will be used for this exercise, rather than modeling fullness. If desired, search for a JPEG image of tall black curtain. Import the image and apply it to the surfaces of the legs. The image's height and width ratio can be adjusted in the Paint Bucket tool's edit window (look for the image of a linked chain). Adjust the image's height-to-width ratio until it fills the leg's surface without appearing tiled. Make each leg and the border their own groups. Alternatively, create the legs and borders and paint them solid black.

Image files can be imported directly into a SketchUp model via the Import… choice on the File drop-down menu. Images can also be imported via the Paint Bucket tool's Entity Info tray. In the upper section of the Materials window, there is an icon that looks like a small box of files with a plus sign. Clicking on this opens the Create Material… window. Check the box for Use Texture Image. This takes you to your folder so you can choose an image file to import. The image will become a tile within the Paint Bucket's menu options for colors and textures in model.

You can also make a single 8′–0″ leg, apply a drapery image, and then copy and paste this leg into position. The Scale tool can then be used to reduce the width of each panel. This will result in the drapery image also narrowing.

6. Place the borders first. To make this easier, turn off the visibility of the venue architecture by Hiding the group. Keep the set group hidden as well. The lineset and batten groups should remain visible (Unhide them if you hid them previously).

    Draw a 20′–0″ vertical guideline at the stage right endpoint of Lineset 1's batten. Move the border by grabbing its upper left corner and align this corner with the upper endpoint of the guideline. If the border is not already vertical and parallel to the batten, Rotate.

    Draw 20′–0″ vertical guidelines at the endpoints of battens 5 and 11. Copy and Move the new borders into place. Erase the three vertical guidelines. Select the three borders and make them a group. Be sure not to include the batten guides or line schedule when selecting the borders (Figure 9-7). Hide the borders before beginning work on the legs.

7. Move the legs by grabbing their lower offstage corners and align the corner with the offstage end of their

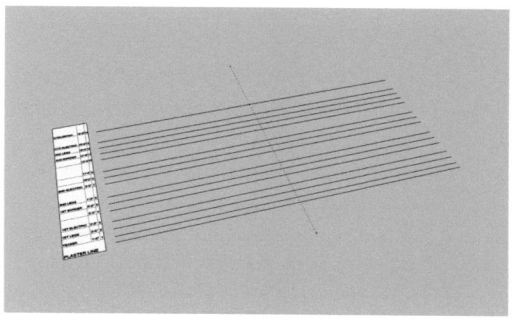

Reestablishing the Center Line Through the Battens

Consructing the Legs and One Border

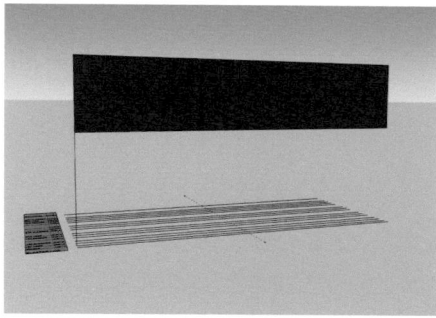

Using a Vertical Line to Position the Teaser

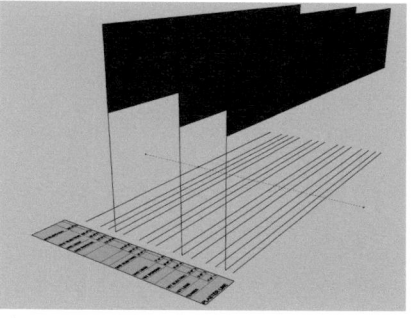

Placing the 1st and 2nd Borders

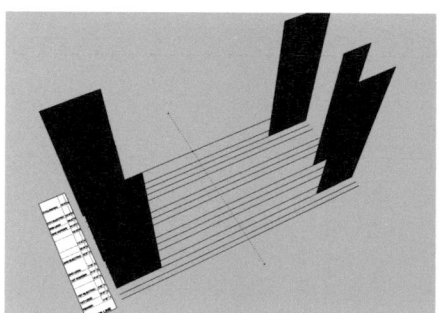

Placing the Legs on Their Linesets

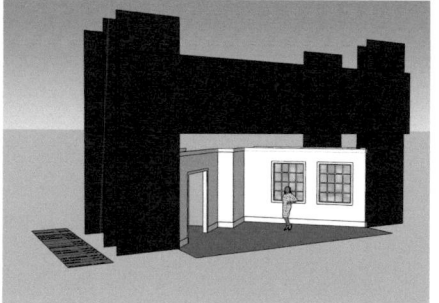

Legs and Borders Seen with the Set (Venue is Hidden)

FIGURE 9-7   *Steps in placing soft goods: constructing panels; positioning the first border/teaser; copying to place the others*

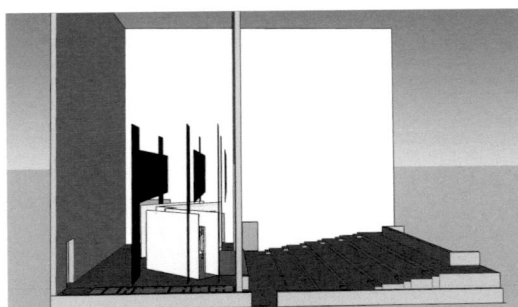

Left View of Set in Venue in Perspective
Projection

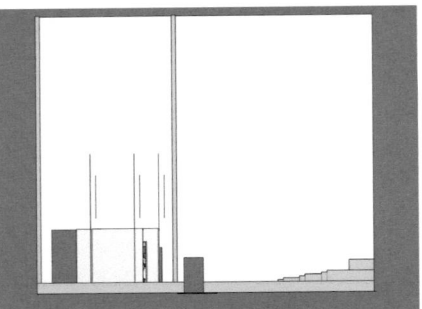

Left View of Set in Venue in Parallel
Projection

FIGURE 9-8   *The model with legs and borders viewed through house left wall; perspective and parallel projections*

assigned batten. If the legs were constructed using their groundplan widths, the onstage edges will automatically be set (Figure 9-7). If you constructed a single leg, draw guidelines to determine their onstage edges and then use the Scale tool to reduce or enlarge their widths.

8. Unhide everything so that the venue architecture, set, legs, and borders are visible. If something was hidden while in Edit Group, you will need to go into Edit Group to Unhide that object. Orbit the model to look at it through the unmodeled stage right wall.

    Since the camera is set for perspective, the legs and borders appear as surfaces, however slender. Most of the set is hidden behind Flats E and A. Switch the camera to parallel projection, left view. The legs and borders now appear as lines rather than surfaces. It's also impossible to see the legs on stage left, as they are directly in line with the stage right legs (Figure 9-8). Return to perspective projection to construct panels for the Main Act Curtain, using measurements from the groundplan. Apply a color or drapery texture and add them to the model.

9. Choose the Cutting Plane tool and click on a surface parallel to the left viewing plane. Drag the cutting plane to the center line. You may need to Orbit to better see the action; return to the standard left view when done (Figure 9-9).

    How much of the set is visible now? Are the legs and borders any more obvious as legs and borders? How much of the venue's stage left wall is visible beyond the set? Which projection system do you think provides more and clearer information? Why?

    Move the cutting plane forward and backward (stage right and stage left) to see what happens to the view of the set. Does moving it to either side of center provide a better view? Is it useful to see the upstage surfaces of certain flats?

    Export 2D graphic JPEGs of these cutting plane options, including one where the cutting plane is indeed positioned along the center line. These will be useful visualization tools while drafting the section. Erase the cutting plane to return to a view of the full model.

Center Line Section in Perspective Projection

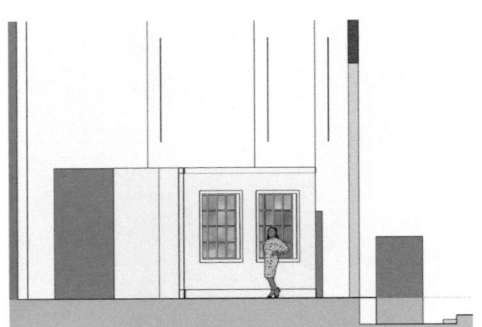

Center Line Section in Parallel Projection

FIGURE 9-9   *The cutting plane at center through the model; perspective and parallel projections*

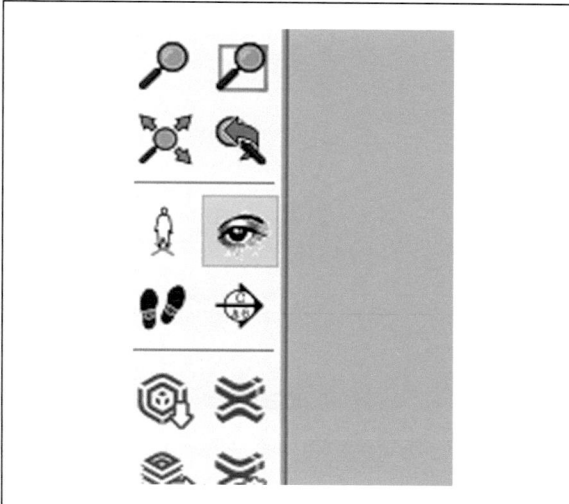

FIGURE 9-10 *Location of the Look Around tool icon; the Position Camera tool icon is next to it*

10. SketchUp features tools that allow you to "walk" around a model and view it as though you were within it. The Look Around tool's icon looks like an eye. It positions the camera view at a stationary point and allows you to swing the viewpoint around as though you were standing at that location. The Position Camera tool's icon looks like a person standing on an X. With this tool, you can drag out a line along which the camera will track (Figure 9-10).

Use the Look Around tool to position your viewpoint in one of the front row sightline seats. Look into the upper areas of the set. Are there gaps in the masking that allow you to see into the backstage areas? When you position the view from the center of the front row and look up toward the borders, do they fully mask each other and the tops of the legs?

Save As this file with a name indicating that it includes the set within the venue, with masking. Since the venue architecture has been substantially altered, consider creating a new venue model file without the set so that it can serve as a base model for subsequent projects. To do so, Hide the masking and set groups. Select the venue group and Copy it. Open a new file and Paste. Give the new file a name to indicate that it is the base venue model. If you move onto the Modeling Battens project, below, wait to create the new base venue model until after completing it.

Project: Modeling Battens

The battens can be modeled so that the soft goods can be seen hanging from pipes. This can be useful for sightline determination. Legs, for instance, typically hang at the sides

of a playing area: will the pipe from which they hang be visible at mid-stage? Hide the set and masking to make this project easier.

Battens are typically 1½″ DIA pipes. Use Push/Pull to extrude a 1½″ circle into a cylinder of the desired length. Drawing the base circle on a rectangular surface helps ensure that a circle aligns with the intended axis. Rotating cylinders can be tricky, so draw the originating circle oriented so that the extruded batten runs in the desired direction. Draw center lines across the cylinder's end surface to aid later alignment. Make this cylinder a group.

When not in use, battens are "gridded." This means they are raised and locked off at their highest possible distance above the deck, right below the grid. In the venue model, the gridded position of the battens will be one foot below the grid: + 49′–0″. The batten array is currently mapped out on the deck. Draw a 49′–0″ long vertical guideline up from one end of the first batten, and then Copy and Move the batten array, dragging the copy directly upward. Affix it to the endpoint of the guideline (Figure 9-11).

Move the cylindrical batten by grabbing the center of its end. Align the center with the endpoint of lineset 1. The line representing lineset 1 is now inside the pipe, running along its central axis. Copy and Move the cylinder, placing the center of each pipe's end at the endpoints of each lineset.

Hide the copy of the lineset array and make the pipes a group. Whenever you wish to move a batten into a lower position, open Edit Group and drop the batten down to the desired trim height (Figure 9-12).

Return to the combined venue/set model file. Select the new pipe array group and open with Edit Group. Move selected pipes into position along the top edges of the legs and borders. If a chandelier has been modeled, move its pipe down to the trim height at which you expect it to play. Exit Edit Group and take a look at the model from the side/section again. Revisit the front of house sightlines.

Save the file.

MODULE 3: DRAFTING THE CENTER LINE SECTION

For both hand drafting and Vectorworks, the section's layout steps are similar. This module walks through the steps via Vectorworks.

The section is a side view of the set; the groundplan is a top view and therefore orthographically diagonal from the side view. These are both sizeable drawings and most drafting tables are not large enough to lay out a miter line for information transfer between the two views. For hand drafting, this means physically turning the groundplan sideways and taping it partway up the drafting table underneath the section's sheet of vellum. Once enough information has been transferred, the groundplan can be removed. If nontransparent

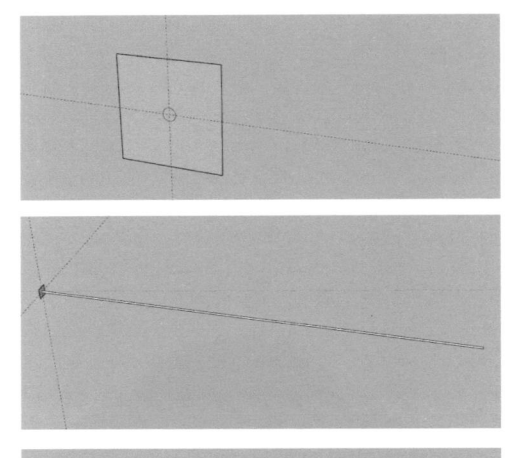

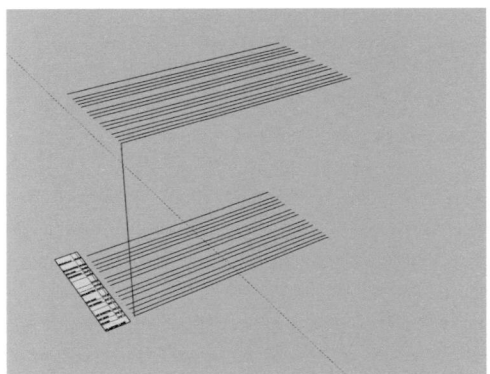

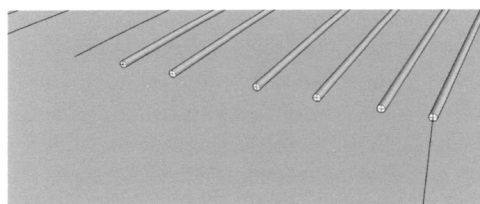

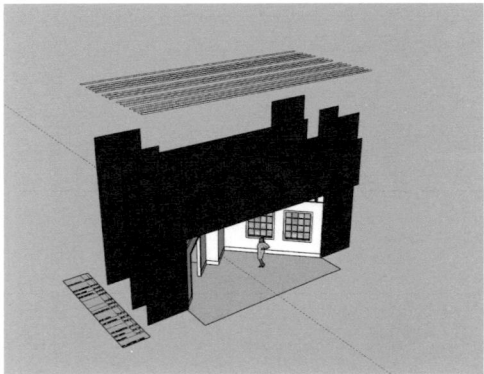

FIGURE 9-11 *Steps of creating an array of battens: lay out end circle; push/pull to length, make group; copy and raise batten guidelines; paste batten over guideline; move and copy*

FIGURE 9-12 *Positioning a pipe against the top of a leg*

bond paper is used, tape the groundplan over the upper third of the section's sheet; remove as soon as possible so that the upper portions of the section can be drawn.

While Vectorworks' infinite drawing space is big enough for a miter line, it's easier to rotate the groundplan and position it above the area in which the section will be drawn. In this exercise, the groundplan will be Copied, Pasted, and Rotated (Figure 9-13).

Students

Open the groundplan constructed in Chapter 8. Create additional layers for this drawing, such as Section Architecture, Section Scenery, Section Dimensions and Notes, and Section Line Schedule. Groundplan Rotated Copy is a useful Layer during the Paste and Rotation phase.

Continue working in ½″ = 1′–0″.

1. First, some drawing space organization. Depending on which section view you've chosen, right or left, that view will be to one side of the to-be-drawn front elevation, sharing a baseline (the zero deck). When the front elevation is addressed in Chapter 10, information from the section will be laterally transferred into it while information from the top view will be transferred vertically.

   In the Guideline layer, drop guidelines from the sides of the groundplan into the area where the elevation will eventually be placed. Be generous with their length.

   The proscenium opening is 18′–0″ high and the elevation will include a certain amount of wall above it. An overall height of + 24′–0″ from the zero deck is a reasonable surmise for the height of the front elevation drawing, since the ceiling height of the auditorium is not yet known. Zone out sufficient white space below

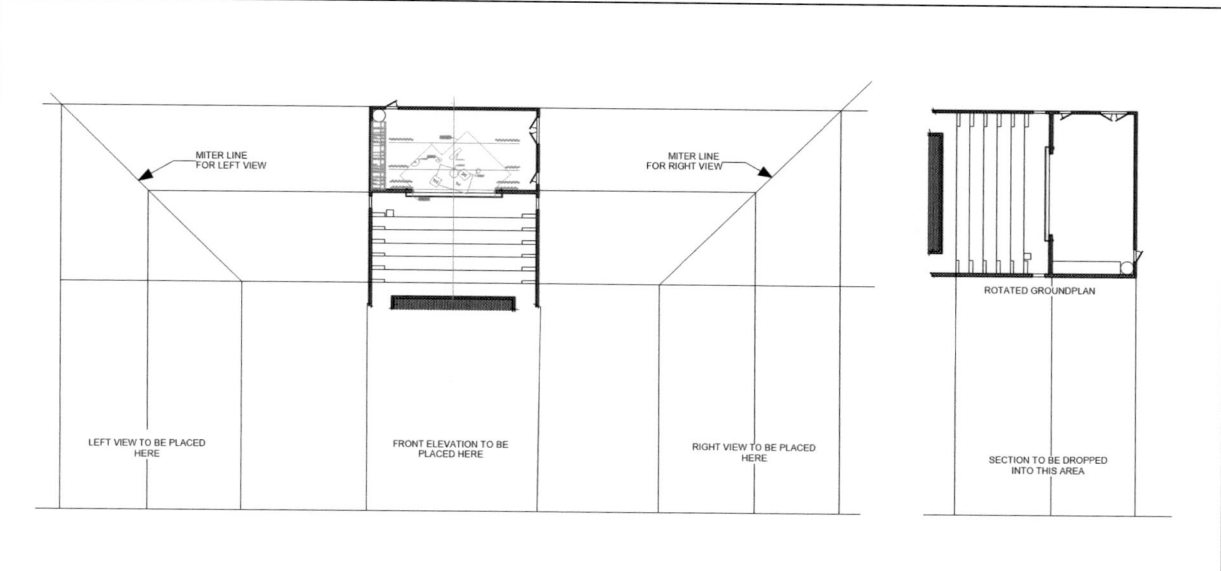

FIGURE 9-13 *Groundplan information transferred via miter lines; the groundplan rotated to drop information directly into the side view*

the groundplan, and then draw a horizontal base line 24'–0" below that. Extend it into the area where the section will be placed. A center line section looking stage left and placed to the left of the front elevation will be generated in this exercise. Leaving a zone of white space to the side of the front elevation area, draw a vertical guideline up from the baseline parallel to the

side of the groundplan. This line will be used to align the groundplan's rotated copy (Figure 9-14).

2. Since the groundplan drawing was constructed in multiple layers, copying the groundplan is not a one-step Copy and Paste process. In the View drop-down menu, double-check that under Layer Options, the Show/Snap to Others option has been chosen. This option enables you to interact with items in

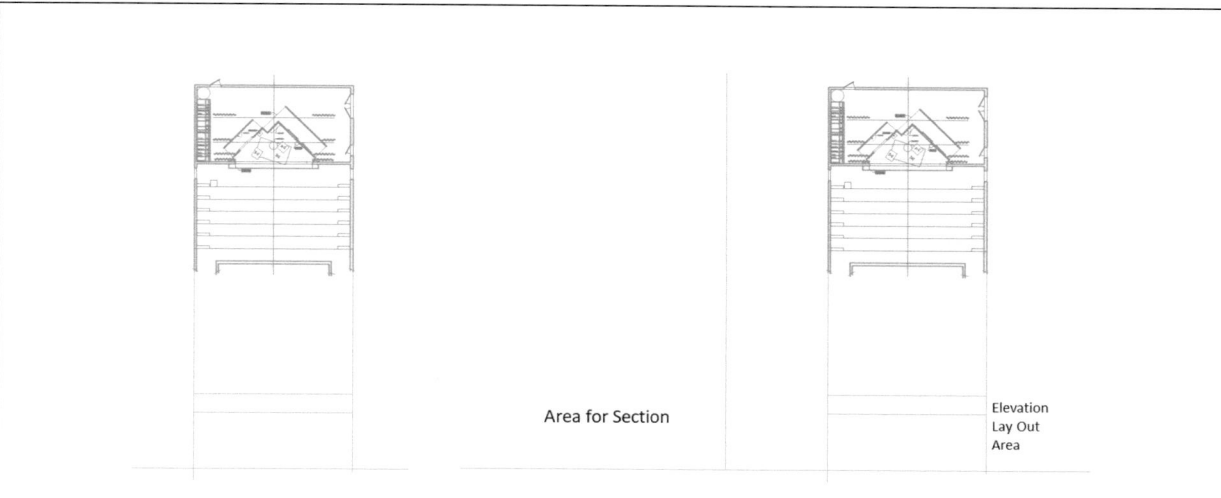

1. Dropping information into front elevation area

2. Extending baseline from elevation area, then placing a vertical line for rear wall of auditorium in section area

FIGURE 9-14 *Section lay out steps: dropping guidelines from the groundplan to place elevation; extending guidelines into section area; drawing a vertical line to align rotated groundplan*

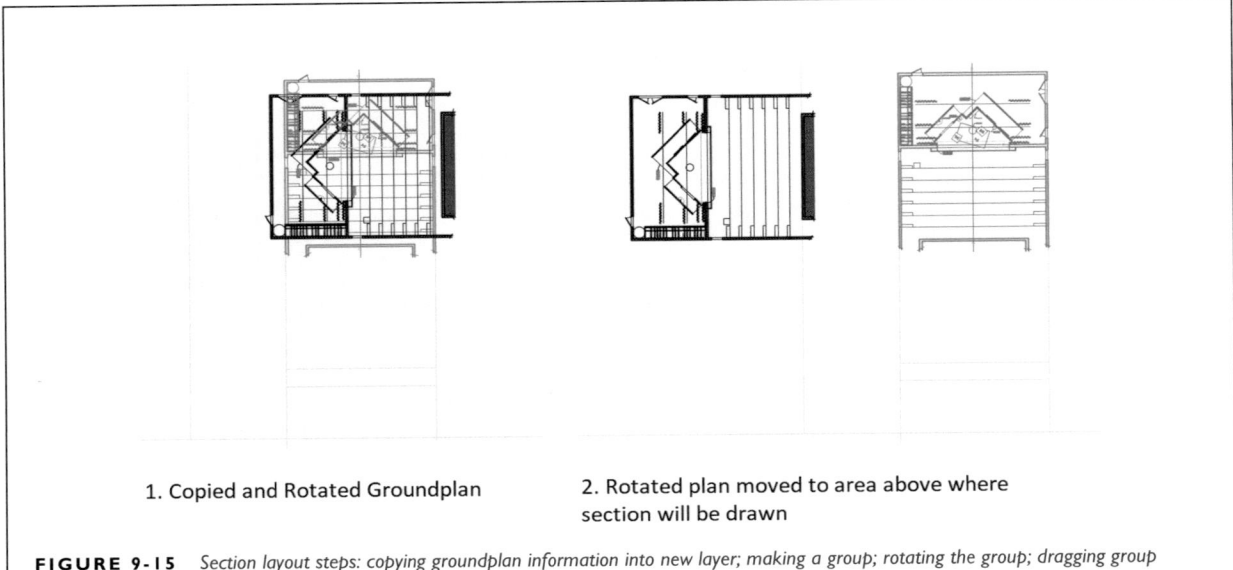

1. Copied and Rotated Groundplan

2. Rotated plan moved to area above where section will be drawn

FIGURE 9-15 *Section layout steps: copying groundplan information into new layer; making a group; rotating the group; dragging group above section area*

non-active layers, making the following steps easier to accomplish.

Go into the design layer for the venue architecture. Select it all. In the Edit drop-down menu, choose Copy. Switch to the Rotated Groundplan Copy design layer and Paste in Place. Copy and Paste in Place all of the required groundplan elements, except the battens, from their separate layers into the Rotated Groundplan Copy layer. Having the battens on the groundplan copy will make it difficult to transfer set information. The line schedule will suffice. When complete, make the drawing a Group. Rotate the group left 90°. Drag it to the left and align the back wall of the auditorium with the vertical guideline (Figure 9-15). This copy is for reference and will not become an actual printed drawing. It may be deleted when no longer needed.

3. Lay out the section's architecture first. In the Guideline layer, drop vertical lines from features in the rotated groundplan until they intersect the baseline. Include the rear wall of the venue, the proscenium (both front and rear edges), the front of the apron, the edges of the seating tiers, steps, doorways, and rear wall of the auditorium.

4. The base line represents the zero deck. Draw a horizontal line for the auditorium floor 2'–0" below the zero deck.

Guidelines will begin to accumulate as additional guidelines are traced over guidelines to outline features with vertical edges. One approach is to create a class just for section guidelines, with a

color different from the usual guideline class. The regular guidelines can be turned off, leaving the section guidelines visible. Another approach is to create a section guideline layer, so that when the Guideline layer is turned off, the section's guidelines remain visible. If neither of these approaches seems convenient, lines can be darkened as you go. Since this is CAD, smudging will not be a problem but there is likely to be a certain amount of snipping and editing as visible lines are placed. For this exercise, a Section Guideline class and layer will be used.

With the Section Guidelines class in the Section Guidelines layer, draw a horizontal line for the auditorium floor between the verticals marking the edge of the apron and the front of the first seating tier. Trace the vertical front edge of the apron. Lay out the side view of the seating tier and steps (Figure 9-16).

5. Using the Section Guideline class, draw a horizontal line over the base line representing the zero deck. Since there is a door in the back wall indicating a continuation of the building, extend the deck's line a bit past the rear wall. This extra bit is not strictly necessary, but helps the section visually "sit" better on the page.

Draw vertical lines for the venue's back wall and the proscenium. Use the Double Line tool, if preferred. The grid, at + 50'–0", is unlikely to fit on the printed sheet, so terminate these walls at + 30'–0". That will still provide adequate space for the full height of the legs and borders, as well as sheet space for the line schedule. Break lines will be added later to indicate the space that was skipped over.

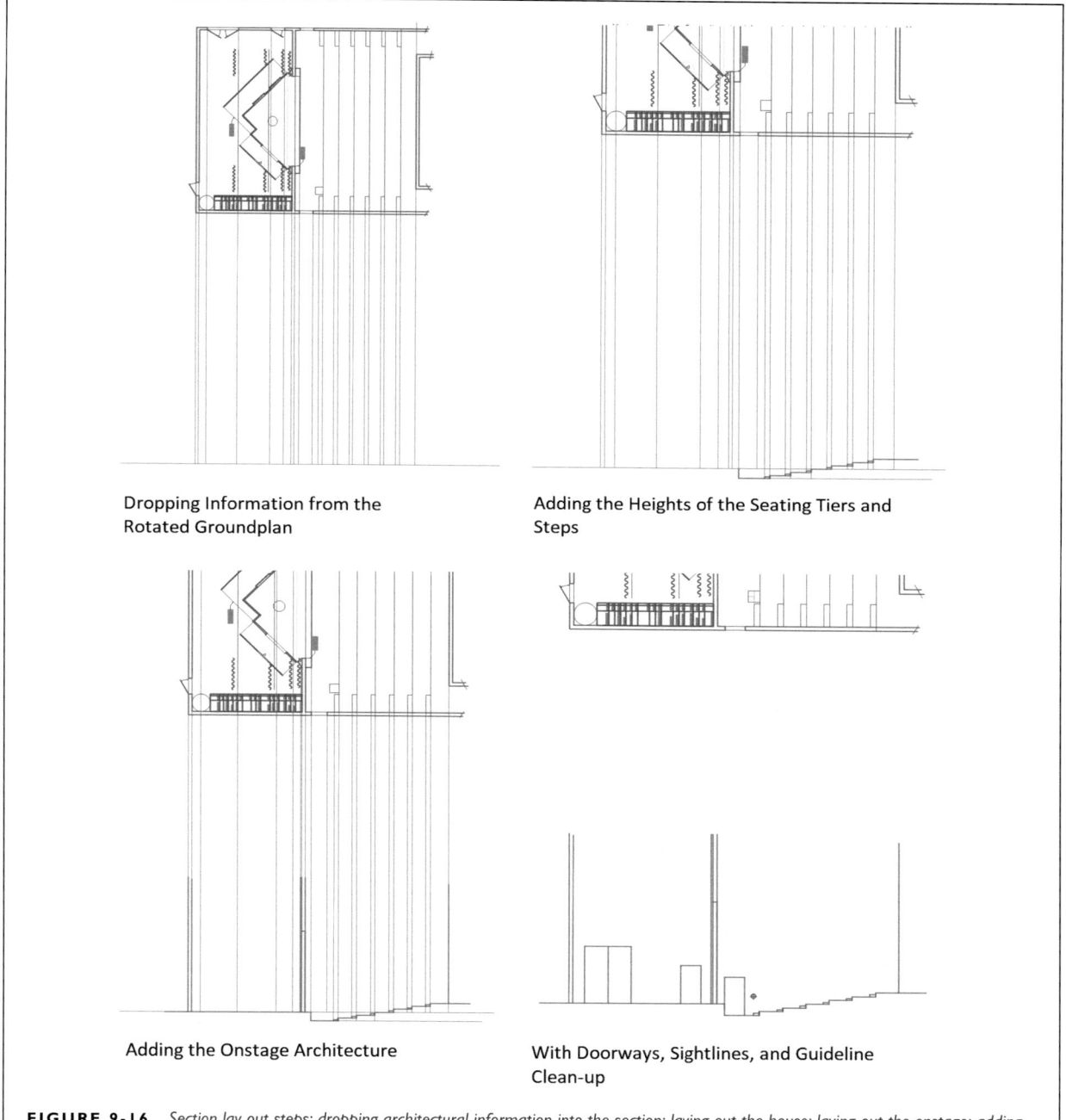

Dropping Information from the
Rotated Groundplan

Adding the Heights of the Seating Tiers and
Steps

Adding the Onstage Architecture

With Doorways, Sightlines, and Guideline
Clean-up

FIGURE 9-16   *Section lay out steps: dropping architectural information into the section; laying out the house; laying out the onstage; adding doorways and sightlines; guideline clean-up*

Add a horizontal line between the proscenium lines to mark the upper edge of the proscenium arch. Add a vertical line to mark the edge of the smoke pocket.

6. Outline the doorways. Drop a guideline for the front row's sightline symbol. On a section, the sightline symbol typically represents a patron sitting in the center of the front row, rather than at the end of a

row. As the rows in this auditorium run straight from side to side, the lateral placement of the sightline symbol remains aligned with its placement on the rotated groundplan. Its height needs to be determined. Find a tape measure, sit in an ordinary chair, and measure the distance from the floor to your eye level. Use that measurement for the sightline symbol's height. Clean up guidelines.

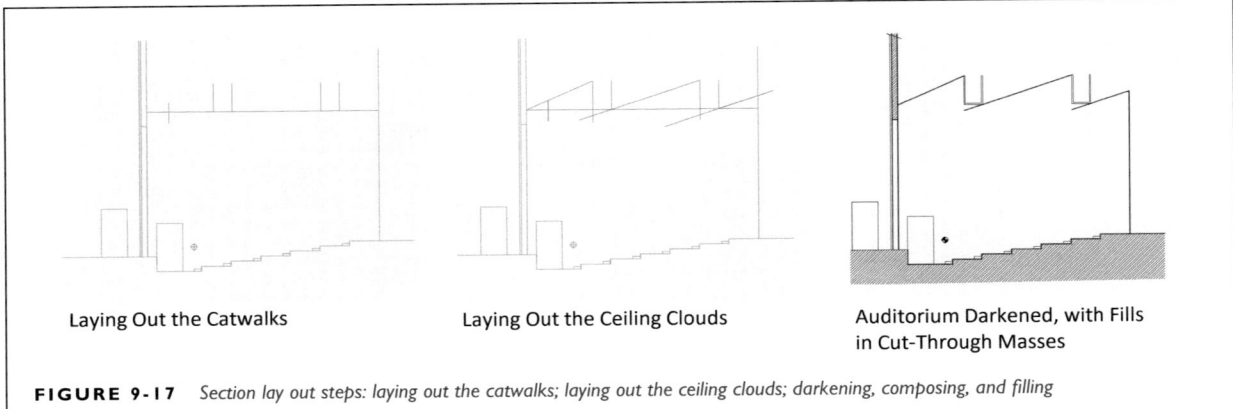

Laying Out the Catwalks   Laying Out the Ceiling Clouds   Auditorium Darkened, with Fills in Cut-Through Masses

FIGURE 9-17 *Section lay out steps: laying out the catwalks; laying out the ceiling clouds; darkening, composing, and filling*

7. The ceiling of the auditorium will now be added. Figure 9-5 includes measurements for the auditorium's ceiling clouds and catwalks. **Cloud** is a term for ceiling sections that "float" between other structures, such as catwalks. A **catwalk** is a permanent suspended walkway, often used to support lighting positions that cannot be accessed from below, as when they span seating areas.

    The catwalk profile is described in a detail drawing in Figure 9-5; 4'–0" high and 2'–6" wide. Catwalk lighting pipes will be added in Chapter 12 (unless you wish to add them now). Lay out the catwalk profiles; use the Double Line tool with a 2" separation to indicate the thickness of the catwalk's structure. Be sure the second line falls within the catwalk outline.

8. Lay out the ceiling clouds by connecting the dots. For the one nearest the proscenium, the start point is + 20'–0" above the deck on the proscenium wall. The endpoint is the top front end of the first catwalk (creating a triangle 9'–0" long and 4'–0" high). The second cloud begins below the front edge of the first catwalk to hide the catwalk's underside from the audience. To find the angle, connect the bottom rear corner of the first catwalk to the top front corner of the second catwalk. Extend the line under the first catwalk until it lines up with its front edge (Figure 9-17). The Line tool's Center mode is useful for extending angled lines. Duplicate the angle for the third cloud, extending the hypotenuse until it terminates it at the back wall of the auditorium.

9. The architecture for the section is now laid out. The line schedule will be placed after the set is drawn so that it does not interfere with keeping track of the guidelines to be dropped for the set.

    Switch to the Section Architecture layer and darken (Figure 9-17). Compose the walls that are cut through and fill with the Cast Iron crosshatch pattern.

Remember, you can use the Rectangle tool or Irregular Polygon tool to create shapes that are automatically composed. Add break lines at the tops of the upstage wall and proscenium wall.

The deck is also a cut-through mass. In hand drafting, crosshatching can fade (whether or not there is a break line) to indicate a continuation of structure. In Vectorworks, composing a shape for crosshatching requires boundaries. You can use an existing line style to outline the shape, or create a new class which has no line style—just a fill (Figure 9-18).

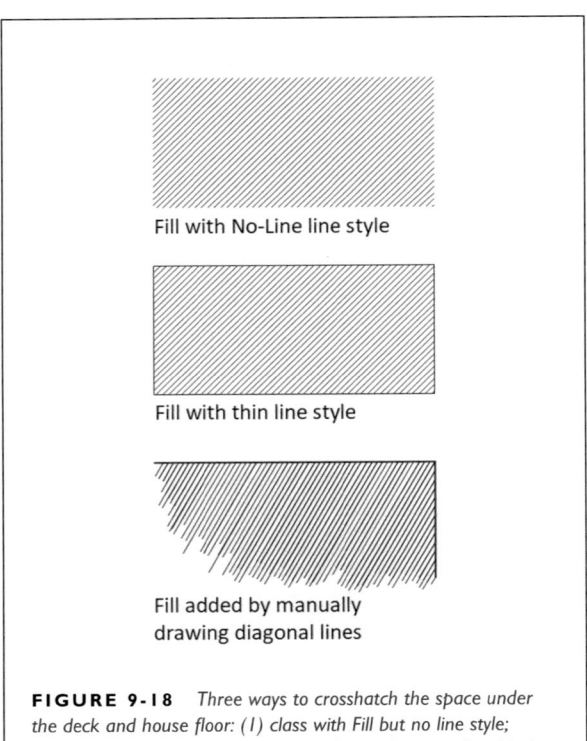

Fill with No-Line line style

Fill with thin line style

Fill added by manually drawing diagonal lines

FIGURE 9-18 *Three ways to crosshatch the space under the deck and house floor: (1) class with Fill but no line style; (2) visible filled outline; (3) partial outline with manually added crosshatching*

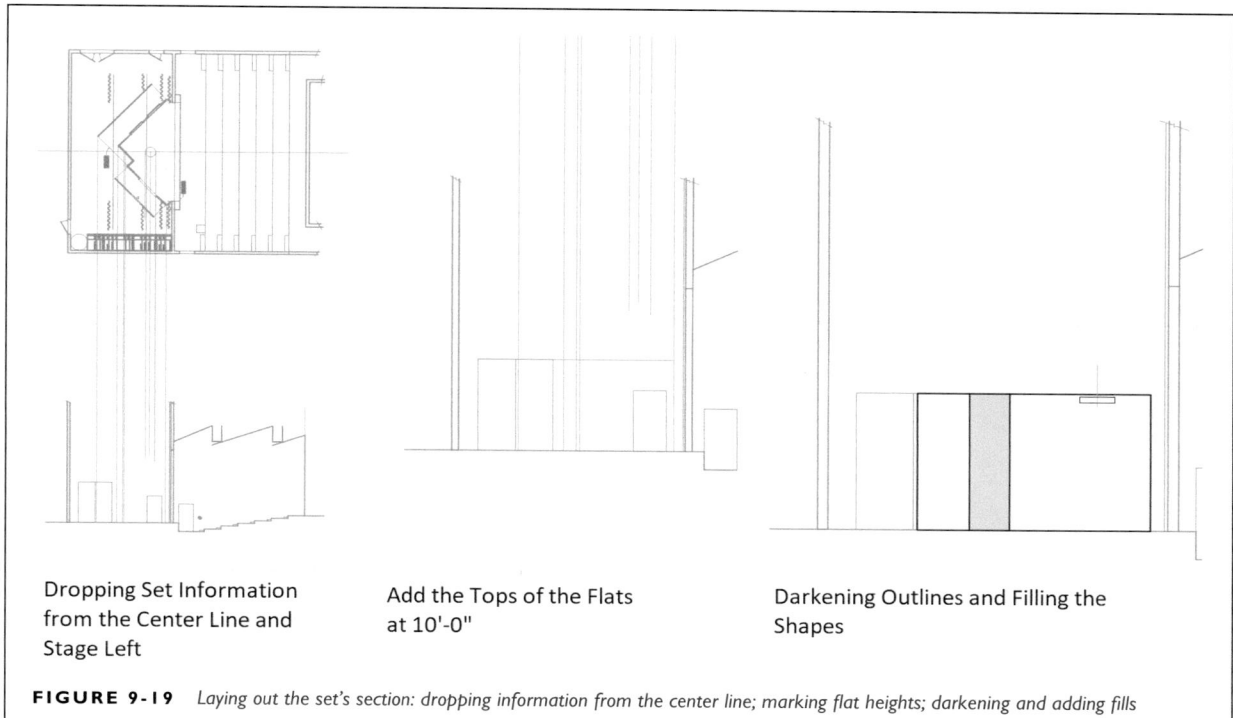

Dropping Set Information
from the Center Line and
Stage Left

Add the Tops of the Flats
at 10'-0"

Darkening Outlines and Filling the
Shapes

FIGURE 9-19   *Laying out the set's section: dropping information from the center line; marking flat heights; darkening and adding fills*

10. Clean up guidelines. As this exercise constructs a section with the cutting plane directly on the center line, the parts of the set to be dropped into the section are those at and above the center line on the rotated groundplan. Start with places where the set intersects the center line: Flat C and the chandelier (Figure 9-18). Remember that the flats are Hollywood style and have thickness. Flat C, where it meets the cutting plane, is a cut-through mass. Drop down the front and rear edges of the sliced area.

Work upward toward stage left of the rotated groundplan. The corner where Flats C and D meet is the next feature encountered. Drop down the downstage end of Flat D and the upstage end of Flat F. Don't drop down the windows, legs, or borders yet; those will be done after the flats are outlined. The end of Flat F will be visible through the downstage window; this will be added after the window is drawn.

11. The set's walls are 10'–0″ tall. Use the Section Guidelines class to draw a horizontal line to mark the top edge of the flats. This should run from the upstage end of Flat F to the downstage end of Flat D (Figure 9-19).

12. Switch to the Section Scenery layer and darken the outline of each flat. Compose the cut-through portion of Flat C and Fill with crosshatching. The outlines of the stage left venue doors will be visible through the flat outlines. Compose the set outlines and give them a white fill to hide the doors beyond. The

example in Figure 9-18 uses a gray fill for the area that represents the upstage side of Flat C, indicating that it is not a scenically treated surface. The chandelier is represented by a temporary rectangle. Clean up guidelines.

13. Take a look at the 2D image of the model's center line section. The windows, with their frames and casing, are seen at an angle. Their height dimensions remain true, but width and depth are foreshortened.

To make the transfer of information easier, a copy of the windows will be pasted in closer proximity to their location in the section. In the Rotated Groundplan layer, select the groundplan and choose Edit Group from the Modify drop-down menu. Select the portion of Flat D containing the windows, make it a group, then copy it. Exit Edit Group. Drop a guideline down from the corner of one of the windows in the rotated groundplan into the section. Paste in Place so a copy of the windows is situated over the groundplan. Select it and drag it down along the guideline until it appears just above the walls in the section. Be sure that it maintains its alignment with the guideline (Figure 9-20). Turn off the visibility of the venue architecture if desired.

Start with the downstage window. Drop guidelines into the section from the corners where the reveals meet the front edge of the flat. The bottom of the window is + 3'–0″ above the deck. The top of the window is at + 8'–0″. Draw those two horizontal lines, but only

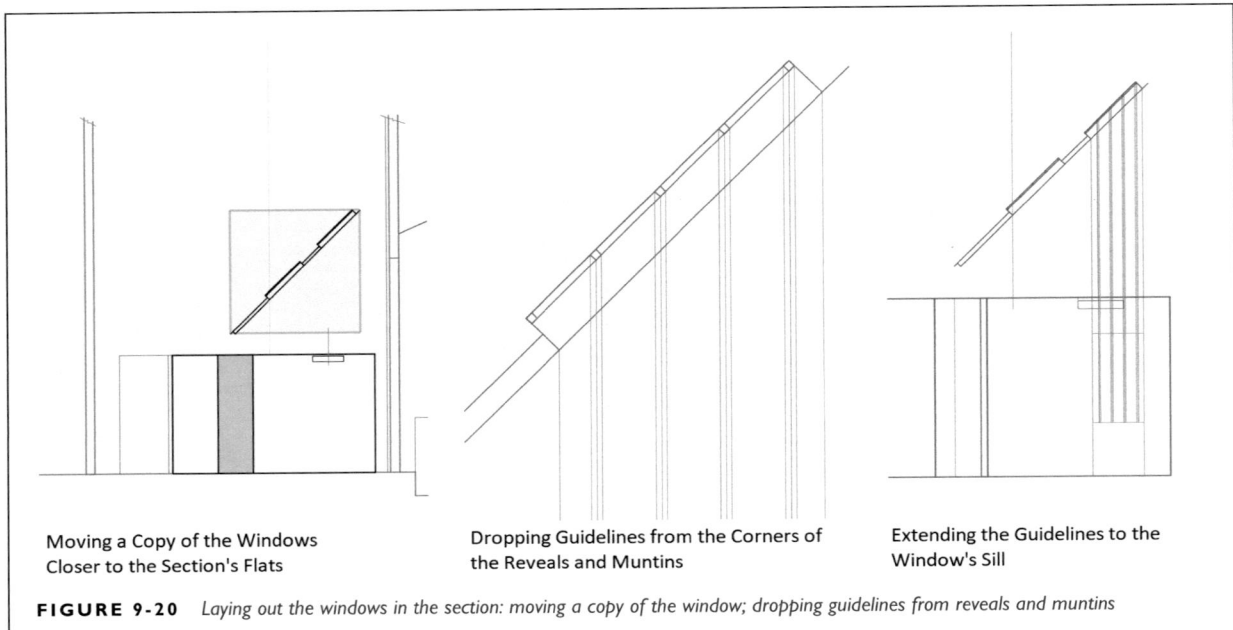

Moving a Copy of the Windows
Closer to the Section's Flats

Dropping Guidelines from the Corners of
the Reveals and Muntins

Extending the Guidelines to the
Window's Sill

FIGURE 9-20   *Laying out the windows in the section: moving a copy of the window; dropping guidelines from reveals and muntins*

between the two vertical lines representing the sides of
the window.

Drop a line from where the stage left reveal meets
the window frame (panes and muntins). The stage
right reveal is hidden behind the flat and is not seen.
Drop lines from the visible corners of the muntins. The
muntin farthest downstage shares an edge with the
reveal, so only two additional lines need to be dropped
for it. The three center muntins will each have three
lines dropped. The muntin farthest upstage (against
the edge of the frame) is hidden behind the flat, and
therefore not seen. The vertical guidelines are very
close together, so drop them down in phases to keep
track of them (Figure 9-20).

14. Consult Figure 8-25 for horizontal muntin heights.
Lay out center guidelines for the horizontals. Use the
Double Line tool to lay out the muntins themselves.
They do not cross the downstage reveal, so stop at that
line (Figure 9-21).

15. Switch to the Section Scenery layer to darken the
window. The outer perimeter of the window is a
major feature, as is the upstage edge of the stage left
reveal, so thicker lines are appropriate. The muntins
are an interior detail, so use a thinner visible line. Pay
attention to which muntin lines are in front and which
are in back. Within each light, the side of a muntin
will be seen. The frame's face is seen as a continuous
surface. When complete, the window frame should
look as it does in the section view of the model.
Drop a guideline from the stage left end of Flat F into

the section. Darken the portions that appear within
the window's lights. If you wish, create a class with
no line style and a light gray fill that can be used to
shade backstage areas. Using this in the window can
help differentiate the surface of Flat F from the space
beyond (Figure 9-21).

16. The same process for drawing the upstage window
can be repeated for this window. However, since the
windows are identical and reside along the same surface
at the same angle to the cutting plane, the first window
can be Copied and Pasted.

Use guidelines to outline the position of the
upstage window. Make the downstage window a
group, and then select it. Hold down Ctrl and drag
the copy into position (or Copy and Paste). However,
since the edge of Flat F and the shading was added to
the downstage window, these features will need to be
deleted from the upstage window since Flat F fills the
whole window (Figure 9-21).

17. No molding was included on the groundplan. However,
the molding's face is seen in the section, which means
molding information must be added to the rotated
groundplan so it can be dropped into the section. At
the downstage end of Flat D, as well as where it is cut
through by the center line, the profile of the crown
molding will be foreshortened, since it is seen at angle
(Figure 9-22).

The crown molding is ¾″ × 5″. On the rotated
groundplan, draw a guideline ¾″ downstage and parallel
to Flats C and D, above the center line. Flat C's crown

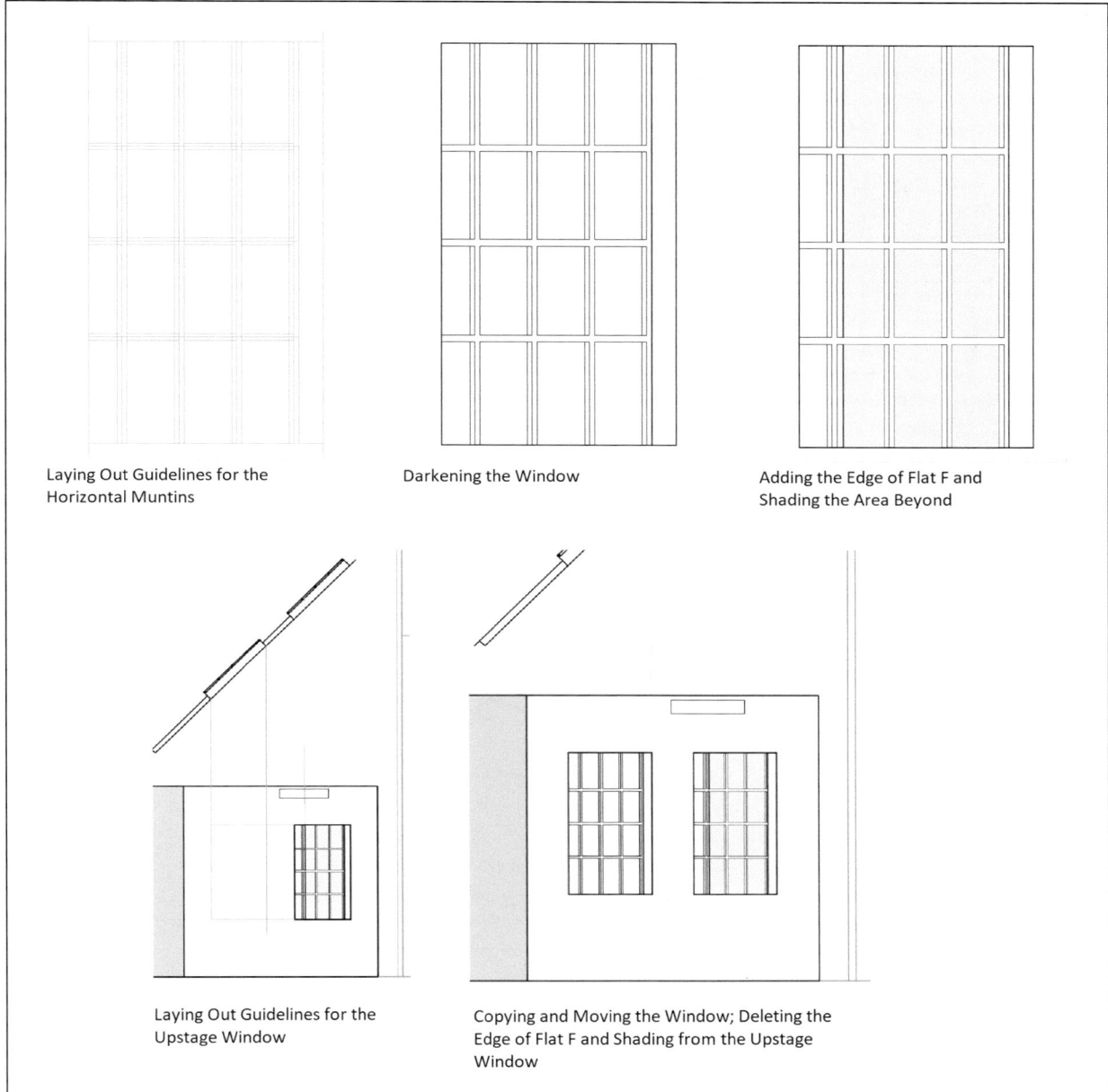

Laying Out Guidelines for the
Horizontal Muntins

Darkening the Window

Adding the Edge of Flat F and
Shading the Area Beyond

Laying Out Guidelines for the
Upstage Window

Copying and Moving the Window; Deleting the
Edge of Flat F and Shading from the Upstage
Window

FIGURE 9-21 *Laying out the window: Adding horizontal guidelines for muntins; darkening the window; adding the visible edge of flat F and shading what lies beyond it; copying the window; deleting flat F's edge and shading from the copy*

molding will intersect the center line at an angle. Flat D's will terminate at the flat's downstage end.

On the section, draw a horizontal guideline 5″ below the top of Flat D. Drop endpoints from the groundplan into the section. The molding sticks out slightly from the downstage end of Flat D. At the top of Flat C, a cut-through section of crown molding will be seen. The cut through area is crosshatched or shaded, depending on which choice will read better; the Pattern menu contains selections for tight

crosshatching. Use the Rectangle tool to draw the crown molding and the cut-through area. Use a white Fill for the molding on Flat D. Send it forward if necessary.

18. The baseboard is ¾″ × 8″. Since it's the same depth as the crown molding, the guidelines added to the rotated groundplan can be used to place the baseboard. Draw a guideline 8″ above the bottom of Flat D. Drop the endpoints from the rotated groundplan or from the crown molding, since the crown molding's face sits in

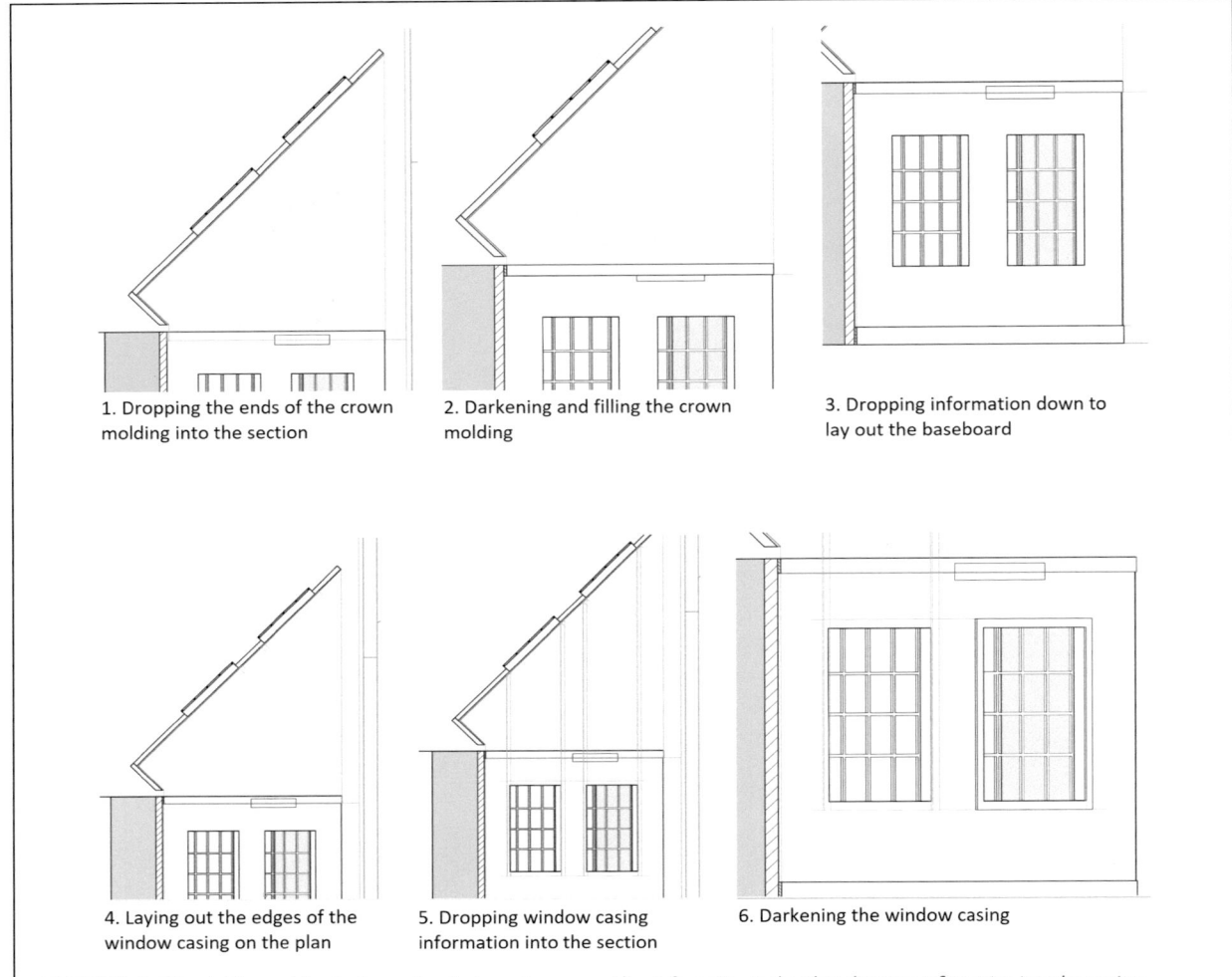

1. Dropping the ends of the crown molding into the section

2. Darkening and filling the crown molding

3. Dropping information down to lay out the baseboard

4. Laying out the edges of the window casing on the plan

5. Dropping window casing information into the section

6. Darkening the window casing

FIGURE 9-22 *Adding molding to the section: laying out crown molding information in the plan; dropping information into the section; darkening crown molding and baseboard; marking window casing information on plan; dropping casing information into the section; darkening*

the same plane as the face of the baseboard. Baseboard cut through at the center line is crosshatched or shaded. Darken and fill. Clean up guidelines.

19. The window casing is ¾″ × 3″. The crown molding and casing have the same thickness, so the guideline already drawn on the plan can also be used for the casing. The inside edge of the casing sits flush against the window opening, so extend the lines of the reveals out to the guideline. Measure the exterior edges of the casing 3″ outward from the reveal edge. The two rectangles created on either side of the window represent cut-through vertical casing. The line across the width of the window opening represents the horizontal edges of the top and bottom casing pieces (Figure 9-22).

     On the section, draw generous horizontal guidelines 3″ above the top and 3″ below the window

openings. Drop guidelines into the section from the casing's endpoints in the rotated groundplan. Note that the upstage vertical guidelines overlap a small portion of the window frame and muntins. Ungroup the windows (or Edit Group) to trim parts of the frame hidden by the casing. Darken.

20. Currently, the upstage side of Flat C is shaded gray. If you wish to include framing detail for the rear of Flat C, draft a construction drawing and use it to transfer framing information into the section. The upstage side of Flat C can also simply be labeled as "Rear of Flat C" without adding construction detail.

21. Before adding soft goods, add the line schedule to the section. Copy the line schedule from the rotated groundplan and paste it into the Section Line Schedule Layer. Drag it down until it is just above the section.

The venue's walls were broken at + 30'–0". Before placing the line schedule, add the upper portion of the stagehouse, well above the break line; the break lines, in effect, skip over about 15' of space. Once the upper sections of the stage house walls have been drawn, drag the line schedule into position. Notice that the lineset numbers and distances are at the top of the line schedule; it will need to be somehow reversed. Select the schedule's group and choose Rotate from the Modify menu. Choose Flip Vertical (Figure 9-24).

On a section where a counterweight system is present, the line schedule is placed horizontally above the grid. An older way to draw the section's line schedule is to angle the rows of text away from the proscenium, typically at 45° or 60°. This may have been done to accommodate the narrow available space at the top of the sheet. This angled format is now seen less and less often. Orientation of text within vertical arrangements varies; some designers orient it to be read from the right, as is standard in the aligned dimensioning system. Others orient it so the lettering aligns with the plaster line, as it is on the groundplan.

22. Drop vertical guidelines for those linesets upon which soft goods will be hung. This is also a good time to double check labels to confirm all hanging items are noted. Remember that the cyclorama is not flown in for this exercise, though it is present on the line schedule. Flown house items that are not used for a production are not usually drawn on the section unless there is a technical or venue requirement, such as keeping track of all the flown items used in a rotating repertory of performances.

23. Before laying out the soft goods, clean up guidelines. The legs have fullness and are 24'–0" high. Borders are 8'–0" without fullness, their bottom edges trimmed at + 12'–0".

24. Drapery without fullness is drawn as a single, unbroken line: the fabric is seen on edge. Drapery with fullness can be drawn as a slender triangle with a depth of couple of inches at the bottom. Some designers use a vertical double line, parallel from top to bottom. If drawn as a triangle, the base is 4" to 6" wide at the deck. A vertical line bisects the triangle from the bottom up to about a third of the leg's height. The bottom is capped with offset curves or straight lines (Figure 9-23).

If scenery obscures the lower portion of a leg, leave a narrow gap between the scenery and the drawn leg to indicate that the leg is not connected to the scenery. If using the triangle method described above, remember that the higher the break, the

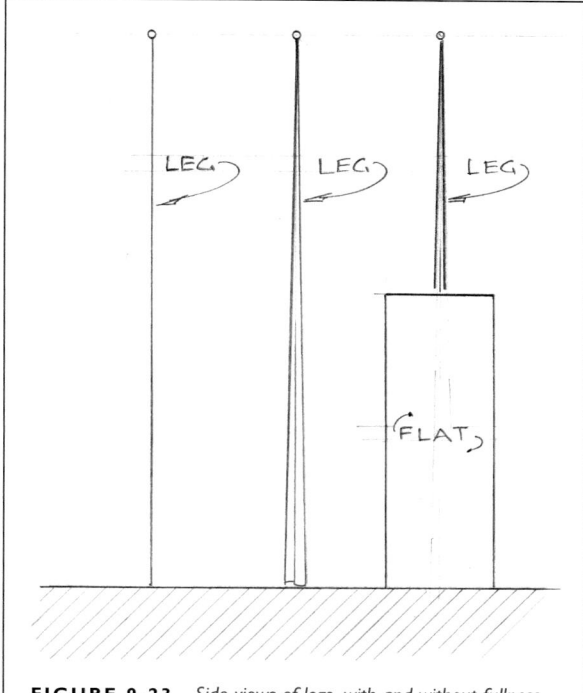

FIGURE 9-23 *Side views of legs, with and without fullness, and seen behind scenery*

narrower the triangle seen. Lay out the full triangle in the guideline layer, then simply darken the portion visible above the scenery. You can also draw the leg elsewhere in the drawing space and make it a group. Drag copies into the section, and then Send them back so that they sit behind filled scenic elements. Add a slender white-filled rectangle with no line style to separate the leg from the edge of the scenery. Sending items to the front or back only works if they are all in the same design layer.

A circle drawn at the top of the soft good indicates the end view of the batten. A thin line may continue upward to the grid to indicate the cable from which the batten is suspended. If the fly space is broken to fit on the sheet, break lines interrupt the cable line.

Create a copy of the leg profile for the Main Act Curtain. Move it into position.

25. Double-check sightlines before finalizing border placement. Lay out the teaser at its intended trim height. Draw a guideline from the front row's sightline symbol to the bottom of the teaser; extend this line well past the teaser. Everything above the angled line is masked by the teaser (Figure 9-24).

Lay out the next border at its intended trim. If the top of the border lies above the sightline, the batten is masked by the teaser. If it falls below the sightline, it will be visible and the trim of the border or teaser needs to be reset. After the trims are adjusted, draw

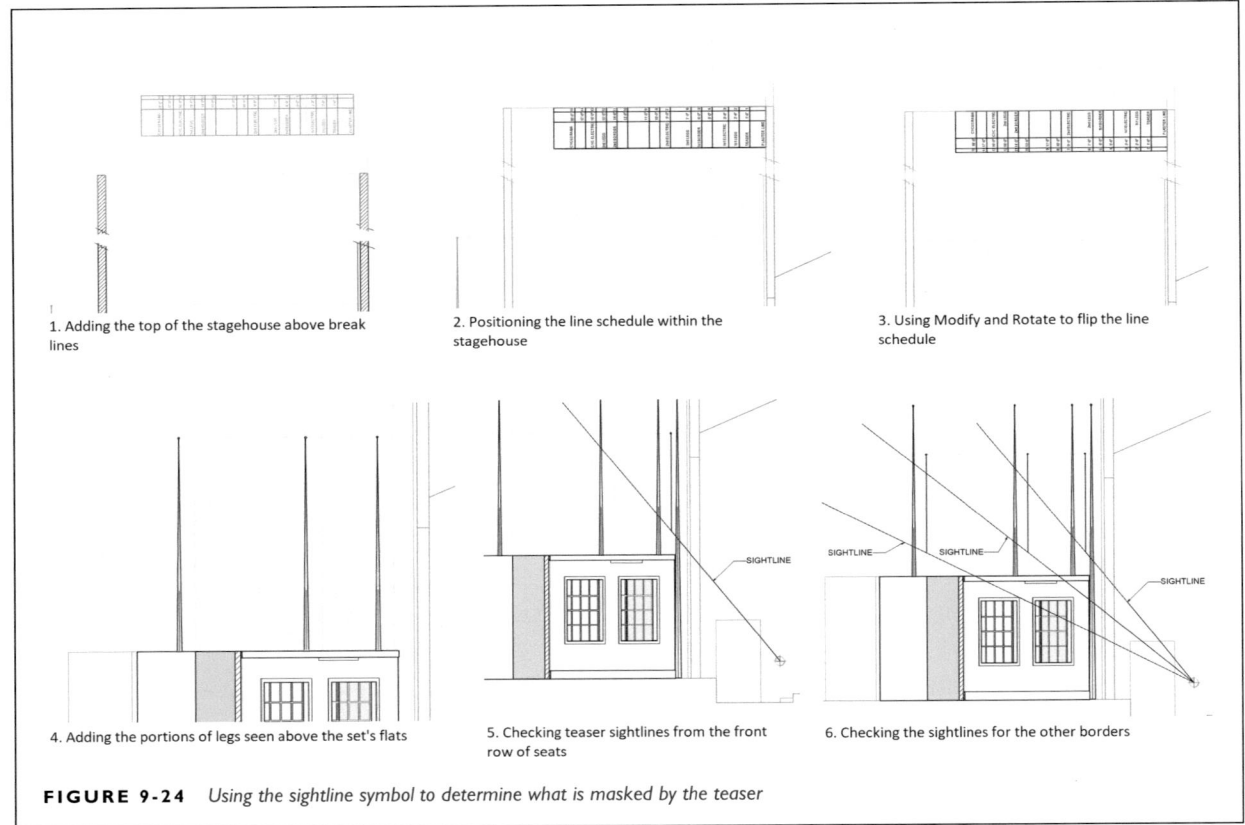

1. Adding the top of the stagehouse above break lines

2. Positioning the line schedule within the stagehouse

3. Using Modify and Rotate to flip the line schedule

4. Adding the portions of legs seen above the set's flats

5. Checking teaser sightlines from the front row of seats

6. Checking the sightlines for the other borders

FIGURE 9-24 *Using the sightline symbol to determine what is masked by the teaser*

a line from the sightline to and past the first border's bottom. Determine whether the top of the second border is masked by the first border. Check whether the legs' battens are masked by the borders. Adjust trims as required.

Originally, the bottom edge of the borders were to play 2′ above the top of the set. Take a look at the upstagemost border and its relationship to the set. How much of the venue's back wall is visible through the gap? Lineset 14 is currently unused; would adding a border to this lineset finish the masking? Would it be preferable to lower the trims of the existing borders? How might lighting be affected if all the borders are lowered even further?

If any trim heights are adjusted or borders added, revise the groundplan accordingly. Empty linesets are not typically drawn, though they are listed in the line schedule.

26. If you've imported a chandelier into the model, look at its parallel projected side view. This 2D graphic can be copied and imported, though a cleaner image may result from importing it into one layer and tracing it in another before inserting it into the drawing. In hand drafting, a scaled image can be placed beneath the vellum and traced. A side view of the chandelier is sufficient; it's not necessary to draw the chandelier's section as that might actually represent it less clearly. Check the trim of the chandelier against the sightlines. Is it low enough to be seen? Is its batten visible? How long does the cable by which it's suspended need to be?

27. From the line schedule, drop guidelines for those linesets indicated as electrics. It's typical to find a trim suggestion on scenic sections, though the lighting designer will determine the final trim heights of lighting positions (often after discussion with the scenic designer, technical director, and sometimes also the sound and projection designers). The batten needs to play high enough to accommodate a lighting unit suspended beneath it. After laying out the sightline, place a lighting symbol above it, then add the batten profile above the symbol (Figure 9-25). When hand drafting, use a lighting template to draw a lighting unit symbol. In Vectorworks, there is a lighting symbol library found in the Spotlight workspace. This library will be discussed in Chapter 12. If unable to include a unit symbol, set the batten at a suggested height and label it as an Electric with trim To Be Determined (TBD).

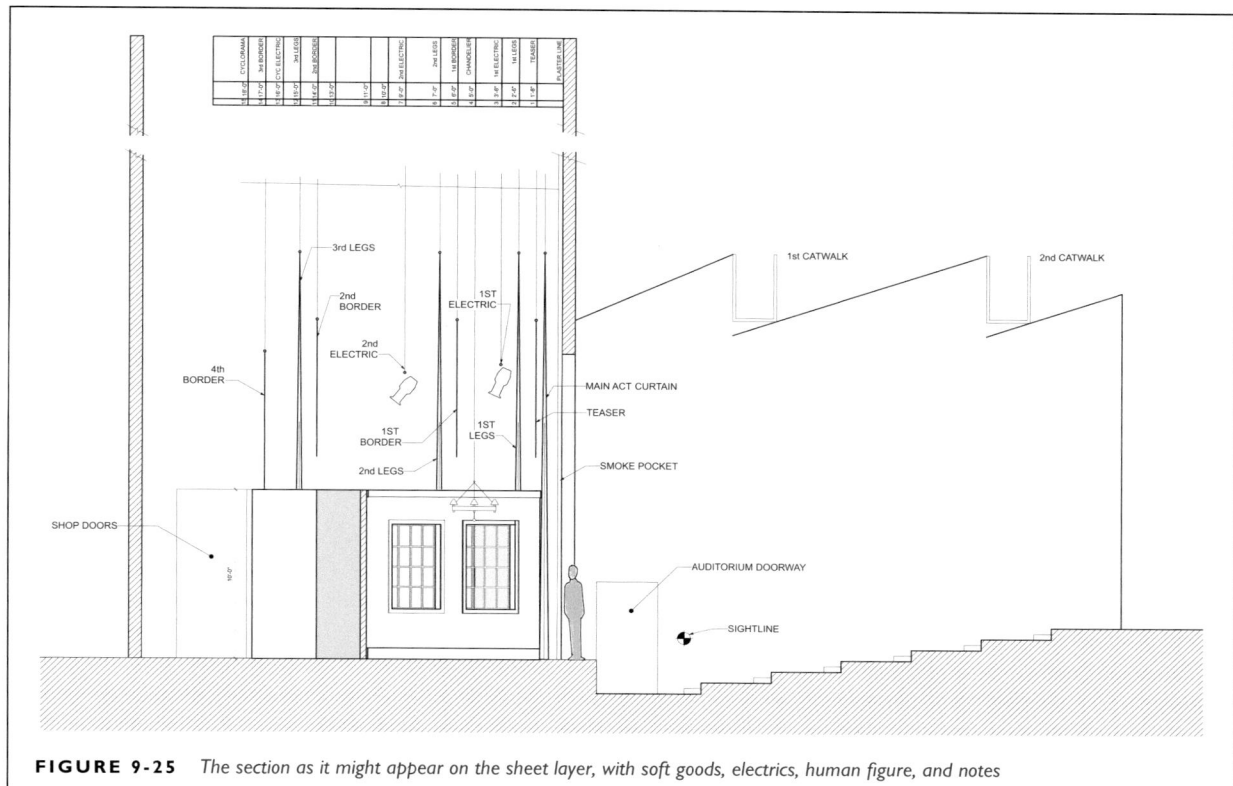

FIGURE 9-25   *The section as it might appear on the sheet layer, with soft goods, electrics, human figure, and notes*

28. Add a human figure standing on the deck. If desired, add chair profiles to each tier in the seating area (Figure 9-25).

29. Switch to the Section Dimensions and Notes Layer. Sections are usually even more lightly dimensioned than groundplans, perhaps including the full heights of major scenic elements. In this exercise, the height of the walls will be dimensioned. Hanging elements are labeled with leader lines; trim heights are often included with the label if they're not noted on the line schedule. Architectural doors are labeled (Shop Door, Hallway, etc.) if visible. If front-of-house (FOH) lighting positions are included, label them. Review the drawing. Add notes and labels for anything else that seems necessary (Figure 9-25).

30. Create a viewport, checkmarking the section layers for visibility. What scale will fit best on an Arch D sheet? What, if anything, has to be cropped to create the sheet layer in ½″ = 1′ –0″? Is shrinking the sheet layer to ¼″ = 1′–0″ a better option than cropping? A scenic section typically features only the onstage portion of the venue unless the scenic environment spills into the house. A lighting section typically requires showing the full venue.

31. Add borders and title block. Fill out title block information. This is Sheet 2. The groundplan is Sheet 1. The total number of sheets is not yet known, though if all sheets have been laid out as rough sketches, an estimate can be made.

32. Save the file. Export the sheet layer as a PDF. Print the PDF.

Homework: Copying an Existing Section

Instructor

This exercise provides further Vectorworks practice, as well as an opportunity to analyze and interpret a section. Distribute printed copies of scenic sections, preferably in ½″ = 1′ –0″. Students should copy the drawing using only the information found on the page; measurements are to be lifted with the scale ruler. Avoid sharing photos or other drawings of the set until complete so that this project focuses on interpreting this drawing. To make this assignment more challenging, the instructor might consider providing sections with missing information so that the students have an opportunity to stretch their interpretive skills even further.

This is also an exercise of paying attention to detail. When evaluating the finished product, look for features, dimensions, labels, etc., that are missing. Ask why the student did not include them. Did they simply miss them? Was the original drawing unclear? Did it seem as though sufficient information was already presented?

If multiple students worked from the same original, compare the range of final products. Discuss similarities and differences.

Students

Draft a Vectorworks copy of the printed section provided to you by the instructor. You can use only the information as found on the drawing; use your scale ruler to lift measurements.

How many layers will be useful and what features will be placed in which layer? Which layer and which feature are the first to be laid out? Do you understand all of the features in the drawing? Are there any mysteries that need to be solved first?

Without a groundplan from which to drop information, what will be the best approach in laying out shapes and features? What details are found on what elements? Are there drawing or graphic options that can add further clarity to the drawing, such as adjustment of line weights, placement of labels, layout of line schedule, etc.?

REVIEW OF CENTER LINE SECTION SYSTEM OF WORKING

- Choose the direction in which to view the scenic environment. Determine the placement of the cutting plane—if not the center line, as close to the center line as possible. Consider what features cross the cutting plane and what masses will be seen as cut through.

- Do rough sketches to aid layout.

- If a base drawing of the venue's section is available, either trace or import it. If it does not face the direction in which you wish to view the space, reverse it. Not all features will be mirrored; consult the venue's groundplan to learn what is seen at the opposite end of the venue.

- Lay out the architecture. Include the seating area and sightlines. If the venue's base drawing includes the line schedule, trace that into the layout as well. Include front of house lighting positions, if present and page space allows.

- Rotate the groundplan, whether hand drafting or in CAD, so that information can be dropped from it into the section. If necessary, copy portions of the rotated groundplan to bring them into closer proximity to the section working space.

- Drop information from the rotated groundplan into the section. Work from the center line toward the direction in which the section looks (stage right or stage left). Add height measurements and establish the outlines of major scenic elements.

- Lay out scenic details within major outlines that can be seen from the cutting plane. If in Vectorworks, darken as you go to keep things straight. Erase guidelines that are no longer needed.

- Lay out hanging features such as legs, borders, act curtain, practicals, electrics, drops, etc. Double-check sightlines to finalize trim heights.

- Lay out dimensions, notes, and labels. Sections tend to feature more labels and notes than dimensioning.

- Darken the drawing.

- Darken borders and title block. Fill in title block information.

NOTES

1. Patricia Woodbridge and Hal Tine, *Designer Drafting and Visualizing for the Entertainment World*, 2nd ed. (New York, NY and London: Focal Press, 2013), p. 85.
2. J. Michael Gillette, *Theatrical Design and Production: An Introduction to Scenic Design and Construction, Lighting, Sound, Costume, and Makeup*, 6th ed. (New York, NY: McGraw-Hill, 2008), p. 130.

CHAPTER 10
THE SCENIC FRONT ELEVATION

The third major orthographic overview of a scenic space is the front elevation. The first module discusses how this drawing may be used. The second module returns to the SketchUp model. The third module constructs a front elevation of the box set developed in the previous chapters, transferring information into it from the groundplan and center line section.

TOPICS AND GOALS

• What is a scenic front elevation?

• Essential elements of a front elevation

• Reading a front elevation

• Adding further detail to a set's model

• Drafting a front elevation

MODULE 1: HOW IS A FRONT ELEVATION USED?

What is a Scenic Front Elevation?

In theatrical usage, "elevation(s)" usually refers to the collection of individual elements' detail drawings. Even though a full stage elevation is not typically offered as part of a standard scenic design package, drafting a full stage elevation is included in this series of exercises to provide further practice in orthographic projection and promote the ability to visualize a scenic environment (Figure 10-1). A full front elevation remains useful when the project is a standalone structure in a larger space such as a convention floor or film studio; the side and front elevations work together to fully describe the structure. If the student designer has yet to master perspective, or does not have access to virtual modeling software, a front elevation can

DOI: 10.4324/9781003154921-11

be a useful stand-in for those images. It can serve as the basis for paint elevations and the lighting designer can use it to consider lighting angles. When the full stage elevation is generated before detail plates, it can inform the content of those drawings.

This drawing comprises the front view of the entire scenic environment as situated within the venue. While it, too, is drawn in parallel projection, it's the most pictorial of the three overview drawings.

For students, generating a scenic front elevation can be a step in establishing the visual relationships of the various scenic elements. Flown items are drawn at their trim heights; a human figure provides a sense of proportion. While the rough groundplan is often generated first, the placement of elements can be informed by decisions made when developing a front view of the structure. As noted earlier, in these exercises the center line section was generated before the elevation because it is a required drawing in the package; constructing the elevation first can provide information that makes for a more efficient section view layout process.

> A **paint elevation** is an elevation of a scenic element with color and texture added to reflect the designer's intentions. To accommodate planning and layout of the painted surfaces in the shop, the paint elevations employ standard orthographic views and are constructed in scale. Paint elevations can be developed through older physical media techniques like watercolor and colored pencil, or generated via paint programs like Adobe Illustrator or Corel Painter. Textures and colors can be imported into SketchUp and Vectorworks, but the two programs offer only limited ways to modify or manipulate them. Create the desired surface quality and *then* import it into the drawing or virtual model.

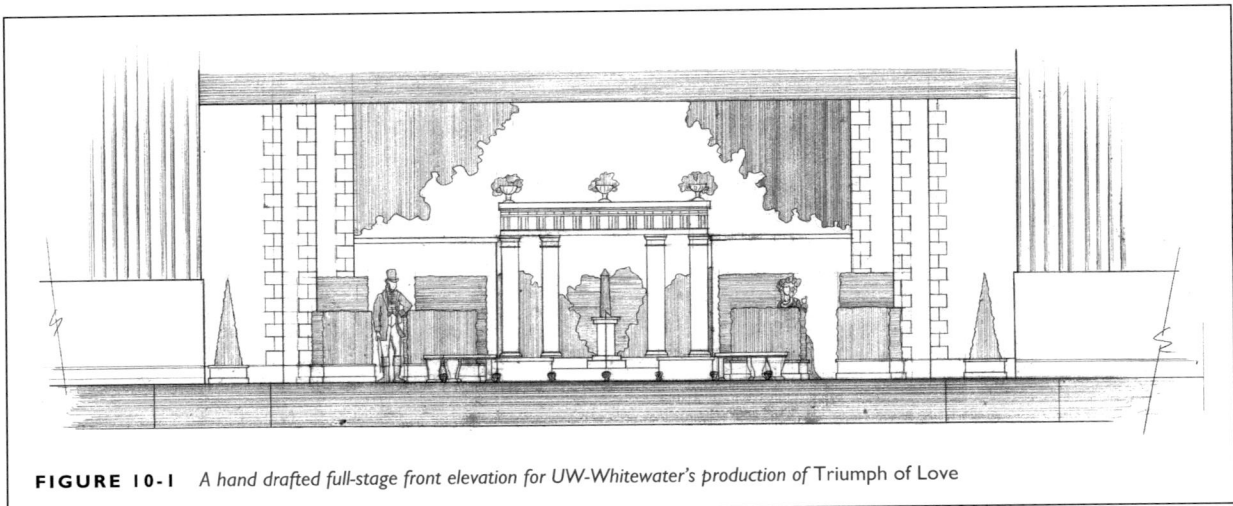

FIGURE 10-1   *A hand drafted full-stage front elevation for UW-Whitewater's production of* Triumph of Love

Parts of a Scenic Elevation

As a more pictorial drawing, a scenic elevation has more variability in its components. How the scenic structure is seen is determined by how it fits within the venue. If in a proscenium theatre, the set will be framed by the proscenium, act curtain, tormentors, and teaser. The front of the apron will be seen from the deck down to the auditorium floor. In a black box or television studio, the lighting grid may be the only architecture that frames the scenic environment. Depending on the venue's width, the view may include seating banks and/or the walls of the space.

The elevation should include all permanent details found on vertical surfaces, such as molding and sconces. Including items like curtains, paintings, and other set dressing is at the discretion of the designer. In Figure 10-2, a piano and dining room table were left off the drawing, as they would obscure the bay window structure at the upstage end of the playing space.

Flown scenery is shown at the height at which it plays during performance. Unless all flown scenery plays at the same time, including it all on an elevation can be confusing. The drawing should represent a single moment in the course of the production. If showing the stage in various configurations is required, draft multiple elevations.

Restrict the elevation to what an audience member would see, given the rules of orthographic projection. Masking blocks backstage areas, battens, and electrics unless it's the designer's intention these should be visible. Drapery with fullness can have just enough drawn texture so that it does not read as a flat surface. Include enough of the proscenium wall or seating area to provide context, particularly if the scenic environment spills beyond a venue's standard performance boundaries. Include a human figure. If the environment features multiple levels, include multiple figures.

Dimensioning is even more limited than what is found on a section: proscenium/grid height, and perhaps the overall

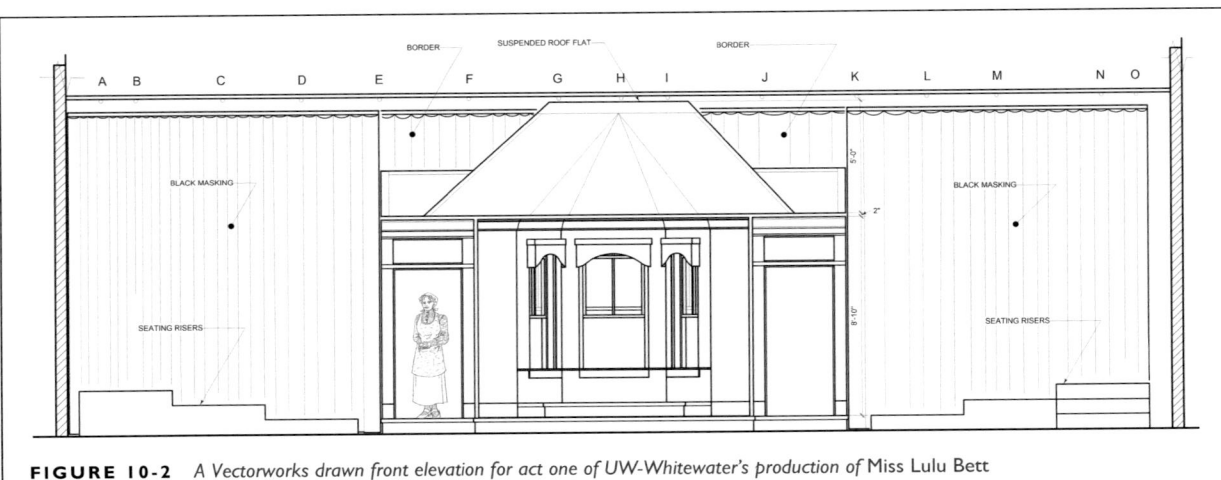

FIGURE 10-2   *A Vectorworks drawn front elevation for act one of UW-Whitewater's production of* Miss Lulu Bett

height of the scenic structure. Labeling, too, is limited, as the features of the drawing should be evident to the reader. Judicious shading can be used to push some objects and surfaces into the background and bring others into the foreground. Resist adding color to shading as that begins the process of turning the drawing into a paint elevation.

MODULE 2: RETURN TO THE BOX SET MODEL

The SketchUp model of the box set within the venue is largely complete, except for set dressing, colors and textures, and desired furniture pieces (Figure 10-3). Open the file and look at the model using parallel projection, a cutting plane situated in the first or second tier of seating, and the standard front view. It will be useful to export a 2D graphic of this view for later reference.

How much of the set is hidden behind the proscenium and masking? How much of Flats E and F are visible through the archway and windows? How much space is visible between the top of the set and bottom of the borders? Since Wall D is at 45° angle to the plaster line, the window frames will look much like they do in the section, only reversed. What interior surfaces of the archway and window openings (the reveals) can be seen?

Exercise: Dressing the Set Model

Instructor

Provide a list of set dressing and furniture pieces for the students to locate in SketchUp's 3D warehouse. If you wish to be more specific about the appearance of items, provide photos and sketches to guide the students' search. This exercise is primarily about exercising aesthetic discretion in using a resource like the 3D warehouse, as well as the finesse required to successfully place these items within a virtual model.

FIGURE 10-3 *The box set model (without set dressing) seen in parallel projection, front standard view, with the cutting plane in the front row*

Students

1. Use the Edit Group function to add colors and textures to the set with the Paint Bucket tool. If color is added to a group without first going into Edit Group, all surfaces of the group will be altered. Add color to surfaces seen by the audience. Add a texture to the floor of the playing space. Add color or texture to the front of the proscenium wall and front of apron. If you are using the model of an actual venue, try to match its colors and textures as closely as possible. Remember that image files of colors and textures can be imported and edited via the Paint Bucket tool.

2. If your instructor has not provided you with one, make a list of furniture and set dressing items that might be found on this set, such as rugs, sofas, wall art, curtains, etc. If you have not already found a chandelier, include that in your list. Go into the SketchUp online 3D warehouse via the File drop-down menu to search for these items (you will need to set up an account in order to do so). Download the selected items into your model. Scale items appropriately and place as desired. Move the human figure back into the playing space. Remember that the more items you import and the more complex the items are, the harder your computer's processor will have to work. The program may begin to pause more often before executing an action.

   Return to parallel projection, front view, reestablishing the cutting plane if needed. Export a 2D graphic of the dressed-up model to use as later reference.

3. SketchUp has an option to view a model monochromatically (in grayscale). In the View menu, there are options for Face Style, among them the Wireframe choice explored in an earlier exercise. Monochrome is another choice. Choose it. Take a look at the front view of the set again. Is it easier to see the outlines and shapes of objects? Export a 2D graphic of the monochrome view for later reference. Take a look at the model in wireframe format. Is this choice at all useful? Why or why not? Explore the Edge Style and Face Style options to see what they do to the model.

4. Save the file. If you wish to keep a basic version of the model as well as the dressed version, Save As with a new file name.

Set dressing will be drafted into the groundplan, section, and elevation as a homework project at the end of Module 3.

MODULE 3: DRAFTING THE ELEVATION

Almost all of the environment's information, minus set dressing, has already been added to the groundplan and section. All that remains is to transfer this information into a front view.

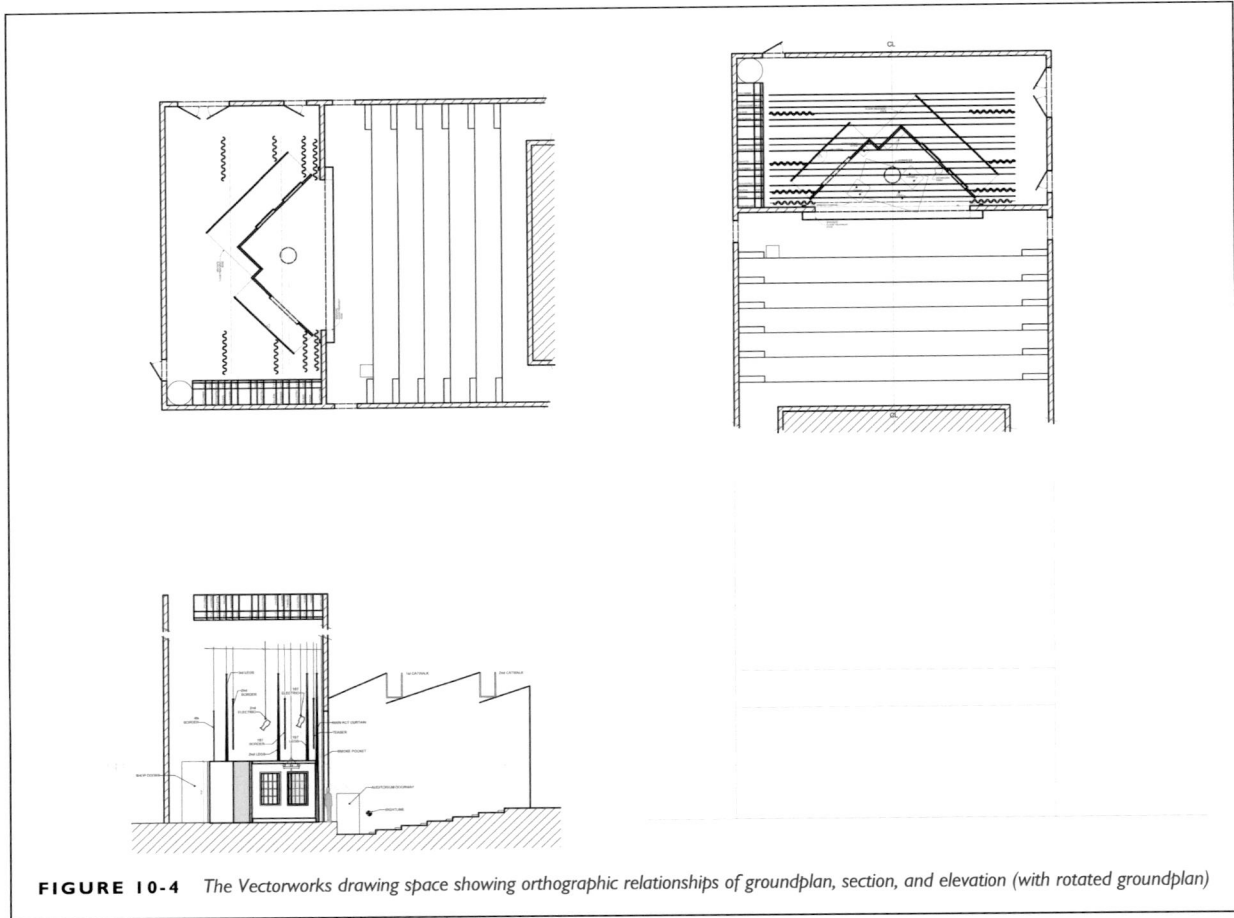

FIGURE 10-4   *The Vectorworks drawing space showing orthographic relationships of groundplan, section, and elevation (with rotated groundplan)*

The front elevation should be situated below the groundplan and to the right of the center line section (Figure 10-4). The rotated groundplan is no longer needed, and its design layer may be either turned off or deleted. Clean up the guideline layer.

The elevation is not a drawing that benefits from being constructed in a number of layers. As was explored while constructing the section, composing and filling elements and sending them to the front or back is often as much organizational complexity as the drawing requires. The purpose of this drawing is to see the set as a whole, so there's little reason (in this exercise, at least) to go beyond adding a layer for the venue and a layer for the set itself.

1. Reestablish the deck baseline and outer architectural edges as needed. Outline the proscenium opening by dropping guidelines from the groundplan and extending the height over from the section. Outline the front of the apron and the lowest floor level of the seating area (Figure 10-5).

2. In the elevation's architecture layer, darken outlines. Since the auditorium's side walls are cut-through, compose and fill them. Include a break line at their upper terminations, if they do not extend the full height of the auditorium (Figure 10-5).

3. Clean up guidelines. From the section, extend a horizontal guideline to mark the top edges of the onstage flats.

    From the groundplan, drop vertical guidelines from all edges and corners of the flats that can be seen within the proscenium opening. If the flats were made a group to copy them into the Rotated Groundplan, go into the Flats layer, make another copy and drag it closer to the elevation so you don't have to extend guidelines through the full depth of the auditorium (Figure 10-6). Clean up guidelines as needed.

    Drop down the placement of the archway and windows. The window heights can be added from your notes directly to the elevation, or transferred from the section. Don't worry about window frame details yet. The archway is 7'–6" high. Remember that the archway and windows are seen at an angle, meaning that upstage reveals will be visible.

4. From your notes or from the section, lay out the top and bottom edges of the baseboard and crown molding. The molding is ¾" thick; so add a guideline to

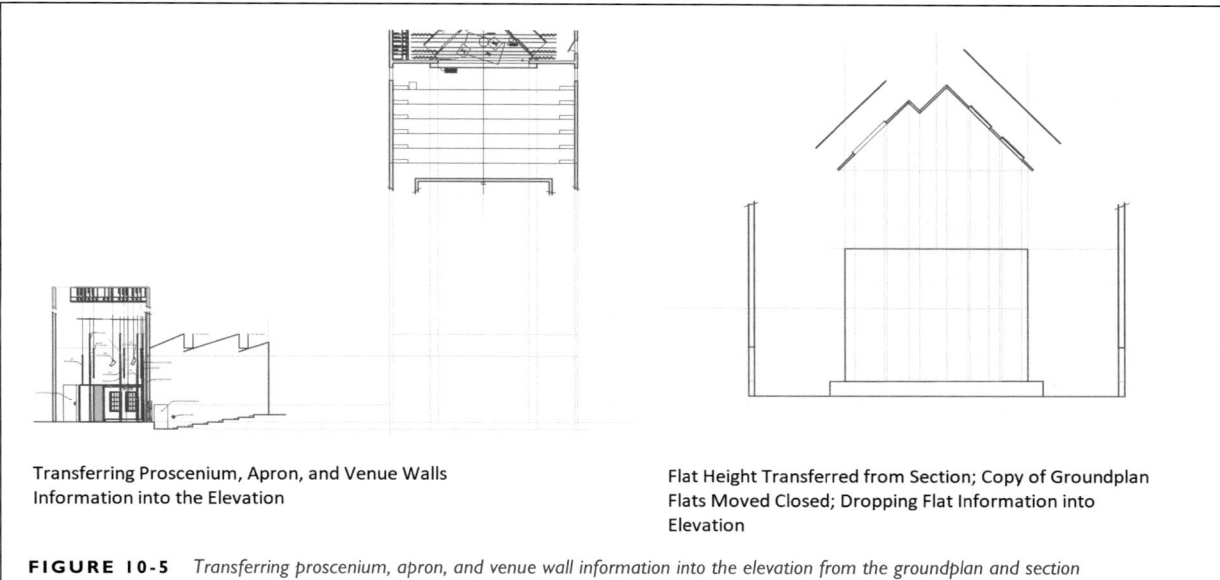

Transferring Proscenium, Apron, and Venue Walls
Information into the Elevation

Flat Height Transferred from Section; Copy of Groundplan
Flats Moved Closed; Dropping Flat Information into
Elevation

FIGURE 10-5 *Transferring proscenium, apron, and venue wall information into the elevation from the groundplan and section*

the groundplan ¾″ downstage of the wall's front edge. Terminate the molding ends to align with the ends of the walls. Mark the window side casing widths (as was done in Figure 9-22) Mark where the baseboard meets the archway along Flat A.

Drop guidelines from the groundplan into the elevation from the corners and endpoints of the molding. The baseboard and crown molding are both 1 × stock, so these points are identical for both, with the exception of the archway. Only drop down those corners and endpoints seen within the proscenium opening. Remember that downstage oriented ends will be seen and represented by a very narrow set of double lines. Since the flats are oriented at 45° to the picture plane, the some of the molding joints may exactly align with corners of the flats.

5. In the front elevation design layer, darken the walls and molding with appropriate visible lines. If you wish, use filled rectangles and Send them to the front or back. Clean up guidelines.

6. Since the walls are oriented at 45° to the proscenium, the window framing can be copied from the section, cleaned up, flipped horizontally, and inserted into the elevation. Select the section's complete window and Copy. In the elevation design layer, Paste the copy into a clear area of the drawing. Trim any lines beyond the window frame's rectangle that were also copied. While the copy is active, open the Rotate option under the main tool bar's Modify drop-down menu and choose flip horizontally. Make the window a group (also under the Modify menu).

Dropping Molding Information from the
Flat Plan Copy into the Elevation

Darkening the Flats and Molding

Copying a Window from the Section, Flipping it Horizontally and
Dragging it into Place; Copying Again for Second Window

FIGURE 10-6 *Dropping molding information into the elevation; darkening the flats and molding; copying and flipping windows from the section*

You'll need two copies of the window, so select the reversed window again, hold down Ctrl, and drag to one side to create a copy. Make the copy a group. Drag each window into position.

7. If you drew a chandelier in the section, Copy and Paste it into the elevation. Extend guidelines from the section to situate it at the correct height. Check to see if the ends of Flats E and F are visible through the archway or windows.

8. From the section, extend a guideline from the bottom edge of the teaser. If any borders play lower than this line because of sightline adjustments, extend those bottom edges over as well. Trim any portions of the chandelier that are hidden behind the teaser (or adjust its trim height). Check onstage edges of the legs to see if any are visible within the proscenium.

9. In the elevation layer, darken the bottom edge(s) of the borders. If adding shading to the soft goods helps clarify the drawing, compose and fill those shapes. When shading, it's often more effective to have the shading lightest in value in the foreground (Figure 10-7).

10. Set dressing and furniture may be added at this point. The 2D graphic exported from SketchUp can be imported into a design layer so that the front elevations of individual items can be traced in another layer. You can also export 2D graphics of each item separately. Trace, Compose, and Fill each object using a white fill. Start by pasting the objects furthest upstage. As you move downstage, Send objects to the front so that they cover parts of the objects upstage of them.

     If you have facility with a paint program, you can import the 2D graphic from SketchUp and clean up the image. Convert the image to grayscale before importing the file into Vectorworks.

11. The venue, set structure, and masking are now complete. Compare to the SketchUp 2D graphic to see if anything is missing; correct as required.

12. Add dimensioning and labels as required. Remember to be minimal.

13. Create a viewport and export into a sheet layer. This is Sheet 3 of the series. Add the border, title block, and title block information. Save the file. Export as a PDF. Print the PDF.

All three primary orthographic overview drawings of this basic box set are now complete. You should now have a reasonable familiarity with not only the appearance of the set and venue, but of the components required to construct

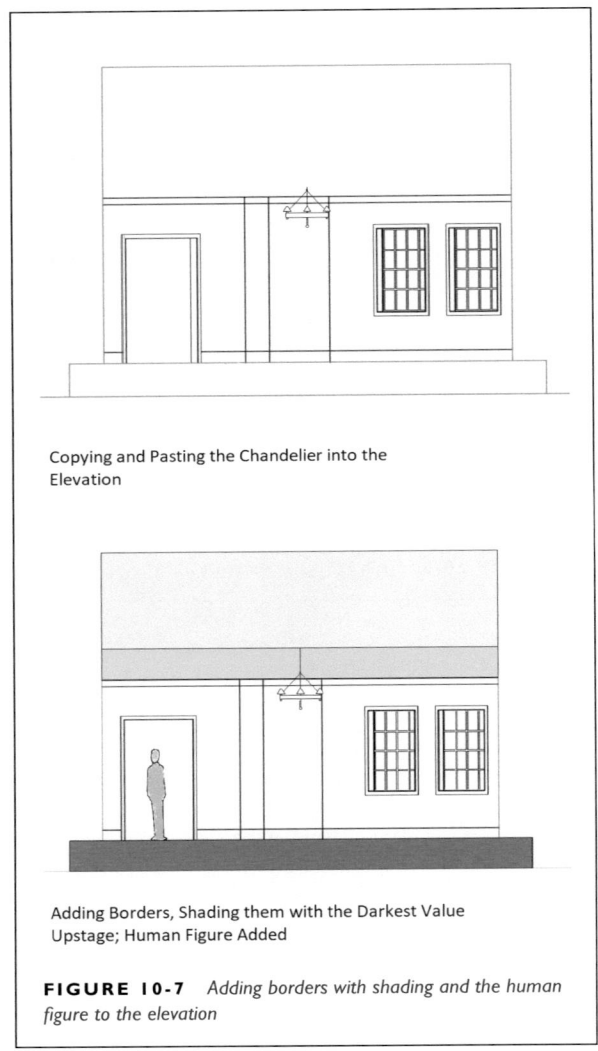

Copying and Pasting the Chandelier into the Elevation

Adding Borders, Shading them with the Darkest Value Upstage; Human Figure Added

FIGURE 10-7   *Adding borders with shading and the human figure to the elevation*

the set. In Chapter 11, the set will be broken apart and detail drawings constructed for all of its parts.

> Once multiple layers are created and it becomes necessary to shuttle back and forth between them, it's easy to lose track of the layer in which a particular object is located. Regardless of the active layer, right click on the object to open the menu and choose either Activate Class or Activate Layer. The drawing will shift to the object's layer or class.

Composite Elevations

When scenic environments feature a continuous series of flats, the designer may construct a **composite elevation** of them. This is another pictorial representation of scenery and useful for describing naturalistic box sets. All of the walls are

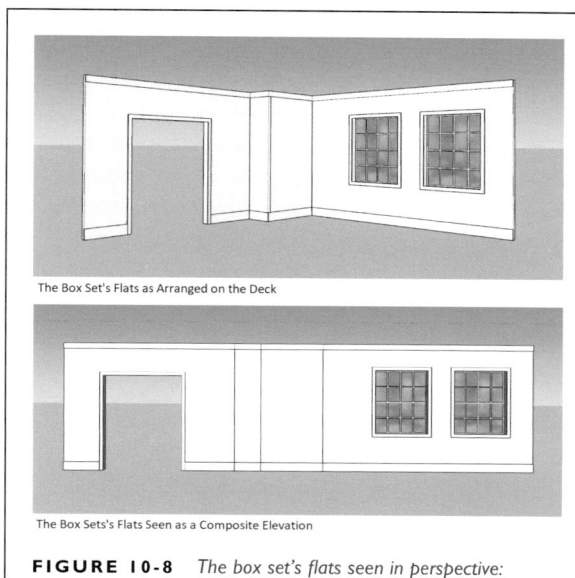

The Box Set's Flats as Arranged on the Deck

The Box Sets's Flats Seen as a Composite Elevation

FIGURE 10-8 *The box set's flats seen in perspective: original configuration and as a composite elevation*

Set dressing and furniture placed directly against walls are often included, as is a human figure. The composite elevation should be as pictorially detailed as possible.

One benefit of a composite section is that each wall, if angled on the stage, is seen without the distortion that arises in a section or full-stage elevation. Detail plates break the wall sequence into individual flats, so a composite can provide an otherwise missing visual overview. It's useful in conversation with other members of the production team and can guide the final dressing of the set after load-in. By printing or photocopying the composite elevation onto cardstock or Bristol board, it can be cut out and folded at the seams to create an "instant" physical model (Figure 10-9). Conversely, a rough composite elevation can be drawn, cut out, folded, and stood up to see what it looks like; the designer can make revisions, then scan and import the rough composite into Vectorworks. This sketch can then be traced, cleaned-up, and adjusted to develop drafted elevations. Detail plates and component elevations will be discussed in depth in Chapter 11.

Exercise: Sketching Individual Flat Elevations

Instructor

To prepare for drawing a composite elevation, the students should construct sketches of the box set's Flats A, B, C, and

drawn as though they inhabit the same plane, parallel to the viewer. The wall sections are joined at their seams to present adjoining walls as a continuous surface (Figure 10-8). This type of elevation is typically drawn in a smaller scale, such as ¼″ = 1′–0″.[1]

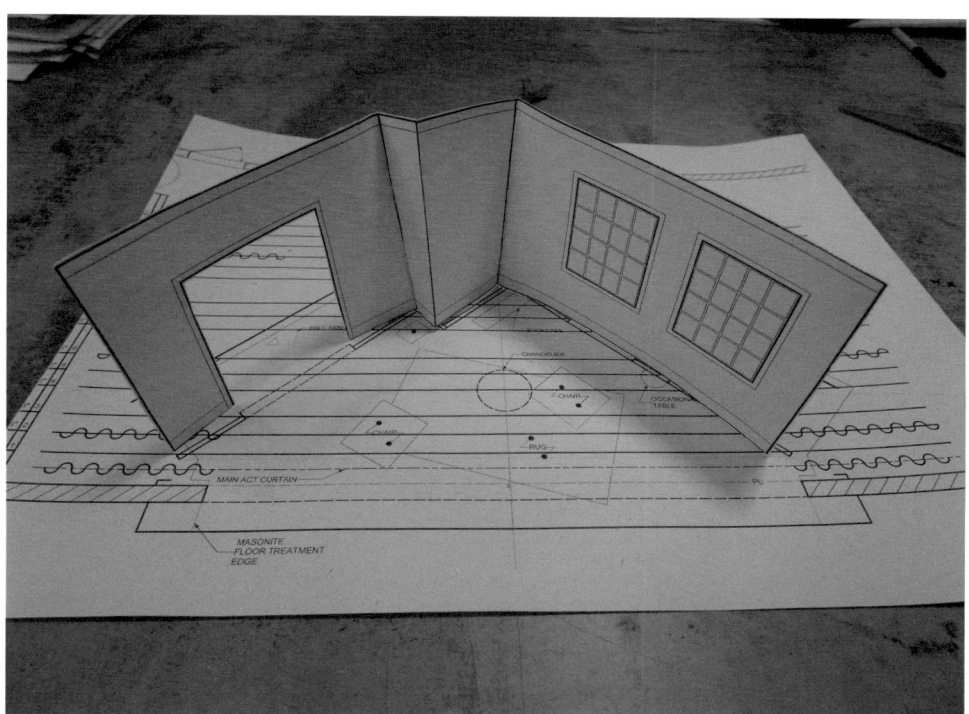

FIGURE 10-9 *Cutting out and folding a composite elevation to create an instant white model*

D. Since undistorted elevations of the box set's flats have not yet been drawn, information will need to be assembled from the groundplan, section, and elevation. The student may sketch the elevation on graph paper, using straightedges. ½″ = 1′–0″ is suggested.

Students

Using information from the box set's groundplan, section, and elevation, as well as the rough sketches accompanying previous exercise, sketch elevations of Flats A, B, C, and D. ½″ = 1′–0″ is suggested; draw on graph paper. Include all details, such as molding and window structure. Note dimensions so they will be available when you draft these flats as a composite elevation.

Exercise: Composite Elevations

Instructor

The students should use the elevation sketches drawn in the previous exercise to draft their composite elevation. This may be done either by hand or in Vectorworks; a printed scale of ½″ = 1′–0″ is suggested. Discuss how molding tracks from one flat to the other and what will be the preferred way to represent it.

Students

Draft the box set's Flats A, B, C, and D as a composite elevation. Include molding and window frame detail. If set dressing has been chosen, include any which hangs upon or sits against the walls. Consult the model and research images to draw set dressing items. Use Arch D paper. If hand drafting, consider overall layout measurements to guide scale selection. In Vectorworks, draft in ½″ = 1′–0″ and select print scale at viewport creation. Include borders and title block. This drawing can be considered as Sheet 4 in the series.

Using information from the groundplan, section, and elevation, construct sketches of each of the four flats before drafting them in a continuous row. Joints between flats are indicated with a vertical line. Molding that wraps around corners may be shown as being continuous, despite the angles and overlaps that would occur if the molding sections (and flat joints) were truly represented. Making molding continuous makes for a more aesthetically cohesive elevation, but one that is less truthful about the construction details. If this elevation is about paint, then continuous is likely fine; if it's about construction, accuracy regarding the molding ends and overlaps will be preferred.

All flats in a composite elevation are shown with their front surfaces parallel to the picture plane. Objects placed on or in front of the flats are also shown in this orientation. For objects that were traced for inclusion in the scenic elevation,

it may be necessary to return to the model to export and trace 2D graphics that view the object from this new position.

All four flats share a common baseline, as well as a top edge. Lay out the perimeter of the composite elevation by adding together the widths of all four flats; divide the large rectangle into its component sections. Lay out openings. No reveals will be visible, since they are perpendicular to the picture plane. Lay out the crown molding and baseboard. Lay out the window frames and muntins. Add set dressing. Review the drawing and darken. Object perimeters should be darkened with thicker lines, interior details with thinner lines.

Exercise: SketchUp Paint Elevations

SketchUp can also be used to construct a composite elevation. In the early stages of model planning, consider how you intend to use the model in later stages. If you think you'll use the model for paint elevations or to add detail images on the drafted sheets, use groups in a way that allows you to easily extract individual features. This may entail building two models: an experimental model in which you work things out, and a presentational model in which individual elements are built in a manner that more closely aligns to construction practice.

1. Open the box set's model. Copy Flats A, B, C, and D. Move the copies off to one side of the model. You can also Paste them into a new file if you wish to create a separate file for details. If the set's walls had been constructed as a single structure, moving the walls and unfolding them into a single plane would require much more editing and revision.

   Since the four flats are constructed individually and each wall is a group, it's a relatively simple process to Rotate each section of wall into alignment. The crown molding is constructed as an independent group and will need to be reestablished in the composite; the baseboard, however, travels with its assigned flat (Figure 10-10).

   Since the flats overlap at the corners, some flats will be too long when reoriented. SketchUp allows masses to reside in the same space, so you can drag one flat into another until their downstage surfaces align properly. This may, though, leave an extra line to one side of the true join. Open Edit Group and then make adjustments so that the seam between the flats is the only visible line.

2. Once the flats are aligned as a continuous surface, draw the crown molding profile at one end and Push/Pull it across the elevation's full extent. Make it a group.

3. Copy and Move any furniture or set dressing that hangs upon or sits directly in front of the flats. It may be necessary to add a floor plane to set furniture upon. When the elevation is viewed in parallel projection from its front, this floor will match the baseline of the flats and disappear from view.

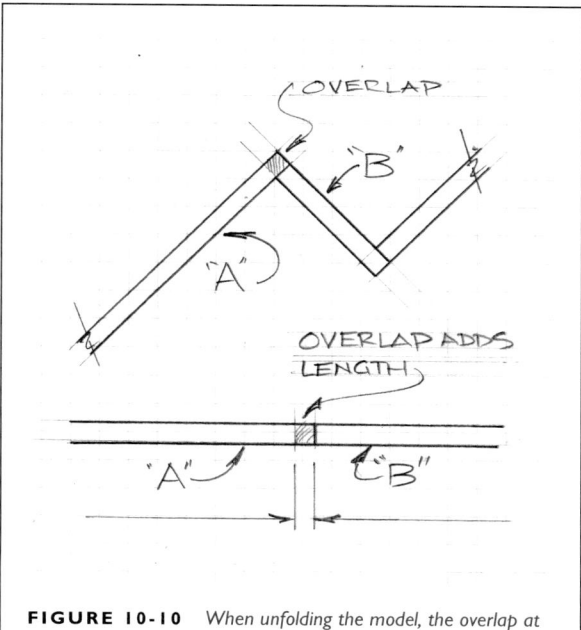

FIGURE 10-10   *When unfolding the model, the overlap at corners must be edited so the composite elevation shows their true widths*

4. Once set dressing has been arranged, export a 2D graphic. If final colors and textures have been applied and the printer's ink colors match the selections, this image can be submitted as a paint elevation (Figure 10-11).

    Paint elevations for individual walls can also be constructed this way. Rather than joining the walls together to make a composite elevation, keep them separate but all facing the same direction in the same plane. Export 2D graphics of each flat. If you have a large format printer or plotter, you can arrange the individual elevations as they would appear on a single sheet of paper Add dimensions to aid paint layout in the shop. Save that array as a graphic, then print on the desired sheet size.

While SketchUp offers a wide range of colors and textures, as well as the ability to import textures and create custom colors, it's difficult to create the shading and nuance required for more advanced paint treatments. You can construct paint elevations elsewhere and then import and paste them onto a virtual model, but that process is more about constructing a detailed presentation model rather than using the experimental model to facilitate in the creations of paint elevations.

Many designers export graphics from SketchUp and Vectorworks into other paint and illustrator programs to become foundation drawings for paint elevations, or as the base line drawing for a perspective illustration. Figure 10-12 shows a plate of paint elevations created by exporting Vectorworks drafted elevations into a paint program. Figure 10-13 shows the front view of a SketchUp model which was exported as a graphic and then used as the foundation drawing for a colored perspective rendering.

Homework: A Different Set in the Venue

Instructor

A set of round platformed areas was drawn and modeled in Chapter 6 (Figure 6-44). In this project, those platforms are used to draw a unit set placed in the venue used for the

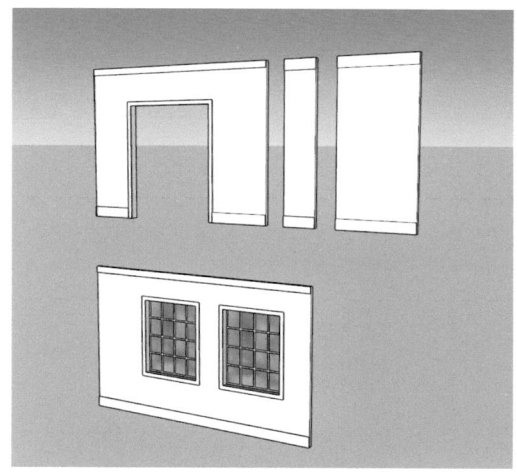

Individual Flats Arrayed for a Sheet of Paint Elevations

The Array of Flats Seen in Front View, Parallel Projection

FIGURE 10-11   *The box set's flats arranged as an array in SketchUp to be exported as a sheet of paint elevations*

FIGURE 10-12 *Vectorworks graphic used as a foundation for a paint elevation drawing for the University of Cincinnati, college-conservatory of music's production of* Guys and Dolls; *designed and drawn by Thomas C. Umfrid. Used by permission of Thomas C. Umfrid*

FIGURE 10-13 *Study for a proposed scenic design for the Hong Kong Academy for Performing Arts' production of* The Magic Flute; *left: 2D graphic of SketchUp model of proposed set; right: Perspective rendering using 2D graphic as foundation drawing; designed and drawn by Thomas C. Umfrid. Used by permission of Thomas C. Umfrid*

previous exercises. A groundplan, center line section, and front elevation should be drafted. Provide a sketch situating the platforming along the venue's reference lines, as well as masking suggestions. Include a list of furniture and set dressing items for a SketchUp 3D warehouse search; you can also provide prepared 2D images and graphics. Make a choice as to what will be seen upstage: cyclorama, traveler, drop, or naked back wall.

Students

Using the round platforming drawn in one of the exercises of Chapter 6 (Figure 6-44), construct a groundplan, section, and elevation. If your instructor provides a placement sketch, use it; if not, determine placement along the center line on your own (don't block the fire curtain). Use sightlines to determine the placement of masking. Your instructor may also provide a list of set dressing items to search for in the 3D warehouse.

If you have not already saved the Vectorworks base drawings as a file without the box set information, open the file and Save As with a new name. Then, delete all of the design layers used for the set and masking. In the Line Schedule layer, delete information specific to the previous project, such as the chandelier. Save again. Alternatively, though less efficient, there are two more options:

a. Go into the design layers and delete the all box set information. Save the file with a new name.

b. Turn off the visibility of the previously used set and masking layers so that multiple sets exist within a single file. Be sure that the names of new layers for the new set clearly differentiate them.

In Vectorworks, the views of the platforms can be copied from the original exercise and pasted into their respective drawings. Be sure their alignment is consistent from drawing to drawing. If you paste the top view into the groundplan first, drop guidelines into the section and elevation with which to align those views.

Use sightlines drawn from the sightline symbols to place stock legs and borders in their optimal locations. Revise the line schedule to reflect the new information. No set walls prevent the audience from seeing the venue's back wall, so a full-stage painted drop, cyclorama, or traveler are options. If the naked back wall is chosen, be sure to include a note so that readers don't think a scenic element was forgotten.

If a set dressing search list was provided, go to the SketchUp online warehouse and download items into a file. Look at the items in parallel projection and export 2D graphics that can be used to create images for the section and elevation. Remember to include a human figure in the section and elevation.

Save the file. Import the drawings into sheet layers. Export as PDFs. Print the PDFs.

REVIEW OF SCENIC FRONT ELEVATION SYSTEM OF WORKING

Figure 10-14 presents a further example of a hand drafted front elevation.

• Transfer information from the groundplan and section into the elevation following the standard orthographic drawing process.

FIGURE 10-14   *Hand drafted front elevation for UW-Whitewater's production of* On The Verge

- Lay out the venue's architectural perimeter.

- Lay out the performance space perimeter, such as the proscenium opening. Remember that downstage masking may hide parts of the scenic environment.

- Lay out the outlines of major structural features, such as walls and platforms. Transfer height information from the section. Transfer width information from the groundplan. Features hidden behind masking, architecture, or other scenic elements are not drawn.

- Lay out structural features within major elements, such as doors, archways, windows, etc.

- Lay out details, such as molding and window frames.

- Lay out remaining masking, including borders.

- Lay out set dressing and furniture that is to be included.

- Lay out minimal dimensioning and labels as required.

- Review the drawing for completeness. Darken the drawing. Shade selected areas, such as masking, if it will bring additional clarity to the drawing.

- Add borders, title block, and title block information.

NOTES

1. Dennis Dorn and Mark Shanda, *Drafting for the Theatre* (Carbondale, IL: Southern Illinois University Press, 1992), p. 175.

CHAPTER 11

DETAIL PLATES

In this chapter, the first module discusses what is found on detail plates and how they are generated as part of the drawing process. In the second module, detail plates for the box set used for Chapters 7, 8, and 9 are discussed and laid out. The final modules discuss the parts, construction, and drafting of a double-hung sash window, a freestanding Palladian window with a window seat, and staircases as examples of further detail drawings.

TOPICS AND GOALS

- Organization of detail plates

- Types and parts of windows

- Parts of stairs and staircases

- Laying out stairs on a groundplan

- Describing staircases, banisters, and balustrades

MODULE 1: WHAT ARE DETAIL PLATES?

In the typical scenic design drawing package, the groundplan, section, and elevation (if included) are followed by detail plates. These drawings break down the scenic environment component by component, starting with major structural elements and progressing toward details (Figure 11-1). Depending on the project, drawings for masking and props may be included. In short, everything that must be fabricated or procured should have a drawing. Even if stock items are available, a drawing identifies, describes, and integrates that item into the whole.

Designers need to understand how the parts of a scenic environment fit together so that construction can fulfill esthetic goals. A designers' drawing package may provide

construction suggestions, but construction decisions are ultimately made via discussion with the technical director and shop technicians and rely upon their knowledge and discretion (Figures 11-2 and 11-3). Still, the deeper a designer's technical knowledge becomes, the more that knowledge can inform the esthetics of the design. Imagining something doesn't mean it *can* be fabricated, or fabricated within the constraints of labor, time, and budget. If a design includes features that "push the envelope," the designer must research the technical solutions that allow the push to be successful. Have conversations with the shop before finalizing the drawings so that the shop isn't tasked with creating the impossible, but rather with following the designer's informed lead in crafting technical solutions that fulfill a clearly defined goal.

Drafting the detail plates generally follows the construction of the three overview drawings. However, preliminary sketches of the components should be in development while the overall look of the environment comes together. Laying out platforming on the groundplan can't be done unless the platforming has been planned. Heights of walls must be determined before the section or elevation is drafted. In the detail plates, features that could not be shown on the groundplan, section, or elevation are now included. It may even be the case that earlier drawings see minor revisions due to discovery as the detail plates are drafted. This is a major reason why a drawing package is typically submitted as a whole, rather than progressively or piecemeal. If construction begins based on the groundplan alone, it may not be possible to integrate alterations and discoveries made while drafting the platform detail plates. Be sure that information is consistent across the drawing package so that, for instance, the groundplan and the platform detail plates do not contradict each other.

Preliminary layout planning of detail plates is almost mandatory when hand drafting. In CAD, this planning seems

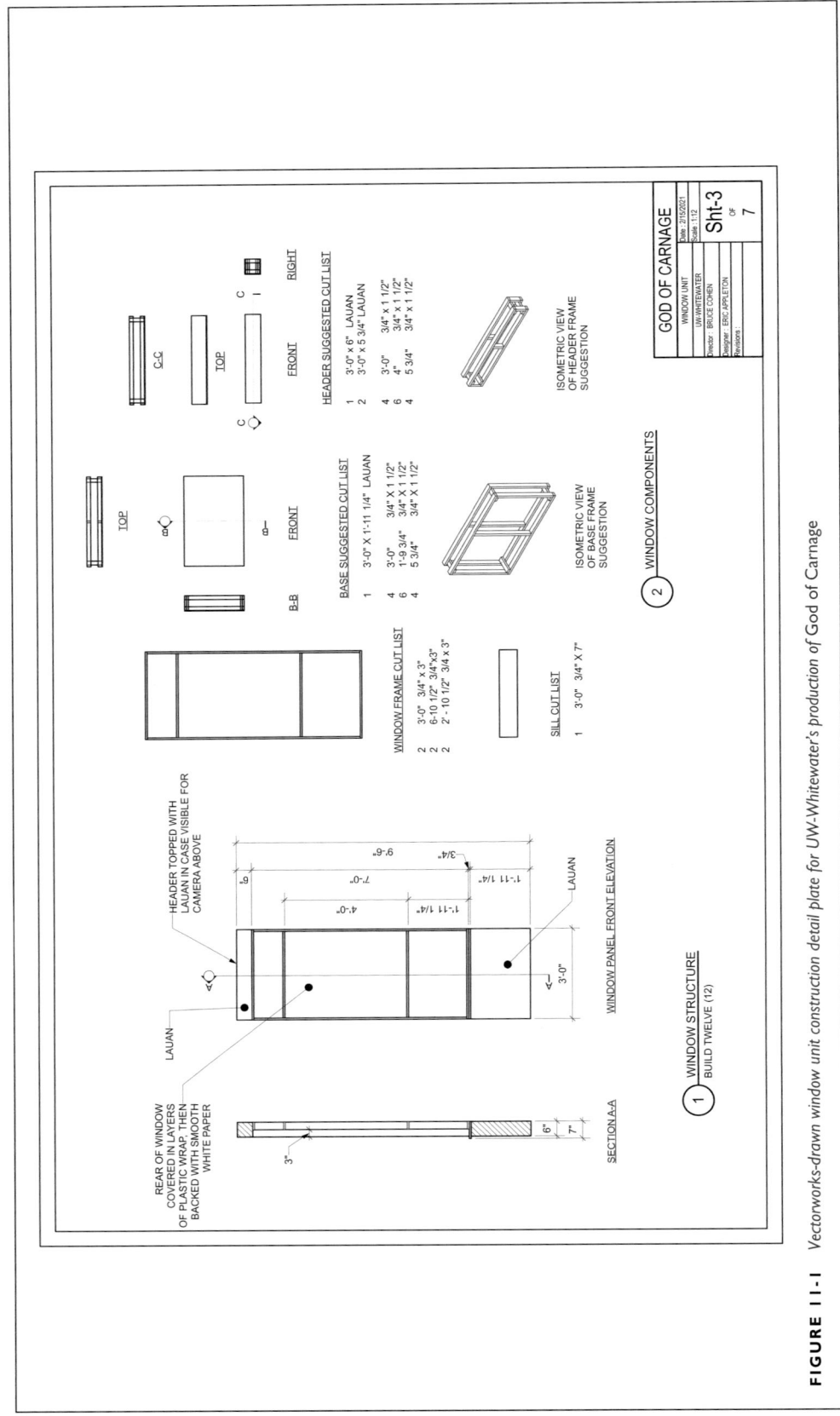

FIGURE 11-1 *Vectorworks-drawn window unit construction detail plate for UW-Whitewater's production of God of Carnage*

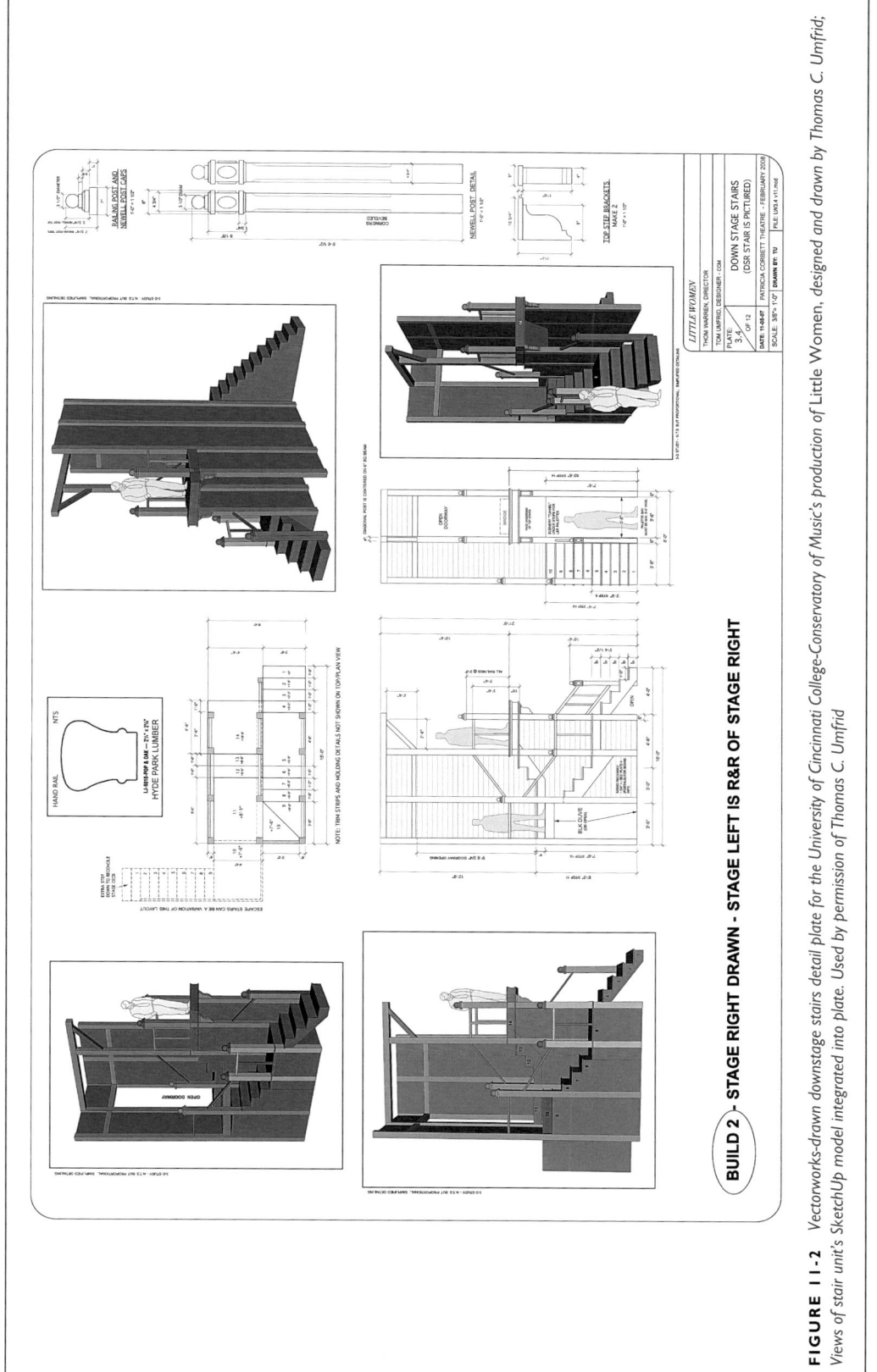

FIGURE 11-2 *Vectorworks-drawn downstage stairs detail plate for the University of Cincinnati College-Conservatory of Music's production of* Little Women, *designed and drawn by Thomas C. Umfrid; Views of stair unit's SketchUp model integrated into plate. Used by permission of Thomas C. Umfrid*

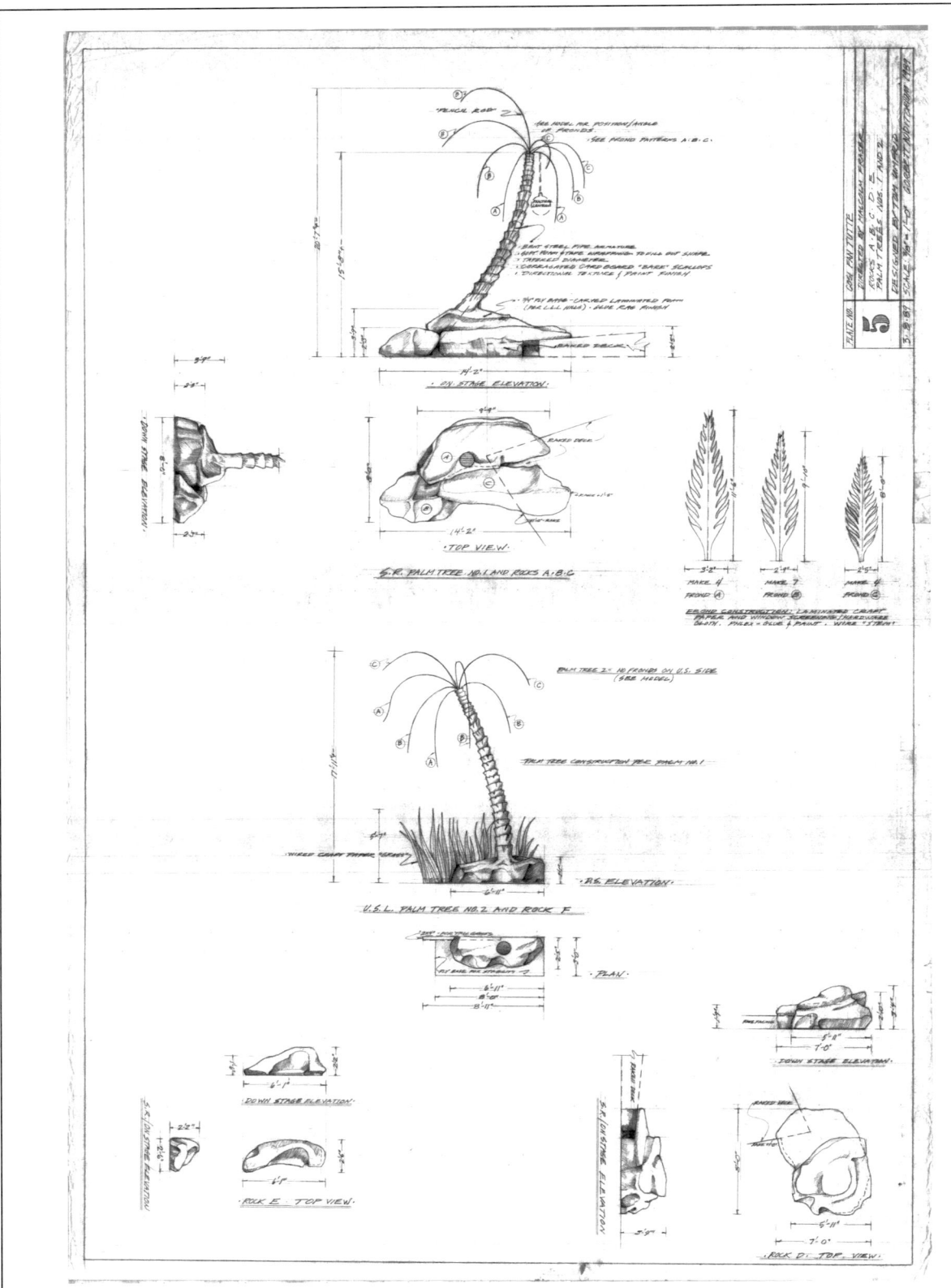

FIGURE 11-3 *Hand-drawn detail plate for rocks A–E, palm trees Nos 1 and 2, for the University of Cincinnati College-Conservatory of Music's production of* Cosi Fan Tutte; *designed and drawn by Thomas C. Umfrid; features organic shapes that are difficult to draft in CAD. Used by permission of Thomas C. Umfrid*

less urgent, as all components can be drafted in the infinite drawing space and then imported into sheet layers one at a time. It is tempting to leave decisions regarding sheet content and layout until the moment of viewport creation; however, this last-minute approach can lead to disjointed page layouts and fragmented information flow when drawings are pieced into place. Keep in mind that sheet content should be thematic, and that white space is not wasted space. When drafting in Vectorwork's infinite space, keep drawings organized and visually related.

Advance detail plate planning has a number of benefits:

- Layout planning helps the draftsperson sort out just what is to be drafted; it's a visual list, and lists are an effective, if not essential, organizational tool.

- It ensures that drawings are not accidentally left out of the package (checking off the items on the list).

- It enables drawings to cross-reference each other as they are drafted: e.g., "For detail, see Drawing 3, Sheet 6."

- Detail plate planning provides information for labeling and indexing, especially when large projects require coded labeling for package navigation.

In detail plates, the individual components of the scenic environment are reoriented so that their surfaces are shown (as far as is possible) parallel to the page's picture plane, not how they may be situated in the overview drawings. The front of a component is most often the surface or side that faces the audience. There are plenty of instances, however, where a different direction offers a clearer view and therefore in a detail drawing becomes the object's front view.

Finally, planning the detail plates is a good point at which to pause and reflect before plunging forward. Breaking the process into achievable steps goes a long way toward preventing paralysis in the face of a large task.

Discussion: Breaking Apart the Box Set

Ideally, rough layout of the detail plates occurs simultaneously with the layout of the introductory drawings. In previous chapters, information about the box set and venue was provided as the project progressed, which prevented pursuing a more global approach in describing the design. That accumulated information will now be used to discuss content and layout of the box set's detail plates.

Scale is taken into account in determining how much content each sheet accommodates. $\frac{1}{2}'' = 1'-0''$ allows smaller details to be more clearly described; $\frac{1}{4}'' = 1'-0''$ allows more content per page. Detail plates are also where larger scales such as $1'' = 1'-0''$ (or even 1:1) are used to describe small components. $3/8'' = 1'-0''$ is becoming more common as a compromise between the resolution of $\frac{1}{2}'' = 1'-0''$ and the compactness of $\frac{1}{4}'' = 1'-0''$. Scale can change from sheet to sheet as well as within the sheet itself, depending on what is being described.

For the box set, if the composite elevation of Flats A, B, C, and D is to be included in the package, subsequent detail plates begin numbering with Sheet 5. This discussion and following exercise include the composite elevation as part of the sequence.

Instructor

Students should have hardcopies of the groundplan, section, and elevation of the box set, as well as 2D graphics of the SketchUp model. A pad of graph paper will aid the layout of each plate. Work through a list and layout of proposed detail content in class. After the class has made decisions and possibly reached consensus, have them read the discussion below. Compare and contrast to see if the author's thoughts matched the class's conclusions. Discuss differences and determine which choices provide the clearest options. Include discussion of construction drawings if desired.

Students

The following steps work through which parts of the box set can be placed on which sheets and in what order. Work through proposed detail plate content in class before reading the discussion below. Have printed copies of the box set groundplan, section, and elevation and model graphics for reference, as well as fresh graph paper to lay out rough sketches of each detail plate. If the piece of graph paper represents an Arch D page, sketch proportionally. Determine how many graph squares are equivalent to one foot; e.g., if two squares equal $1'-0''$, then one square equals $6''$.

For the first sheet, sketch a version in proportional $\frac{1}{2}'' = 1'-0''$ and another in proportional $\frac{1}{4}'' = 1'-0''$ to determine which scale provides sufficient description within the page's space limitations.

After in-class conversation, read through the following to see if the author's thoughts matched class consensus.

Discussion of Detail Plate Layout Possibilities

The box set features no platformed areas. However, the floor of the set is created by laying painted sheets of Masonite on the venue's deck. Sheet 5 would be a diagram of the Masonite covered area, with notes regarding paint finishes and indicating wall footprints. If no composite elevation was drawn, this would be Sheet 4, instead.

Six flats total are used. Four flats form the interior walls of the set. Another is used for the hallway, and the last one creates a garden vista seen through the windows. Grouping flats by purpose is good practice, so Sheet 6 might feature

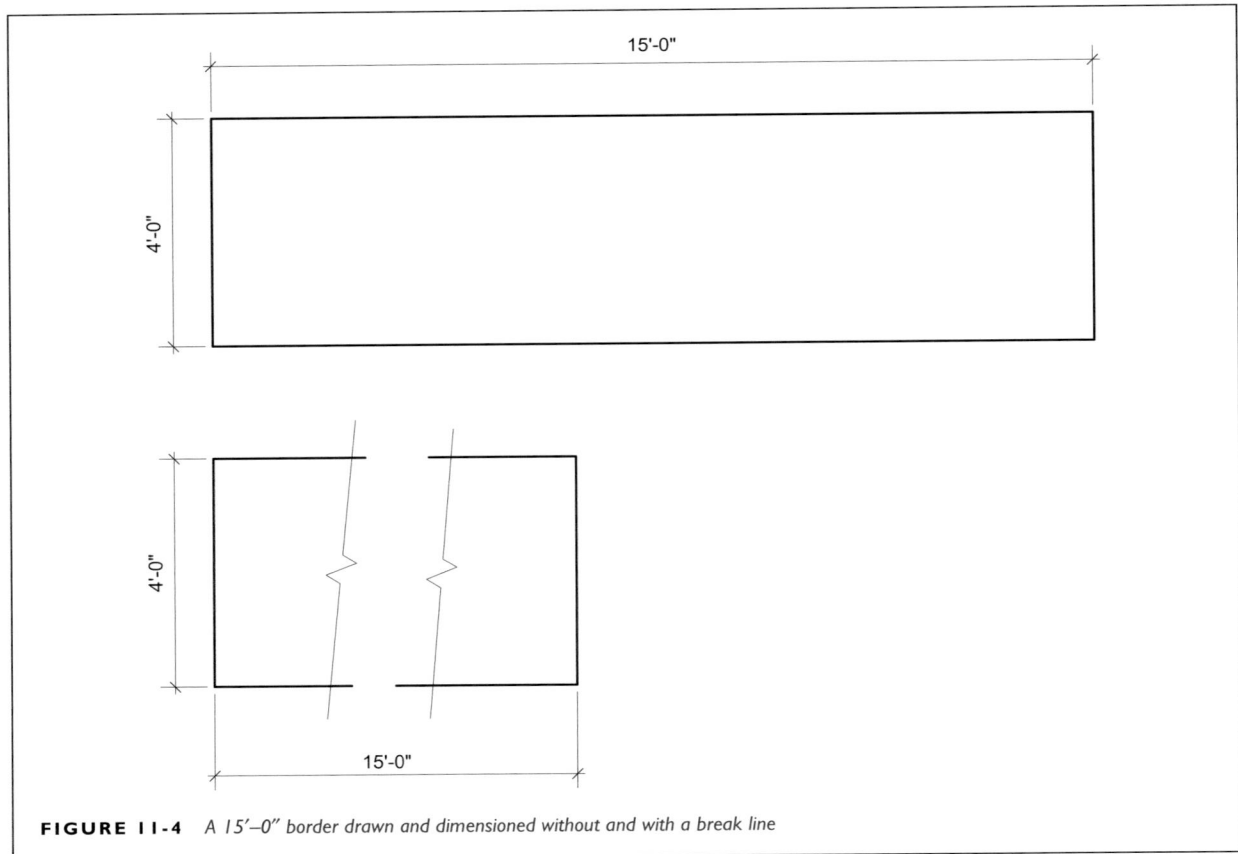

FIGURE 11-4 *A 15'–0" border drawn and dimensioned without and with a break line*

the elevations of Flats A, B, C, and D, along with any sections, top, and side views required.

On the other hand, Flat D features two windows with reasonably complex frames; it might be best to place Flat D on its own sheet (Sheet 7) with an enlarged detail drawing of the window frames on the same page. All the flats feature molding, which also needs to be described. Molding profiles might best fit on the plate with Flat D. A note on the Flats A, B, C sheet will direct the reader to see Sheet 7 for molding details.

Flat E sports a sconce. Flat F is a flat painted surface. It's unlikely that either flat will require sections or views other than a front elevation, especially as they are both Standard construction. Notes regarding jacks to support the flats are sufficient. Both of these flats should fit on Sheet 8. A sconce research image and information should be included.

These four sheets should be enough to accommodate the constructed elements of the set. It was noted earlier that the legs and borders would be pulled from the venue's stock; in that case, a drawing is less likely to be necessary. However, for this exercise, a drawing of the soft goods will be included. There are three borders and six legs. They are all rather sizable, so there are a couple of ways to approach their drawings.

The borders are all 38'–0" long by 8'–0" tall. All three are identical; therefore, only one needs to be drawn, with a note that three will be required. There is no interior detail,

so the border can be drawn in ¼" = 1'–0" scale, which should allow it to easily fit on an Arch D sheet. Top and side views are not necessary. If the length of the border as drawn does not fit on the sheet, break lines can be used to remove the center and shorten the drawing; be sure that dimensioning indicates the border's full length (Figure 11-4). In Vectorworks, turn off Show Dim Value and enter the correct value on the drawing using the Text tool.

The legs are all 24'–0" tall, but vary in width depending on their placement. Since they are not all identical, it's necessary to label them individually. Since there are no interior details and the soft goods can be drawn in a smaller scale, they should all fit on the same sheet (Sheet 9).

When an object is to be duplicated in a series uniform distances, the **Duplicate Array** command is useful. Draw an object and Select it. In the Edit drop-down menu choose Duplicate Array… The Duplicate Array window will open, offering choices for number of copies and the distance between copies (Cartesian Offset). You can choose whether to place the copies linearly, in rectangular or circular arrays, and even resize or rotate the copies. This command is very useful when an object or group of objects repeat with regular spacing across a drawing.

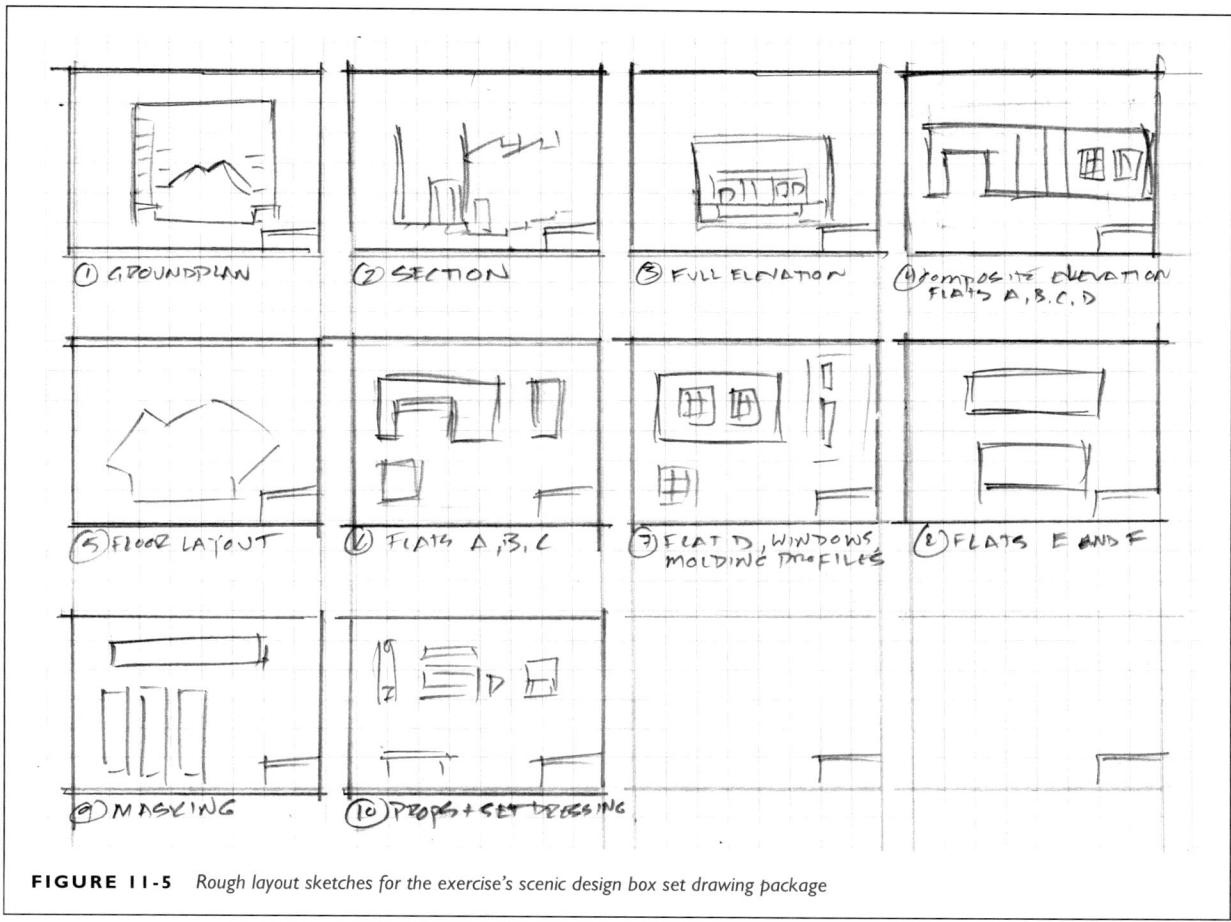

FIGURE 11-5   *Rough layout sketches for the exercise's scenic design box set drawing package*

The final drawing in the series laid out in Figure 11-5 (Plate 10) includes research images and information on furniture, the chandelier, and large set dressing pieces.

Working drawings are generated from the drawings included in the scenic design package. The number of sheets and their respective content depend upon how the technical director breaks up the project for the shop. If teams build different items simultaneously, placing multiple construction drawings on the same sheet is not efficient. The white space of construction drawings is likely to be marked up with notes, lists, and math as they are used; keeping these notes as part of the drawing can be useful when projects are handed off between technicians or if as-built drawings are required at the completion of the project.

As-Built Drawings

Another type of construction drawing is the **as-built drawing**. When working on a project that is expected to have many years of use, as-built drawings provide information on how the structure was actually built. In most projects, there is a certain amount of construction adjustment during the build, as well as necessary on-site alteration. These drawings aid long-term maintenance and safe removal when

the structure becomes obsolete. If as-built drawings are required by the client, documentation of all alterations must be documented as they occur. The drawings are then revised accordingly as part of the project wrap-up.[1]

Exercise: Layout and Drafting of Detail Plates

Instructor

This exercise walks through the layout and content of each of the detail plates as discussed previously. The following steps tend toward the general, as the student should by now have enough information for self-guidance. Steps are written to reflect Vectorworks usage, though they remain similar for hand drafting. Decide whether the students are to draft construction drawings as well as, or instead of, design drawings.

Students

Using the rough detail plate layout sketches generated in the previous discussion, draft a full set of detail plates for the box set.

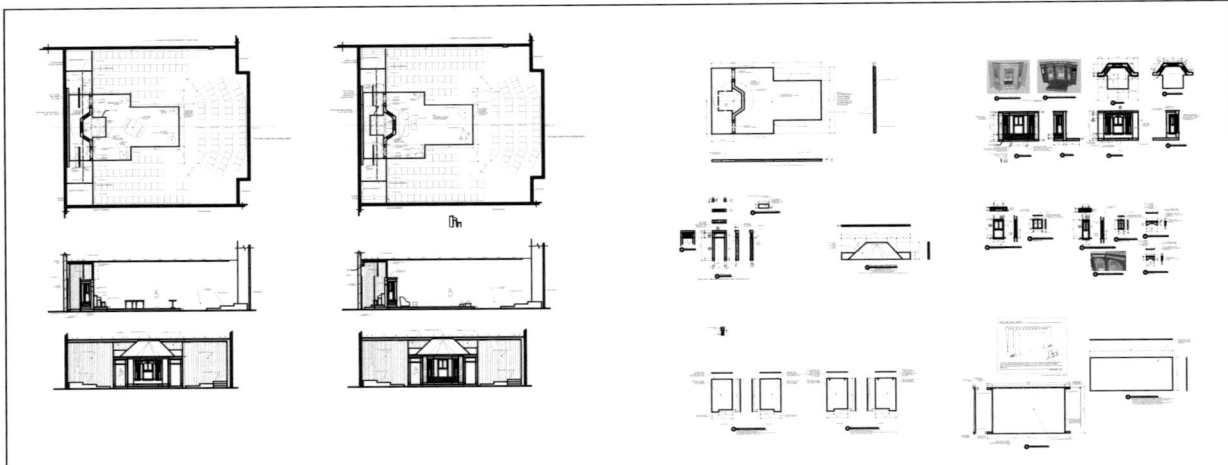

FIGURE 11-6 *Drawings for UW-Whitewater's production of* Miss Lulu Bett *as laid out in Vectorworks' infinite working space; content grouped by theme; elements within a drawing area share baselines*

When working in Vectorworks, keep the drawing space organized so that drawings can be easily located as more are added. Placing each new drawing along a common baseline is an easy way to do this, as this will also reflect the order in which they are found in the package's sheet layers. Design layer creation may reflect the overall content of a sheet or each individual component. See Figure 11-6 for an example of a set of drawings laid out in Vectorworks.

Construction drawing suggestions are included as each sheet is discussed.

Sheet 5: Floor Layout

To construct this drawing, trace the floor perimeter from the groundplan. In Vectorworks, create a new design layer. Trace the playing space perimeter in the new layer. Include the footprint of flats as dashed or colored lines. This provides reference points while indicating that the flats are not part of the deck information being drawn. Note where the Masonite sheets will meet the proscenium and apron edges.

Since this sheet contains only the floor covering, there are two options before creating the sheet's viewport:

a. Select the tracing and drag it to a new, open location in the drawing space.

b. Turn off the visibility of all layers other than the floor outline layer either before creating the viewport or via the viewport creation window's Layers button.

Dragging the outline to a new location presents it in the infinite drawing space as its own drawing; you are less likely to lose track of it.

Include dimensions and notes as to floor treatment if finishes have been selected. If the treatment is a regular pattern,

such as wood planks or tile, light, thin lines (gray) may be used to suggest the pattern on a selected patch of the floor. This is easier and more effective in hand drafting, as the lines fade to indicate continuation across the rest of the surface (Figure 11-7).

The construction drawing indicates placement of individual Masonite sheets. Trace the floor outline again in a new layer. Indicate individual sheets marked with thin lines from corner to corner. Label individual sheets. Include dimensions and angles. Include a center line to mark overall alignment with the venue and rest of the set.

Sheet 6: Flats A, B, and C

Use a common baseline to begin the layout of the front elevations. All flats are of the same height, so their tops can also be marked with a common horizontal line.

Lay out the perimeter of Flat A's front elevation. Determine cutting planes. Lay out top and side views and sections as required; remember to add the reveals. Lay out molding. If in Vectorworks, darken the drawing. Add dimensioning, notes, and labels.

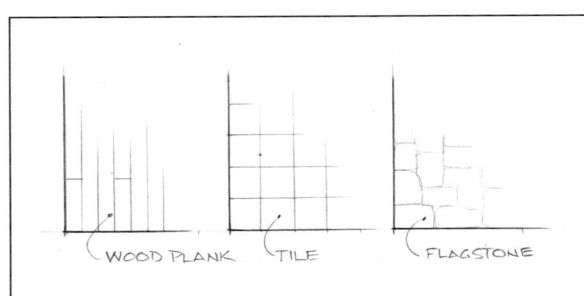

FIGURE 11-7 *Using light lines to suggest a floor treatment on a hand-drawn drawing*

Move farther along the common baseline to lay out Flat B's front elevation. Lay out side and top views; the top view may not be necessary since the crown molding obscures everything below it. No section is necessary as there are no interior features to be described. Lay out molding. If in Vectorworks, darken. Add dimensioning, notes, labels.

Repeat for Flat C.

If hand drafting, darken the full sheet of drawings after everything is laid out. On all three drawings, a note should refer the reader to the molding detail drawing to be included on Sheet 7.

Construction drawings feature the rear elevation of each flat drawn to describe its Hollywood framing. Flat A will likely require a section; Flats B and C will not. Remember Flat A's reveals. Maintain the requested archway measurements. If the opening's dimensions shift because of reveal placement, this will impact precutting and installation of casing and baseboard; carpenters may have to cut twice (or worse, start again with fresh lumber) because drawn dimensions are no longer accurate.

Remember that the flats butt together to create corners. Remember each flat's thickness and how the flat is oriented at joints to prevent adding or losing flat width. Consider an overlap of lauan at the corner so that the sides of flats don't create unsightly seams running parallel to corners (Figure 11-8). Framing decisions for one flat may affect measurements for adjacent flats; be sure to keep track of them.

Include a lauan sheet placement front elevation for each flat.

Sheet 7: Flat D, Window Detail, Molding Detail

Lay out the perimeter of the front elevation. Extend guidelines to the top and side view areas. Determine cutting planes. Remember window reveals. Lay out molding and window details. Transfer information into the top and side views and any sections. Dimension, note, and label. Note that window details are found elsewhere on the plate. Dimension the opening, not the placement of muntins, as this will get crowded unless you're working in a very large scale.

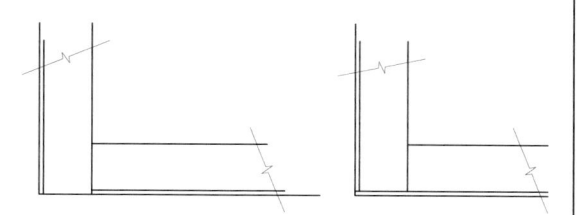

FIGURE 11-8 *Left: Hollywood flats with side of flat visible at corner; Right: Using a lauan overlap to make the joint less apparent, other flat must have ¼" subtracted from its width*

Lay out the window frame elsewhere on the sheet. If lumber stock is indicated with a note, it's likely not necessary to include top and side views. Dimension overall height and width and to the centers of muntins. Provide width of frame and muntins. Note what material is to be used as "glass," such as Plexiglas. Draw the frame once; note that two are to be built. When the viewport is created, you may select a larger scale. If this is the case, be sure that the scale of each drawing appears on the sheet as part of its label.

Lay out the molding profiles for the crown molding, casing, and baseboard. Arrange the profiles so they form a block of information that can be easily placed on the sheet to one side of Flat D (if space permits). In Vectorworks, since this is drawn in the same scale as the rest of the drawing it will look tiny—scale, chosen at viewport creation, will enlarge the drawing for the sheet layer. When hand drafting, draw the profiles in the selected larger scale. If you have switched the molding from simple 1x to more complex shapes, include inventory and purchase information along with the profiles.

The construction drawing includes the rear elevation, framed Hollywood style. Consult the drawing for Flat C to double-check any length adjustments made to accommodate corner joints. Determine cutting planes and sections as needed. Include front elevations with lauan sheet placement. Include a window frame detail drawing. If this is exactly the same as appears on the design drawing, it may be copied and pasted, or a second viewport of the original can be created and placed on the construction drawing sheet.

Sheet 8: Flats E and F

As these flats are the same height, they can share top and bottom guidelines. Lay out the perimeter of Flat E. Lay out the perimeter of Flat F. If hand drafting, determine if they fit side by side on the sheet, or are better placed one above the other. Lay out molding. Flats E and F are soft-covered (muslin) Standard flats; note the surface covering.

Even though Flat E features a sconce, it may not be necessary to include a top or side view unless there is something about the sconce that requires additional description. When dimensioning, include the height and lateral placement of the sconce where it is attached to the wall. If you wish, look at research and model images to draw the sconce as pictorially as possible; this makes it less easy for a reader to miss the sconce. Provide a research image and purchase information to one side. A note on the flat should direct the reader to the research image.

Since these are both Standard flats, they require some sort of structure or bracing to keep them upright. Include a note that upstage jack placement is left to the discretion of the technical director or shop.

Jacks are independent braces that are attached to the back of a flat and anchored to the floor. In some venues, the deck cannot be drilled into or otherwise damaged; in those cases, weights are used to stabilize the jack. There are different styles of shop-built jacks, but they all generally consist of an L-shaped frame with an angled member running from bottom to top. An older traditional jack style is the **stage brace**. This is an extendable rod that hooks into a metal cleat near the top of the flat and whose base is fastened to the deck with a **stage screw.**

The construction drawings feature rear elevations that describe framing. When determining interior stile and toggle placement for Flat E, be sure that a member is placed to support the sconce. Also consider jack placement when determining framing: how many jacks will be necessary to keep the flat upright without sagging or leaning?

Side and top views are not necessary unless sconce attachment requires extra description. Cornerblocks and straps may be drawn if the printed scale is big enough; otherwise, include a note. If stock jacks are available, note how many will be needed; otherwise, include a drawing showing jack construction and how many are to be built. If stage braces are used, indicate appropriate hardware and its placement. USITT has recommended symbols for this hardware.

Sheet 9: Soft Goods

How these are drawn will determine how they are laid out on the sheet. Consider whether break lines are desirable. Dimension fully.

The borders have no fullness and can be drawn as simple rectangles. If these are to be constructed, include notes about the webbing and grommets typically found along the upper edge, as well as the fabric to be used (in short, research soft good construction). Note how many borders are to be procured or constructed.

The legs have fullness; remember to note this. If the leg is to be constructed, include information about the webbing, grommets, and fabric to be used, as well as the percentage of fullness to be sewn into the pleating. Note how many legs of each width are to be procured or fabricated.

Sheet 10: Props and Furniture

This sheet provides an array of images and diagrams of the set pieces and furniture. If items are to be built, provide orthographic as well as photographic or model image description. Consult your model to determine what items are to be placed on this sheet. Arrange the various items in rows on the sheet and label them to match their labels on the groundplan. Be sure images are large enough to provide sufficient detail. Note each drawing's scale in its label; photos and model images should be noted as "Image Not to Scale."

Review and Finalization

Review the full series of drawings. Be sure that notes cross-reference correct sheet and drawing location. Revise earlier drawings as required. As viewports are created and imported into sheet layers, create labels as needed and enter drawing and sheet information. Once all sheet layers have been created, review them as well. Corrections made in the design layer automatically appear on the sheet layer if the correction is made within the viewport crop.

Construction drawings may be incorporated into the full series, if desired; adjust series numbering as necessary. They may also be created as their own sheet series. The drawings can still be cross-referenced; for example, "For Flat A Construction, See Construction Package Sheet 5" and "For Flat A Design Drawing, See Design Package Sheet 6."

Homework: Detail Plates for an Existing Scenic Structure

Instructor

Provide the groundplan, section, and elevation of a realized design. Examine the drawings in class, and spend some time discussing what detail plates are required. The students should then develop a series of rough layout sketches for the detail plates. If time allows, they should then draft them. The three overview drawings will likely not contain all the information required; let the students know that they are allowed to make design decisions based upon evidence found in the three plates.

Construction drawings may be included as part of the project. This will likely require multiple weeks to complete. Depending on the complexity of the selected structure; it might be useful as a final project.

When the projects are complete, compare, contrast, and discuss the rough drawing package layouts in class. Did everyone come to the same conclusions regarding the various components? What information was missing, or elusive? What differences in sheet layouts were there? Were any sheet layouts more effective or more efficient than others? If drafted, have the students present their detail plates. Did the students treat any of the structures differently than the original draftsperson who drew them? Do any of the student drawings show advantages over the original? How does the original present information in a way that improves upon the students' efforts?

Students

Your instructor should provide a groundplan, section, elevation for a realized design. Discuss them as a class to determine how many detail plates will be required, and what structures might appear on each. If the three overview drawing don't contain sufficient information required to develop a particular detail drawing, extrapolate from the information provided and make your own decisions.

Prepare rough sketches each plates' layout. Try to be as detailed as possible; include layout for dimensions and notes, as well as the placement of breakout detail drawings. If assigned to do so, draft the series.

When the completed project is brought into class and discussed, consider the similarities and differences between your work and your classmates. Did you miss anything? Did they miss something? Is there a particular layout that works very well? Was information missing from the original documents that would have been good to have? Which structures did you have particular difficulties with and why do you think that was the case? Which structures were particularly easy to draw, and why was that the case? For structures in which not enough information was provided in the three given drawings, how similar or different were your extrapolations from those of your classmates?

MODULE 2: MORE ABOUT WINDOWS

The box set used for previous projects features two fixed windows with multiple panes and muntins. They are relatively simple windows, and the applicable groundplan graphic was introduced in Chapter 8. For most styles of windows, their groundplan representation can be developed from that chapter's graphic examples. Design and construction drawings require more detail and further knowledge of window parts

and assembly, particularly if the window is expected to be functional.

Window Vocabulary

Listed below are some broad categories of window types. First, a **sash** is a frame holding a pane or panes of glass (or other window material). How this frame moves determines the type of window. A **fixed** window has no moving parts.

- **Single-** and **double-hung** windows feature sashes that slide up and down along tracks in the window frame.

- **Casement** windows feature sashes that swing from side hinges. They typically open outward. **French doors** are large versions of casement windows.

- **Sliding** windows feature a sash or sashes that move laterally in the frame along tracks.

- **Awning** windows have a sash that is hinged at the top so that the bottom of the sash swings open.

- **Hopper** windows have a sash hinged at the bottom so the top of the sash swings open.

- A **Pivoting** window features a sash that rotates open around a vertical pivot point rather than a hinge attached to one side of the sash.

Any of these types of windows can be incorporated into a wide variety of shapes and styles. Keep in mind that regardless of the type of window, the depth of the reveal reflects the thickness of the wall; older buildings tend to have deeper window sills because their walls are often thicker.

Within a window and its frame are standard parts (Figure 11-9). Some of the terms listed below have already

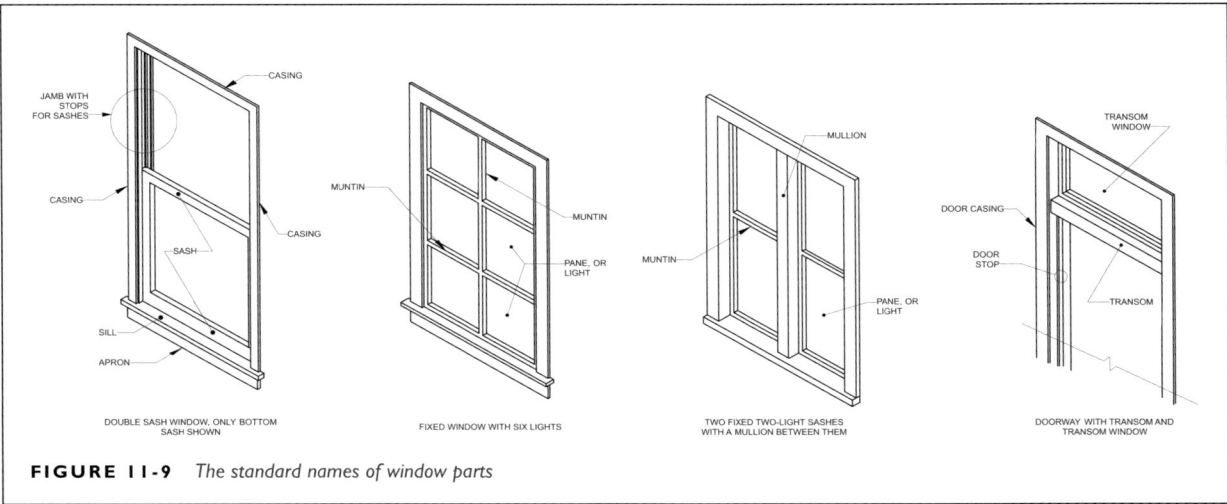

FIGURE 11-9 *The standard names of window parts*

been introduced. They are defined again here to add further context.

- A section of glass or other translucent material is called a **pane** or **light**.

- The frame holding the panes of glass is the **sash.** The sash is comprised of stiles and rails.

- The side vertical framing members are called **jambs**, as they are with doors.

- The top framing member is the **soffit**. Soffit is a term used for the flat, horizontal undersides of several architectural features.

- The bottom framing member is the **sill**. This often extends slightly beyond the window opening and can interrupt the casing.

- **Stops** and tracks may be found on these framing members, depending on the type of window.

- Multiple panes are supported and separated by **muntins** within the frame (the window version of toggles).

- Vertical architectural members that separate/divide adjacent window frames are called **mullions**.

- A horizontal member that separates adjacent windows is a **transom**. When a window is placed directly above a door, is it called the transom window.

- Many windows feature a piece of molding flat against the wall beneath the sill. This is called an **apron**.

- Molding surrounding a window is called **casing.**

Take a look at windows in your residence or school. Are they all the same type? Can you identify what category they fall into? Do they have all the parts listed above, or are they missing anything? Is there a part that was not listed above? If there is casing and an apron, examine how they relate to each other. Does the apron extend to the edge of the casing, or go beyond it? Does the apron extend the full width of the sill, or does it extend beyond the casing?

The second photo in Figure 11-10 depicts the lower corner of a mid-twentieth-century casement window and the relationship between apron, casing, and sill. A crank handle opens the sash. A metal-edged window screen sits within the sash frame. The sill extends beyond the edge of the apron and casing and protrudes into the room about 1″. The apron and casing have the same profile, though the ends of the apron are slightly rounded. The casing is set back about ¼″ from the interior edge of the jamb. The stop is comprised of strips of ovolo profile molding. In the photo of the whole window, a mullion is seen to separate the halves. Paying attention to details can sell a realistic window in an intimate performance environment, especially in film or television where the audience may see close-ups of the scenery.

If a high degree of realism is required, be sure to understand the exterior of the window as well as the interior, and vice versa. If the viewer can see through the window, edges and portions of the exterior side of the window opening may be visible.

If budget allows and a high degree of realism is required, prefabricated windows can be purchased from suppliers. However, the flats supporting them will likely need to be more substantial as these items are meant for architectural construction and have particular installation requirements. Prefab windows and doors still need to be drawn and described as part of the larger scenic piece, with their purchase information, specifications, and catalog image included on the sheet.

FIGURE 11-10 *Photograph of a casement window and its details; Examining the relationship between the apron, sill, and casing*

Drafting different types and styles of windows follows much the same process as outlined in previous chapters. As with doors, the designer must research the style of window to be used. Including photos will clarify details and the "feel" of the window even if the photo is not an exact match for the window to be constructed (be sure to include a note indicating that difference). Layout begins with the perimeter and works its way inward to the details.

Exercise: Drafting a Double-Hung Window

Instructor

In this exercise a double-hung sash window will be drafted. The window will not be operational, though the lower sash will be fixed in an open position. The window is horizontally centered in a 10′–0″ by 10′–0″ hard-covered Hollywood flat with 6″ reveals. For construction drawings, frame the flat with 1×4 and cover with lauan. The front of the flat represents a room's interior; include baseboard and crown molding. Molding uses 1× at various widths. If desired, assign more complex molding. Window hardware will be included for appearance rather than operation.

The steps of this exercise are described for Vectorworks, though the hand drafting process is similar; darkening occurs at the end of overall layout to prevent smudging. Dimensions are provided in Figure 11-11.

Students

For this drawing of a sash window in a Hollywood flat, ½″ = 1′–0″ is suggested for the design drawing. 1″ = 1′–0″ is suggested for the construction drawing. Use Arch D paper. Dimension, label, and note fully. Include borders and title block. Include molding profile details and purchase information for required hardware and pane material.

Figure 11-11 features two detail photos describing the bottom and top corners of the molding. There is a lock where the sashes meet and another piece of hardware on the left side of the upper sash that prevents the lower sash from rising above this point. The exterior storm window will be ignored for this project, and other features will be simplified. The lower sash is to be fixed in a raised position, so tracks along the jambs will not be included, though tracks would contribute to the realism of the window even if the window is not operational.

In Figure 11-11, sash detail drawings are on the right. The horizontal section shows the two sashes, cut-through where they overlap within the frame. The lower sash is raised 6″ above the sill. The upper sash is flush with the flat's rear. The lower sash's rear surface aligns with the upper sash's front surface. The lock has two halves: a turning part on the

lower sash's upper rail that twists into the catch on upper sash's lower rail. While the window in the research photo does not have handles, two handles will be added to the lower rail of the lower sash. Research appropriate locking hardware and handles.

For most double-hung windows, the top rail of the lower sash and the bottom rail of the upper sash overlap completely. When the height of the two sashes is added together, they will equal a distance greater than the opening's height. Remember this when drawing the sashes in the opening or in the detail drawing.

For this exercise, the pane material is inserted in the center of the sashes rather than attached to the sashes' rear surfaces. **Glazing** approaches are discussed below. Glazing refers to the act of installing glass panes, as well as the materials used to do so. A glazier is a person who installs glass.

As per usual, work from the outside in: perimeter toward detail. The flat and window are collections of regular geometric shapes. Consider what those shapes are and how they fit together as you begin the layout process.

1. To begin the front elevation, lay out the flat's perimeter, then lay out the window's vertical center line and a horizontal line marking the top surface of the sill. Extend the sill guideline into the area in which a side view or section will be drawn. Extend the center line up into the area where the top view or section (or both) will be drawn.

2. In the front elevation, lay out the perimeter of the window opening. Lay out the casing, sill, and apron around the opening. Note that not all pieces of casing are the same width.

3. Lay out baseboard and crown molding.

4. Lay out the sashes within the window opening.

5. Using your research, lay out the location of the hardware. If they cannot be drawn convincingly, note the placement of centers and/or edges to guide their installation.

6. Determine the placement of cutting planes. Lay them out.

7. In Vectorworks, go ahead and darken the drawing if this will prevent later guideline confusion. Clean up guidelines.

8. Extend fresh guidelines into the top view area. If constructing both a horizontal section and a top view, lay out the views one at a time. Remember that guidelines for the horizontal section refer to features at and below the cutting plane. Lay out details for the top view and/or horizontal section.

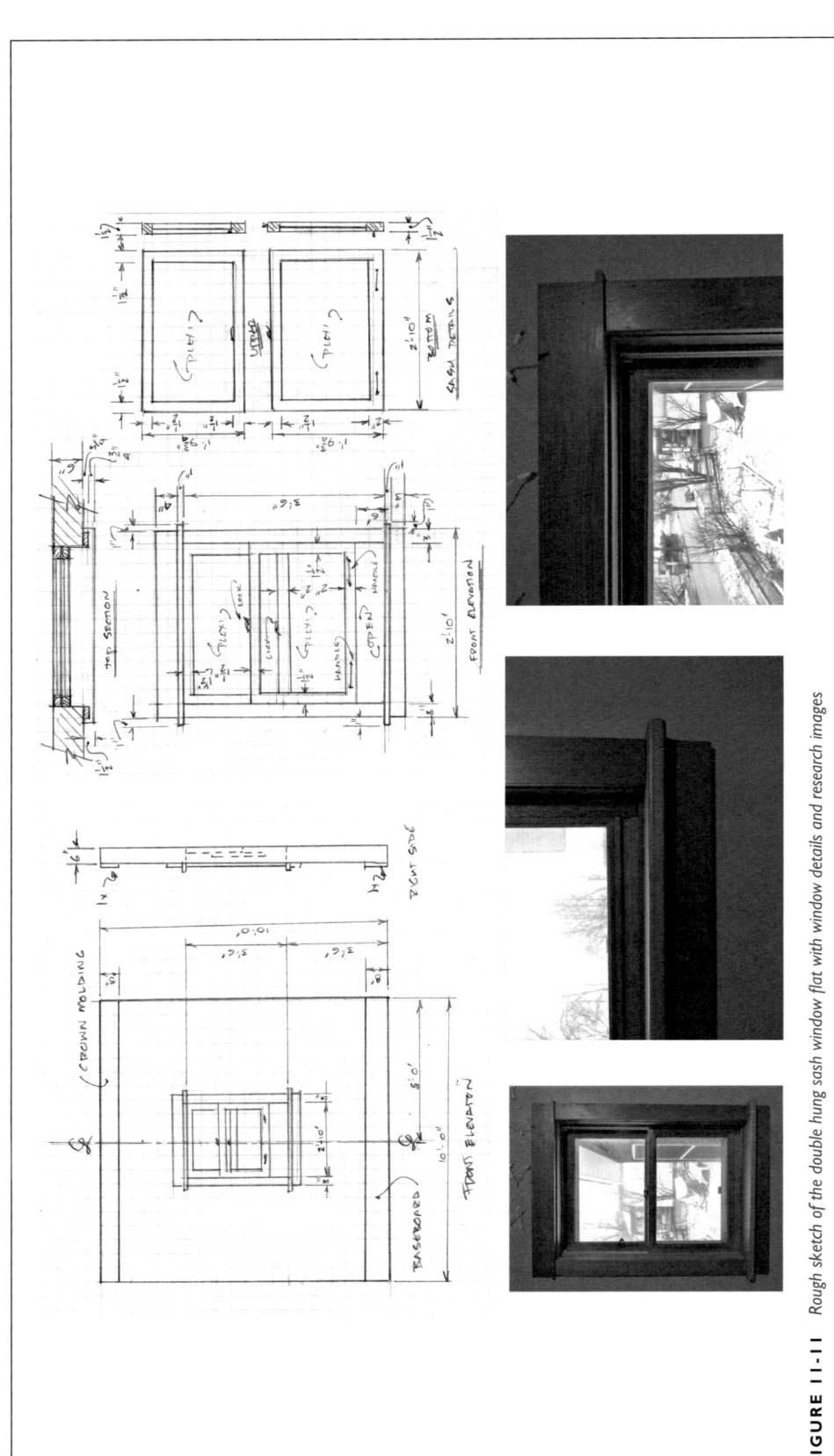

FIGURE 11-11 Rough sketch of the double hung sash window flat with window details and research images

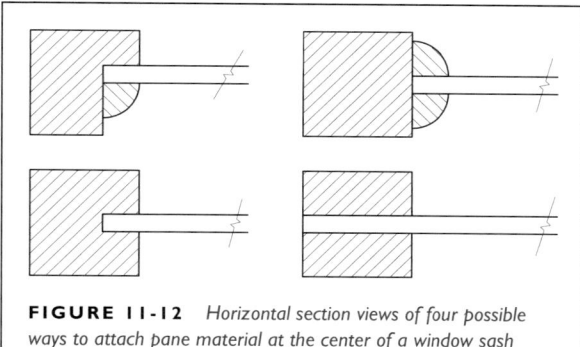

FIGURE 11-12   *Horizontal section views of four possible ways to attach pane material at the center of a window sash*

The glazing approach can be described as a detail drawing. Provide a note in the main view referring the reader to it. Figure 11-12 presents four possible options.

The first option has an inset into which the pane is placed; it is then secured with a strip of quarter-round molding (caulk is another option). The second shows the pane held front and back with molding. The third inserts the pane into a groove. The fourth sandwiches the pane between two frames. The degree of realism, proximity of audience, and need to reclaim materials at strike will drive the glazing choice. Even if the designer has no preference, there should be a conversation with the shop to ensure that the appearance of the window meets expectations (at which point, the construction drawing will detail the glazing approach).

If a glazing approach using strips of molding is chosen, add lines to represent them within the sashes in the front elevation.

9. Darken the top view and/or section. Clean up guidelines.

10. Extend fresh guidelines into the side view area. Add a miter line to transfer information from the top view, if necessary. Again, if constructing both a side view and a vertical section, lay them out one at a time; a second miter line will be required if both side view and vertical sections are drawn. Remember that the lower sash is raised; draw it accordingly.

11. Darken the side view/section. Clean up guidelines.

12. Lay out the sash detail drawings. Darken.

13. Lay out the molding profiles. Darken.

14. Import hardware imagery and information.

15. Lay out dimensions, notes, and labels. Darken the drawing (if you have not already done so).

16. Create viewport(s). Import drawing(s) into a sheet layer. Remember that detail drawings should appear in a larger scale. Add border and title block. Fill in title block information.

For the construction drawing, use the guidelines outlined in earlier chapters to lay out the framing. Remember that the flat is Hollywood style, using 1 × 4. Be sure to consider the thickness of the reveals when laying out the window opening. Include a front elevation describing lauan layout.

When building windows, the author prefers to use a single piece of wood for the sill, notching the ends and sliding it into the bottom of the opening. A single-piece sill is more durable and usually better looking than a sill created by tacking a narrow strip of wood to the flat's face. It's less likely to rip off when leaned upon or snagged, with fewer fastener heads and dimples to hide. It does, however, take more time to cut and install. If a single-piece sill is used, the bottom of the window framing must be adjusted so that the top surface of the sill is at the designated height.

Exercise: SketchUp Model of the Double-Hung Window

Instructor

In this exercise, a model of the flat housing the double-hung sash window is constructed. Decide which degree of construction detail the modeling approach will employ, such as building from a parts list. If that approach is chosen, provide the parts list. Otherwise, students should use the drawings generated in the previous exercise for reference.

Students

Model the double-hung window flat in SketchUp, using the construction approach chosen by your instructor. Your instructor should provide a parts list if that approach is designated; otherwise, use your drawings as reference. It is suggested that the sashes be constructed off to one side and then inserted into the frame. Be sure the window panes have appropriate thickness. Select an appropriate glass texture for the panes from the SketchUp texture library. Look for hardware in the online warehouse.

Exercise: Drafting a Free-Standing Palladian Window with Window Seat

Palladian windows are named for the Italian late-Renaissance architect Andrea Palladio (1508–1580). An arch flanked by two rectangular openings was one of his signature building features. This style of opening became popular in England during the Neoclassical revivals of the eighteenth century, and continues to be featured in neoclassically influenced Western architecture.

FIGURE 11-13 *SketchUp model of the Palladian window with window seat; human figure included for scale*

Instructor

Figures 11-13 and 11-14 provide information for the window seat with window to be drawn. Students should correct any erroneous or missing information they discover during lay out. Provide and discuss images of Palladian windows to give the project further context; students should then feel more confident if they need to make adjustments and corrections. Choose whether students should draft design or construction drawings (or both).

The steps are described for Vectorworks, though once again, hand drafting follows similar steps with the primary exception being when to darken the drawing. Allow the students to select the printed scale for an Arch D sheet, and whether they feel the need to break the drawing up into multiple plates.

Students

After reviewing Figures 11-13 and 11-14, choose which scale to use and whether you wish to include all the views and details on a single Arch D sheet, or break them out across multiple sheets.

This version of a Palladian window is fixed. Pane material is attached to the upstage side of the framing; a single sheet may cover multiple pane openings. The cornice at the top of each side light extends slightly into the space of the central window. The window framing and muntins are all 1″ × 1″. Research what materials and approaches can be used to create the curved

surfaces to make recommendations, especially for construction drawings. The construction drawing may include a view in which pane material placement is laid out to ensure that seams are hidden by muntins or mullions.

Two sections are included in the rough sketch. The cutting plane of the vertical section cutting plane is the window's vertical center line. It may be desirable to include a second vertical section through one of the side lights. The cutting plane of the horizontal section runs horizontally through the middle panes (second up from the bottom) to show the mullions and sill.

Remember that a rough sketch is just that: a rough sketch. It is a preliminary attempt to describe the window in scale with measurements, and may include error. As you lay out the drawing, make corrective decisions if you find misalignment or measurements that don't add up. Refer to the model image in Figure 11-13 if the rough sketch doesn't provide clear enough information.

Determine how many views and detail drawings are required. In Vectorworks, create necessary design layers.

1. This window features a semi-circle, so placing its center is the best place to begin. It is useful to lay out the curves for an object of this size and complexity rather than waiting for the darkening phase, especially as there are a number of concentric semicircles (and in Vectorworks, there's no worry about compass control). Lay out the arcs' center lines.

2. Using the center lines, lay out semi-circles for all components. Measure down the center line to mark the top of the bench (5′–6″). Extend endpoints down to the bench top to begin layout of the parts of the central window. Since the outermost arc terminates at the top of the lights' upper molding, those lines only need to extend down about 1′–0″ (Figure 11-15).

3. The lights' openings are 1′–0″ wide, flanking the central window. Lay out their widths, as well as the 3″ wide mullions. Lay out the molding that caps the top of the lights.

4. The sill sits within the 5′–6″ between the bench seat and the arcs' horizontal center line. If you lay the sill out below the window perimeter, you'll add ¾″ to the overall height of the unit. Lay out the perimeter of the bench. Lay out the nosing and baseboard. These both wrap around the sides of the bench and extend ¾″ beyond the sides of the bench itself.

5. Lay out the interiors of the central window and side lights. Remember that the center of the muntins is the measurement given in the rough sketch (Figure 11-16).

6. Lay out cutting plane lines for any planned sections.

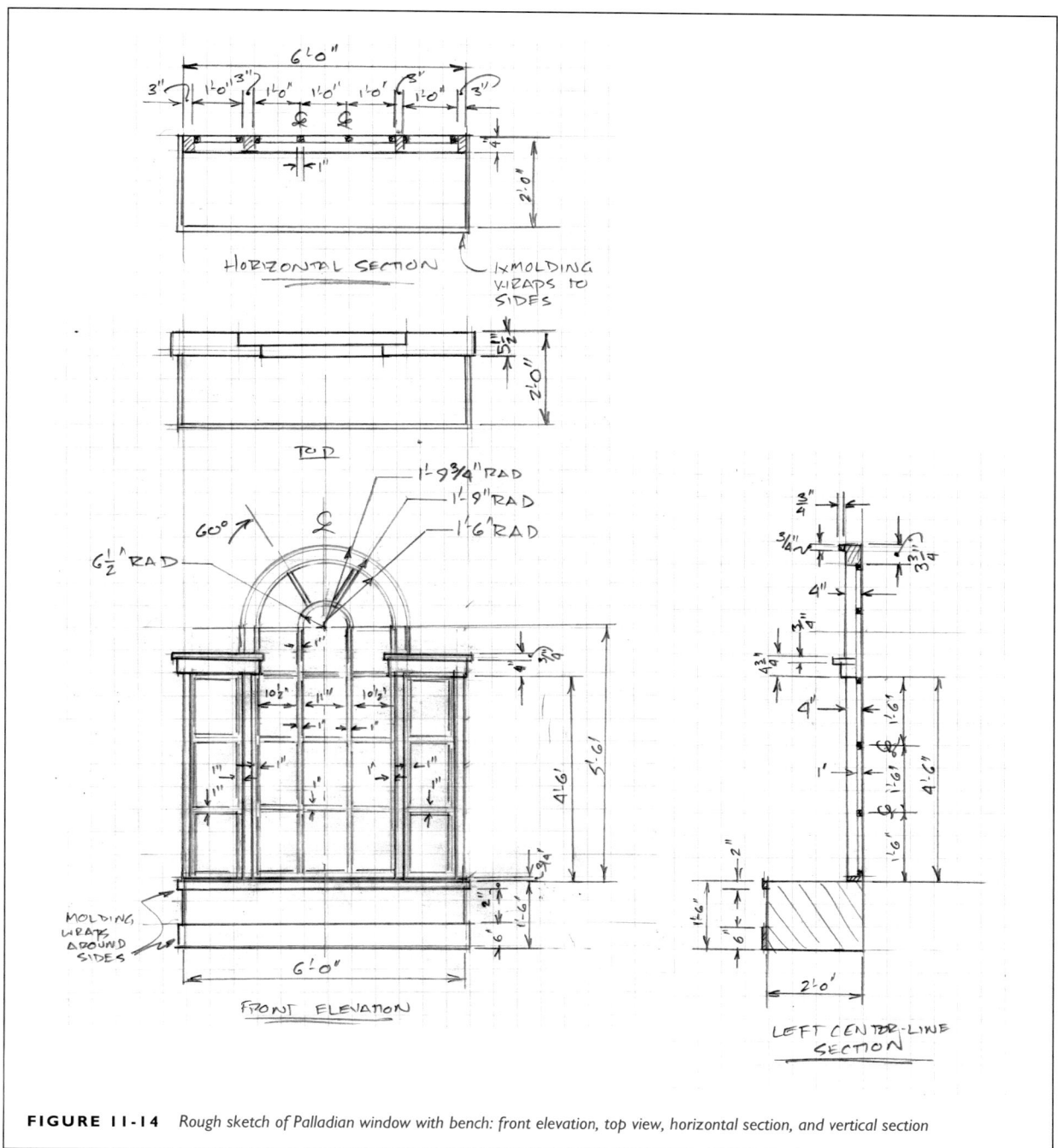

FIGURE 11-14   *Rough sketch of Palladian window with bench: front elevation, top view, horizontal section, and vertical section*

7. Since there a fair number of guidelines have been drawn for this front elevation, darken the elevation before extending guidelines into the top and side views/sections. Use heavier line weights for major outlines and perimeters, and medium or thin line weights for interior details. Clean up guidelines.

8. Extend fresh guidelines into the top view/section area. Decide which will be drawn first and which will be placed closest to the front elevation: the top view or the horizontal section. Lay it out. After darkening, lay out the remaining view or section. Darken. Clean up guidelines.

9. Extend guidelines into the side view/section area. Decide which will be drawn first: the side view or the vertical section. Lay it out. Use a miter line to move information if necessary. After darkening, lay out the remaining side view or section. Darken. Clean up guidelines.

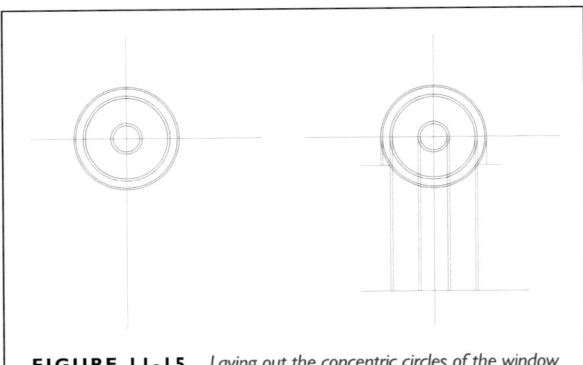

FIGURE 11-15 *Laying out the concentric circles of the window frame, jamb, and casing; extending lines down to the bench seat*

10. Lay out detail drawings, such molding profiles. Are there any other small elements that require their own detail drawing? Darken.

11. Lay out dimensions, notes, and labels. Don't forget to note what pane material will fill the openings (only one pane needs to be noted as such; e.g., "1/8" clear acrylic sheet in all openings, see detail drawing for layout," or, "1/8" clear styrene TYP").

12. Finish darkening the drawing.

13. Create viewport(s). Import to a sheet layer. If multiple sheets are used, be sure to add notes on the first sheet to direct the reader.

14. Add borders, title block, and title block information.

On the construction drawing, take into consideration the thickness of the facing lauan when calculating the length of the bench's framing members. If this is not subtracted from the frame, the additional thickness adds width and depth to the bench—not only will the nosing and baseboard be longer than expected, but the additional width may throw off the placement of adjoining scenic pieces.

Exercise: Model the Palladian Window in SketchUp

Instructors

This exercise uses the drawing drafted in the previous exercise to model the Palladian window and bench. Decide what level of construction detail the modeling approach should employ, such as assembling the window unit from a parts list. If a parts list is to be used, provide that list.

Students

Construct a model of the Palladian window using the approach chosen by the instructor. Use your completed drawing for reference. If a parts list has been provided, construct each element as a group, then assemble. It is suggested that the window frames, with muntins and pane material, be modeled to one side and then inserted into the openings.

Use the Cutting Plane tool to section the model as indicated on your drawing. Compare the drawing and the model. Do they match? Why or why not? Does something need to be corrected? Do any features read more clearly in the model's section than on the drawing so that you now better understand what you were drafting?

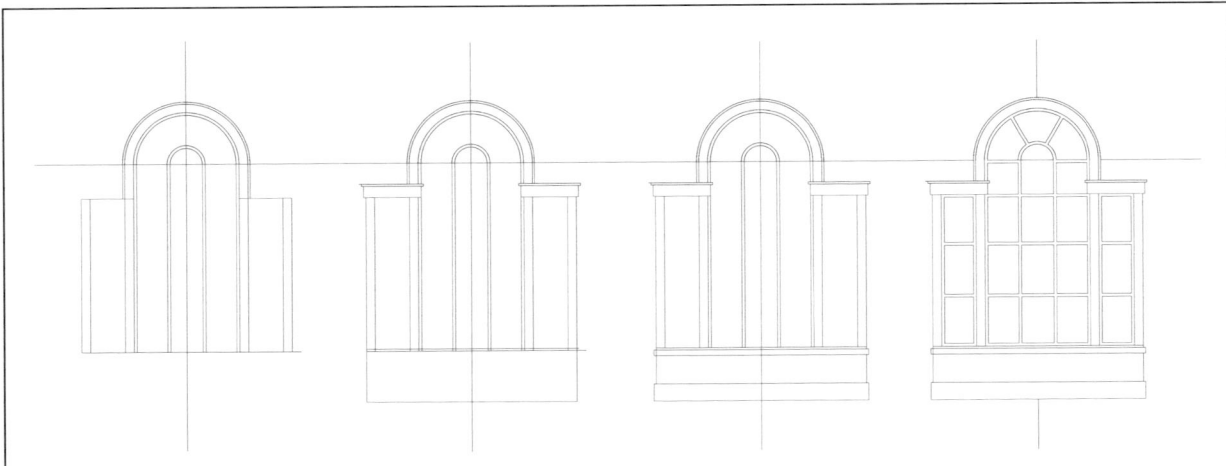

FIGURE 11-16 *Laying out Palladian window seat: (1) laying out the lights; (2) adding sill, bench, and cornice over the lights; (3) adding bench molding; (4) laying out muntins*

MODULE 3: STAIRS, BANISTERS, AND BALUSTRADES

Stairs and staircases provide access to raised platforms. They can range from the purely functional step unit to a grand, sweeping staircase. If there are more than three or four steps, a railing of some kind is usually required for safety.

Staircase Vocabulary

There are three basic parts to a staircase or stair unit, as shown in Figure 11-17:

a. **Stringers**. These are the main supports that run the length of the staircase. These are typically beefy planks cut from 2×8 or 2×10, though small portable step units may have stringers cut from ¾″ plywood. Stairs three feet and narrower typically require only two; one on either side. The number of stringers also depends on the structural strength of the stair treads and the materials used for construction. For wooden staircases, the wider the stairs become, the more stringers are required to support the middle. Modern steel staircases may have a single stringer running up the center. A stringer supporting the interior of the staircase and not seen by the audience is also called a **carriage**.[2]

b. **Treads**. These are planks laid horizontally across the stringers upon which the performer steps. The depth of the tread is referred to as the **run**.

c. **Risers**. These are planks set vertically to close the face of each step. They provide further structural stability, preventing the tread from flexing when it's stepped upon. They also mask interior structure. Depending on the style of the staircase (such as an ultramodern cantilevered flight of stairs), they are not always included. The height of the step is called the **rise.**

Together, the rise and run describe the slope of the staircase. The run should be deep enough to set a human foot fully upon it, typically 11″ to 12″. The rise should provide an easy and comfortable step up, typically 7″ to 8″ inches. Rises higher than 9″ and shallower than 6″ tend to be more difficult to ascend and descend. Exceeding a slope of 45° becomes particularly dangerous;[3] *The Backstage Handbook* lists slopes between 50° and 65° and between 7.5° and 20°s as "accident prone ranges."[4] Very steep slopes (from 65° to 75°) transition a flight of stairs into becoming a **ship's ladder** (Figure 11-17). If a slope less than 7.5° is required, a ramp is recommended.

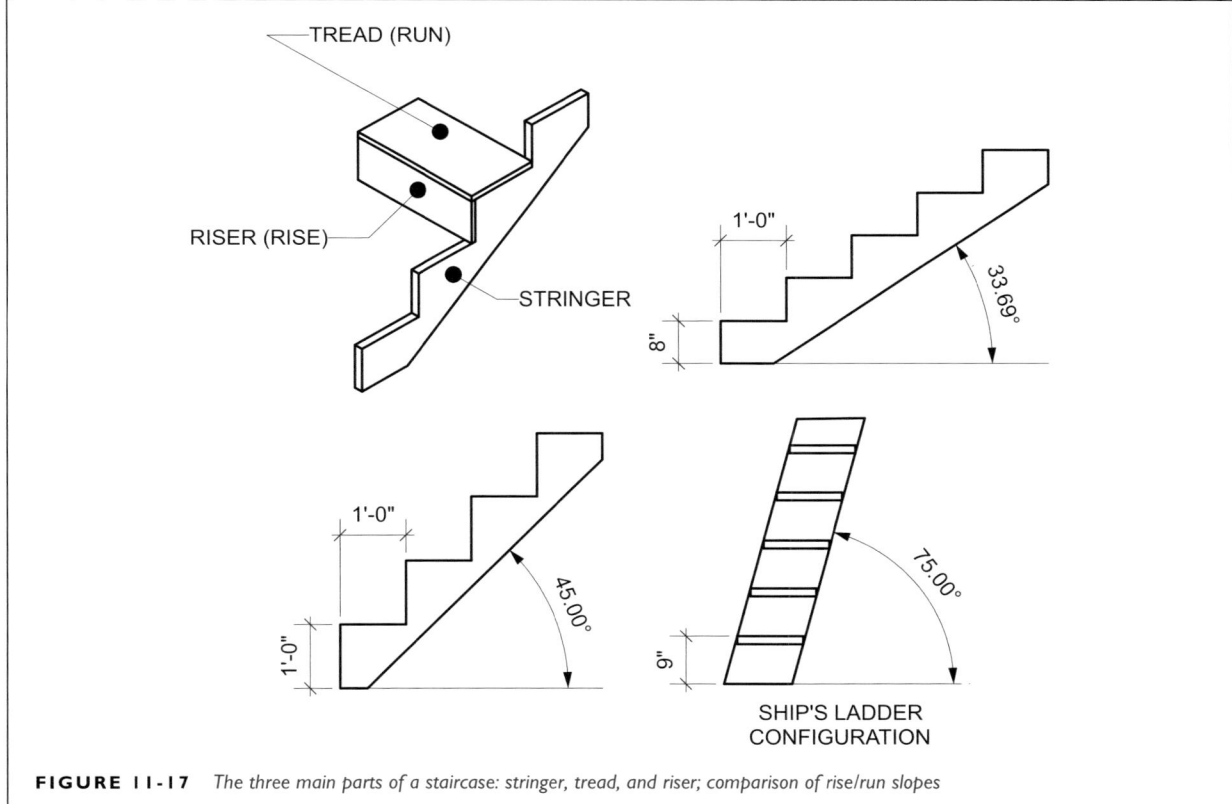

FIGURE 11-17   *The three main parts of a staircase: stringer, tread, and riser; comparison of rise/run slopes*

Typically, the rise and run are consistent across the full height and width of the staircase; this is an important safety consideration. Variation from the norm requires special attention and extra rehearsal time for those using the staircase.

> Among other things, Actor's Equity Association (AEA) employs safety recommendations for stair configurations based upon ANSI standards. These recommendations can be found on the Entertainment Services and Technology Association (ESTA) website (www.esta.org). Further AEA recommended practices and standards can also be found at: www.actorsequity.org/resources/ Producers/safe-and-sanitary/technical-standards/.

Further stair vocabulary (Figure 11-18):

- **Flight**: A section of staircase; "a flight of stairs."

- **Landing**: A small platform used to change a staircase's direction or where a staircase terminates.

- **Handrail**: A rail placed above and along the stairs such that a climber can use it for support as they climb or descend. The top of the handrail is typically 2′–6″ to 2′–8″ above the lip of each stair. It is typically 2′–10″ to 3′–0″ above a flat platform (such as a landing).

- **Baluster**, or **spindle**: A narrow post that supports the handrail. The base may be set into a tread, or supported by a **baserail** or a **shoerail**.

- **Banister**: Collective term for a hand rail and balusters as part of a staircase.

- **Balustrade**: Collective term for a handrail and balusters as part of a porch, landing, or other non-staircase area.

- **Baserail**: Parallels the handrail just above the treads to provide a bottom anchor for balusters.

- **Closed Stringer**: When the treads and risers are attached to the side of the stringer, rather than atop the stringer, the stringer is said to be "closed." When the treads sit atop the stringer, the stringer is said to be "open."

- **Shoe rail**: Closed stringers may be capped with a shoe rail. It may provide a bottom anchor for balusters.

- **Newel Post**: A large post at the top and bottom of flights to which the hand (and baserail) are attached.

- **Newel Post Cap**: A decorative element attached to the top of the newel post. May include a **finial**, or, decorative knob.

- **Nosing**: The forward lip of the tread often projects slightly over the riser; this can be done by adding depth to the tread itself or attaching a strip of molding.

Since staircases often have many repeated components (treads, balusters, etc.) and spacings, remember that using the abbreviation TYP (for typical) reduces the need to dimension each and every identical element.

Discussion and Exercise: Staircase Math

Stairs tend to take up more space than many student designers expect; this is a case where thinking and sketching in scale prevents heartache later on.

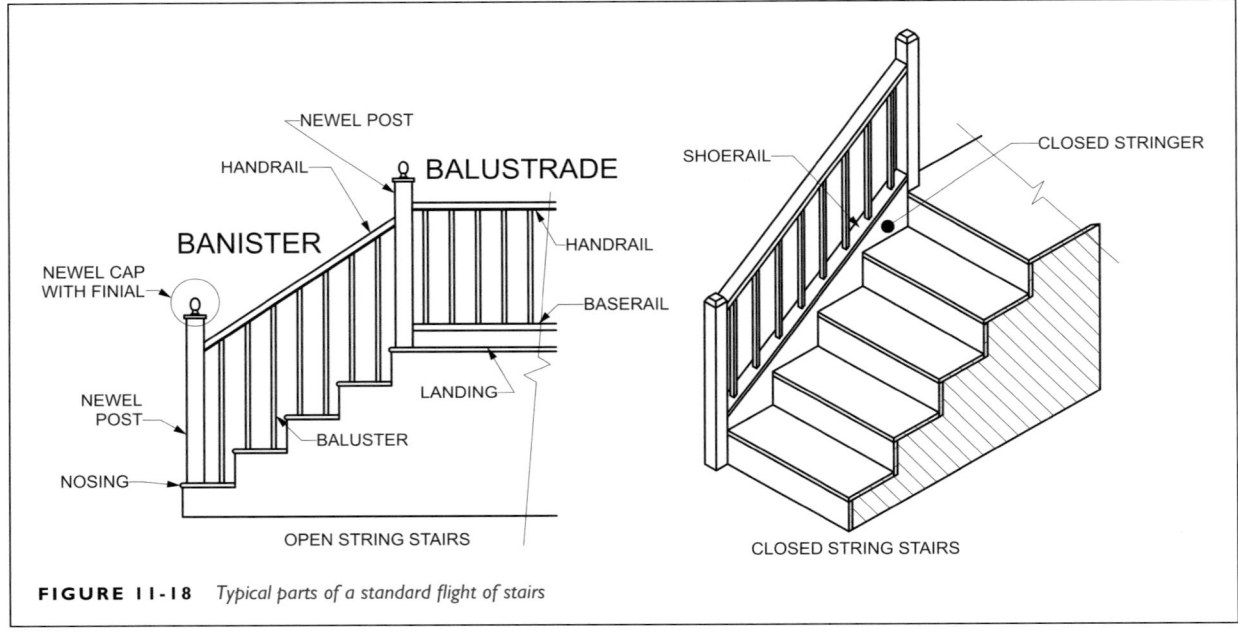

FIGURE 11-18 *Typical parts of a standard flight of stairs*

To determine how many stairs are required, it's necessary to know the height of the platform to which the flight ascends and to know the rise and run of the intended staircase. While simple division can determine the number of steps needed, it's often not until the run is laid out that just how much space the staircase requires is truly understood. After calculating the number of treads needed, the height of the platform—or the rise of the stairs—may need to be adjusted.

For example, a platform is to be + 63″. The designer wishes the rise to be 8″ and the run to be 12″. 63 ÷ 8 = 7.875; that's 7 increments of 8″ plus some inches left over. 8 times 7 equals 56. 63 minus 56 equals 7. That means a final step 1″ lower than the others. All the rises could be changed to 7.875″, but that's likely to frustrate the shop. Perhaps 1″ could be added to the height of the platform, making it + 64″. Of course, the designer has to double-check everything else in the design affected by the change in platform height. If the rise is changed to 7″, then 63 ÷ 7 equals 9, resulting in 9 steps rather than the expected 8.

If the platform's height is changed to + 64″ and the flight is comprised of eight steps, the staircase has a footprint 8′–0″ long. If the stairs are changed to have a rise of 7″, the staircase now has a footprint 9′–0″ long. Can this extra foot be accommodated on the groundplan? If the tread is made less than 12″, can the footprint be reduced to 8′–0″? How steep and/or safe is this new staircase expected to be?

Instructor

Supply a list of platform heights coupled with rise and run measurements. Students may slightly adjust the rise and run measurements, as well as the platform heights. Review the calculations and adjustments generated by the first item on the list with the class as a whole. The students should finish working through the list on their own, drafting profiles of their final determinations in scale. Compare and discuss findings.

Students

For the platform list provided, calculate the number of stairs required and work out the length of the staircase. What compromises and adjustments does each set of measurements require? Draft the profiles of the staircases and check the angle of their slopes. Which are the steepest? Which are the shallowest? Did any set of measurements not work unless major adjustments were made?

Discussion and Exercise: Staircase Layout

Instructor

To explore how the overall length of a staircase impacts a scenic environment, prepare one or two scale groundplans

(Figure 11-19). The groundplan generated in Chapter 8 can be altered, replacing the archway with a doorway, landing, and staircase. Provide several possible landing heights and rise/run combinations. The students must then determine how many stairs are required, as well as the orientation of the staircase, for each designated landing height and rise/run combination. Groundplans should include upstage platforming so students can lay out backstage escape stairs. Provide enough copies of each groundplan so that each landing measurement can be drawn separately.

Doing this exercise in both ½″ = 1′–0″ and ¼″ = 1′–0″ will help students further develop thinking in scale and therefore be less likely to underestimate the amount of space stairs require.

After completion, discuss the solutions presented. Consider how each staircase configuration may or may not affect sightlines and blocking choices, both impediments and opportunities.

Students

Using the landing height and rise/run combinations provided, determine how many stairs are required to access the landing from the zero deck, as well as the best placement for the entire footprint of the staircase. If necessary, the staircase may be interrupted by further landings. For each set of landing/rise/run measurements, lay out the staircase's footprint on a groundplan in scale.

Performers must be able to exit the stage once they have ascended the stairs. Platforming must be placed upstage of the entrance and **escape stairs** supplied. For each onstage staircase configuration, determine where escape stairs can be placed, making sure the flights don't run into masking, the fly rail, or other scenery. If sightline symbols are indicated on the drawing provided, see how long the upstage platform must be so that the audience does not see a performer begin their descent down the escape stairs.

Discussion and Exercise: Stairs for a Unit Set

Being consistent in rise/run throughout a scenic environment provides a level of safety for performers, as they don't have to change their stair-climbing gait in different parts of the environment. There are always exceptions, but safety must go hand in hand with esthetic goals.

Instructor

Prepare a rough scale groundplan featuring an array of interlocking platformed areas (Figure 11-20). Provide a wish list of platform heights; be sure a few are of reasonably substantial elevation. Provide expected rise/run measurements and an overall width of tread to be kept consistent throughout the set.

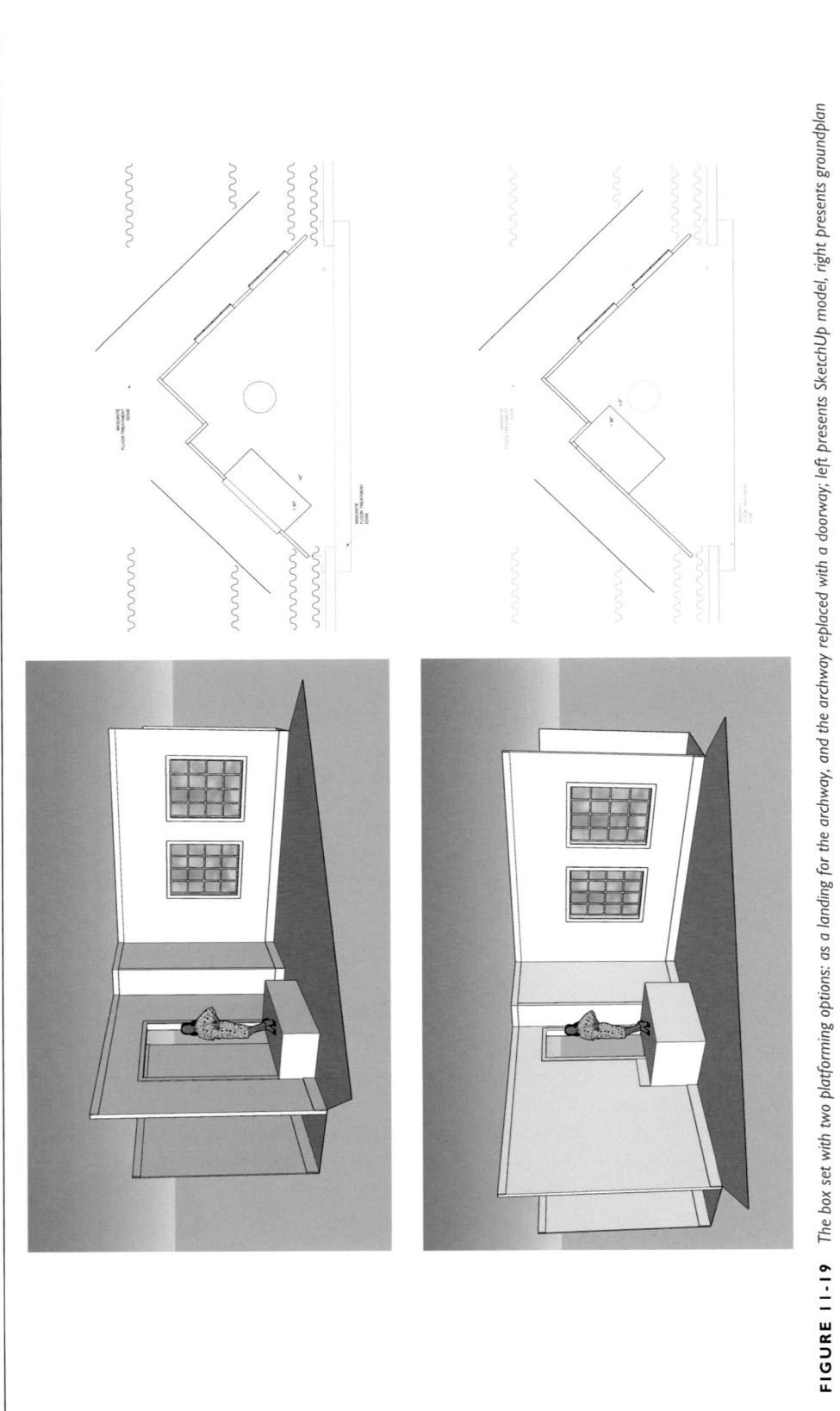

FIGURE 11-19 *The box set with two platforming options: as a landing for the archway, and the archway replaced with a doorway; left presents SketchUp model, right presents groundplan*

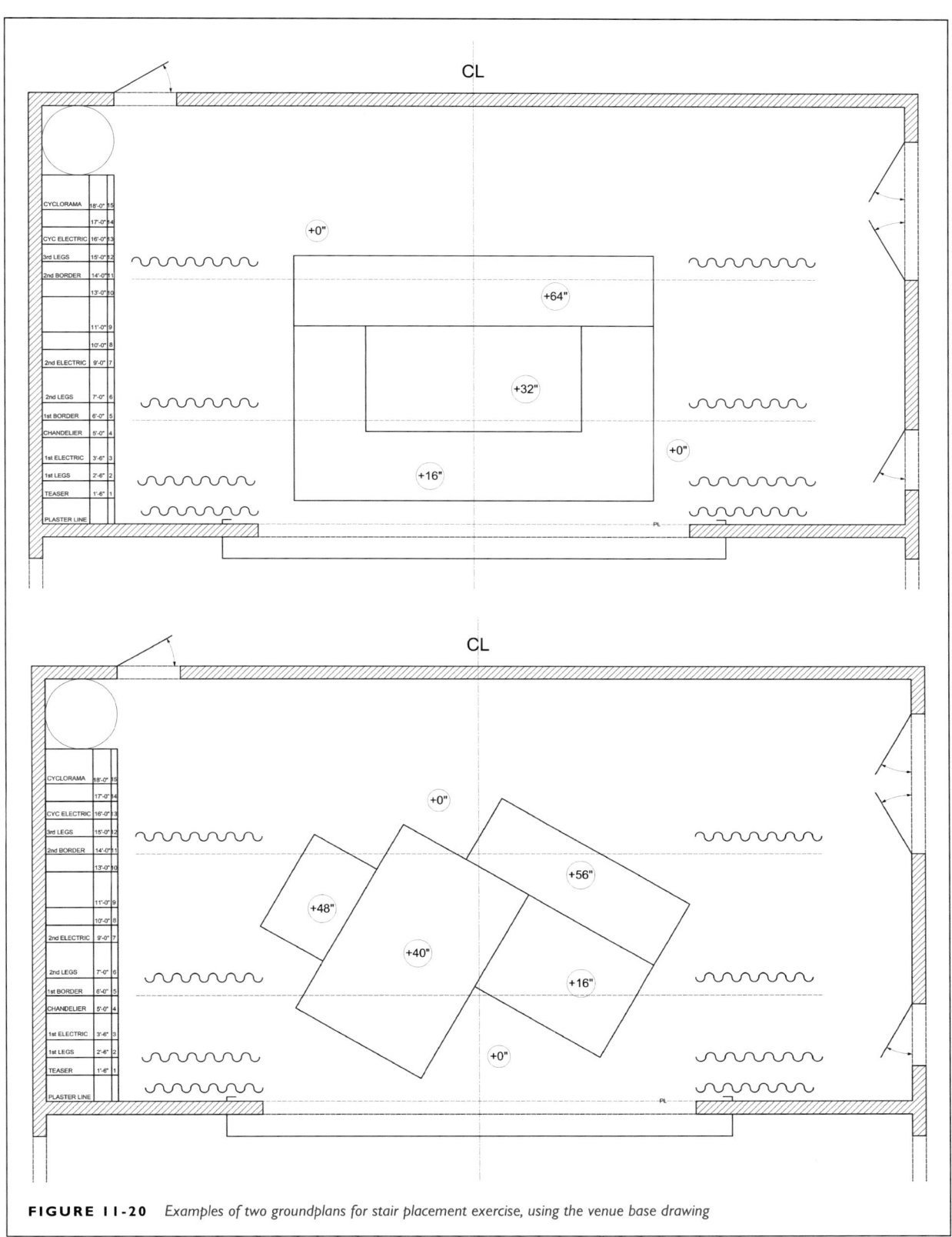

FIGURE 11-20   *Examples of two groundplans for stair placement exercise, using the venue base drawing*

Students

Using the groundplan(s) provided by your instructor, compare the given rise/run measurements against platform heights. Are adjustments to platform heights required to ensure that the rise/run remains consistent across the whole of the set? Which platforms require height adjustments? How might these height adjustments affect the overall look of the environment?

Lay out sets of stairs connecting the platforms. Which locations seem best? How much platform area is converted to stair footprint? Remember that stairs are not obstacles and can be seen as opportunities for blocking, providing further levels and spatial relationships between performers.

Are escape stairs needed? Where are those best placed? How far offstage do they extend?

Exercise: Stairs for a Unit Set with Curved Platforms

Instructor

Supply rough groundplans featuring round platformed areas of various heights. Prepare corresponding wish lists of platform heights, rise/runs, and suggested flight width (Figure 11-21). Each onstage stair unit should follow the curve of the platform to which it ascends in order to create steps that appear from above as concentric arcs. Be sure the centers of platforms are indicated with center lines so that concentric stair outlines can be placed with relative ease. On the other hand, finding the platform centers can be an added geometric construction exercise.

After students are done, compare and contrast their work. What solutions were common? How much space did the footprint of each set of stairs take up on each level? Discuss sightlines and blocking and how the stairs affected them.

Students

For each groundplan provided by your instructor, calculate rise/runs and the number of required steps, adjust platform height accordingly, and place steps to access the platformed areas. Include escape stairs, if required. Each onstage step unit should follow the curve of the platform to which it ascends in order to create steps that appear from above as concentric arcs. You have esthetic freedom in the placement of steps, though they should all be visible to the audience. In scale, lay out the proposed steps on each groundplan. How much floor space is lost to the run of the steps? How wide are the steps, and do you think that width will be safe for performers? Where did you place escape stairs?

Why would it be useful to access that particular platform via escape stairs? Compare and discuss your decisions with your classmates. Was there a wide variety of choices, or did the platform shapes drive everyone toward similar configurations?

Exercise: Modeling Platformed Areas and Stairs in SketchUp

Select one or two of the unit set configurations from the previous exercises and model them in SketchUp. Add nosing (size and style to be student's choice) to the edges of all platforms and stair units. Add color and texture to all surfaces.

Copy the platforming into the empty venue model and add necessary masking and upstage soft goods (such as cyclorama, drop, or black traveler). Remember to include a human figure.

Exercise: Drafting Step Units

A **step unit** is a term generally applied to shorter stock sets of stairs, typically of four or fewer treads. They are meant to reasonably portable.

Instructor

Two drawings will be constructed, one for each step unit in Figure 11-22. The rise is 8″, the run is 12″. Each unit has three treads. They are constructed with ¾″ plywood. The first unit has flush edges around each step. The second step unit features treads that extend ¾″ on three sides, creating built-in nosing. Students may select the scale in which they feel the object is best described. Construction drawings may be included, expanding the exercise to four drawings (though design and construction drawings for each step unit may be placed on the same sheet, space allowing). Overall dimensions are provided by the rough sketch in Figure 11-22.

Students

Construct drawings of the orthographic views of the two step units in Figure 11-22. Draw each unit on its own sheet. If construction drawings are assigned, the design and construction drawing for each step unit may share a sheet, if space allows. Choose an appropriate scale and sheet size. The first step unit has treads flush to the stringers and risers. The treads of the second unit protrude slightly to create built-in nosing. Include borders and title block.

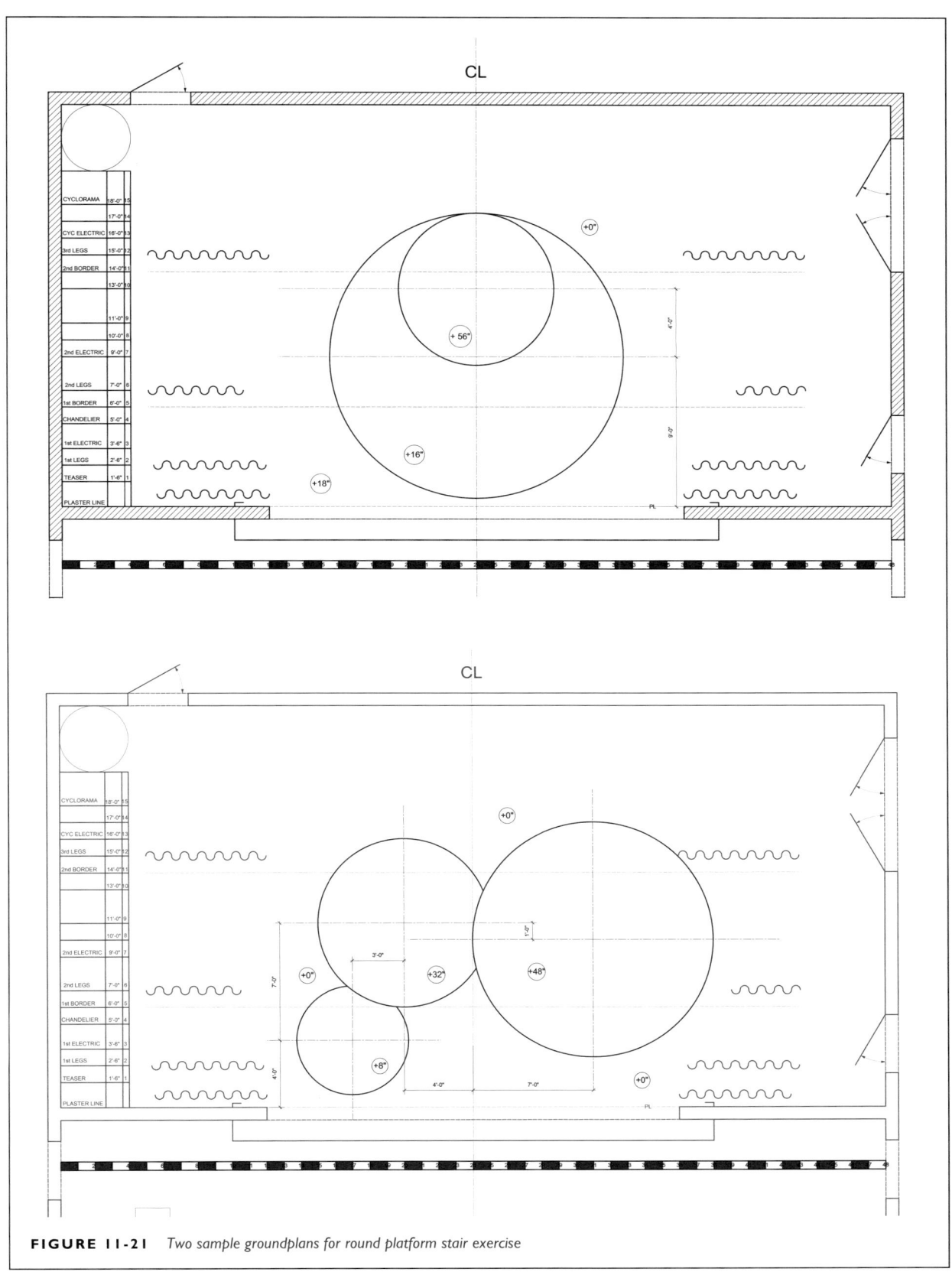

FIGURE 11-21   *Two sample groundplans for round platform stair exercise*

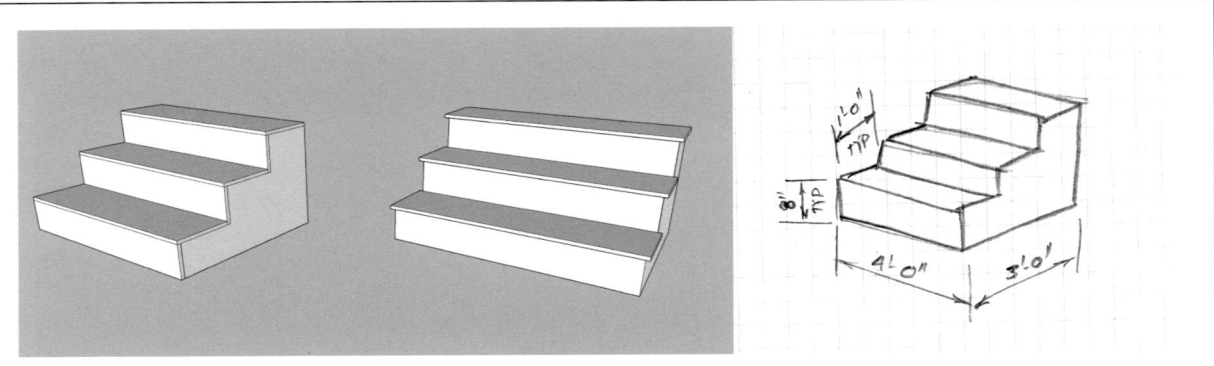

FIGURE 11-22   *SketchUp models of the two step units, one without nosing, one with ¾″ nosing; Rough sketch of step unit with measurements*

Step Unit #1

Design Drawing

As a design drawing, interior structure is not shown unless the designer needs to specify a particular construction approach. For this exercise, no hidden lines are required.

1. Do layout math to situate the views.

2. Lay out the front view. Extend guidelines into the areas for the top and side views.

3. Lay out the top view. Lay out the side view.

4. Lay out dimensions, labels, and notes. Be sure to note that the step unit is constructed with ¾″ plywood.

5. Darken the drawing. Create viewport(s). Import to a sheet layer. Add borders, title block, and title block information.

Construction Drawing

Consider the parts of the stair units. Three treads and three risers are required. The span of the treads is 4′–0″, so it's a good idea to add a center stringer, or carriage.

The placement of the risers determines some of the stringer dimensions. Typically, a riser's top edge sits under the front edge of the tread. The lower edge of the riser sits behind the back edge of the tread; the rear of the tread butts against the riser and is fastened to it.

At the top of the flight, the riser's thickness must be accommodated by the length of the stringer edge upon which the top tread sits. The units in this exercise use ¾″ plywood for all components, so ¾″ must be subtracted from length of the stringer's top step. This begins an offset so that the ¾″ lost by the riser at the rear is made up by the ¾″ added by the riser at the front. If the stringer is cut so that all of its steps are 1′–0″ deep, the unit as a whole will end up ¾″ too long (Figure 11-23).

To keep the stringers from splaying apart at the rear of the unit, a toggle is placed across the rear opening. It's good to have one at the bottom and one at the top. The top one will also support the rear of the top tread. Since a carriage is included, notching the carriage where it meets the toggle allows the toggle to be fastened securely to the carriage. This also allows the toggle to remain a single piece of wood

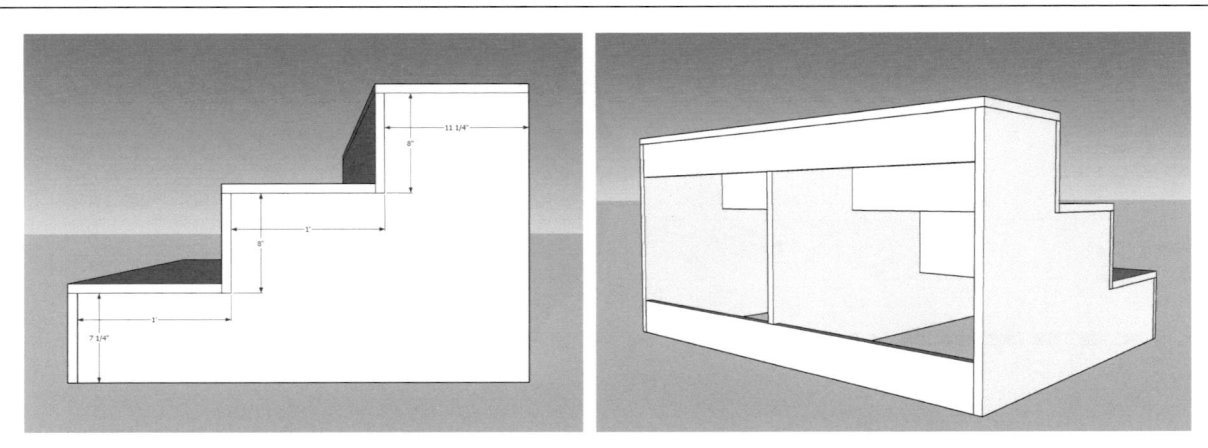

FIGURE 11-23   *SketchUp models showing: 1) stringer measurements; 2) location of toggles and carriage*

rather than two separate pieces on either side of the carriage (Figure 11-23).

A detail drawing of the carriage may be necessary to clearly describe the notches.

The treads and risers extend the full width of the step unit. Their ends will be seen in the side view. Include hidden lines for all internal edges. Consider whether a section is necessary to describe interior structure more clearly. A rear view may be required to describe toggles.

Step Unit #2

Follow the same process and construction approach for Step Unit #2, except add a ¾″ overhang on the front and both sides of each tread. How does dimensioning layout shift with this added feature? Are hidden lines necessary in the design drawing to adequately describe the overhang, especially in the top view? Does enlarging the scale to 1″ = 1′–0″ aid in understanding the meaning of the hidden lines?

Homework: Modeling Two Step Units in SketchUp

Students

Construct two models for each step unit. The first model will use the blunt approach: treat the step unit as a single mass, Push/Pulling it into shape. For the second model, use your drawings to create a parts list to construct each component; use them to assemble the stair unit.

Exercise: Drafting a Stair Unit with Banister and Landing

The photo of the exterior staircase in Figure 11-24 serves as the research image for this exercise. The newel posts are simple, the balusters are supported by a baserail, and lattice is inserted below the stringer and beneath the landing. The treads hang over the edge of the stringer. The landing balustrade in the photo is a different style than the banister; that will be changed to match. The newel posts at the top and bottom of the banister will be stylistically merged so that all newel posts are identical.

To keep the structure simple for this exercise, the handrail is 2×4. The balusters are 1″ × 1″. The base rail will be eliminated so that balusters sit on the treads. The newel posts are all 4″ × 4″, with the addition of a cap made of layers of 1×. All treads and risers are ¾″ plywood. Treads have a ¾″ lip on their front and audience-facing edge. The rise is 8″, the run is 1′–0″. The whole structure is faced with lattice, held within frames made of 2″ wide strips of ½″ plywood. The landing edge is beefed up with a fascia and

FIGURE 11-24 *Research photograph of a staircase with banister*

nosing to make the platform look thicker than it really is. The legs holding up the landing will be 4″ × 4″. The lower newel post sits on the first step, in line with the riser's face (not flush with the tread overhang edge). The upper newel post sits on the landing and lines up with the landing's leg.

Figure 11-25 presents a model of the revised staircase to be drafted. Figure 11-27 presents a rough sketch with most of the measurements.

The first step is to break the structure down into components to figure out what the major outlines and masses will be. The structure pictured in Figure 11-25 has six major exterior components:

a. the staircase (stringers, tread, risers)

b. the landing (platform, fascia, legs)

c. the banister (handrail, balustrades, newel posts)

d. the landing's front balustrade (handrail, base rail, balusters, newel posts)

e. the landing's left balustrade (handrail base rail balusters, newel posts)

f. lattice panels (three of them)

By considering these elements separately during the planning stage the structure reveals itself as a much more approachable object. This list also begins the process of determining which elements require breakout detail drawings,

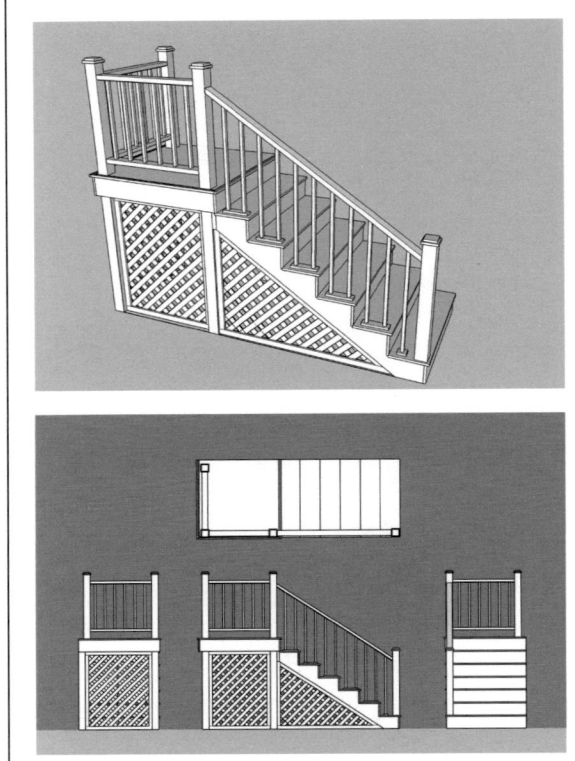

FIGURE 11-25 *SketchUp model of staircase for exercise: top, model in perspective; below, copies arranged orthographically in parallel projection*

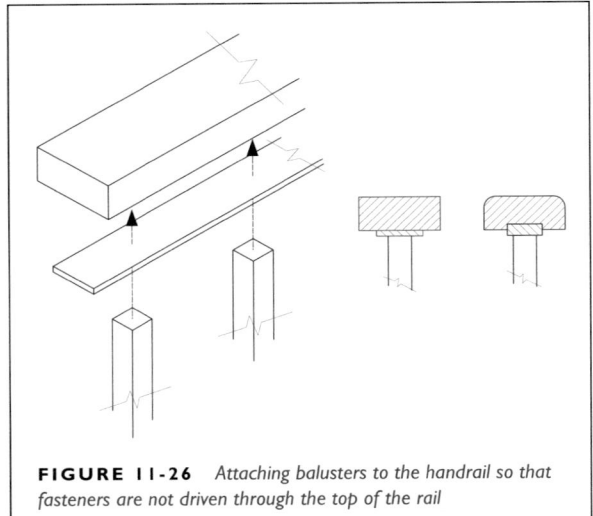

FIGURE 11-26 *Attaching balusters to the handrail so that fasteners are not driven through the top of the rail*

Discussion and Exercise: Laying out the Staircase Drawing

This discussion will not provide detailed instructions. Rather, a series of rough sketches and discussion points are presented for both the instructor and students. Take a moment to do some rough layout work based on Figure 11-25 before looking at the sketch in Figure 11-27 and reading through the following discussion. Situate views, place detail drawings. Not all dimensions are given in Figure 11-27; some will need to be extrapolated during the layout process, particularly the height of the balusters and the angle of the banister's handrail.

To begin laying out the design drawing, start with the largest mass of the object: the staircase and landing. Every other feature of the object is attached to this underlying structure. Think of the views creating an exchange of information; some structures may be a little easier to lay out in a particular view. If you can't visualize the front view of a feature, try working it out in the top or side view and then transfer the discoveries into the front view.

Because this is a reasonably complex object, a number of detail rough sketches are required before drafting can commence. The overview begins the process; subsequent detail sketches fill in the gaps so that drafting can proceed more confidently. Detail information is generated by the integration of research and production needs with information about typical staircase construction, adjusted to fit research images, and then further adjusted to fulfill the aesthetic needs of the scenic environment.

Figure 11-28 describes how the balusters, handrail, and stair treads/risers relate to each other. The balusters are paired on small 1× plates which are fastened to each tread. The tops of the balusters are fastened to the ½″ plywood strip, which is in turn fastened to the underside of the 2×4 handrail. This arrangement is most apparent in the front

and how dimensioning will work in cramped areas such as the railing balustrades. The standard orthographic views will do a good job of describing this rectilinear structure. Both side views will likely be desired, as the view from the right shows the stairs, and the view from the left shows the landing-end balustrade and lattice. It is useful to include breakout detail drawings for the banister and two balustrades, the newel posts (build four), and the lattice frames. Construction drawings will have to open up the staircase to show interior structure; carriage, legs, bracing, etc.

There are multiple ways to attach balusters to handrails; the simplest and most brutal way is to drive long fasteners through the top of a wooden handrail down into the balusters. Metal banisters typically feature balusters welded to the handrail. Most wooden handrails consist of two parts: the handrail itself, and a strip that runs along the underside of the handrail. Fasteners are driven through the top of this strip into the tops of the balusters. The strip is then fastened from below into the underside of the handrail (Figure 11-26). This way, no fastener heads mar the finish. This strip is often seated in a channel along the underside the handrail. For this exercise, a 2″ wide strip of ½″ plywood will be used, secured to the underside of the 2×4 handrail. A detail drawing depicting the banister/baluster connection will be useful.

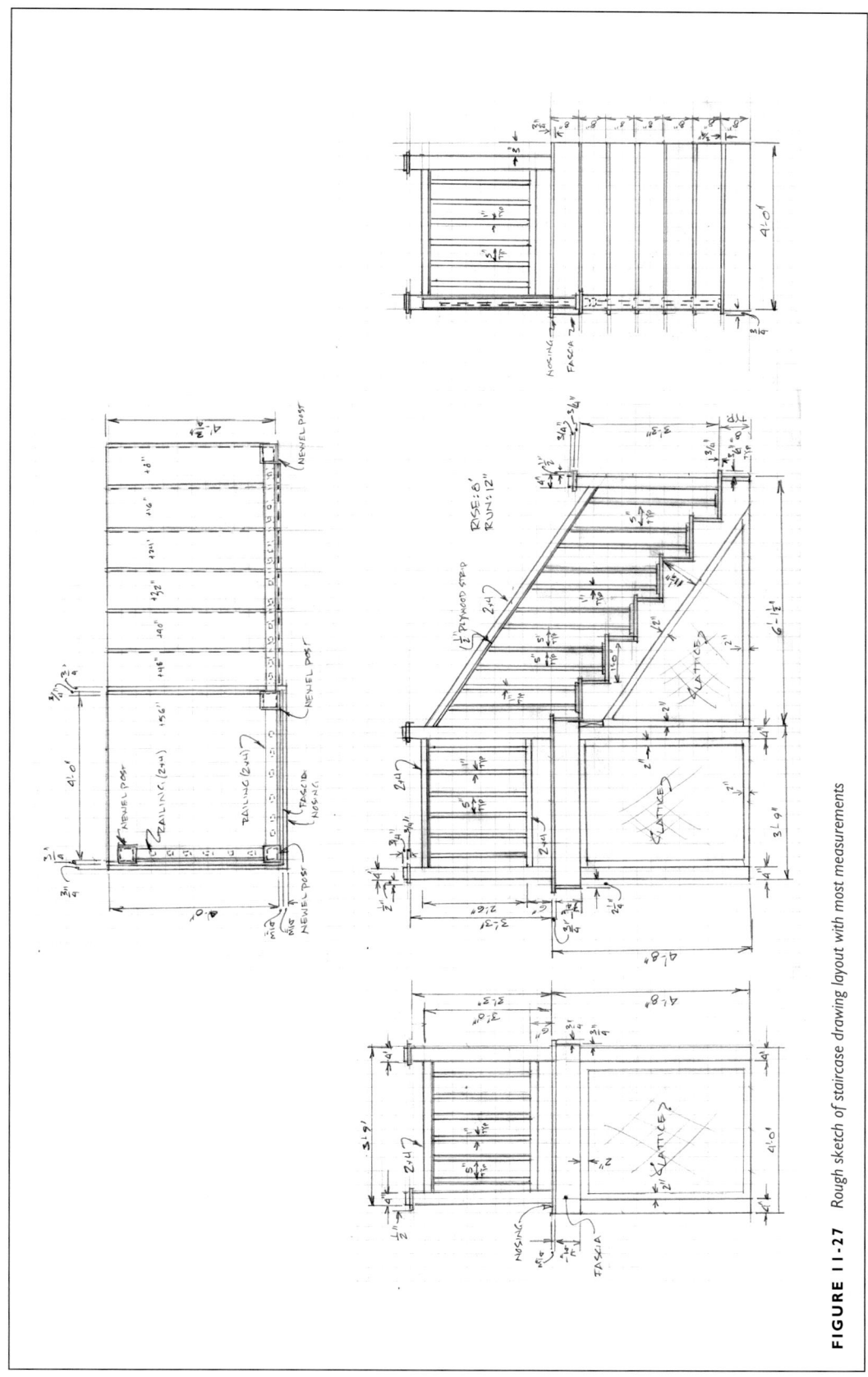

FIGURE 11-27 *Rough sketch of staircase drawing layout with most measurements*

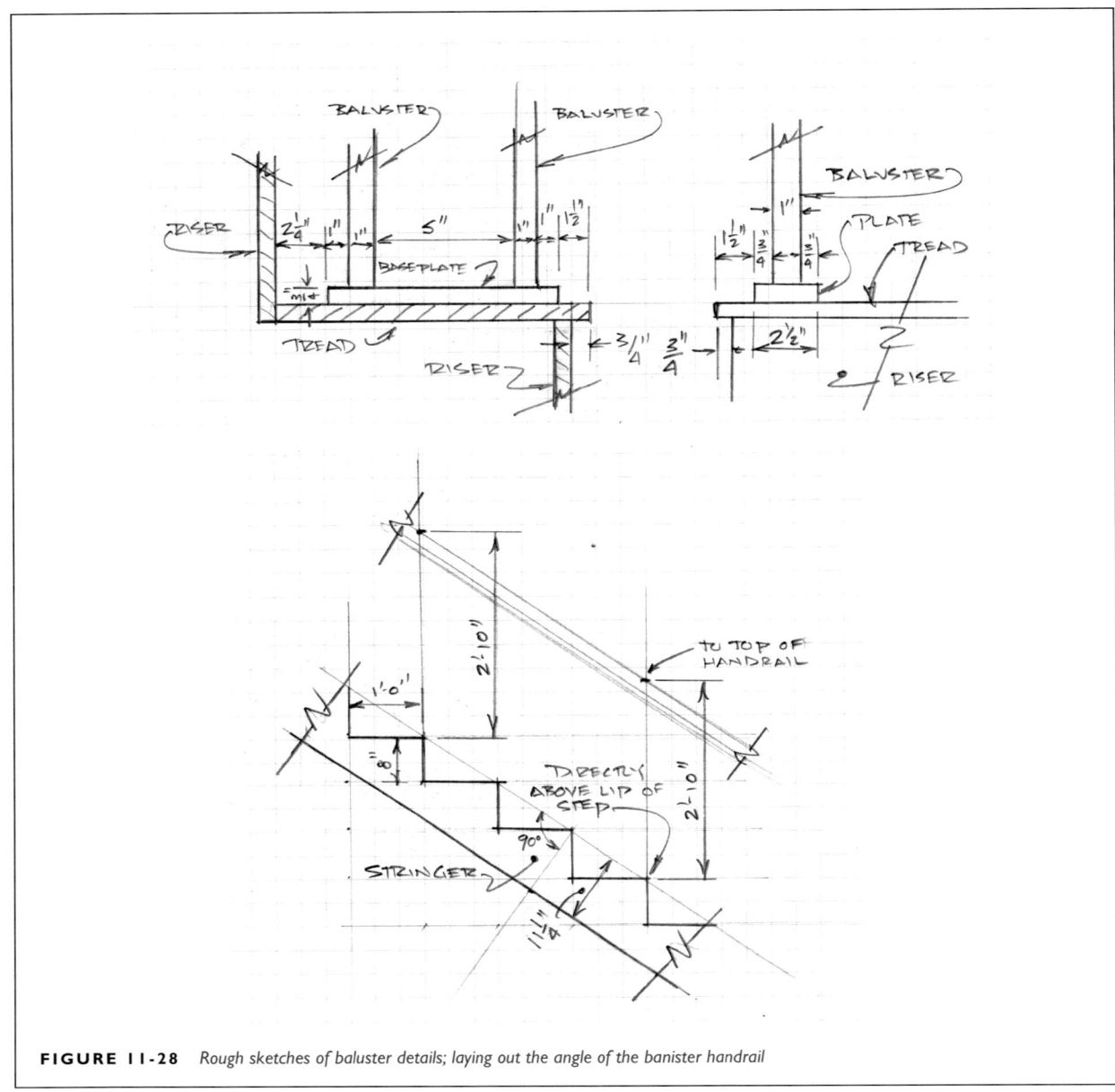

FIGURE 11-28 *Rough sketches of baluster details; laying out the angle of the banister handrail*

view of the staircase. In the side and top views, the balusters are hidden behind or beneath the handrail and newel posts; many overlapping hidden lines in a very narrow space would be needed to include the balusters in those views.

To find the angle of the handrail, the easiest method is to measure up from the face of the steps and connect the dots. Figure 11-28 indicates that the height is to be measured from the top of the tread to the top of the handrail (2′–10″). Lay out the handrail before drawing the tread nosing to ensure that the measurements line up with the face of the riser rather than the forward edge of the nosing. This angle determines the height at which the handrail attaches to the newel posts.

After the handrail and baluster base plates are laid out, the balusters themselves can be added. In each pair, the forward baluster will be shorter than the rear one. After laying out the first pair, copy and paste to lay out the remaining pairs (if in Vectorworks). The first step (the one with the newel post) contains a single baluster; its placement should match the rear baluster seen in the pairs. Shorten the base plate accordingly.

The front and side landing balustrades are identical and share a corner newel post. If you draw construction elevations of both the front and side balustrades, you'll have to decide whether both drawings include the shared newel post or whether to leave it out of one view to avoid

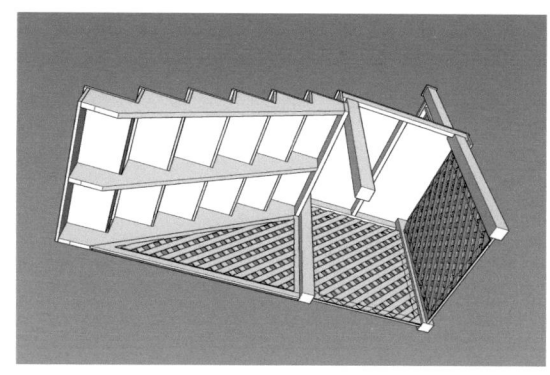

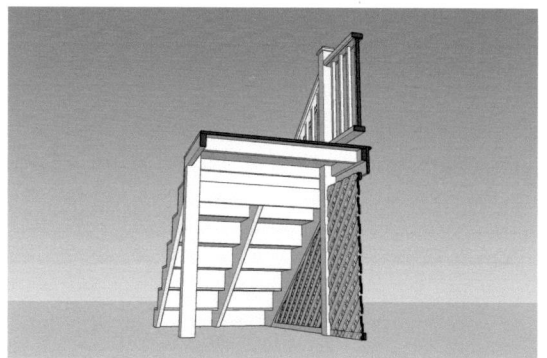

FIGURE 11-29 *The SketchUp model viewed from beneath (upper) and as a vertical section (lower), revealing the notched leg configuration choice*

accidental replication. The top view, though, should make clear how many newel posts are required in which positions.

All balusters are 5″ apart. For the landing balustrades, count the number of spaces between the balusters and multiply by 5. Then add the total width of all the balusters. This will give you the overall width of the set of balusters. Subtract this from the distance between the newel posts, then divide by two. The final number is the distance of the first baluster from the newel post, and will center the array of balusters between the newel posts.

The landing's legs are 4″ × 4″ (not 4×4). The platform is 4′-0″ by 4′-0″, with fascia and nosing attached to its side. If a stock wooden platform built with 2×4 framing is used, how the legs are attached becomes a further construction consideration. The span between the outside edges of the legs should be 4′-0″. If the legs are built and attached against the inner surface of the platform framing, they are set back 1½″. One solution is to notch the top of the leg so the 2×4 frame sits in the notch, allowing the face of the leg to be flush with the outer face of the platform frame (Figure 11-29).

The placement of the landing legs determines the sizes of the frames holding the lattice. Similarly, the width of

the stringers determine the size of the lattice frame beneath the steps. In this case, the stringers are cut from 2×12s, which means their actual size is 1½″ × 11¼″.

To lay out the width of the stringer in the front view (Figure 11-28):

1. Lay out the rise and run for the full length of the flight
2. Draw a line connecting the lip of the steps
3. Draw a line perpendicular to the angled line (use the two-triangle technique if drafting by hand)
4. Measure 11¼″ along the perpendicular line
5. Draw a line parallel to the first angled line through the 11¼″ measurement

Once the perimeters of the lattice openings are laid out, you'll have determined the overall size of each lattice frame. For this project, the frame will be made from 2″ wide strips of ½″ plywood, fastened to the face of the lattice. To specify the lattice, go to a lumber supply website and search for wooden lattice (it's typically available in 4′-0″ × 8′-0″ sheets); look for a specification for the width of the lattice's strips.

Lattice is constructed of strips lying in one direction with perpendicular strips laid atop them; it's not a basket weave where the strips are interwoven (Figure 11-30). To draw lattice:

1. Lay out diagonal guidelines in one direction
2. Lay out diagonal guidelines in the other direction
3. Darken one direction of lines
4. In Vectorworks, create a class for visible lines with a light gray fill. Use the Rectangle tool and the Irregular Polygon tool to outline the spaces between slats. If hand drafting, darken alternate strips of the lines running in the other direction; shade spaces if desired.
5. If in Vectorworks, turn off guidelines and restablish perimeter lines as required.

Shading the gaps helps snap the lattice into resolution and show the reader that those are indeed open spaces. You can also search the Hatch libraries for a lattice pattern; be sure that it matches the proportions of the drawing.

Another consideration is how much of the interior structure is to be seen through the lattice. If you wish to hide the interior structure, include a note that a fabric like black duvetyn should be attached to the interior of the lattice frames. If seeing the structure is acceptable, you might specify a paint treatment such as "paint all visible interior structure black," or "paint interior structure to match weathered exterior finish."

While the design drawing generally shows the staircase and landing as a single object with breakout detail drawings, the construction drawings will explode the structure to

   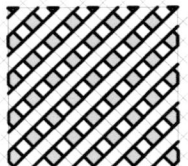 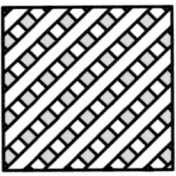

Use Double Line Tool to Lay Out Diagonal Guidelines

Lay Out Diagonals in Other Direction

Darken Lines in One Direction

Use a Class with Thick Lines and Light Gray Fill to Draw Alternating Squares

Turn Off Guidelines and Reestablish Perimeter

FIGURE 11-30 *Constructing lattice in Vectorworks*

describe the individual construction components. Remember that all the dimensions given for the individual components must allow them to fit together properly. This where building a model can be very useful. Use the given dimensions to ensure that the parts fit as they are intended to.

The staircase requires support. The upper step should be bolted to the landing structure; one approach is to install a 2×4 or 2×6 under the back edge of the top tread and between the stringers, then fasten this to the landing's legs. If the upper portion of the staircase cannot be securely fastened to the landing structure, the top of the stairs must be supported by its own legs. It's not a bad idea to also place a leg at the stair's midpoint. Tying the legs together with stretchers to prevent prevent splaying is also a good strategy. Don't rely on the lattice panels to perform the function of a stretcher; they are cosmetic. The width of the stairs makes including a center carriage is a wise choice.

What other breakout detail drawings are necessary for the construction drawings? From which views are interior structural features best described? How many cutting planes and section views are needed?

REVIEW OF DETAIL PLATE SYSTEM OF WORKING

1. Do the research. While a large amount of research has certainly already gone into the esthetics and appearance of the overall design, this is the time to double down on technical information, regardless of whether it's a design or construction drawing. Determine materials, preferred construction methods, finishes, etc., before starting to draft. Determine which photos and model images can be included to provide visual reference for fabrication.

2. Plan the detail plates. Work large to small, perimeter to detail. Group sheets by item and theme. Work out indexing from sheet to sheet so the reader knows where to look for subset detail drawings.

3. Sketch the detail plates so that you know exactly what you intend to place on each one, and where. Without a plan in place it's difficult to know what to change, how to change it, and how to keep track of the changes; you can't change your mind if you haven't already made a choice. Make as many small detail sketches as necessary to figure out a structure. Determine what section views may be required. Determine what detail breakout drawings may be required.

4. As you lay out the various scenic components, leave adequate white space for dimensions and notes. Start drafting by laying out perimeters and center lines, then work your way through major features toward details. Allow views to inform each other; as you lay out one view, transfer pertinent information into the other orthographically related views. If you can't figure something out in one view, attack a different view and transfer that discovered information into the trouble view.

5. When in doubt, note and label.

NOTES

1. John Blurton, *Scenery: Drafting and Construction for theatres, museums, exhibitions, and trade shows* (New York: Theatre Arts, 2001), p. 29.
2. Patricia Woodbridge and HalTine, *Designer Drafting and Visualizing for the Entertainment World*. 2nd ed. (New York and London: Focal Press, 2013), p. 215.
3. John Holloway, *Illustrated Theatre Production Guide*. 2nd ed. (New York and London: Focal Press, 2010), p. 312.
4. Paul Carter, *Backstage Handbook: an Illustrated Almanac of Technical Information*. 3rd ed. (Louisville, KY: Broadway Press, 1994), p. 237.

CHAPTER 12

DRAWINGS FOR LIGHTING DESIGN

This chapter discusses the drawings generated by the lighting designer for use by the electricians.

The first module discusses a typical package of drawings and reviews vocabulary, as stage lighting is a jargon-rich field with multiple terms applying to the same objects. The second module outlines a typical light plot layout process for proscenium and black box (studio) style venues. The third module discusses lighting sections and the detail plates that are constructed for positions such as booms, ladders, and set mounts.

TOPICS AND GOALS

- Vocabulary terms

- Lighting graphics and symbols

- Reading a light plot

- Drafting the light plot

- The lighting section

- Detail plates (booms, ladders, etc.)

MODULE 1: LIGHTING VOCABULARY AND STANDARD GRAPHICS

The light plot is the primary technical drawing generated by the lighting designer in the execution of a lighting design. As such, it is a distillation of esthetic and technical choices. The plot alone gives the reader very little sense of what the design will ultimately look like. Even an architectural or display lighting design is not fully revealed until the lighting units are turned on, focused, and their intensities adjusted not only to the qualities of the object being lit but also to

the qualities of all the other lights being used. Live events are lit for the human eye, but in film and television camera settings and film stock determine the final appearance of the lighting design. 3D computer-generated renderings can come close to showing the client what the lighting will eventually look like, but this virtual model is still displayed on a 2D screen.

Unlike the more pictorial drawings constructed for scenic design and construction, light plots and their associated drawings are schematic diagrams. Symbols represent lighting units; an array of information is associated with each symbol. The light plot is a geographic description of the lighting rig and further information is located in a database/spreadsheet called a **hook-up**. The plot and the hook-up are cross-referenced tools and are to be used as such. Professional theatrical light plots from the early to mid-twentieth century often feature only the unit symbol accompanied by the unit number and channel. This is just enough information for the electrician to find that particular unit in the hook-up. Contemporary light plots typically place more information around the symbol, but keep in mind that the plot becomes less legible as more and more information is arrayed around each symbol (Figures 12-1 and 12-2).

The light plot is drawn so the electricians can hang, circuit, and troubleshoot the rig. It is used during the focus call to keep track of which units have been focused and will be marked up on site with copious notes. While the plot is revised and updated until opening, at the focus call the designer typically shifts to the hook-up as their primary onsite reference tool. In preparation for cue writing, the hook-up is condensed into a **magic** (or **cheat**) **sheet,** which is a handy one- or two- page graphic index of control. The plot comes out again when it's necessary to add, remove, troubleshoot, or resituate a lighting unit.

DOI: 10.4324/9781003154921-13

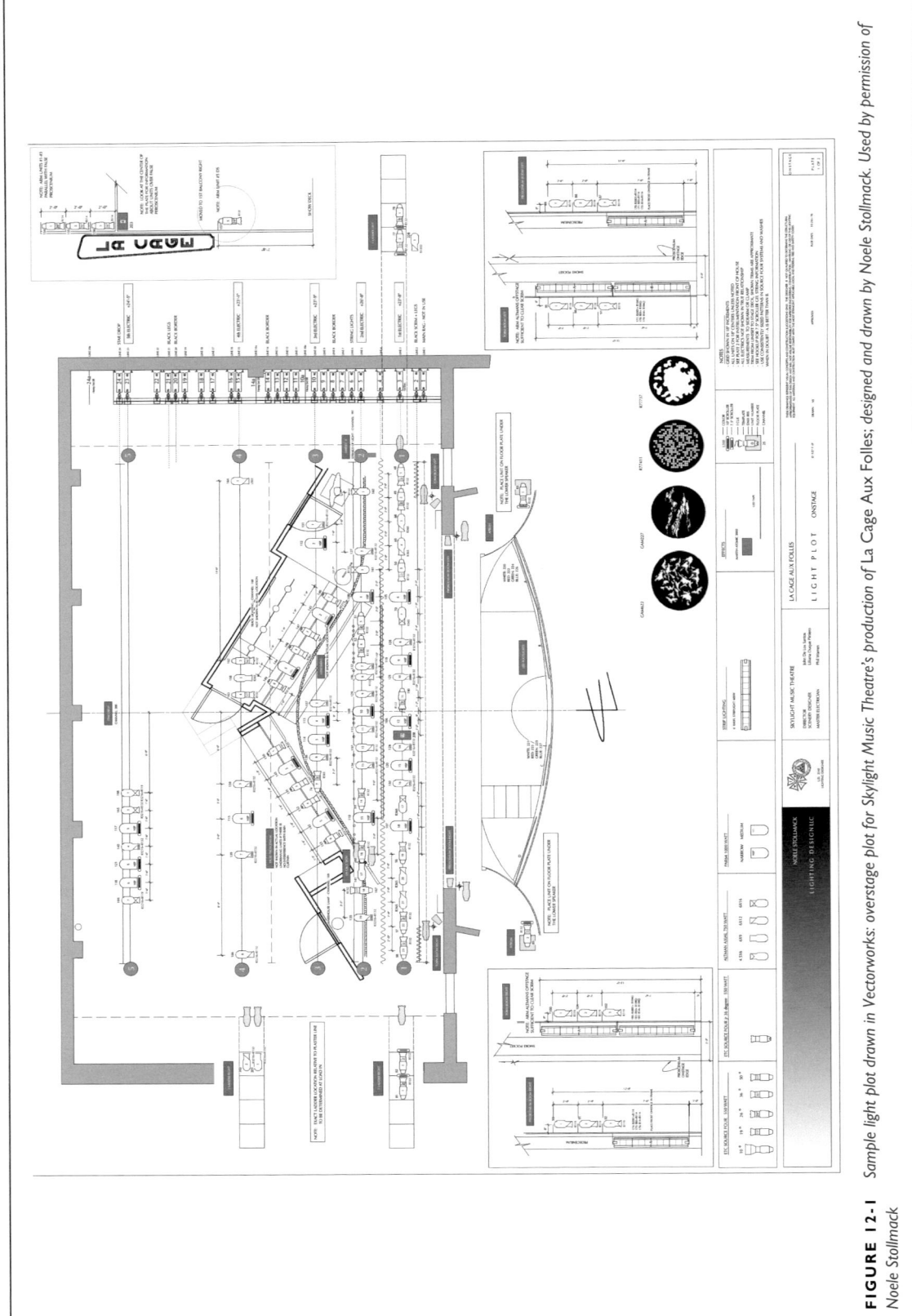

FIGURE 12-1 *Sample light plot drawn in Vectorworks: overstage plot for Skylight Music Theatre's production of La Cage Aux Folles; designed and drawn by Noele Stollmack. Used by permission of Noele Stollmack*

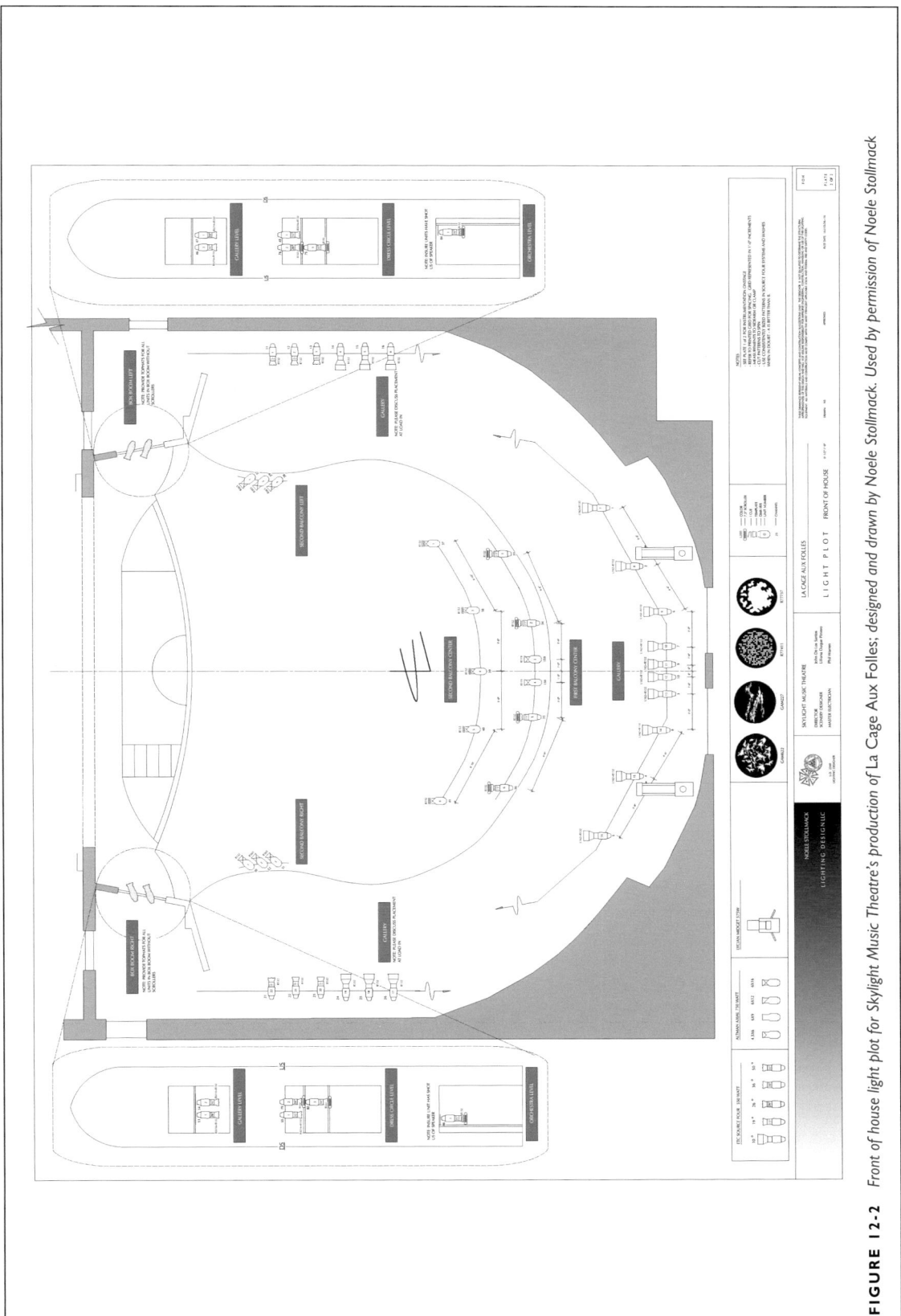

FIGURE 12-2   *Front of house light plot for Skylight Music Theatre's production of La Cage Aux Folles; designed and drawn by Noele Stollmack. Used by permission of Noele Stollmack*

The technical director integrates lighting information with that of the scenic, sound, and projection designers to create a global understanding of the electrical and data needs of the production. At opening, as the designer moves onto other projects, copies of the paperwork are filed with the venue for troubleshooting, maintenance, and archival purposes. Dance in particular requires extensive lighting record keeping, as pieces are remounted from season to season with perhaps gaps of years as pieces move in and out of the repertory. Stage managers should know how to read a light plot, as in many small venues they and their team also fill the role of house electrician.

A typical lighting drawing package contains a light plot, a lighting section, and detail drawings as needed. The plot is a top view of the entire venue with lighting positions and instrumentation taking visual precedence. The lighting section is a center line section of the full venue, indicating the heights and trims of lighting positions as well as sightlines. If a production has complex scenic movement, multiple lighting sections may be required. Detail drawings include anything that does not fit onto the plot and section, such as set mounts and booms as well as information pertaining to accessories or equipment.

As noted previously, the plot is a working drawing and should be printed/drawn in ½″ = 1′–0″ to allow the electricians to easily read the information and have room to add notes as the hang and focus progress. Due to printing from CAD files, 3/8″ = 1′–0″ is becoming more common and at least allows more legibility than ¼″ = 1′–0″. If the whole of the printed plot does not fit on a single Arch D sheet, the venue should be broken into sections. The division most typical for a proscenium venue is the overstage and the front of house (FOH). If possible, procure a plot from a previous production to see a typical plot layout for that venue.

During the hang, a copy of the plot may be cut apart or printed out position by position. Because these strips are sometimes stiffened by taping or gluing them to cardboard, they're often called **cardboards**. The plot can then be divvied up so that teams can take relevant portions to the areas in which they are working. Information discovered while working is written on the cardboard and returned to the master electrician, who collates the information on a master plot and hook-up.

Touring productions and other venues that require all equipment be rented or purchased may require further diagrams and drawings for the bidding process; however, most of this information is collected database style and used to create the **shop order**. The shop order and the preliminary light plot are generated in tandem, as bids typically must go out to vendors long before rehearsals begin or technical activities commence.

Lighting Vocabulary

The devices that emit light are referred to by a variety of terms across the entertainment industry. Technicians often refer to them interchangeably as a **unit, instrument,** **fixture, luminaire,** or **light**. A category of unit might be known by the manufacturer's brand name (Source Four, Leko, Klieglight, Vari-lite, Shakespeare) as well as by the type of lens/reflector configuration that differentiates one category of device from another (ellipsoidal, Fresnel, PAR). The term "unit" will be used for the rest of this chapter to refer to the piece of equipment that emits light, just to keep things straightforward.

Until the end of the twentieth century, the vast majority of lighting units used an incandescent light source—a **lamp**. An incandescent lamp is a filament enclosed in a glass envelope (bulb). Electricity runs through the filament and resistance creates heat and light. The lamp is mounted in front of a **reflector**, which collects the light coming off the rear of the lamp and reflects it back through the lens(es) at the front of the unit. An incandescent unit with these three basic components (lamp, reflector, lens) that is clamped to a lighting position and aimed at a single point on the stage throughout a production is called a **conventional**. The only variable that can be controlled during the course of a performance is the lamp's brightness (**intensity**). Only one **channel** of control is required to operate a conventional. The flow of electricity sent to the lamp is controlled by a **dimmer**, which sends out the amount of electricity as directed by the **control console** (light board, **desk**) (Figure 12.3). The channel's setting on the console tells the dimmer how bright to make the lamp's output: "Channel 1 at 75%" means the unit controlled by Channel 1 is at 75% of its full intensity.

Matching a channel to a dimmer is called **patching**. In analog days, this was a physical act. The cable or wire that a unit is plugged into is called a **circuit** (abbreviated "ckt"). Circuit is also a verb: one circuits a unit by plugging it into a circuit. Circuiting is the phase of the hang during which all of the units are plugged in. The circuit runs to the dimmer

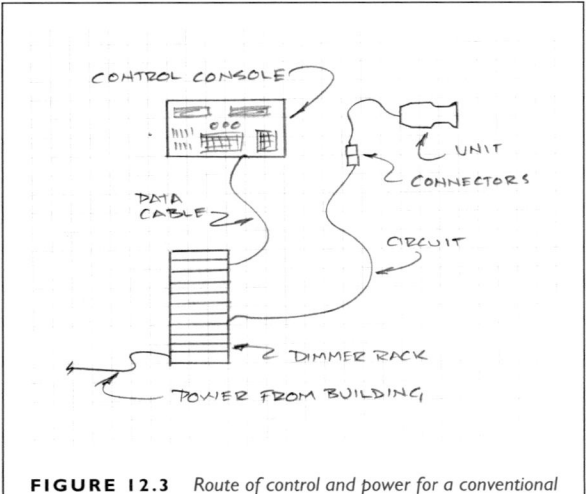

FIGURE 12.3   *Route of control and power for a conventional unit*

rack and is connected to a dimmer. Circuits and dimmers are numbered to keep track of them; for example, Unit #4 on the 1st Electric may be plugged into circuit 43, patched into dimmer 8, and controlled by channel 15.

When circuits are not hardwired or plugged directly into the dimmer of choice, a mechanical interface called a **patch panel** is used, though they are becoming more and more rare. Some of them looked like old telephone switchboards where you pulled out the termination of the circuit and plugged it into a receptacle that led to the dimmer. With digital control consoles, newer venues are typically hardwired 1-to-1; circuits are permanently placed and numbered, and wires run through raceways and conduit to dimmers with the same number assignment. Assigning dimmers to channels then occurs within the lighting console's digital **softpatch**.

Units that have motorized or electronic features or attributes are generally called **automated**, **intelligent,** or **movers**. These attributes include movement (pan, tilt), color changing (via LED, glass filters, or scrollers), intensity, irises, and gobo choices. Many automated units use high-intensity discharge lamps. The intensity of these lamps cannot be controlled by varying the amount of electricity sent through them but is dimmed by shutters or louvers that are opened and closed.

Each attribute requires its own channel of control. A simple LED unit with an array of red, green, blue, and white LEDs requires at least five channels of control: one for each color plus overall intensity. Power for the unit is not controlled through a dimmer (except as an on/off switch); all units are powered up when the system is powered up (Figure 12-4). Attributes are then controlled directly through a data line to the console.

Gobos are also called **templates** and **patterns**. Television and film have a similar accessory called a **cucaloris,** or cookie. A cucaloris, strictly speaking, is a perforated sheet of plywood placed in front of a unit to create textured light. Gobos began as small perforated metal plates dropped into an ellipsoidal reflector spotlight to create textures and patterns. They are now also made in glass and plastic, with printed color images. "Gobo" will be the preferred term for this chapter.

Because a number of channels are required to control automated fixtures, the units have **addresses**. An address is the channel at which the attributes of that unit begin. For example, the LED unit mentioned above needs five channels. If you have three units you wish to control, the first unit might have an address of 1, meaning that the attributes of that unit will respond to signals from the console sent to channels 1 through 5. The second unit's address would be 6 (channels 6 through 10); the third's address would be 11

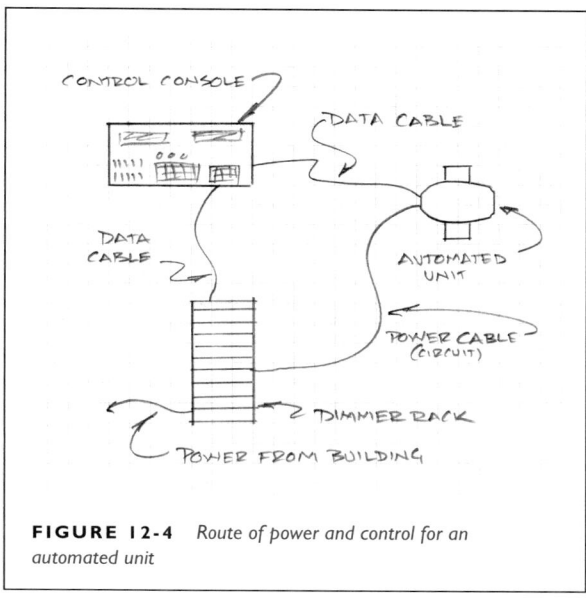

FIGURE 12-4   *Route of power and control for an automated unit*

(channels 11 through 15). Dimmers also have addresses. A dimmer pack is addressed so that each individual dimmer within the pack knows by what channel it will be controlled. Dimmer packs typically come in increments of 6 or 12, with 48 and 96 being the larger rack sizes. For a series of 4 small six-dimmer packs, their addresses would be 1, 7, 13, and 19 respectively.

The control console may have a library of manufacturer settings that can be assigned to the channel as specific units are softpatched. This allows the operator to control the unit via one channel with a subset menu of attributes instead of working with multiple channels at the same time. The unit's address may be 756, taking up 12 channels, but it can be patched as channel 1 with 12 subset functions.

Most lighting equipment speaks **DMX**, a software language. DMX is broken into sets of 512 channels (DMX-512). The first set of 512 channels is the first **universe**. The second, from 513 to 1024, is the second universe, and so on. As data networks become more complex, it's important to know in which universe a unit or device is addressed so that channel assignments can be made smoothly.

Basic Unit Categories and Their Symbols

Among conventional units, there are a handful of basic categories in common usage:

1. **Ellipsoidal Reflector Spotlight**. These units house an ellipsoidal-shaped reflector with the lamp placed at one of the ellipsis's centers (ellipses have two centers). Light is reflected forward through lenses, which are housed in a **lens train**, or **barrel**. Adjusting the barrel changes the quality of the beam of light. The second

center of the ellipse is where accessories such as irises and gobos are slotted (the **gate**). These units are also called **ERSs** and **Lekos**. Leko is a shortened form of Lekolite, which was the brand name of the first ellipsoidal units designed by Levy and Kook.

2. **Fresnel**. These units set a lamp in front of a reflector that used to be a segment of a sphere but has become more dish-shaped over time. The light passes through a Fresnel lens, named for Augustin-Jean Fresnel, a French physicist and optical engineer from the nineteenth century. The lamp and reflector travel together on a sled, their distance from the lens changing the size of the beam of light emitted.

3. **PAR**. PAR stands for parabolic aluminized reflector. Older forms of these units are little more than cylindrical cans into which a PAR lamp is placed (a PAR-can). The old-style lamp is a self-contained combination of reflector, lens, and internal lamp. The beam size is changed by swapping out the whole lamp for one with a different lens (WFL = wide flood, MFL = medium flood, NSP = narrow spot, etc.). The beam is oval shaped, and in older units, the lamp is rotated by reaching into the back of the unit, grabbing the ceramic base and socket and twisting (also called "spinning the bottle"). These were favored units for rock concerts because of their simplicity and punch. Newer versions contain a lamp and reflector mounted within the housing of the unit, with interchangeable lenses. PAR sizes are named by the number of 1/8″ increments in the lens diameter. A PAR64's lens is 64 1/8″ increments across.

4. **Cyc Lights**. Cyc is short for cyclorama. The shape of these units' reflectors is such that when set at the top or bottom of a drop or cyc, the light spreads a considerable distance across its surface. To light a full drop or cyc, multiple units are required, set at regular intervals across its distance; lighting the surface evenly usually requires both a top and bottom row. These are often housed in strips or arrays since to mix color with conventional units, multiple units are required.

5. **Strip Lights**. These are series of lamps laid out in a long rectangular housing, also intended to light surfaces such as drops and cycloramas. They are also often used as footlights. In conventional strip lights, there are typically multiple circuits (usually three) to allow for color mixing.

The symbols used for these categories of units are reasonably consistent (Figure 12-5). A symbol of a certain shape will likely correspond to one of these unit types. Not every lighting unit has a corresponding standard symbol, though most manufacturers now release CAD symbols for their latest products. If you need to fabricate a symbol, stick as close as possible to the base shape of the unit

type. Any electrician should be able to read any plot, and standardized symbols make that possible. USITT has a roster of standard symbol recommendations, and most lighting templates and CAD libraries are well labeled.

The Legend

Every plot contains a **legend**, which graphically explains what information is placed where around the unit symbol (Figure 12-6). The amount of information offered depends on the size and needs of the production. Every symbol requires at least enough information so that it can be easily located in the hook-up.

1. **Unit number**: Every unit has an individual name, and that is the Position name and the Unit Number. Every position starts its numbering at 1 and continues the numeric sequence until the end of the position is reached. The unit number is placed in the center of the symbol, in the main body of the unit (Figure 12-6).

2. **Channel:** The number by which the console will control the unit's attributes.

3. **Beam and accessory information:** Different categories of units have different features. ERSs come in a variety of beam spreads—the width of the beam emitted. Contemporary ERSs measure this in degrees (50, 36, 26, 19, 10, etc.), and there is a standard format for designating beam spread within the ERS symbol (Figure 12-7). ERSs may also have gate accessories like gobos and irises. PARs will include beam designation, lamp axis orientation, and wattage. The legend indicates where this information is to be found within the symbol; the description of individual beam spread symbols is placed in the key.

   There are standard symbols for different types of gate accessories, though there is variation. A gobo may be represented by a filled-in circle, a filled-in triangle, or a T. The author prefers a T because it is instantly recognizable. An iris may be designated by an unfilled circle or an I. The legend indicates where an accessory's symbol is located in relation to the unit symbol. Further information describing the accessory symbol is located in the key. Details and specifics about the accessory, such as a gobo inventory number or colors making up a scroller string, are included in the hook-up.

4. **Color**: This is written at the front of the symbol (or as close as possible to the front), and is the color media (**gel**) manufacturer's letter and number designation of that color (e.g., R25, L213, AP2280, Gam268). If no color media is used, the unit is noted **NC** (N/C), for No Color. This shows the reader that lack of color media is intentional. LED units may be noted as LED, since color is assigned through the console.

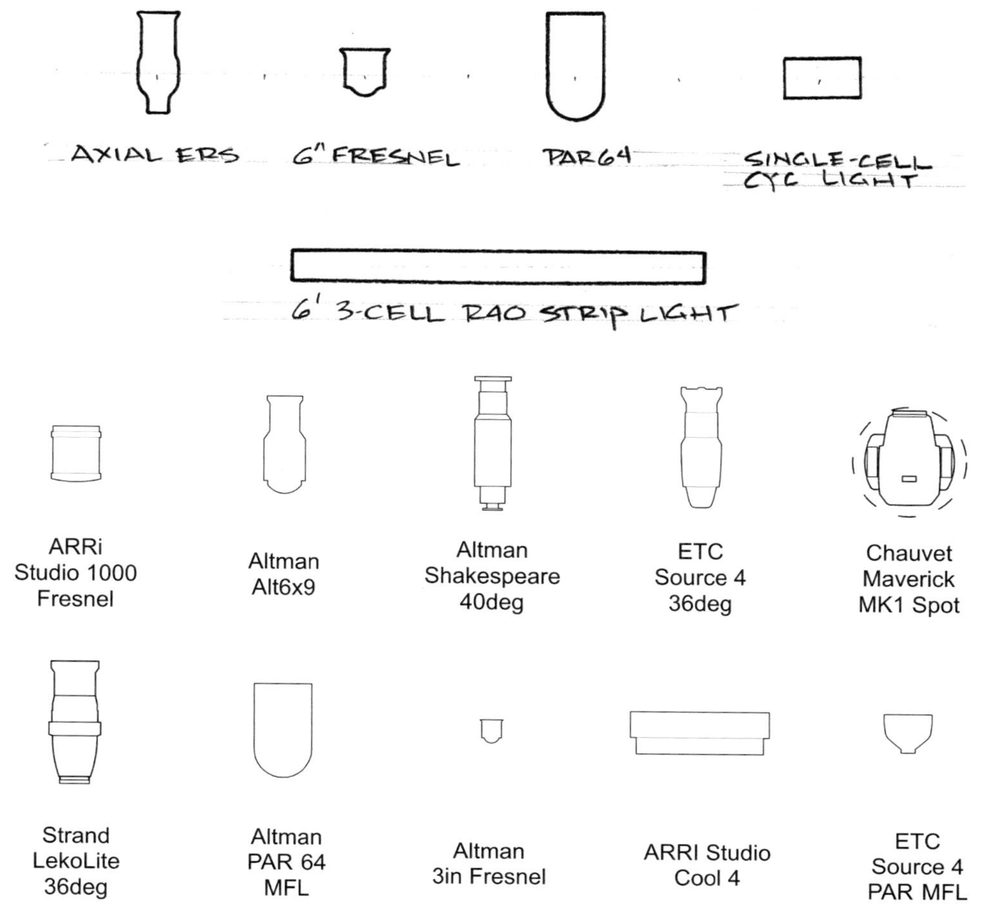

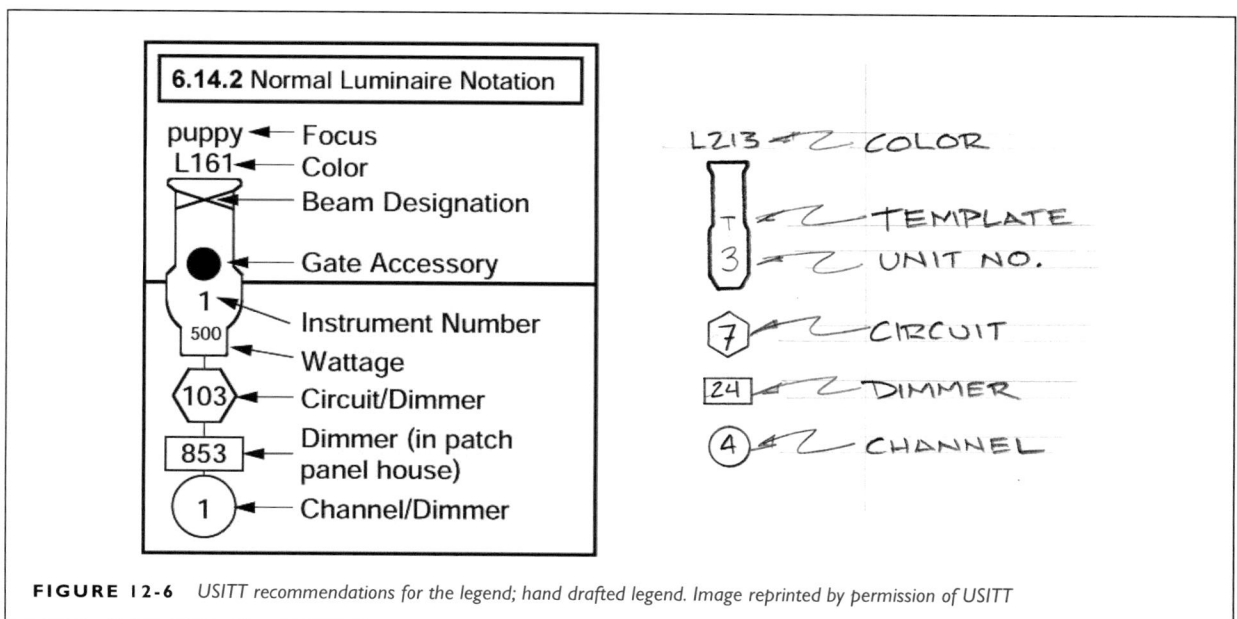

FIGURE 12-5   *Base symbols for ERS, Fresnel, PAR, cyc light, and striplight; hand drafted with template and via Vectorworks library*

FIGURE 12-6   *USITT recommendations for the legend; hand drafted legend. Image reprinted by permission of USITT*

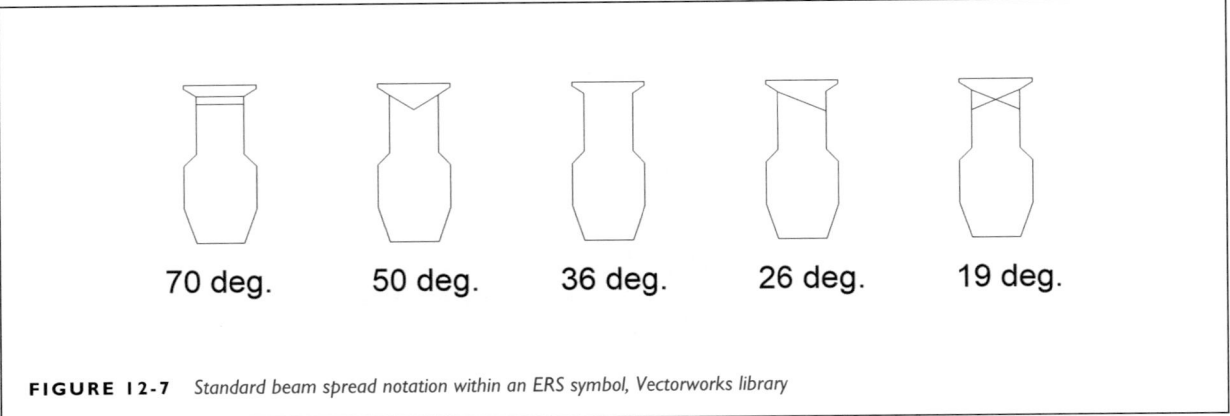

FIGURE 12-7 *Standard beam spread notation within an ERS symbol, Vectorworks library*

5. **Dimmer/Address:** This number is placed in a rectangle between the unit and the channel. Depending on the venue and working practices of the crew, the dimmer/address information may be generated as the hang commences and filled in by the electricians. If the production is such that dimmer/address assignments must be planned before the load-in, it is included in the paperwork.

6. **Circuit**: The circuit is written in a hexagon placed between the unit and the dimmer/address rectangle. Again, depending on the venue and working practices of the crew, the circuit information may be generated as the hang commences and filled in by the electricians unless production needs require it to be determined before the load-in.

   If the dimmer/address and circuit information are generated during the hang, the master electrician must be sure that this information is added to the permanent paperwork. Since this information is likely to be penciled onto the plot as it is discovered, the designer may choose to include empty shapes for these numbers so that notation can remain as neat and ordered as possible.

7. **Focus**: Some designers include the area of the stage where the unit is to be focused in a second tier beyond the color note. However, it's more text that can clutter up the plot and make it less readable. Focus information should be found in the hook-up, which is the document the designer refers to throughout the focus call. Detailed **focus charts** are drawn during the focus call to describe the final placement of the beam. Focus charts are important for touring shows, productions that run a long time, or productions expected to be remounted at a future date.

The Symbol as Drawn on the Position

Units are drawn with the line of their position (pipe) meeting the sides of the symbol. This indicates that the unit is to be clamped to that pipe. Units that are situated below the position, as a set mount or on the ground, are drawn as a crosshatched outline. If a pipe crosses the outline, the symbol is interrupted by the pipe to indicate that the unit sits well below it (Figure 12-8).

Placing the symbol ahead or behind the pipe usually means that you are asking the electricians to **yoke** the unit out. Yoking is the practice of loosely clamping the unit to the pipe and pushing it forward (or backward) and then locking it into place so that the unit no longer hangs directly below the pipe. A unit clamped so it sits above the pipe is **overhung** (or, "roostered"). When yoking or overhanging, the pipe must be secured so that it does not rotate because of the unit's offset weight. This practice is most often found in nontraditional venues with irregular lighting positions.

In pre-CAD days it was standard practice to orient the unit symbol in one of four cardinal directions: to the front, rear, right, or left. During the hang, the electricians aimed the unit in one of these four general directions. The unit was not pointed in its final direction until focus. With CAD, it has become common to point the symbol in the direction of its

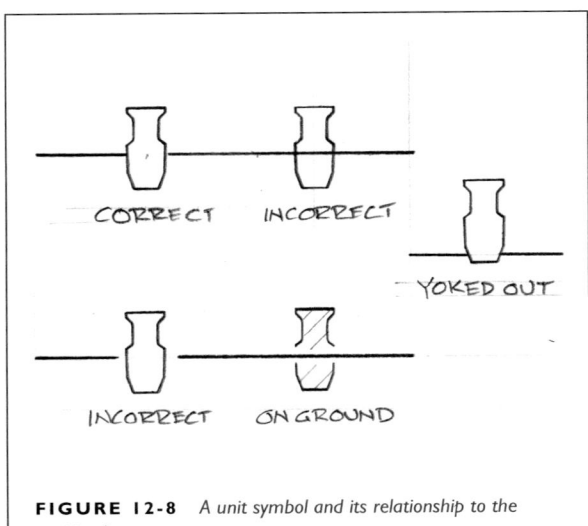

FIGURE 12-8 *A unit symbol and its relationship to the position line*

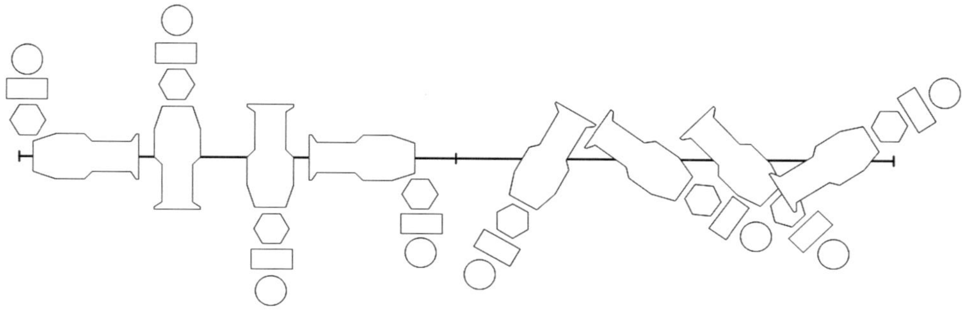

FIGURE 12-9   *Units drawn to cardinal directions; units drawn angled in their focus orientations*

ultimate focus. While this practice may allow an instructor to better understand a student's intentions when reviewing a plot, it also means that the information arranged around the symbol runs the risk of overlapping and interrupting the information of its neighboring units (Figure 12-9). While Vectorworks allows you to move text boxes around, much time can be lost finessing something that could have been done when the symbols are first placed. Of course, there are times when angling symbols is unavoidable or even preferred, such as when a lighting position itself is angled or contoured to an irregular space.

Currently, many students are encouraged to construct the light plot as a CAD 3D model. This can be useful for visualizing the focus of individual units or working out beam angles in a complex scenic environment. However, a 2D view of the 3D model is not the same as a drafted plot or section. What may be clear in a model may not translate directly or well into the 2D graphics recommended for a working drawing.

Essential Elements of a Light Plot

As with scenic groundplans and sections, there are certain elements that are common to light plots, regardless of venue or production.

Reference Lines

The center line; the plaster line if present.

Key

A list of all unit types used in the design, matching symbol to description (type, manufacturer, wattage, beam spread, etc. [Figure 12-10]). Some designers include all unit types in the venue's permanent inventory, in case something is added or replaced during the load-in and tech process. For pertinent unit types, include any notation associated with the unit type, such as lamp orientation or lamp/lens type. Include symbols for all accessories, such as scrollers, barndoors, and top hats.

Vectorworks has an Instrument Summary tool located in the Lighting Tool Set. This tool creates a key and legend that can be situated on the plot (Figure 12-11). The key is populated by all lighting unit types found on the drawing. After choosing the icon, click twice to affix it. It will take a moment to gather information. A window will then appear, allowing you to select from various options.

Legend

A diagram of all the information to be associated with a unit symbol, such as channel, unit number, color, etc.

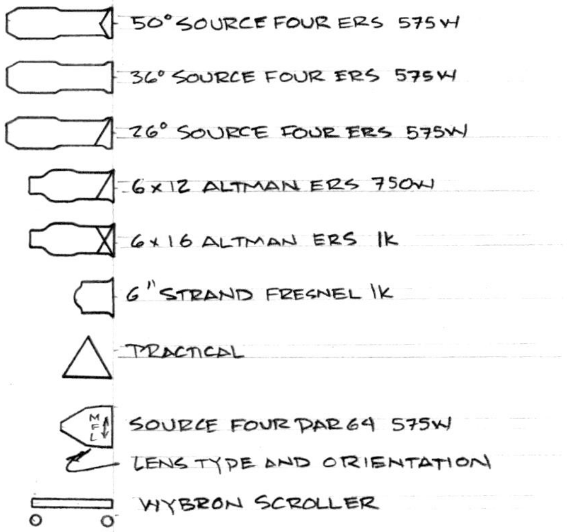

FIGURE 12-10   *A hand drafted key featuring symbols and their descriptions*

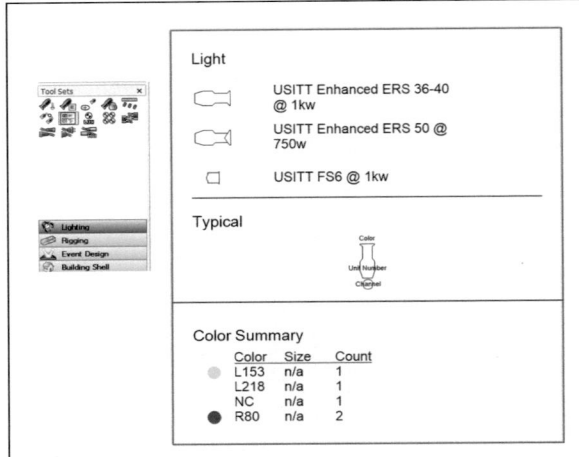

FIGURE 12-11   *The Instrument Summary tool icon and the summary it produces*

Lighting Positions

All permanent and created locations in the venue where lighting equipment is to be placed (hung). This includes overstage electrics, light bridges, dead hung pipes, ceiling catwalks, trusses, box booms, and balcony rails. For black box and studio venues, the full pipe grid is drawn. Created positions include such structures as onstage booms, ladders, units mounted to scenic structures, and trusses suspended from winches. Even if no units are hung on a position, it is useful to include it on the plot—something might be added during tech week.

Onstage positions are numbered from the plaster line, with #1 closest to it. FOH positions are also numbered from the plaster line. Position labels often include the number of circuits available and/or required at that position, as well as information on data network nodes and connections. Trim heights may be included. Some designers list the position's unit inventory and required number of circuits.

Lighting Units

Each unit represented by an appropriate, standardized symbol. Each symbol is accompanied by its legend information, including channel, unit number, color, etc.

Line Schedule (if present)

All battens that have lighting equipment on them are considered Electrics. Include trim heights, and if possible, the estimated weight of all equipment on the batten. All information on the scenic groundplan's line schedule is transferred into the light plot's line schedule.

Venue Architecture

The lighting positions take precedence, but as much venue architecture as possible is included to help readers understand the placement of the positions. Sightline information is included, as is the smoke pocket (if present). The edge of the playing space, whether apron lip or painted line, should be included. Any obstructions special to the venue are indicated, such as HVAC ductwork, steel beams, support posts, etc.

Scenic Structures

These are indicated with thin or light-colored lines interrupted by the lighting information as required (so the scenery appears to be situated below the lighting positions). This is generally restricted to outlines of the most important features—just enough to let the electrics crew know if a unit is to one side of a wall or hung directly over a piece of furniture.

Practicals (if present)

Practicals are denoted with equilateral triangles and labels.

Masking

Placement of masking is essential to determining sightlines and beam angles. Label all legs, borders, and other masking elements.

Notes

There is often a block of notes. These notes include things like "All units on 18″ centers unless indicated," "Final boom placement TBD by designer at focus," "No changes without designer's written approval," etc.

Dimensioning

Lighting units are typically hung on 18″ centers to ensure that they do not bump into their neighbors when focused. One way to dimension positions is to place tick marks every 18″ along empty stretches. For long gaps or irregular placement, include dimensioning from center axis to center axis (C-clamp to C-clamp) of the units on either end of the gap (Figure 12-12). Booms and other vertical positions include height measurements from the deck to the unit or sidearm. A scale bar is often included, especially if the size of the printed plot has yet to be determined, or it is possible it may be reduced.

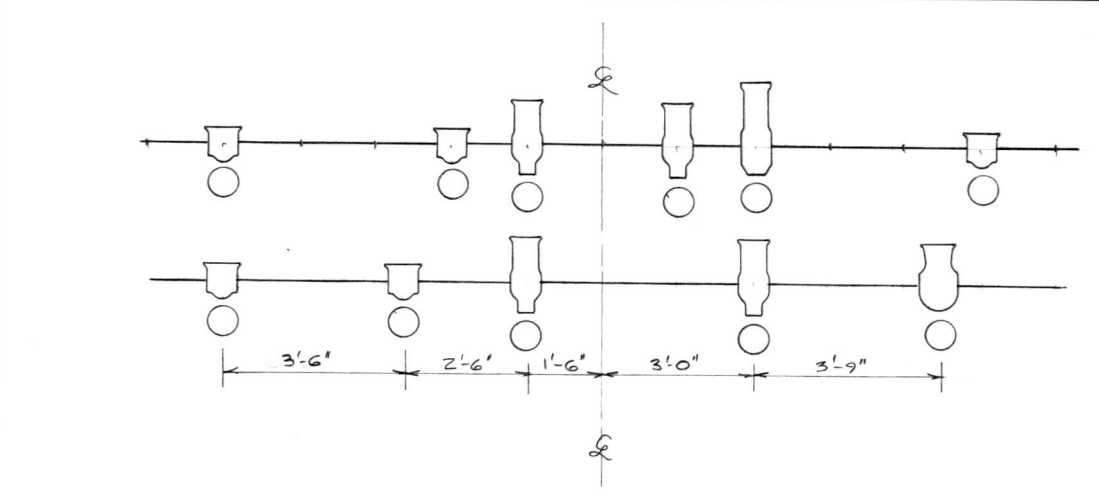

FIGURE 12-12 *Two ways to dimension positions: if tick marks at 18" centers are used, written dimensions are generally not needed; Include dimensions if tick marks are not present or if spacing is irregular*

Title Block

As with all drafting, the name of the designer, production company, name of venue, scale, liability disclaimer, etc. are included. For lighting, the name of the master electrician is included. If the designer is a member of United Scenic Artists, there should be space for the union stamp.

Discussion and Exercise: Reading a Light Plot

Instructor

Provide three to four sample light plots for the class to examine and discuss as a group. See Figures 12-13 and 12-14, as well as Figures 12-15 and 12-16, for examples. Reduced copies can augment the examination of a full-sized or projected version. The questions below can be used to guide the discussion. If possible, bring in some lighting units so students can compare their profiles to their symbols.

If the light plots represent designs in an accessible venue, take a field trip and identify the lighting positions, matching them to their placement on the plot.

Students

Examine the light plots provided by the instructor. Where is your eye drawn first, and why do you think that is? Does there seem to be any sort of visual starting point? What information would be most useful to know right away in order to read the plot? Does knowing what information is most useful first change where you look first? The array of lighting instruments scattered across the sheet should all be drawn in the same line weight and style, as they are the most important element of the drawing. This can make it a bit difficult to find a visual starting point, especially if you have never encountered a light plot before.

Where are the key and legend located? Are important notes easily located? Does the drawing seem cramped and complicated or is there enough white space to allow the eye to travel easily from feature to feature? Is there a center line?

How easy is it to read the information around the symbols? Does legend information overlap or bump into neighboring symbols or legend information? Is spacing clearly indicated? Are there tick marks or dimensioning? Is it easy to tell which lines represent lighting positions, and which lines represent scenic or architectural structures? Are all the positions clearly labeled? How easily can you find a given unit, such as "1st Electric Unit #4," or "Pipe N, Unit # 8?" Is there information that appears to be missing or placed in a manner contrary to the guidelines outlined earlier? Do you think it was a mistake or a choice? If a choice, why do you think the designer or draftsperson made that choice? Is it clearer than what has been recommended?

MODULE 2: DRAFTING THE LIGHT PLOT

For novice designers, there can often be an intellectual conflation of drafting the light plot with the development of the design. This can lead to determining and mapping out the various systems and angles as the plot is being drafted, which complicates both processes. If a student is designing as they draft the light plot unit by unit, or worse, focus point by focus point, they should step back and rethink their process. A **rough**, or preliminary, **plot** should be assembled before drafting begins (Figure 12-17). The design process itself is beyond the scope of this book; however,

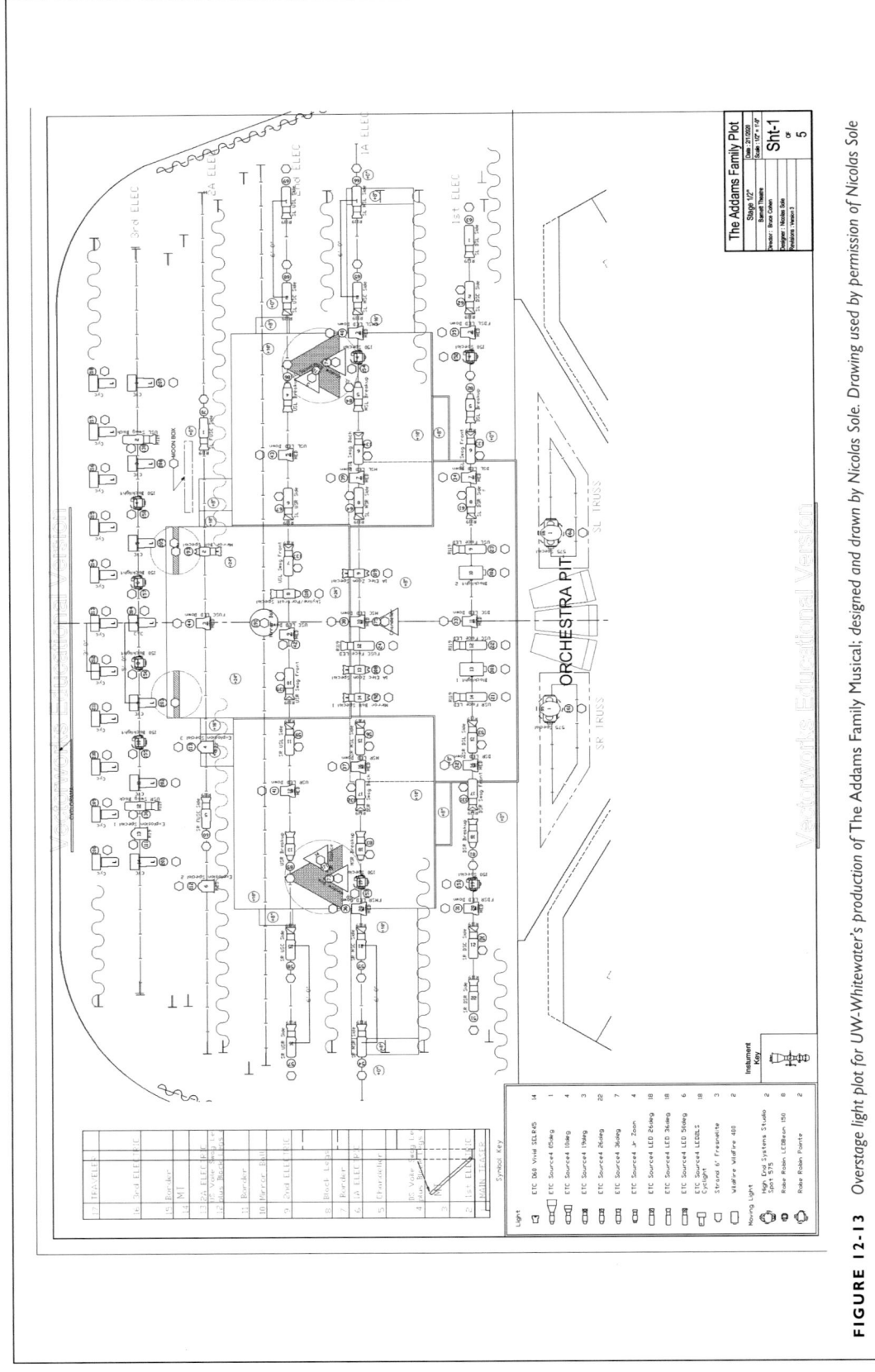

FIGURE 12-13 Overstage light plot for UW-Whitewater's production of The Addams Family Musical; designed and drawn by Nicolas Sole. Drawing used by permission of Nicolas Sole.

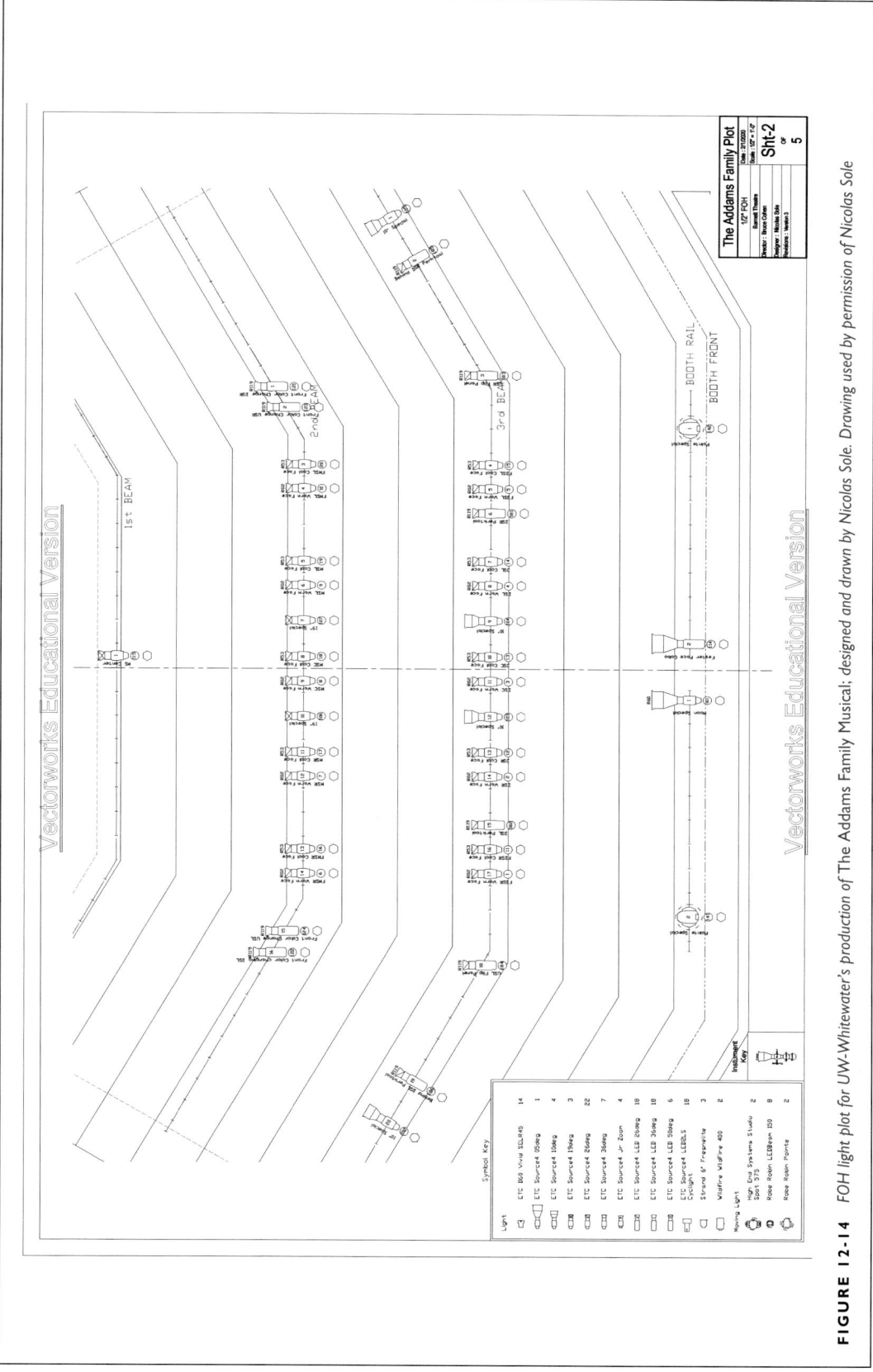

FIGURE 12-14 FOH light plot for UW-Whitewater's production of The Addams Family Musical; designed and drawn by Nicolas Sole. Drawing used by permission of Nicolas Sole.

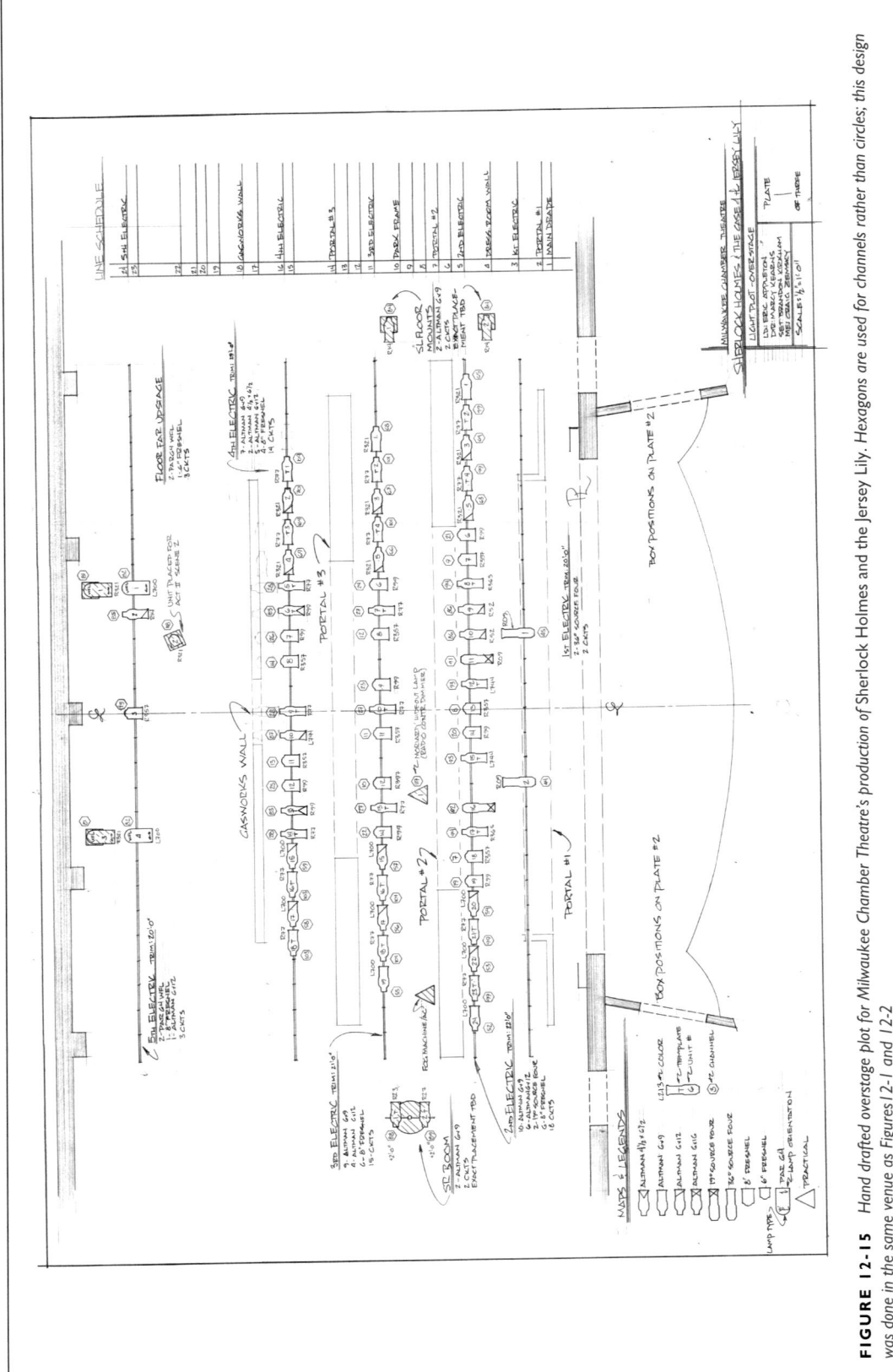

FIGURE 12-15 *Hand drafted overstage plot for Milwaukee Chamber Theatre's production of Sherlock Holmes and the Jersey Lily. Hexagons are used for channels rather than circles; this design was done in the same venue as Figures 12-1 and 12-2*

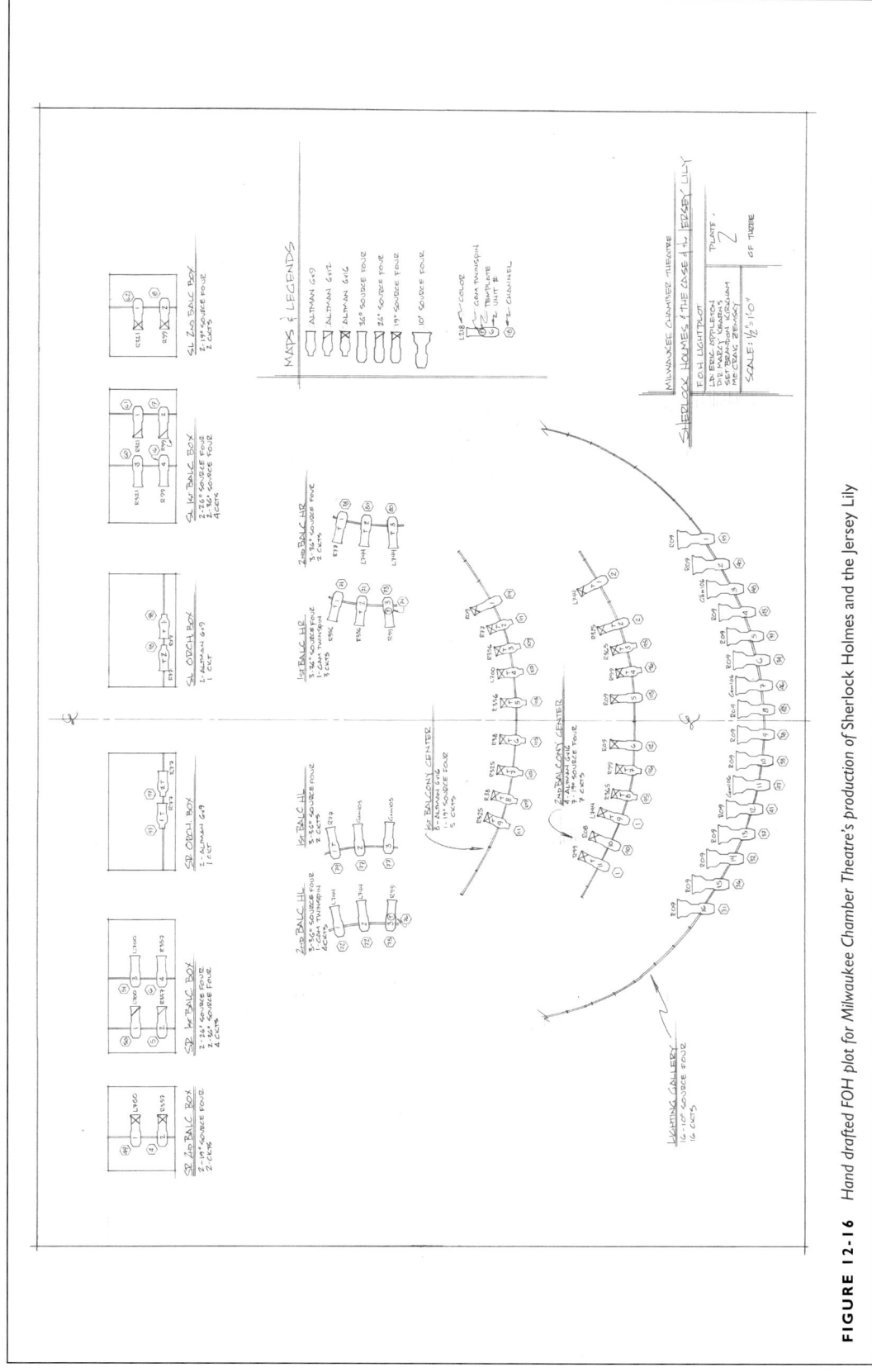

FIGURE 12-16   *Hand drafted FOH plot for Milwaukee Chamber Theatre's production of Sherlock Holmes and the Jersey Lily*

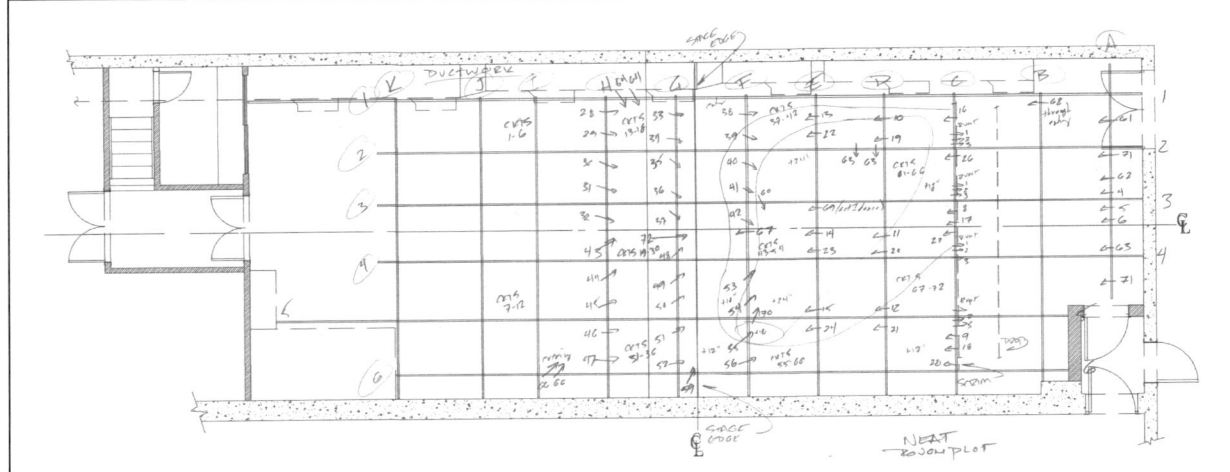

FIGURE 12-17 *Rough plot for lighting design of Renaissance Theatreworks's production of* Neat; *base architectural drawing author unknown*

the author feels that any lighting design process must aim toward and pause at the construction of a rough plot before going any further.

A rough plot uses the venue's base drawing to map out, in pencil, the locations and channels of all the lighting units intended to be used for the design. Since the notation is rougher and simpler than an actual plot and is only for the designer's preliminary work, ¼″ = 1′–0″ works well; the draftsperson can see the whole of the venue and all of its lighting positions in a reasonable size at the same time. This is generally not possible on a computer monitor.

To prepare the lighting positions, use a ruler to lay out tick marks at 18″ increments if they do not already have centers marked. Work outward from each position's center. As you lay out the rough plot, place the units on tick marks whenever possible. Venues with pipe grids and other non-proscenium positions may not easily accommodate themselves to a rigorous agenda of 18″ centers; try to be regular in your spacing and allow pivot space for focusing. Regular spacing also facilitates a speedier hang.

The rough plot is developed in tandem with the rough hook-up and is the step in which the placement of the units listed on the rough hook-up is finally determined. It's also the phase in which it's determined whether all the desired units fit on a position and whether they can be arranged on that position as desired. The hook-up should already contain most of the information previously generated during the design process: channel numbers, purposes and focus areas, unit types, color, gobo choices, etc. Revision and compromise should largely happen on the rough plot, not on the final plot as it is being drafted (even though the final plot will be revised throughout load-in and technical rehearsals).

The rough plot requires little more than an arrow to represent the unit and the channel number as designated in the hook-up. Colored pencils are useful in visually identifying individual systems in order to better understand how their parts are scattered around a venue. 3D models of the venue, as well as work in a light lab, can help the draftsperson determine the optimal placement of units. Remember that at this stage, the model feeds information into the rough plot—it is not a substitute for the rough plot.

Exercise: Laying out a Position Using a Rough Plot

In this exercise, a short First Electric is drawn. In placing unit symbols, remember to work from the center line out to the ends of the pipe. The center of the performance area is usually the most important lit area. If units are laid out starting from one end of the position and an error or miscalculation occurs, critical units may end up placed where their effectiveness is diminished. In the Vectorworks portion of this exercise, the Spotlight workspace will introduced. Spotlight is the location in which the majority of tools for entertainment industry use are found.

Instructor

The students should do this exercise by hand, first. They will need ½″ = 1′–0″ lighting templates, appropriately sized paper, and their usual other drafting equipment. A softer pencil such as 2H or H is suggested for darkening the lighting symbols. Figures 12-18 and 12-19 provide the rough plot and hook-up information. Distribute copies of the USITT lighting graphics recommendations so symbol examples can be referenced. This exercise will be repeated in Vectorworks.

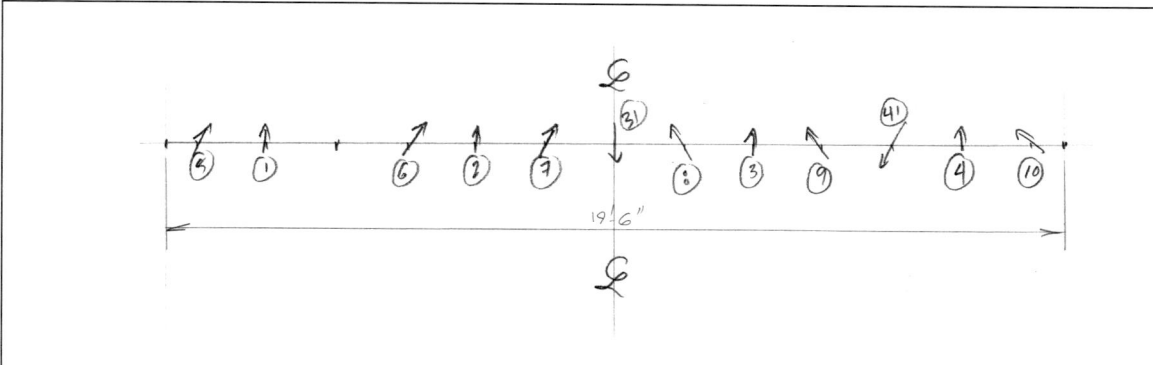

FIGURE 12-18   *Rough sketch for plot exercise*

CUN	PURPOSE	UNIT TYPE/WATTAGE		COLOR + TEMPLATE
1	NIGHT TONALITY DSR	6" FRESNEL	1K	R82
2	NIGHT TONALITY DSRC	6" FRESNEL	1K	R82
3	NIGHT TONALITY DSLC	6" FRESNEL	1K	R82
4	NIGHT TONALITY DSL	6" FRESNEL	1K	R82
5	MOONLIGHT USR ↗	36° ERS	575W	R62 + LEAF BRKUP
6	MOONLIGHT USC ↗	36° ERS	575W	R62 + LEAF BRKUP
7	MOONLIGHT USL ↗	36° ERS	575W	R62 + LEAF BRKUP
8	NIGHT SHADOW FILL USR ↗	36° ERS	575W	L714
9	NIGHT SHADOW FILL USC ↗	36° ERS	575W	L714
10	NIGHT SHADOW FILL USL ↗	36° ERS	575W	L714
31	WANDA BACK SPECIAL ↓	19° ERS	575W	R60
41	ROCKING CHAIR BACK ↓	26° ERS	575W	NC

FIGURE 12-19   *Rough hook-up for rough plot exercise*

Students

Using the information provided in Figures 12-18 and 12-19, draft the lighting position by hand, using a ½″ = 1′–0″ lighting template. No borders or title block are required for this exercise. Include a key and legend. The rough sketch shows unit placement, direction, and channel. The rough hook-up includes columns for channel, purpose/focus, unit type and wattage, and color and template/gobo. Color is specified, but the gobo is not—just that there is one. Include empty container shapes for dimmer and circuit if directed by the instructor.

1. Lay out the lighting position (First Electric). The position is 19′–6″ long. Draw outlines to place the key and legend on the sheet.

2. Lightly mark 18″ centers along position, working outward from the center line. The tick marks that end up within symbols will remain light and gray. The ones without symbols placed atop them will be reinforced during the darkening phase.

3. Read through the rough hook-up to see what unit types are required. Two unit types are noted: ERSs and Fresnels. The ERSs have three different beam spreads: 36°, 26°, and 19°. The Fresnels are 6″ Fresnels (the diameter of the lens is 6″). You may choose the instrument manufacturer and brand unless a list is provided by your instructor. Consult USITT lighting graphic recommendations to determine the best symbols to use. Many contemporary lighting templates are reasonably manufacturer-specific with their symbol outlines.

4. Lay out the key's content, especially if this will help you remember which symbols you are using. Remember that there should be ERS symbols for every beam spread used. The key can also be filled in as one of the final steps.

5. Since you already have all the position's information, unit symbols may be darkened right away. As with circle templates, keep the pencil vertical. Lighting symbols have more nooks and crannies than circles, so you may have to break the symbols into a couple of strokes. Start with the unit closest to the center line.

   Place the template so that the tick mark is centered in the body of the unit (Figure 12-20). While the arrows on the rough plot may point in the direction where the designer intends to focus the unit (because that's what the designer was thinking about as they laid it out), for this exercise orient the unit symbol in one of four cardinal directions: front, back, right, or left.

6. To keep legend information tidy, add guidelines to place each tier (Figure 12-21). While the symbols are

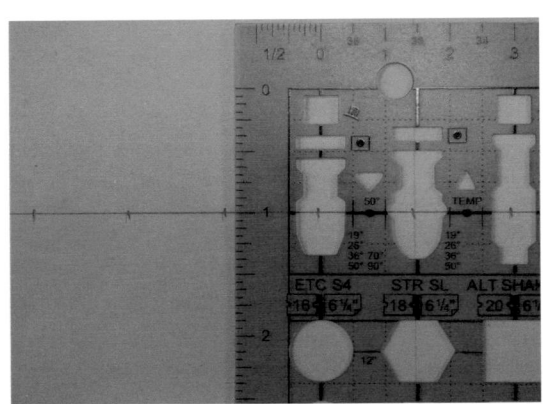

FIGURE 12-20 *Centering the lighting template on the tick mark*

centered vertically on the position, they are not all the same length; Fresnels are shorter than ERSs, for instance. As much as possible, the information should all be a uniform distance from the symbol rather than a uniform distance from the position. Once guidelines are placed, draw the empty information shapes in their correct locations. Letter the information. For this exercise, only the channel and color are given; the electricians will fill in dimmer and circuit information as they circuit and patch. The unit number will be determined in the next step.

7. Number the units starting from stage left (the right-hand side of the sheet). Units are always numbered from stage left to stage right along a horizontal/lateral position, and from top to bottom on vertical positions. Positions perpendicular to the plaster line or stage are typically numbered outward from the proscenium or stage edge. Once the units are numbered, this information can be added to the rough hook-up.

8. Darken the position with either a thick line or double thin lines (Figure 12-22).

9. Label the position and add the inventory below. How many circuits do you think are needed?

10. Finalize; darken the key and legend.

Exercise: Drafting from the Rough Plot in Vectorworks

Setup is required before a plot can be drawn in Vectorworks. This exercise uses the same information used in the previous exercise. It will be drafted in landscape, in ½″ = 1′–0″. Borders and title block are not required. A key and legend will be included.

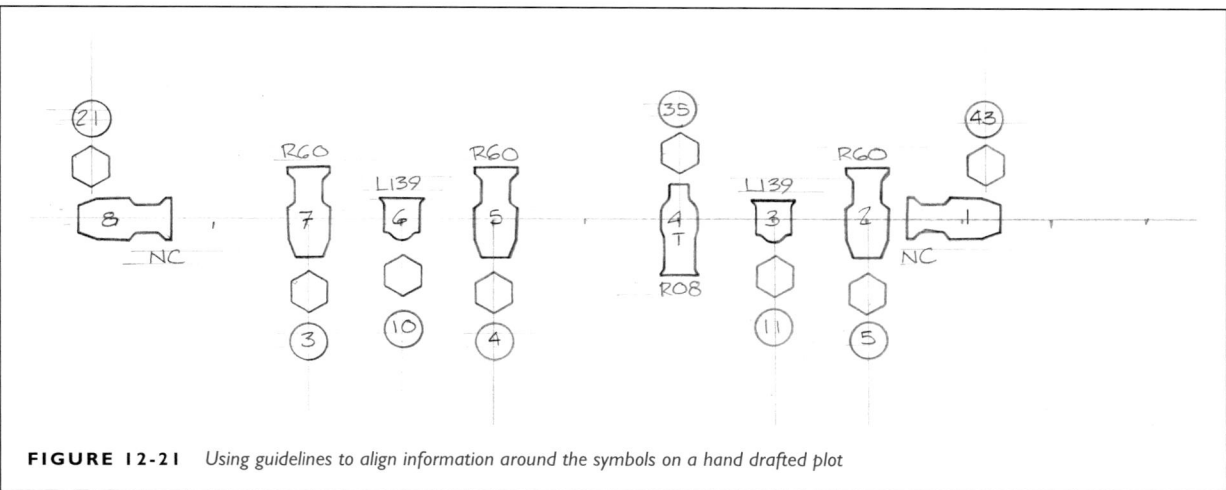

FIGURE 12-21   *Using guidelines to align information around the symbols on a hand drafted plot*

1. Open your base file. In Page Setup, switch to landscape orientation. Switch from Vectorworks' Fundamentals workspace to the Spotlight workspace via the Tools drop-down menu. Choose Workspaces. Choose Spotlight. The screen with go through a moment of reconfiguration. A new assortment of tool sets will appear in the lower left corner of the screen. Some icons will migrate or replicate in the Basic tool set, such as the Dimension and Tape Measure tools (Figure 12-23).

2. Create design layers for guidelines, lighting positions, and lighting units.

3. In the lighting position layer, draw the pipe with a thick visible line. If you wish to lay out a guideline first, do so in the guideline layer. In hand drafting, the position is darkened after the symbols are placed; in Vectorworks, the position is darkened first. In Vectorworks, the position is not just a line but a method of record-keeping; symbols placed on a position are automatically associated with it. The lighting symbols include a Fill, which obscures the pipe's line as the symbols are placed.

4. If you've explored the Spotlight tool sets, you may have noticed a tool in the Rigging tool set called the Lighting Pipe tool. When this tool is selected, clicking the cursor affixes the position's startpoint. Dragging the cursor to another point and clicking creates a position segment. You can drag out and click to create as many segments in whichever direction you desire. Double-clicking ends the line and turns the series of segments into a single lighting position.

    For this exercise, however, you'll draw a line first and then turn it into a Lighting Pipe. Select the 19′–6″ line just drawn. Right-clicking the mouse opens a long menu;

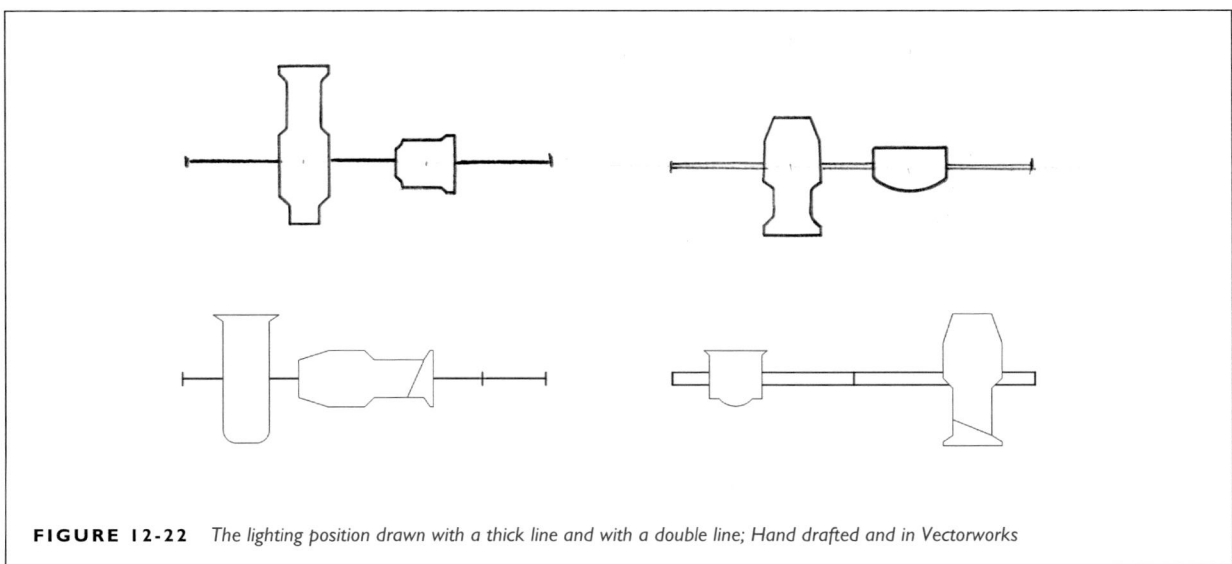

FIGURE 12-22   *The lighting position drawn with a thick line and with a double line; Hand drafted and in Vectorworks*

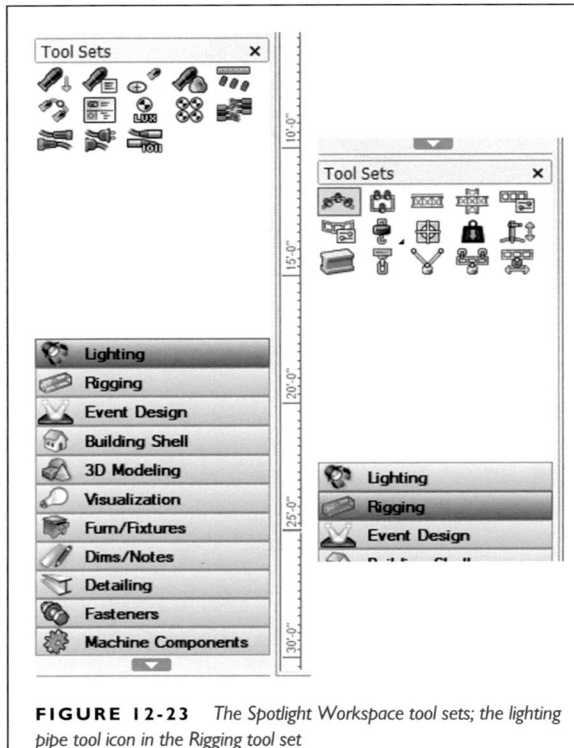

FIGURE 12-23 *The Spotlight Workspace tool sets; the lighting pipe tool icon in the Rigging tool set*

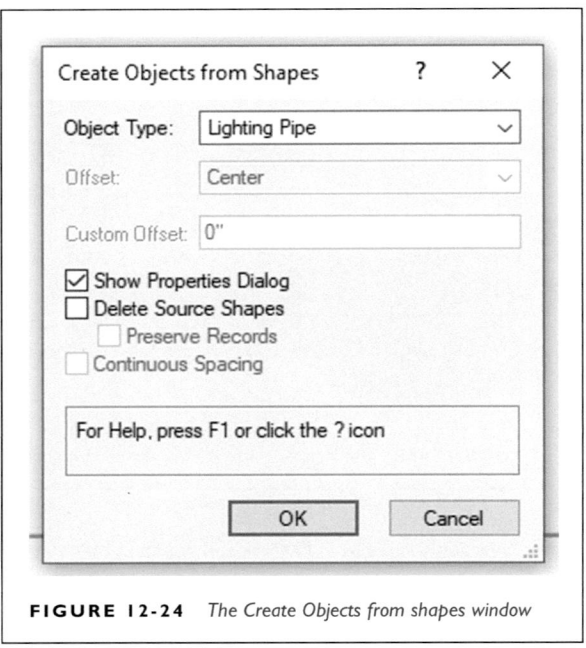

FIGURE 12-24 *The Create Objects from shapes window*

if you get the shorter menu, close it, then Select the line again and right-click. Open Create Objects from Shapes… Object Type offers a drop-down menu (Figure 12-24). Open it and choose Lighting Pipe. When you click OK, an Object Info window for a Lighting Pipe will open (Figure 12-25). This is where you can enter the pipe diameter, thick or double line for the pipe, arrangement of tick marks, etc. (Figure 12-25).

5. If these are not already the default settings, choose:

> Draw Pipe (checkmark)
> Line Type: Single
> End Marker: Both
> Display ticks as: Tick Marks
> Centers: 1'6"
> Centered on: Total Length
> Origin: On Center

If you wish to, take a few moments and explore the settings, clicking OK after each change to see how the selected line is altered. Undo after each iteration to return to the default settings. Once you've finalized the look of the position, click OK.

6. Now, whenever the pipe is selected, the Lighting Pipe navigation window appears on the right side of the screen. Select the pipe, and in the navigation window,

find the Position Name text box. Name the pipe First Electric (Figure 12-26).

7. Switch to the Lighting Unit layer. In the Lighting tool set, find the Lighting Instrument tool. This tool places lighting unit symbols on the pipe. Vectorworks will create and add classes for lighting symbols. They can be edited through the Class Organization window.

Choose the Lighting tool, then click the cursor at the intersection of a tick mark and the pipe. A dashed outline will appear, along with a Select Object window. The outline will rotate around its center as you move the mouse. Orient it so it points straight up. Click on the Select an Object to Insert menu, and a library of lighting units will open (Figure 12-27). Scroll through them until you find USITT Enhanced ERS 36-40. Double-click on it. The Select Object window will now show this symbol, which also appears on the pipe.

The Lighting Device Object Info window will appear, allowing you to enter information about this particular unit (Figure 12-27).

The Edit button in the Object Info window allows you to change symbols after they have been placed.

Notice that the body of the symbol is filled and interrupts the pipe. Information now needs to be arrayed around the symbol. This is set up through Vectorworks' **Label Legend** manager.

8. Open the Spotlight drop-down menu on the main toolbar. Choose Label Legend. Open Label Legend Manager from the selections that appear. The Label Legend Manager window will open. Choosing Add… opens the Add New Legend window (Figure 12-28).

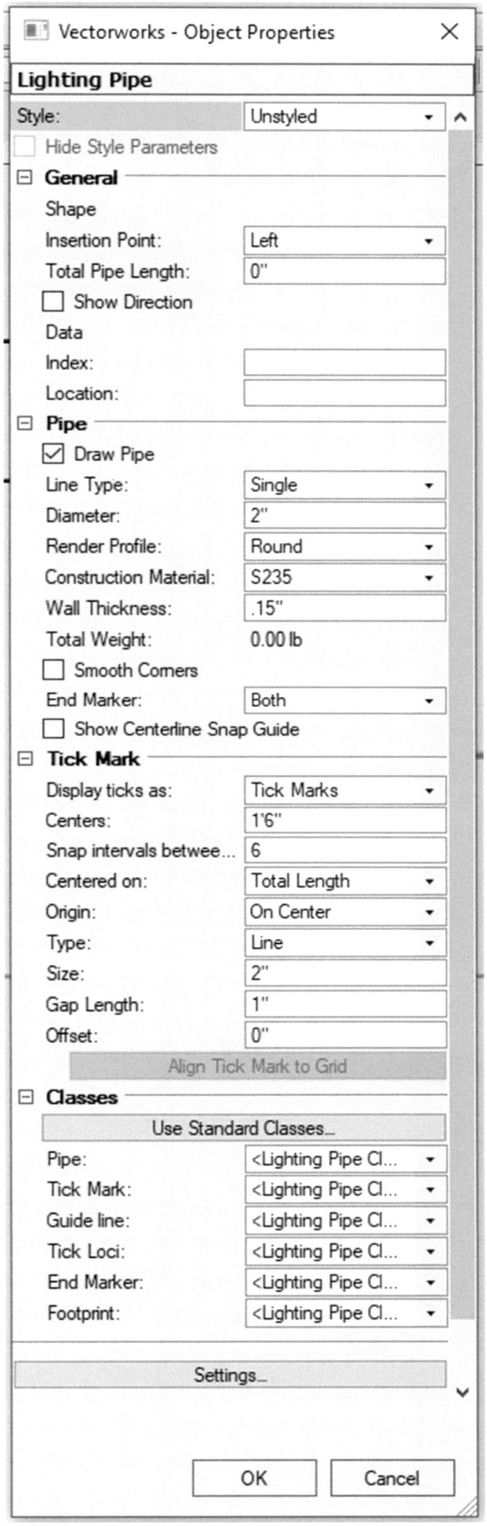

FIGURE 12-25   *The Object Properties window for a Lighting Pipe*

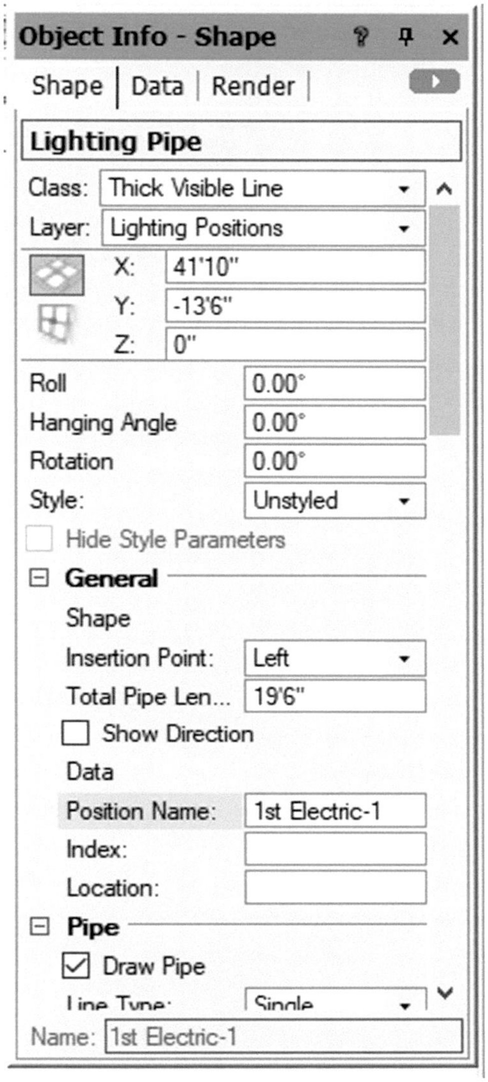

FIGURE 12-26   *Naming the pipe via the navigation window*

The Label Legend positions the information arrayed around the symbol. The Attributes menu lists the types of information available to be positioned around a symbol. The Use Legend Symbol shows the symbol with which the array of attributes will be associated; if it is not USITT Enhanced ERS 36-40, open the menu and select it. Click the Use… column to choose the attributes you wish to use. For now, select Channel, Color, and Unit Number, just to keep things simple.

The column named Container Type allows you to select the shape that will house the information. Clicking on the space rotates you through the various options. Do not choose a container for unit number or color. Choose a circle for Channel.

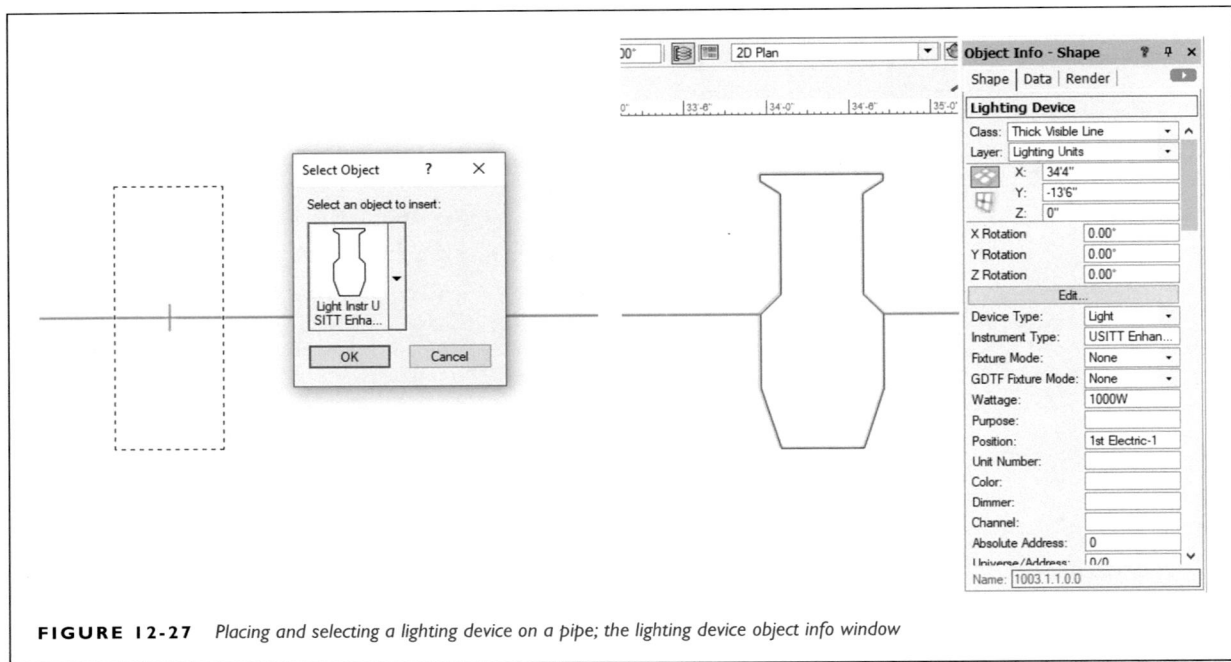

FIGURE 12-27 *Placing and selecting a lighting device on a pipe; the lighting device object info window*

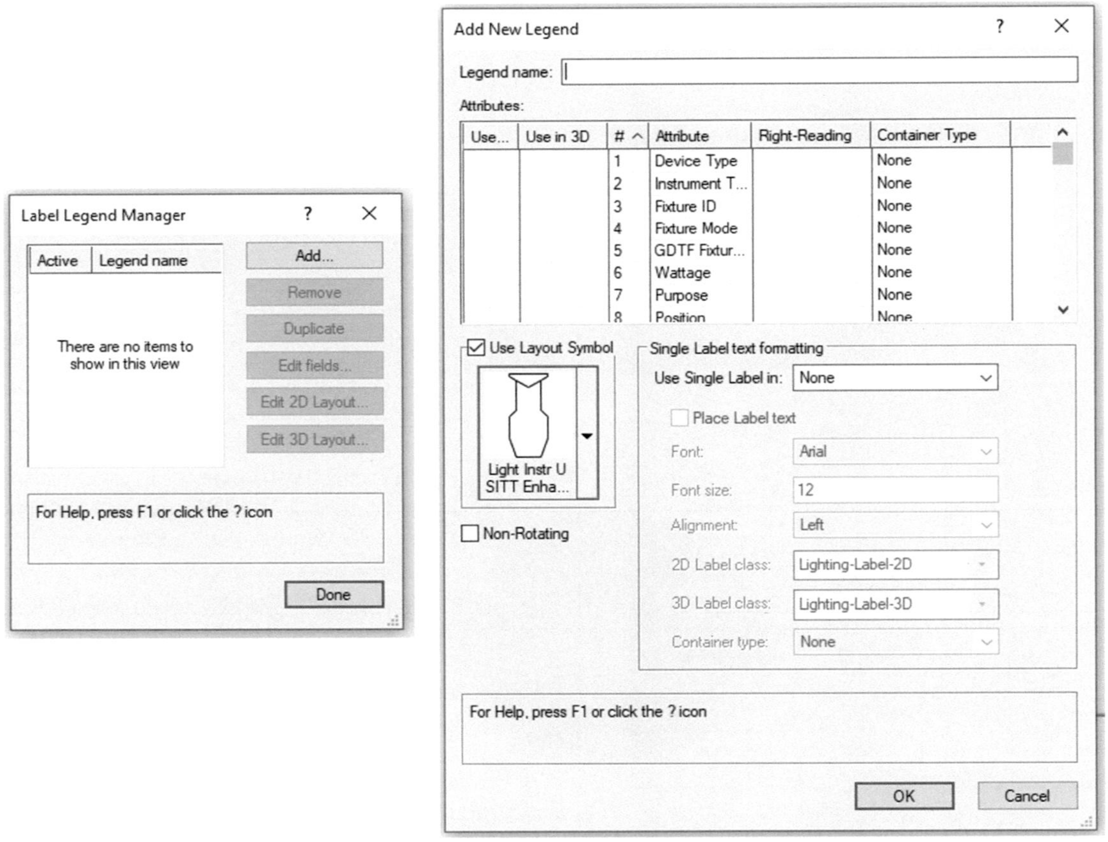

FIGURE 12-28 *The Label Legend Manager and add new legend windows*

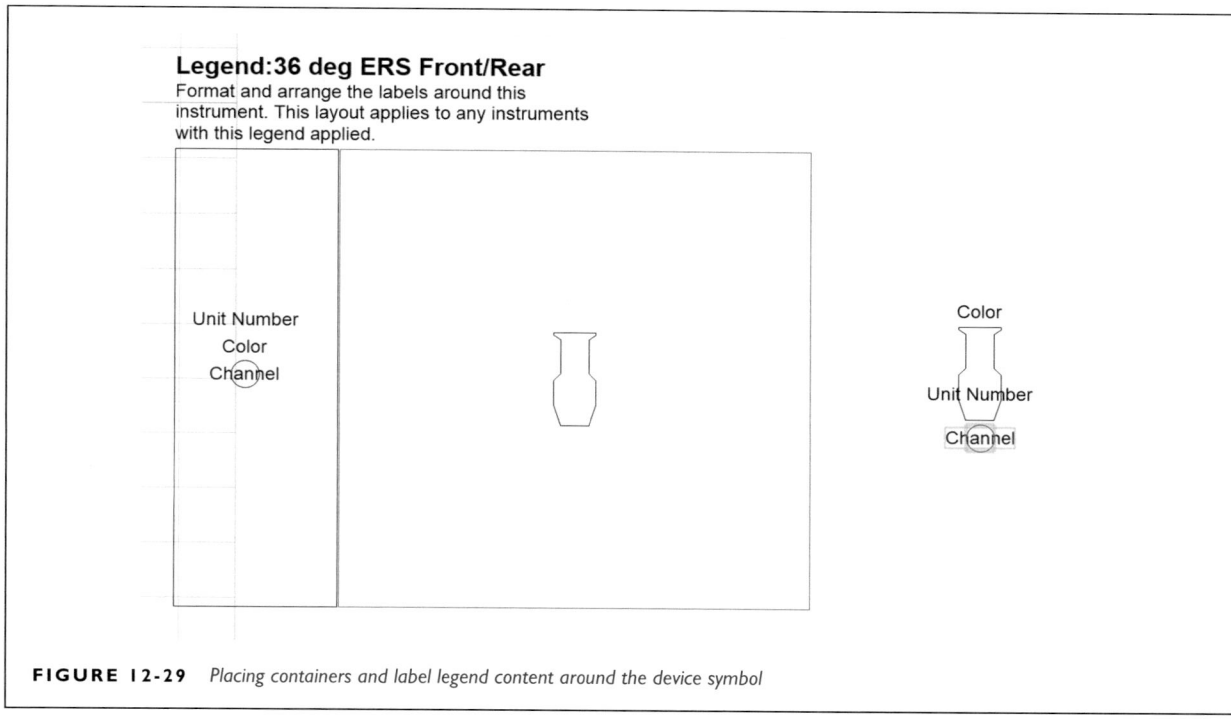

FIGURE 12-29 *Placing containers and label legend content around the device symbol*

For this exercise, the symbols will be pointed in one of the four cardinal directions. This symbol's legend array will be arranged around the unit for when the symbol faces upstage (a front light). Name the legend ERS Front/Rear. Later on, you'll construct legends for units facing to the right and left. Click OK to return to the Label Legend Manager window.

9. In the Label Legend Manager window, click in the right-hand column to activate the legend. A checkmark will appear. Click on Edit 2D Layout. The screen will shift to show the unit with the labels off to one side.

    Select and drag each legend component to its standard location around the unit symbol. Align centers with the axis of the unit. As each component is highlighted, its object info will appear in a navigation window, allowing you to change font size, etc. When the legend info array is set up to your satisfaction, click on the Exit Symbol box in the upper right corner of the screen. If you don't see it, it may be hidden behind a navigation window; move the window to reveal it (Figure 12-29).

10. When you return to the lighting unit layer, add a new unit to the pipe. It should automatically carry the legend array with it. While the symbol is active (selected), an Object Info window will open, allowing you to add information to the legend, such as color and channel (Figure 12-30).

11. This label legend can be applied to other types of units. After clicking to affix a unit, click on Edit in the navigation window. Change the symbol to a new unit type. In the Lighting Device window, open the Use Legend menu to choose from among legends already created. That arrangement of attribute information will appear around the new unit symbol on the drawing (Figure 12-31).

    You can also select a unit and assign a label legend to it by opening the Legend Label Manager from the Spotlight drop-down menu, then choosing Assign Legend to Instruments… A window appears listing all the legends you have created. Choose the desired legend for the highlighted unit symbol.

12. Since all of the units on this exercise's rough plot face forward or to the rear, this one label legend should suffice. Use the Lighting Instrument tool to place unit symbols in the appropriate locations. Assign the ERS Front/Rear label legend to each symbol. Add channel and color information via each unit's Object Info window. Once all the units have been placed, add their unit numbers by returning to each symbol's Object Info window.

13. Label the position and write the inventory below the label. Use the Instrument Summary tool to place a key and legend on the drawing.

14. Import into a sheet layer (be sure to check all applicable layers in the Create Viewport window). Add a border, title block, and title block information.

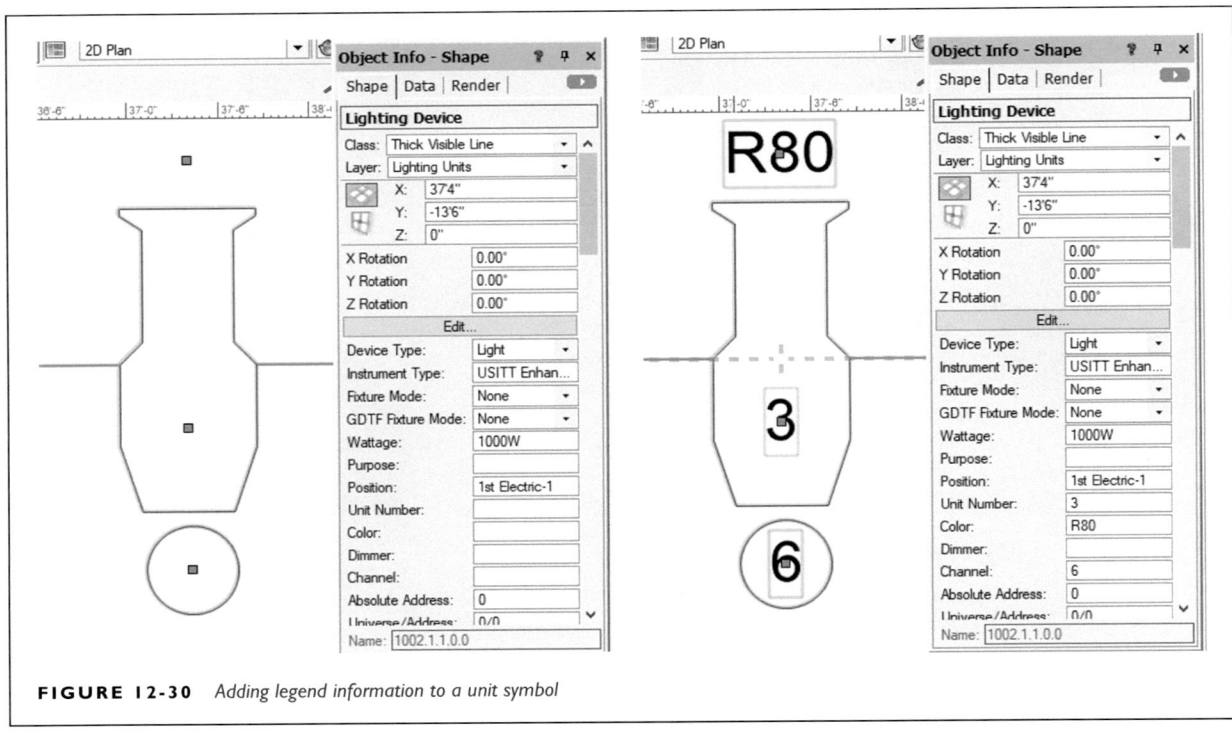

FIGURE 12-30   *Adding legend information to a unit symbol*

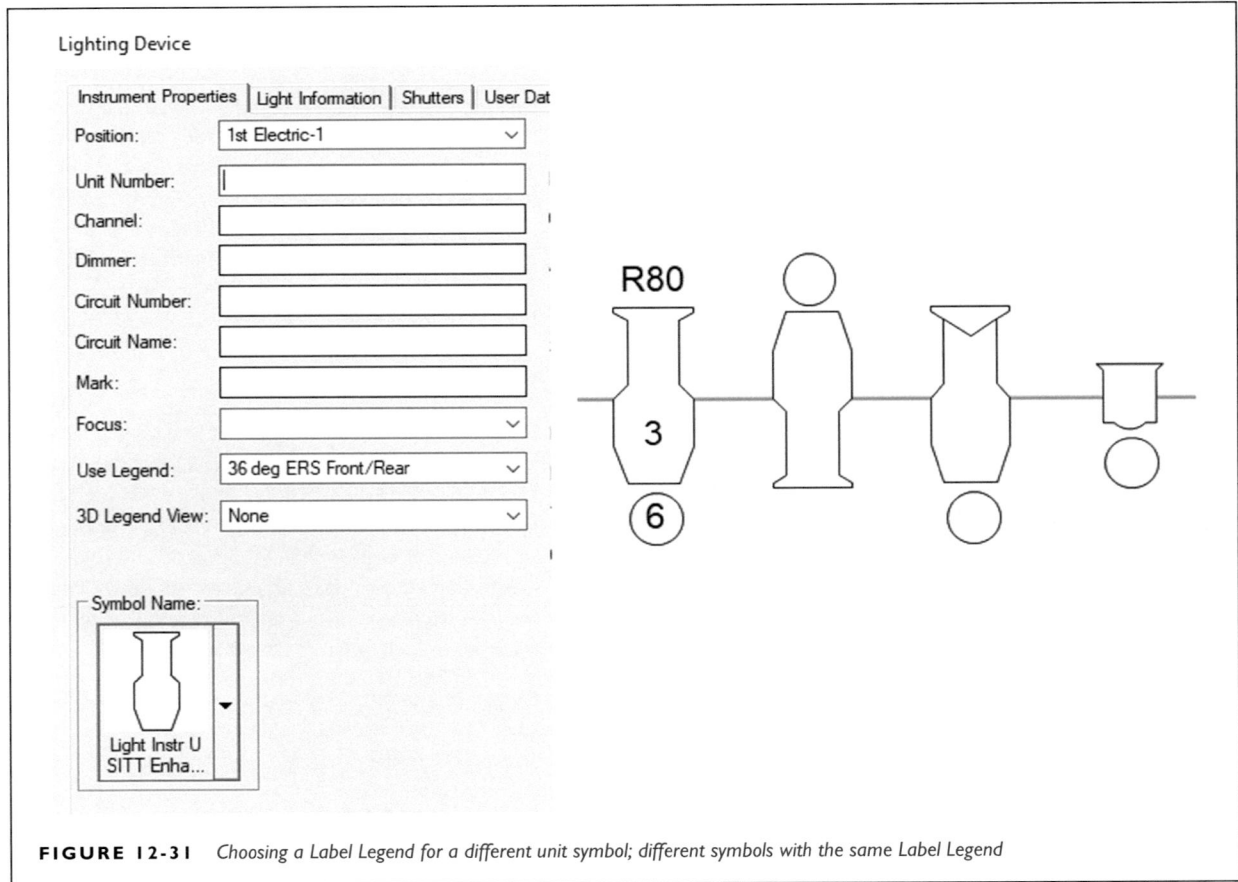

FIGURE 12-31   *Choosing a Label Legend for a different unit symbol; different symbols with the same Label Legend*

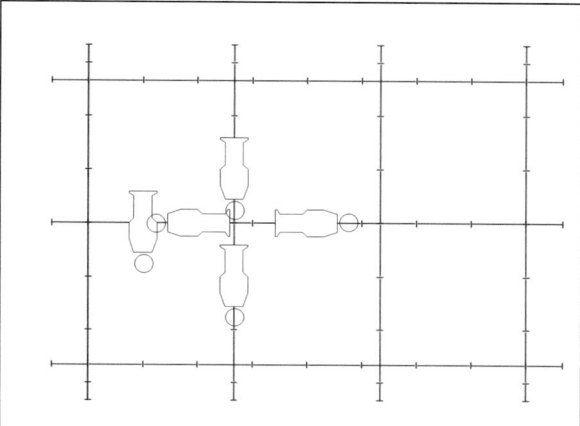

FIGURE 12-32 *Units on a pipe grid, their legend information overlapping other units and pipes*

Label Legends for Right and Left Facing Units

Figure 12-32 shows what happens when the Front/Rear label legend is the only legend used for units aligned with the pipe. First, legend information sits right on top of the position itself, diminishing readability and taking up pipe space. Second, legend information overlaps nearby units. Sideways facing units need a legend that places the legend information next to the unit symbol, rather than directly before or behind it. Legend information can be rotated in the Edit 2D Layout window by Selecting the attribute, right-clicking to open the menu, and choosing one of the Rotate options. This can help in some situations, but

multiple label legends are usually required for optimal organization.

In Figure 12-33, note that the center unit's information is also sideways to the reader. This is permissible, but not optimal. It may be necessary to create yet another label legend. Figure 12-34 shows a second version of a right/left facing unit's label legend. Because of the symbol's close proximity to other symbols, the color is connected to the symbol with a leader line to make it clear that that information is associated with that symbol (done as a step after adding legend information to that symbol).

If you simply need to swap legend information from one side of the symbol to another, the unit's Object Info window has options. You can find them by scrolling down though the various options until you reach Flip Front & Back 2D Legend (Figure 12-35).

> Lightwright, developed by John McKernon, is currently the lighting design industry's major database software. It can be used as a standalone spreadsheet program or linked to Vectorworks. It also links with the ETC EOS family of control consoles. Further information can be found at www.mckernon.com and www.lightwright.com.

Drawing Units Placed below the Grid or Rig

Any number and type of lighting units may be placed below the permanent overhead lighting rig or grid. These may be units mounted directly to scenic pieces, mounted

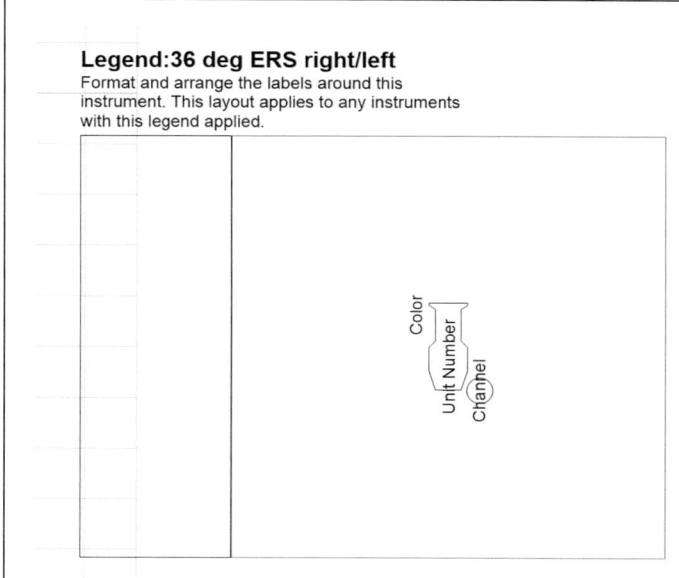

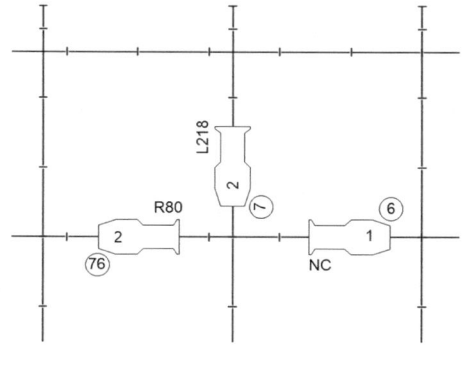

FIGURE 12-33 *Making a Label Legend for right- and left-facing units*

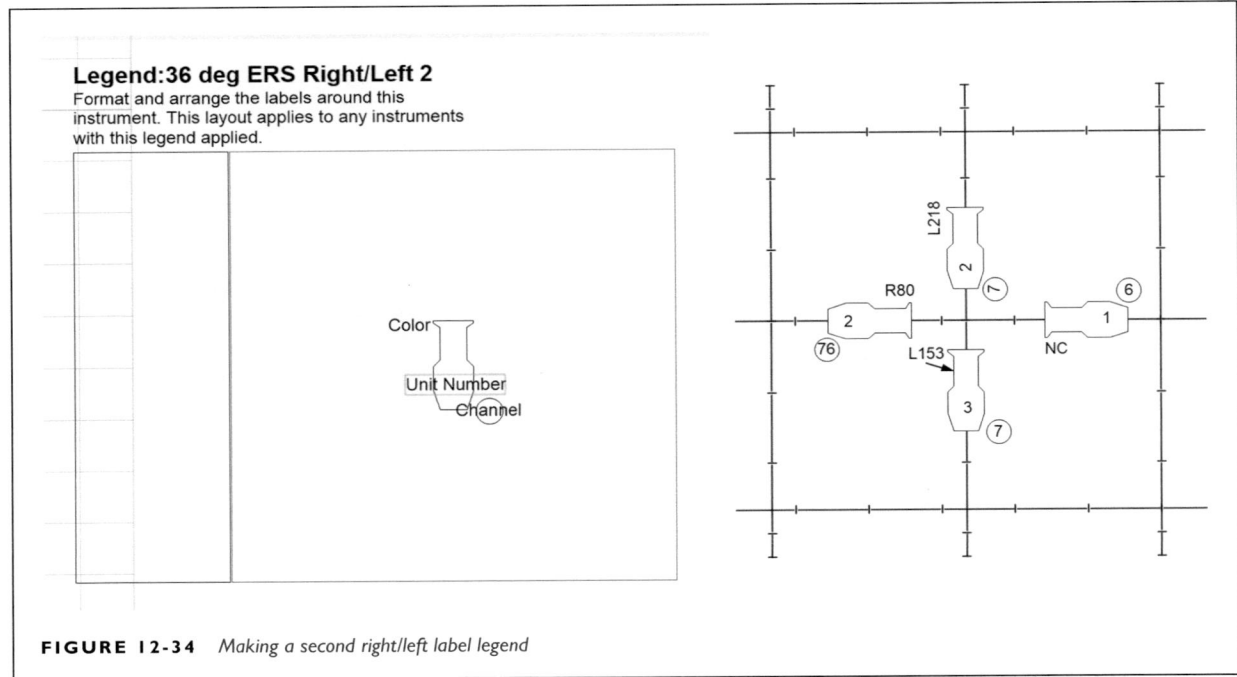

FIGURE 12-34 *Making a second right/left label legend*

to booms and stands (a unit on a stand that moves around is a **rover**), mounted to floor plates (floor mounts, pancakes), and installed in troughs, such as a permanent footlight array. Cycloramas are often lit from above and below; a line of lighting units at the base of cyc or drop is called a **groundrow**, which is also the name of the piece of scenery that typically masks them from view.

Units below the venue's rig/grid are typically crosshatched. Booms and other positions that carry multiple units are shown by their footprints on the deck with elevations that explode outward or are shown on other sheets. If a position has an elevation associated with it, no legend information is placed around the crosshatched footprint.

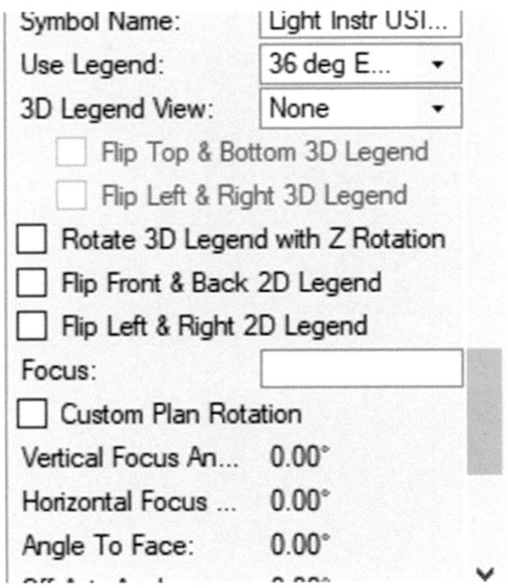

If the unit does not have an elevation or detail drawing elsewhere, such as a single floor-mounted Fresnel, legend information is arrayed around the crosshatched symbol. The crosshatching is interrupted to provide clear space for the unit number and any other legend information located within the symbol's outline. As much as possible, the mounting context of the unit should be graphically provided. If the unit is attached to a floor mount, the plate is drawn around the unit symbol. Practicals (designated by a triangle) are placed on pieces of indicated furniture or aligned against walls (Figure 12-37).

In Vectorworks, you can't select a lighting symbol and simply change the Fill. The simplest way to crosshatch symbols in Vectorworks is to trace the symbol in a new layer and then compose and fill the new shape. If this is done, consider placing an additional unit symbol in a detail area with legend information; otherwise, the unit will not be counted by Vectorworks' Instrument Summary tool. When the crosshatched unit is placed on a lighting position, the position will be overlaid on the symbol when the viewport is created (Figure 12.38). A narrow borderless white rectangle can be overlaid atop the unit to provide further visual separation between the unit and the pipe. Creating a design layer for floor mounts and other below-grid items is suggested.

If you construct a crosshatched lighting unit, it can be saved as a symbol. Select the object. In the Modify menu, select Create Symbol… The Create Symbol window will

FIGURE 12-35 *Object Info options to flip legend information around a symbol*

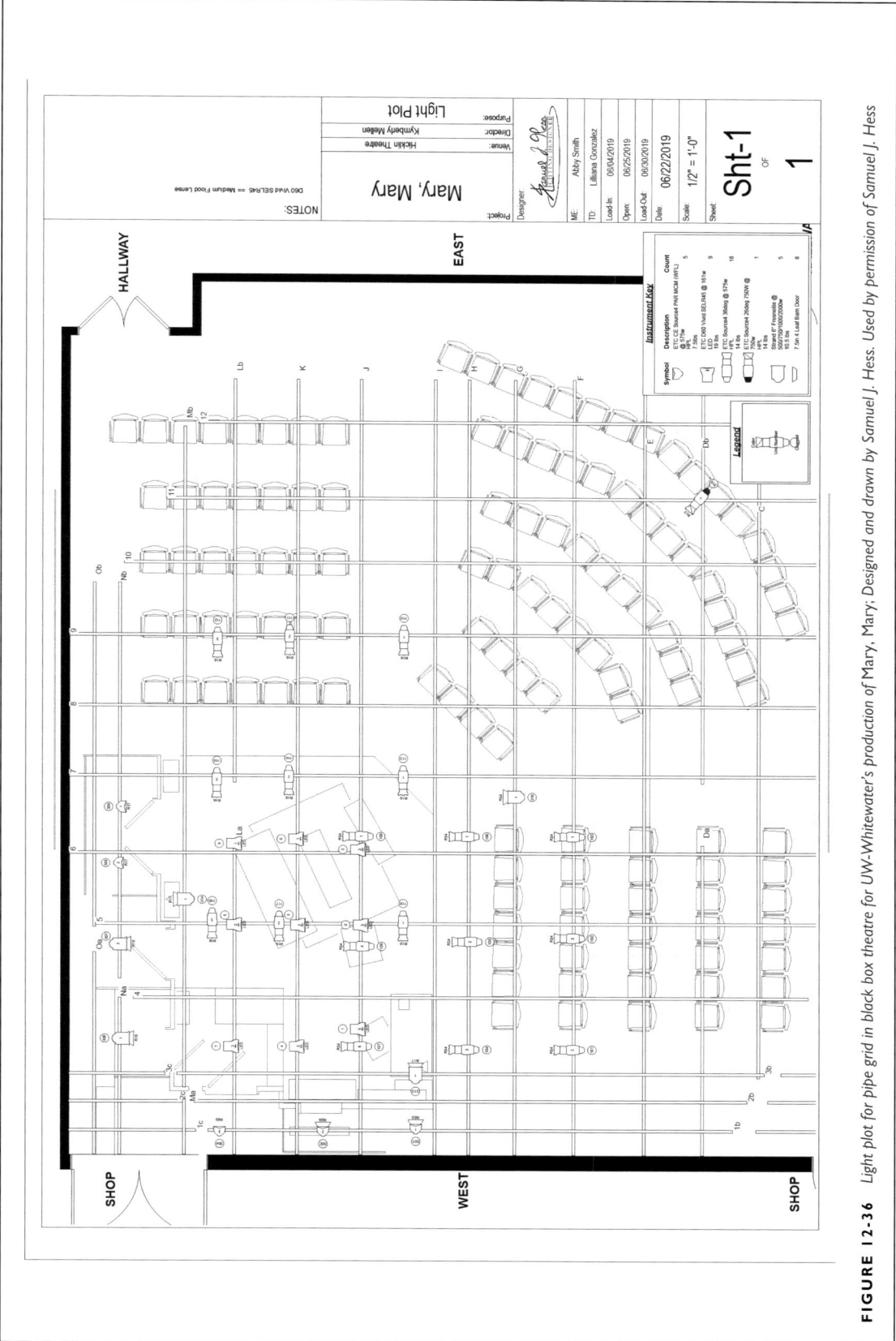

FIGURE 12-36 Light plot for pipe grid in black box theatre for UW-Whitewater's production of Mary, Mary; Designed and drawn by Samuel J. Hess. Used by permission of Samuel J. Hess

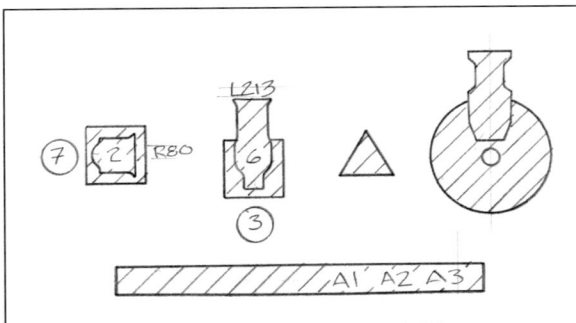

FIGURE 12-37 *Hand drafted crosshatched floor mounts, practical, boom footprint, and groundrow striplight*

appear, asking you to name the symbol and assign other attributes to it. Your symbol can then be added to a folder in the Resource Manager (Window > Palettes > Resource Manager). When you want to use the symbol, open the file in the Resource Manager, click on it and then click on the drawing to place it in the desired location (more about symbol creation in Chapter 13).

Exercise: Drafting Vertical Positions

At its simplest, a **boom** is a pipe screwed into a heavy base that sits on the deck. Lighting units are either clamped directly to the pipe, or hung from **sidearms** which in turn are clamped to the pipe. A sidearm is a short length of pipe clamped horizontally to the boom and minimizes unwanted angling when the unit is panned or tilted. On stage, booms are most often used to provide positions for sidelight, particularly in dance production. Booms mounted to the upstage side of the proscenium wall on either side of the opening are called **torms.** In a studio space, pipes mounted to walls may be named for the compass direction of the wall to which they are mounted to (e.g., North #1, South #1, etc.).

Ladders are runged rectangular structures that provide hanging positions for multiple units, typically in pairs on each rung. They are often used to create sidelight positions when there is no space on the floor for booms, such as a venue with extremely limited wingspace or productions with a lot of scenery that moves on and off stage. A **tail down** is a vertical pipe clamped to a batten. It serves much the same purpose as a ladder.

A **box boom** is a position found along the side walls of an auditorium or studio space. In the early days of theatrical electrification when lighting units began to be placed in the FOH, booms were placed in the box seats; hence the name box boom. Box booms are found in many configurations; their nomenclature is often specific to the venue. They are numbered in order, moving away from the proscenium.

Vertical positions are numbered in order, away from the plaster line or proscenium opening. Tail downs are typically named and numbered for the existing position to which they are mounted (e.g., First Electric SL Tail Down).

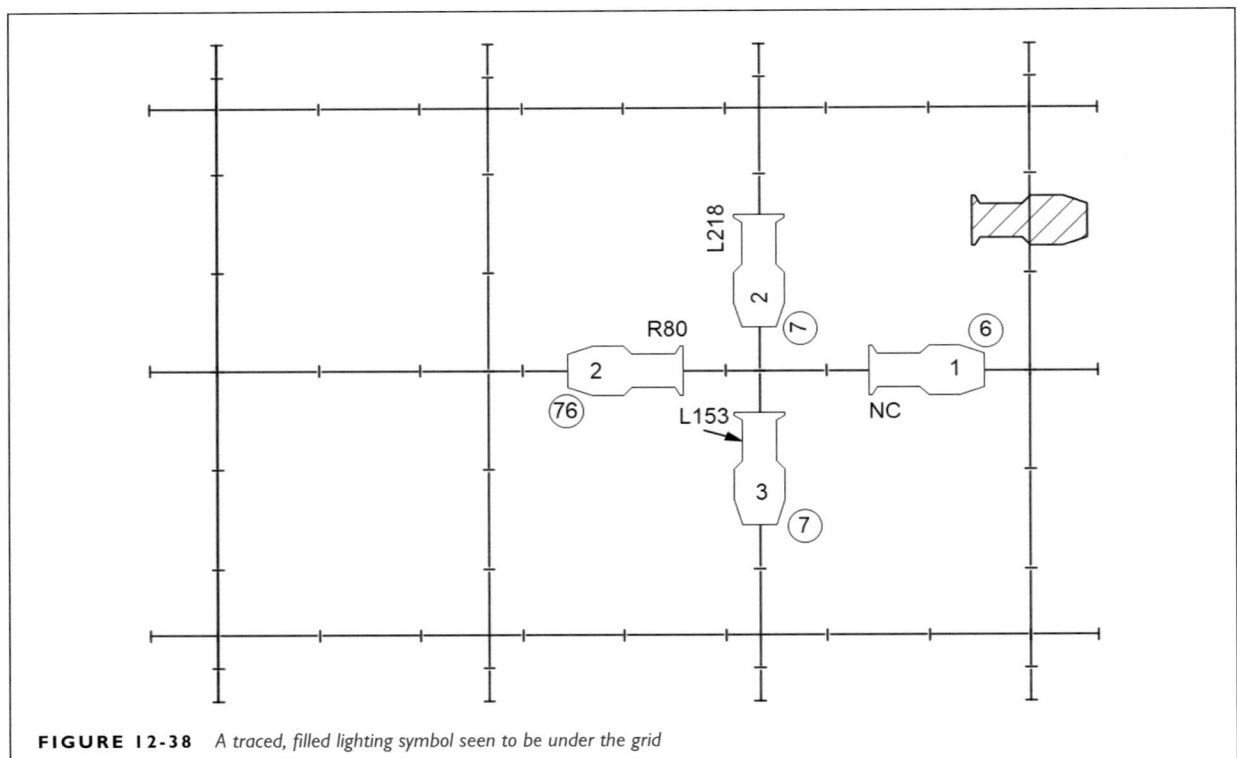

FIGURE 12-38 *A traced, filled lighting symbol seen to be under the grid*

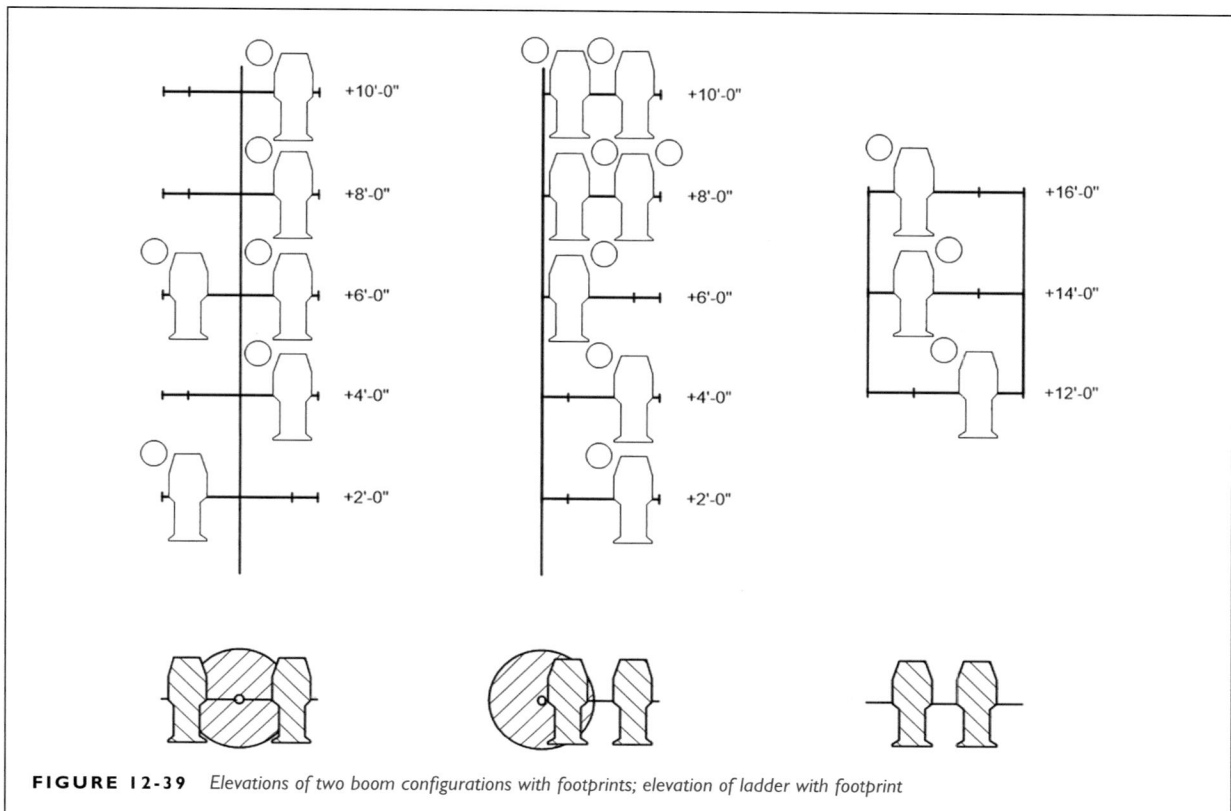

FIGURE 12-39 *Elevations of two boom configurations with footprints; elevation of ladder with footprint*

Elevations describing vertical lighting positions are still part of the schematic agenda of the light plot. Lighting units are depicted with the same symbols used on the plot with the same arrays of legend information. In Vectorworks, new label legends need to be created to accommodate the unique unit placement of vertical positions.

The footprints of vertical positions are situated on the plot. Since they are typically below the grid or rig, they are crosshatched. If the boom is screwed into a base, the base is also crosshatched. The position is labeled, but no legend information is arrayed around the units (Figures 12-39 and 12-40).

If page space allows, the boom elevation is shown exploded outward from the base location. All legend information is included around the symbols in the exploded view, as well as the heights of all units from the deck (e.g., + 8'–0") (Figure 12-41). If the pipe has a long stretch between units, the pipe may be broken with a break line. If the page space does not allow, the boom elevation is drawn on a detail sheet. A note at the footprint's location directs the reader to the correct sheet. For onstage vertical positions, it's generally best to congregate them either all on the overstage

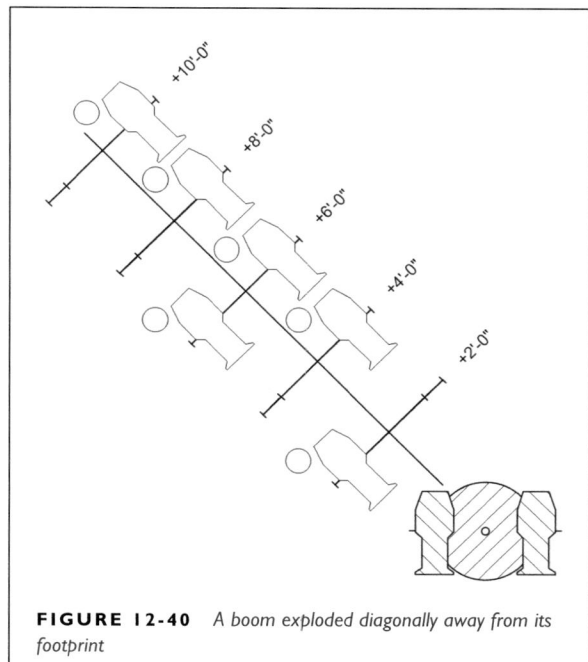

FIGURE 12-40 *A boom exploded diagonally away from its footprint*

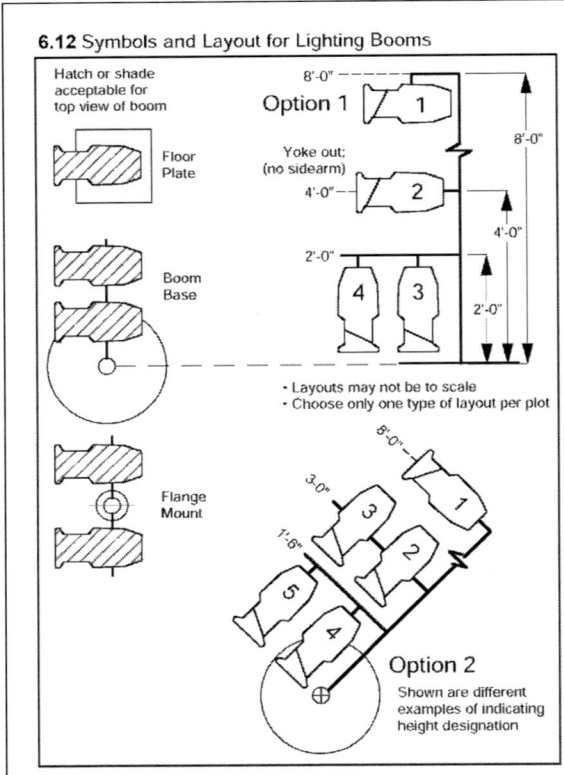

FIGURE 12-41   *USITT recommendations for boom notation. Reprinted by permission of USITT*

plot or all on a detail plate, rather than splitting them between sheets. This way, a team can be handed a single sheet of the plot when instructed to work on the booms.

The height indicates the placement of the sidearm clamp (or unit C-clamp) on the boom; 2'–0" is typical vertical spacing. For ladders, note the height of each rung. For vertical pipes in the FOH, such as those mounted on walls or in recessed slots, the height from the bottom of the position itself may be preferred.

Exercise: Drafting Booms and Ladders

Instructor

Have the students do this exercise by hand first. When complete, do it again in Vectorworks. Draft in ½" = 1'–0". Arch D paper is suggested. The footprint is to be included below the elevation. The box boom position is attached to the auditorium wall. Pipes and sidearms may be drawn with either a single thick line or thin double lines.

As added practice in thinking in scale, after the Vectorworks version is complete, have the students create the sheet layer in 3/8" = 1'–0" and 1/4" = 1'–0". Compare

to the ½" = 1'–0" version and discuss whether this reduced scale maintains or sacrifices the drawing's legibility.

Students

Using the information distributed by your instructor (or seen in Figure 12-42 and Tables 12-1 through 12-4), draft four vertical lighting positions in ½" = 1'–0" on Arch D paper. Draft them by hand first, then in Vectorworks. All four positions should fit on a single sheet. Include the crosshatched footprint below the elevation. Include all legend information and sidearm heights. Include labels and inventory for each position. Include a key and legend for the page as a whole.

1. Determine the placement of the positions on the sheet. As much as possible, seat the booms on the same baseline.

2. Lay out the verticals. Mark the height of the pipe itself, as well as the placement of the sidearms. For ladders, lay out the outline of the structure.

3. Lay out horizontal guidelines for each sidearm at the indicated heights.

4. Lay out the position's footprint, aligned with the boom or ladder.

5. It is useful to draw vertical guidelines to align units in neat columns (unless the design requires otherwise). Measure out along one sidearm (or ladder rung) to place the vertical axis of the unit symbol. Be sure the unit is far enough away from the boom to allow pan and tilt. Extend these vertical guidelines the full height of the position. Use them to place the unit symbols on the footprint.

6. Draw the units. Darken as they are placed.

7. Place legend information so that it is clearly associated with the correct symbol. Information should not be laid on top of or interrupt the lines of the sidearms or booms.

8. Darken the boom and sidearms. Add height information. Crosshatch the footprint. Add labels and notes as required. Draw the key and legend.

The steps are similar in Vectorworks. However, using the lighting pipe creation options on a boom with multiple sidearms creates each sidearm as its own position. This is not unacceptable, as they can be named Boom #1 Sidearm #1, Boom #1 Sidearm #2, etc. Unit numbers can be assigned so that the two units on Sidearm #1 are units 1 and 2, and the units on Sidearm #2 are 3 and 4. This is not so much a plot issue as it is a hook-up and paperwork maintenance issue. The **Instrument Schedule** takes the hook-up

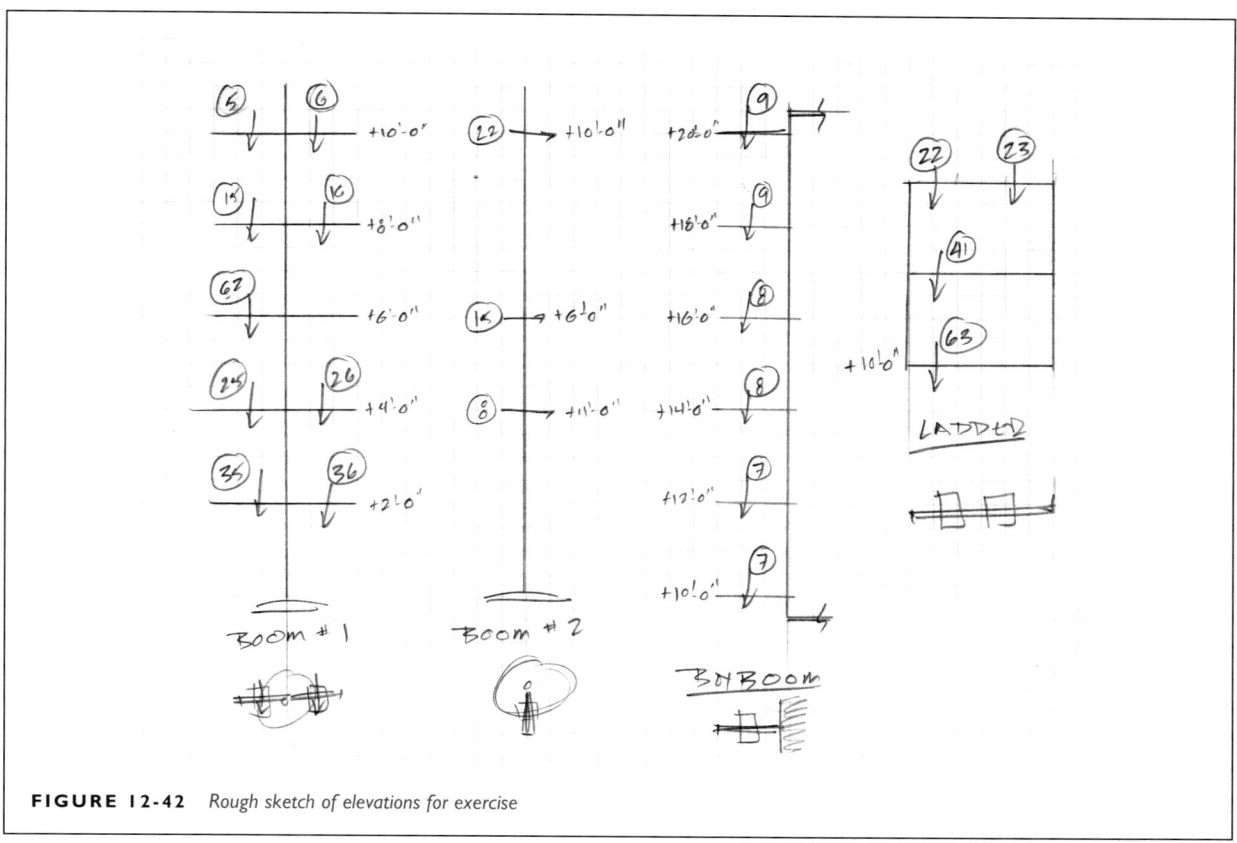

FIGURE 12-42   *Rough sketch of elevations for exercise*

TABLE 12-1   *Information for exercise: Boom #1*

Chan	Purpose	Unit Type/Wattage	Color/Template
5	Gold High Side SR	50° ERS 575w	R318
6	Gold High Side SL	36° ERS 575w	R318
15	Red Side SR	50° ERS 575w	R26
16	Red Side SL	36° ERS 575w	R26
25	Blue Side SR	50° ERS 575w	L183
26	Blue Side SL	36° ERS 575w	L183
35	NC Side SR	50° ERS 575w	NC
36	NC Side SL	36° ERS 575w	NC
62	Center Dot Special	26° ERS 575w	NC T = dots

TABLE 12-2   *Information for exercise: Boom #2*

Chan	Purpose	Unit Type/Wattage	Color/Template
8	Mid-Afternoon Window	PAR 64 WFL 1k	L218
15	Dusk Window	PAR 64 WFL 1k	R99
22	Pond Reflection	50° ERS 575w	R362 T = wavy breakup

TABLE 12-3 *Information for exercise: Box Boom*

Chan	Purpose	Unit Type/Wattage	Color/Template
9	Day Front SL	19° ERS 575w	R08
9	Day Front SR	26° ERS 575w	R08
8	Evening Front SL	19° ERS 575w	AP2120
8	Evening Front SR	26° ERS 575w	AP 2120
7	Dusk Front SL	26° ERS 575w	Gam101
7	Dusk Front SR	36° ERS 575w	Gam101

TABLE 12-4 *Information for exercise: Ladder*

Chan	Purpose	Unit Type/Wattage	Color/Template
22	Hallway side near	36° ERS 575w	R08
23	Hallway side far	26° ERS 575w	R08
41	Window on wall	50° ERS 575w	R60 T = window
63	Sconce support	36° ERS 575w	L108

information and organizes it by position; breaking up a single boom into six different positions means generating that many more tables in the document. On the other hand, that kind of granular breakdown of positions may be of benefit for complex productions.

When constructing the footprints in Vectorworks, construct the crosshatched shapes as a "stack."

Start by drawing and filling the shape on the bottom, such as the boom base. Draw guidelines over the filled shape to help place other shapes, such as the pipe at the center of the base. Copy and paste copies of traced unit symbols and place them last. If a shape disappears behind another shape, open the mouse's right click menu and Send the shape to the front. Finally, draw lines for the sidearms. Be sure all of the required layers are selected when the viewport is created. Remember to turn off guideline visibility (Figure 12-43).

Other FOH Positions

As well as having lighting positions that span the auditorium ceiling, such as coves and catwalks, many venues have FOH positions that are stacked on top of each other. The venue shown in Figure 12-2 features box booms on either side of the apron with three levels of hanging positions. In the house itself, there are two tiers of Balcony Right and Balcony Left; the designer has drawn only the second level as the first level is left unused. There are also two levels of Balcony Center, and then above that, the Gallery. While their relative locations can be graphically described on the section, their drawn locations on the plot must be adjusted.

Also in Figure 12-2, the First and Second Balcony Center positions are placed one in front of another, with the Gallery at the rear. This space compression can also be applied to

ceiling catwalks, narrowing the space between them so that they all fit on the same sheet. Compressing space on the drawing should be used with caution when positions are not permanent to the venue. Whenever a position is graphically adjusted, there should be a note stating that it is not drawn in its actual location.

Exercise: Adding Lighting Positions to the Venue Base Drawings

Instructor

This exercise adds lighting positions to the venue's plan to create a lighting base drawing. If you have been using a real venue for these exercises, continue using it. A field trip to and survey of the space will be helpful in visualizing the positions.

If hand drafted, the lighting base drawing should be ½″ = 1′–0″. The whole of the venue will not fit on a single Arch D sheet; the additions to the auditorium can be drafted onto a fresh sheet, with the proscenium and apron situated at the top of the page.

Once the lighting base drawing is complete, the students can construct a light plot as a new drawing/file. Provide a rough plot and rough hook-up information so they can lay out the units. Consider placing a few booms on stage. If the rough plot is generated with the box set in mind, booms can be situated to shoot light down the hallways or perhaps to get units just above Flat F to shoot through the windows at a shallow angle.

In Vectorworks, two main viewports will be created: one for the overstage and one for the FOH. These should be imported onto separate sheets. The box booms may be

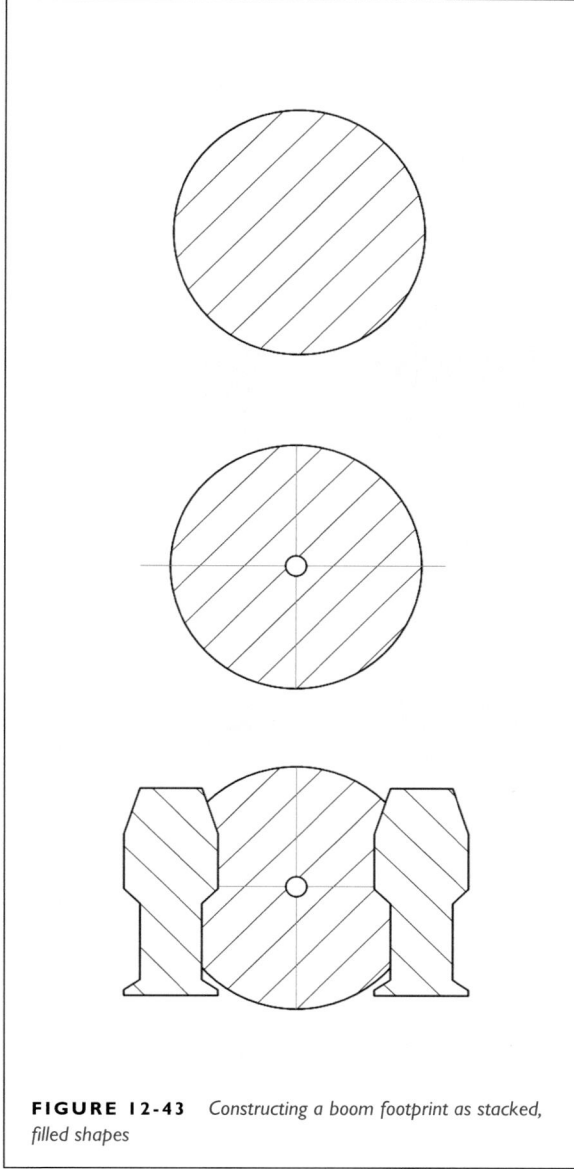

FIGURE 12-43 *Constructing a boom footprint as stacked, filled shapes*

broken out into a detail sheet, along with any booms whose exploded elevations do not easily fit on the overstage.

Students

Lighting positions will be added to the venue's base drawings. This creates a lighting base drawing that can be used for subsequent light plot projects. Figure 12-44 presents information for the FOH lighting positions.

If you are hand drafting this project, draft the auditorium directly onto a fresh second sheet. Situate the proscenium and apron at the top of the sheet to serve as a visual connection between this sheet and the overstage sheet. Box boom elevations should be included on the sheet as

well, outside the walls of the auditorium; if there is not enough space, create a base drawing detail plate for them.

1. Open the venue file. If you saved a version without the box set, open that. Create a lighting position design layer.

2. Eight lighting positions will be added to the auditorium's groundplan: two catwalks, an apron pipe, a center rail, and two box booms on house right and house left.

3. Use the center line to measure the distances of the Apron Pipe, Catwalk #1, and Catwalk #2 from the proscenium. The catwalks themselves may be shaded or colored; they can also be placed in the venue architecture layer. Lay out lengths from the center line to the right and to the left. The Catwalk Pipes are mounted 1'–0" away from their respective catwalks on the proscenium-facing side. The Center Rail is attached to auditorium's rear wall, the pipe 1'–0" away from it. The box booms are mounted 1'–0" away from the walls; they have 2'–0" sidearms clamped to them that extend perpendicularly from the wall (Figure 12-46).

4. The elevations of the box booms will be drawn outside of the auditorium walls, placed so that they are in visual proximity to their plan positions. Figure 12-45 shows the dimensions of the box booms. The sidearms should point toward the center line.

5. In the lighting position layer, darken in the Apron Pipe, Catwalks #1 and #2 (the pipes, not the catwalks themselves), and the Center Rail. Select each line in turn and turn them into lighting pipes with tick marks on 18" centers. Enter each position's names via its object info navigation window. The line schedule notes dedicated electrics: #1, #2, and the Cyc Electric. Trace these battens and turn them into lighting pipes. If a production requires other battens to become electrics, all the designer needs to do is trace over that batten and convert it into a lighting pipe.

6. The box boom footprints will not be turned into lighting pipes. They can simply be darkened. Turn their elevations into lighting pipes and name them accordingly.

7. Label the positions. Be sure that the box booms are labeled both on the plan and at their elevations. Note sidearm heights.

8. Save the drawing as Lighting Base Groundplan.

It should look much like Figure 12-46:

A light plot can now be drafted on this finished drawing. Save as a new file before drafting.

The steps remain the same: using a rough plot, start at the rear of the venue and lay out units position by position, working your way to the rear of the auditorium. For those

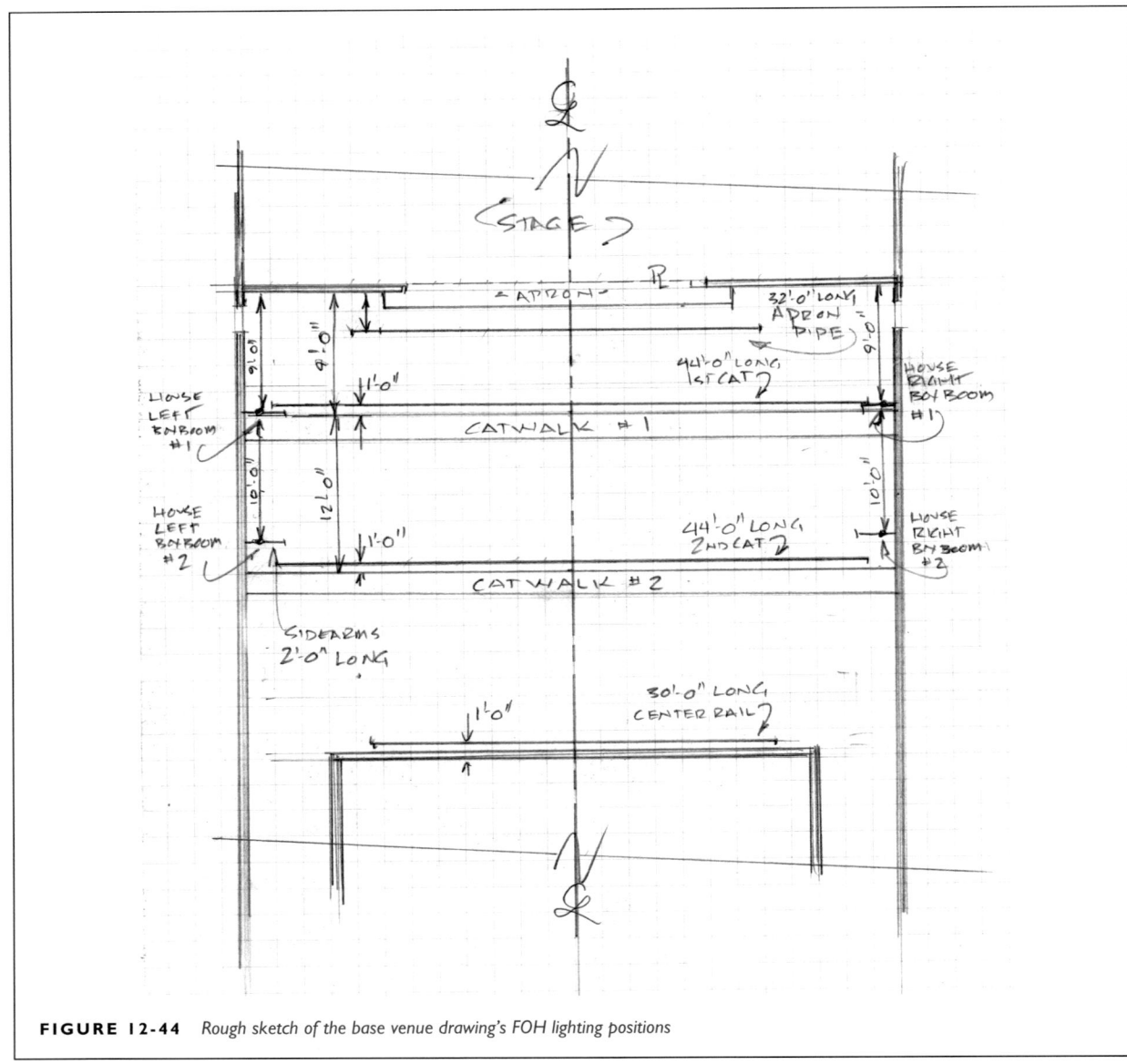

FIGURE 12-44 *Rough sketch of the base venue drawing's FOH lighting positions*

positions that straddle the center, work from the center out. For vertical positions, work from the top down.

If your instructor provides a rough plot based on the box set, include the outlines of major scenic elements and all masking. Trace them into a fresh layer using a colored line style to prevent the scenery from visually interfering with the lighting information; turn off the visibility of the original scenery layers.

When the light plot is printed, use two viewports: one for the overstage and one for the FOH. Place them on separate sheets. Box booms may be placed on a detail plate. Any onstage booms whose elevations do not fit on that sheet can also be moved to the detail plate.

MODULE 3: LIGHTING SECTIONS

There are two types of lighting sections. The first is the rough version used by the designer to determine optimal lighting angles and the placement of moveable positions. It is usually drawn in pencil directly onto a hardcopy of the scenic section and no one but the designer is likely to see it.

The second version presents the finalized lighting positions as they relate to the scenic environment and other technical installations, such as sound (Figure 12-47). The unit symbols do not include legend information. For those positions crossing the center line, the unit on center or closest to center is drawn in profile. Vertical positions with units aimed toward center may show lens-end views of the units. Yokes

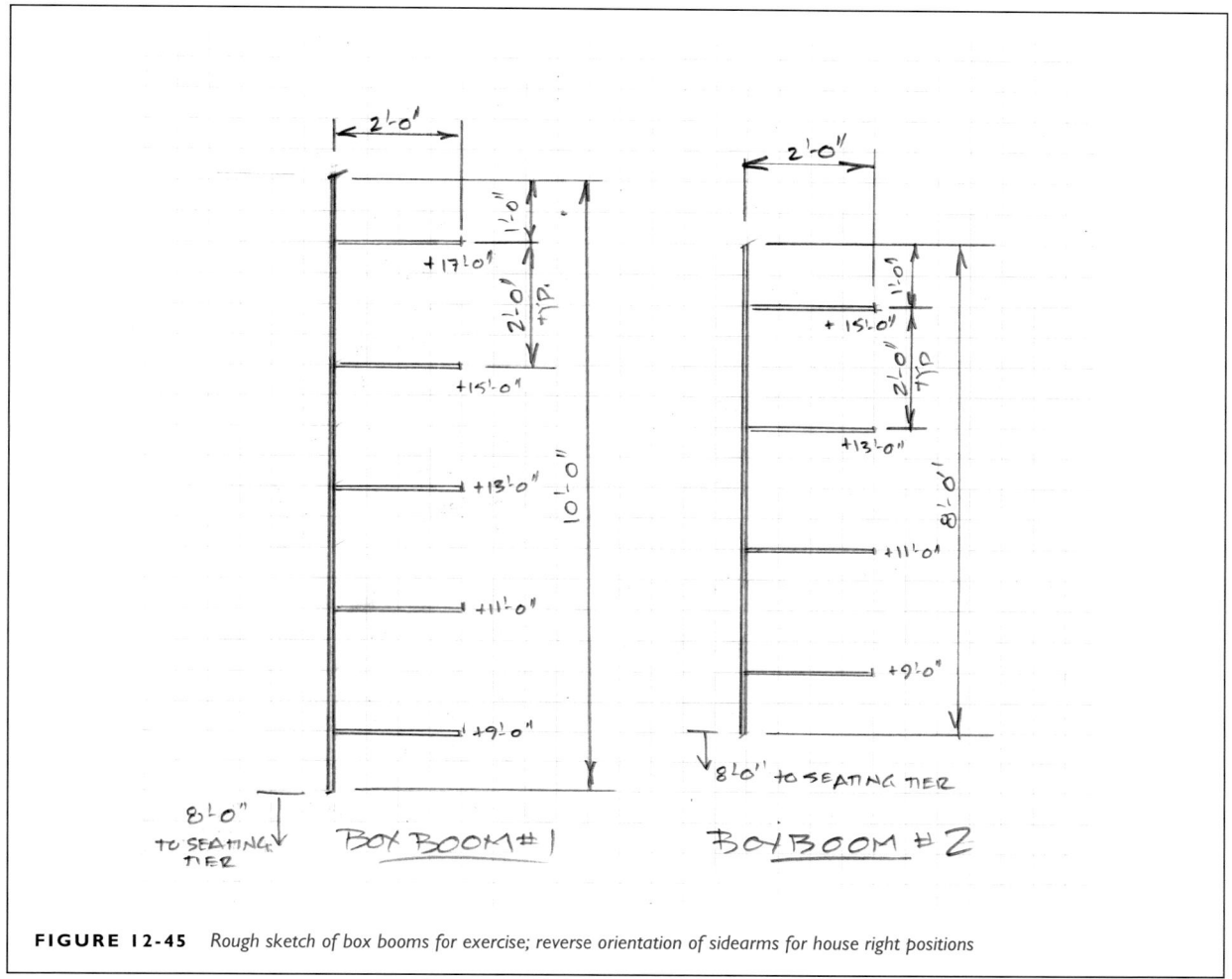

FIGURE 12-45 *Rough sketch of box booms for exercise; reverse orientation of sidearms for house right positions*

and clamps may be drawn, or included if the CAD library contains unit symbols with those features (Figure 12-48).

In Vectorworks, a profile unit symbol can be created by tracing over the library unit symbol and adding a yoke and clamp to the drawing. Select the drawing and choose Create Symbol… from the Modify drop-down menu. The symbol will be stored in a Reference Manager file.

It has become a trend to model the plot in 3D, creating the section via a center line cutting plane. Since the units are typically depicted pictorially with those in the foreground overlapping those in the background, the effect is often one of visual confusion. This kind of modeling (to the author's mind at least) does not present much in the way of increased efficiency or clarity for the technicians (Figure 12-49). The 3D model should be edited to present only the units closest to the center line to avoid overlapping symbols.

If the final section is drawn as part of the same file as the 2D plot, all the unit symbols on both drawings will be counted and added to the Instrument Summary. This means that symbols on the section that are duplicates of units on

the plot will be added to the summary, even if no legend information is added to the symbol. While the Instrument Summary can be customized to not include instrument counts, this miscount is more of an issue when Vectorworks exchanges data with programs like Lightwright to generate hook-ups and instrument schedules.

The rough section should be generated by the designer before the rough plot is created. The best unit for the job is selected by examining the location of the lighting position and the beam spread required to adequately illuminate desired portions of the playing space. A 36° ERS hung 40 feet away from the stage will cover a larger area than a 26°, or a 19°. How much of the playing space to be lit by a single unit will determine its type and hanging position (Figure 12-50).

The other major function of the rough section is to determine the trim heights of variable positions, such as ladders, booms, and electrics over the stage. The scenic designer is likely to have determined the trim heights of overhead masking; the lighting designer must now determine whether those trim heights allow them to do what they

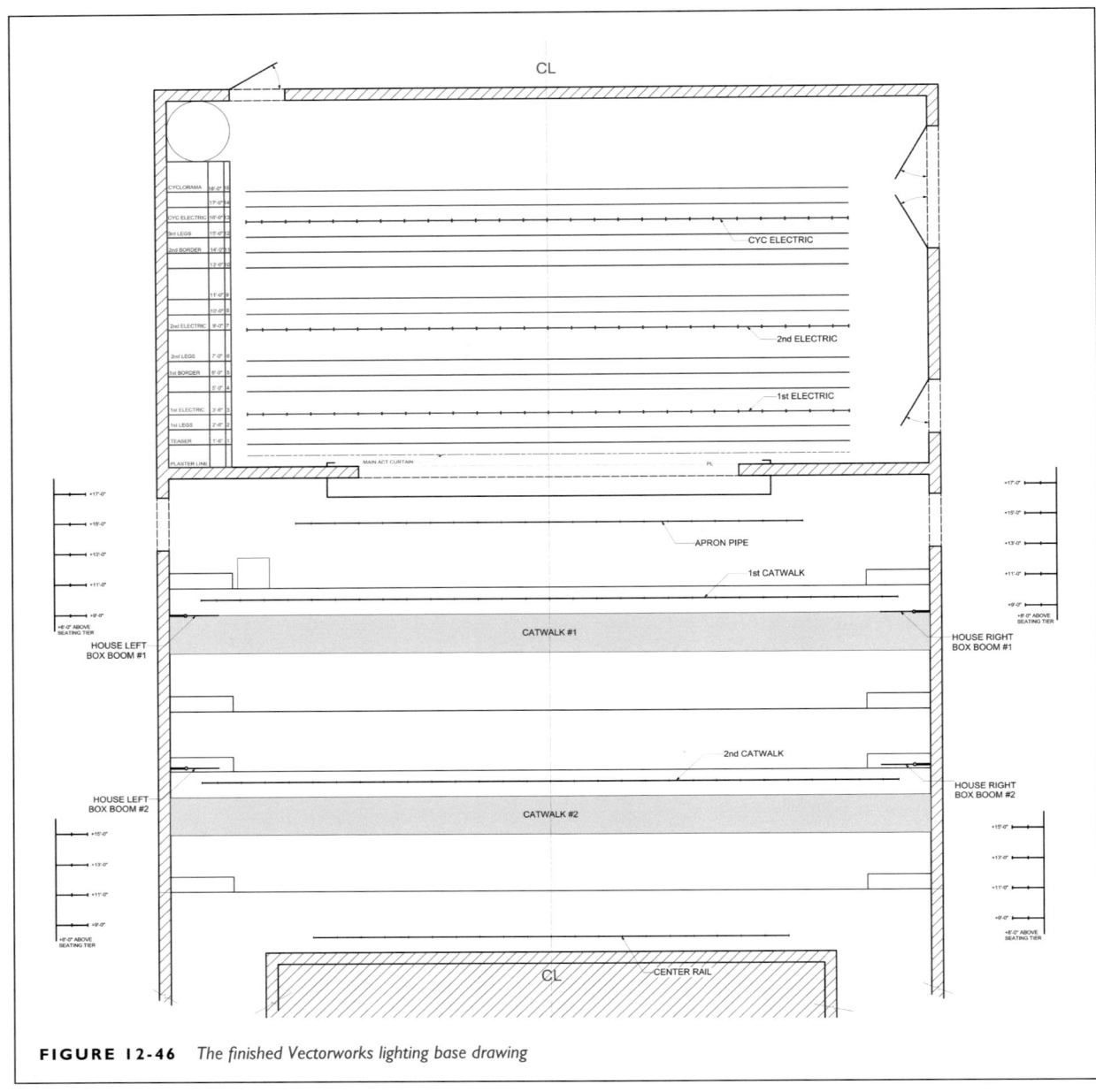

FIGURE 12-46 *The finished Vectorworks lighting base drawing*

need to do and adjust the suggested trim heights to final trim heights as dictated by lighting needs and discussion with the rest of the production team. Dance performances are one type of production where the lighting designer may be the first on the production team to designate masking trims.

The drafted section presents these finalized decisions to the production team (see Figure 12-51 for an example). The architecture and scenic elements are traced from the scenic section. Electrics are drawn at final trim heights, booms and other vertical positions are presented as elevations integrated into the scenic section. Onstage boom description is particularly important for dance production, as the sidelight positions share the same exits and entrances as the dancers. Some designers also lay out the sightlines and beam angles with dashed or lightly colored lines.

Adding Box Boom Positions to the Base Venue Section

On the venue's base section, the catwalk pipes, box booms, and center rail will be added to construct a lighting section base drawing. Consult Chapter 9 (Figure 9-5) for the auditorium's architectural dimensions and the placement of the catwalks.

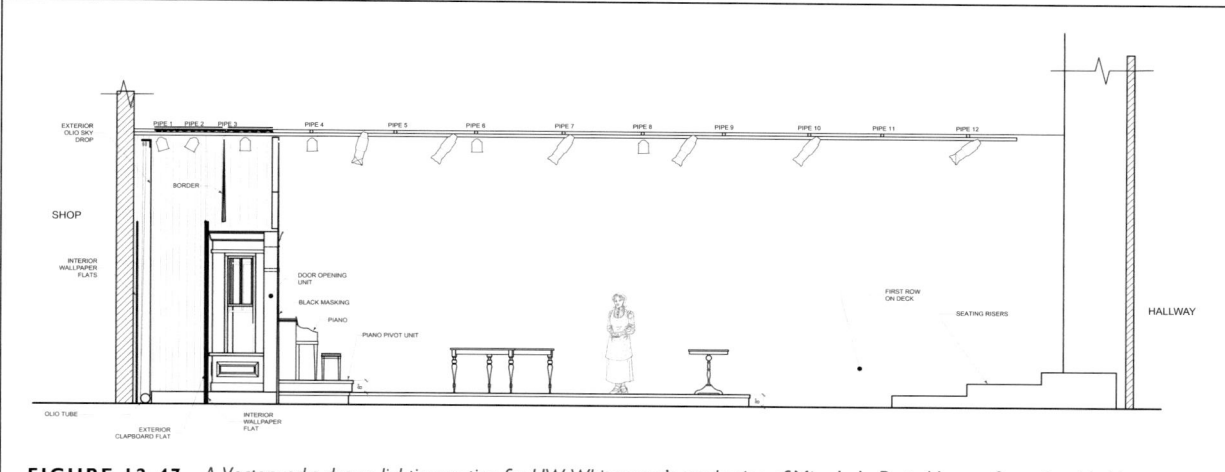

FIGURE 12-47 *A Vectorworks drawn lighting section for UW-Whitewater's production of* Miss Lulu Bett. *Human figure provided by* Marshall Anderson

1. Open your Vectorworks venue base drawing file. Create a layer for the lighting section. Turn off the visibility of other layers, such as the scenic section and soft goods.

2. Figure 12-52 presents height information for the FOH lighting positions. The catwalks themselves are cut through, with the pipe attached by a triangle of bracing. The Apron Pipe is suspended from the ceiling and therefore accessible only via a personnel lift or A-frame ladder.

3. Measure up 2'–0" from the top of the proscenium opening and extend a horizontal guideline out across the auditorium. The Apron Pipe hangs 3'–0" away from the proscenium wall (Figure 12-52). Include a line to represent the aircraft cable by which the pipe is suspended.

4. The catwalk pipes are 1'–0" above and 1'–0" away from the forward bottom edge of the catwalk itself (Figure 12-52). Include a triangle of bracing to connect the pipe to the catwalk.

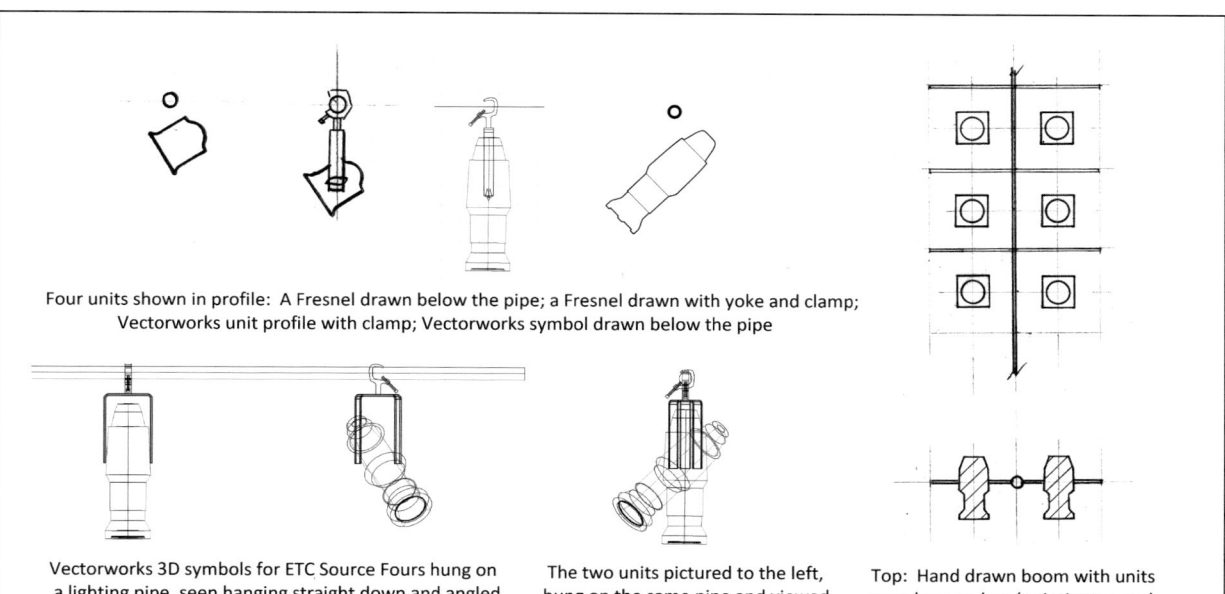

Four units shown in profile: A Fresnel drawn below the pipe; a Fresnel drawn with yoke and clamp; Vectorworks unit profile with clamp; Vectorworks symbol drawn below the pipe

Vectorworks 3D symbols for ETC Source Fours hung on a lighting pipe, seen hanging straight down and angled

The two units pictured to the left, hung on the same pipe and viewed from the right

Top: Hand drawn boom with units seen lens-end on (pointing toward center); Bottom: Footprint of boom

FIGURE 12-48 *Drawing Unit Profiles for the Section: (1) hand-drawn Fresnel below pipe profile; (2) hand-drawn Fresnel with yoke and C-clamp; (3) Vectorworks 3D symbol for ETC Source Four ERS, seen in side view; (4) 2D Source Four symbol drawn below pipe profile; (5) Source Four 3D symbols hung on pipe in front view, with same units seen overlapping in side view; (6) hand-drawn boom seen lens-end from center*

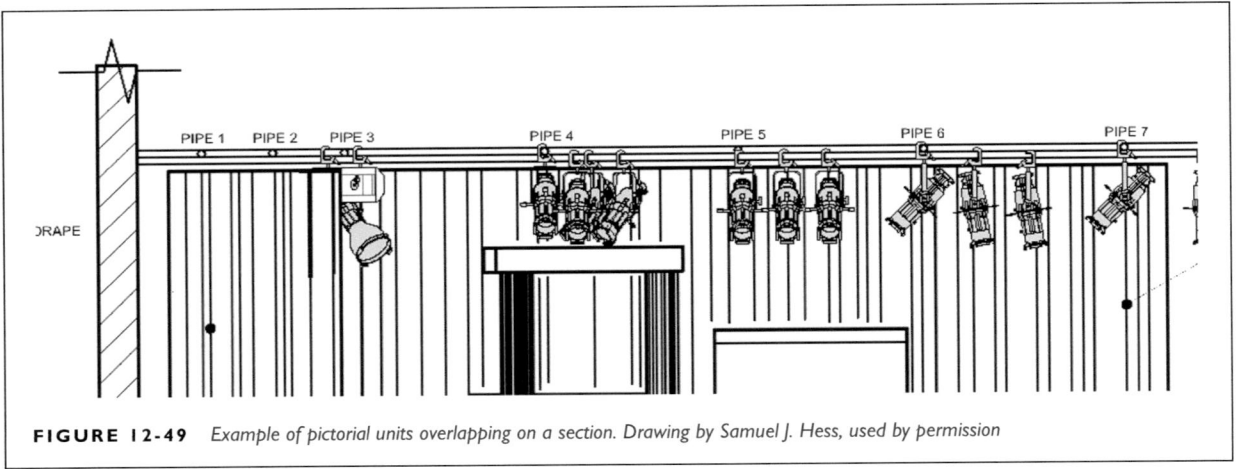

FIGURE 12-49 *Example of pictorial units overlapping on a section. Drawing by Samuel J. Hess, used by permission*

5. The Center Rail on the back wall is 12′–0″ above the level of the final seating tier. It too has bracing attaching it to the wall.

6. Box Boom #1 is 9′–0″ from the proscenium wall. Box Boom #2 is 12′–0″ beyond Box Boom #1. Even though the sidearms extend into the auditorium, they can be drawn in profile as circles, allowing unit profiles to be placed. Use 1 1/2″ circles to indicate the ends of the sidearms. Box booms can be drawn with either a thick line, or double line. Consult Figure 12-45 for the box booms' dimensions and height above their respective seating tiers.

7. Draw the overstage Electrics, dropping their positions from the line schedule.

8. Label all positions. When used to construct the lighting section, the designer will select electrics and battens and draw them at their intended trim heights. The finished lighting section base drawing should look like Figure 12-53.

Homework: Drafting a Lighting Section

Instructor

Using the information provided in the earlier plot drafting exercise, the students should construct a section for this plot. Trim heights for electrics should be double checked from the sightline symbol. All masking should be included. Scenic features may be traced with thin or colored lines. Discuss

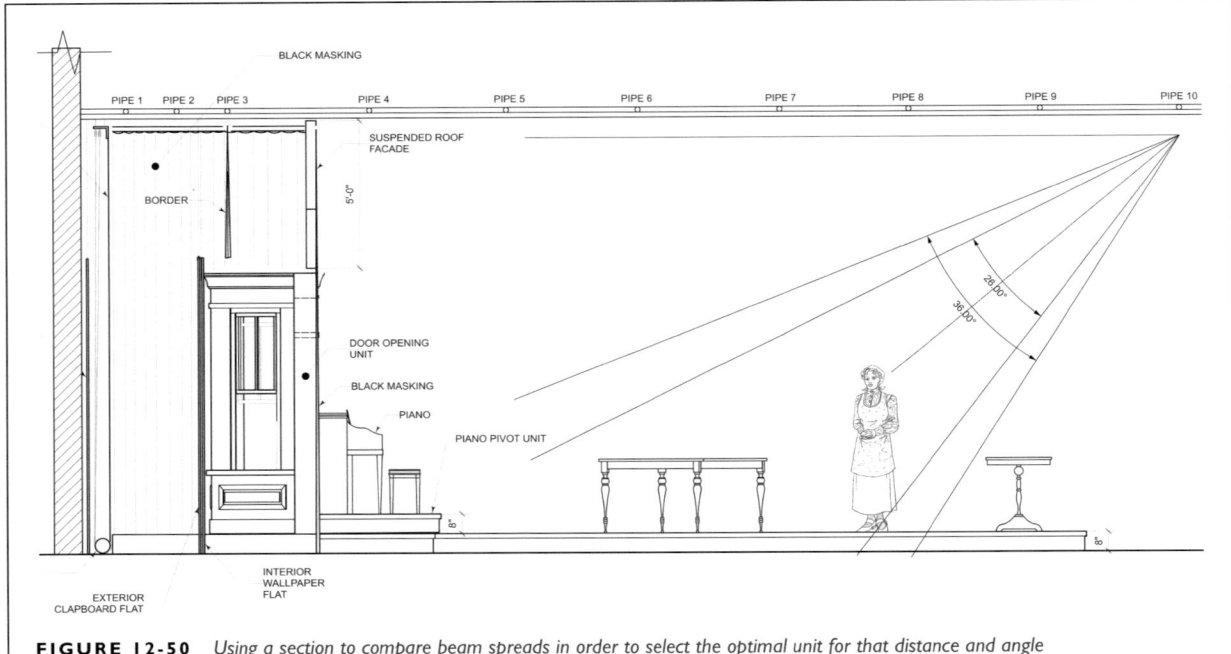

FIGURE 12-50 *Using a section to compare beam spreads in order to select the optimal unit for that distance and angle*

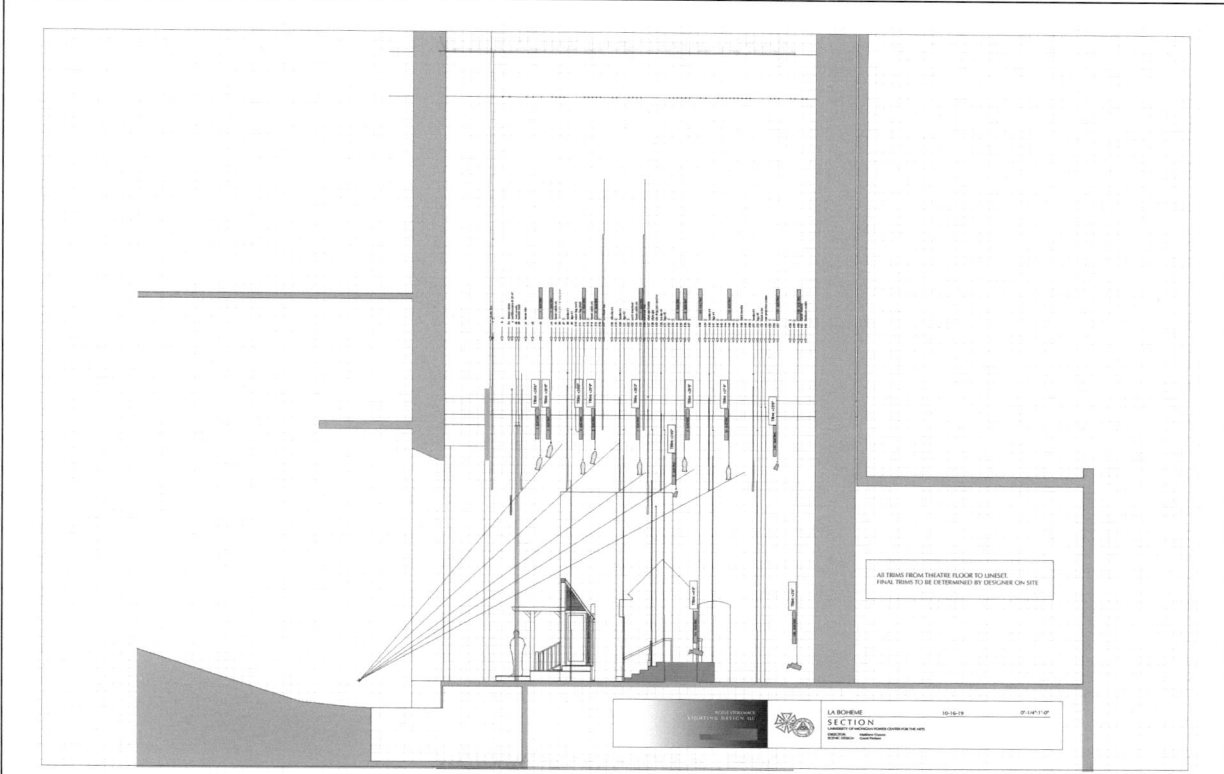

FIGURE 12-51   *Section drawn in Vectorworks for the University of Michigan's production of La Boheme; designed and drawn by Noele Stollmack; sightlines included; reference grid retained in lieu of scale bar. Used by permission of Noele Stollmack*

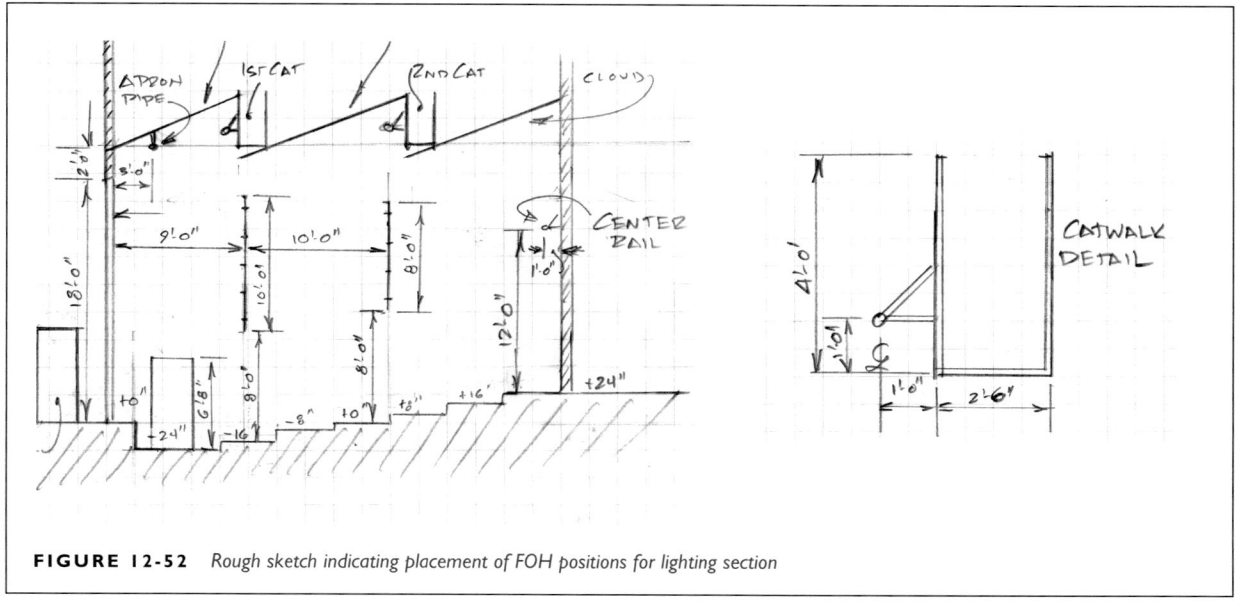

FIGURE 12-52   *Rough sketch indicating placement of FOH positions for lighting section*

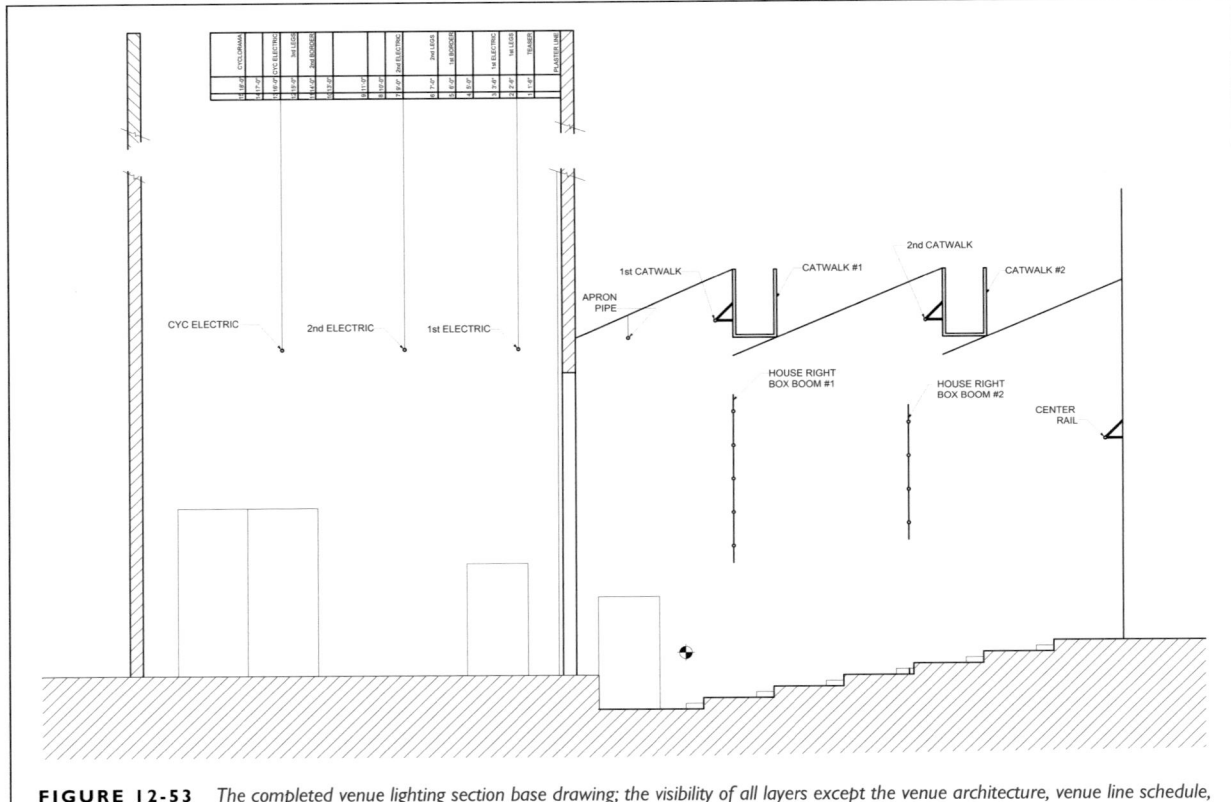

FIGURE 12-53 *The completed venue lighting section base drawing; the visibility of all layers except the venue architecture, venue line schedule, and the lighting section have been turned off*

preferred unit profiles for booms and box booms. A key is to be included, especially as the profile symbol of a unit may not match its plan view symbol.

Students

Draft a section for the plot drafted earlier, using the same information. Include onstage booms and FOH vertical positions. Include all masking. Scenic elements may be traced in thin or colored lines. Preferred unit profiles for booms should be discussed and chosen in class discussion. Include labels for all positions and trim heights where required. No legend information is arrayed around the unit symbols. Include a key.

If boom detail plates were generated during the plot drafting phase, renumber the sheets so that the section is Sheet 2 of the series. Revise notes on the plot directing the reader to the boom elevations.

REVIEW OF LIGHTING DRAWINGS SYSTEM OF WORKING

1. Procure venue base drawings and scenic groundplan, section, and other drawings as required. Survey the space to acquire additional information. Check with venue staff regarding venue-specific position names.

2. Construct the rough section to determine angles and unit types. Construct the rough plot in pencil on a ¼″ = 1′–0″ print of the whole venue so you can see the relationship of all the positions to the venue and the scenic environment.

3. If hand drafting, trace architectural and scenic elements from the base drawing in ½″ = 1′–0″. Break the drawing into multiple sheets, if necessary. Trace all lighting positions. Space between positions may be condensed to fit on the page, if necessary. If so, be sure that a note indicates that the position is not shown in its true location. If using Vectorworks, review the lighting layers of the base drawing file.

4. Prepare positions by lightly marking 18″ centers on all positions (if possible, and if not already automatically done by the CAD program). Work from the position's center out to its ends wherever possible. Work outward from the scenic center line wherever possible. For studio spaces with pipe grids, consider drawing a grid of guidelines by which to align units.

5. Consider where to place the key and legend and lightly outline the required area. Lay out the line schedule (if present).

6. Begin with the position at the top of the sheet (or rear of the venue). Draw darkened units along the position, working from the center toward its ends. As much as possible, point the symbols in one of the four cardinal directions (oriented to the position).

7. In Vectorworks, legend information can be added as each symbol is placed; it can also be added once units have been added to the whole position. If hand drafting, draw empty shapes (hexagon, etc.) to position information around each symbol. Avoid crowding neighbor units. Use small leader lines to connect information shapes to symbols if necessary. Lay out lettering guidelines for color. Fill in the information for each symbol. Once all units are complete, darken the position between symbols. Reinforce 18″ center tick marks between symbols. Add dimensions as required.

8. Repeat the process for all other positions, working down the sheet or from the rear of the venue. For venues with a rectilinear pipe grid, fill in all positions aligned in one direction, and then all positions aligned in the other.

9. Units or positions below the rig/grid, as well as boom and ladder footprints, should be indicated with a crosshatched symbol. If space allows, lay out the exploded view of booms and other vertical positions on the plot itself. If not, include a note to direct the reader to the pertinent detail sheet.

10. Once all positions have been drawn and darkened, add position labels (and inventories, if desired) and notes.

11. Darken architecture. Interrupt lines where they intersect lighting information. Darken the line schedule; fill in line schedule information. Include sightline symbols.

12. Use thin black lines to darken scenic outlines (or use a colored line, if in Vectorworks). Interrupt lines where they intersect lighting information. Include masking, such as legs and borders.

13. Draw the key and legend. Add other required production information and notes.

14. Darken the borders and title block. Fill in title block information.

15. Repeat this process for the section. Trace architecture and scenic elements. Trace the line schedule and lighting positions. Situate the key. For positions that cross or are perpendicular to the venue center line, draw the unit found on center or closest to center as that position's representative unit. Positions seen head-on, or otherwise fully seen, should include their full complement of units. These symbols do not require legend information. Label positions, and include trim heights as needed. Draw all masking.

16. If detail sheets are required, lay out the positions' elevations. If required for clarity, include the position's footprint (crosshatched). Outline the key and legend. Draw darkened unit symbols and their information arrays. Darken positions. Include height information. Label positions, noting on which sheet the position's location can be found. Add other notes as required. Darken the key and legend. Add border, title block, and title block information.

17. Review the drawing; check information against the hook-up. Revise as needed.

CHAPTER 13

DRAFTING FOR SOUND DESIGN

The first module of this chapter introduces the types of drawings typically generated for sound designs and how they are used, as well as introductory sound vocabulary. The second module discusses speaker placement plans and sections. The third module discusses the layout of signal flow diagrams and rack drawings.

TOPICS AND GOALS

- Types of sound design drawings and how they are used

- Sound vocabulary

- Constructing symbols for schematic layouts

- Signal flow diagrams

- Speaker placement drawings

- Rack drawings

MODULE 1: TYPES OF DRAWINGS AND SOUND VOCABULARY

Sound designer Beth Lake notes that it's important for any sound designer to remember that what they think of as standard procedure may not be the other person's standard procedure. Paperwork and documentation form a bridge between variants in practice and technology, and what is assumed to be "standard practice" must be clearly and fully apparent in the paperwork—never assumed. Practitioners from the concert world, the theater world, and the architectural world may have very different ways of talking about the same thing.[1]

One of the challenging and interesting things about sound design is that it is aggregated from many other practices, each with its own canon of work and research.[2] It has seen major technological changes through the end of the twentieth and into the twenty-first century, particularly the shift from analog to digital technologies. There's a heightened need for designers to understand engineering and engineers to understand design. In many respects, sound is even more ephemeral than light and its path from source to audience more intangible, especially in digital systems that have further reduced the physicality of sound technology. Clear and well-ordered documentation is essential in describing systems.

Sound design generates two broad categories of drawings: ones that describe the placement of equipment (speaker plots, rack drawings) and ones that describe the movement of information (**signal**) between equipment. Video and projection design generate drawings similar to those created for sound design.

Depending on the venue, client, and project, the scope of these drawings can vary. For commercial and permanent projects, every last piece of equipment must be specified so that bids can go out to vendors. In a not-for-profit storefront theater, the sound designer may be working from a laptop connected directly into an in-house system comprised of off-the-shelf home entertainment components of unknown age. Commercial and touring projects require rack drawings; a storefront theater might not even have a rack. If speakers and speaker arrays (or, "clusters") are in fixed positions, there is little need for a speaker plot unless the budget allows for additional speakers. A permanent, invariable system is often referred to as a **PA (public address) system,** especially in non-theatrical venues.

DOI: 10.4324/9781003154921-14

As with most design fields, there are two primary and complementary halves of a sound design:[3]

1. Content of the design (aesthetics)

2. Engineering the playback system (technology)

Developing content is akin to any other design process: understanding the production requirements (whether it be product presentation, play, ballet, or dining experience), research, making lists of what sound is to be heard at which point in the experience (cue lists), and seeking out and/or building audio recordings.

Content is then often broken into two realms: that of **input** and that of **output**. Knowing what goes in and wanting a particular result when it comes out determines what is required in the middle. Many sound designers employ a variant of this "building from the ends to the middle" approach.[4] If a ghost needs to sound ghostly, some sort of effects processor will likely be needed. If there is live music, what device will balance the sound among the musicians' microphones? If a truck is to sound as though it's barreling through the audience, how can the sound be shifted from one speaker to another?

Delivering the content can be broken into three phases:

1. Using signal mapping to determine the system requirements: playback devices, microphones, the number of speakers, mixing consoles, amplification, processors, etc.

2. Tuning the system/ringing it out. This is akin to the focus call of a lighting design, adjusting and refining the system so that it does what the designer needs it to do. Time alignment is a major part of this, as a signal moves through wire more quickly than sound waves move through air; e.g., delay may be added to speakers positioned closer to the audience than those further away so the whole audience receives the sound at the same time.[5]

3. Tech process. This involves the setting of levels, editing cues, setting timings, etc.—in short, all of the activities during which the sound design is integrated with the other elements of the production in real time.

Drafting comes into play mostly in the first phase, but it is used and updated throughout. The basic drawings of a sound design package are:

- Plan (top view of venue with speaker/equipment placement)

- Section (side view of venue with speaker/equipment placement)

- Signal flow (schematic diagrams of input/output, can also be created as a spreadsheet)

- Rack drawings (all the components required for the audio racks)

3D acoustic modeling is an important tool for most entertainment sound designers.[6] Since this is typically a step in the design process, its documentation is not usually shared with other members of the production team unless required by the client.[7] Much modeling software is proprietary and supports only its brand of speakers/processors.[8] For these reasons, acoustic modeling drawings lie outside the scope of this book.

Plan and Section speaker placement drawings are built upon the architectural base drawings as well as those provided by the scenic and lighting designers. Having accurate drawings for the auditorium is critical. Often, to begin acoustic modeling as soon as possible, the sound designer will begin working with the architectural base drawings long before the scenic and lighting drawings are available.[9] If the sound designer develops a speaker placement drawing early, the lighting designer can take this information into account when developing their light plot. Alternately, the sound designer might draft speaker placement directly onto a copy of the light plot.

As with lighting design, a venue's resources and budget determine what equipment is available to the designer. Most non-profit and regional venues have made investments in in-house systems, but are somewhere within a 10- to 20-year system life cycle. A designer might work on a project with the latest equipment and then follow it up with work in a different venue that last renovated their system in the 1990s.[10]

Many sound equipment manufacturers issue CAD symbols for their equipment. The draftsperson must usually do some editing to get the symbol to function as intended on the drawing; the manufacturer may include layers of information relevant to their industry, but not to yours.[11] CAD symbols can also be built from scratch, creating customized graphics best suited to the designer's working preferences.

Recent efforts in addressing recommendations and standards resulted in the 2021 release of the *USITT/TSDCA Sound Documentation Recommended Practice* document, a major updating of USITT's *Sound Graphics Recommended Practice* of 2008. The new document notes that it is "not a comprehensive collection of graphic symbols, terms, and necessary paperwork, but is a flexible framework of document forms and practices for practical communication. The information which designers and engineers must communicate can be represented in many forms; this document will identify and describe several of the most common."[12] Overall sound drafting practices are similar to those previously discussed. The major difference, as with lighting, is understanding the content that is to be drafted and understanding the format in which that content is typically presented.

TDSCA stands for Theatrical Sound Designers and Composers Association. Their website and public resources can be found at www.tsdca.org.

Sound Vocabulary

Sound follows a path (Figure 13-1). At its simplest technical level, sound is created, travels through the air as waves, and is received by a listener. If you are adding sound to a live production by use of sound effect machines, such as a piece of canvas stretched over a rotating drum to imitate wind, the sound travels directly (more or less) from device to listener. The art of creating live sound effects is called **Foley**, and its practitioners Foley artists. Foley is named for sound effect artist Jack Foley, and remains a technique by which sound is added to film and television in post-production.

Most sound design involves the conversion of acoustic sound waves into an analog signal, which may be further converted into a digital signal. The signal is processed and converted back into acoustic sound waves. Any device that converts one form of energy into another (e.g., acoustic wave to analog signal) is called a **transducer**. Microphones (mics) and loudspeakers (speakers) are transducers.

Signal going into a device is **input**. Signal coming out of a device is **output**. Sound waves hitting a microphone's diaphragm are input; the signal traveling down the wires is output. Signal heading into a speaker is input; the sound emanating from the speaker's diaphragm into the air is output. System diagrams are typically drawn with the sound origination devices (input) on the left and the sound dissemination devices (output) on the right.

Processing live sound during a performance, such as from a musical instrument or a human voice, is called **reinforcement.** A component that plays recorded sound, such as a CD player or MP3 player, is a **playback device**. Any system where recorded sound (whether vinyl album,

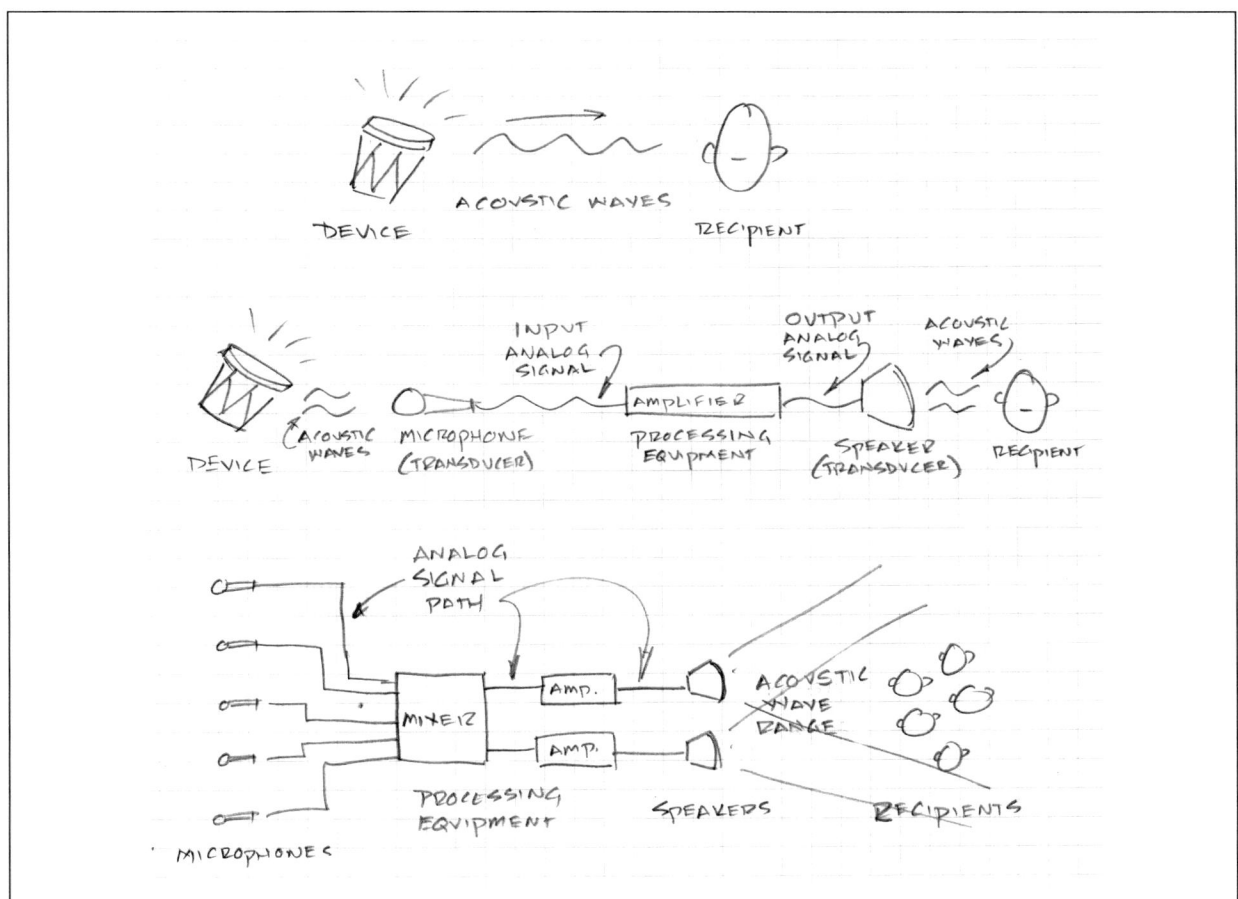

FIGURE 13-1 *Simple schematics of signal flow: device to ear; device to microphone to speaker to ear; multiple mics sending signal through processing equipment to multiple speakers*

magnetic tape, or digital file) is played back at any given time can be called a **playback system.**

Most systems include the integration of a **mixer**. A mixer accepts multiple inputs from a variety of sources and allows the user to send those inputs to selected outputs. If six microphones send signals to the mixer, it can combine those six signals and send the combined signal out to four speakers. The mixer can also send the signal from a single microphone to a single speaker. **Amplifiers** are located between the mixer and the loudspeaker to boost the signal to a strength strong enough to drive the speaker. A **preamplifier** (**preamp**) may be located between the microphone or playback device and the mixer; it boosts the signal to make it more usable by the mixer.

Processing devices such as amps and preamps are typically stacked in a **rack**. Racks are metal storage structures that hold sound equipment and are typically the point at which the system is connected to the main power supply.

Students often find the digital world daunting if they haven't already worked through analog systems and their components. Even with digital signal flow, you still need to know where signals originate, how they are processed, and their intended destinations within the software.[13] USITT recommends creating digital patch paperwork when working with signal flow systems such as AES67 or Dante. On some of the signal flow diagrams later in this chapter, there are symbols indicating that a signal is being digitally processed; what happens within that digital processor symbol is described elsewhere in the paperwork.

> Audinate, the company that developed and sells the Dante Audio Networking System, offers certification courses. The company website can be found at www.audinate.com.

Constructing Symbols with Records in Vectorworks

As with lighting, sound uses symbols to represent pieces of equipment. Information about the equipment—manufacturer name, type, signal inputs and outputs—is compiled on spreadsheets and other database record keeping. Sound signal mapping gets very complicated very quickly, and it's important to have clear, updated records. For sound drafting, the three commonly used categories of symbols are:

- Speakers and Microphones
- Signal flow components (device blocks)
- Rack components

In Vectorworks, the Spotlight libraries are already populated with lighting symbols. Typically, a sound designer must collect and file an assortment of manufacturer symbols or construct them from scratch (Figure 13-2). Equipment symbols range

from the pictorial to the schematic depending on designer preference and whether the appearance of front, top, side, or 3D views of the device is important. Signal flow and rack component symbols tend toward simplicity, such as subdivided rectangles of various standard sizes. Color coding is common on sound drawings, as this helps to differentiate between types of signal and categories of equipment.

Sound drawings typically feature text placed within a symbol's partitions: equipment name, number of inputs and outputs, etc. In Vectorworks, text can be entered via the symbol's Object Info navigation window. Certainly, you could draft the symbol (without using Create Symbol), save it as a group, and then use the Text tool to add required information. For smaller systems, this may be perfectly adequate. Creating a library of fillable symbols should be a tool to ease your workload and make future drafting more efficient.

Since the signal flow drawing is schematic, and the rack drawing also tends toward the schematic, these drawings tend to be constructed in 1:1. The printed scale is determined by considering how well the drawing will fit on the selected page size and remain clear and readable. For sound schematics, using classes to assign attributes to symbol outlines, subdivisions, and text is often preferred.

Exercise: Constructing a Symbol in Vectorworks

Instructor

The following steps construct a basic record keeping symbol in Vectorworks.[14] This exercise constructs a symbol as might be used for a rack drawing equipment symbol: a rectangle with two fillable text spaces (Figure 13-3). There are three main steps to the process:

1. Drafting the symbol
2. Creating the record format and fields
3. Connecting fields to the symbol

Students

As with any drafting project, when building symbols, sketch what you intend to include in the symbol's layout before drafting. This symbol will be a rectangle with two fillable text spaces as might be found on a rack drawing.

1. Open a new Vectorworks file. This will become a Sound Base File. The first layer may remain Design Layer-1. Create a guideline layer.
2. Create three classes: guidelines, symbol outlines, and text. Outline:

   > Line Style: Thick (0.50 mm)
   > Fill: Light Gray
   > Text: Un-styled

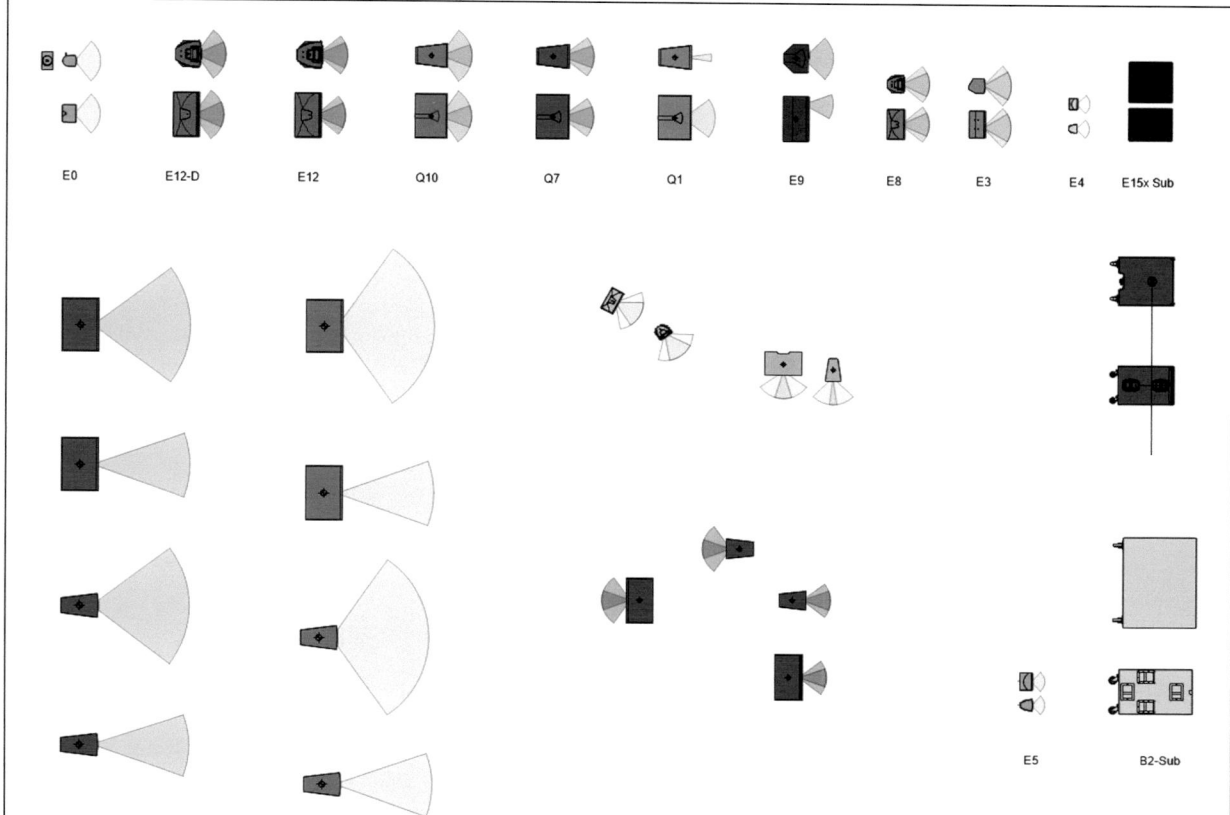

E0   E12-D   E12   Q10   Q7   Q1   E9   E8   E3   E4   E15x Sub

E5   B2-Sub

FIGURE 13-2   *A library of custom-drawn speaker and subwoofer symbols; shaded fans suggest range of speaker coverage; original drawing is color coded; drawn by Ben Truppin-Brown. Used by permission of Ben Truppin-Brown*

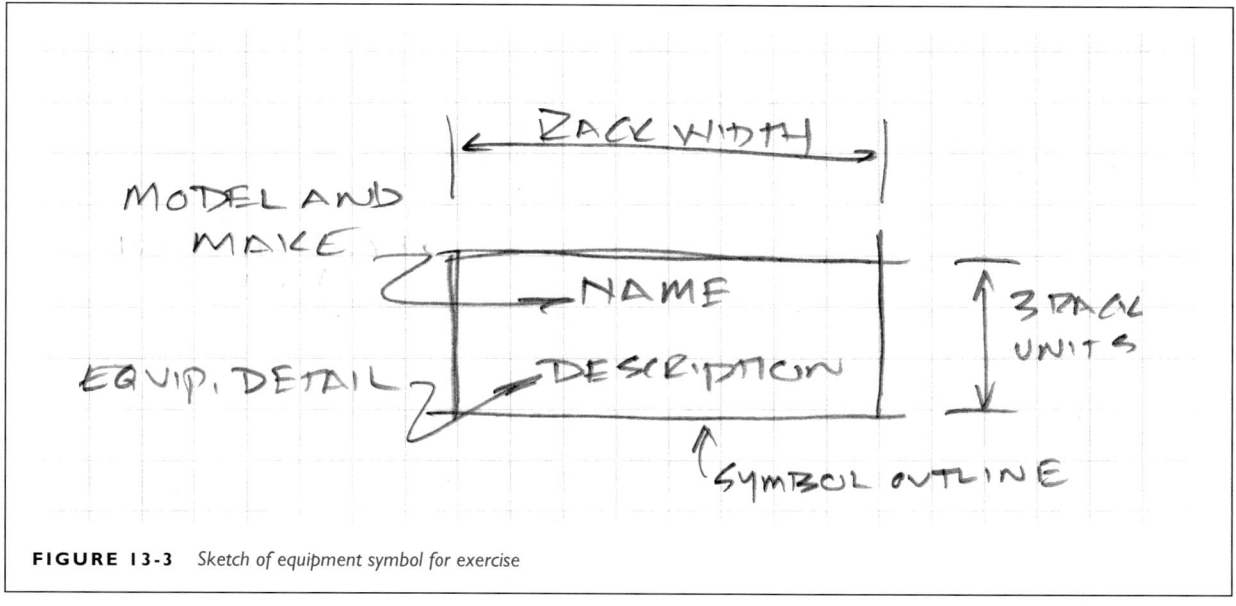

FIGURE 13-3   *Sketch of equipment symbol for exercise*

Text:

>   Line Style: Solid (0.50 mm)
>   Fill: None
>   Text: Un-styled

After creating these classes, click OK to return to the drawing. On the main toolbar, choose Text to open its drop-down menu. Choose Format Text. The Format Text window will open, listing all the font choices available. As this symbol will be drawn in 1:1, the font size needs to be proportional. Change the font size to 48pts. At the top of the window is a Save button. Click on it; a window asking you to name the font style will appear. Name it Symbol Font (Figure 13-4).

OK back to the drawing, and then open the Class organization window. Open the Text style.

When you open the selection menu for Text Style, Symbol Text will appear in the menu. Choose it and OK back to the drawing. Now, whenever you use the Text class, the letters will be drawn in 48pt. As you develop your library of symbols, set up a selection of desired font sizes and styles ahead of time so you can choose them during class creation.

3. In the guideline layer, lay out a rectangle 1'–7" wide by 5¼" high. This represents a piece of equipment that will be slotted into a rack. Add guidelines at the vertical and horizontal axes to divide the rectangle into four areas. These internal guidelines will be used to center the text. Switch to the design layer. Darken the rectangle with the Symbol Outline class. The guidelines will disappear behind the Fill. Select the surface and Send to Back.

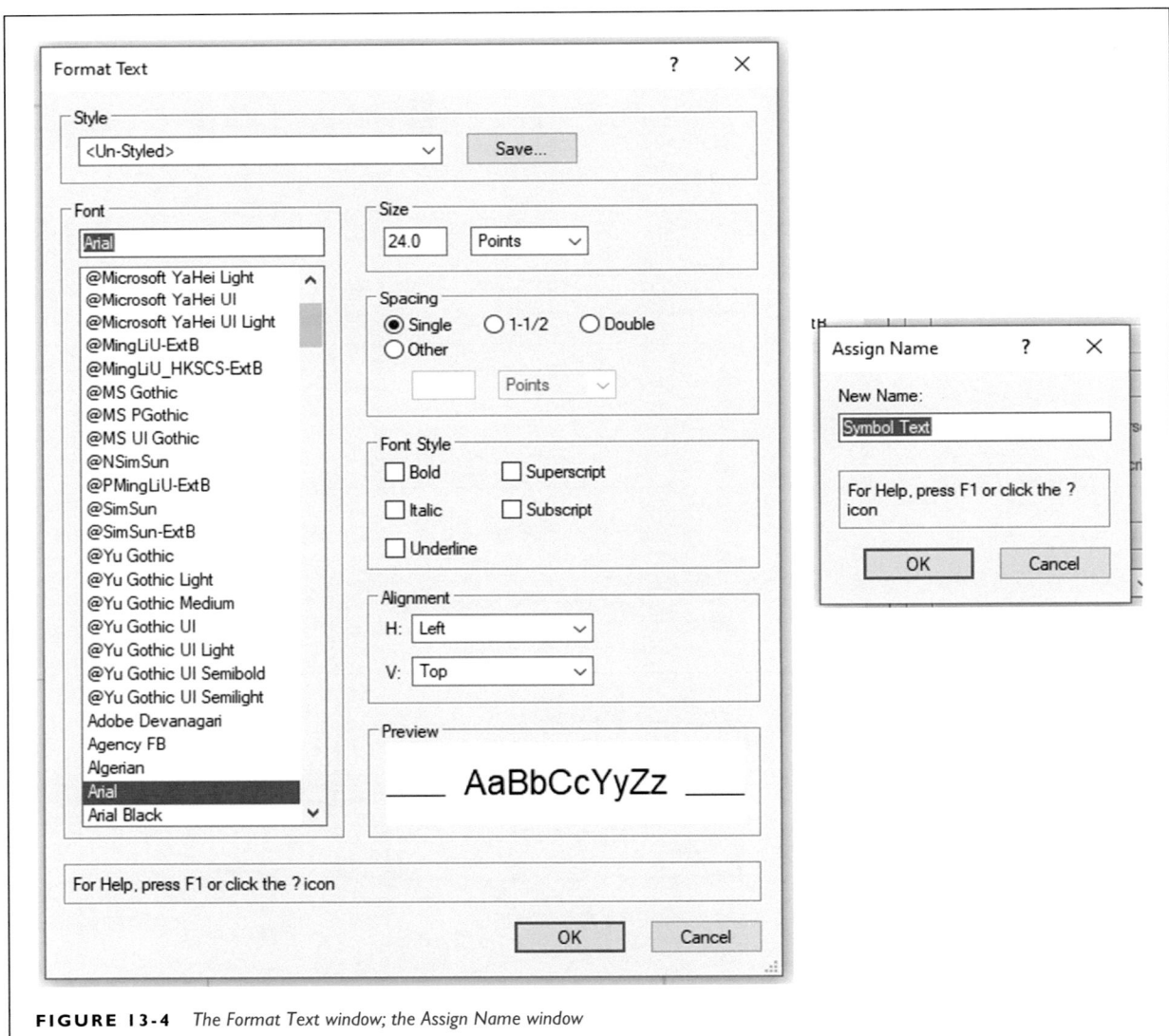

FIGURE 13-4   *The Format Text window; the Assign Name window*

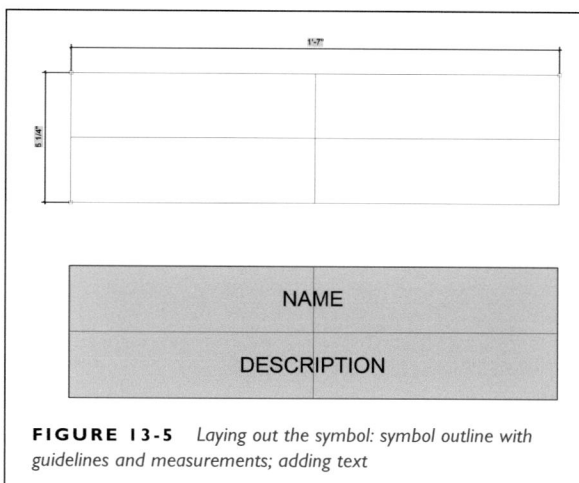

FIGURE 13-5 *Laying out the symbol: symbol outline with guidelines and measurements; adding text*

4. Switch to the Text class. Write NAME and center the words in the upper section. Write DESCRIPTION, centering the words on the lower intersection. Turn off visibility for guidelines (Figure 13-5).

5. Select the whole object. Open the Modify menu from the upper toolbar. Choose Create Symbol… to open the Create Symbol window. Name the symbol 3 RU Rack Equipment Symbol. RU refers to rack units, a standard rack spacing increment. Checkmark the boxes next to Next Mouse Click, Leave Instance in Place, and Change 2D Objects for Layer Place. Click OK to return to the drawing, and then click on the symbol to open a different Create Symbol window. This window allows you to send it to a location in the Resource Manager library.

If you click OK, the symbol will be sent to the Untitled folder. If you click on New Folder, you will be asked to name it. When you click OK, the new folder will be placed within the Untitled folder. Choose New Folder and name it My Symbols (Figure 13-6).

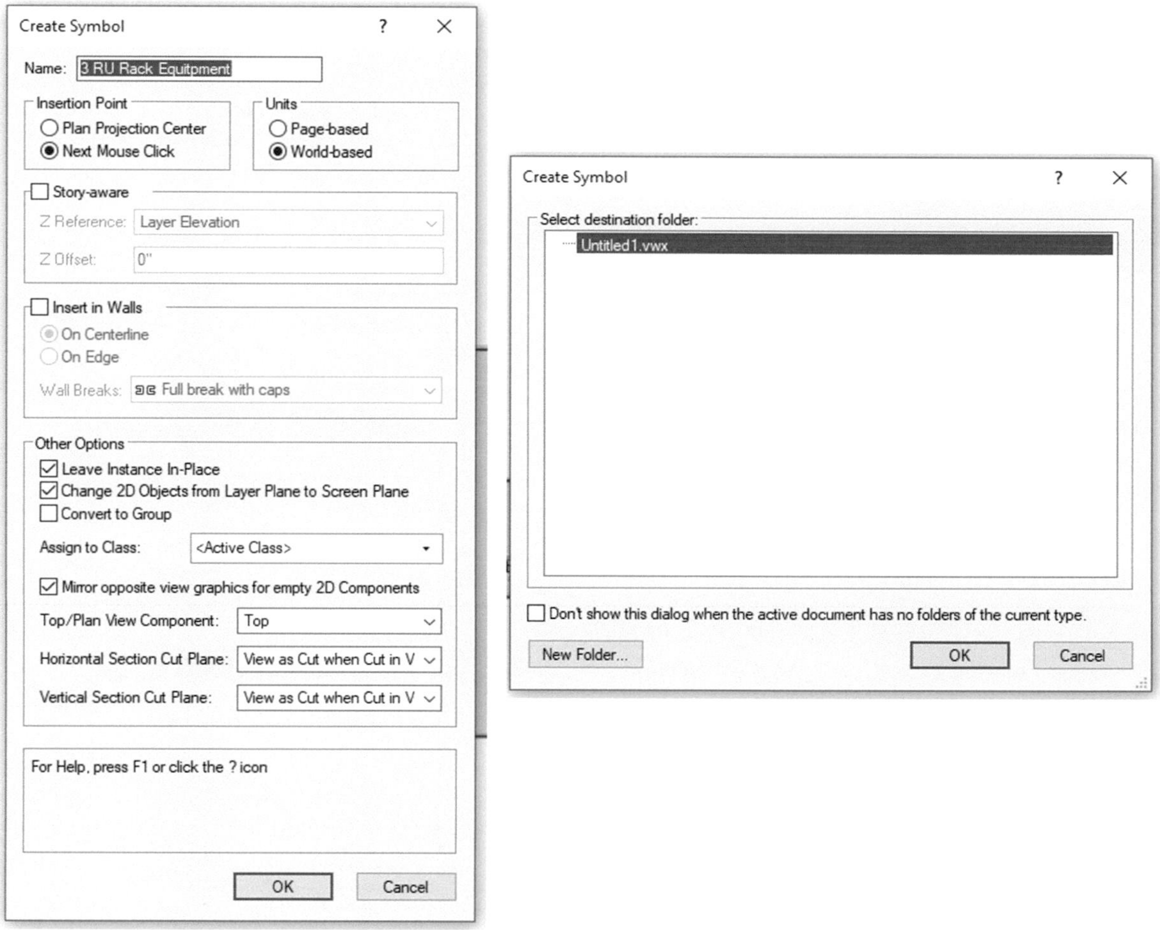

FIGURE 13-6 *The create symbol and create symbol destination folder windows*

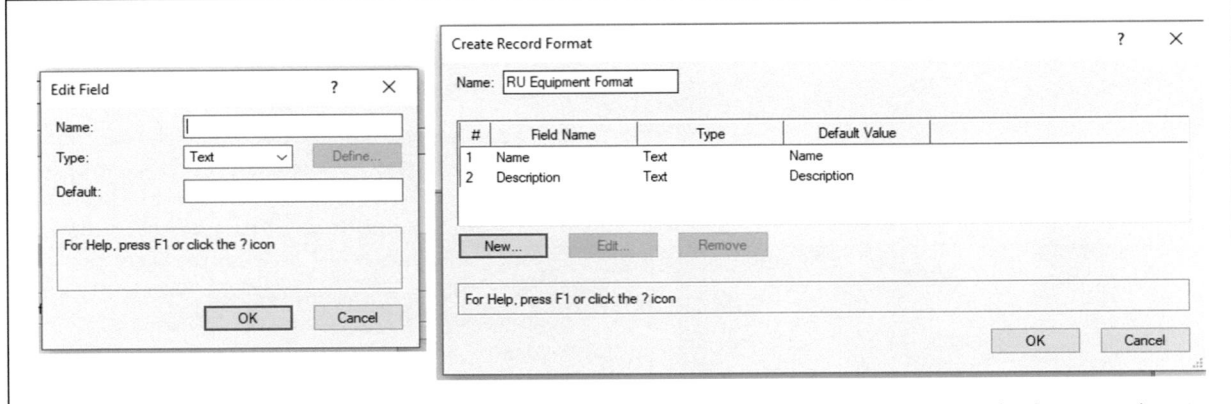

FIGURE 13-7   *The Edit Field window opens when New is clicked in the create Record Format window; the exercise's information is shown in the create Record Format window*

6. The symbol now needs to have records associated with it. The first step is to create a Records Format folder in your My Symbols folder. Open the Resource Manager (Windows > Palettes > Resource Manager). The Resource Manager window collapses when the cursor is moved away from it, leaving just its header on the screen; move the cursor over the header to expand it.

    Right-click on the My Symbol file name to open a menu. Choose New Resource in Untitled1. Its menu will open. Choose Record Format. The Create Record Format window will open.

    Name the format Rack Equipment Format. There are currently no Field Names listed. Click on New to open the Edit Field window. "Name" refers to the title of the text box that will appear in the symbol's Object Info navigation window. "Type" refers to the type of object within the symbol with which the record will be associated. "Default" refers to the text that will appear when the symbol is first placed and will be replaced by what is typed into the Navigation window's options.

    The upper text of this exercise's symbol is the name of the piece of equipment. The lower text is the equipment's description. Click on New to open the Edit Field window. Enter Name for Name, choose Text, and enter Name for Default. Click OK, and then New to create a second field. Enter Description for Name, choose Text, and enter Description for Default. OK back to the drawing (Figure 13-7).

7. Now these fields must be attached to the text in the symbol. Right click on the symbol and choose Edit from the menu that appears. The Edit Symbol window appears. Click the selection for 2D Components, and switch Double Click: to Edit the 2D Components. Click Edit. The appearance of the screen will shift. Click the cursor outside the symbol to deactivate everything that

might be currently selected. Double click on the feature to be linked to a record, in this case NAME. The color of box around the text will change.

On the main toolbar, open the Tools drop-down menu. Choose Records, and then Link Text to Record…. The Choose Field window will appear (Figure 13-8). Click on the desired Format in the upper window, and the Name Field in the lower window. This will associate the Name field with the symbol's name location. Click OK, and then repeat the process for the

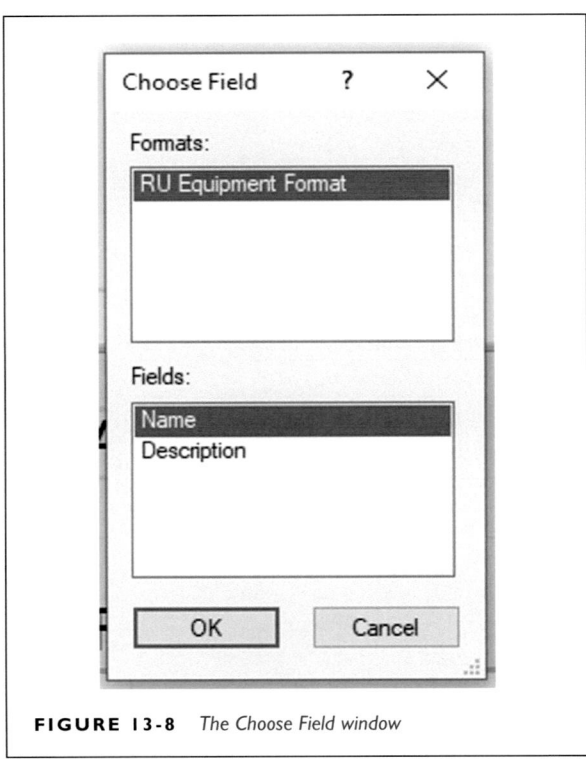

FIGURE 13-8   *The Choose Field window*

Description, choosing the Description option from the lower window.

OK back to the drawing, and then exit the Symbol Editing screen by clicking on the exit button in the upper right corner of the drawing space (move the upper navigation window if you do not see it; the button may be hidden behind it).

8. When you wish to use this symbol, open the Resource Manager, open the My Symbol Folder, and double click on the symbol to select it. Return to the drawing space and double-click to place it.

9. To change the text, open the Data tab in the Object Info navigation window (Figure 13-9). The names of the fields with their default text will be listed in the lower section of the window (you may need to expand or scroll down the window to find these options). After typing in the name or description, hit Enter. The new text will appear within the symbol.

Many of the symbols used for sound drawings are variants of what was created in this exercise. The symbol may be broken into columns and rows, each with its own record-associated text. Internal divisions may be color coded. Information within any symbol may be written with a variety

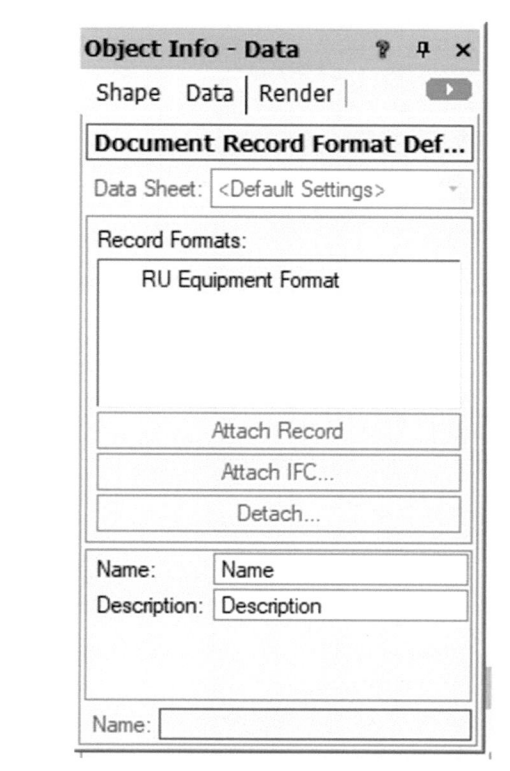

FIGURE 13-9 *The symbol's Object Info navigation window; open the data tab to enter field information*

of font sizes depending on the category of the information: names, description input/output numbers, etc. As symbols are created for various projects and drawings, sort them into separate folders; e.g., one for speaker symbols, one for rack equipment, one for signal flow diagrams.

Make symbol creation part of your preparation process. Save new symbols into your sound base file before you begin a new drawing.

MODULE 2: SOUND PLOTS AND SECTIONS

Generally, sound plots and sections situate equipment symbols onto existing base, scenic, or lighting drawings. As with light plots, laying out rough sound plots and sections in pencil on base drawing hardcopies helps finalize the drawing before drafting commences.

Speaker Plots

Information for each speaker symbol is provided through callout boxes connected to the symbol with leader lines. The sound information should have visual precedence on the sheet, so other information can be drawn with lighter value lines and colors. Often, the equipment symbols are filled with a color to make them more visible. Notation within the callout box includes location, equipment name, and hanging and placement information and guidance. Unlike on light plots, equipment sitting on the deck or below the grid is not crosshatched.

Figures 13-10–13-12 present sound plots in three different venues. Figure 13-11 features some acoustic modeling in the form of colored fans indicating the speaker's lateral coverage. Figure 13-12 indicates the speaker coverage with lines. In Figure 13-10, callout boxes are connected to the speaker symbol with a leader line. They are colored red to help them stand out from the underdrawing. In Figure 13-11, the callout boxes are placed in clear proximity to the symbol; the symbols themselves are color coded. In Figure 13-12, the speakers are numbered, referring the reader to accompanying paperwork. On Figures 13-10 and 13-11, keys are not included, as all the speaker's information is contained within the associated callout box. Figure 13-12 includes a key since the reader is referred beyond the drawing for further speaker details.

Information may also be broken apart onto separate sheets, such as in Figures 13-13 through 13-16. Figure 13-13 drawing presents onstage speaker locations and Figure 13-11 presents onstage microphone placement. In Figure 13-14, note that the symbol used to represent microphones is the same filled-quadrant circle used to denote sightlines in scenic and lighting drawings.

Full venue views are not the only form a sound plot may take. Equipment may be placed on or within scenic elements and detail drawings indicate their placement. Figure 13-15

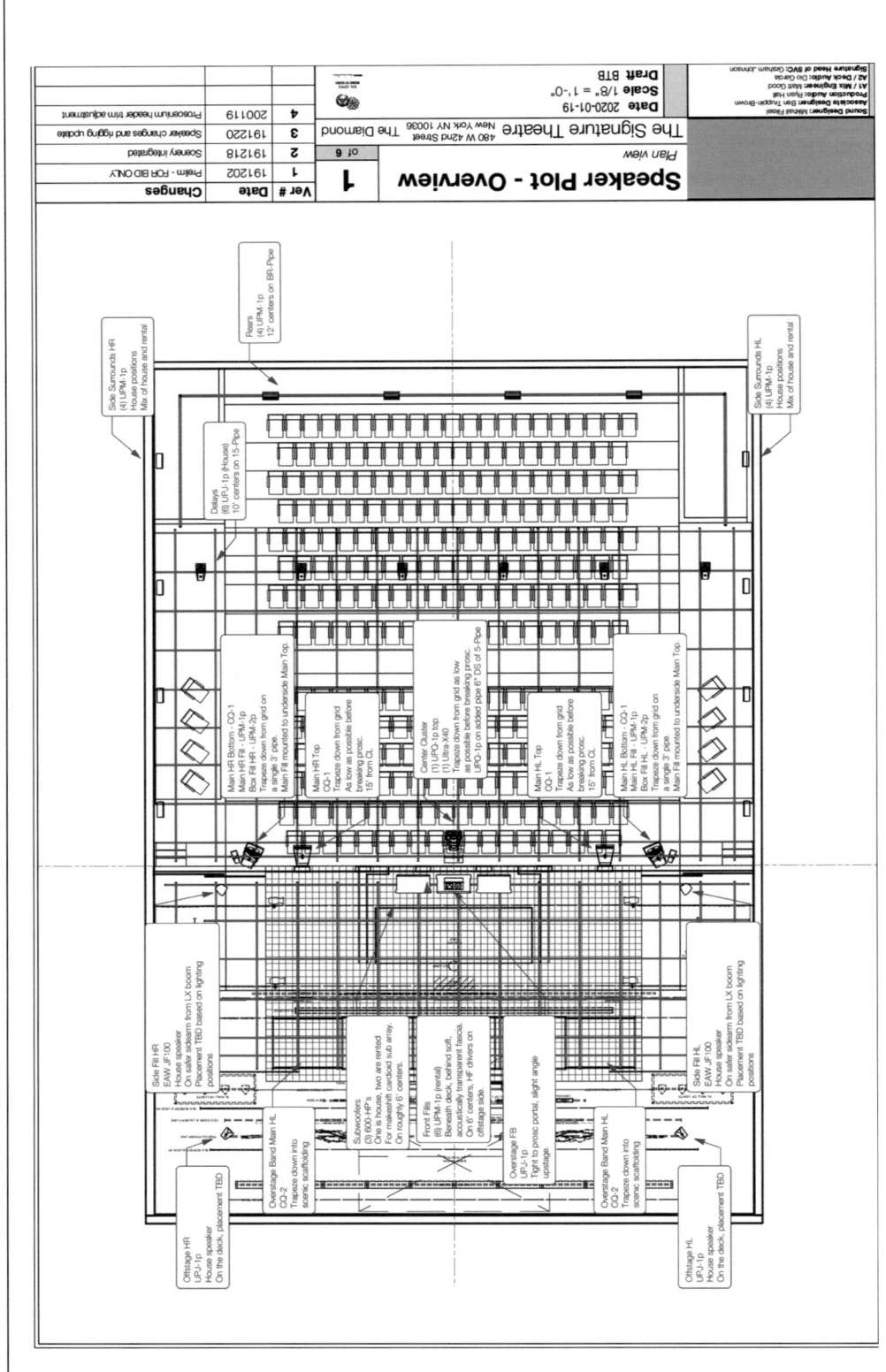

FIGURE 13-10 *Speaker plot overview for Signature Theatre's production Cambodian Rock Band, sound design by Mikhail Fiksel, drawing by Ben Truppin-Brown, associate sound designer. Used by permission of Ben Truppin-Brown and Mikhail Fiksel*

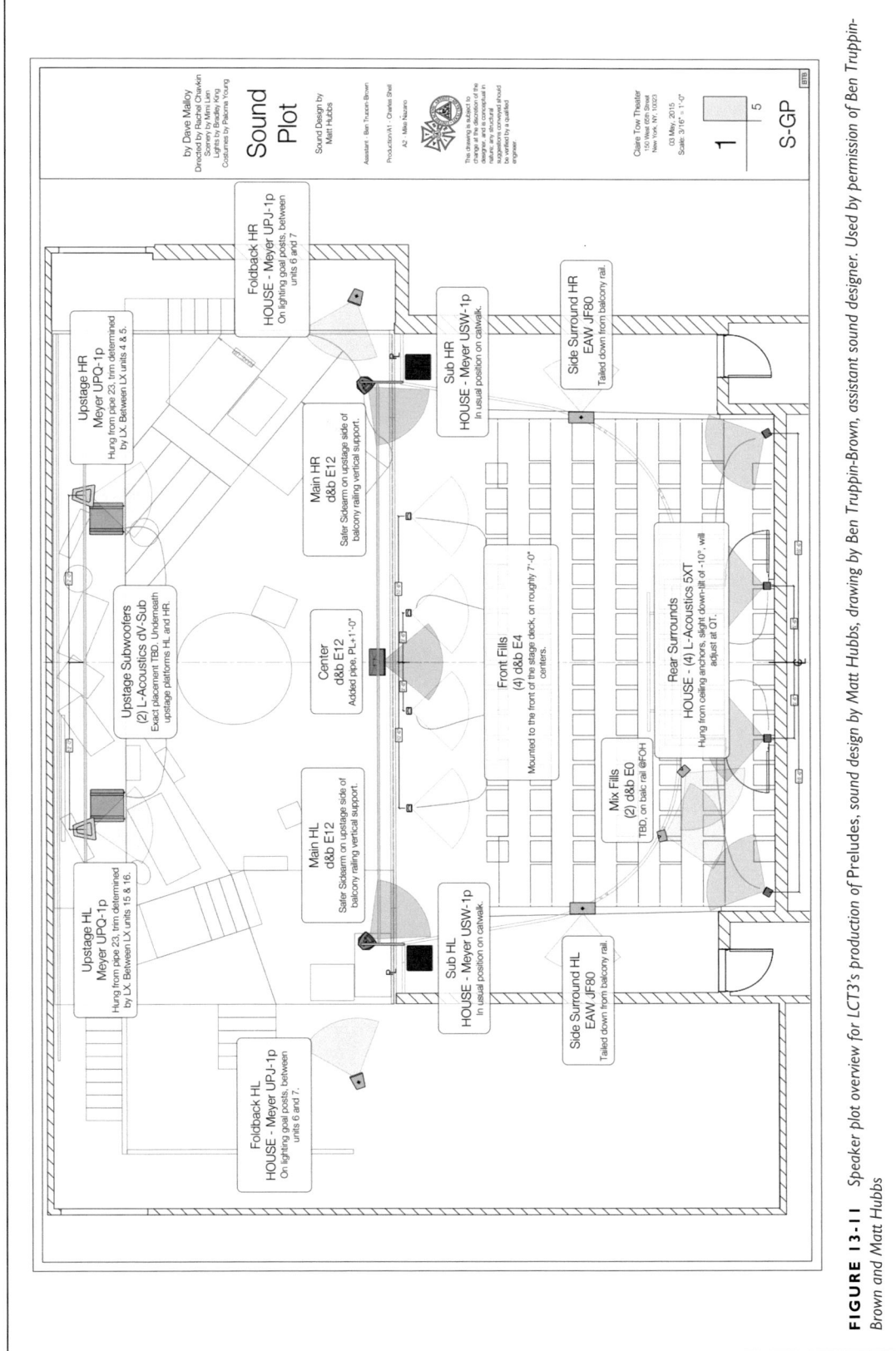

FIGURE 13-11 *Speaker plot overview for LCT3's production of Preludes, sound design by Matt Hubbs, drawing by Ben Truppin-Brown, assistant sound designer. Used by permission of Ben Truppin-Brown and Matt Hubbs*

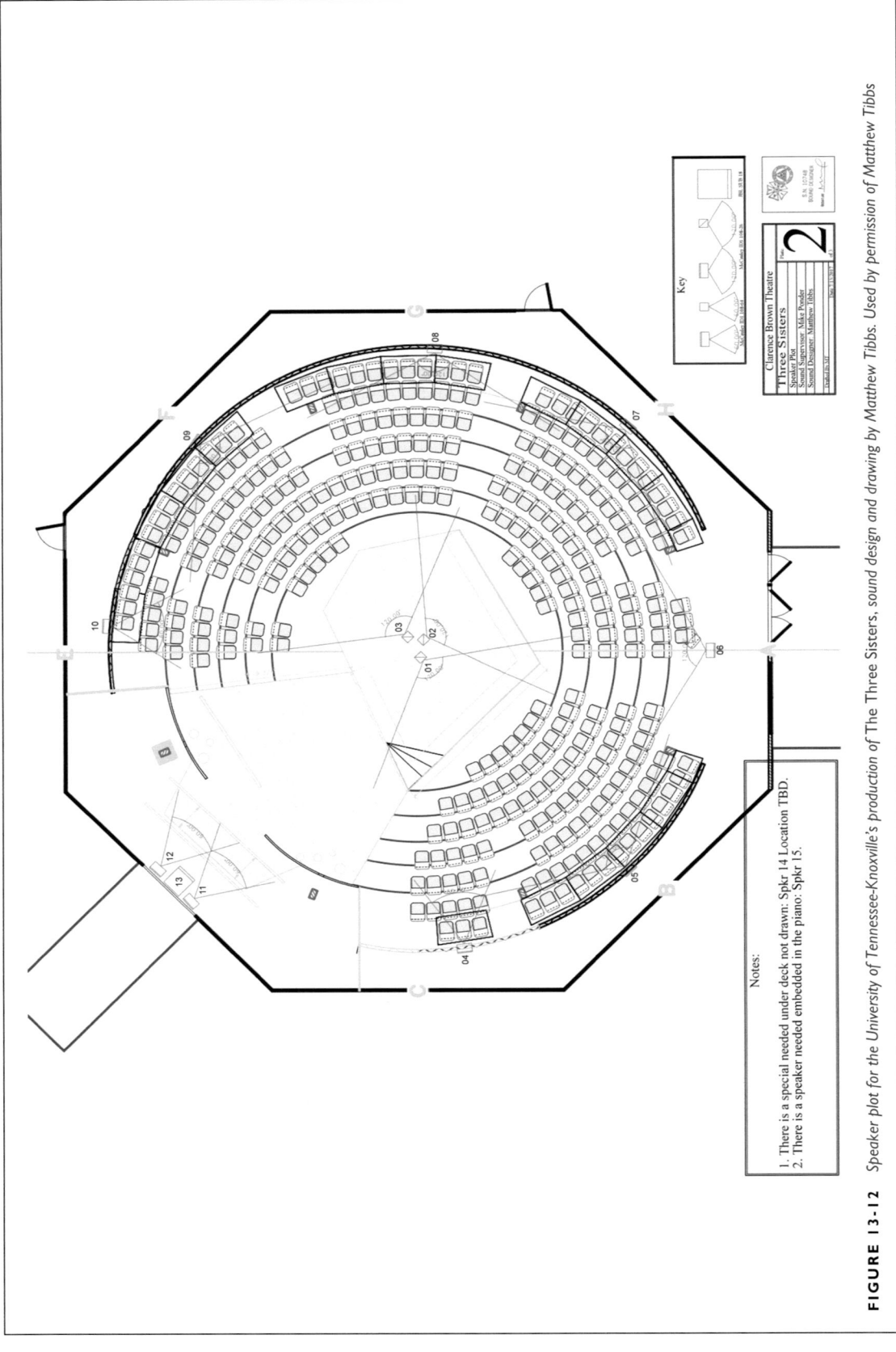

Notes:

1. There is a special needed under deck not drawn: Spkr 14 Location TBD.
2. There is a speaker needed embedded in the piano: Spkr 15.

FIGURE 13-12 *Speaker plot for the University of Tennessee-Knoxville's production of The Three Sisters, sound design and drawing by Matthew Tibbs. Used by permission of Matthew Tibbs*

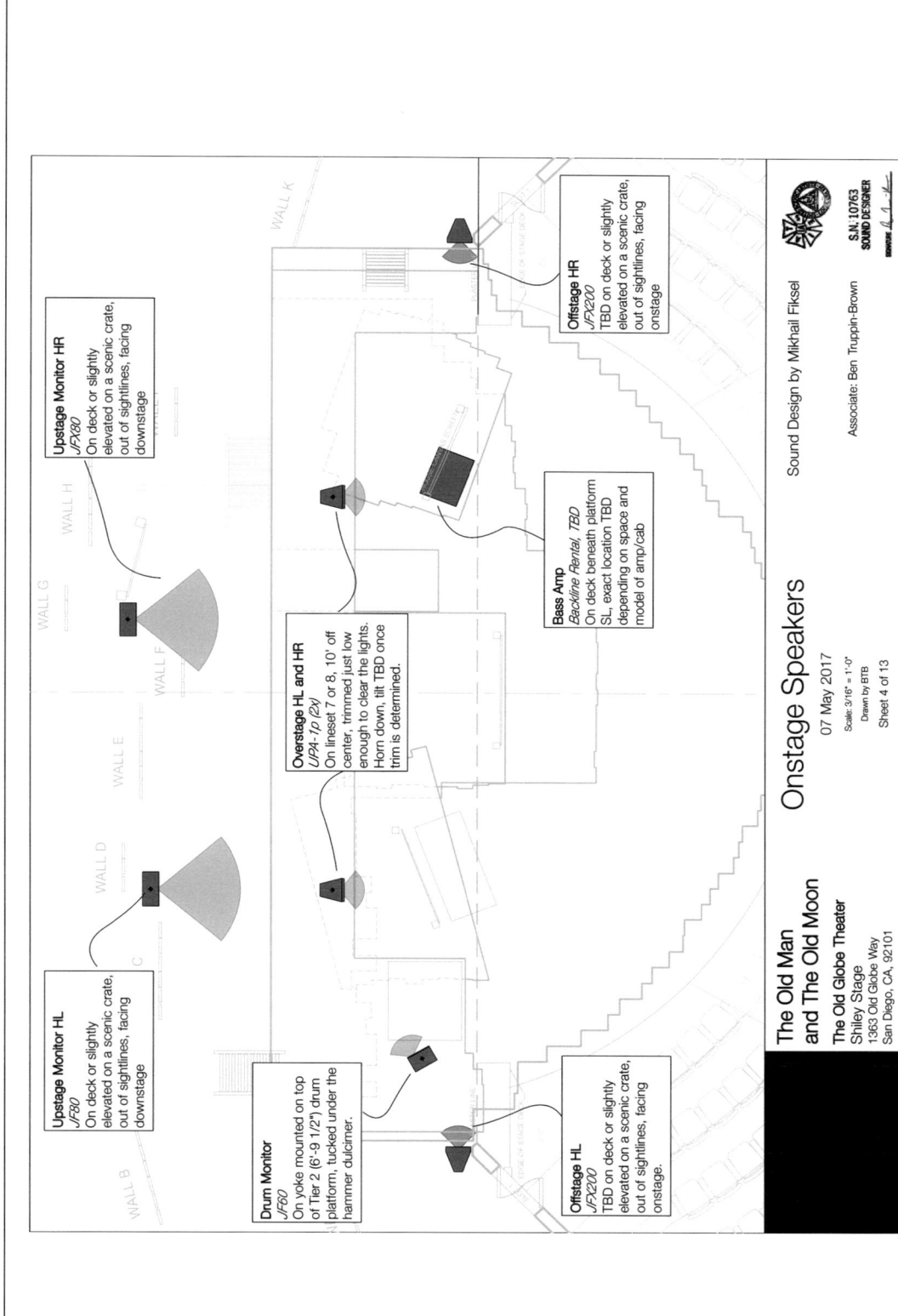

FIGURE 13-13 *Drawing for onstage speaker placement for Pigpen Theatre Company's production of The Old Man and the Old Moon, sound design by Mikhail Fiksel, drawing by Ben Truppin-Brown, associate sound designer. Used by permission of Ben Truppin-Brown and Mikhail Fiksel*

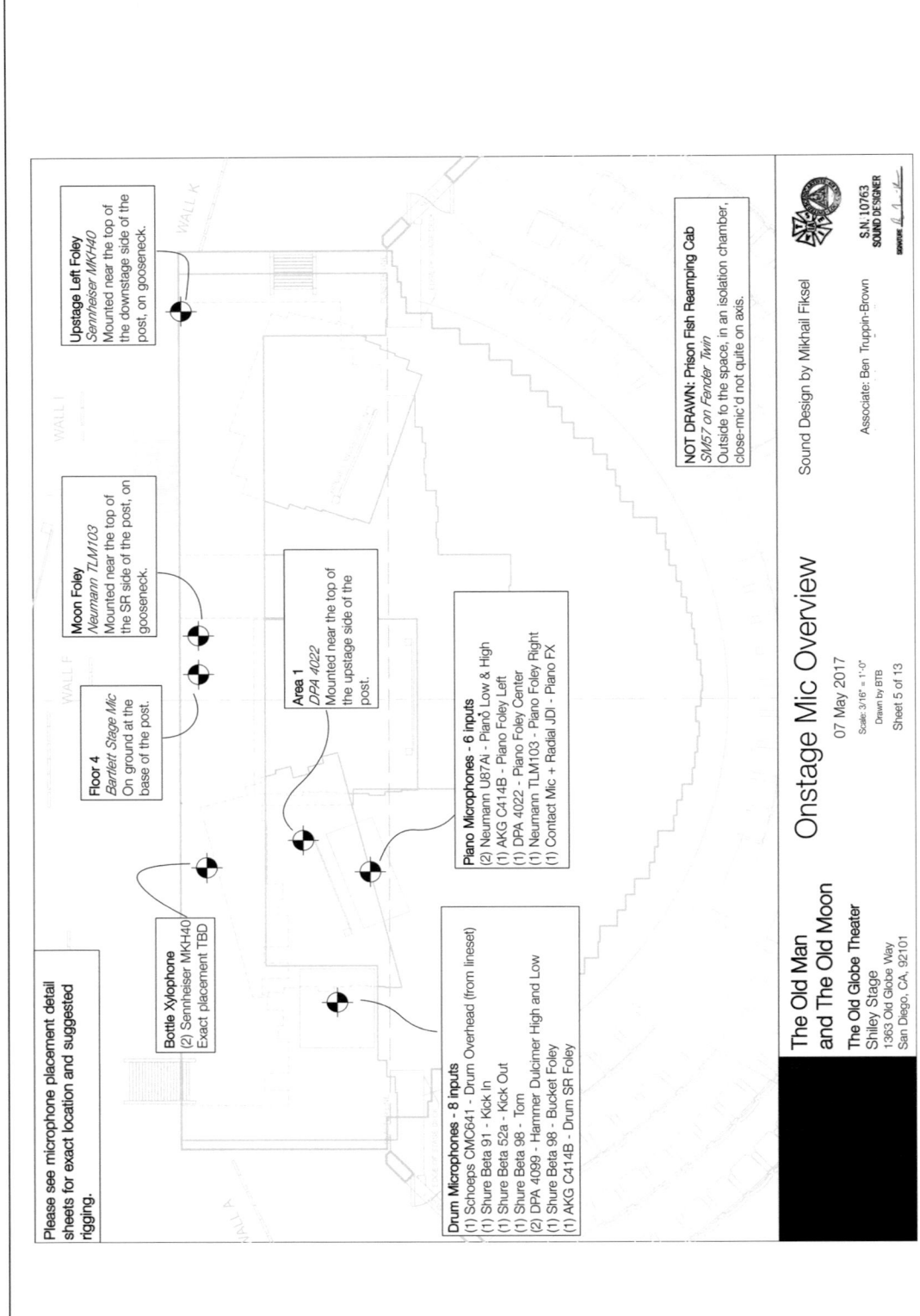

FIGURE 13-14 *Drawing for onstage mic overview for Pigpen Theatre Company's production of The Old Man and the Old Moon, sound design by Mikhail Fiksel, drawing by Ben Truppin-Brown, associate sound designer. Used by permission of Ben Truppin-Brown and Mikhail Fiksel*

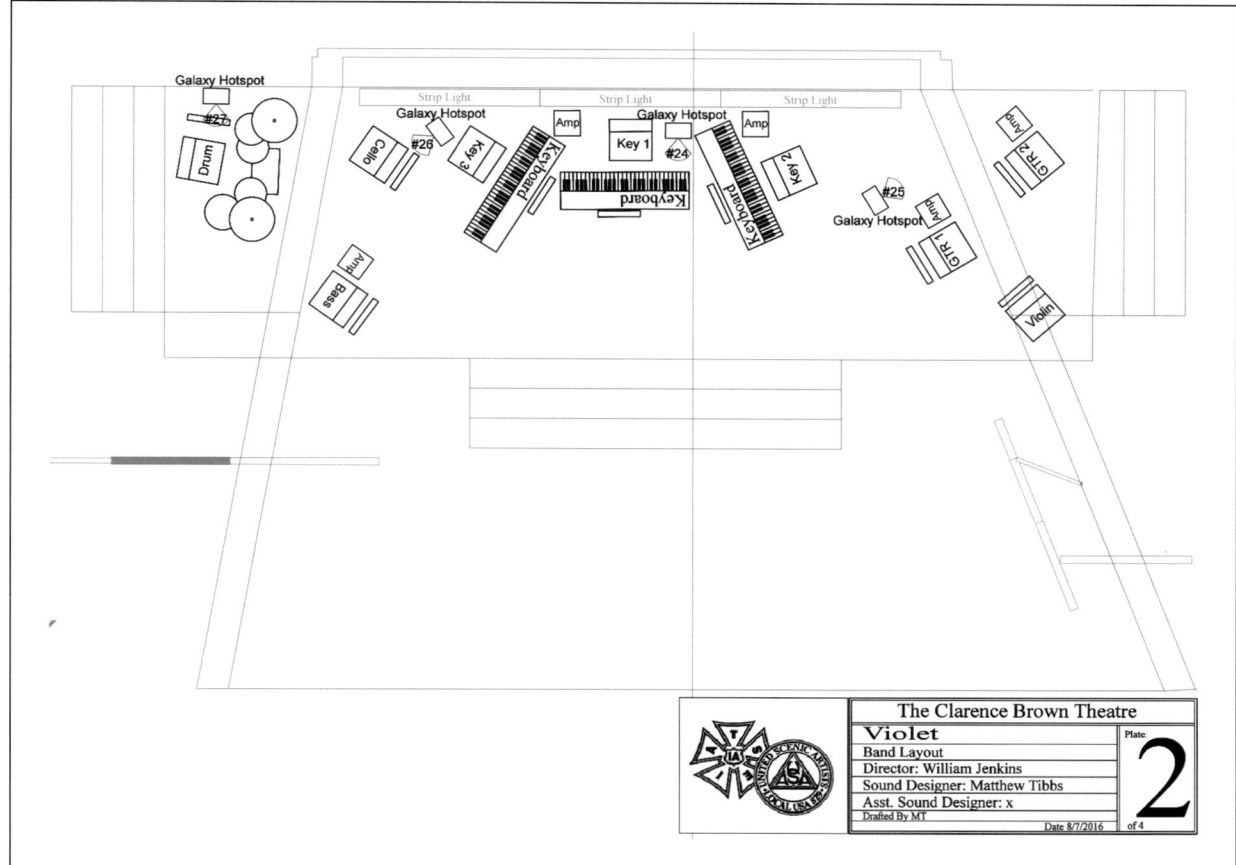

FIGURE 13-15 *Band layout drawing for the University of Tennessee-Knoxville's production of* Violet, *sound designed and drawn by Matthew Tibbs. Used by permission of Matthew Tibbs*

presents equipment on a band platform, with instrument and amplifier placement.

Detail drawings may include photographs. This is especially useful in touring situations where each locale has a fresh crew for each load-in and strike, as well as for repertory performance and long-term system maintenance. This drawing is added to the package after load-in, as the photos can't be taken until the equipment is actually set up. A section of Figure 13-16 features a plan of the stage with reference numbers indicating the placement of each piece of equipment. The numbers are associated with photos of each piece of equipment as the designer wishes it to be placed, along with descriptions.

Sections

Sections are constructed in a similar manner. Symbols are placed on base, scenic, or lighting drawings and callout boxes or labels provide information about each symbol. Since the plot provides the bulk of details for each symbol, the section's information may be limited to include only location and equipment name (enough information to clearly cross-reference other paperwork). Sections may also include

detailed height information to place equipment. Figure 13-17 presents a section with shaded fans representing acoustic coverage. Figure 13-18 indicates coverage with lines.

Project: Drafting Speaker Plots and Sections

Instructor

If a venue with a permanent sound system is available to the class, lead the students on a field trip to locate the space's speakers and microphones (if present). If equipment is accessible, have the students measure their locations. Measure the speakers themselves, taking note of make and model; otherwise, provide this information. If you wish students to create visualizations of the speakers' acoustic spread, have the students research the manufacturer's specification sheets for each type of speaker found.

If a venue is not available, use a copy of the venue drawing from a previous chapter to provide the students with a rough plot and section with speaker placement indicated. Include a list of speaker makes and models, as well as their locations/purposes.

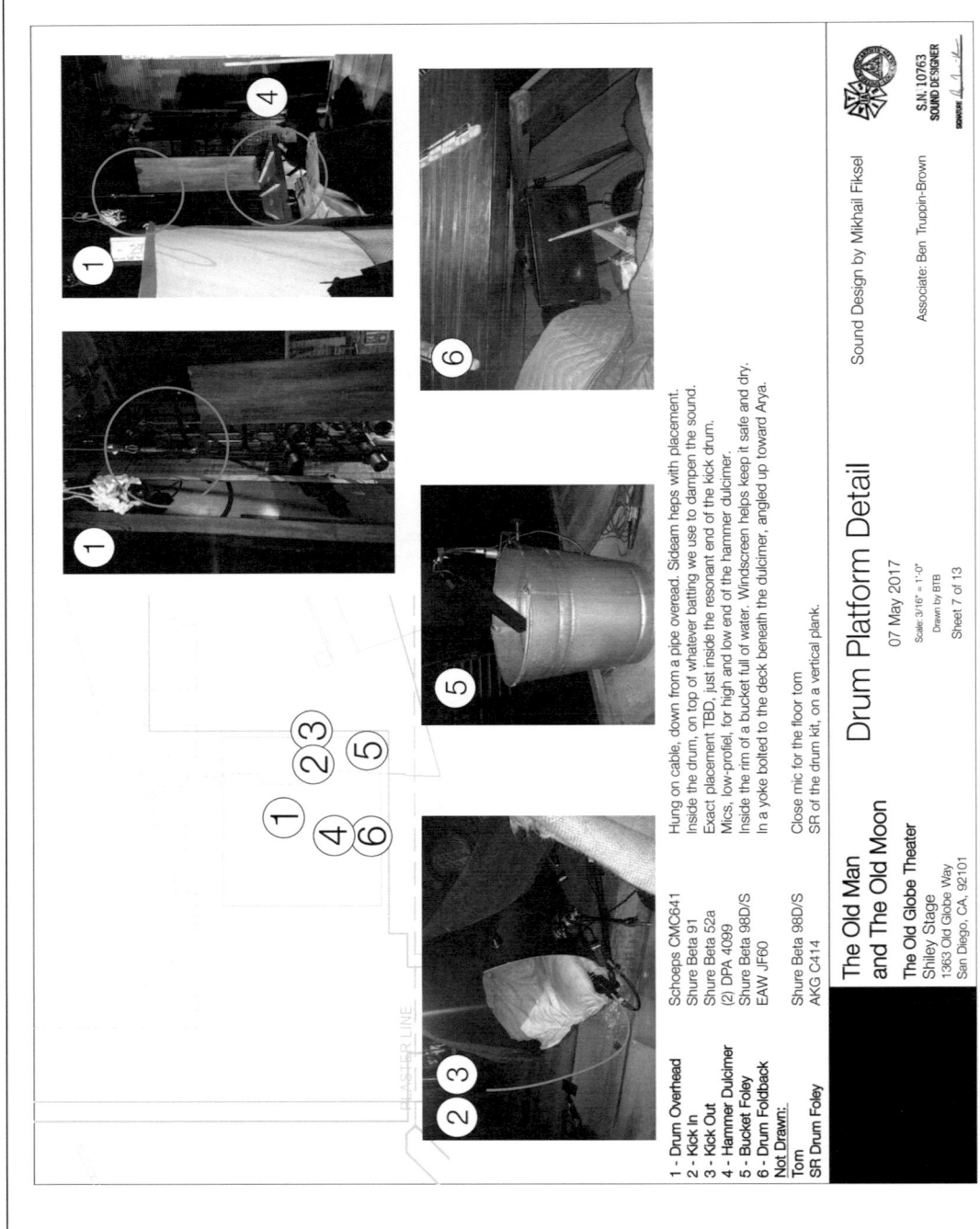

1 - Drum Overhead   Schoeps CMC641   Hung on cable, down from a pipe overead. Sidebeam helps with placement.
2 - Kick In   Shure Beta 91   Inside the drum, on top of whatever batting we use to dampen the sound.
3 - Kick Out   Shure Beta 52a   Exact placement TBD, just inside the resonant end of the kick drum.
4 - Hammer Dulcimer   (2) DPA 4099   Mics, low-profiel, for high and low end of the hammer dulcimer.
5 - Bucket Foley   Shure Beta 98D/S   Inside the rim of a bucket full of water. Windscreen helps keep it safe and dry.
6 - Drum Foldback   EAW JF60   In a yoke bolted to the deck beneath the dulcimer, angled up toward Arya.

Not Drawn:
Tom   Shure Beta 98D/S   Close mic for the floor tom
SR Drum Foley   AKG C414   SR of the drum kit, on a vertical plank.

The Old Man and The Old Moon

The Old Globe Theater
Shiley Stage
1363 Old Globe Way
San Diego, CA, 92101

Drum Platform Detail

07 May 2017

Scale: 3/16" = 1'-0"
Drawn by BTB
Sheet 7 of 13

Sound Design by Mikhail Fiksel

Associate: Ben Truppin-Brown

S.N. 10763
SOUND DESIGNER

FIGURE 13-16 *Drum platform detail drawing for Pigpen Theatre Company's production of The Old Man and the Old Moon, sound designed by Mikhail Fiksel, drawing by Ben Truppin-Brown, associate sound designer. Used by permission of Ben Truppin-Brown and Mikhail Fiksel*

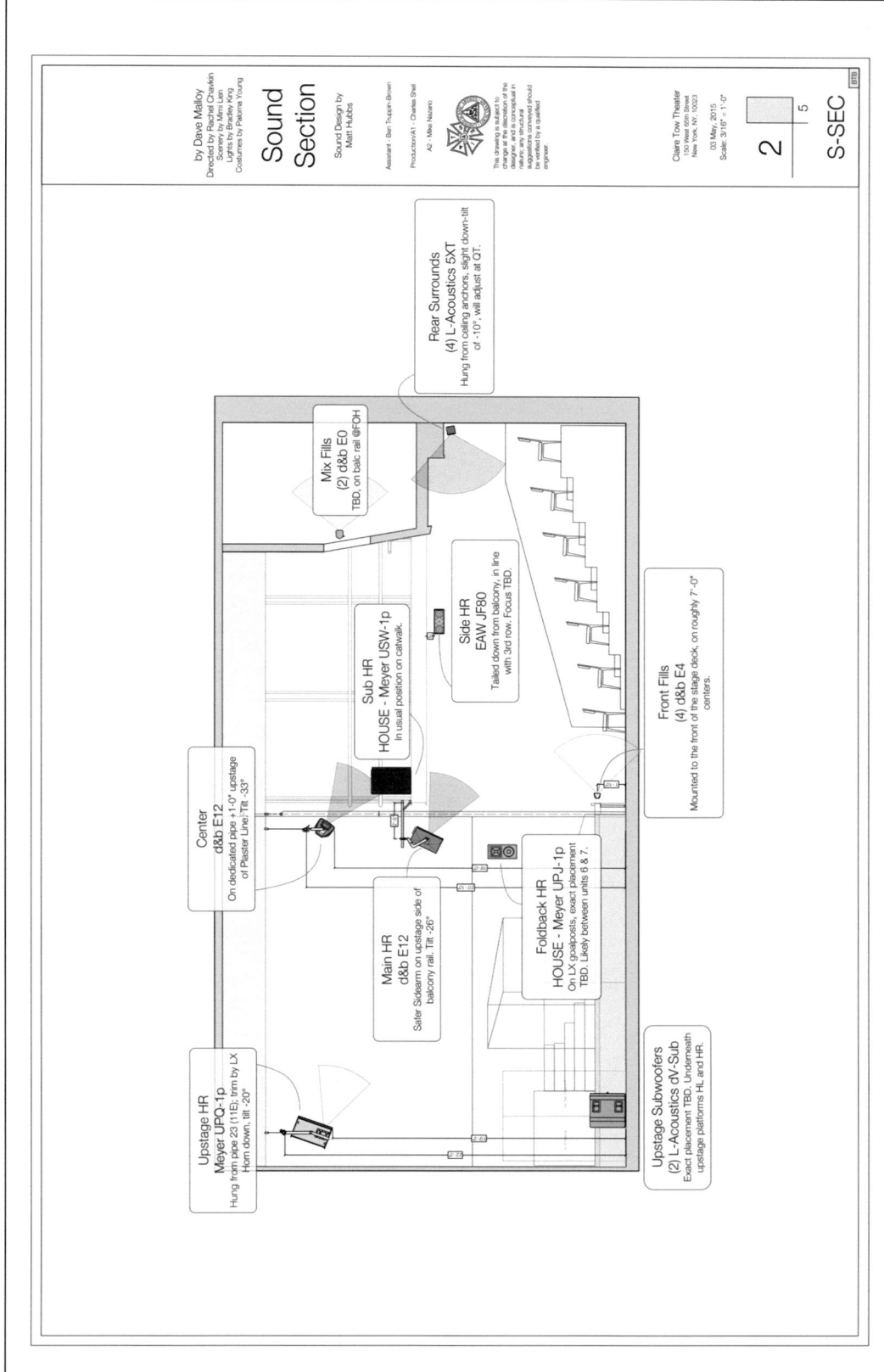

FIGURE 13-17 *Sound section drawing for LCT3's production of Preludes, sound design by Matt Hubbs, drawing by Ben Truppin-Brown, assistant. Used by permission of Ben Truppin-Brown and Matt Hubbs*

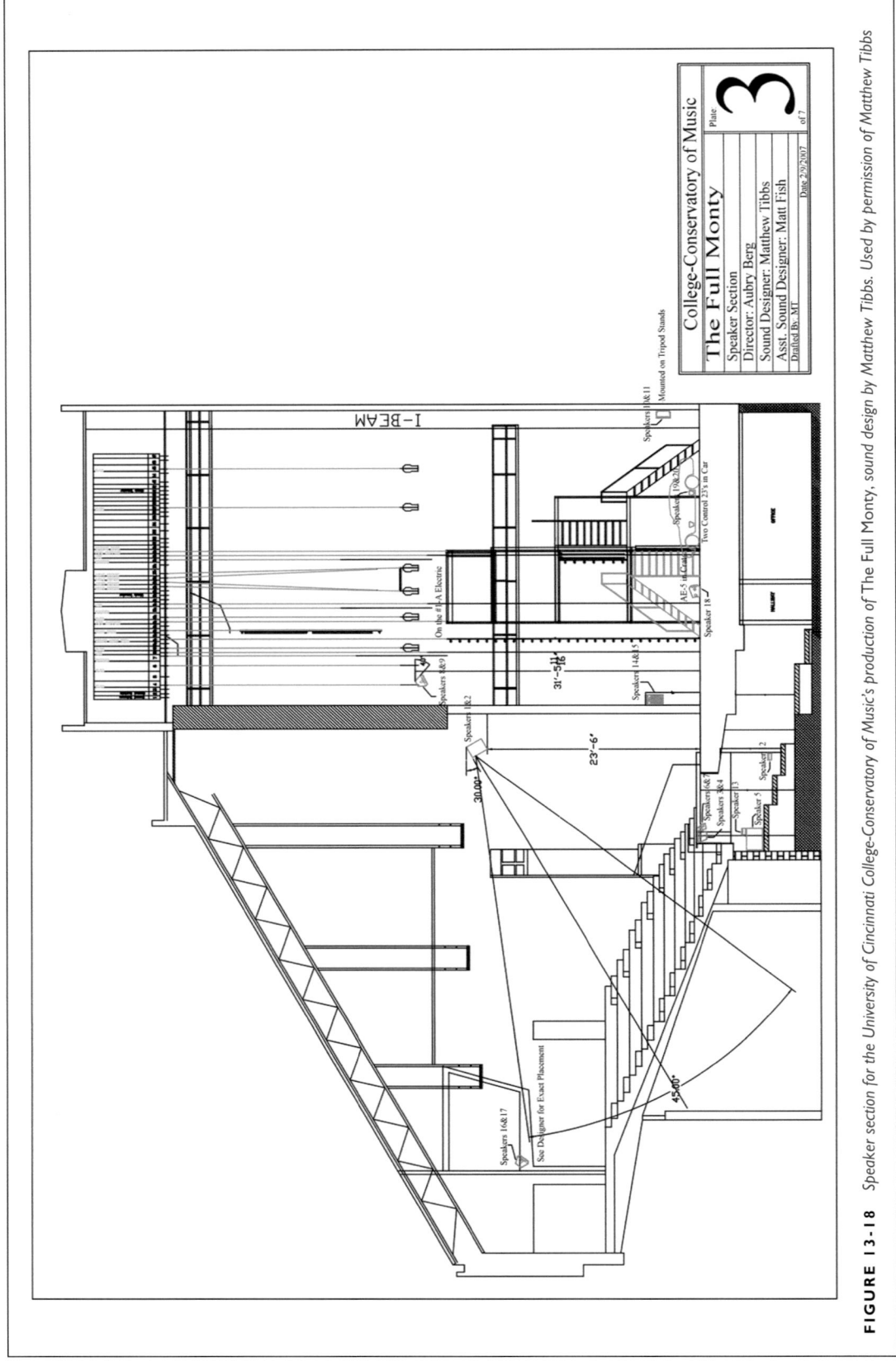

FIGURE 13-18 *Speaker section for the University of Cincinnati College-Conservatory of Music's production of The Full Monty, sound design by Matthew Tibbs. Used by permission of Matthew Tibbs*

The students should use this information to create their own Vectorworks symbols to be used on a sound plot and section. The speaker symbols may be either simple geometric shapes, or drawn to reflect the desired view of the speaker. Text information should include make and model (name), and location/purpose (description). Once symbols are created, use them to draft a sound plot and section using the venue base drawings.

Students

Using speaker information either provided by your instructor or gathered from a survey of an available venue, draft a speaker plot and section using the venue base drawings from previous exercise.

In Vectorworks, create symbols for each type of speaker. Text information should include the make and model and the location and purpose. Create a top-view symbol for the plot and a side-view symbol for the section, unless the speakers are totally symmetrical.

On the venue base drawing, create new layers for speaker placement: one for the groundplan, and one for the section. Place the symbols into position. Add callout boxes with each speaker's information. Be sure the callout boxes have a Fill so that they take visual precedence over the drawing below. If desired, adjust the line styles of the architecture to boost the

sound information's presence. Layers can be grayed out when creating a sheet layer; when selecting layers in Create Viewport, click on the third column. A gray eye will appear, indicating that that layer will be shown in gray on the sheet layer.

Include a key on a separate design layer. Unlike in Spotlight, there is no tool to aggregate sound symbols into a key. Draw the key manually and make it a group. For future projects, the key can be edited and expanded with new symbols. As a group, it can be readily copied and pasted into new files.

Create a viewport for the plot and for the section, placing each drawing on its own sheet layer Add borders, title blocks, and title block information. Save as PDFs and print.

MODULE 3: SIGNAL FLOW DIAGRAMS AND RACK DRAWINGS

Signal Flow Diagrams

As with all drafting, planning the signal flow diagram with pencil and paper beforehand will expedite layout and finalization. As a schematic, this diagram is not drawn to a scale; rather, the draftsperson needs to ensure that all text is a readable size, and that there is enough space that paths can be laid out clearly.

Figure 13-19 presents a signal flow diagram that tracks signal from a mic and computer (on the left), through the audio interface (MOTU 16A) and amplifiers to the speakers

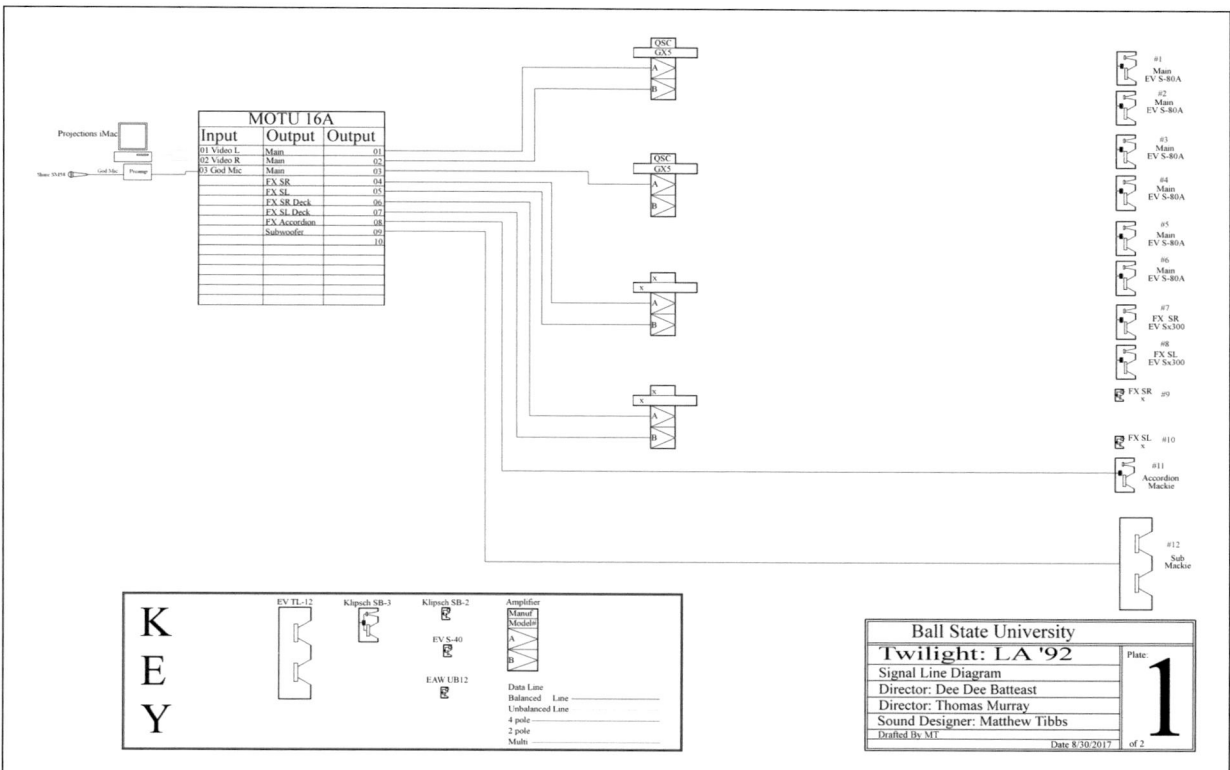

FIGURE 13-19 *Signal flow diagram for Ball State University's production of* Twilight: Los Angeles 1992, *sound designed and drawn by Matthew Tibbs. Used by permission of Matthew Tibbs*

(right). The mixer is represented by a rectangle called a **device block**. A device block can represent any type of processing component and contains:

1. a header for the device make and model above the block
2. a header for the device Type/Usage within the block
3. a column for inputs on the left
4. a column for outputs on the right

In Figure 13-19, the device block for the audio interface includes a column for the name and purpose of the output as well as one for the output's number. Device blocks may also include sections for data networks and power supplies, should those be part of the component. All inputs and outputs should be included in the device block, regardless of whether they are used as part of the design. Figure 13-20 presents the USITT recommendation for a standard device block graphic. Figure 13.21 presents an assortment of custom-drawn symbols.

Layout begins by gently zoning the sheet so that each device the signal passes through is met in order from left to right. Components of a single purpose, such as speakers, should all be located in their own column as much as possible. Parallel lines are drawn to connect inputs to outputs so that the signal's path is easily discernable. USITT recommends that path lines have rounded corners when they change direction to visually distinguish device paths from device blocks, though some designers note that rounded corners are harder to edit.[16] Information about the path (such as cable type, length, connectors, and patching) may be included as text associated with a path line. In complicated systems, paths may need to jump each other or split to indicate that the signal goes to multiple components. For very large systems, information may need to jump the sheets themselves, terminating at the right side of the first sheet and being picked up on the left side of the second.

Other symbols on signal flow diagrams include those for microphones, loudspeakers, antennas, and wireless signals, though unique input devices may be presented with an icon or pictures. A small lightning bolt representing a wireless signal can replace a connective path line. The generic symbols presented in Figure 13-22 are sometimes replaced with more pictorial images, such as when there are a variety of speaker types. Figure 13-23 features images of a computer and a guitar as inputs. A key listing all symbols used must be included. Paths can be color coded to differentiate between digital and analog signals. Device blocks can be color coded to differentiate various categories of equipment. The original drawing for Figure 13-24 features colored device blocks and signal paths.

Exercise: Constructing Stock Symbols

Three symbols will be constructed for this exercise: a microphone, an amplifier device block, and a loudspeaker. Work in 1:1. Sketch out the symbols and consider what classes will facilitate building the symbols; for example, should a microphone symbol be drawn with a microphone class or with a symbol outline class? How many text formats will you need? For the microphone and speaker symbols pictured in Figure 13-22, two text formats are used—a third if the number within the symbol is considered a separate format.

For this exercise, classes will be created for the features of the symbol, rather than the type of symbol. The style of device block symbol will be based upon the ones used for the amplifiers seen in Figure 13-24. See Figure 13-25 for sketches of the three symbols to be constructed for this exercise.

1. Begin with setup for the microphone symbol. Create classes for Guidelines and Symbol Outline. See Figure 13-22 for fonts and proportional text sizes. The scale is 1:1; the diameter of the mic symbol circle is ¾". Before setting text format font size, draw a circle

Make & Model
Type/Usage (#)

Input Group Info			Output Group Info	
1	Source 1		Destination 1	**1**
2	Source 2		Destination 2	**2**
3				**3**
4				**4**
5				**5**
6				**6**
7				**7**
8				**8**
9				**9**
10				**10**
11				**11**
12				**12**
13				**13**
14				**14**
15				**15**
16				**16**

FIGURE 13-20 *USITT recommended sample device block.[15] Reprinted by permission of USITT*

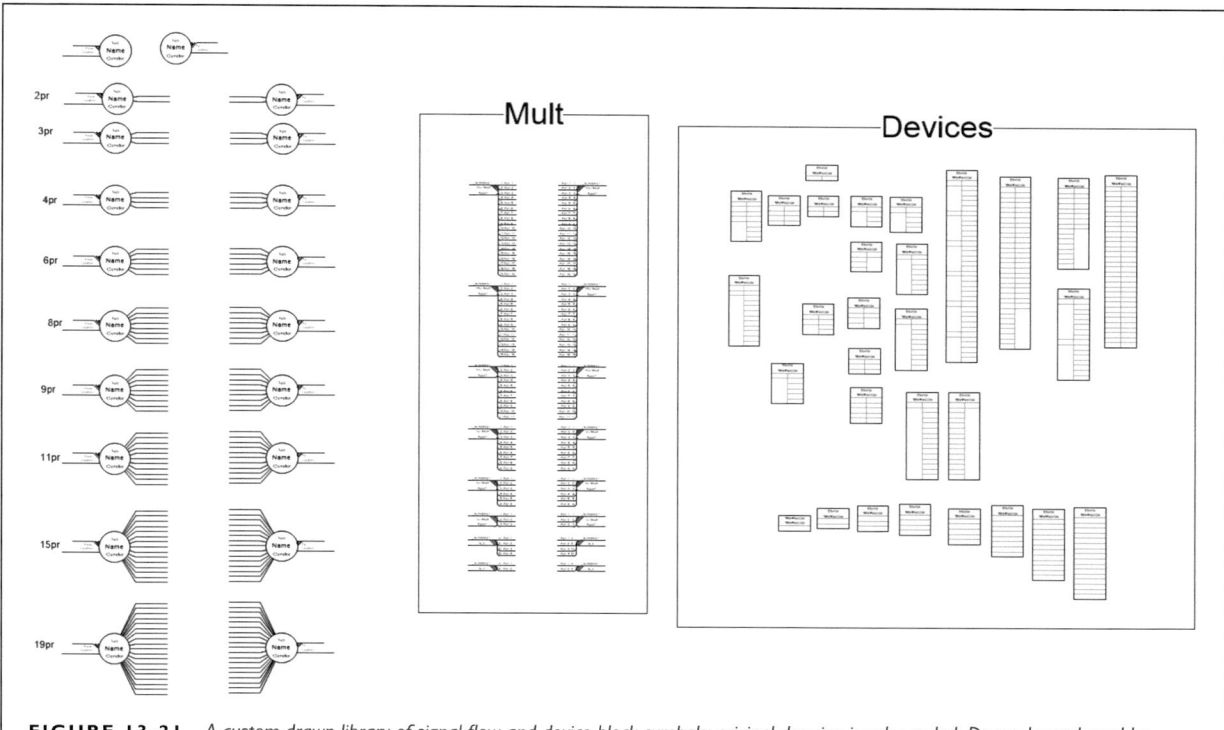

FIGURE 13-21   *A custom-drawn library of signal flow and device block symbols; original drawing is color coded. Drawn by and used by permission of Ben Truppin-Brown*

and use the text tool to determine the appropriate font sizes (Figure 13-26). When creating the text formats, choose alignment to the right (justify right). Create text formats for Description, Make and Model, and Numeral.

2. Draft the microphone symbol. Place Text in desired locations. You may wish to add a few extra spaces to text boxes so that when it is filled in with longer words, the word is not cut off by a too-short text

box. Select the whole array and Create Symbol… (Modify menu). Click on the symbol to open the Resource Manager. Create a New Folder; name it Sound Symbols.

3. While in the Resource Manager, right click on the name of the file and choose New Resource in Untitled1, and then choose Record Format. Create three fields associated with the Text Formats: Description, Make and Model, and Number.

4. OK back to the drawing space. Select the symbol and right click to choose Edit. Select a piece of text. While active, open Tools in the main toolbar, the Records, then Link Text to Record. Choose the field to be associated with the text. Repeat for each piece of text. Close the Edit window.

5. Go to the Resource Manager to file the microphone symbol you just built. Double-click to select it and then return to the drawing space. Double-click to affix it. Open the Data Tab in the Object Info navigation window and enter relevant information.

6. The process is similar for the loudspeaker symbol. Make the overall width of the symbol 7/8″, and overall height ¾″. Create text formats for speaker description and speaker make and model. You may

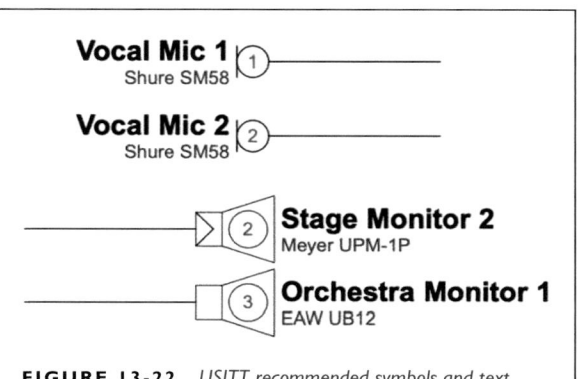

FIGURE 13-22   *USITT recommended symbols and text placement for microphones and speakers; Self-powered speakers are indicated with the addition of a small triangle.[17] Reprinted by permission of USITT*

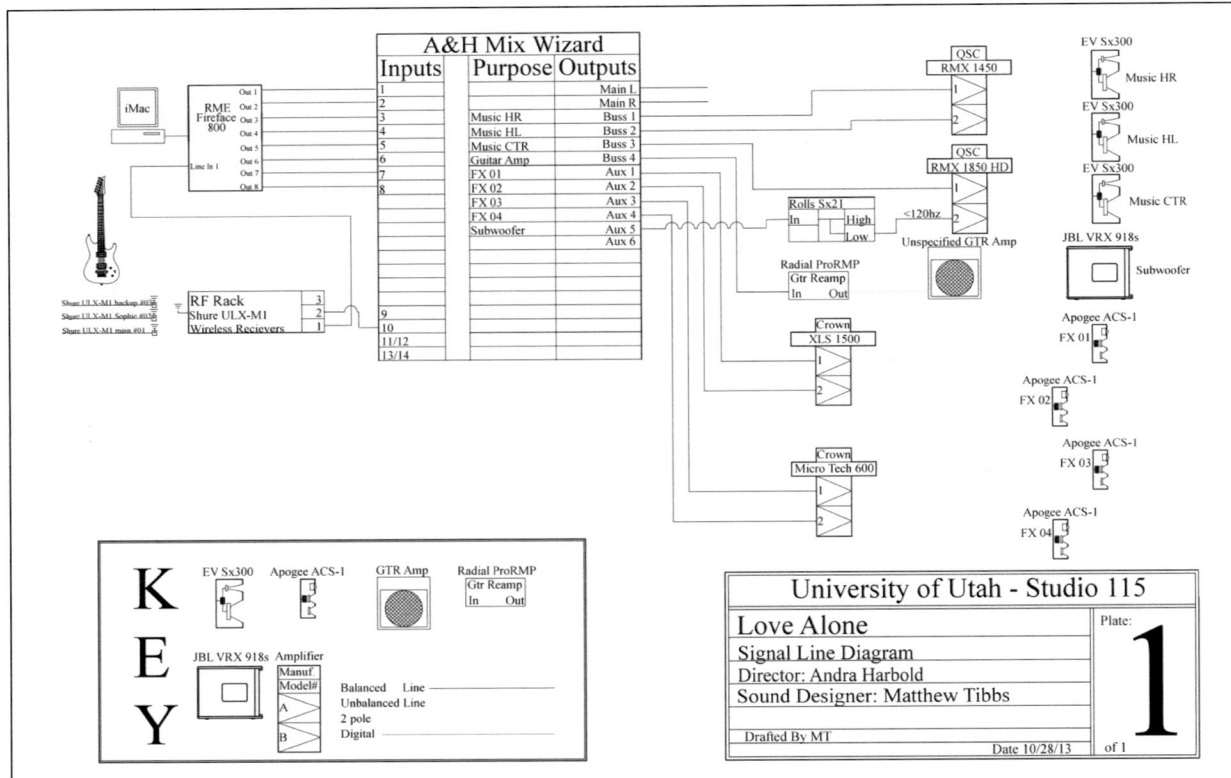

FIGURE 13-23 *Signal flow diagram for the University of Utah's production of* Love Alone, *sound design and drawing by Matthew Tibbs. Used by permission of Matthew Tibbs*

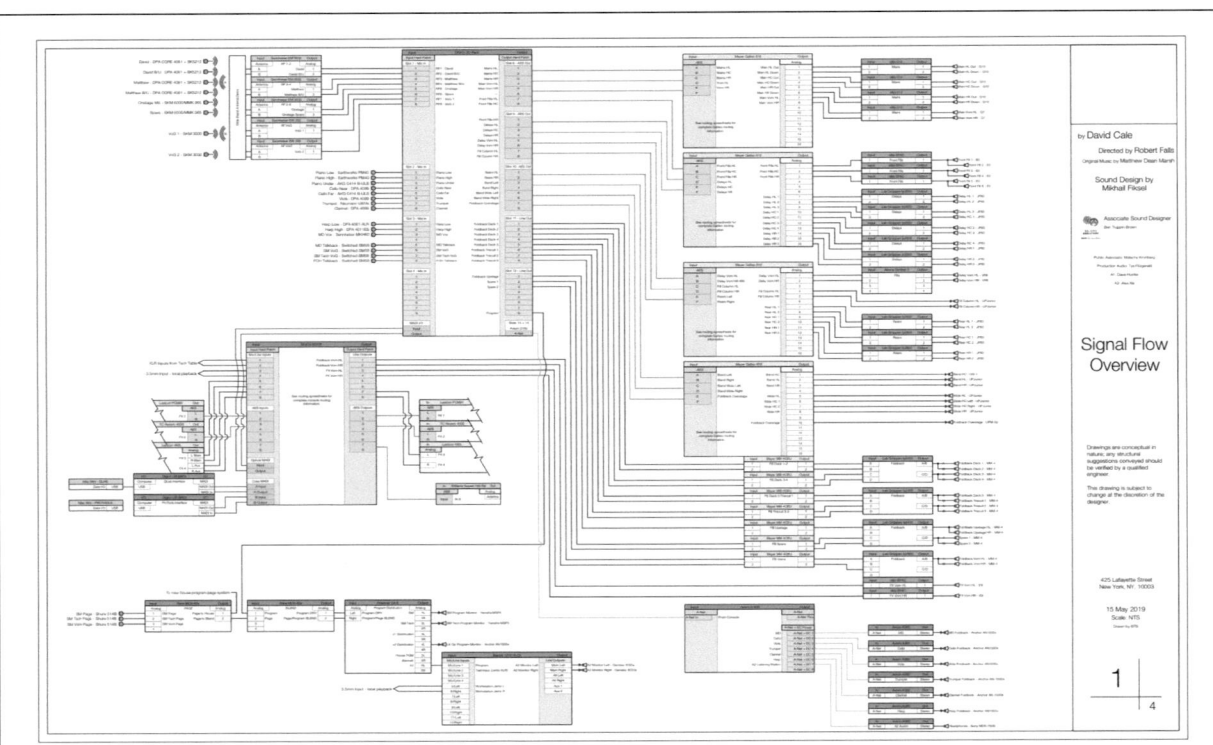

FIGURE 13-24 *Signal flow overview for the Public Theatre's production of* We're Only Alive a Short Amount of Time, *sound design by Mikhail Fiksel, drawing by Ben Truppin-Brown, associate sound designer. Used by permission of Ben Truppin-Brown and Mikhail Fiksel*

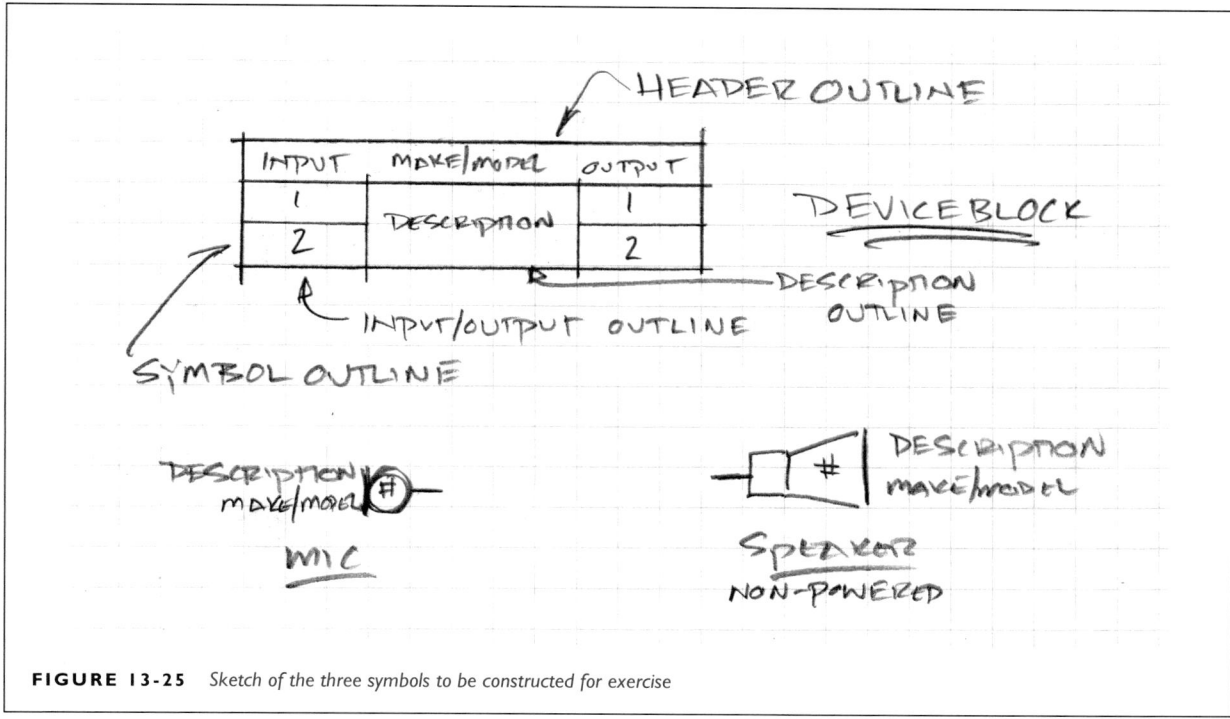

FIGURE 13-25 *Sketch of the three symbols to be constructed for exercise*

be able to use the previous created number format. Since the speakers are placed at the right-hand side of the signal flow diagram, justify to the left when creating the text format.

7. Create the symbol. Create a new Record Format named for the speaker symbols. Assign text formats to the fields. Return to the drawing space and Edit the symbol. Link Text to Records.

8. The device block symbol can be a little more complicated, depending on color coding and the number of font sizes. Each box holding text can be a separate class and therefore coded a different color. Draw boxes with the rectangle tool, starting with the overall block outline. Add boxes for each type of information in turn. Place text last. Send filled boxes and text to the front or back to maintain visibility. The words Input and Output do not need to have a text format associated with them, as they are permanent and constant headers in the symbol. The numbers within the Input and Output columns are also permanent and constant.

Drawing each box as a rectangle preserves the snapping point found along the midpoint of their sides; this makes attaching signal path lines in regular increments easier. Figure 13-27 presents a visualization of possible back to front organization. Make each tier of the device block 1″ high. The total width of the block is 1′—1″. The Input and Output columns are 3″ wide.

FIGURE 13-26 *Determining font sizes for microphone symbol*

20pt   3/4" DIA

DESCRIPTION
make and model

14pt

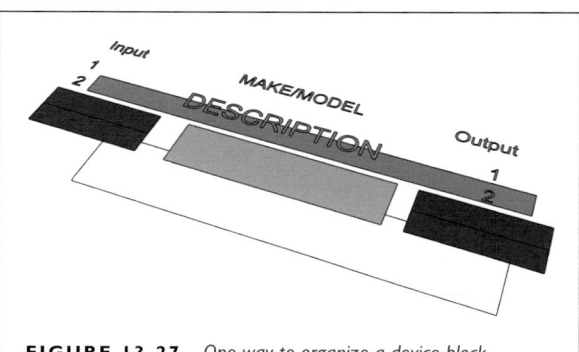

FIGURE 13-27 *One way to organize a device block symbol's filled rectangles, back to front*

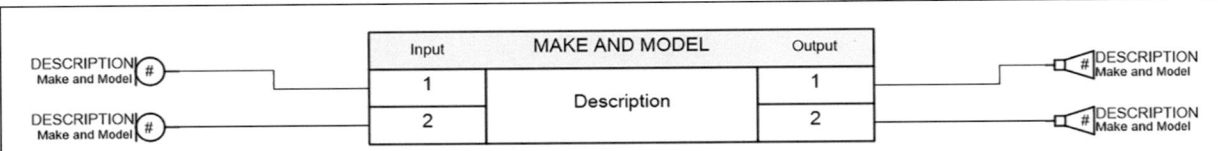

FIGURE 13-28   *A simple signal flow drawing using the symbols created in the exercise; The device block represents an all-in-one combo mixer/amplifier such as a musician might use*

9.   Again, create your desired text formats. Draw the symbol, add text. Create as a symbol. Create a Record Folder for Device Block. Edit the Symbol, and then Link Text to Record.

10.   Once you've completed these three symbols, save your sound base file. Save As to create a new file in which to draft a simple signal flow diagram. Place two microphones connected to the inputs of the device block, and two speakers at the outputs. Use ½″ spacing for the signal path lines. If you have model and make information, add it to each symbol. If you wish to color code the signal paths, create appropriate line style classes (Figure 13-28).

11.   This drawing will likely fit on an 8½″ × 11″ sheet of paper, landscape orientation. Create a viewport, select 1:1 for scale. Add borders, title block, and title block information. Export as a PDF and print.

Homework: Drafting a Signal Flow Diagram

Figure 13-29 presents a simple signal flow diagram, drafted in AutoCAD rather than in Vectorworks. Draft your own copy of this drawing. Note that the speaker symbols are drawn to represent the shape of the actual speakers and a different device block format is used for the amplifiers. You may duplicate these symbols or replace them with variations of the device block created in the previous exercise. Note that the drawing was originally created in 2010, before the 2021 USITT recommendations were issued. The inputs are all

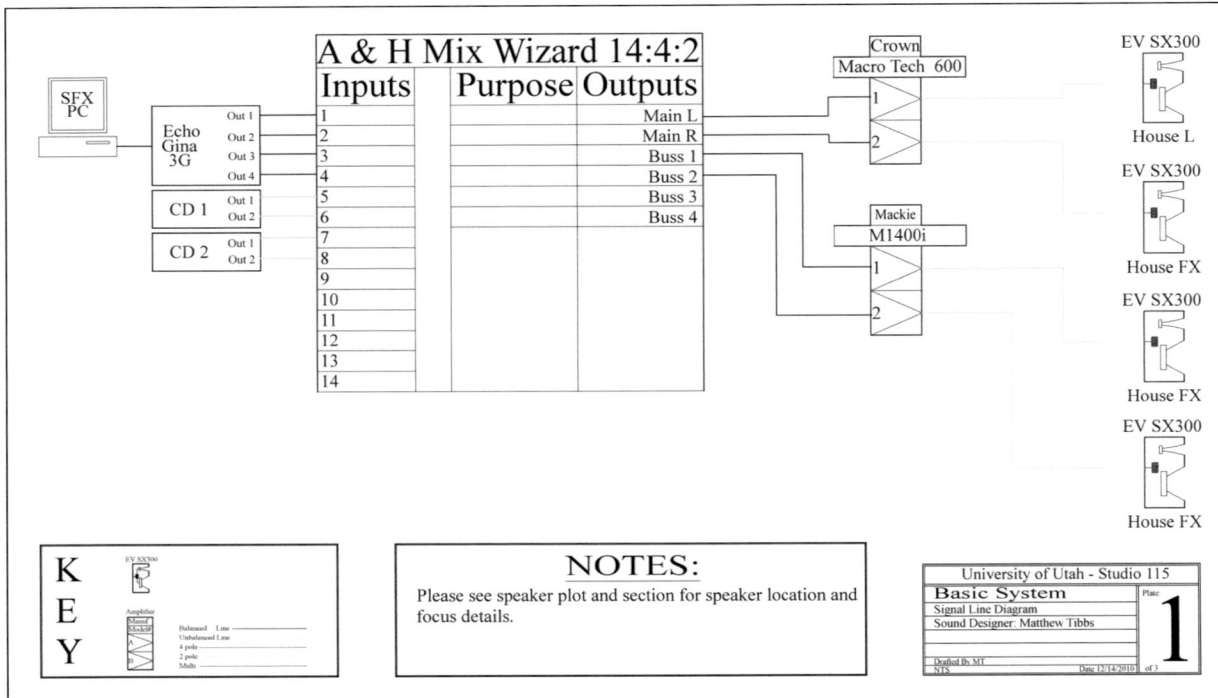

FIGURE 13-29   *Basic signal flow diagram for the University of Utah's Studio 115, designed and drawn by Matthew Tibbs. Used by permission of Matthew Tibbs*

playback devices, including a PC running a sound software program called SFX.

After creating the required symbols and saving them to your base sound file, Save As to open a new file for this drawing. Use guidelines to lay out zones on an Arch D reference grid to help align each equipment type.

Start placing equipment from the center and work outward. Set the mixer in place, then the playback equipment to the left. Set the amps in place on the right, and then the speakers. Use path lines spaced in ½″ increments to connect inputs and outputs. Adjust placement of equipment if anything becomes crowded.

Add a key and any notes regarding color coding, if used. When creating a viewport, consider whether 1:1 is an adequate printed size, or whether the drawing would benefit from being enlarged or reduced. On the sheet layer, add borders, title block, and title block information. Note of scale is not required, as this is a schematic diagram not drawn to scale (or note on drawing: Not to Scale).

Rack Drawings

Rack drawings are typically drawn when a system requires custom configuration. For venues with permanent systems, the designer is unlikely to generate one. Rack drawings are drawn or printed in scale; however, the scale simply needs to be one in which all labeling is a readable size (Figure 13-30).

There are two types of rack drawings: indicative and representative. Indicative ones present the little more than the outline of pieces of equipment in their rack locations (Figures 13-31 and 13-32). Representative ones are more pictorial, using detailed images to represent each component.[18]

FIGURE 13-30 *The rack in UW-Whitewater's Hicklin Theatre*

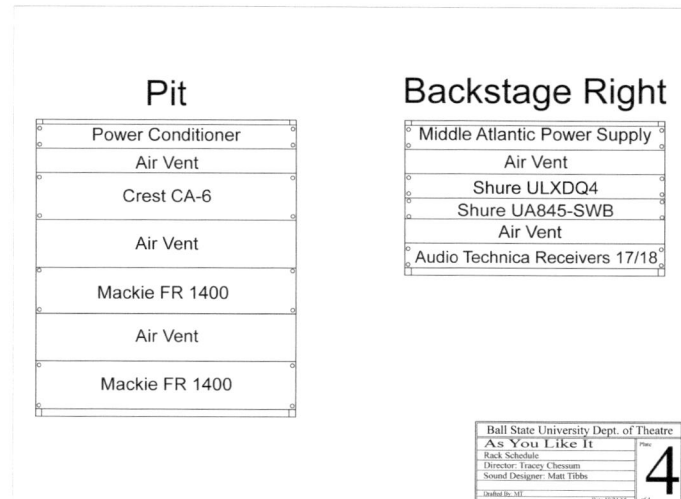

FIGURE 13-31 *A rack schedule drawing for Ball State University's production of* As You Like It, *sound design and drawing by Matthew Tibbs; Note spaces allowed for air vents. Used by permission of Matthew Tibbs*

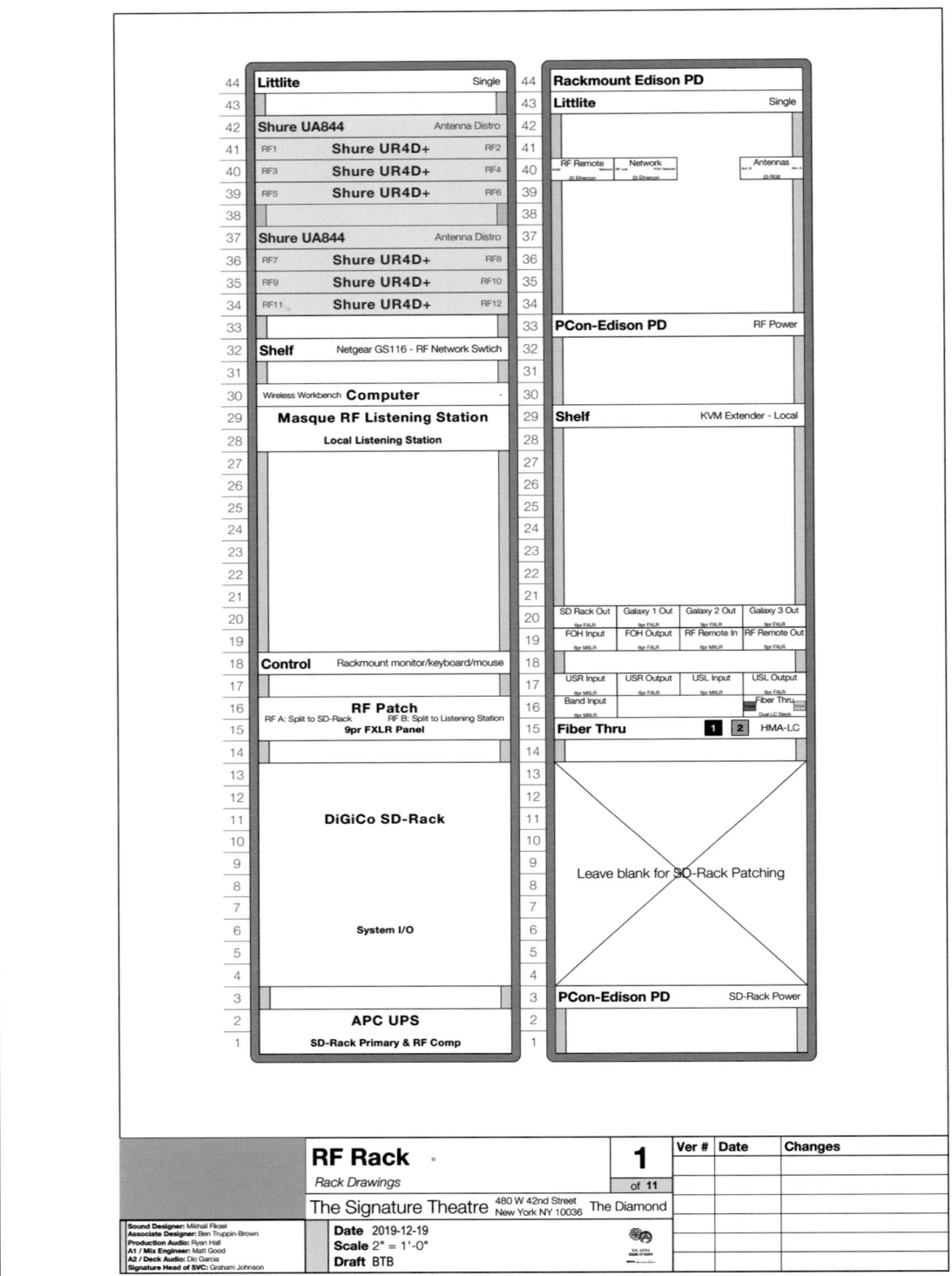

FIGURE 13-32 *RF rack drawing for the Signature Theatre's production of* Cambodian Rock Band, *sound design by Mikhail Fiksel, drawing by Ben Truppin-Brown, associate sound designer. Used by permission of Ben Truppin-Brown and Mikhail Fiksel*

Since commercial racks are standardized, stock device blocks can be used. Equipment is typically slotted into the rack and affixed into place via regularly spaced screw holes; the increments in which these holes are placed are called **Rack Units** (RUs). A standard RU is 1¾″.[19] Each RU is numbered, starting from the bottom of the rack. Each rack receives its own set of RU numbers, always starting with 1. For very small racks, some designers may omit the numbers if further documentation does not require that information.

When creating rack device block symbols, they vary in height by RUs: 1 RU, 2RU, 3RU, etc. Non-standard sized equipment requires custom device blocks. The make and model and any required description is placed within the block. Depending on the equipment, input and output information may be found within blocks. For representative drawings, a key is included, as the component's image has little to no space for text. In most cases, a front elevation is sufficient. A rear elevation may be included if some components require further explication.

Exercise: Drafting a Rack Drawing

Instructor

If a venue is available to the class, lead a field trip to the rack location. Have the students measure the rack and the heights of the components. Make a list of the makes and models of the equipment, as well as any additional descriptive information.

If a rack is not accessible, consult a sound supply company's website and look for rack specifications. Compile a list of equipment that might be found in a relatively simple rack.

Students

Using information provided by your instructor or from a survey of a sound system, construct an indicative rack drawing. All symbols should be drawn in 1:1.

Review the list and measurements of the equipment found in the rack to determine how many device blocks of what height are required. Create symbols as required (see the exercise that included Figure 13-5). Create a symbol for the rack itself. Include RU increments in a column along one side so that device blocks can be easily aligned with them.

Insert device block symbols in their appropriate locations. Add make and model and any descriptive information to each symbol. If the label does not fit within the block, use leader lines to place the label next to the rack.

When complete, create a viewport and consider the scale of the sheet layer. Unlike the signal flow diagram, this is a scale drawing. Enlarge or reduce as required to ensure readability, and note the scale in the title block.

Add border, title block, and title block information. Save the file. Export as PDF, then print.

NOTES

1. Beth Lake, interview, 12/29/20.
2. Matthew Tibbs, email correspondence, 1/4/21.
3. Josh Schmidt and Ben Truppin-Brown, interview, 12/7/20
4. Schmidt and Truppin-Brown, 12/7/20; Lake, 12/29/20; Matthew Tibbs, interview, 12/23/20
5. Tibbs, 1/4/21
6. Schmidt and Truppin-Brown, 12/7/20
7. Lake, 12/29/20
8. Tibbs, 1/4/21
9. Lake, 12/29/20
10. Tibbs, 1/4/21
11. Matthew Tibbs, interview, 12/23/20
12. USITT Education Commission, *USITT/TSDCA Sound Documentation Recommended Practice* (Syracuse, NY: United States Institution for Theatre Technology, 2020), p. 3
13. Tibbs, 12/23/2020
14. Ben Truppin-Brown, interview, 1/22/21
15. USITT Education Commission, p. 12.
16. Truppin-Brown, 1/22/21
17. USITT Education Commission, p. 20.
18. USITT Education Commission, p. 48.
19. USITT Education Commission, p. 48.

BIBLIOGRAPHY

Allen, Edward. *Fundamentals of Building Construction; Materials and Methods.* 2nd edition. John Wiley & Sons, New York. 1990.

Allen, Kevin Lee. *Vectorworks for Entertainment Design: Using Vectorworks to Design and Document Scenery, Lighting, and Sound.* Focal Press, New York and London. 2015.

Bay, Howard. *Stage Design.* Drama Book Specialists/ Publishers, New York. 1974.

Bentham, Frederick. *Stage Lighting.* Sir Isaac Pitman & Sons, London. 1950.

Blurton, John. *Scenery: Drafting and Construction for theatres, museums, exhibitions, and trade shows.* Theatre Arts, New York. 2001.

Boy Scouts of America. *Merit Badge Series: Drafting.* Boy Scouts of America, New Brunswick, NJ. 1965.

Brewster, Karen and Shafer, Melissa. *Fundamentals of Theatrical Design.* Allworth Press, New York. 2011.

Carter, Paul. *Backstage Handbook: an Illustrated Almanac of Technical Information.* 3rd edition. Broadway Press, Louisville, KY. 1994.

Ching, Francis D. K. *Architectural Graphics.* 4th edition. John Wiley & Sons, New York. 2003.

Cunningham, Glen. *Stage Lighting Revealed: A Design and Execution Handbook.* Waveland Press, Long Grove, IL. 1993.

Dorn, Dennis and Shanda, Mark. *Drafting for the Theatre.* Southern Illinois University Press, Carbondale, IL. 1992.

Essig, Linda. *Lighting and the Design Idea.* Thomson Wadsworth, Belmont, CA. 2005.

French, Thomas E. and Svensen, Carl L. *Mechanical Drawing.* 7th edition. McGraw-Hill, New York. 1968.

Giesecke, Frederick E., Mitchell, Alva and Spender, Henry Cecil. *Technical Drawing.* 4th edition. The MacMillan Company, New York. 1958.

Gillette, J. Michael. *Theatrical Design and Production: An Introduction to Scenic Design and Construction, Lighting, Sound, Costume, and Makeup.* 6th edition. McGraw-Hill, New York. 2008.

Gillette, J. Michael and McNamara, Michael. *Designing with Light: An Introduction to Stage Lighting,* 6th edition. McGraw-Hill, New York. 2014.

Hayes, David. *Light on the Subject: Stage Lighting for Directors and Actors – and the Rest of Us,* 5th edition. Limelight Editions, New York. 1998.

Helvenston, Harold. *Scenery: A Manual of Scene Design.* Stanford University Press, Stanford, CA. 1931.

Holloway, John. *Illustrated Theatre Production Guide.* 2nd edition. Focal Press, New York and London. 2010.

Hopgood, Jeromy. *Dance Production: Design & Technology.* Focal Press, New York and London. 2016.

Kaye, Donna and Lebrecht, James. *Sound and Music for the Theatre: The Art and Technique of Design.* Focal Press, New York and London. 2009.

Lake, Beth. *Interview.* 12/29/20.

Loar, Josh. *The Sound System Design Primer.* Routledge, New York and London. 2019.

Malloy, Kaoime E. *The Art of Theatrical Design: Elements of Visual Composition, Methods, and Practice.* Focal Press, New York and London. 2015.

McHugh, Robert C. *Working Drawing Handbook: A Guide for Architects and Builders.* 2nd edition. Van Nostrand Reinhold Company, New York. 1982.

Mielziner, Jo. *Designing for the Theatre: A Memoir and a Portfolio.* Bramhall House, New York. 1965.

Palmer, Richard H. *The Lighting Art: The Aesthetics of Stage Lighting Design.* 2nd edition. Prentice-Hall, Englewood Cliffs, NJ. 1994.

Parker, Oren. *Sceno-graphic Techniques.* Carnegie-Mellon University, Pittsburgh, PA. 1969.

Parker, Oren, Smith, Harvey K. and Wolf, Craig R. *Scene Design and Stage Lighting.* 5th edition. Holt, Rinehart, and Winston, New York. 1985.

Pecktal, Lynn. *Designing and Painting for the Theatre.* Holt, Rinehart, and Winston, New York. 1975.

Petroski, Henry. *The Pencil: A History of Design and Circumstance.* Alfred A. Knopf, New York. 2006.

Pilbrow, Richard. *Stage Lighting Design: The Art, the Craft, the Life.* Design Press, New York. 1997.

Ramsey, Charles George, and Sleeper, Harold Reeve. *Architectural Graphic Standards for Architects, Engineers, Decorators, Builders, and Draftsmen,* 4th edition. John Wiley & Sons, New York. 1953.

Raoul, Bill. *Stock Scenery Construction Handbook.* Broadway Press, Shelter Island, NY. 1990.

Rose, Rich. *Drafting Scenery for Theatre, Film, and Television.* Betterway Books, Cincinnati, OH. 1990.

Rose, Rich. *Drawing Scenery for Theatre, Film, and Television*. Betterway Books, Cincinnati, OH. 1994.

Rosenthal, Jean and Wertenbaker, Lael. *The Magic of Light: The Craft and Career of Jean Rosenthal, Pioneer in Lighting for the Modern Stage*. Theatre Arts Books, New York. 1972.

Rowe, Clare. P. *Drawing & Rendering for the Theatre: A Practical Course for Scenic, Costume, and Lighting Designers*. Focal Press, New York and London. 2007.

Schmidt, Joshua, and Truppin-Brown, Ben. *Interview*. 12/7/2020.

Shelley, Steven Louis. *A Practical Guide to Stage Lighting*. 3rd edition. Focal Press, New York and London. 2014.

Smith, Fran Kellogg, and Bertolone, Fred. J. *Bringing Interiors to Light: The Principles and Practices of Lighting Design*. Whitney Library of Design/Watson Guptill Publications, New York. 1986.

Thorne, Gary. *Stage Design: A Practical Guide*. The Crowood Press, Ramsbury. 1999.

Tibbs, Matthew. *Personal Interview*, 12/23/2020

Tibbs, Matthew. *Email Correspondence*. 12/13/20 to 2/1/21.

Truppin-Brown, Ben. *Interview*. 1/22/21

Umfrid, Thomas C. *Email Correspondence*. 5/20 to 3/21.

USITT Education Commission. *Recommended Practice for Theatrical Lighting Design Graphics*. United States Institution for Theatre Technology, Syracuse, NY. 2006.

USITT Education Commission. *USITT Scenic Design and Technical Production Graphic Standards*. United States Institution for Theatre Technology, Syracuse, NY. 1992.

USITT Education Commission. *USITT Graphic Recommended Best Practices*. United States Institution for Theatre Technology, Syracuse, NY. 2020.

USITT Education Commission. *USITT/TSDCA Sound Documentation Recommended Practice*. United States Institution for Theatre Technology, Syracuse, NY. 2020.

USITT Sound Commission. *Sound Graphics Recommended Practice: A Project of the USITT Sound Commission*. United States Institution for Theatre Technology, Syracuse, NY. 2008.

Watson, Lee. *Lighting Design Handbook*. McGraw-Hill, New York. 1990.

Wolfe, Welby B. *Materials of the Scene: An Introduction to Technical Theatre*. Harper & Row, Publishers, New York. 1977.

Woodbridge, Patricia and Tine, Hal. *Designer Drafting and Visualizing for the Entertainment World*. 2nd edition. Focal Press, New York and London. 2013.

Yee, Rendow. *Architectural Drawing: A Visual Compendium of Types and Methods*. John Wiley & Sons, New York. 1997.

INDEX

Note: Italicized *page numbers refer to figures,* **bold** *page numbers refer to tables*

act curtain 182
Actor's Equity Association (AEA) 280
actual lumber size 141; *see also* nominal
    lumber size
acute angle 36
addresses 297
adjacent views, with missing information
    77–78
American National Standards Institute
    (ANSI) 2
American Society of Mechanical Engineers
    (ASME) 2
amplifiers 338
angles 36; dimensioning 106, *108*
Angular Dimension tool (Vectorworks) 106
apron 174, 272
Arc tool (Vectorworks) 160
architectural drawing 3–4
architectural scale ruler 6–7, *7;* reading
    20–21
archway 153
arcs: defined 36; meeting lines 70–71
arena 174; *see also* in-the-round
as-built drawings 267
Audinate 338; *see also* Dante Audio
    Networking System
AutoCAD 11
automated lamps 297
auxiliary views 49, 81–97; constructing
    96–97; dimensioning with 112–114;
    drawing 85, 95–96; hand drafting
    85–86; inclined 82; oblique 82;
    placement of *83;* two-triangle technique
    83–84, *84*
axonomic projection 47

Backstage Handbook (Carter) 138
baluster 280
balustrade 280
banister 280, 287–288
barrel 297
bars 174
base drawings 173–174, 185
baseboard 157
baserail 280
battens 174, 235
Bay, Howard 139–140
beam and accessory information 298

bluelines 2, *4; see also* Wisconsin State Capitol
blueprint 2, *3; see also* Brisbois House
booms 320; drafting 322; notations *322*
Border tool (Vectorworks) 61, 87, *133*
borders 21–22, 27–28, *29*
boundaries 21–22
bow compass 7
box boom 320; adding to venue base
    section 328–330; rough sketch *321*
box set: detail plates 265–267; dressing
    the model 251; modeling in SketchUp
    210–218
Break Line tool (Vectorworks) 178
break lines 178
Brisbois House *3; see also* blueprint
Broadway style flat 141
brush 10, *11*
butt joints 141

cabinet 47
Callout Box tool (Vectoworks) 105
Callout tool (Vectorworks) 106
caps 163
cardboards 296
carriage 279
Carter, Paul 138
casement window 207, 271
casing 157, 272
catwalk 240
cavalier 47
center line section 227–248; adding height
    to venue model 231–235; copying
    an existing section 247–248; defined
    227–228; drafting 235–247; elements of
    228–229; system of working 248
center lines 101–102, 194
chair rail 157
channel 296, 298
cheat sheet 293; *see also* magic sheet
circle templates 7, *9*
Circle tool (SketchUp) 69–70
Circle tool (Vectorworks) 67–69
circles 36, 66–67; drawing in SketchUp
    69–70; drawing in Vectorworks 68–69;
    drawing with template 67–68
circuit 296, 300
circumference 36
clarity 5

Class Name 32
Class Organization window (Vectorworks)
    31–35
classes 31
closed stringer 280
cloud 240
color notation 298
composed objects 66
composite elevations 254–255
construction drawing 154–155
construction lines 24–26
control console 296
conventional incandescent lamp 296
cookie 297; *see also* cucaloris
corner blocks 141
corner braces 140
cornice 157
crosshatching 93; adding to shapes 93–95;
    section views 93
crossover 174
crown molding 157
cubes: building in SketchUp 54–59; fully
    dimensioned orthographic views of
    114–125
cucaloris 297; *see also* cookie
cut list 150
cutting plane 89
cutting plane lines 90–92, *91*
Cutting Plane tool (SketchUp) 218, 234, 278
cyclorama 298

Dante Audio Networking System 338;
    see also Audinate
dead-hung 174
deck 174
design drawing 154
design layer 22
desk 296
detail plates 261–292; benefits of 265; box
    set 265–267; as-built drawings 267;
    defined 261; for existing scenic structure
    270–271; flats 268–270; floor layout 268;
    layout and drafting of 267–270; layout
    possibilities 265–267; molding detail
    269; props and furniture 270; for rocks
    264; soft goods 270; stairs and staircases
    279–292; system of working 292;
    window detail 269; windows 271–278

device block 354

diagonal views, with missing information 81

diameter 36

dimension lines 31; defined 39, 99–100; terminators 102

dimensioning 99–114; angles 106, *108*; with auxiliary views and hidden lines 112–114; center lines 101–102; defined 99; dimension lines 102; extension lines 100–101; format systems 99–100; hand drafting 106–107; and labels 198; leader lines 105–106; lighting units 302; measurement notation 102–104; in SketchUp 111–112; in Vectorworks 104–105

dimensions: choice of 108; placement of 108–111

dimmer 296

dimmer/address 300

Dims/Notes tool (Vectorworks) *62, 63,* 126

DMX 297

door flat with details 157–162

Door tool (Vectorworks) *179*

doorknob 156

doors 156

doorways 153, *206*

Double Line tool (Vectorworks) 177, 238, 240, 242

double prime 19

double-hung windows 271, *274*; drafting 273–275; SketchUp model of 275

downstage 174

drafting: booms and ladders 322–324; center line section 235–247; checklist 97–98; defined 1; double-hung windows 273–275; equipment and materials 5–11; flat with opening and reveal 153–154; light plot 303–308; lighting sections 330–332; objects 76–77; Palladian window with window seat 275–278; rack drawings 361; reader 5; scenic front elevation 251–254; sections 349–353; series of flats 151–152; signal flow diagrams *358,* 358–359; speaker plots 349–353; stair unit with banister and landing 287–288; standard style rectangular flat 142; units 284; venue plan in Vectorworks 174–180; vertical positions 320–322

drafting board/table 7, *8*; working edge 22–23

drafting projects, evaluating 12–13

draperies 208

drawing: auxiliary views 95–96; door flat with details 157–162; groundplans graphics 202–209; platformed area 169–170; platforms 164; stepped platform 167–169

Drawing Label too (Vectorworks) 126

drawing package 136–138

drawing space 21–37; borders 21–22, 27–28, *29*; boundaries 21–22; construction lines/guidelines 24–26; exercise 23–24; setting up 22–24

drawings: creating PDF of 45; printing 44; transporting 28

drop 207

dry erase pad 9

DWG file 187

electric batten 174

elevations 49

ellipsoidal reflector spotlight (ERSs) 297–298

Entertainment Services and Technology Association (ESTA) 280

equilateral triangle 36

equipment and materials 5–11; architectural scale ruler 6–7, *7*; bow compass 7; brush 10, *11*; circle templates 7, *9*; drafting board/table 7, *8*; dry erase pad 9; eraser shield 9; erasers 9; French curve 7; lighting symbol drafting template 7, *10*; paper 11; pencils 8–9; protractor 7; sharpeners 9; software 5–6; tape 10, *11*; triangles 7, *8*; T-square 7, *8*; vellum 11

Erase tool (SketchUp) 56, 70

eraser shield 9

erasers 9

escape stairs 281

existing plan, copying 185–186

explode 212

extension lines 99, 100–101

facing 164

false proscenium 182

Fill attribute 32

fills, adding to shapes 93–95

film style flat 141

finial 280

fixed window 271

fixture 296

flats 139–152; construction 140; defined 139–140; detail plates 268–270; drafting checklist 171; drafting series of 151–152; modeling in SketchUp 146–151; with opening and reveal 153–156; parts of 140

flight 280

floor layout 268

floor treatment 198

fly rail 174

focus 300

focus charts 300

FOH positions 324

Foley 337

Follow Me tool (SketchUp) 167, *168*

fractions 17

French, Thomas E. 135

French curve 7

French doors 271

Fresnel 298

front elevation 249–260; composite elevations 254–255, 256; definition of 249; drafting 251–254; paint elevation 249, 256–257; parts of 250–251; sketching individual flat elevations 255–256; system of working 259–260; venue 257–259

front-of-house (FOH) 174

full section 90

fullness 208

furniture 270

gate 298

gel 298

general notes 125

geometric construction 37, 66–67

glass box concept *48,* 48

glazing 273–275

gobo 297; *see also* patterns; templates

grand drape 182

grand valance 182

graphic language 1

groundplans 193–226; checklist 185; components of 194–198; creating rough sketch 209–210, *211*; defined 193; drawing graphics for 202–209; extrapolating from image 225–226; graphics 202–209; laying out 218–225; master 194; reading 198–202; system of working 226; using 193–194

groundrow 318

group 74

guidelines 24–26, 31, 39

half-straps 141

hand drafting: auxiliary views 85–86; crosshatching 93–95; dimensioning 106–107; setting up drawing space for 23–24; standard style rectangular flat 142–144

handrail 280

hanging plot 194

hanging time 198

hidden lines 31, 39; dimensioning with 112–114; rules for 60–66

hinges 156

hollow core door 156

Hollywood style flat 141; modeling 151

hook-up 293

Hooley's Opera House 2; *see also* lithograph

hopper windows 271

house 174

house left 174

house right 174

human figure 144–145

identity labels 125

image files 187

incandescent lamp 296

inclines: described 82; drafting in Vectorworks 86–87; irregular 87–88; modeling 84–85

individual drawings 135–136

input 336, 337

instrument 296

instrument schedule 322

Instrument Summary tool (Vectorworks) 301, *302,* 318

intelligent lamps 297

intensity 296

International Alliance of Theatrical Stage Employees (IATSE) 129

International Organization for Standardization (ISO) 3
in-the-round 174; *see also* arena
irregular inclines 87–88
irregular polygon 36
Irregular Polygon tool (Vectorworks) 291
isometric projection 47, *47*
item placement 198

jacks 270
jamb 152
joists 152
JPEG file 187

key 301
keystones 141
knee walls 163

label legends 310–315, *316*
labels 125–127, 198
ladders 320; drafting 322
Lake, Beth 335
lamp 296
landing 280, 287–288
lauan 140
Leader Line Simple tool (Vectorworks) 106
leader lines 100, 105–106
legends 298–300, 301
Leko 298
lens train 297
lettering 15–18; exercise 17–18; goals
        15–16; guidelines 16; homework 18;
        spacing 17
light 272
light plot *294*; drafting 303–308; elements of
        301–303; reading 303
lighting 293–334; adding positions to venue
        base drawings 324–326; categories
        297–298; cyclorama 298; ellipsoidal
        reflector spotlight 297–298; Fresnel
        298; label legends 310–315, *316*;
        legends 298–300; parabolic aluminized
        reflector 298; strip lights 298; symbol
        as drawn on the position 300–301;
        system of working 332–333; units
        placed below the grid or rig 317–320;
        vocabulary 296–297
lighting bridge 174
Lighting Instrument tool (Vectorworks) 315
Lighting Pipe tool (Vectorworks) 311
lighting positions 302
lighting sections 326–332; drafting 330–332;
        types of 326–327
lighting symbol drafting template 7, *10*
lighting units 302
Lightwright 317, 327
Line attribute 32
line schedule 174, 181–185, 198, 302
line styles: choosing appropriate 59–66;
        familiarization 39–47; hand drafting
        45–46
Line tool (SketchUp) 56, 73, 75
Line tool (Vectorworks) 35, 41, 42, 66, 219
line type 26–27
line weight 26

lines: arcs meeting 70–71; center 101–102,
        194; constraining 35; construction
        24–26; cutting plane 90–92, *91*;
        defined 36; dimension 31, 39, 99–100,
        102; drawing in Vectorworks 35–36;
        extension 99, 100–101; hidden 31, 39,
        60–66, 112–114; leader 100, 105–106;
        miter 79–81; parallel 36; perpendicular
        36; phantom 175, 202, 207; reference
        301; snapping 35; styles 31; thick visible
        31, 39; thin visible 175; types 31; visible
        25, 26; weights 31; witness 99
lineset 174
lithograph, *2*; *see also* Hooley's Opera House
local notes 125
locking rail 174
Look Around tool (SketchUp) 235
lumber sizes 141
luminaire 296

magic sheet 293; *see also* cheat sheet
main act 182
main rag 182
masking 174, 302
master groundplans 194
McKernon, J. 317
measurement notation 100; aligned system
        104; placement 102–104; unidirectional
        system 104
measurements: exercise 21; notation 18–19;
        writing 19
mechanical drawing 3–4
Mechanical Drawing (French and Svensen)
        135
mechanical perspective 46, *47*
medium visible lines 31, 39; *see also* lines
miter lines 79–81; *see also* reflection line
mixer 338
modeling: battens 235; box set in SketchUp
        210–218; flat with opening and
        reveal 155–156; Hollywood style
        flat construction 151; incline 84–85;
        Palladian window in SketchUp 278;
        platformed area 167; tent-shaped
        objects 88–90; venue in SketchUp
        186–191
molding 152; crown 157; details 269;
        nomenclature 157
Move tool (SketchUp) 74, 75
movers 297
mullions 272
multi-view orthographic projection 48;
        auxiliary view 49; correct and incorrect
        alignment *50*; glass box concept 48;
        section view 49
multi-view projection 47
muntins 205, 272
muslin 140

NC (N/C) 298
New Sheet Layer window *130*
newel post 280
newel post cap 280
nominal lumber size 141; *see also* actual
        lumber size

nosing 164, 280
notes 125–127, 302

Object Info window (Vectorworks) *92*
objects: composed 66; decomposing 66;
        drafting 76–77; group 74; tent-shaped
        88–90
oblique 82
oblique projection 47, *47*
obtuse angle 36
Offset tool (SketchUp) 149–150, 214–215
off-stage 174
offstage storage of scenery 198
ogee curve 160
onstage microphone *348*
onstage speaker placement *347*
orthographic projection 47
orthographic visualization 51–53; using
        SketchUp 73–76
output 336, 337
Ozalid Streamliner 2

page 137
Paint Bucket tool (SketchUp) 147, *148*, 232,
        251
paint elevations 49; defined 249; SketchUp
        256–257
Paint tool (Vectorworks) 149
Palladian windows: drafting 275–278;
        modeling in SketchUp 278; with
        window seat 275–278
Palladio, Andrea 275
Pan tool (SketchUp) 55
Pan tool (Vectorworks) 31, *32*
pane 272, *275*
panel door 156
paper 11
parabolic aluminized reflector (PAR) 298
parallel lines 36
parallel projection 47, 57
parallelogram 36
patch panel 297
patterns 297; *see also* gobo; templates
PDF file 187
Pen attribute 32
pencils 8–9
perpendicular lines 36
phantom lines 175, 202, 207; *see also* lines
picture rail 157
pipe 174
pit 174
pivoting windows 271
plan 49
plane of projection 46
plaster line 174, 175, 194; *see also*
        groundplans
plates 137, 163
platformed area: construction drawing of
        169–170; modeling 167
platforms 162–170; defined 162; drafting
        checklist 171; drawing 164; modeling
        167; parts of 162–164; stepped
        167–169
playback device 337
playback system 338

plinth 157
PNG file 187
point 36
polygon 36
Position Camera tool (SketchUp) 235
practicals 198, 302
preamplifiers 338
prime 19
printing drawings 44
profiles 152, 157
projection: axonomic 47; isometric 47, 47;
    multi-view 47; multi-view orthographic
    48–49, 50; oblique 47, 47; orthographic
    47; parallel 47; plane of 46
projection systems 46–53; defined 46;
    orthographic visualization exercise
    51–53
projectors 46
props 270
proscenium 174
proscenium arch 174
protractor 7
Protractor tool (SketchUp) 212, 213
public address (PA) system 335
Push/Pull tool (SketchUp) 73

quadrilateral 36

rack drawings 359–360, 359–361
rack units 361
racks 338
radius 36
rails 140
rear elevation 140, 142
Rectangle tool (SketchUp) 73
Rectangle tool (Vectorworks) 66, 240, 291
reference guidelines 175
reference lines 301
Reference Marker tool (Vectorworks) 225
reflection line 81; see also miter lines
reflector 296
reinforcement (sound) 337
reveals 152
Rigging tool (Vectorworks) 311
right angle 36
right triangle 36
rise 279
risers 279
Rotate tool (SketchUp) 74
rough plot: drafting in Vectorworks 310–315;
    laying out position using 308–310
rover 318
ruler, reading 19
running in 279

sash 271, 272, 275
scabs 141
Scale Bar tool (Vectorworks) 180, 182
scale bars 180, 198
Scale tool (SketchUp) 233
scales: measurements 21; reading scale ruler
    20–21
scenic front elevation 249–260; composite
    elevations 254–255, 256; definition of

249; drafting 251–254; paint elevation
    249, 256–257; parts of 250–251;
    sketching individual flat elevations
    255–256; system of working 259–260;
    venue 257–259
scenic structure 270, 302
Section Plane tool (SketchUp) 89, 90
section views 49; crosshatching 93; and
    irregular inclines 87–88
Section-Elevation Marker tool (SketchUp) 92
Section-Elevation Marker tool (Vectorworks)
    90, 92
sections 349, 351, 352; drafting 349–353
Selection tool (Vectorworks) 31
set line 210
shapes, adding crosshatching and fills to
    93–95
sharpeners 9
sheet 137
sheet layers 22, 127–128
shift plots 194
ship's ladder 279
shoe rail 280
shop orders 296
show curtain 182
show portal 182
sidearms 320
sightline symbols 179, 194; see also
    groundplans
sightlines 174
signal flow diagrams 353, 353–354, 356;
    drafting 358, 358–359
sill 272
single-hung windows 271
single-stroke commercial gothic 15
SketchUp 5, 6; adding height to venue
    model 231–235; building four cubes
    in 54–59; constructing paint elevations
    in 256–257; dimensioning in 111–112;
    drafting objects 76–77; drawing circles
    in 69–70; modeling an incline in 84–85;
    modeling box set in 210–218; modeling
    double-hung windows in 275; modeling
    flat in 146–151; modeling Palladian
    windows in 278; modeling platformed
    areas and stairs in 284; modeling tent-
    shaped objects in 88–90; modeling two
    step units in 287; modeling venue in
    186–191; visualizing orthographic views
    with 73–76; wireframe view 76
smoke pocket 174
soft goods 208, 270
softpatch 297
solid core 156
sound design 335–361; constructing symbols
    in Vectorworks 338–343; content of
    336; rack drawings 359–360, 359–361;
    sections 349, 351, 352; signal flow
    diagrams 353, 353–354, 356; speaker
    plots 343–349, 344, 345, 346; stock
    symbols 354–358, 357; vocabulary
    337–338
speaker plots 343–349, 344, 345, 346;
    drafting 349–353

speaker sections 352
spindle 280
Spotlight tool (Vectorworks) 311
Spotlight Workspace toolsets (Vectorworks)
    312
Stage Design (Bay) 139
stage left 174
stage right 174
stage style flat 141
staircases: laying out drawing 288–292;
    layout 281; parts of 279, 279;
    vocabulary 279
stairs: escape 281; modeling in SketchUp
    284; parts of 279, 280; placement 283;
    stair unit with banister and landing
    287–288; step units 284–287; for unit
    set 281–284; for unit set with curved
    platforms 284; vocabulary 280
standard style flat 141
step units 284; construction drawing
    286–287; design drawing 286; modeling
    in SketchUp 287
stiles 140
stock symbols 354–358, 357
stops 272
straightedges 24, 26–27
straps 141
stretchers 163
strike plate 156
stringers 279
strip lights 298
stud walls 163
studio style flat 141
Svensen, Carl 135
sweeps 140

tail down 320
tape 10, 11
Tape Measure tool (SketchUp) 55, 56, 212
Tape Measure tool (Vectorworks) 62
teaser 182
television style flat 141
templates 297; see also gobo; patterns
terminals 11
terminators 102
Text tool (Vectorworks) 43, 223, 225, 318
thick visible lines 31, 39; see also lines
Thickness attribute 32
thin visible lines 175; see also lines
thresholds 156
thrust 174
Title Block Border Settings window
    (Vectorworks) 65
title blocks 127–132, 198, 303
titles 125–127
toggles 140
torm position 182
tormentor 182
torms 320
traditional style flat 141; hand drafting
    142–144; in Vectorworks 145–146
transducer 337
transom 272
travel 182

treads 279
Triangle tool (Vectorworks) 220
triangles 7, *8*, 36
trim 152
trim heights 181
Trim tool (Vectorworks) 41, 42, 69
T-square 7, *8*
2D Plan 31
2D Polygon tool (Vectorworks)
    178, 220
two-triangle technique 83–84, *84*

Undo (Vectorworks) 41
unit 296
unit number 298
United Scenic Artists (USA) 129
United States Institute for Theatre
    Technology (USITT) 3
United States National CAD Standard 3
universe 297
upstage 174
USITT 1992 Graphic Standard 3

Vectorworks 5; classes 31; constructing
    symbols in 338–343; crosshatching in
    93–95; dimensioning in 104–105; draft-
    ing a standard style rectangular flat in
    145–146; drafting from the rough plot
    in 310–315; drafting simple incline in
    86–87; drafting venue plan in 174–180;
    drawing circles in 68–69; drawing fully
    dimensioned orthographic views of
    four cubes in 114–125; drawing lines
    in 35–36; geometric construction in
    66–67; Layer Organization window
    30; layers 28; line style familiarization
    40–44; line styles 60–66; page naviga-
    tion 30–31; Pan tool 31, *32*; Rectangle
    tool 66; Selection tool 31; setting up
    base file in 28–30; setting up classes
    31–35; Zoom selection 30
vellum 2, 11
venue: adding box boom positions to
    base section 328–330; adding lighting
    positions to base drawings 324–326;

architecture 194; model, adding height
    to 231–235; modeling in SketchUp
    186–191; *see also* groundplans
Vertical American Standard Alphabet 15, *16*
Viewport Layer Properties window *130*
viewports 127–128, *131*
Visibility 32
visible lines 25, 26; *see also* lines

windows *206*, 271–278; details 269; double-
    hung 273–275, *274*; fixed 271; Palladian
    275–278; parts, standard names 271
wings 174
wireframe view (SketchUp) 76
Wisconsin State Capitol *4*; *see also*
    bluelines
witness lines 99
working edge 22–23

zero deck 164
zones 63
Zoom tool (SketchUp) 55